Lecture Notes in Computer Science 3876

Commenced Publication in 1973
Founding and Former Series Editors:
Gerhard Goos, Juris Hartmanis, and Jan van Leeuwen

Shai Halevi   Tal Rabin (Eds.)

Theory of Cryptography

Third Theory of Cryptography Conference, TCC 2006
New York, NY, USA, March 4-7, 2006
Proceedings

 Springer

Volume Editors

Shai Halevi
Tal Rabin
IBM T.J. Watson Research Center
19 Skyline Drive, Hawthorne, NY 10532, USA
E-mail: shaih@alum.mit.edu; talr@watson.ibm.com

Library of Congress Control Number: 2006921180

CR Subject Classification (1998): E.3, F.2.1-2, C.2.0, G, D.4.6, K.4.1, K.4.3, K.6.5

LNCS Sublibrary: SL 4 – Security and Cryptology

ISSN	0302-9743
ISBN-10	3-540-32731-2 Springer Berlin Heidelberg New York
ISBN-13	978-3-540-32731-8 Springer Berlin Heidelberg New York

Springer is a part of Springer Science+Business Media

springer.com

© International Association for Cryptologic Research 2006
Printed in Germany

Typesetting: Camera-ready by author, data conversion by Scientific Publishing Services, Chennai, India
Printed on acid-free paper   SPIN: 11681878   06/3142   5 4 3 2 1 0

Preface

TCC 2006 was the third Theory of Cryptography Conference, which was held at Columbia University in Manhattan, New York, March 4-7, 2006. TCC 2006 was sponsored by the International Association for Cryptologic Research (IACR) and organized in cooperation with the Computer Science Department of Columbia University. The local arrangements chair was Tal Malkin.

The Program Committee, consisting of 13 members, received 91 submissions and selected for publication 31 of these submissions. The quality of the submissions was very high, and the selection process was a challenging one. The proceedings consist of the revised versions of these 31 papers. Revisions were not checked as to their contents, and the authors bear full responsibility for the contents of their papers. In addition to the 31 accepted papers, the program included two tutorials: A tutorial on "Black-Box Separation Results" by Omer Reingold and a tutorial on "Non-Black-Box Techniques" by Boaz Barak. The conference featured a rump session for informal short presentations of new results, chaired by Charlie Rackoff and boosted by Tequilas!

We are in debt to the many people who contributed to the success of TCC 2006, and we apologize for those whom we have forgotten to mention. First and foremost we thank the authors who submitted their papers to TCC 2006; a conference is only as good as the submissions that it receives. The Program Committee members made a concentrated effort during the short review period contributing their time, knowledge, expertise and taste, and for that we are extremely grateful. We also thank the large number of external reviewers who assisted the committee in the review process.

A heartfelt thanks goes to our local arrangements chair Tal Malkin and her assistant Sophie Majewski for facilitating the communication with Columbia University. Their hard work made the local arrangements an effortless process for us. We also thank Angelos D. Keromytis, Michael Locasto, and Angelos Stavrou for giving us a web server at Columbia University on which to host the TCC work and helping us manage it. We also want to thank IBM for their generous donation of our time and the financial support for students attending TCC.

This was the first year that TCC was sponsored by the IACR. Several people at the IACR helped us navigate this new terrain, in particular Andy Clark, Helena Handschuh and Kevin McCurley. We also benefited from advice from members of the TCC Steering Committee, including Mihir Bellare, Ivan Damgård, Oded Goldreich and Moni Naor. Additional help came from the organizers of last year's TCC: Shafi Goldwasser, Joe Kilian and Joanne Talbot-Hanley, and the people at Springer, in particular Alfred Hofmann, Ingrid Beyer and Anna Kramer.

And last but not least, thanks to our group members Ran Canetti, Rosario Gennaro, Hugo Krawczyk and Masa Abe for all their support (emotional and otherwise).

December 2005

Shai Halevi and Tal Rabin
TCC 2006 Program Co-chairs

External Reviewers

Michel Abdalla	Danny Harnik	Thomas Brochmann
Masayuki Abe	Alejandro Hevia	Pedersen
Jesús F. Almansa	Nick Howgrave-Graham	Krzysztof Pietrzak
Benny Applebaum	Yuval Ishai	Benny Pinkas
Boaz Barak	Oleg Izmerly	Bartosz Przydatek
Mihir Bellare	Stanisław Jarecki	Oded Regev
Alexandra Boldyreva	Yael Tauman Kalai	Omer Reingold
Dan Boneh	Joe Kilian	Leonid Reyzin
Xavier Boyen	Eike Kiltz	Tom Ristenpart
Jan Camenisch	Tadayoshi Kohno	Ron Rivest
Ran Canetti	Chiu-Yuen Koo	Louis Salvail
Melissa Chase	Hugo Krawczyk	Hovav Shacham
Richard Cleve	Gunnar Kreitz	Tom Shrimpton
Ivan Damgård	Homin Lee	Alice Silverberg
Anupam Datta	Arjen Lenstra	Jessica Staddon
Ante Derek	Anna Lysyanskaya	Tamir Tassa
Yevgeniy Dodis	Phillip MacKenzie	Mårten Trolin
Cynthia Dwork	Stephen Miller	Wim van Dam
Ariel Elbaz	Sara Miner	Salil Vadhan
Marc Fischlin	Anton Mityagin	Vinod Vaikuntanathan
Matthias Fitzi	Tal Mor	Emanuele Viola
Ariel Gabizon	Ruggero Morselli	Andrew Wan
Rosario Gennaro	Steven Myers	Bogdan Warinschi
Craig Gentry	Gregory Neven	Hoeteck Wee
Kristian Gjøsteen	Damian Niwiński	Douglas Wikström
Mikael Goldmann	Shien Jin Ong	
Venkat Guruswami	Saurabh Panjwani	

TCC 2006

The Third Theory of Cryptography Conference

Columbia University, New York, NY, USA
March 4-7, 2006

Sponsored by the *International Association for Cryptologic Research*

Organized in cooperation with the *Computer Science Department,
Columbia University*

General and Program Co-chairs

Shai Halevi and Tal Rabin, IBM T.J. Watson Research Center

Program Committee

Stefan Dziembowski	Warsaw University
Johan Håstad	Royal Institute of Technology
Jonathan Katz	University of Maryland, College Park
Eyal Kushilevitz	Technion Israel Institute of Technology
Yehuda Lindell	Bar-Ilan University
Tal Malkin	Columbia University
Daniele Micciancio	University of California, San Diego
John C. Mitchell	Stanford University
Chanathip Namprempre	Thammasat University
Jesper Buus Nielsen	University of Århus
Manoj Prabhakaran	University of Illinois, Urbana-Champaign
Adam Smith	Weizmann Institute of Science
Luca Trevisan	University of California, Berkeley

TCC Steering Committee

Mihir Bellare	University of California, San Diego
Ivan Damgård	University of Århus
Oded Goldreich	Weizmann Institute of Science
Shafi Goldwasser	MIT
Johan Håstad	Royal Institute of Technology
Russell Impagliazzo	University of California, San Diego
Ueli Maurer	ETH Zurich
Silvio Micali	MIT
Moni Naor	Weizmann Institute of Science
Tatsuaki Okamoto	NTT Labs

Table of Contents

Zero-Knowledge

Primitives

Assumptions and Models

The Bounded-Retrieval Model

Privacy

Secret Sharing and Multi-party Computation (I)

Universally-Composible Security

One-Way Functions and Friends

Secret Sharing and Multi-party Computation (II)

Pseudo-Random Functions and Encryption

Concurrent Zero Knowledge
Without Complexity Assumptions*

Daniele Micciancio[1,**], Shien Jin Ong[2,***], Amit Sahai[3,†], and Salil Vadhan[2,‡]

[1] University of California, San Diego, La Jolla CA 92093, USA
daniele@cs.ucsd.edu
[2] Harvard University, Cambridge MA 02138, USA
{shienjin, salil}@eecs.harvard.edu
[3] University of California, Los Angeles, Los Angeles CA 90095, USA
sahai@cs.ucla.edu

Abstract. We provide *unconditional* constructions of *concurrent* statistical zero-knowledge proofs for a variety of non-trivial problems (not known to have probabilistic polynomial-time algorithms). The problems include Graph Isomorphism, Graph Nonisomorphism, Quadratic Residuosity, Quadratic Nonresiduosity, a restricted version of Statistical Difference, and approximate versions of the (**coNP** forms of the) Shortest Vector Problem and Closest Vector Problem in lattices.

For some of the problems, such as Graph Isomorphism and Quadratic Residuosity, the proof systems have provers that can be implemented in polynomial time (given an **NP** witness) and have $\tilde{O}(\log n)$ rounds, which is known to be essentially optimal for black-box simulation.

To the best of our knowledge, these are the first constructions of concurrent zero-knowledge proofs in the plain, asynchronous model (i.e., without setup or timing assumptions) that do not require complexity assumptions (such as the existence of one-way functions).

1 Introduction

In the two decades since their introduction [2], zero-knowledge proofs have taken on a central role in the study of cryptographic protocols, both as a basic building block for more complex protocols and as a testbed for understanding important new issues such as composability (e.g., [3]) and concurrency (e.g., [4]). The "classic" constructions of zero-knowledge proofs came primarily in two flavors. First, there were direct constructions of zero-knowledge proofs for specific problems, such as QUADRATIC RESIDUOSITY [2] and GRAPH ISOMORPHISM [5]. Second, there were general constructions of zero-knowledge proofs for entire classes of

---

* A full version of this paper is available [1].
** Supported by NSF grant 0313241 and an Alfred P. Sloan Research Fellowship.
*** Supported by ONR grant N00014-04-1-0478.
† Supported by NSF ITR and Cybertrust programs, an equipment grant from Intel, and an Alfred P. Sloan Foundation Fellowship.
‡ Supported by NSF grants CNS-0430336 and CCR-0205423.

S. Halevi and T. Rabin (Eds.): TCC 2006, LNCS 3876, pp. 1–20, 2006.
© Springer-Verlag Berlin Heidelberg 2006

problems, such as all of **NP** [5].[1] Both types of results have played an important role in the development of the field.

The general results of the second type show the wide applicability of zero knowledge, and are often crucial in establishing general feasibility results for other cryptographic problems, such as secure multiparty computation [8,5] and CCA-secure public-key encryption [9, 10, 11]. However, they typically are too inefficient to be used in practice. The specific results of the first type are often much more efficient, and are therefore used in (or inspire) the construction of other efficient cryptographic protocols, e.g., identification schemes [12] and again CCA-secure public-key encryption [13, 14, 15]. Moreover, the specific constructions typically do not require any unproven complexity assumptions (such as the existence of one-way functions), and yield a higher security guarantee (such as *statistical* zero-knowledge proofs).[2] The fact that the proof systems are unconditional is also of conceptual interest, because they illustrate the nontriviality of the notion of zero knowledge even to those who are unfamiliar with (or who do not believe in the existence of) one-way functions.[3]

Concurrent zero knowledge. In recent years, a substantial effort has been devoted to understanding the security of cryptographic protocols when many executions are occurring concurrently (with adversarial scheduling). As usual, zero-knowledge proofs led the way in this effort, with early investigations of concurrency for relaxations of zero knowledge dating back to Feige's thesis [22], and the recent interest being sparked by the work of Dwork, Naor, and Sahai [4], which first defined the notion of concurrent zero knowledge. Research on concurrent zero knowledge has been very fruitful, with a sequence of works leading to essentially tight upper and lower bounds on round complexity for black-box simulation [23, 24, 25, 26, 27, 28], and partly motivating the first non-black-box-simulation zero-knowledge proof [29]. However, these works are primarily of the *second* flavor mentioned in the first paragraph. That is, they are general feasibility results, giving protocols for all of **NP**. As a result, these protocols are fairly inefficient (in terms of computation and communication), rely on unproven complexity assumptions, and only yield computational zero knowledge (or, alternatively, computational soundness).

There have been a couple of works attempting to overcome these deficiencies. Di Crescenzo [30] gave unconditional constructions of concurrent zero-knowledge

---

[1] See the textbook [6] and survey [7] by Oded Goldreich for a thorough introduction to zero-knowledge proofs.

[2] Of course, this partition into two types of zero-knowledge protocols is not a precise one. For example, there are some efficient zero-knowledge proofs for specific problems that use complexity assumptions (e.g., [16] and there are some general results that are unconditional (e.g., [17, 18, 19]).

[3] It should be noted that the results of [20,21] show that the existence of a zero-knowledge proof for a problem outside **BPP** implies some weak form of one-way function. Still, appreciating something like the perfect zero-knowledge proof system for GRAPH ISOMORPHISM [5] only requires believing that there is no *worst-case* polynomial-time algorithm for GRAPH ISOMORPHISM, as opposed to appreciating notions of average-case complexity as needed for standard one-way functions.

proofs in various timing models. That is, his protocols assume that the honest parties have some synchronization and may employ delays in the protocol, and thus do not work in the standard, asynchronous model (and indeed he states such a strengthening as an open problem). Micciancio and Petrank [31] gave an efficient (in terms of computation and communication) transformation from honest-verifier zero-knowledge proofs to concurrent zero-knowledge proofs. However, their transformation relies on the Decisional Diffie–Hellman assumption, and yields only computational zero knowledge.

Our Results. We give the first unconditional constructions of concurrent zero-knowledge proofs in the standard, asynchronous model. Our proof systems are statistical zero knowledge and statistically sound (i.e. they are interactive proofs, not arguments [32]). Specifically, our constructions fall into two categories:

1. Efficient proof systems for certain problems in **NP**, including QUADRATIC RESIDUOSITY, GRAPH ISOMORPHISM and a restricted form of quadratic non-residuosity for Blum integers, which we call BLUM QUADRATIC NONRESIDUOSITY. These proof systems all have prover strategies that can be implemented in polynomial time given an **NP** witness and have $\tilde{O}(\log n)$ rounds, which is essentially optimal for black-box simulation [27].

2. Inefficient proof systems for other problems, some of which are not known to be in **NP**. These include QUADRATIC NONRESIDUOSITY, GRAPH NON-ISOMORPHISM, the approximate versions of the complements of the CLOSEST VECTOR PROBLEM and SHORTEST VECTOR PROBLEM in lattices, and a restricted version of STATISTICAL DIFFERENCE (the unrestricted version is complete for statistical zero knowledge [33]). These proof systems have a polynomial number of rounds, and do not have polynomial-time prover strategies. These deficiencies arise from the fact that our construction begins with a public-coin, honest-verifier zero-knowledge proof for the problem at hand, and the only such proofs known for the problems listed here have a polynomial number of rounds and an inefficient prover strategy.

Techniques. One of the main tools for constructing zero-knowledge proofs are commitment schemes, and indeed the only use of complexity assumptions in the construction of zero-knowledge proofs for all of **NP** [5] is to obtain a commitment scheme (used by the prover to commit to the **NP** witness, encoded as, e.g., a 3-coloring of a graph). Our results rely on a relaxed notion of commitment, called an *instance-dependent commitment scheme,*[4] which is implicit in [35] and formally defined in [36,34,19]. Roughly speaking, for a language L (or, more generally, a promise problem), a instance-dependent commitment scheme for L is a commitment protocol where the sender and receiver algorithms also depend on the instance x. The security requirements of the protocol are relaxed so that the hiding property is only required when $x \in L$, and the binding property is only required when $x \notin L$ (or vice-versa).

---

[4] Previous works [34,19] have referred to this as "problem-dependent" commitment scheme, but this new terminology of "instance-dependent" seems more accurate.

As observed in [36], many natural problems, such as GRAPH ISOMORPHISM and QUADRATIC RESIDUOSITY, have simple, unconditional instance-dependent commitment schemes. This is useful because in many constructions of zero-knowledge proofs (such as that of [5]), the hiding property of the commitment scheme is only used to establish the zero-knowledge property and the binding property of the commitment scheme is only used to establish soundness. Since, by definition, the zero-knowledge property is only required when the input x is in the language, and the soundness condition is only required when x is not in the language, it suffices to use a instance-dependent commitment scheme. Specifically, if a language $L \in \mathbf{NP}$ (or even $L \in \mathbf{IP}$) has a instance-dependent commitment scheme, then L has a zero-knowledge proof [36] (see also [34,19]).

Existing constructions of *concurrent* zero-knowledge proofs [24,27,28] also rely on commitment schemes (and this is the only complexity assumption used). Thus it is natural to try to use instance-dependent commitments to construct them. However, these protocols use commitments not only from the prover to the verifier, but also from the verifier to the prover. Naturally, for the latter type of commitments, the roles of the hiding and binding property are reversed from the above — the hiding property is used to prove soundness and the binding property is used to prove (concurrent) zero knowledge. Thus, it seems that we need not only a instance-dependent commitment as above, but also one where the security properties are reversed (i.e. binding when $x \in L$, and hiding when $x \notin L$).

Our first observation is that actually we only need to implement the commitment schemes from the verifier to the prover. This is because the concurrent zero-knowledge proof system of Prabhakaran, Rosen and Sahai [28] is constructed by a general compiler that converts *any* public-coin zero-knowledge proof into a concurrent zero-knowledge proof, and this compiler only uses commitments from the verifier to the prover. (Intuitively, the verifier commits to its messages in an initial "preamble" stage, which is designed so as to allow concurrent simulation.) Since all the problems we study are unconditionally known to have public-coin zero-knowledge proofs, we only need to implement the compiler. So we are left with the task finding instance-dependent commitments that are binding when $x \in L$ and hiding when $x \notin L$. Thus, for the rest of the paper, we use this as our definition of instance-dependent commitment.

This idea works directly for some problems, such as GRAPH NONISOMORPHISM and QUADRATIC NONRESIDUOSITY. For these problems, we have instance-dependent commitments with the desired security properties, and thus we can directly use these commitments in the compiler of [28]. Unfortunately, for the complement problems, such as GRAPH ISOMORPHISM and QUADRATIC RESIDUOSITY, we only know of instance-dependent commitments that are hiding when $x \in L$, and binding when $x \notin L$.

Thus, for some of our results, we utilize a more sophisticated variant of instance-dependent commitments, due to Bellare, Micali, and Ostrovsky [35]. Specifically, they construct something like a instance-dependent commitment scheme for the GRAPH ISOMORPHISM problem, but both the hiding and binding properties are non-standard. For example, the binding property is as follows: they show that if $x \in L$ and the sender can open a commitment in two different ways, then it is possible for the sender to extract an **NP** witness for $x \in L$. Thus we call these *witness-binding commitments*. Intuitively, when we use such commitments, we prove concurrent zero knowledge by the following case analysis: either the verifier is bound to its commitments, in which case we can simulate our proof system as in [28], *or* the simulator can extract a witness, in which case it can be simulated by running the honest prover strategy. In reality, however, the analysis does not break into such a simple case analysis, because the verifier may break the commitment scheme in the middle of the protocol. Thus we require that, in such a case, an already-begun simulation can be "continued" once we are given an **NP** witness. Fortunately, the classic (stand-alone) proof systems for GRAPH ISOMORPHISM and QUADRATIC RESIDUOSITY turn out to have the needed "witness-completable simulation" property.

An additional contribution of our paper is to provide abstractions and generalizations of all of the above tools that allow them to be combined in a modular way, and may facilitate their use in other settings. First, we show how the "preamble" of the Prabhakaran–Rosen–Sahai concurrent zero-knowledge proof system [28] can be viewed as a way to transform any commitment scheme into one that is "concurrently extractable," in the sense that we are able to simulate the concurrent execution of many sessions between an adversarial sender and the honest receiver in a way that allows us to extract the commitments of the sender in every session. This may be useful in constructing other concurrently secure protocols (not just proof systems). Second, we provide general definitions of witness-binding commitment schemes as well as witness-completable zero-knowledge proofs as possessed by GRAPH ISOMORPHISM and QUADRATIC RESIDUOSITY and as discussed above.

Perspective. The recent works of Micciancio and Vadhan [34] and Vadhan [19] hypothesized that every problem that has a statistical (resp., computational) zero-knowledge proof has a instance-dependent commitment scheme.[5] There are several pieces of evidence pointing to this possibility:

1. A restricted form of a complete problem for statistical zero knowledge has a instance-dependent commitment scheme [34].
2. If instance-dependent commitments exist for all problems with statistical zero-knowledge proofs, then instance-dependent commitments exist for all of problems with (general, computational) zero-knowledge proofs [19].
3. Every problem that has (general, computational) zero-knowledge proofs also has inefficient instance-dependent commitments. These commitments are in-

---

[5] Actually, the works of [34] and [19] refer to instance-dependent commitments where the hiding property holds on YES instances and the binding property on NO instances, which is opposite of what we use. For statistical zero knowledge, this does not matter because the class of problems having statistical zero-knowledge proofs is closed under complement [17]. But for computational zero knowledge, it means that outline presented here might yield a concurrent zero-knowledge *argument* system rather than a proof system.

efficient in the sense that the sender algorithm is not polynomial-time computable [19]. Unfortunately we cannot use these commitments in our protocols in this paper, because our verifier plays the role of the sender.

If the above hypothesis turns out to be true, then our work suggests that we should be able prove that *any* problem that has a zero-knowledge proof has a concurrent zero-knowledge protocol: simply plug the hypothesized instance-dependent commitment scheme into our constructions. (We do not claim this as a theorem because in this paper, we restrict our attention to instance-dependent commitment schemes that are noninteractive and perfectly binding for simplicity, but the hypothesis mentioned above make no such restriction.)

Outline. Section 2 details some nonstandard notations that are used in this paper. In Sect. 3, we abstract the preamble stage in the Prabhakaran-Rosen-Sahai concurrent zero-knowledge protocol [28, Sect. 3.1], showing how it transforms any noninteractive commitment scheme into one satisfying a desirable extraction property. In Sect. 4, we apply this transformation to *instance-dependent* commitments, and thereby obtaining some of our concurrent zero-knowledge proofs. In Sect. 5, we extend this transformation to problems with *witness-binding* commitments, and thereby obtaining concurrent zero-knowledge proofs for QUADRATIC RESIDUOSITY and GRAPH ISOMORPHISM. Many details and proofs are contained in the full version of this paper [1].

2   Preliminaries

For the most part, we use standard notations found in the theoretical cryptography and complexity theory literature. In the next few paragraphs, we highlight several nonstandard notations used.

Transcript and output of interactive protocols. For an interactive protocol (A, B), let $\langle A, B \rangle(x)$ denote the random variable representing the output of B after interaction with A on common input x. In addition, let $\mathrm{view}_B^A(x)$ denote the random variable representing the content of the random tape of B together with the messages received by B from A during the interaction on common input x.

Committed-verifier zero knowledge. Prabhakaran, Rosen and Sahai [28], in their works on concurrent zero knowledge, showed that adding a $\widetilde{O}(\log n)$-round preamble to a specific form of zero-knowledge protocol (the Hamiltonicity protocol) results in a concurrent zero-knowledge proof system, assuming the existence of a collection of claw-free functions. Alon Rosen, in his PhD thesis, noted that the preamble can be added to a more general form of zero-knowledge protocol, which he informally defines as *challenge-response zero knowledge* [37, Sect. 4.8.1]. We formalize this notion and call it *committed-verifier zero knowledge*.

Definition 1 (committed-verifier zero knowledge). *A committed-verifier V_m, where $m = (m_1, m_2, \ldots, m_k)$, is a deterministic verifier that always sends m_i as its i-th round message.*

An interactive proof (P, V) *for (promise) problem* Π *is* perfect (resp., statistical, computational) committed-verifier zero knowledge (CVZK) *if there exists a probabilistic polynomial-time simulator* S *such that for all committed verifier* V_m, *the ensembles* $\{\text{view}^P_{V_m}(x)\}_{x \in \Pi_Y}$ *and* $\{S(x, m)\}_{x \in \Pi_Y}$ *are perfectly (resp., statistically, computationally) indistinguishable.*

This CVZK property is closely related to notion of *honest-verifier zero knowledge* (HVZK) in that any CVZK protocol is also trivially HVZK. Conversely, any *public-coin* HVZK protocol can be converted into a *public-coin* CVZK protocol by allowing the prover to send random coins m' before the verifier's public-coin message m, and making the prover respond to $m' \oplus m$ (instead of just m).

Lemma 2. *Promise problem* Π *has* public-coin *(perfect/statistical/computational) CVZK proofs if and only if it has* public-coin *(perfect/statistical/computational) HVZK proofs.*

3 Concurrently-Extractable Commitment Scheme

3.1 Overview

A key component in our concurrent zero-knowledge protocols is a commitment scheme with a *concurrent extractability* property. We call this scheme *concurrently-extractable commitment (CEC) scheme.* The notion of concurrent extractability informally means that we are able to simulate the concurrent execution of many sessions between an adversarial sender and the honest receiver in a way that allows us to extract the commitments of the sender in every session.

This notion of concurrent extractability is inspired by the rewinding and simulation strategy of the Prabhakaran-Rosen-Sahai (PRS) [28] concurrent zero-knowledge protocol. The PRS protocol essentially consists of two stages, the preamble (first) stage and the main (second) stage [28, Sect. 3.1]. The concurrent zero knowledge feature of the protocol comes from the preamble stage, in which the verifier is required to commit to the messages that it will use in the main stage. Our goal in this section is to modularize the PRS protocol by abstracting this key feature (preamble stage) that allows for concurrent security.

3.2 Definitions

Standard commitment schemes. A standard (interactive) commitment scheme typically consists of a sender S, a receiver R and a verification algorithm Verify. A message bit $m \in \{0, 1\}$ is given as private input to S, and the common input to both is 1^n, where n is the security parameter. After the interaction $(S(m), R)(1^n)$, R outputs a commitment string c and S outputs a decommitment pair (m, d). (Without loss of generality, we can assume that c is R's view of the interaction and d is S's coin tosses.) The verification algorithm Verify checks that (m, d) is a valid decommitment of c by accepting if it is, and rejecting otherwise.

Commitment schemes with partial verification. To extend standard commitments to concurrently extractable ones, we require an additional verification procedure denoted as Partial-Verify, which is needed for the special binding property (see Definition 6).

Definition 3. *A commitment scheme with partial verification consists of probabilistic polynomial-time algorithms $(S, R, \mathsf{Verify}, \mathsf{Partial\text{-}Verify})$ such that the following conditions hold.*

1. *After the interaction $(S(m), R)(1^n)$, R outputs a commitment string c and S outputs a decommitment pair (m, d).*
2. *For all $(c, (m, d)) \leftarrow (S(m), R)(1^n)$, we have that $\mathsf{Verify}(c, m, d) = 1$.*
3. *For all c, m and d, $\mathsf{Verify}(c, m, d) = 1$ implies $\mathsf{Partial\text{-}Verify}(c, m, d) = 1$.*

A decommitment (m, d) to c with $\mathsf{Verify}(c, m, d) = 1$ is called a *full decommitment*, whereas if we have only that $\mathsf{Partial\text{-}Verify}(c, m, d) = 1$, it is called a *partial decommitment*. Note that a standard commitment scheme is a special case of the above definition by imposing $\mathsf{Partial\text{-}Verify} = \mathsf{Verify}$.

Remark 4. Our above notion of a commitment scheme with partial verification shares some similarities with *mercurial commitments*, a notion recently defined in [38]. For our notion, we have a single kind of commit phase that has two kinds of decommitments, a full decommitment and a partial decommitment. For mercurial commitments, the *hard commitments* correspond to our single commit phase, and thus has two kinds of decommitments; standard decommitments and *tease*. Standard decommitments and tease correspond to full decommitments and partial decommitments, respectively. Mercurial commitments also have a notion of *soft commitments* (that cannot be opened with standard decommitments, but can be teased to any value), which we do not require. Mercurial commitments were defined as a primitive for constructing *zero-knowledge sets* [39].

Statistical hiding and perfect binding. Definition 3 only refers to the syntax of a commitment scheme, and does not impose any security requirements (e.g., hiding and binding). For that, we have the following two definitions.

Definition 5 (hiding). *A commitment scheme with partial verification $(S, R, \mathsf{Verify}, \mathsf{Partial\text{-}Verify})$ is statistically hiding if for every adversarial receiver R^*, the ensembles $\{\langle S(0), R^* \rangle(1^n)\}_{n \in \mathbb{N}}$ and $\{\langle S(1), R^* \rangle(1^n)\}_{n \in \mathbb{N}}$ are statistically indistinguishable.*

The above definition is restricted to statistically hiding since for the purposes of our paper, we will only need to consider statistically hiding commitments. It is straightforward to extend Definition 5 to encompass perfect and computational hiding. Next, we define the *perfectly binding* property for commitment schemes with partial verification. This perfectly binding notion will be used throughout Sect. 4.

Definition 6 (binding). *A commitment scheme with partial verification* $(S, R, \mathsf{Verify}, \mathsf{Partial\text{-}Verify})$ *is perfectly binding if for every commitment c, there do not exist decommitments (m, d) and (m', d') such that $m \neq m'$ and* $\mathsf{Verify}(c, m, d) = \mathsf{Partial\text{-}Verify}(c, m', d') = 1$.

Intuitively the above definition says that a partial decommitment of c to a message m is a proof that c can only be full decommitted to m. Also, observe that Definition 6 implies that the scheme is binding with respect to Verify alone. That is, there do not exist c, (m, d) and (m', d') with $m \neq m'$ and $\mathsf{Verify}(c, m, d) = \mathsf{Verify}(c, m', d') = 1$. But the scheme need not be binding with respect to $\mathsf{Partial\text{-}Verify}$ alone. Hence, the binding property specified in Definition 6 is a natural extension of the binding property of standard commitments (where $\mathsf{Partial\text{-}Verify} = \mathsf{Verify}$).

Concurrent simulatability with extractability. The commitment scheme with partial verification (as in Definition 3) will be used as a building block for our concurrent zero-knowledge protocols in Sects. 4 and 5. For these concurrent zero-knowledge protocols, the prover P and adversarial verifier V^* will play the role of the receiver R and concurrent adversarial sender $\widehat{S}$, respectively. Therefore, we will need to simulate the concurrent interaction between R and $\widehat{S}$, but it turns out this alone is not sufficient. We will also need the simulator to determine partial decommitments of $\widehat{S}$ in every completed session that it has simulated. This property is called *concurrent extractability*, a notion we formalize next.

Definition 7. *A commitment scheme with partial verification* $(S, R, \mathsf{Verify}, \mathsf{Partial\text{-}Verify})$ *is concurrently extractable if there exists a probabilistic polynomial-time simulator* Sim *such that for every $Q \leq \mathrm{poly}(n)$, and for every concurrent adversary $\widehat{S}$ that executes at most Q concurrent sessions, we have:*

1. *(Statistical simulation) The output of* $\mathsf{Sim}^{\widehat{S}}(1^n, 1^Q)$ *is statistically indistinguishable to the output of $\widehat{S}$ in the concurrent interaction* $\langle R, \widehat{S} \rangle (1^n)$.

2. *(Concurrent extractability) Whenever* Sim *queries $\widehat{S}$ on a transcript T, for every completed session s in T with a commitment $c[s]$, it provides partial decommitment $(m[s], d[s])$ such that* $\mathsf{Partial\text{-}Verify}(m[s], c[s], d[s]) = 1$.

For short, we call this a concurrently-extractable *commitment scheme. Also,* Sim *is called the* concurrently-extracting simulator.

Note that we require that the concurrent extractability property hold for all adversaries $\widehat{S}$, even computationally unbounded ones. The only limitation on $\widehat{S}$ is that it executes at most polynomial sessions, which is a natural restriction since it is infeasible to simulate a superpolynomial number of sessions in polynomial time. In addition, the simulator is only required to provide partial decommitments for every completed session. This suffices because a valid partial decommitment (m, d) of a commitment c effectively binds it to the message m if we insist on a full decommitment later on (see Definition 6).

3.3   Construction of Concurrently-Extractable Commitments

A circuit $\mathsf{Com}\colon \{0,1\} \times \{0,1\}^n \to \{0,1\}^n$ can be viewed as a generic (noninteractive) commitment scheme, with n being the security parameter. The commitment to a message bit m is $\mathsf{Com}(m;r)$, where $r \leftarrow \{0,1\}^n$ is a uniformly chosen random key. Likewise, the decommitment of c to a bit m is a pair (m,r) such that $c = \mathsf{Com}(m;r)$. Note that this definition only refers to the syntax of a commitment scheme and does not impose any security requirements (i.e., hiding and binding).

The next lemma states that we can transform any generic commitment scheme $\mathsf{Com}\colon \{0,1\} \times \{0,1\}^n \to \{0,1\}^n$ into a new scheme with the concurrent extractability property. This new scheme is essentially the preamble stage of the PRS concurrent ZK protocol [28], with the sender (verifier) using Com to commit in the $\widetilde{O}(\log n)$ rounds of interaction, and the receiver (prover) just sending random coins.

Lemma 8. *For any generic noninteractive commitment scheme* $\mathsf{Com}\colon \{0,1\} \times \{0,1\}^n \to \{0,1\}^n$, *there is a* concurrently-extractable *commitment scheme* $C_{\mathsf{Com}} = (S_{\mathsf{Com}}, R_{\mathsf{Com}}, \mathsf{Verify}_{\mathsf{Com}}, \mathsf{Partial\text{-}Verify}_{\mathsf{Com}})$ *(taking the circuit* Com *as auxiliary input), such that:*

1. *If* Com *is perfectly binding, then* C_{Com} *is perfectly binding.*
2. *If* Com *is statistically hiding, then* C_{Com} *is statistically hiding.*
3. *$(S_{\mathsf{Com}}, R_{\mathsf{Com}})$ has $\widetilde{O}(\log n)$ rounds of interaction.*

We denote $\mathsf{CEC\text{-}Sim}_{\mathsf{Com}}$ *as the concurrently-extracting simulator for* C_{Com}.

Committing to multi-bit messages. The concurrently-extractable commitment scheme obtained from Lemma 8 is for a single-bit message; to commit to a ℓ-bit message, we independently repeat the scheme ℓ times in parallel. It is important to note that even if we do so, all the properties required in Definition 7 still hold. (Concurrent extractability follows because parallel repetition is a special case of concurrent interaction.) Later in Sect. 4, it will be more convenient to think of $\widehat{S}$ as committing to an ℓ-bit message per session, rather than ℓ senders committing to a single-bit message each.

Finally, when S commits to multi-bit messages, it can full-decommit in multiple steps, one for each committed bit. This is because the full decommitment for each bit of the message is independent of the others.

4   Unconditional Concurrent Zero-Knowledge Proofs for Problems with Instance-Dependent Commitments

In this section, we demonstrate a generic technique for transforming certain standalone public-coin zero-knowledge protocols into concurrent zero-knowledge protocols. In doing so, we construct unconditional concurrent zero-knowledge proofs for

non-trivial problems like QUADRATIC NONRESIDUOSITY, GRAPH NONISOMOR-
PHISM, a variant of STATISTICAL DIFFERENCE and approximate lattice problems.

The main tool used in the transformation is a *instance-dependent commitment
scheme*, formally defined in Definition 9. Later in Sect. 5, we demonstrated
a modified transformation that works for certain problems possessing *witness-
binding commitments*.

4.1   Instance-Dependent Commitments

In order to prevent the adversarial verifier from deviating widely from the original
protocol specification, the previous constructions of concurrent zero-knowledge
protocols require the verifier to commit to certain messages in advance [23,25,28].
While these commitments can be constructed from one-way functions [40,41],
proving the existence of one-way functions remains a major open problem in
complexity theory.

To achieve concurrent security without relying on unproven assumptions, we
observe that the standard verifier's commitments used in [28] can be replaced
by *instance-dependent commitments* [36] (cf., [34]). A instance-dependent commitment, roughly speaking, is a commitment protocol that takes the problem
instance x as an additional input, is binding on the YES instances ($x \in \Pi_Y$), and
is hiding on the NO instances ($x \in \Pi_N$). Standard commitments, by contrast,
are required to always be both hiding and binding regardless of the problem
instance.

Because the hiding and binding properties of instance-dependent commitments depend on the problem instance, we can construct instance-dependent
commitments that are both perfectly binding (on the YES instances) and statistically hiding (on the NO instances).[6] We give a simplified, noninteractive
definition of instance-dependent commitments that suffices for our applications
in this section.

Definition 9. (noninteractive instance-dependent commitment) *Promise
problem* $\Pi = (\Pi_Y, \Pi_N)$ *has a instance-dependent commitment if there exists a
polynomial-time algorithm* PD-Com *such that the following holds.*

1. *Algorithm* PD-Com *takes as input the problem instance* x, *a bit* b, *and a
   random key* r, *and produces a commitment* $c = $ PD-Com$_x(b; r)$. *The running
   time of* PD-Com *is bounded by a polynomial in* $|x|$, *hence without loss of
   generality we can assume that* $|c| = |r| = \mathrm{poly}(|x|)$.
2. *(perfectly binding on* YES *instances) For all* $x \in \Pi_Y$, *the distributions*
   PD-Com$_x(0)$ *and* PD-Com$_x(1)$ *have disjoint supports. That is, there does
   not exist strings* r *and* r' *such that* PD-Com$_x(0; r) = $ PD-Com$_x(1; r')$.
3. *(statistically hiding on* NO *instances) For all* $x \in \Pi_N$, *the commitments to
   0 and 1 are statistically indistinguishable. In other words, the distributions*
   PD-Com$_x(0)$ *and* PD-Com$_x(1)$ *are statistically indistinguishable (w.r.t.* $|x|$,
   the length of the instance).

---

[6] By contrast, standard commitments *cannot* be both statistically binding and statistically hiding.

The commitment c can be decommitted to by sending the committed bit b and random key r. Since both parties have access to the problem instance x, this decommitment can be verified by checking that $c = \mathsf{PD\text{-}Com}_x(b; r)$.

4.2   Main Results

Before presenting the our unconditional concurrent zero-knowledge protocol, we state our main results for this section.

Theorem 10. *If promise problem Π has a public-coin CVZK proof system (P_0, V_0) (in the sense of Definition 1) and also a instance-dependent commitment, then Π has a proof system (P, V) with the following properties:*

1. *If (P_0, V_0) is statistical (resp., computational) zero knowledge, then (P, V) is concurrent statistical (resp., computational) zero knowledge.*
2. *Prover P is black-box simulatable in* strict *polynomial time.*
3. *The round complexity of (P, V) increases only by an additive factor of $\widetilde{O}(\log n)$, with n being the security parameter, compared to the original protocol (P_0, V_0).*
4. *The completeness of (P, V) is exactly the same as that of (P_0, V_0), while the soundness error increases by only a negligible additive term (as a function of n).*
5. *The prover strategy P can be implemented in probabilistic polynomial-time with oracle access to P_0. In particular, if P_0 is efficient, so is P.*

We provide an outline of the proof of Theorem 10 in Sects. 4.3 and 4.4. Several natural problems that Theorem 10 applies to are listed below.

Corollary 11. *The following problems have concurrent statistical zero-knowledge proofs:*

- *The statistical difference problem $\mathrm{SD}^1_{1/2}$.*
- *The languages* QUADRATIC NONRESIDUOSITY *and* GRAPH NONISOMORPHISM.
- *The lattice problems* CO-GAPCVP$_\gamma$ *and* CO-GAPSVP$_\gamma$*, for $\gamma = \Omega(\sqrt{(n/\log n)})$.*

Proof. All the problems listed—$\mathrm{SD}^1_{1/2}$, QUADRATIC NONRESIDUOSITY, GRAPH NONISOMORPHISM, CO-GAPCVP$_\gamma$ and CO-GAPSVP$_\gamma$, for $\gamma = \Omega(\sqrt{(n/\log n)})$ —have honest-verifier statistical zero-knowledge proofs [2,5,42,33], which can be made public-coin by [17]. In addition, they all have instance-dependent commitments [36,34].

The above corollary does not guarantee a polynomial-time prover strategy (with auxiliary input) nor round efficiency. The reason is that the public-coin honest-verifier zero-knowledge proof systems known for these problems do not have a polynomial-time prover nor a subpolynomial number of rounds. For BLUM

QUADRATIC NONRESIDUOSITY,[7] however, we can start with the noninteractive statistical zero-knowledge proof[8] of [43], whose prover is polynomial time (given the factorization of the modulus), and obtain the following:

Corollary 12. *The language* BLUM QUADRATIC NONRESIDUOSITY *has a concurrent statistical zero-knowledge proof systems with* $\widetilde{O}(\log n)$ *rounds and a prover that can be implemented in polynomial time given the factorization of the input modulus.*

We note that we do not expect to obtain efficient provers for GRAPH NONISOMORPHISM or $\mathrm{SD}^1_{1/2}$, since these problems are not known to be in **NP** (or **MA**), which is a prerequisite for an efficient-prover proof system. However, QUADRATIC NONRESIDUOSITY is in **NP** (the factorization of the input is a witness), as are CO-GAPCVP$_\gamma$ and CO-GAPSVP$_\gamma$ for larger approximation factors $\gamma = \Omega(\sqrt{n})$ [44], so we could hope to obtain an efficient prover. The bottleneck is finding *public-coin* honest-verifier zero-knowledge proofs with a polynomial-time prover for these problems.

4.3   Our Concurrent Zero-Knowledge Protocol

A high-level description of our unconditional concurrent zero-knowledge protocol is as follows: We begin with a public-coin CVZK protocol. We make it concurrent zero knowledge by forcing the verifier to commit in advance to its (public-coin) messages in the CVZK protocol using concurrently-extractable commitments C_{Com} provided for by Lemma 8. However, C_{Com} still requires a generic noninteractive commitment scheme Com; for this, we plug-in the instance-dependent commitment scheme $\mathsf{PD\text{-}Com}_x$.

Now, let us formally describe our concurrent zero-knowledge protocol. Let (P_0, V_0) be a public-coin CVZK proof system for Π with $q(|x|)$ rounds on common input x. Denote the messages sent by V_0 in the protocol as $m = (m_1, \ldots, m_q)$, and let $\ell \stackrel{\text{def}}{=} |m|$ be the verifier-to-prover communication complexity. Let $\mathsf{PD\text{-}Com}_x \colon \{0,1\} \times \{0,1\}^n \to \{0,1\}^n$, where $n = \mathrm{poly}(|x|)$, be a instance-dependent commitment for Π.

From Protocol 13 and Lemma 8, we can easily derive the prover efficiency, round complexity and completeness claims of Theorem 10. For soundness, observe that since $\mathsf{PD\text{-}Com}_x$ is statistically hiding, $C_{\mathsf{PD\text{-}Com}_x}$ is also statistically hiding (by Lemma 8). Hence, the soundness of Protocol 13 only decreases by a negligible amount because a cheating prover will not know the committed messages of the verifier until the verifier decommits to m_t (in round t of the main stage).

We show that Protocol 13 is concurrent zero-knowledge by highlighting the main ideas behind its simulation in the next subsection. The full description of our concurrent zero-knowledge protocol (P, V) is next.

---

[7] The problem BLUM QUADRATIC NONRESIDUOSITY is a variant of quadratic residuosity restricted to Blum integers.

[8] Noninteractive zero knowledge implies (in fact is equivalent to) 2-round public-coin honest-verifier zero knowledge since the honest verifier just sends the common random string in the first round, and the prover sends the single-message proof in the second round.

Protocol 13 *Our unconditional concurrent zero-knowledge protocol (P, V) for problem Π with instance-dependent commitments.*

Input: Instance x of Π.

Preamble stage (using instance-dependent commitments)

> Let $C_{\mathsf{PD\text{-}Com}_x} = (S_x, R_x, \mathsf{Verify}_x, \mathsf{Partial\text{-}Verify}_x)$ be the concurrently-extractable commitment scheme provided for by Lemma 8 by substituting $\mathsf{Com} = \mathsf{PD\text{-}Com}_x$.

V : Select a random message $m = (m_1, \ldots, m_q) \leftarrow \{0, 1\}^\ell$.

$V \to P$: Send the message `"start session"`.

$V \leftrightarrow P$: Run the following instance-dependent CEC schemes $(S_x(m_1), R_x)(1^n), \cdots, (S_x(m_q), R_x)(1^n)$ in parallel, with the verifier V acting as S_x and the prover P as R_x.

> Let the output of R_x be the commitments $(c_1, \ldots, c_q)$, and be the output of S_x be the decommitments $((m_1, d_1), \ldots, (m_q, d_q))$. Note that neither P nor V sends the outputs of R_x or S_x to the other party at this stage.

Main stage (stand-alone zero-knowledge protocol)

$V \to P$: Send the message `"start main stage"`.

P : Select randomness $r_{P_0} \leftarrow \{0, 1\}^*$ for the original prover P_0.

> For $t = 1, \ldots, q$, do the following:

$V \to P$: Decommit to m_t by sending full decommitment (m_t, d_t) of c_t.

$P \to V$: Verify the decommitment received is valid by checking if $\mathsf{Verify}(c_t, m_t, d_t) = 1$. If so, answer as the original prover P_0 would, that is, send $\pi_t = P_0(x, m_1, \ldots, m_t; r_{P_0})$. Otherwise, halt and abort.

> Verifier V accepts if the original verifier V_0 accepts on $(m_1, \pi_1, \ldots, m_q, \pi_q)$.

4.4   Our Simulator

Observe that the prover's strategy can be broken into two parts, P_{pre} and P_{main}, denoting the preamble stage and main stage, respectively. Both P_{pre} and P_{main} use independent randomness. The simulation procedure for our concurrent zero-knowledge protocol (Protocol 13) is broken into three main steps.

1. First, we analyze the concurrent interaction of P and V^* in the context of concurrently-extractable commitment schemes (provided for by Lemma 8, substituting $\mathsf{Com} = \mathsf{PD\text{-}Com}_x$). To do so, we define a new adversarial sender

$\widehat{S}$ that takes V^* and P_{main} as oracles and only returns the preamble messages of V^*. The preamble stage prover P_{pre} acts as the honest receiver R_x. By Definition 7 and Lemma 8, we can simulate the output of $\widehat{S}$ (after interaction with P_{pre}), while having the additional property of being able to extract the commitments.

By virtue of the way we defined $\widehat{S}$, its output after concurrently interacting with P_{pre} is equivalent to the output of V^* after concurrently interacting with P. Nevertheless, this simulation is inefficient because $\widehat{S}$ uses an oracle for P_{main}.

2. Since we can extract partial decommitments, we are able to determine the verifier's main stage messages in advance.[9] Hence, we can replace the adaptive queries to P_{main} by a single query made to a new oracle, called $\mathcal{O}_P$, at the start of each main stage.

3. However, $\mathcal{O}_P$ is still not an efficiently implementable oracle. In the final step, we replace oracle $\mathcal{O}_P$ with a committed-verifier zero knowledge (CVZK) simulator S_{CVZK} to obtain an efficient simulation strategy.

5 Unconditional Concurrent Zero-Knowledge Proofs for Problems with Witness-Binding Commitments

Here we extend the techniques in Sect. 4 to obtain unconditional concurrent statistical zero-knowledge proofs for certain problems like QUADRATIC RESIDUOSITY and GRAPH ISOMORPHISM. These problems are not known to have instance-dependent commitments (in the sense of Definition 9), but have a variant of instance-dependent commitments called *witness-binding commitments* (see Sect. 5.1). Informally, these commitments are not guaranteed to be perfectly binding but breaking the binding property of these commitments is as hard as finding a witness.

Using these witness-binding commitments, we proceed to transform them into ones with the concurrently extractability property. (In Sect. 3.3 we did a similar transformation for standard instance-dependent commitments.) Our concurrent zero-knowledge protocol combines the witness-binding concurrently-extractable commitments with an underlying stand-alone ZK protocol.

Recall that in Sect. 4, we required the stand-alone protocol to be committed-verifier zero knowledge (CVZK), as in Definition 1. However, since we are using only witness-binding commitments, we require the underlying stand-alone protocol to have a stronger property that we call *witness-completable* CVZK (see Sect. 5.2). The additional witness-completable property, informally stated, gives our simulator the ability to complete the simulation even when the verifier sends a message different from its committed one, if we provide our simulator with a valid witness at that time. This is important because the binding property of witness-binding commitments can be broken, but if that is the case, the simulator can obtain a witness that it can use to complete the simulation.

---

[9] The binding property in the sense of Definition 6 allows us to determine the committed message in any valid full decommitment by just knowing a partial decommitment.

5.1 Witness-Binding Commitments

Based on the techniques used in Sect. 4, the first natural step towards constructing concurrent zero-knowledge protocols would be to construct instance-dependent
commitments. Consider the naive commitment scheme for GRAPH ISOMORPHISM
specified as follows: Let (G_0, G_1) be an instance of the problem. To commit to bit b,
send a random isomorphic copy of G_b. This commitment is perfectly hiding on the
YES instances (when $G_0 \cong G_1$) and perfectly binding on the NO instances (when
$G_0 \not\cong G_1$). However, this is exactly the opposite of what we require in a instance-
dependent commitment (see Definition 9). In fact, every problem satisfying Definition 9 is in **coNP**, but GRAPH ISOMORPHISM is not known to be in **coNP**.

Protocol 14 *Witness-binding commitment scheme for* GRAPH
ISOMORPHISM *(implicit in [35]).*

To commit to bit b using problem instance (G_0, G_1), proceed as
follows.

Index generation stage
    $R \to S$: Let H_1 be a random isomorphic copy of G_0, and send
        H_1. That is, $H_1 = \sigma(G_0)$ for a random permutation
        σ of the vertices of G_0. In addition, both parties set
        $H_0 = G_0$.

Commitment stage
    $S \to R$: To commit to bit b, send F, a random isomorphic
        copy of H_b.

Decommitment stage
    $S \to R$: To decommit, send b together with the isomorphism
        between H_b and F.

Verification stage
    After the decommitment stage, the receiver R_x proves that H_1,
    sent in the index generation stage, is isomorphic to G_0 by sending
    the isomorphism σ between G_0 and H_1.

To overcome this apparent difficulty, the above commitment scheme (Protocol 14) makes use of additional index generation and verification stages to
do instance-dependent commitments. It can be shown that this witness-binding
commitment scheme is perfectly hiding on every instance (in particular the NO
instances) if H_1 is generated correctly, that is if $H_1 \cong G_0$. On the YES instances,
the scheme is "computationally binding" in that breaking the scheme is as hard
as finding an **NP**-witness (an isomorphism between G_0 and G_1). More precisely,
we can extract the witness if we use a *simulated* index generation stage, where H_1
is taken to be a random isomorphic copy of G_1 (which is distributed identically
to the actual index generation).

This scheme can be generalized to a number of other **NP** languages, and a formal definition capturing the notion of witness-binding commitments is in the full version of this paper [1]. In addition, we note that QUADRATIC RESIDUOS-ITY has a similarly structured witness-binding commitment scheme (based on Protocol 14 and its 3-round perfect zero-knowledge proof system [2]).

5.2   Witness-Completable CVZK

Recall that witness-completable CVZK (wCVZK) is a strengthening of the notion of CVZK (Definition 1) in that our simulator, when given a valid witness, must have the ability to complete the simulation even when the verifier sends a message different from its committed one. The formal definition of wCVZK is the full version of this paper [1].

The 3-round perfect zero-knowledge protocols for both QUADRATIC RESIDU-OSITY [2] and GRAPH ISOMORPHISM [5] turns out to have the witness-completable property, as desired.

5.3   Main Results

Our main result for this section can be summarized in a very similar manner as Theorem 10 in Sect. 4.2. The main differences are (1) the promise problem Π needs to have a witness-binding commitment scheme and a 3-round, public-coin, wCVZK proof system (instead of instance-dependent commitment scheme and CVZK proof system), and (2) our new simulation runs in *expected* polynomial time instead of strict polynomial time. With that, we obtain the following theorem.

Theorem 15. *Both languages* GRAPH ISOMORPHISM *and* QUADRATIC RESID-UOSITY *have concurrent statistical zero-knowledge proof systems with* $\widetilde{O}(\log n)$ *rounds and efficient provers. The simulator for both protocols runs in* expected *polynomial time.*

Note that the round complexity of $\widetilde{O}(\log n)$ for the concurrent zero-knowledge protocols of both GRAPH ISOMORPHISM and QUADRATIC RESIDUOSITY is essentially optimal for black-box simulation [27].

5.4   Our Modified Concurrent Zero-Knowledge Protocol

Since we are dealing with witness-binding commitments, we have to modify Protocol 13 in Sect. 4.3. Our modified concurrent zero-knowledge protocol is similar in structure with the main difference being that instead of just the preamble stage and the main stage, it also an index generation stage before the preamble stage and a verification stage after the main stage (for implementing the corresponding stages of the witness-binding commitment scheme). The full description of our modified protocol is in the full version of this paper [1].

5.5 Our Simulator

Recall the three main steps of the simulation procedure in Sect. 4.4.

1. Analyze the concurrent interaction of P and V^* in the context of the concurrently-extractable commitment schemes. Specifically, define a new adversarial sender $\widehat{S}$ that takes V^* and P_{main} as oracles and only returns the preamble messages of V^*, and simulate its interaction while extracting its commitments.
2. Replace the adaptive queries to P_{main} by a single query made to a new oracle, called $\mathcal{O}_P$, at the start of each main stage.
3. Replace oracle $\mathcal{O}_P$ with a CVZK simulator S_{CVZK} to obtain an efficient simulation strategy.

For the simulation of our modified concurrent zero-knowledge protocol, we keep Step 1 the same, but in Step 2 observe that the prover responses provided by $\mathcal{O}_P$ depends on the witness w given the to prover. Hence, we denote it more precisely as $\mathcal{O}_{P(w)}$. In Step 3, we simulate the answers from $\mathcal{O}_{P(w)}$ with our wCVZK simulator. However, our wCVZK simulator needs a witness w in order to continue the simulation when the verifier's V^* response does not match the expectation of our simulator.

This can only happen if V^* breaks the binding of the witness-binding commitment. And when that happens, our simulator is able to obtain a witness w, which it can then feed to the wCVZK simulator to continue the simulation. Actually, a subtlety is that the witness-binding commitment allows us to extract a witness only if we *simulate* the index generation stage, whereas here we need to run the actual index generation in order to complete the verification stage. Thus, if needed, we run a separate offline process to extract a witness, and this is what causes our simulator to run in expected polynomial time. For details, see the full version of this paper [1].

Acknowledgements. We thank Alexander Healy, Manoj Prabhakaran and Alon Rosen for helpful discussions.

References

1. Micciancio, D., Ong, S.J., Sahai, A., Vadhan, S.: Concurrent zero knowledge without complexity assumptions. Technical Report 05-093, Electronic Colloquium on Computational Complexity (2005)
   http://eccc.uni-trier.de/eccc-reports/2005/TR05-093/.
2. Goldwasser, S., Micali, S., Rackoff, C.: The knowledge complexity of interactive proof systems. SIAM Journal on Computing **18**(1) (1989) 186–208
3. Goldreich, O., Krawczyk, H.: On the composition of zero-knowledge proof systems. SIAM Journal on Computing **25**(1) (1996) 169–192
4. Dwork, C., Naor, M., Sahai, A.: Concurrent zero-knowledge. In: Proc. 30th STOC. (1998) 409–418

5. Goldreich, O., Micali, S., Wigderson, A.: Proofs that yield nothing but their validity or all languages in NP have zero-knowledge proof systems. Journal of the ACM **38**(1) (1991) 691–729
6. Goldreich, O.: Foundations of cryptography. Volume 1. Cambridge University Press, Cambridge, UK (2001)
7. Goldreich, O.: Zero-knowledge twenty years after its invention. http://www.wisdom.weizmann.ac.il/~oded/zk-tut02.html (2002)
8. Yao, A.C.: How to generate and exchange secrets. In: Proc. 27th FOCS. (1986) 162–167
9. Naor, M., Yung, M.: Public-key cryptosystems provably secure against chosen ciphertext attack. In: Proc. 22nd STOC. (1990) 427–437
10. Dolev, D., Dwork, C., Naor, M.: Nonmalleable cryptography. SIAM Journal on Computing **30**(2) (2001) 391–437
11. Sahai, A.: Non-malleable non-interactive zero knowledge and adaptive chosen-ciphertext security. In: Proc. 40th FOCS. (1999) 543–553
12. Feige, U., Fiat, A., Shamir, A.: Zero-knowledge proofs of identity. Journal of Cryptology **1**(2) (1988) 77–94
13. Cramer, R., Shoup, V.: A practical public key cryptosystem provably secure against adaptive chosen ciphertext attack. In: Proc. CRYPTO '98. (1998) 13–25
14. Elkind, E., Sahai, A.: A unified methodology for constructing public-key encryption schemes secure against adaptive chosen-ciphertext attack. Cryptology ePrint Archive, Report 2002/042 (2002) http://eprint.iacr.org/.
15. Cramer, R., Shoup, V.: Design and analysis of practical public-key encryption schemes secure against adaptive chosen ciphertext attack. SIAM Journal on Computing **33**(1) (2004) 167–226
16. Gennaro, R., Micciancio, D., Rabin, T.: An efficient non-interactive statistical zero-knowledge proof system for quasi-safe prime products. In: Proc. of the 5th ACM Conference on Computer and Communications Security. (1998) 67–72
17. Okamoto, T.: On relationships between statistical zero-knowledge proofs. Journal of Computer and System Sciences **60**(1) (2000) 47–108
18. Goldreich, O., Sahai, A., Vadhan, S.: Honest-verifier statistical zero-knowledge equals general statistical zero-knowledge. In: Proc. 30th STOC. (1998) 399–408
19. Vadhan, S.: An unconditional study of computational zero knowledge. In: Proc. 45th STOC. (2004) 176–185
20. Ostrovsky, R.: One-way functions, hard on average problems, and statistical zero-knowledge proofs. In: Proceedings of the Sixth Annual Structure in Complexity Theory Conference. (1991)
21. Ostrovsky, R., Wigderson, A.: One-way functions are essential for non-trivial zero-knowledge. In: Second Israel Symposium on Theory of Computing Systems. (1993) 3–17
22. Feige, U.: Alternative models for zero knowledge interactive proofs. PhD thesis, Weizmann Institute of Science, Israel (1990)
23. Richardson, R., Kilian, J.: On the concurrent composition of zero-knowledge proofs. In: Proc. EUROCRYPT '99. (1999) 415–431
24. Kilian, J., Petrank, E., Rackoff, C.: Lower bounds for zero knowledge on the Internet. In: Proc. 39th FOCS. (1998) 484–492
25. Kilian, J., Petrank, E.: Concurrent and resettable zero-knowledge in poly-logarithm rounds. In: Proc. 33rd STOC. (2001) 560–569
26. Rosen, A.: A note on the round-complexity of concurrent zero-knowledge. In: Proc. CRYPTO '00. (2000) 451–468

27. Canetti, R., Kilian, J., Petrank, E., Rosen, R.: Black-box concurrent zero-knowledge requires (almost) logarithmically many rounds. SIAM Journal on Computing **32**(1) (2003) 1–47
28. Prabhakaran, M., Rosen, A., Sahai, A.: Concurrent zero knowledge with logarithmic round-complexity. In: Proc. 43rd FOCS. (2002) 366–375
29. Barak, B.: How to go beyond the black-box simulation barrier. In: Proc. 42nd FOCS. (2001) 106–115
30. Di Crescenzo, G.: Removing complexity assumptions from concurrent zero-knowledge proofs. In: Proc. 6th COCOON. (2000) 426–435
31. Micciancio, D., Petrank, E.: Simulatable commitments and efficient concurrent zero-knowledge. In: Proc. EUROCRYPT '03. (2003) 140–159
32. Brassard, G., Chaum, D., Crepeau, C.: Minimum disclosure proofs of knowledge. Journal of Computer and System Sciences **37**(2) (1988) 156–189
33. Sahai, A., Vadhan, S.: A complete problem for statistical zero knowledge. Journal of the ACM **50**(2) (2003)
34. Micciancio, D., Vadhan, S.: Statistical zero-knowledge proofs with efficient provers: lattice problems and more. In: Proc. CRYPTO '03. (2003) 282–298
35. Bellare, M., Micali, S., Ostrovsky, R.: Perfect zero-knowledge in constant rounds. In: Proc. 22nd STOC. (1990) 482–493
36. Itoh, T., Ohta, Y., Shizuya, H.: A language-dependent cryptographic primitive. Journal of Cryptology **10**(1) (1997) 37–49
37. Rosen, A.: The Round-Complexity of Black-Box Concurrent Zero-Knowledge. PhD thesis, Weizmann Institute of Science, Israel (2003)
38. Chase, M., Healy, A., Lysyanskaya, A., Malkin, T., Reyzin, L.: Mercurial commitments with applications to zero-knowledge sets. In: Proc. EUROCRYPT '05. (2005) 422–439
39. Micali, S., Rabin, M.O., Kilian, J.: Zero-knowledge sets. In: Proc. 44th FOCS. (2003) 80–91
40. Naor, M.: Bit commitment using pseudorandomness. Journal of Cryptology **4**(2) (1991) 151–158
41. Hastad, J., Impagliazzo, R., Levin, L.A., Luby, M.: A pseudorandom generator from any one-way function. SIAM Journal on Computing **28**(4) (1999) 1364–1396
42. Goldreich, O., Goldwasser, S.: On the limits of nonapproximability of lattice problems. Journal of Computer and System Sciences **60**(3) (2000) 540–563
43. Blum, M., De Santis, A., Micali, S., Persiano, G.: Noninteractive zero-knowledge. SIAM Journal on Computing **20**(6) (1991) 1084–1118
44. Aharonov, D., Regev, O.: Lattice problems in NP $\cap$ coNP. In: Proc. 45th FOCS. (2004) 362–371

Interactive Zero-Knowledge with Restricted Random Oracles

Moti Yung[1] and Yunlei Zhao[2]

[1] RSA Laboratories and Department of Computer Science,
Columbia University, New York, NY, USA
moti@cs.columbia.edu
[2] Software School, School of Information Science and Engineering,
Fudan University, Shanghai 200433, China
ylzhao@fudan.edu.cn

Abstract. We investigate the design and proofs of zero-knowledge (ZK) interactive systems under what we call the "restricted random oracle model" which restrains the usage of the oracle in the protocol design to that of collapsing protocol rounds a la Fiat-Shamir heuristics, and limits the oracle programmability in the security proofs. We analyze subtleties resulting from the involvement of random oracles in the interactive setting and derive our methodology. Then we investigate the Feige-Shamir 4-round ZK argument for $\mathcal{NP}$ in this model: First we show that a 2-round protocol is possible for a very interesting set of languages; we then show that while the original protocol is not concurrently secure in the public-key model, a modified protocol in our model is, in fact, concurrently secure in the bare public-key model. We point at applications and implications of this fact. Of possible independent interest is a concurrent attack against the Feige-Shamir ZK in the public-key model (for which it was not originally designed).

1 Introduction

The basic random oracle (RO) methodology was originally introduced in [2] as an idealization and abstraction of the Fiat-Shamir heuristics [14] that transforms Σ-protocols (i.e., 3-round public-coin special honest verifier zero-knowledge SHVZK protocols) into signature schemes. The methodology was used to achieve non-interactive schemes (signatures, public-key encryption and non-interactive zero-knowledge NIZK). However, nowadays more and more complicated *interactive* protocols are developed employing random oracle (one example is the recent direct anonymous attestation (DAA) scheme [3] developed for industrial use).

In this work, we show that the design of *interactive* schemes with advanced security notions using the normal random oracle methodology is more subtle than is typically believed. The subtlety lies in the programmability of the RO in security proofs. Namely, in security proofs the simulator defines output of query values and in fact "programs" the RO on any query, in particular *any query it chooses*. We further investigate the usages of RO in protocol designs.

S. Halevi and T. Rabin (Eds.): TCC 2006, LNCS 3876, pp. 21–40, 2006.
© Springer-Verlag Berlin Heidelberg 2006

For a protocol developed in the RO model, we consider the sensitivity of the security of the protocol when the RO is replaced (realized) by practical (hash) functions, showing that different usages lead to different sensitivities (say, to future cryptanalysis of the function).

We attempt to minimize the usage of the ROs both in **protocol designs** and in **security proofs**. This is naturally desired property. In our approach we impose the following restrictions:

- We limit the usage of ROs in security proofs by letting the honest player and the simulator (who plays the role of the honest player) use a *non-programmable* RO.
- Furthermore, the non-programmable RO could be replaced by *any* hash function without compromising the security of the *honest* player.
- Finally, we limit the usage of ROs in protocol designs to collapsing Σ-protocols just as in the original Fiat-Shamir methodology.

We note that on one hand, it is desirable to minimize using the truly random function property of real hash functions in protocol designs (due to its idealized nature and its unrealization within certain constructions, e.g., [5, 6, 22, 18, 1]). On the other hand, we justify the original Fiat-Shamir methodology by the simple (yet, we believe important) observation that even very weak hash functions (clearly not collision-resistant and not pseudorandom) can be used to collapse Σ-protocols into non-interactive ones with remarkable security guarantee (not ZK but nevertheless a useful property that can be employed).

In this work, we refer to protocols, which are developed with limiting ROs in security proofs and protocols designs according to our approach (which we motivate herein), as *protocols with restricted ROs*.

Although our approach of restricted ROs is seemingly very limiting, it turns out to be still very powerful for achieving practical *interactive* cryptographic schemes with a (seemingly) better balance between "(idealized) provable" security and implementation efficiency. In particular, we show that a Σ_{OR}-based practical implementation (without $\mathcal{NP}$-reductions) of the Feige-Shamir 4-round ZK arguments (the version that appears in [13]) can give us *generic yet practical* 2-round (that is optimal) ZK systems with restricted ROs.

We then investigate the *concurrent security* of the Feige-Shamir ZK protocol. We show that (perhaps surprisingly) the Feige-Shamir ZK protocol is, in general, *not* concurrently secure in public-key settings (when users possess public keys), by identifying a concurrent attack. Though it may look quite natural to run the Feige-Shamir ZK protocol in public-key models when it is used in practice and perhaps to do it concurrently (even though the protocol was not designed for concurrency), our attack shows that this intuition is wrong. Fortunately, the Feige-Shamir ZK protocol is concurrently secure with restricted ROs in the bare public-key (BPK) model. In this process, we also identify and clarify complications and subtleties in dealing with concurrent adversaries in the setting of *interactive* zero-knowledge with restricted ROs.

We remark that the 2-round ZK systems with restricted ROs (with or without registered public-keys) can be used to construct more complicated interactive systems with restricted ROs. It can also be used to transform a large number of existing interactive schemes, which are developed originally with the normal random oracle methodology, into schemes with restricted ROs, by adding *at most one additional round* but with seemingly stronger security guarantees.

Finally, as part of this work, two constructions are given which are of independent interest and may be worthy of further explorations: a one-round witness-hiding (WH) protocol for DLP, and a concurrent attack against the Feige-Shamir ZK when it is run in the new setting of public-key models.

2   Preliminaries

In this section we review the major cryptographic tools used, and present a key observation on non-interactive Σ_{OR}-protocols with ROs.

Definition 1 (Σ-protocol [7]). *A 3-round public-coin protocol $\langle P, V \rangle$ is said to be a Σ-protocol for a relation R if the following hold:*

- *Completeness. If P, V follow the protocol, the verifier always accepts.*
- *Special soundness. From any common input x of length n and any pair of accepting conversations on input x, (a, e, z) and (a, e', z') where $e \neq e'$, one can efficiently computes w such that $(x, w) \in R$. Here a, e, z stand for the first, the second and the third message respectively and e is assumed to be a string of length t (that is polynomially related to n) selected uniformly at random in $\{0, 1\}^t$.*
- *Special honest verifier zero-knowledge (SHVZK). There exists a probabilistic polynomial-time (PPT) simulator S, which on input x and a random challenge string e, outputs an accepting conversation of the form (a, e, z), with the same probability distribution as the real conversation between the honest P, V on input x.*

Σ-protocols have been proved to be a very powerful cryptographic tool and are widely used. Many basic Σ-protocols have been developed, and the following are Σ-protocol examples for DLP and RSA.

Σ-Protocol for DLP [24]. The following is a Σ-protocol $\langle P, V \rangle$ proposed by Schnorr [24] for proving the knowledge of discrete logarithm, w, for a common input of the form (p, q, g, h) such that $h = g^w \bmod p$, where on a security parameter n, p is a uniformly selected n-bit prime such that $q = (p - 1)/2$ is also a prime, g is an element in $\mathbf{Z}_p^*$ of order q. It is also actually the first efficient Σ-protocol proposed in the literature.

- P chooses r at random in $\mathbf{Z}_q$ and sends $a = g^r \bmod p$ to V.
- V chooses a challenge e at random in $\mathbf{Z}_{2^t}$ and sends it to P. Here, t is fixed such that $2^t < q$.
- P sends $z = r + ew \bmod q$ to V, who checks that $g^z = ah^e \bmod p$, that p, q are prime and that g, h have order q, and accepts iff this is the case.

Σ-Protocol for RSA [19]. Let n be an RSA modulus and q be a prime. Assume we are given some element $y \in Z_n^*$, and P knows an element w such that $w^q = y$ mod n. The following protocol is a Σ-protocol for proving the knowledge of q-th roots modulo n.

- P chooses r at random in Z_n^* and sends $a = r^q \mod n$ to V.
- V chooses a challenge e at random in Z_{2^t} and sends it to P. Here, t is fixed such that $2^t < q$.
- P sends $z = rw^e \mod n$ to V, who checks that $z^q = ay^e \mod n$, that q is a prime, that $gcd(a, n) = gcd(y, n) = 1$, and accepts iff this is the case.

The OR-proof of Σ-protocols [8]. One basic construction with Σ-protocols allows a prover to show that given two inputs x_0, x_1, it knows a w such that either $(x_0, w) \in R_0$ or $(x_1, w) \in R_1$, but without revealing which is the case. Specifically, given two Σ-protocols $\langle P_b, V_b \rangle$ for R_b, $b \in \{0, 1\}$, with random challenges of (without loss of generality) the same length t, consider the following protocol $\langle P, V \rangle$, which we call Σ_{OR}. The common input of $\langle P, V \rangle$ is (x_0, x_1) and P has a private input w such that $(x_b, w) \in R_b$.

- P computes the first message a_b in $\langle P_b, V_b \rangle$, using x_b, w as private inputs. P chooses e_{1-b} at random, runs the SHVZK simulator of $\langle P_{1-b}, V_{1-b} \rangle$ on input (x_{1-b}, e_{1-b}), and let $(a_{1-b}, e_{1-b}, z_{1-b})$ be the output. P finally sends a_0, a_1 to V.
- V chooses a random t-bit string s and sends it to P.
- P sets $e_b = s \oplus e_{1-b}$ and computes the answer z_b to challenge e_b using (x_b, a_b, e_b, w) as input. He sends (e_0, z_0, e_1, z_1) to V.
- V checks that $s = e_0 \oplus e_1$ and that conversations (a_0, e_0, z_o), (a_1, e_1, z_1) are accepting conversations with respect to inputs x_0, x_1, respectively.

Theorem 1. [8] *The protocol Σ_{OR} above is a Σ-protocol for R_{OR}, where $R_{OR} = \{((x_0, x_1), w) | (x_0, w) \in R_0$ or $(x_1, w) \in R_1\}$. Moreover, for any malicious verifier V^*, the probability distribution of conversations between P and V^*, where w is such that $(x_b, w) \in R_b$, is independent of b. That is, Σ_{OR} is perfectly witness indistinguishable (WI).*

Given access to a random oracle (RO) $\mathcal{O}$, we can transform a Σ-protocol into a non-interactive protocol, which in this work we call non-interactive Σ-protocol in the RO model. On a common input x, an auxiliary input aux and a private witness w, the prover then generates the first message a, queries $\mathcal{O}$ with (x, a, aux) to get the challenge e and then computes the answer z. The proof is then (a, z, aux). To verify such a proof, query $\mathcal{O}$ with (x, a, aux) to get e and then run the verifier of the original Σ-protocol. The transformed non-interactive protocol is *zero-knowledge proof of knowledge* in the random oracle model [2, 25]. The key observation here (which is simple yet powerful) is the following claim:

Claim 1. *The non-interactive Σ_{OR}-protocol in the RO model remains witness-indistinguishable even if the random oracle is replaced by any real hash function.*

Proof. Note that the WI property of Σ_{OR} is with respect to *any* malicious verifier. In particular, the WI property holds for a specific malicious verifier that uses a real hash function to generate the challenge in Round-2.

3 Restricted ROs: Motivation and Discussions

In this section, we first provide some motivating examples along with discussions and comments. Then, in light of the motivating examples and discussions, we give some desirable principles for limiting the uses of ROs in security proofs and protocol designs that are naturally derived from the motivation.

3.1 Motivating Examples and Discussions

A motivating example for limiting the uses of RO in security proofs. Consider the following protocol depicted in Figure-1.

We note that the zero-knowledge property of the protocol of Figure-1 can be easily shown *in the normal random oracle model*, where the ZK simulator simulates (programs) the RO (i.e, the simulator defines the outputs of the random oracle on any queries, in particular *any queries it chooses*). The proof is omitted here due to space limitation. But we argue that this protocol (though simulatable (in the programmable oracle case, which is what the proof shows) is not

---

Common input. An element $x \in L$ of length n, where L is an $\mathcal{NP}$-language that admits Σ-protocols.

P's private input. A witness w for $x \in L$.

Random oracle. A random oracle denoted $\mathcal{O}$.

Round-1. The verifier V selects a OWF f_V that admits Σ-protocols, randomly selects two elements in the domain of f_V, x_V^0 and x_V^1, computes $y_V^0 = f_V(x_V^0)$ and $y_V^1 = f_V(x_V^1)$, randomly selects a bit b from $\{0,1\}$, sends to the prover a non-interactive Σ_{OR}-proof on (y_V^0, y_V^1), denoted $\pi_V = (y_V^0, y_V^1, a_V, e_V, z_V, aux_V)$, that it knows either the preimage of y_V^0 or the preimage of y_V^1. The witness used by V in forming π_V is x_V^b. The random challenge e_V is generated by querying $\mathcal{O}$ with $(x, y_V^0, y_V^1, a_V, aux_V)$ where aux_V is the auxiliary information of V that possibly includes a time stamp.

Round-2. The prover P first checks the validity of π_V and aborts if it is not valid. Otherwise, P randomly select a bit b' from $\{0,1\}$, sends back a non-interactive Σ_{OR}-proof on (x, y_V^0, y_V^1, b'), denoted $\pi_P = (a_P, e_P, z_P, b')$, that it knows either a witness for $x \in L$ or the preimage of $y_V^{b'}$. The witness used by P is its private input w. The random challenge e_P is generated by querying the random oracle with (x, a_P, b', π_V).

Verifier's Decision. The verifier checks the validity of π_P and accepts if it is valid, otherwise it rejects.

Fig. 1. An insecure ZK protocol in the normal random oracle model

really ZK even in the random oracle model. Observe the following attack: On an input y_{V^*} that the adversary V^* does not know the preimage of under the OWF f_{V^*} (selected by V^* in Round-1), the adversary V^* additionally randomly selects a x'_{V^*} from the domain of f_{V^*} and computes $y'_{V^*} = f_{V^*}(x'_{V^*})$. Then, V^* randomly selects a bit b from $\{0, 1\}$, sets $y^b_{V^*}$ to be y'_{V^*} and $y^{1-b}_{V^*}$ to be y_{V^*}. Finally, V^* sends to the honest prover a non-interactive Σ_{OR}-proof on $(y^0_{V^*}, y^1_{V^*})$ that it knows either the preimage of $y^0_{V^*}$ or the preimage $y^1_{V^*}$ by using x'_{V^*} as the witness. Then according to the perfect witness indistinguishability of Round-1, with probability $1/2$ the honest prover will select $b' = 1 - b$ in Round-2. That is, with probability $1/2$ V^* will get a non-interactive Σ_{OR}-proof for showing the knowledge of either the witness for $x \in L$ or the preimage of $y^{1-b}_{V^*} = y_{V^*}$, a knowledge of such a witness cannot be generated by V^* alone without interacting with P. In other words, the honest prover divulges (seemingly significantly) valuable "knowledge" to the above malicious verifier.

Note that the above protocol is a very simple interactive protocol in the random oracle model, and so we can easily identify the above simple attack. When we construct much more complicated *interactive* schemes in the random oracle model, the security analyses might be much more complicated and subtle. In light of the above attack, we believe that for interactive protocols in the random oracle model, letting the ZK simulator define the random outputs of the random oracle on queries *it chooses* may be too artificial to reflect the real power of the malicious verifier even in the random oracle model.

According to the above arguments, for proving the security of *interactive* protocols in the random oracle model we should restrict the power of the simulator in defining the random outputs of the random oracle. Also note that in a much more complicated interactive scheme in the random oracle model, where both the prover and the verifier prove using the non-interactive Σ-protocols, the provers may actually be the verifier of the high-level complicated interactive protocol.

Comment. Note that it is easy to check that the protocol depicted in Figure-1 is not zero-knowledge if the simulator (who plays the role of the honest prover) uses a non-programmable random oracle.

Motivating examples for limiting the uses of RO in protocol designs
Random oracles have been employed in many ways. Consider, for example, the following commitment scheme in the random oracle model employed in [23]: To commit to a message m, the commitment sender randomly picks a random string r and sends $RO(x, r)$. The security (both binding and hiding) can be easily checked in the random oracle model. But in practice when the random oracle RO is replaced by real practical hash functions, the security properties of this commitment scheme are (critically) sensitive to the realization. The zero-knowledge protocol with ROs developed in [23] critically uses the above commitment scheme. This means that for a large complicated cryptographic system built with the ZK protocol of [23] as a building block, the security of the *whole* system will also be (critically) dependent on the assumed random-function property of the underlying hash functions. This means it will be critically sensitive to

the realizations of the underlying practical hash functions used, which we would like to avoid in certain critical settings (e.g., if the realization is cryptanalyzed in the future and the binding property is lost, say).

So, what uses of random oracle are less sensitive to the case that the hash function realizing it is cryptanalyzed (perhaps in the future)? We justify the original Fiat-Shamir methodology by showing that even very weak hash functions can be used to collapse Σ-protocols into non-interactive ones with remarkable security guarantee. Consider the following one-round witness hiding (WH) protocol $\langle P, V \rangle$ for DLP.

Common input. (p, q, g, h), where on a security parameter n, p is a uniformly selected n-bit prime such that $q = (p - 1)/2$ is also a prime, g and h are elements in $\mathbf{Z}_p^*$ of order q.

P's private input. w such that $h = g^w \bmod p$.

The protocol. P chooses r at random in $\mathbf{Z}_q$, computes $a = g^r \bmod p$. If a is an even number then let $e = \frac{1}{2}a \bmod p$ and if a is an odd number then let $e = \frac{1}{2}(a - 1) \bmod p$ (this guarantees that $e \in \mathbf{Z}_q$). Then P computes $z = r + ew \bmod p$. Finally P sends (a, z) to V.

V's decision. V computes e from a and checks that $g^z = ah^e \bmod p$, that p, q are prime and that g, h, a have order q, and accepts iff this is the case.

The above protocol can be viewed as the non-interactive version of Schnorr's Σ-protocol for DLP [24] when the random oracle is replaced by the following hash function H: for any strings x, y in $\{0, 1\}^*$ and $e \in \mathbf{Z}_q$, $H(xe0) = H(ye1) = e$. Clearly this hash function is not collision-resistant and not pseudorandom. But this extremely weak hash function still provides remarkable security guarantee for the above transformed non-interactive protocol.

We first note that under the DLP hardness assumption, the above non-interactive protocol is witness hiding (WH) for DLP. Specifically, suppose with non-negligible probability a PPT adversary can produce w from (a, z), then the adversary can also compute $\log_g(a)$ for a random a in $\mathbf{Z}_p^*$ of order q, which violates the DLP hardness assumption.

Now, we consider the soundness. We want to argue that if a malicious prover P^* does not know w, then it should not give the correct pair (a, z) such that $g^z = ah^{\frac{1}{2}a}$ for even a, or $g^z = ah^{\frac{1}{2}(a-1)}$ for odd a. Suppose P^* does not know w but can successfully produce (a, z), then it must be the case that a is a hard instance of DLP and P^* does not know $\log_g(a)$ (since otherwise P^* can compute w from $\log_g a$), which seems infeasible. In particular, based on the following specifically tailored but seemingly hard (and reasonable) assumption the soundness of the above non-interactive protocol holds. (*Note that the WH property does not rely on the new assumption.*)

Hardness assumption. Given (p, q, g, h) of the above form, no PPT algorithm A can with non-negligible probability produce a pair (a, z) such that $g^z = ah^{\frac{1}{2}a}$ for even a or $g^z = ah^{\frac{1}{2}(a-1)}$ for odd a, where $a \in \mathbf{Z}_p^*$ of order q and $z \in \mathbf{Z}_q$. (Note that this assumption implies that a is a hard instance of $\mathbf{Z}_p^*$ and the producer A itself also does not know $\log_g a$.)

Summary. For the security of cryptographic schemes proved with ROs, what may be lost in real world when ROs are replaced by real practical hash functions? The above motivating examples and discussions show that it depends on both, the uses of ROs in security proofs and the uses of ROs in protocol design.

3.2   Principles for Restricting Uses of ROs in Security Proofs and Protocol Designs

In light of the above motivating examples and discussions, we introduce restrictions on uses of RO in *interactive* protocols. We describe the principles in the two-party case, but extensions to the multi-party case are immediate. For a two-party interactive scheme (with restricted ROs), there are two random oracles: $\mathcal{O}_{\mathcal{P}}$ for the prover and $\mathcal{O}_{\mathcal{V}}$ for the verifier. The uses of the ROs in security proofs and protocol design are limited in the following way:

1. For proving prover's security properties (i.e., zero-knowledge), the random oracle $\mathcal{O}_{\mathcal{P}}$ used by the simulator (which plays the role of the honest prover) is *non-programmable* (and the adversary can access $\mathcal{O}_{\mathcal{P}}$ for verifying messages). The random oracle $\mathcal{O}_{\mathcal{V}^*}$ used by the adversary V^* (malicious verifiers) is *programmable* (and the simulator can access $\mathcal{O}_{\mathcal{V}^*}$ for verifying messages). Similarly, for proving verifier's security properties (i.e., soundness), the simulator (playing the role of the honest verifier) uses the *non-programmable* random oracle $\mathcal{O}_{\mathcal{V}}$. The adversary P^* (malicious provers) uses a *programmable* RO $\mathcal{O}_{\mathcal{P}^*}$. Note that this essentially requires that in security proofs the simulator can only define the outputs of the *programmable* random oracle on queries made *by the adversary* in question.

2. Furthermore, the non-programmable random oracles, $\mathcal{O}_{\mathcal{P}}$ and $\mathcal{O}_{\mathcal{V}}$, can be replaced by *any* real (i.e., hash) function without compromising the security of the *honest* players P and V respectively. This requirement essentially says that the restricted random oracle model could be viewed as a "hybrid" between the normal random oracle model and the standard model with real (hash) functions, in the sense that the malicious player still lives in the idealized random oracle world but the honest player could live in real hash function world. In other words, if you are *honest* (e.g. a *trusted authority*) then you could use any real hash function in generating messages from you without compromising your security. Note that, in practice many interactive schemes in the random oracle model (e.g. the DAA protocol of [3]) involve (possibly quite complicated) interactive setup/join protocols between users and *a trusted authority*.

3. As in the original Fiat-Shamir methodology, the random oracles are used only to collapse Σ-protocols into non-interactive ones. This requirement reduces the dependency of the security of the protocol upon the idealized random function property of the realizations of the ROs (e.g., it is not used as a long term commitment, which is naturally desirable in certain critical settings as discussed above).

In the rest of this work, we refer to protocols which are developed with limiting the uses of ROs according to the above principles as *protocols with restricted ROs*.

Remark. In the above description, we only give the general principles of limiting the uses of ROs. When it comes to formally and exactly defining certain cryptographic primitives (e.g., ZK) with restricted ROs, we need to formally specify (and embed) the above general principles in the specific cryptographic primitives. In particular, the next section provides the formal definition of ZK with restricted ROs.

Comparisons with the work of [23]. We have noted recently the related work of Pass [23] who nicely treats the issue of deniability. [23] observed that non-interactive zero-knowledge in the random oracle model [2] does not preserve deniability and presented a new definition of ZK, named deniable ZK, in the RO model, and constructed a 2-round deniable zero-knowledge protocol from any Σ-protocol.

The approach taken by Pass in [23] for defining and constructing deniable ZK with ROs amounts to the following non-programmable random oracle methodology: all players (including the ZK simulator) access a unique *non-programmable* random oracle. The non-programmable RO methodology is also investigated by Nielsen in the non-committing encryption setting [22]. Below, we provide detailed comparisons between our approach and Pass's non-programmable random oracle methodology. We believe we have some noticeable advantages, though one should welcome various methodologies in this subtle area.

- In Pass's approach, all players access a unique non-programmable RO. But in our approach, there are a pair of ROs: one is non-programmable RO through which messages from the honest player and the simulator (who plays the role of the honest prover) are generated; and one is a programmable RO through which the messages from the adversary in question are generated. Furthermore, the non-programmable RO could be replaced by any real hash function without compromising the honest player's security, a property we do not know how to achieve with Pass's approach.
- We attempted to develop efficient protocols for important cryptographic languages (e.g., DLP and RSA); the efficient protocols are a central reason for which we employ the random oracle idealization, to start with. Solutions for ZK protocols with Pass's approach seems intrinsically inefficient, due to the cut-and-choose technique used, which leads to blow-ups in both computational complexity and communication complexity. Specifically, if the 2-round ZK protocol of [23] is from Schnorr's Σ-protocol for DLP, then on a security parameter n the prover needs to perform $8n$ modular exponentiations and the verifier needs to perform $10n$ exponentiations. For communication complexity, there are about $12n^2$ bits exchanged in total. This inefficiency may violate the spirit of RO protocols as was noted by Pass, who, in fact, suggested as an urgent open problem to find more efficient constructions of zero-knowledge with ROs with the approach. In comparison, our schemes

are generic yet practical solutions with restricted ROs and go through only 9 modular exponentiations at each player's side in the BPK model (in the plain model the verifier needs 11 exponentiations), with the DLP as an example.
- In our approach, we further restrict the uses of ROs in protocol designs by limiting the uses of ROs only to collapsing Σ-protocols, while Pass's approach does not. In fact, the ZK protocol with ROs developed in [23] critically uses the commitment scheme $c = RO(m, r)$. This by itself seems fine, but there is the concern (since we deal with realized idealized objects) that a realization can be broken implying weakened commitment which reveals knowledge. We attempt a design where security properties (WI) for honest parties remain intact even if the realization turns out to be weak.

4   *Generic Yet Practical* Round-Optimal Zero-Knowledge with Restricted ROs

Here, we provide the definition of ZK with restricted ROs, show its round-complexity lower-bound, and then present a generic yet practical round-optimal zero-knowledge argument with restricted ROs for any $\mathcal{NP}$-language that admits Σ-protocols (which includes many important languages most relevant to cryptography).

Definition 2 (zero-knowledge argument with restricted ROs). *A pair of interactive machines, $\langle P, V \rangle$, is called a zero-knowledge argument with restricted ROs for a language $L \in \mathcal{NP}$ (with $\mathcal{NP}$-relation R_L), if both machines are polynomial-time and the following conditions hold:*

Completeness. *For any $x \in L$ and its $\mathcal{NP}$-witness w, and any auxiliary input $aux_V \in \{0,1\}^*$, it holds that*

$$\Pr[\langle P^{(\mathcal{O}_P, \mathcal{O}_V)}(w), V^{(\mathcal{O}_V, \mathcal{O}_P)}(aux_V)\rangle(x) = 1] = 1$$

where $\mathcal{O}_P$ and $\mathcal{O}_V$ are two random variables uniformly distributed in $\{0,1\}^{poly(|x|)} \rightarrow \{0,1\}^{poly(|x|)}$. The random oracles $\mathcal{O}_P$ and $\mathcal{O}_V$ are used in the following way: random oracles are used only to collapse Σ-protocols (namely, deriving the challenges in such protocols and running them non-interactively), where each player only access its designated random oracle for generating messages from it (and the second random oracle is only accessed for verifying messages from its counterpart).

Computational soundness. *For any $x \notin L$, any PPT interactive machine P^*, and any auxiliary input $aux_{P^*} \in \{0,1\}^*$ and $aux_V \in \{0,1\}^*$, it holds that:*

$$\Pr[\langle P^{*(\mathcal{O}_{P^*}, \mathcal{O}_V)}(aux_{P^*}), V^{(\mathcal{O}_V, \mathcal{O}_{P^*})}(aux_V)\rangle(x) = 1] \leq \varepsilon(|x|)$$

where $\varepsilon(\cdot)$ is a negligible function, and $\mathcal{O}_{P^*}$ and $\mathcal{O}_V$ are random variables uniformly distributed in $\{0,1\}^{poly(|x|)} \rightarrow \{0,1\}^{poly(|x|)}$. Furthermore, the soundness condition holds even if $\mathcal{O}_V$ is arbitrarily (rather than uniformly) distributed from $\{0,1\}^{poly(|x|)} \rightarrow \{0,1\}^{poly(|x|)}$ (i.e., $\mathcal{O}_V$ can be any function).

(Black-box) zero-knowledge. *There exists an expected polynomial-time simulator S such that for every PPT verifier V^*, any $aux_{V^*} \in \{0,1\}^*$, any sufficiently long $x \in L$, the following two ensembles are computationally indistinguishable (where the distinguishing gap is a function in $|x|$):*

- *$\{(\mathcal{O}_{V^*}, \mathcal{O}_P, view_{V^*(\mathcal{O}_{V^*}, \mathcal{O}_P)(aux_{V^*})}^{P(\mathcal{O}_P, \mathcal{O}_{V^*})(w)}(x))\}_{x \in L, aux_{V^*} \in \{0,1\}^*}$ for arbitrary w such that $(x, w) \in R_L$.*
- *$\{(\mathcal{O}_P, S^{\mathcal{O}_P}(x, aux_{V^*}))\}_{x \in L, aux_{V^*} \in \{0,1\}^*}$*

where $\mathcal{O}_P$ and $\mathcal{O}_{V^*}$ are random variables uniformly distributed in $\{0,1\}^{poly(|x|)} \to \{0,1\}^{poly(|x|)}$, and $view_{V^*(\mathcal{O}_{V^*}, \mathcal{O}_P)(aux_{V^*})}^{P(\mathcal{O}_P, \mathcal{O}_{V^*})(w)}(x)$ is a random variable describing V^*'s state and all messages exchanged during a joint computation between P and V^* on common input x when P has w as its auxiliary input and accesses $(\mathcal{O}_P, \mathcal{O}_{V^*})$, and V has aux_{V^*} as its auxiliary input and accesses $(\mathcal{O}_{V^*}, \mathcal{O}_P)$. Furthermore, the zero-knowledge condition holds even when $\mathcal{O}_P$ is taken arbitrarily (rather than uniformly) from $\{0,1\}^{poly(|x|)} \to \{0,1\}^{poly(|x|)}$ (i.e., $\mathcal{O}_P$ can be any function).

Comment. Note that in the definition of (black-box) ZK with restricted ROs, the RO $\mathcal{O}_P$ is given (pre-specified) in the above two probability ensembles, which means that S cannot "program" $\mathcal{O}_P$. But, for the RO $\mathcal{O}_{V^*}$, S is allowed to "program" and output a "simulation" of $\mathcal{O}_{V^*}$ in its simulation. Note that the random oracle is actually an infinite object, S thus, for this purpose, has a special fill out function. We refer the reader to [2] for a more formal treatment of "programming" ROs.

We next show, by the Goldreich-Krawczyk technique [17], the impossibility of non-interactive black-box zero-knowledge arguments with restricted ROs for non-trivial languages. Specifically, we give the following proposition (whose proof is omitted here due to space limitation):

Proposition 1. *Suppose an $\mathcal{NP}$-language L admits a one-round black-box zero-knowledge argument with restricted ROs, then $L \in \mathcal{BPP}$.*

Finally, we show that a protocol based on the Feige-Shamir 4-round ZK argument for $\mathcal{NP}$ renders a generic yet practical 2-round (i.e., optimal) ZK argument with restricted ROs for any language that admits Σ-protocols. The protocol is depicted in Figure-2.

Comment. At a high level, the protocol depicted in Figure-2 can be viewed as a Σ_{OR}-based implementation of the Feige-Shamir 4-round ZK arguments for $\mathcal{NP}$ (the version appearing in [13]) in the RO model. The construction of [13] is a plausible $\mathcal{NP}$-solution and goes through general (inefficient) $\mathcal{NP}$-reductions, whereas here (for this section) we emphasize the fact that the protocol works directly for any language that admits Σ-protocols (a large set that includes, in particular, both DLP and RSA). If the underlying Σ-protocols for the language are practical, then the transformed protocols are also practical. With Σ-protocol for DLP as an example, our scheme goes through 9 modular exponentiations at the prover side and 11 exponentiations at the verifier side.

Common input. An element $x \in L$ of length n, where L is an $\mathcal{NP}$-language that admits Σ-protocols.

P's private input. A witness w for $x \in L$.

Random oracles. There are two random oracles $\mathcal{O}_P$ and $\mathcal{O}_V$: $\mathcal{O}_P$ is used by the prover for generating non-interactive Σ-proofs and $\mathcal{O}_V$ is used by the verifier for generating non-interactive Σ-proofs.

Round-1. The verifier V selects a OWF f_V that admits Σ-protocols, randomly selects two elements in the domain of f_V, x_V^0 and x_V^1, computes $y_V^0 = f_V(x_V^0)$ and $y_V^1 = f_V(x_V^1)$, randomly selects a bit b from $\{0,1\}$, sends to the prover a non-interactive Σ_{OR}-proof on (y_V^0, y_V^1), denoted $\pi_V = (y_V^0, y_V^1, a_V, e_V, z_V, aux_V)$, that it knows either the preimage of y_V^0 or the preimage of y_V^1. The witness used by V in forming π_V is x_V^b. The random challenge e_V is generated by querying $\mathcal{O}_V$ with $(x, y_V^0, y_V^1, a_V, aux_V)$, where aux_V is the auxiliary information of V that possibly includes a time-stamp.

Round-2. The prover P first checks the validity of π_V and aborts if it is not valid. Otherwise, P sends back a non-interactive Σ_{OR}-proof on (x, y_V^0, y_V^1), denoted $\pi_P = (a_P, e_P, z_P)$, that it knows either the witness for $x \in L$ or the preimage of either y_V^0 or y_V^1. The witness used by P is its private input (i.e., the $\mathcal{NP}$-witness w). The random challenge e_P is generated by querying $\mathcal{O}_P$ with (x, a_P, π_V).

Verifier's Decision. The verifier checks the validity of π_P and accepts if it is valid, otherwise it rejects.

Fig. 2. The generic yet practical 2-round ZK arguments with restricted ROs

Theorem 2. *Let f_V be any one-way function that admits Σ-protocol and L be a language that admits Σ-protocols, the protocol depicted in Figure-2 is a 2-round ZK argument with restricted ROs for L.*

Proof (sketch). Intuitively, P proves $x \in L$ only after it is convinced that the verifier does know the preimage of either y_V^0 or y_V^1 (this means that the verifier can also generate the second-round message *by itself*), so the ZK property of the protocol should hold. In more details, for a malicious verifier V^* who accesses the programmable random oracle $\mathcal{O}_{V^*}$, the zero-knowledge simulator S first extracts $x_{V^*}^b$ (the witness used by V^* in generating the Round-1 message) by redefining the random oracle $\mathcal{O}_{V^*}$ on queries made by V^*, then using $x_{V^*}^b$ as the witness S generates a simulated Round-2 message *through its fixed random oracle $\mathcal{O}_P$*. By the perfect WI property of Round-2, the simulated transcript is indistinguishable from the real transcript. Furthermore, since in security proof we only need the WI property of the non-interactive Σ_{OR}-protocol of Round-2, the fixed random oracle $\mathcal{O}_P$ can be replaced by *any real function*, as the WI property of Round-2 does hold with respect to any function (*see Claim 1*). That is, the honest prover's security (i.e., ZK) holds even when the honest prover uses any real function (say cryptographic hash based one) in generating non-interactive Σ_{OR}-protocols (of Round-2).

For proving soundness, however, one may argue that seeing the Σ_{OR}-proof π_V sent by the honest verifier in Round-1 (*which could be generated through any real function*) may help a malicious prover P^* to give a false Σ_{OR}-proof π_{P^*} in Round-2. What save us here are the key-pair technique (originally introduced in the Public-Key Encryption setting by Naor and Yung [21]) and the witness indistinguishability of the Σ_{OR}-proof even with real functions. In more details, suppose a malicious prover P^* can successfully convince of a false statement $x \notin L$ with a non-negligible probability, then we show a PPT algorithm E that will break the one-wayness of f_V. Specifically, on an input y_V E runs P^* as a subroutine and works as follows: E randomly selects x'_V from the domain of f_V, computes $y'_V = f_V(x'_V)$. Then, E randomly selects a bit b from $\{0,1\}$, sets y^b_V be y'_V and y^{1-b}_V be the input y_V. Finally, by using x'_V as the witness E sends to P^* the Round-1 message (generated through the fixed random oracle $\mathcal{O}_V$), claiming that it knows the preimage of either y^0_V or y^1_V. After receiving a successful Round-2 message from P^* that is generated through the programmable RO $\mathcal{O}_{P^*}$, by rewinding P^* and redefining $\mathcal{O}_{P^*}$ E will extract a preimage of either y^0_V or y^1_V (as we assume $x \notin L$). Then by the perfect WI property of Round-1, with probability $1/2$ (conditioned on P^* successfully giving the Round-2 message), the extracted value will be the preimage of $y^{1-b}_V = y_V$, which violates the one-wayness of f_V. Again, in the above security analysis, we only need the WI property of the non-interactive Σ_{OR} of Round-1, and so the fixed random oracle $\mathcal{O}_V$ can be replaced by *any real function*. That is, the honest verifier's security (i.e., soundness) holds even when the honest verifier uses any function in generating the non-interactive Σ_{OR}-protocols of Round-1.

5   Concurrent Security of the Feige-Shamir ZK with Registered Public-Keys and with Restricted ROs

Dealing with concurrent adversaries in the interactive random oracle model turns out to be much more complicated and subtle. The reason is that for messages sent by an adversary we need to rewind the adversary and redefine the outputs of the programmable random oracles to extract the witnesses used by the adversary. But for a concurrent adversary, it can make both concurrent interleaving interactions with the honest player instances and concurrent interleaving oracle queries (across multiple existing sessions). We thus risk an exponential blow-up when tracking back through the interleaving interactions or the interleaving oracle queries across multiple sessions, in the sense that previous simulation efforts (interaction rewinding-s or random oracle redefining-s) will become void. This phenomenon is first observed by Dwork, Naor and Sahai [12] for *inter-active* protocols in the standard model in dealing with adversaries that make concurrent interleaving interactions with honest player instances, and observed also by Shoup and Gennaro [25] for *non-interactive* schemes in the random oracle model in dealing with adversaries that make concurrent interleaving oracle queries across multiple sessions in the context of threshold decryption.

   To avoid the exponential blow-up in dealing with concurrent adversaries, several computational models have been proposed: the timing model [12, 16], the preprocessing model [10], the common reference string model [9], and the bare public-key model [4].

The bare public-key (BPK) model was introduced by Canetti, Goldreich, Goldwasser and Micali [4] to achieve round-efficient resettable zero-knowledge (rZK) that is a generalization and strengthening of the notion of concurrent zero-knowledge [12]. A protocol in the BPK model simply assumes that all verifiers have deposited a public key in a public file before any interaction takes place among the users[1]. to all users at all times. Note that an adversary may deposit many (possibly invalid or fake) public keys in it, particularly, without even knowing corresponding secret keys or whether such exist. That is, no trusted third party is assumed in the BPK model. What is essentially guaranteed by the BPK model is only a limitation on the number of different identities that a potential adversary may assume and there are no other assurances. The adversary, in turn, may try to impersonate any user registered in the public-file, but it cannot act on behalf of a non-registered user. The BPK model is thus very simple, and it is, in fact, a weaker version of the frequently used public-key infrastructure (PKI) model (recall that PKI underlies any public-key cryptosystem or any digital signature scheme). Despite its apparent simplicity, the BPK model turns out to be quite powerful in dealing with concurrent adversaries and stronger resetting adversaries.

Soundness in public-key models is more subtle than in the standard model [20]. In public-key models, a verifier V has a secret key SK, corresponding to its public-key PK. A malicious prover P^* could potentially gain some knowledge about SK from an interaction with the verifier. This gained knowledge may help him to convince the verifier of a false theorem in another interaction. Micali and Reyzin [20] showed that under standard intractability assumptions there are four distinct meaningful notions of soundness, i.e., from weaker to stronger, one-time, sequential, concurrent and resettable soundness. In this paper we focus on concurrent soundness which roughly means, for zero-knowledge protocols, that a malicious prover P^* cannot convince the honest verifier V of a false statement even when P^* is allowed multiple interleaving interactions with V.

Due to space limitation, the definitions of concurrent ZK and concurrent soundness in the BPK model with restricted ROs are omitted here, and will be presented in the full version of this work.

5.1   The Feige-Shamir ZK Protocol Is *Not* Secure in the Public-Key Setting

Next we show that the Feige-Shamir ZK protocol [13] is, in general, *not* concurrently secure in the public setting (indeed it was not designed for that more modern setting). Specifically, we show a concurrent attack against the Σ_{OR}-based implementation of the Feige-Shamir ZK protocol in the public-key setting (which may be of independent interest).

Consider the version of the protocol depicted in Figure-2 when the pair (y_V^0, y_V^1) is published as the verifier's public-key (i.e., fixed once and for all sessions) and ROs are removed (i.e., random challenges e_V and e_P are not obtained

---

[1] The BPK model does allow dynamic key registrations (see [4]).

any longer by querying the ROs, but sent by the prover and the verifier respectively). We remark that, at a first glance, *it is quite natural* for the verifier to publish (y_V^0, y_V^1) as its public-key when the Feige-Shamir ZK protocol (especially its Σ_{OR}-based implementation) is used in practice. But, the following attack shows that this intuition is wrong.

Let L (wlog, the $\mathcal{NP}$-complete language Directed Hamiltonian Cycle DHC) be a language that admits Σ-protocols. We show how a malicious prover P^* can convince an honest verifier V (with public-key (y_V^0, y_V^1)) of a false statement "$x \in L$" while $x \notin L$, by concurrently interacting two sessions with V. The message schedule of P^* in the two sessions is specified as follows.

1. P^* interacts with V in the first session and works just as the honest prover does in Phase-1. When P^* moves into Phase-2 of the first session and needs to send V the first-round message, denoted by a_P, of the Σ_{OR}-protocol of Phase-2 of this session on common input (x, y_V^0, y_V^1), P^* suspends the first session and does the following:
   - It first runs the SHVZK simulator (of the underlying Σ-protocol for L) on x to get a simulated conversation, denoted by (a_x, e_x, z_x), for the false statement "$x \in L$".
   - Then, P^* initiates a second session with V; After receiving the first-round message, denoted by a_V', of the Σ_{OR}-protocol of Phase-1 of the second session on common input (y_V^0, y_V^1) (i.e., V's public-key) , P^* sets $a_P = (a_x, a_V')$ and suspends the second session.
2. Now, P^* continues the execution of the first session, and sends $a_P = (a_x, a_V')$ to V as the first-round message of the Σ_{OR}-protocol of Phase-2 of the first session.
3. P^* Runs V further in the first session. After receiving the second-round message of Phase-2 of the first session, denoted by e_P (i.e., the random challenge from V), P^* sets $e_V' = e_P \oplus e_x$ and suspends the first session again.
4. P^* continues the execution of the second session, and sends $e_V' = e_P \oplus e_x$ to V as its random challenge in the second-round of the Σ_{OR}-protocol of Phase-1 of the second session. After receiving the third-round message of Phase-1 of the second session, denoted by z_V', P^* sets $z_P = ((e_x, z_x), (e_V', z_V'))$ and suspends the second session again.
5. P^* continues the execution of the first session again, sending the value $z_P = ((e_x, z_x), (e_V', z_V'))$ to V as the last-round message of the first session.

Note that (a_x, e_x, z_x) is an accepting conversation for showing "$x \in L$", (a_V', e_V', z_V') is an accepting conversation for showing the knowledge of the preimage of either y_V^0 or y_V^1, and furthermore $e_P = e_x \oplus e_V'$. According to the description of Σ_{OR} (presented in Section 2), this means that, from the viewpoint of V, (a_P, e_P, z_P) is an accepting conversation on common input (x, y_V^0, y_V^1) of the Σ_{OR}-protocol of Phase-2 of the first-session, and thus P^* successfully convinced V of a false statement in the first session. We remark that, in general, the above attack also enables P^* to convince V of a true statement $x \in L$ without knowing any $\mathcal{NP}$-witness for $x \in L$.

5.2   The Feige-Shamir ZK Is Concurrently Secure in the BPK Model with Restricted ROs

As shown, the Feige-Shamir ZK is not concurrently secure in public-key model, but we next show that it is still concurrently secure in the BPK model with restricted ROs (which may make it useful in certain applications which it was not originally designed for).

Specifically, consider the following modified version of the protocol depicted in Figure-2 in the BPK model: there is a key generation phase before any interaction takes place among the users, in which each verifier V_i registers $(y_{V_i}^0, y_{V_i}^1)$ in a public-key file F, where $y_{V_i}^0 = f_{V_i}(x_{V_i}^0)$, $y_{V_i}^1 = f_{V_i}(x_{V_i}^1)$ and f_{V_i} is a OWF that admits Σ-protocols. For a bit b randomly chosen from $\{0, 1\}$, V_i keeps $x_{V_i}^b$ in secret as its secret-key while discarding $x_{V_i}^{(1-b)}$. Then in Round-1 of the modified protocol, by using $x_{V_i}^b$ as the witness, V_i sends a non-interactive Σ_{OR}-proof that it knows the preimage of either $y_{V_i}^0$ or $y_{V_i}^1$. Round-2 remains unchanged.

Theorem 3. *Under any one-way functions that admit Σ-protocols, the above modified protocol is a generic yet practical 2-round concurrently sound concurrent ZK argument with restricted ROs in the BPK model for any language that admits Σ-protocols.*

Below, we present the high-level proof overview of Theorem 3 and identify some complications and subtleties of dealing with concurrent adversaries for *interactive* schemes with ROs.

The simulation procedure for concurrent zero-knowledge is similar to the simulation procedure for resettable zero-knowledge presented in [4]. Specifically, for any concurrent adversary V^* that has as its output a public-key file of the form $F = \{(y_{V_1^*}^0, y_{V_1^*}^1), (y_{V_2^*}^0, y_{V_2^*}^1), \cdots, (y_{V_q^*}^0, y_{V_q^*}^1)\}$, the zero-knowledge simulator S runs V^* as a subroutine and works in at most $q + 1$ phases. In each phase, S either successfully gets a simulated transcript or "breaks" a new public-key $(y_{V_i^*}^0, y_{V_i^*}^1)$, $1 \le i \le q$, in the sense that S can extract the corresponding secret-key $x_{V_i^*}^b$. In this process, we identify that dealing with concurrent adversaries for interactive schemes in the random oracle model actually amounts to dealing with resetting adversaries in the standard model. Specifically, in dealing with these resetting adversaries for proving resettable zero-knowledge in the standard model for the sake of extracting the witness used by a malicious resetting verifier in one session (for facilitating the successful simulation), we normally need to rewind the adversary and change the random challenge that has been sent with respect to some message of that session (e.g. the first message of a Σ-protocol), and give back, in turn, a different random challenge. But, the random challenge to be changed in that session may have been "defined" in a previous session, we may thus need to rewind the adversary in a previous session in which the random challenge is defined *for the first time*. Similarly, in dealing with concurrent adversaries for proving concurrent zero-knowledge in the random oracle model, to extract the witness used by the adversary in forming the Round-1 message in one session, we need to redefine the random output of the programmable random oracle. But the random output of the programmable random oracle used in that session may be obtained by the adversary by querying the random oracle in a previous session, and thus we need to rewind the adversary in the previous session where it made the oracle query in question *for the first time*. We remark that in the proof of concurrent ZK we only need the WI property of the non-interactive Σ_{OR}-proofs (of Round-2) generated through the non-programmable RO $\mathcal{O}_\mathcal{P}$ (that does hold even when $\mathcal{O}_\mathcal{P}$ is replaced by any function of a proper size). This means that the honest prover's security (i.e., concurrent ZK) holds even when the honest prover uses any function in generating the non-interactive Σ_{OR}-protocols of Round-2.

For concurrent soundness, assume a PPT q-concurrent adversary P^* can successfully convince V with public-key (y_V^0, y_V^1) of a false statement with non-negligible probability p in one of the q concurrent sessions, then we will construct an algorithm E that on an input y in the range of f_V outputs the preimage of y with non-negligible probability $\frac{p^2}{2q}$ in expected polynomial-time, which violates the one-wayness of f_V.

Algorithm E on an input y, first randomly selects an element x' in the domain of f_V, computes $y' = f_V(x')$, randomly selects a bit b from $\{0,1\}$, sets y_b be y' and y_{1-b} be y, publishes (y_0, y_1) as its public-key while keeping x' privately as the corresponding secret-key. Then, E randomly chooses i from $\{1, 2, \cdots, q\}$, and runs P^* by playing the role of the honest verifier (with (y_0, y_1) as its public-key and x' as its secret-key) in any session other than the i-th session. In the i-th session on a common input x_i, suppose P^* successfully gives a Round-2 message, denoted by $(a_{P^*}^{(i)}, e_{P^*}^{(i)}, z_{P^*}^{(i)})$, with respect to a Round-1 message, denoted by $\pi_V^{(i)}$, sent by E in the first-round of the i-th session, where $e_{P^*}^{(i)}$ is the random oracle answer given by E to P^* on a query of the form $(x_i, a_{P^*}, \pi_V^{(i)})$ to the programmable random oracle $\mathcal{O}_{\mathcal{P}^*}$. Then E rewinds P^* to the point that P^* just made the oracle query $(x_i, a_{P^*}, \pi_V^{(i)})$, gives back a new random oracle answer $e_{P^*}^{(i)'}$ and runs P^* from the above rewinding point and on. We stress that in the above process all Round-1 messages from E to P^* are generated through the fixed random oracle $\mathcal{O}_V$.

Since we assume that P^* can, with probability p, convince V of a false statement in one of the q concurrent sessions, then conditioned on E correctly guessing the value i, it is easy to see that with probability p^2 E will extract an $\mathcal{NP}$-witness for $x_i \in L$ or a preimage of either y_0 or y_1, which is guaranteed by the special soundness of the Σ_{OR} protocol. Since we further assume that $x_i \notin L$ and E randomly guesses i from $\{1, \cdots, q\}$, we conclude that with probability $\frac{p^2}{q}$ E will output the preimage of either y_0 or y_1. Furthermore, according to the perfect WI property of Σ_{OR}-protocol, we know that with probability $\frac{p^2}{2q}$ E will output a preimage of $y = y_{1-b}$, which violates the one-wayness of f_V. Note that in the above proof we only need the WI property of the non-interactive Σ-proofs (of Round-1) generated through the non-programmable RO $\mathcal{O}_V$ (that does hold even when $\mathcal{O}_V$ is replaced by any properly sized function), which means that the honest verifier's security (i.e., concurrent soundness) holds even for this type of $\mathcal{O}_V$.

Comment. At a first glance, it seems that the above proof procedure for concurrent soundness can also be applicable to the Feige-Shamir ZK protocol in the public-key model *without ROs* (that is however, as we have shown, not concurrently secure). This subtle point needs further elaboration: The WI property is only guaranteed to be concurrently composable when the *same* protocol is composed concurrently. But in our case, the concurrent adversary P^* actually also runs WI protocols to V (or E) with the player role reversed with respect to the WI protocols from V (or E) to P^*. In general, in this case the concurrent WI property of the Σ_{OR}-protocols from V (or E) to P^* is not guaranteed. This is also the very reason why the Feige-Shamir ZK protocol is not concurrently secure in the public-key model *without ROs*, as shown by our concurrent attack. In contrast, in the (Σ_{OR}-based implementation of) Feige-Shamir ZK protocol in the BPK model *with restricted ROs*, the important fact is that we are, both, working in the random oracle model and the WI (i.e., Σ_{OR}) protocols are *non-interactive*. In more details, suppose in this case the preimage extracted by E is dependent on the witness used by E, then we can show a PPT algorithm E' that violates the WI property of *non-interactive* Σ_{OR}-protocols as follows. For a PPT concurrent adversary P^* and the honest verifier V with public-key (y_0, y_1) who actually is a non-interactive Σ_{OR}-prover on (y_0, y_1) with random challenges generated through the fixed random oracle $\mathcal{O}_V$, E' runs P^* as a subroutine and interacts with V. E' works just as E does but with the following modifications: Whenever E' needs to send a non-interactive Σ_{OR}-proof in Round-1 of a session, E' just interacts with V to get such a proof and sends it to P^*. Note that E' never redefines the fixed random oracle $\mathcal{O}_V$. Clearly, E' can violate the WI property of the *non-interactive* Σ_{OR}-proofs received from V if the extracted preimage (from P^*) is dependent on the witness used by V.

6   A Note on the Applications of the 2-Round ZK with Restricted ROs

The notion of zero-knowledge plays a central role in modern cryptography and we are now at the point where more and more complicated *interactive* schemes with random oracle methodologies are under development (including ones for industrial use). Thus, we expect that the generic yet practical 2-round ZK protocols with restricted ROs (with or without registered public-keys) can be used as a building block in constructing more complicated *interactive* schemes provably secure with restricted ROs.

In particular, we note that the 2-round ZK protocols with restricted ROs can be used to transform a large number of (but not necessarily all) interactive schemes (and non-interactive systems with interactive setup/join protocols like PKI, group signatures or e-cash) developed originally in the normal random oracle model, which use the random oracle only to collapse Σ-protocols, into schemes with provable security using restricted ROs, paying in efficiency *at most* one extra round, but with seemingly more sound provable security guarantees. The idea is to replace each non-interactive NIZK in the original interactive scheme (developed in the normal interactive RO model) by our 2-round ZK protocols with restricted ROs. The key observation here is that all 2-round ZK protocols with one party playing the role of the prover can share the same Round-1 non-interactive Σ_{OR}-protocol sent by its counterpart. This way, the non-interactive nature of the NIZK-protocols in the original interactive systems can be preserved at the price of at most one additional initiating round on top of the protocol. This general transformation, along with detailed discussions, will be presented in the full version of this work.

Acknowledgments. We are grateful to Yehuda Lindell for referring us to [13] and for valuable discussions and suggestions. We thank the anonymous referees of TCC'06 for valuable and detailed comments and suggestions.

References

1. M. Bellare, A. Boldyreva and A. Palacio. An Uninstantiable Random-Oracle-Model Scheme for a Hybrid-Encryption Problem In *C. Cachin and J. Camenisch (Ed.): Advances in Cryptology-Proceedings of EUROCRYPT 2004, LNCS 3027*, pages 171-188. Springer-Verlag, 2004.
2. M. Bellare and P. Rogaway. Random Oracles are Practical: A Paradigm for Designing Efficient Protocols. In*ACM Conference on Computer and Communications Security*, pages 62-73, 1993.
3. E. Brickell, J. Camenisch and L. Chen. Direct Anonymous Attestation. ACM's CCS 2004.
4. R. Canetti, O. Goldreich, S. Goldwasser and S. Micali. Resettable Zero-Knowledge. In *ACM Symposium on Theory of Computing*, pages 235-244, 2000.
5. R. Canetti, O. Goldreich and S. Halevi. The Random Oracle Methodology, Revisited. In *ACM Symposium on Theory of Computing*, pages 209-218, 1998.
6. R. Canetti, O. Goldreich and S. Halevi. On the Random-Oracle Methodology as Applied to Length-Restricted Signature Schemes. In *1st Theory of Cryptography Conference (TCC), LNCS 2951* , pages 40-57, Springer-Verlag, 2004.
7. R. Cramer. Modular Design of Secure, yet Practical Cryptographic Protocols, PhD Thesis, University of Amsterdam, 1996.
8. R. Cramer, I. Damgard and B. Schoenmakers. Proofs of Partial Knowledge and Simplified Design of Witness Hiding Protocols. In *Y. Desmedt (Ed.): Advances in Cryptology-Proceedings of CRYPTO 1994, LNCS 839*, pages 174-187. Springer-Verlag, 1994.
9. I. Damgard. Efficient Concurrent Zero-Knowledge in the Auxiliary String Model. In *B. Preneel (Ed.): Advances in Cryptology-Proceedings of EUROCRYPT 2000, LNCS 1807*, pages 418-430. Springer-Verlag, 2000.
10. G. Di Crescenzo and R. Ostrovsky. On Concurrent Zero-Knowledge with Pre-Processing. In *M. J. Wiener (Ed.): Advances in Cryptology-Proceedings of CRYPTO 1999, LNCS 1666*, pages 485-502. Springer-Verlag, 1999.
11. D. Dolev, C. Dwork and M. Naor. Non-Malleable Cryptography. *SIAM Journal on Computing*, 30(2): 391-437, 2000. Preliminary version appears in STOC'91.
12. C. Dwork, M. Naor and A. Sahai. Concurrent Zero-Knowledge. In *ACM Symposium on Theory of Computing*, pages 409-418, 1998. Full version to appear in Journal of the ACM.

13. U. Feige. Alternative Models for Zero-Knowledge Interactive Proofs. Ph.D. Thesis, Department of Computer Science and Applied Mathematics, Weizmann Institute of Science, Rehovot, Israel, 1990.

14. A. Fiat and A. Shamir. How to Prove Yourself: Practical Solutions to Identification and Signature Problems. In *A. Odlyzko (Ed.): Advances in Cryptology-Proceedings of CRYPTO'86, LNCS 263*, pages 186-194. Springer-Verlag, 1986.

15. U. Feige and Shamir. Zero-Knowledge Proofs of Knowledge in Two Rounds. In *G. Brassard (Ed.): Advances in Cryptology-Proceedings of CRYPTO 1989, LNCS 435*, pages 526-544. Springer-Verlag, 1989.

16. O. Goldreich. Concurrent Zero-Knowledge with Timing, Revisited. In *ACM Symposium on Theory of Computing*, pages 332-340, 2002.

17. O. Goldreich and H. Krawczyk. On the Composition of Zero-Knowledge Proof Systems. *SIAM Journal on Computing*, 25(1): 169-192, 1996.

18. S. Goldwasser and Y. Tauman. On the (In)security of the Fiat-Shamir Paradigm. In *IEEE Symposium on Foundations of Computer Science*, pages 102-115, 2003.

19. L. Guillou and J. J. Quisquater. A Practical Zero-Knowledge Protocol Fitted to Security Microprocessor Minimizing both Transmission and Memory. In *C. G. Gnther (Ed.): Advances in Cryptology-Proceedings of EUROCRYPT 1988, LNCS 330* , pages 123-128, Springer-Verlag, 1988.

20. S. Micali and L. Reyzin. Soundness in the Public-Key Model. In *J. Kilian (Ed.): Advances in Cryptology-Proceedings of CRYPTO 2001, LNCS 2139*, pages 542–565. Springer-Verlag, 2001.

21. M. Naor and M. Yung. Public-Key Cryptosystems Provably Secure Against Chosen Ciphertext Attacks. In *ACM Symposium on Theory of Computing*, pages 427-437, 1990.

22. Jesper Buus Nielsen. Separating Random Oracle Proofs from Complexity Theoretic Proofs: The Non-Committing Encryption Case. In *M. Yung (Ed.): Advances in Cryptology-Proceedings of CRYPTO 2002, LNCS 2442*, pages 111-126, Springer-Verlag, 2002.

23. R. Pass. On Deniabililty in the Common Reference String and Random Oracle Models. In*D. Boneh (Ed.): Advances in Cryptology-Proceedings of CRYPTO 2003, LNCS 2729*, pages 316-337, Springer-Verlag 2003.

24. C. Schnorr. Efficient Signature Generation by Smart Cards. *Journal of Cryptology*, 4(3): 24, 1991.

25. V. Shoup and R. Gennaro. Securing Threshold Cryptosystems Against Chosen Ciphertext Attack. *Journal of Cryptology*, 15(2): 75-96, 2002.

Non-interactive Zero-Knowledge from Homomorphic Encryption

Ivan Damgård[1], Nelly Fazio[2,*], and Antonio Nicolosi[2,*]

[1] Aarhus University, Denmark[**]
ivan@brics.dk
[2] Courant Institute of Mathematical Sciences, New York University, NY, USA
{fazio, nicolosi}@cs.nyu.edu

Abstract. We propose a method for compiling a class of Σ-protocols (3-move public-coin protocols) into non-interactive zero-knowledge arguments. The method is based on homomorphic encryption and does not use random oracles. It only requires that a private/public key pair is set up for the verifier. The method applies to all known discrete-log based Σ-protocols. As applications, we obtain non-interactive threshold RSA without random oracles, and non-interactive zero-knowledge for NP more efficiently than by previous methods.

1 Introduction

In a zero-knowledge proof system, a prover convinces a verifier via an interactive protocol that some statement is true *i.e.*, a given word x is in some given language L. The verifier must learn nothing beyond the fact that the assertion is valid. Zero-knowledge is an extremely useful notion and has found innumerable applications.

One efficient variant is known as Σ-protocols, which are three-move protocols where conversations are tuples of the form (a, e, z) and e is a random challenge sent by the verifier. A large number of such protocols are known for languages based on discrete logarithm problems, such as Schnorr's protocol [16] and many of its variants, *e.g.*, for proving that two discrete logs are equal [4]. This last variant is useful, for instance, in threshold RSA protocols [17], where a set of servers hold shares of a private RSA key, and clients can request them to apply the private key to a given input. The Σ-protocol is used here by the servers to prove that they follow the protocol.

One well-known technique for making Σ-protocols non-interactive is the Fiat-Shamir heuristic [11], where e is computed by the prover himself as a hash of the statement proved and the first message a. In the random oracle model, where the hash function is replaced by a random function, this can be shown to work.

[*] Research conducted while visiting BRICS.
[**] Supported by BRICS, Basic Research in Computer Science, Center of the Danish National Research Foundation, and FICS, Foundations in Cryptography and Security, funded by the Danish Research Council.

S. Halevi and T. Rabin (Eds.): TCC 2006, LNCS 3876, pp. 41–59, 2006.

However, it is not in general possible to instantiate the random oracle with a concrete function and have the security properties preserved (*cf.* [12]). In other words, a proof in the random oracle model does not guarantee security in the real world.

Cramer and Damgård [7] suggest a different type of proof for equality of discrete logarithms in the *secret-key zero-knowledge* model, where prover and verifier are assumed to be given private, but correlated secret keys initially. These proofs can be applied to build non-interactive threshold RSA protocols without random oracles, but unfortunately, it is required that *every* client using the system must have keys for the proofs set up with *every* server. This seems quite impractical in many cases, due to the large amount of interaction and secure memory needed to set up and manage these keys. Moreover, [7] does not include any protocols for more general statements (such as NP-hard problems).

In this paper, we present a technique to compile a class of Σ-protocols into efficient non-interactive protocols, in the registered public-key model [1]. This model includes a trusted functionality for setting up a private/public key pair individually for each player (in fact, we only need this for the verifiers). Hence, unlike [7], the key setup is not tied to a particular prover/verifier pair: it can be implemented, for instance, by having the verifier send her public key to a trusted "certification authority" who will sign the key, once the verifier proves knowledge of her private key. Now, any prover who trusts the authority to only certify a key after ensuring that the verifier knows her private key, can safely (*i.e.,* in zero-knowledge) give non-interactive proofs to the verifier.

Our technique requires homomorphic public-key encryption such as Paillier's cryptosystem [15], and it preserves the communication complexity of the original protocol up to a constant factor. This is in contrast to the NIZK construction of Barak et al. [1] for the registered public-key model, which provides a much less efficient transformation from CCA-encryption and ZAP's [10].

The zero-knowledge property of our protocols is unconditional, whereas the soundness is based on an assumption akin in spirit to "complexity leveraging" [3]. More precisely, we assume that, by choosing large enough keys for the cryptosystem, the problem of breaking it can be made much harder than the problem underlying the Σ-protocol (for a particular meaning of "much harder" that we formalize in the paper).

An immediate consequence of our results is non-interactive threshold RSA and discrete-log based cryptosystems without random oracles, and assuming only that each client has a registered key pair. In the context of threshold cryptography where keys must be set up initially anyway, this does not seem like a demanding assumption. Our protocols are as efficient as the best known previous solutions (that required random oracles) up to a constant factor.

Another consequence is efficient non-interactive zero-knowledge arguments for circuit satisfiability, and hence for NP (in the registered public-key model). Namely, the prover commits to his satisfying assignment using a bit-commitment scheme for which appropriate efficient Σ-protocols exist. Then, using well-known techniques, for instance from [6], he could prove via a Σ-protocol that the committed bits satisfy the circuit. Compiling this protocol using our technique leads to the desired non-interactive protocol, whose communication complexity is essentially $O(ks_c)$ bits, where s_c is the size of the circuit, and k is the security parameter. This compares favorably to the solution of Kilian and Petrank [14] in the common random string model, which have complexity $O(k^2 s_c)$, even when using similar algebraic assumptions as we do here. Recently, Groth et al. [13] proposed non-interactive zero-knowledge proofs for NP in the common reference string model based on a specific assumption on bilinear groups, and with the same communication complexity as our protocol. This result is incomparable to ours: [13] uses a more conventional setup assumption and does not need a complexity-leveraging type of cryptographic assumption. On the other hand, it needs to assume that the statement shown by the prover is chosen independently from the reference string; in our model, the prover may see the verifier's public key first and then attempt to prove any theorem of his choice.

2 Preliminaries

We start by introducing some concepts and assumptions that will be useful later.

2.1 Problem Generators and a Complexity Assumption

A *problem generator* $\mathcal{G}$ is a pair $\mathcal{G} = \langle G, g \rangle$, where G is a probabilistic polynomial-time algorithm and $g : \{0,1\}^* \rightarrow \{0,1\}^*$ is an arbitrary (and possibly non-efficiently computable) function. On input 1^k, algorithm G outputs a string u, which we call an *instance*; we refer to $g(u)$ as the *solution* to u, and we require that $g(u)$ has length polynomial in k. For instance, u might be the concatenation of a public key and a ciphertext while $g(u)$ is the corresponding plaintext. We will only be considering problems with unique solutions, since that is all we need in this paper.

We will say that a probabilistic algorithm A *breaks* $\mathcal{G} = \langle G, g \rangle$ *on instances of size* k, if setting $u \xleftarrow{r} G(1^k)$ and $y \xleftarrow{r} A(1^k, u)$, results in $y = g(u)$ with non-negligible probability. We will be looking at the running time of A as a function only of its first argument 1^k; notice that A will not always be restricted to time polynomial in k.

We define that a probabilistic algorithm A *completely breaks* $\mathcal{G} = \langle G, g \rangle$ *on instances of size* k by considering the same experiment: Set $u \xleftarrow{r} G(1^k)$ and $y \xleftarrow{r} A(1^k, u)$; however, this time we demand that there exists a polynomial P such that, except with negligible probability (over the random choices of G), and for all large enough k, we have $Pr(y = g(u)| u) \geq 1/P(k)$, where the last probability is only over the random choices of A. In other words, A should be able to solve (almost) any instance u with good probability.

Definition 1. *Consider two problem generators $\mathcal{G}$ and $\mathcal{H}$ and let f be a polynomial. We say that $\mathcal{H}$ is f-harder than $\mathcal{G}$ if there exists a probabilistic algorithm A running in time $T(k)$ such that A completely breaks $\mathcal{G}$ on instances of size k, but no algorithm running in time $O(T(k) + poly(k))$ breaks $\mathcal{H}$ on instances of size $f(k)k$ or larger.*

In other words, completely breaking $\mathcal{G}$ on instances of size k requires time $T(k)$, but given a similar amount of time, there is no significant chance to break $\mathcal{H}$—where, however, the $\mathcal{H}$-instances to be solved have size at least $f(k)k$. Note that if $T(k)$ is polynomial in k, then $O(T(k) + poly(k)) = poly(k)$ and the definition amounts to say that $\mathcal{H}$ generates instances that are hard in the usual sense. But if $T(k)$ is superpolynomial, more is required about the hardness of $\mathcal{H}$-instances—essentially that the complexity of breaking $\mathcal{H}$ grows "fast enough" with the security parameter k.

For problem generators $\mathcal{F}, \mathcal{G}$, we will say that $\mathcal{F}$ is *as easy as* $\mathcal{G}$, if there exists an algorithm that completely breaks $\mathcal{F}$ on instances of size k in time polynomial in k, plus a constant number of oracle calls to any algorithm that completely breaks $\mathcal{G}$. The lemma below now follows trivially from the above definitions:

Lemma 1. *Let $\mathcal{F}, \mathcal{G}, \mathcal{H}$ be problem generators. If $\mathcal{F}$ is as easy as $\mathcal{G}$ and $\mathcal{H}$ is f-harder than $\mathcal{G}$, then $\mathcal{H}$ is also f-harder than $\mathcal{F}$.*

As an example, consider the following problem generator $\mathcal{G}_{dlog} = \langle G_{dlog}, g_{dlog} \rangle$: on input 1^k, G_{dlog} outputs an instance $u \doteq (p, p', g, h)$, where p, p' are primes, p' is k-bit long, $p = 2p' + 1$, g is an element of Z_p^* of order p' and $h = g^w \bmod p$, for some $w \in Z_{p'}$. In this case, the solution is $g_{dlog}(u) \doteq w$.

As another example, let $\mathcal{H}_{Paillier} = \langle H_{Paillier}, h_{Paillier} \rangle$, where $H_{Paillier}(1^k)$ outputs a k-bit RSA modulus n along with $c \doteq (1 + n)^w r^n \bmod n^2$ (*i.e.*, c is a Paillier encryption of w), where w is chosen in some given interval. Here, $h_{Paillier}(n, c) \doteq w$ is the solution. We can then make the following:

Assumption 1. *$\mathcal{H}_{Paillier}$ is 2-harder than $\mathcal{G}_{dlog}$.*

To discuss why this might be a reasonable assumption, note that no method is known to break one-way security of Paillier encryption other than factoring the modulus n. Furthermore, state of the art is (and has been for several years) that discrete log and factoring are of similar complexity for moduli of the same size. Moreover, with the current best known attacks (based on the number field sieve), doubling the modulus length has a dramatic effect on the expected time to solve the problem. Indeed, this is the reason why 1024-bit moduli are currently considered secure, even though 512-bit moduli can be broken in practice. It would therefore be very surprising, if it turned out to be possible to factor $2k$-bit numbers using only the time we need to find k-bit discrete logs. Note that if we had chosen a constant larger than 2 in Assumption 1, the assumption would be weaker, but all our results would remain essentially the same. We could even have used a polynomial f of degree ≥ 1, but then our compilation would be less efficient.

Definition 1 calls for an algorithm that completely breaks $\mathcal{G}$, and we will need this for technical reasons in the following. This makes Assumption 1 stronger than if we had only asked for one that breaks $\mathcal{G}$ in the ordinary sense. However, in the concrete case based on discrete logs, this makes no difference, as far as current state of the art is concerned: The best known attack on the discrete logarithm problem modulo p is the index calculus algorithm which works for all prime moduli, and has complexity that only depends on the size of the modulus. Furthermore, the discrete-log problem is random self-reducible and hence an algorithm solving a random instance modulo p with probability ϵ can solve any *fixed* instance modulo p with the same probability. In other words, the best known attack on the discrete log problem does in fact break it completely in our sense (albeit in superpolynomial time, of course).

2.2 Σ-Protocols

Consider the following protocol (adapted from [4]), which we will call $\mathcal{P}_{eqdlog}$:

Prover P and Verifier V get as common input $x \doteq (p, p', g_1, g_2, h_1, h_2)$, where p, p' are prime, p' is k-bit long, $p = 2p' + 1$, $g_1 \in Z_p^*$ has order p', $g_2, h_1, h_2 \in \langle g_1 \rangle$ and $h_1 = g_1^w \bmod p$, $h_2 = g_2^w \bmod p$, for some $w \in Z_{p'}$. P gets w as private input.

1. P chooses a random $3k$-bit integer r and sends $a \doteq (a_1, a_2)$ to V, where $a_1 \doteq g_1^r \bmod p, a_2 \doteq g_2^r \bmod p$;
2. V chooses e at random in $Z_{p'}$ and sends it to P;
3. P sends $z \doteq r + ew$ to V who checks that $g_1^z = a_1 h_1^e \bmod p, g_2^z = a_2 h_2^e \bmod p$.

Define the relation R_{dlog} as the set of pairs (x, w) as specified above, and $L_{R_{dlog}} \doteq \{x |\ \exists w : (x, w) \in R_{dlog}\}$. It is easy to see that the protocol above is an interactive proof system for membership in $L_{R_{dlog}}$, that is, it proves to the verifier that $\log_{g_1}(h_1) = \log_{g_2}(h_2)$.

In general, we define a Σ-protocol [5] for a relation R to be an interactive proof systems $\mathcal{P}$ for $L_R \doteq \{x |\ \exists w : (x, w) \in R\}$ with conversations of the form (a, e, z) and with the following additional properties:

Relaxed Special Soundness. Consider an input $x \notin L_R$, and any a. We say that a value of e is *good* if there exists z such that $x, (a, e, z)$ would be accepted by the verifier. The requirement now is that for any pair $x \notin L_R, a$, at most one good e exists.

Special Honest-Verifier Zero-Knowledge. There exists a probabilistic polynomial time simulator which on input x, e outputs a conversation (a, e, z) with distribution statistically indistinguishable from conversations between P and V, for the given statement $x \in L_R$ and challenge e.

Usually, one considers Σ-protocols for R, which have the standard Special Soundness property, namely that from $x \in L_R$ and accepting conversations $(a, e, z), (a, e', z')$ where $e \neq e'$, we can efficiently compute w such that $(x, w) \in R$. This clearly implies Relaxed Special Soundness, which is all we will need here.

The properties are straightforward to verify for the example protocol $\mathcal{P}_{eqdlog}$. In addition, $\mathcal{P}_{eqdlog}$ is an example of what we call a Σ-*protocol with linear answer*.

Definition 2. *A Σ-protocol with linear answer is a Σ-protocol where the prover's final message z is a sequence of integers, $z = (z_1, \ldots, z_m)$, where $z_j = u_j + v_j e$, and where u_j, v_j are integers that can be computed efficiently from x, P's random coins and his private input w.*

For a relation R to be useful, it is typically necessary that one can efficiently generate pairs $(x, w) \in R$ from a security parameter 1^k. We say that x is a k-instance, and we will assume that R comes with a polynomial ℓ_x such that k-instances have length $\ell_x(k)$.

Finally, we point out a consequence of Relaxed Special Soundness which will be important in the following: Let us consider any probabilistic polynomial-time algorithm $G_\mathcal{P}$ that, given a security parameter 1^k, generates a pair (x, a) where x has length $\ell_x(k)$. This defines a problem generator $\mathcal{G}_\mathcal{P} = \langle G_\mathcal{P}, g_\mathcal{P} \rangle$ in the sense of Section 2.1, where (x, a) is the problem instance and the solution function $g_\mathcal{P}$ is defined as follows: If $x \notin L_R$ and there exists a good e for (x, a), this e-value is the solution (which is unique by relaxed special soundness). These are the interesting instances. In all other cases (i.e., if $x \in L_R$ or if there is no good e for (x, a)), we define the solution to be $g_\mathcal{P}(x, a) \doteq 0^k$ (just to ensure that there is an answer for any instance). We call any such problem generator $\mathcal{G}_\mathcal{P}$ a *fake-proof generator* for $\mathcal{P}$.

For the example protocol $\mathcal{P}_{eqdlog}$, it is straightforward to verify that we can find the solution to any instance (x, a) by computing a constant number of discrete logarithms mod p. Therefore, any fake-proof generator for $\mathcal{P}_{eqdlog}$ is as easy as $\mathcal{G}_{dlog}$, and so by Lemma 1, we get

Proposition 1. *Under Assumption 1, $\mathcal{H}_{Paillier}$ is 2-harder than any fake-proof generator for $\mathcal{P}_{eqdlog}$.*

2.3   Homomorphic Encryption

A public-key cryptosystem is as usual defined by algorithms E, D for encryption and decryption and a key generation algorithm KG. The key generation receives 1^k as input and outputs a pair of private and public key (sk, pk). We will consider systems where plaintexts are integers from some interval $[0, n-1]$ where n can be computed from pk. Given plaintext a and random coins r, the ciphertext is $E_{pk}(a; r)$, and we require, of course, that $a = D_{sk}(E_{pk}(a; r))$.

We will be looking at systems that are *homomorphic*, in the following sense: the set of ciphertexts is an Abelian group, where the group operation is easy to compute given the public key. Furthermore, for any a, b, r_a, r_b it holds that $E_{pk}(a; r_a) \cdot E_{pk}(b; r_b) = E_{pk}((a+b) \bmod n; s)$ for some s. We will assume throughout that n is a k-bit number. Note that by multiplying $E_{pk}(a; r)$ by a random encryption of 0, one obtains a random and independently distributed encryption of a; we denote such operation with randomize$(E_{pk}(a; r))$.

A typical example of homomorphic encryption is Paillier's cryptosystem, where pk is a k-bit RSA modulus n, and sk is the factorization of n. Here, $E_{pk}(a; r) \doteq (1 + n)^a r^n \bmod n^2$, where r is uniformly chosen in Z_n^*.

2.4   The Registered Public-Key Model

Below we briefly review the registered public-key model (introduced in [1]), focusing on the aspects that we will need in the following. We refer the reader to [1] for the original description of the model and its relation to other setup assumptions (e.g., the common random string model).

Let $KS(1^k)$ (for *Key Setup*) be a probabilistic polynomial-time algorithm which, on input a security parameter 1^k, outputs a private/public key pair. We write $KS(1^k; r)$ to denote the execution of KS using r as random coins.

The registered public-key model [1] features a trusted functionality F_{reg}^{KS}, which the parties can invoke to register their key pairs and to retrieve other parties' public keys. Key registration takes place by having the registrant privately sending F_{reg}^{KS} the random coins r that she used to create her key pair. F_{reg}^{KS} will then run $KS(1^k; r)$, store the resulting public key along with the identity of the registrant, and later give the public key to anyone who asks for it. Note that this in particular means that to register a public key one needs to know the corresponding private key. Note also that one need not have registered a public key of his own to ask F_{reg}^{KS} for somebody else's public key.

2.5   Non-interactive Zero-Knowledge with Key Setup

Below we present a stand-alone definition of *Non-Interactive Zero-Knowledge* in the registered public-key model.[1]

Let $KS(1^k)$ be the key setup for the key-registration functionality F_{reg}^{KS}, and let R be a relation for which one can efficiently generate pairs $(x, w) \in R$ from a security parameter 1^k. A non-interactive system for R with key setup KS is a pair of efficient algorithms (P, V), where:

- $P(1^k, x, w, pk_V)$ is a probabilistic algorithm run by the prover. It takes as input a k-instance x and w such that $(x, w) \in R$, along with the verifier's public key pk_V, which the prover obtains from F_{reg}^{KS}. It outputs a string π as a non-interactive zero-knowledge proof that $x \in L_R$;
- $V(1^k, x, \pi, sk_V)$ is a deterministic 0/1-valued algorithm run by the verifier, satisfying the following *correctness* property: for all k-instances x and w such that $(x, w) \in R$, it holds that:

$$\Pr[V(1^k, x, \pi, sk_V) = 1 \mid (sk_V, pk_V) \xleftarrow{r} KS(1^k); \pi \xleftarrow{r} P(1^k, x, w, pk_V)] = 1$$

where the probability is over the random coins of KS and P;

The system is *zero-knowledge* if there exists a probabilistic polynomial-time algorithm M, such that for all k-instances x and w such that $(x, w) \in R$, the following two ensembles are indistinguishable:

VERIFIER'S KEY PAIR, REAL PROOF:

$$\{(sk_V, pk_V, \pi) \mid (sk_V, pk_V) \xleftarrow{r} KS(1^k); \pi \xleftarrow{r} P(1^k, x, w, pk_V)\}$$

VERIFIER'S KEY PAIR, SIMULATED PROOF:

$$\{(sk_V, pk_V, \pi) \mid (sk_V, pk_V) \xleftarrow{r} KS(1^k); \pi \xleftarrow{r} M(1^k, x, pk_V, sk_V)\}$$

---

[1] We only consider the setting where the key setup is required just for the verifier, as that is all we need in this paper. Adapting the definition to the case in which provers also have private/public key pair is straightforward; we omit the details.

As usual, depending on the quality of the indistinguishability of the above ensembles, one obtains computational, statistical or perfect zero-knowledge.

To define soundness, we consider a probabilistic polynomial-time adversary $\tilde{P}$ who plays the following game:

- Execute $(sk_V, pk_V) \xleftarrow{r} KS(1^k)$ and give pk_V to $\tilde{P}$.
- Repeat until $\tilde{P}$ stops: $\tilde{P}$ outputs x, π and receives $V(1^k, x, \pi, sk_V)$.

We say that $\tilde{P}$ wins if he produces at least one x, π that V accepts, where $x \notin L_R$. The protocol is *sound* if any $\tilde{P}$ wins with probability negligible in k. We say that the system is sound for a particular number of proofs $m(k)$ if the game always stops after at most $m(k)$ proofs are generated.

3   A Compilation Technique

In this section, we assume we are given a relation R and a Σ-protocol $\mathcal{P}$ for R with linear answer. When running the protocol on input (x, w), where x is a k-instance, we let $\ell_x(k)$ be the bit-length of x, $\ell_e(k)$ be the bit-length of the verifier's challenge, and $\ell_z(k)$ be the maximal bit-length of a component in the prover's answer z i.e., $z = (z_1, \ldots, z_m)$ and $\ell_z(k) \doteq \max(len(z_1), \ldots, len(z_m))$. We also use a homomorphic cryptosystem with key generation algorithm KG.

Our compilation technique works in the registered public-key model of [1] (*cf.* also Section 2.4). Specifically, we assume that each player acting as verifier has initially registered a private/public key pair with the trusted functionality F_{reg}^{KS}, using the following key setup algorithm:

$KS(1^k)$ (Key setup for the Verifier):
Set $(sk, pk) \xleftarrow{r} KG(1^{k'})$ where we choose $k' \doteq \max(f(k)k, \ell_z(k) + 1)$, and where $f(k)$ is a polynomial specified in Theorem 2 below. Choose a challenge e as V would do in the given Σ-protocol (that is, e will be a $\ell_e(k)$-bit string), and set c to be a random (homomorphic) encryption of e under pk. The public key is now (pk, c) and the private key is (sk, e).

In Section 6, we discuss how our key setup functionality F_{reg}^{KS} can be implemented efficiently in a standard PKI setting.

Note that the algorithm $KS(1^k)$ for the verifier's key setup can also be thought of as defining a problem generator, where (pk, c) is the problem instance, and e is the solution. We will call this problem generator $\mathcal{H}_{KG}$ in the following. It will be identical to $\mathcal{H}_{Paillier}$ if we use Paillier encryption.

To understand the compilation technique itself, note that because the Σ-protocol is with linear answer, it is possible to execute the prover's side of the protocol given only an encryption of the challenge e. Namely, the prover starts by computing his first message a. Then, if the answer z is supposed to contain $z_j = u_j + v_j e$, the prover will be able (by linearity) to derive the values of u_j, v_j from x, his private input w and the random coins used to create a. At this point, the prover can compute $E_{pk}(z_j)$ as $E_{pk}(u_j) \cdot c^{v_j}$. This can be decrypted and then checked as usual by V.

Now, soundness of any Σ-protocol is based on the fact that a cheating prover has to generate the first message a without knowing what the challenge is. Since, in this case, the prover is only given an encryption of the challenge, we might hope that soundness would still hold. More specifically, if the prover can, for a false statement x, come up with a first message a, and encrypted responses that the verifier would accept, then relaxed special soundness implies that x, a uniquely determines the challenge e that the verifier encrypted. If the complexity of finding e from x, a is much smaller than the complexity of breaking the verifier's cryptosystem, this gives a contradiction, as formalized below. On the other hand, zero-knowledge simulation is easy if the challenge is known to V, and the key setup exactly guarantees that V knows the challenge.

A more detailed description of the compiled protocol follows. Our construction is designed to give proofs for instances x of length up to $\ell_x(k)$. It is in general understood that the verifier will reject immediately if x is longer than $\ell_x(k)$ or if the proof is in any other way obviously malformed.

Protocol compile($\mathcal{P}$)

1. Given a k-instance x, w to prove, P gets V's public key (pk, c) from F_{reg}^{KS} and computes the first message a in a proof according to $\mathcal{P}$. Let the final message z be of the form $(u_1 + v_1 e, \ldots, u_m + v_m e)$; then, for $i = 1, \ldots, m$, P computes $c_i \xleftarrow{r} \text{randomize}(E_{pk}(u_j) \cdot c^{v_j})$. P sends x, π to V, where $\pi \doteq (a, (c_1, \ldots, c_m))$.
2. On input x and a proof $\pi \doteq (a, (c_1, \ldots, c_m))$, V sets $z_i' \leftarrow D_{sk}(c_i)$, and then verifies that $x, (a, e, (z_1', \ldots, z_m'))$ would be accepted by the verifier of protocol $\mathcal{P}$, and accepts or rejects accordingly.

Theorem 1. compile($\mathcal{P}$) *is complete and statistical zero-knowledge (in the registered public-key model).*

Proof. Completeness is clear by inspection. In particular, $D_{sk}(c_i)$ equals the correct value $z_i \doteq u_i + v_i e$, since the fact that $k' > \ell_z(k)$ ensures that $z_i < n$.

As for zero-knowledge, the simulator M will as usual interact with V and attempt to emulate the view V would see in real life. In particular, M will receive the string V sends initially (namely, the random coins r intended for F_{reg}^{KS}). This allows M to generate V's private key, and in particular the e-value inside c. Now, to simulate a proof for $x \in L_R$, M will use the special honest-verifier simulator for $\mathcal{P}$ on input x, e to generate $(a, e, z) = (a, e, (z_1, \ldots, z_m))$. It then outputs $x, (a, (E_{pk}(z_1), \ldots, E_{pk}(z_m)))$. The only difference between this simulation and real proofs is that the values $a, z_1, \ldots, z_m$ are generated by the prover in $\mathcal{P}$ in real proofs, while in M's output they are simulated. The theorem now follows from special honest-verifier zero-knowledge of $\mathcal{P}$. $\square$

Theorem 2. *Let $\mathcal{P}$ be a Σ-protocol with linear answer, and $\mathcal{H}_{KG}$ be the problem generator associated with the key setup for the verifier. Assume that $\mathcal{H}_{KG}$ is f-harder than any fake-proof generator $\mathcal{G}_\mathcal{P}$ for $\mathcal{P}$, and that the verifier's public key for the homomorphic encryption scheme is generated with security parameter $1^{k'}$, where $k' \doteq \max(f(k)k, \ell_z(k) + 1)$. Then compile($\mathcal{P}$) is sound for provers generating $O(\log k)$ proofs.*

Proof. Assume we have a probabilistic polynomial-time cheating prover $\tilde{P}$ contradicting the conclusion of the theorem. At a high level, our proof will proceed as follows: first, we describe how to use $\tilde{P}$ to obtain a fake-proof generator $\tilde{\mathcal{G}}_\mathcal{P} = \langle \tilde{G}_\mathcal{P}, g_\mathcal{P} \rangle$ for $\mathcal{P}$; then, using $\tilde{P}$ and any algorithm A that completely breaks $\tilde{\mathcal{G}}_\mathcal{P}$, we will show how to construct an algorithm A' breaking $\mathcal{H}_{KG}$ on instances of size $k' \geq f(k)k$, in time comparable to A's. This will contradict the assumption that $\mathcal{H}_{KG}$ is f-harder than any fake-proof generator for $\mathcal{P}$.

Consider the algorithm $\tilde{G}_\mathcal{P}$ which, on input 1^k, starts by generating a public key (pk, c) for the verifier according to the protocol (*i.e.*, (pk, c) was produced by $KS(1^{k'})$). Then, $\tilde{G}_\mathcal{P}$ runs $\tilde{P}$ on (pk, c), and whenever $\tilde{P}$ outputs a statement/proof pair, $\tilde{G}_\mathcal{P}$ replies with a random bit to represent the verifier's reaction to each proof. Once $\tilde{P}$ halts, $\tilde{G}_\mathcal{P}$ chooses uniformly one of the statement/proof pairs generated by $\tilde{P}$ (it will be of the form $x, (a, (c_1, \ldots, c_m)))$, and outputs (x, a).

Note that with probability $1/poly(k)$, all the bits that $\tilde{G}_\mathcal{P}$ sends to $\tilde{P}$ are identical to what the verifier would have sent. Hence, the fact that $\tilde{P}$ is a successful cheating prover implies that, with non-negligible probability, one of the statement/proof pairs $x, (a, (c_1, \ldots, c_m))$ generated by $\tilde{P}$ is such that $x \notin L_R$, yet the verifier would accept. Given that there is such a proof, there is at least a $1/(\log k)$ probability that $\tilde{G}_\mathcal{P}$ chooses this proof to generate its output. In conclusion, with overall non-negligible probability, $\tilde{G}_\mathcal{P}$ outputs $x \notin L_R, a$ for which exactly one good e exists. This value of e must be identical to the plaintext inside c since the verifier would accept the corresponding proof.

Algorithm $\tilde{G}_\mathcal{P}$ defines a fake-proof generator $\tilde{\mathcal{G}}_\mathcal{P} = \langle \tilde{G}_\mathcal{P}, g_\mathcal{P} \rangle$ for $\mathcal{P}$ (where, as in Section 2.2, $g_\mathcal{P}(x, a)$ is the good e-value if one exists and $x \notin L_R$, and 0^k otherwise). Hence, the assumption that $\mathcal{H}_{KG}$ is f-harder than any fake-proof generator for $\mathcal{P}$ implies in particular that $\mathcal{H}_{KG}$ is f-harder than $\tilde{\mathcal{G}}_\mathcal{P}$.

Let A be a probabilistic algorithm that breaks $\tilde{\mathcal{G}}_\mathcal{P}$ completely in time $T(k)$, and consider the following algorithm A' to break $\mathcal{H}_{KG}$. On input a k'-instance (pk, c) for $\mathcal{H}_{KG}$ (*i.e.*, (pk, c) was produced by $KS(1^{k'})$) A' invokes $\tilde{P}$ on (pk, c) and interacts with it according to the exact same strategy that we described above for $\tilde{G}_\mathcal{P}$. At the end of such interaction, A' will obtain a pair (x, a): at this point, A' runs A on (x, a), and outputs the value e returned by A.

By the above analysis of $\tilde{G}_\mathcal{P}$ and the fact that A breaks $\tilde{\mathcal{G}}_\mathcal{P}$ completely, we see that A' returns the plaintext encrypted inside c with non-negligible probability. Since A' runs in time $T(k) + poly(k)$ and $k' \geq f(k)k$, this contradicts the assumption that $\mathcal{H}_{KG}$ is f-harder than $\tilde{\mathcal{G}}_\mathcal{P}$. $\square$

For the example protocol $\mathcal{P}_{eqdlog}$, the above theorem and Proposition 1 imply the following:

Corollary 1. *Suppose we construct* compile($\mathcal{P}_{eqdlog}$) *using Paillier encryption with security parameter* $1^{k'}$, *where* $k' \doteq \max(2k, \ell_z(k) + 1)$. *Then, under Assumption 1,* compile($\mathcal{P}_{eqdlog}$) *is sound for provers generating* $O(\log k)$ *proofs. Moreover, its communication and computational complexity are a constant factor times those of* $\mathcal{P}_{eqdlog}$.

While the restriction to a logarithmic number of proofs may seem like a serious one, there are in fact many applications where this result is good enough. The point is that our reduction only fails for polynomially-many proofs because we assume that the prover learns whether the verifier accepts each individual proof. However, when a zero-knowledge protocol is used as a tool in a larger construction, the prover often does not get this information, and thus in such cases, it is enough that soundness holds for a single proof. The application to threshold RSA in the next section is an example of this.

Moreover, we believe that compile($\mathcal{P}_{eqdlog}$) is in fact sound, even for an arbitrary polynomial number of proofs. We can show this under a stronger non-standard assumption: we report the details in Appendix A.

4   Threshold RSA

Our technique can be used in most known threshold RSA- or discrete-log-based cryptosystems to obtain efficient solutions not relying on random oracles. As a concrete example, we consider here Shoup's threshold RSA protocol [17].

In this construction, a trusted dealer generates an RSA modulus $N = pq$, where $p = 2p' + 1, q = 2q' + 1$ and p', q' are k-bit primes. In addition, the dealer publishes an element $v \in Z_N^*$ of order $p'q'$, and sets up a secret sharing of the private exponent. Each server S_i in the protocol privately receives a share s_i (which is a number modulo $p'q'$). Finally, the dealer publishes the value $v_i \doteq v^{s_i} \bmod N$ for each server.

When the system is operational, a client may send an input α to be signed to all servers. Each server S_i in the protocol produces an element β_i which is guaranteed to be in the subgroup of Z_N^* of order $p'q'$ (because it is a square of another element). Server S_i then sends β_i to the client, claiming that $\beta_i = \alpha^{s_i} \bmod N$. Assuming that the majority of the servers are honest, the client can reconstruct the desired signature, as long as he does not accept any incorrect β_i's. Each server must therefore prove to the client that β_i was correctly formed.

The following Σ-protocol $\mathcal{P}_{dlmodN}$ can be used as the basis for a solution:

1. S_i chooses a random $4k$-bit integer r and sends $a \doteq (a_1, a_2)$ to V, where $a_1 \doteq v^r \bmod N, a_2 \doteq \alpha^r \bmod N$.
2. The verifier chooses a random $(k - 1)$-bit string e and sends it to P.
3. S_i sends $z \doteq r + es_i$ to the verifier who checks that $v^z = a_1 v_i^e \bmod N, \alpha^z = a_2 \beta_i^e \bmod N$.

Assuming that N, v are generated by the trusted dealer as described, it follows from the arguments given in [17] that this is a Σ-protocol for proving that $\log_v(v_i) = \log_\alpha(\beta_i)$. Indeed, the non-interactive solution proposed in [17] is simply the Fiat-Shamir heuristic applied to this protocol.

We propose to apply instead our compilation technique based on Paillier encryption to get a non-interactive solution. This leads to:

Theorem 3. *Under Assumption 1, there exists a non-interactive threshold RSA scheme, secure in the registered public-key model. Its communication and computational complexity are the same as in Shoup's scheme, up to a constant factor.*

Proof. The protocol given above has the right properties for applying the compilation technique, the only exception being a small technical issue with soundness: the protocol has relaxed special soundness only for inputs where N, v are correctly formed, while our definition requires it for all inputs. However, we can simply instruct the verifier to reject all inputs not containing the N, v generated by the dealer. This will force a cheating prover to only use inputs for which relaxed special soundness holds, and the proof of Theorem 2 then goes through in the same way as before.

To apply Theorem 2, we need to show that $\mathcal{H}_{Paillier}$ is $f(k)$-harder (for some $f(k)$) than any fake-proof generator $\mathcal{G}_{dlmodN} = \langle G_{dlmodN}, g_{dlmodN} \rangle$ for $\mathcal{P}_{dlmodN}$. To this end, observe that when we argue soundness, we may assume that the factors p, q of N are known, since soundness is based only on security of the (independently chosen) Paillier public key, specified by the verifier's key pair. Now, instances for a fake-proof generator $\mathcal{G}_{dlmodN}$ have the form $(x, a) \doteq ((N, v, v_i, \alpha, \beta_i), (a_1, a_2))$, whereas the solution $g_{dlmodN}(x, a)$ typically is the only e-value that S_i can answer (unless either there is no such e-value, or the theorem x is true, in which cases $g_{dlmodN}(x, a) \doteq 0^k$). Since p, q are known, we can reduce everything modulo p and q and the Chinese remainder theorem now implies that we can find the solution by computing a constant number of discrete logarithms mod p and q. Consequently, any fake-proof generator for $\mathcal{P}_{dlmodN}$ is as easy as $\mathcal{G}_{dlog}$; therefore Assumption 1 implies that compile($\mathcal{P}_{dlmodN}$) is sound against provers giving $O(\log k)$ proofs (though, as we will see below, we only need soundness for provers giving a single proof).

To prove that the RSA protocol is secure, we must first show that an adversary corrupting at most half the servers learns nothing from the protocol, except for the RSA signatures that the protocol is supposed to produce. This follows from zero-knowledge of compile($\mathcal{P}_{dlmodN}$) and the simulator given in [17].

Second, we must show that no probabilistic polynomial-time adversary can make an honest client fail to output a correct RSA signature (even on input messages chosen by the adversary). We will show that existence of an adversary *Adv* doing this with non-negligible probability contradicts soundness of compile($\mathcal{P}_{dlmodN}$), namely we construct from *Adv* a prover that cheats the client on a single proof with non-negligible probability.

For this, we will execute the dealer's algorithm to set up the RSA key and give shares of the private key to the adversary for those servers he wants to corrupt. We also give him the public key of the client we want to attack. Assume *Adv* chooses a maximum of i_{max} input messages for the client before halting. We pick i at random in $[1, i_{max}]$ and hope that the i-th message is the first where *Adv* is successful in cheating the client. Since i_{max} is polynomial in k, our guess is correct with $1/poly(k)$ probability. Assuming our guess is correct, we can perfectly simulate what *Adv* sees for any previous message m_j, $j < i$: for the actions of honest servers, we can simply follow the protocol (as we know the private RSA key and all its shares); as for the client, for $j < i$ he will just output a correct RSA signature on m_j, which we can also compute.

Since (assuming a correct guess of i) we can perfectly simulate Adv's view up to message m_i, there is a non-negligible probability that Adv successfully cheats the client when he tries to get a signature on m_i. But for this to happen, Adv must fool the client into accepting an incorrect share, which can only occur if Adv produced (for at least one of the corrupt servers) an acceptable proof for an incorrect statement. Thus, we choose at random one of the corrupt servers and output its statement and proof. This is clearly a successful cheating prover. □

5 The OR-Construction and NIZKs for NP

5.1 Closure Under OR-Construction

A construction that is widely used in designing efficient Σ-protocols is the so-called OR-construction [8]. Given Σ-protocols Σ_l and Σ_r for relations R_l and R_r, the OR-construction yields a Σ-protocol Σ_{OR} for the following relation R_{OR}:

$$((x_l, x_r), (w_l, w_r)) \in R_{OR} \Leftrightarrow ((x_l, w_l) \in R_l \vee (x_r, w_r) \in R_r).$$

The OR-construction is based on executing the two protocols for relations R_l, R_r in parallel, where the prover derives the two challenges from a single value chosen by the verifier. In our case, we do all computations on challenges over the integers, which means that some details of the standard construction have to be modified slightly; this is covered in Appendix B.

An attractive feature of the compilation technique proposed in Section 3 is that if it is applicable to both Σ_l and Σ_r, then it is also applicable to the composed protocol Σ_{OR}. In other words:

Theorem 4. *The class of Σ-protocols that can be made non-interactive using our homomorphic-encryption-based technique is closed under OR-construction.*

Proof. Let Σ_l and Σ_r be Σ-protocols with linear answer. The theorem amounts to proving that the Σ-protocol Σ_{OR} resulting from the OR-construction also features a "linear answer," and so we can apply the compiler from Section 3. Now, valid conversations of Σ_{OR} (*cf.* Appendix B) have the form:

$$((a_l, a_r), e, (e_l, z_l, e_r, z_r)),$$

where (e_l, e_r) is a "split" for e, that is, $e = e_l - e_r$ and either e_l or e_r was chosen randomly by the prover when preparing (a_l, a_r). Hence, e_l and e_r are clearly linear; moreover, since both Σ_l and Σ_r have linear answer, z_l and z_r are also linear, and the theorem follows. □

5.2 Non-interactive Bit Commitments

We now describe a non-interactive bit-commitment scheme for the registered public-key model, along with non-interactive protocols to prove boolean relations among committed bits.

Consider the Σ-protocol $\mathcal{P}_{eqdlog}$ for equality of discrete logarithms described in Section 2.2. Applying the OR-construction to two instances of $\mathcal{P}_{eqdlog}$ yields a Σ-protocol $\mathcal{P}_{1out2}$ for proving that one out of two pairs of discrete logarithms is equal. In other words, $\mathcal{P}_{1out2}$ is a proof system for statements of the form $x \doteq (p, p', g_1, g_2^0, g_2^1, h_1, h_2)$, where p, p' are prime, $p = 2p' + 1$, $g_1, g_2^0, g_2^1 \in \mathbb{Z}_p^*$ have order p', $h_1 = g_1^w \bmod p$ (for some $w \in \mathbb{Z}_{p'}$) and either $h_2 = (g_2^0)^w \bmod p$ or $h_2 = (g_2^1)^w \bmod p$.

To commit to a bit b, the prover picks p, p', g_1, g_2^0, g_2^1 as described above,[2] randomly selects $w \in \mathbb{Z}_{p'}$ and computes $h_1 = g_1^w \bmod p$, $h_2 = (g_2^b)^w \bmod p$. At this point, the prover uses $\mathsf{compile}(\mathcal{P}_{1out2})$ to prove (non-interactively) that the statement $x \doteq (p, p', g_1, g_2^0, g_2^1, h_1, h_2)$ is well-formed. The commitment then consists of x along with such NIZK, though in the following we will often refer to x by itself as the commitment to keep the discussion simpler.

To open the commitment to b, it suffices to show that $\log_{g_1} h_1 = \log_{g_2^b} h_2$, which the prover can do non-interactively via the protocol $\mathsf{compile}(\mathcal{P}_{eqdlog})$.

Now, suppose that we want to show that three bits b_1, b_2, b_f (hidden within commitments x_1, x_2, x_f, respectively) satisfy $b_f = f(b_1, b_2)$, for some binary boolean function f. Proving such relation amounts to prove that (x_1, x_2, x_f) can be opened either to $(0, 0, f(0, 0))$, or to $(0, 1, f(0, 1))$, or to $(1, 0, f(1, 0))$, or to $(1, 1, f(1, 1))$. But this is just the disjunction of statements that can each be proven using three instances of $\mathcal{P}_{eqdlog}$; hence, applying the OR-construction we get a Σ-protocol Σ_f that can be made non-interactive as described in Section 3.

5.3   NIZK for Circuit Satisfiability

The discrete-logarithm-based non-interactive bit-commitment scheme from Section 5.2 can be used, in conjunction with the approach of [6], to obtain efficient non-interactive zero-knowledge arguments for Circuit Satisfiability, and hence for any NP language.

To show that a given circuit is satisfiable, the prover P commits to his satisfying assignment and to all intermediate bits resulting form the computation of the circuit, and sends all these non-interactive bit-commitments to the verifier V. Additionally, P non-interactively opens the output bit to 1, and prove non-interactively to the verifier that the commitments to the inputs and the output of each gate of the circuit are consistent.

Upon receiving such non-interactive proof, V checks that all the commitments are well-formed, that the output of the circuit actually opens to 1, and that the proof of consistency of each gate is correct, and if so, V accepts P's proof.

Notice that the length of such non-interactive proof is proportional to the circuit's size and to the security parameter 1^k, and is thus "linear" in the sense of the "Linear Zero-Knowledge" of [6], whereas previous constructions [14] in the common random string model are quadratic in this regard, even under specific number-theoretic assumptions.

---

[2] As a matter of efficiency, we notice that, when committing to many bits, the prover can safely reuse the values p, p', g_1, g_2^0 and g_2^1.

6   Implementing the Key Setup

The compilation technique of Section 3 works in the registered public-key setting. In this model, each verifier V registers her public key by sending the random coins used to generate her private key/public key pair to a trusted functionality. This is exploited in the proof of Theorem 1 to enable the simulator M to derive the private key of the verifier, and to ensure the validity of the public key.

Of course, such a functionality can always be implemented using a more standard PKI with a certification authority CA, and generic zero-knowledge techniques. The verifier sends her public key to the CA and proves in zero-knowledge that she knows a set of random coins that, using the given key-generation algorithm, leads to the public key she sent.

This will be very inefficient in general. But in fact, taking a closer look at the simulation for the case where the verifier uses Paillier encryption, one can see that all that is needed is knowledge of the challenge value e and of the RSA modulus n, plus assurance that e lies in the proper interval and that n is well-formed. (Knowledge of the factorization of n, in particular, is not required.) In our case, it is enough to know that n is the product of two distinct primes and that n is relatively prime to $\phi(n)$. Hence, registration of the verifier's key pair for the key setup from Section 3 can be efficiently implemented by having V and CA engage in the following protocol:

Step 0: V sends her public key (n, c) to CA;
Step 1: V proves to CA that n is well-formed;
Step 2: V proves knowledge of the plaintext e hidden within c; and that this value e lies in the specified interval.

All the above steps can be efficiently realized leveraging known tools from the literature [18, 2, 9]. In particular, Step 1 can be carried out by first using the protocol of van de Graaf and Peralta [18], by which one can show that $n = p^i q^j$ where $p \equiv q \equiv 3 \bmod 4$ and i, j are odd. Then one can use the following folklore trick: the verifier chooses a random element in Z_n^*, and the prover proves in zero-knowledge that it has an n-th root mod n. This will always be the case if $\gcd(n, \phi(n)) = 1$, but fails with constant probability otherwise. As for Step 2, one can first use an integer commitment scheme (like the one of Damgård and Fujisaki [9]) to create a commitment Com to e, and then prove knowledge of the value committed within Com (e.g., using the protocol in Section 4.1 of [9]). Then, using standard techniques, it is possible to show that the commitment Com and the ciphertext c hide the same value e. For completeness, in Appendix C we sketch a simple Σ-protocol to achieve this. Finally, Boudot's efficient proof of membership in intervals [2] allows the prover to prove that the e contained in Com lies in the required range.

Acknowledgement. We thank the anonymous referees for useful advise on improving the presentation.

References

1. B. Barak, R. Canetti, J. B. Nielsen, and R. Pass. Universally Composable Protocols with Relaxed Set-Up Assumptions. In *Proceedings of the 45th Annual IEEE Symposium on Foundations of Computer Science (FOCS'04)*, pages 186–195. IEEE Computer Society, 2004.
2. F. Boudot. Efficient Proofs that a Commited Number Lies in an Interval. In *Advances in Cryptology—EUROCRYPT '00*, volume 1807 of *LNCS*, pages 431–444. Springer, 2000.
3. R. Canetti, O. Goldreich, S. Goldwasser, and S. Micali. Resettable zero-knowledge. In *STOC'99*, pages 235–244. ACM Press, 1999.
4. D. Chaum and T. P. Pedersen. Wallet databases with observers. In *Advances in Cryptology—CRYPTO '92*, volume Volume 740 of *LNCS*, pages 89–105. Springer, 1992.
5. R. Cramer. *Modular Design of Secure yet Practical Cryptographic Protocols*. PhD thesis, CWI and University of Amsterdam, 1996.
6. R. Cramer and I. Damgård. Linear Zero-Knowledge—A Note on Efficient Zero-Knowledge Proofs and Arguments. In *Proceedings of the 29th Annual ACM Symposium on Theory of Computing*, pages 436–445. ACM Press, 1997.
7. R. Cramer and I. Damgård. Secret-Key Zero-Knowledge. In *Theory of Cryptography—TCC '04*, pages 223–237. Springer-Verlag, 2004. LNCS 2951.
8. R. Cramer, I. Damgård, and B. Schoenmakers. Proofs of Partial Knowledge and Simplified Design of Witness Hiding Protocols. In *Advances in Cryptology—CRYPTO '94*, pages 174–187. Springer, 1994. LNCS 839.
9. I. Damgård and E. Fujisaki. A Statistically-Hiding Integer Commitment Scheme Based on Groups with Hidden Order. In *Advances in Cryptology—ASIACRYPT '02*, pages 125–142. Springer, 2002. LNCS 2501.
10. C. Dwork and M. Naor. Zaps and Their Applications. In *Proceedings of the 41st Annual IEEE Symposium on Foundations of Computer Science (FOCS'00)*, pages 283–293. IEEE Computer Society, 2000.
11. A. Fiat and A. Shamir. How to Prove Yourself: Practical Solutions to Identification and Signature Problems. In *Advances in Cryptology—Crypto'86*, volume 263 of *LNCS*, pages 186–194, Berlin, 1987. Springer.
12. S. Goldwasser and Y. Tauman Kalai. On the (In)security of the Fiat-Shamir Paradigm. In *FOCS '03*, pages 102–115. IEEE Computer Society, 2003.
13. J. Groth, R. Ostrovsky, and A. Sahai. Perfect Non-Interactive Zero Knowledge for NP. http://eprint.iacr.org/2005/290, 2005.
14. Joe Kilian and Erez Petrank. An Efficient Non-interactive Zero-Knowledge Proof System for NP with General Assumptions. *J. Cryptology*, 11(1):1–27, 1998.
15. P. Paillier. Public Key Cryptosystems Based on Composite Degree Rediduosity Classes. In *Advances in Cryptology—EUROCRYPT '99*, pages 223–238. Springer, 1999. LNCS 1592.
16. C. Schnorr. Efficient Signature Generation by Smart Cards. *Journal of Cryptology*, 4(3):161–174, 1991.
17. V. Shoup. Practical Threshold Signatures. In *Advances in Cryptology—EUROCRYPT '00*, pages 207–220. Springer, 2000. LNCS 1807.
18. J. van de Graaf and R. Peralta. A Simple and Secure Way to Show Validity of Your Public Key. In *Advances in Cryptology—CRYPTO '87*, volume 293 of *LNCS*, pages 128–134. Springer, 1988.

A   Unbounded Soundness of compile($\mathcal{P}_{eqdlog}$)

Below we sketch an argument showing soundness of compile($\mathcal{P}_{eqdlog}$) for provers generating any polynomial number of NIZKs, assuming Paillier cryptosystem is used for the homomorphic encryption. Throughout, all exponentiations are meant modulo p, unless noted otherwise.

Recall that valid statements for the protocol $\mathcal{P}_{eqdlog}$ have the form $x \doteq (p, p', g_1, g_2, g_1^w, g_2^w)$, where w is the secret input to the prover. Using w, an honest prover computes his proof as $\sigma \doteq ((g_1^r, g_2^r), E_n(r) \cdot c^w \bmod n^2)$, where c is the encrypted challenge and n is the verifier's modulus.

Our argument works in a "generic model" for the homomorphic encryption scheme: namely, we assume that whenever the prover outputs a proof $\sigma \doteq ((a_1, a_2), \bar{c})$, the ciphertext $\bar{c}$ is specified as a pair of integers (u, v) such that $\bar{c} = E_n(u) \cdot c^v \bmod n^2$.

Theorem 5. *Suppose we construct* compile($\mathcal{P}_{eqdlog}$) *using Paillier encryption. Then under Assumption 1,* compile($\mathcal{P}_{eqdlog}$) *is unboundedly sound for provers using the homomorphic properties of Paillier encryption in a black-box fashion.*

Proof. Let A be an algorithm completely breaking k-instances of $\mathcal{G}_{dlog}$ in time $T(k)$, and assume we have a cheating prover $\tilde{P}$ contradicting the conclusion of the theorem. We show how to use A and $\tilde{P}$ to construct an algorithm A' that breaks $2k$-instances of $\mathcal{H}_{Paillier}$ in time $O(T(k) + poly(k))$, contradicting Assumption 1.

On input (n, c), A' starts by executing $\tilde{P}$ on the same values n, c. During its execution, $\tilde{P}$ produces several statement/proof pairs x, σ to which A' ought to reply with a bit representing the verifier's reaction. Let $x \doteq (p, p', g_1, g_2, h_1, h_2)$ and $\sigma \doteq ((a_1, a_2), (u, v))$; then A' replies with 1 if and only if the following relations hold:

$$h_1 = g_1^v \bmod p, \quad h_2 = g_2^v \bmod p, \quad a_1 = g_1^u \bmod p, \quad a_2 = g_2^u \bmod p \quad (\sharp)$$

When $\tilde{P}$ stops running, A' picks at random a statement/proof pair x, σ among those produced by $\tilde{P}$. Then, calling A twice, A' can compute $w_1 \doteq \log_{g_1} h_1$ and $w_2 \doteq \log_{g_2} h_2$, thus being able to decide whether x is a valid statement or not. In either case, A' can recover e from σ with either one or two more calls to A:

A-1. if x is a false statement (*i.e.*, $w_1 \neq w_2$), then A' invokes A to learn $r_1 \doteq \log_{g_1} a_1$, $r_2 = \log_{g_2} a_2$, and computes $e \doteq (r_2 - r_1)(w_1 - w_2)^{-1} \bmod p$;

A-2. if x is a valid statement (*i.e.*, $w_1 = w_2 = w$), then A' invokes A to learn $r \doteq \log_{g_1} a_1$, and computes $e \doteq (r - u)(v - w)^{-1} \bmod p$ (if $w = v$, then A' aborts).

The running time of A' is clearly $O(T(k) + poly(k))$. We now argue about its success probability. In the analysis, we use the term "funny" proof to refer to a proof σ for a true statement x that was not obtained according to the protocol, yet it passes the verifier's test. In our "generic model," given the fact that $\mathcal{P}_{eqdlog}$

admits at most one valid answer z for any given x, a, e, a funny proofs satisfies $(u,v) \neq (\log_{g_1} a_1, \log_{g_1} h_1)$, but $u + v \cdot e = \log_{g_1} a_1 + \log_{g_1} h_1 \cdot e$.

The view that $\tilde{P}$ sees within the simulation put on by A' deviates from what $\tilde{P}$ would see in a real interaction with the verifier only after $\tilde{P}$ produces a statement/proof pair x, σ for which either of the following two cases occurs:

B-1. x is a false statement, but σ passes the verifier's test (whereas according to the test ($\sharp$), A' always rejects σ in such case);

B-2. x is a true statement, but σ is a "funny" proof (notice that the test ($\sharp$) ensures that A' rejects all proofs not created according to the protocol).

Observe that since $\tilde{P}$ is a successful cheating prover, then at least one of the above cases will occur with non-negligible probability. Let i^* be the index of the first such occurrence. With $1/poly(k)$ probability, the random statement/proof pair chosen by A' will be exactly the i^*-th pair. Conditioning on such event, the simulation of the verifier's answers up to that point is perfect, and moreover:

C-1. if i^* corresponds to a false statement (case B-1. above), then the fact that σ passes the verifier's condition, along with relaxed special-soundness, implies that the value e computed by A' according to case A-1. is indeed correct;

C-2. if i^* corresponds to a "funny" proof (case B-2. above), then the fact that σ passes the verifier's condition implies that the value e computed by A' according to case A-2. is correct. (Notice that σ being a "funny" proof excludes the possibility of aborting in case A-2.)

In conclusion, with non-negligible probability, A' outputs the correct solution e to the $2k$-instance n, c in time $O(T(k) + poly(k))$, contradicting the assumption that $\mathcal{H}_{Paillier}$ is 2-harder than $\mathcal{G}_{dlog}$. $\qquad\square$

B   The OR-Construction of [8]

The OR-construction [8] derives a Σ-protocol Σ_{OR} from Σ_l and Σ_r by allowing the prover to "split" the challenge $e \in [0, 2^k[$ into two parts $e_l, e_r \in [0, 2^{2k}[$ as he wishes, as long as $e_l - e_r = e$. This enables the prover to "simulate" the false part of the statement, while actually carrying out the proof for the part which is true. More in details, conversations in the OR-construction have the form:

$$((a_l, a_r), e, (e_l, z_l, e_r, z_r)),$$

where an honest prover P constructs his flows differently depending on whether P holds a valid witness w_l for x_l, or a valid w_r for x_r.

In the first case, P picks a random e_r from $[0, 2^{2k}[$ and uses the simulator for Σ_r to obtain an accepting conversation (a_r, e_r, z_r) for x_r. Then, P selects a_l according to Σ_l and sends (a_l, a_r) to V. When P receives e, he sets $e_l \doteq e_r + e$ and computes z_l with respect to x_l, a_l, e_l and the witness w_l, according to Σ_l.

The second case is completely analogous, except that P sets $e_r \doteq e_l - e$.

As for the verification condition, V checks that (a_l, e_l, z_l), (a_r, e_r, z_r) are accepting conversations respectively for x_l and x_r, and that (e_l, e_r) is a valid "split" for e, that is, $e = e_l - e_r$.

Observe that choosing e_l and e_r to be k bits longer than e ensures that the joint distribution of (e_l, e_r) does not reveal (to the verifier) information about whether P had a valid witness for the "left" or for the "right" part of Σ^{OR}. Indeed, given any fixed value of e in $[0, 2^k[$, the statistical distance between the two marginal distributions on (e_l, e_r) induced by the experiments described below is clearly negligible in k:

"Left" distribution: randomly choose e_r from $[0, 2^{2k}[$, and set $e_l \doteq e_r + e$;
"Right" distribution: randomly choose e_l from $[0, 2^{2k}[$, and set $e_r \doteq e_l - e$.

C   An Efficient Sub-protocol for the Key Setup

Let n be the verifier's modulus, e be her secret k-bit challenge, and c be a random Paillier encryption of e under n, namely $c \leftarrow (1+n)^e r^n \bmod n^2$, for some random $r \in \mathbb{Z}_n^*$. Recall that in the integer commitment scheme of [9], a commitment Com to e has the form $Com \doteq G^e H^s \bmod N$, where N is the product of two k-bit strong primes, G, H are generators of the subgroup of quadratic residues modulo N, and s is a $2k$-bit randomizer.

For binding, it is important that the verifier does not know neither the factorization of N nor the discrete log of H base G. In our setting, this can be enforced by having CA choosing G, H and N. Afterward, V can prove to CA that c and Com hide the same value via the following protocol:

1. V randomly selects $\hat{e} \in [0, 2^{3k}[$, $\hat{s} \in [0, 2^{4k}[$, $\hat{r} \in \mathbb{Z}_n^*$, and sends CA the values $\widehat{Com} \leftarrow G^{\hat{e}} H^{\hat{s}} \bmod N$ and $\hat{c} \leftarrow (1+n)^{\hat{e}} \hat{r}^n \bmod n^2$;
2. CA replies with a random $(k-1)$-bit challenge t;
3. V computes $\tilde{e} \leftarrow \hat{e} + et$ and $\tilde{s} \leftarrow \hat{s} + st$ (over the integers), and $\tilde{r} \leftarrow \hat{r} \cdot r^t \bmod n$, and sends $\tilde{e}, \tilde{r}, \tilde{s}$;
4. CA checks that $G^{\tilde{e}} H^{\tilde{s}} \overset{?}{=} \widehat{Com} \cdot Com^t \bmod N$ and $(1+n)^{\tilde{e}} \tilde{r}^n \overset{?}{=} \hat{c} \cdot c^t \bmod n^2$.

It is easy to check the usual properties of this protocol; in particular since t is chosen so that it is less than each prime factor in N, ability to answer more then one challenge unconditionally implies that the values hidden within Com and c are the same.

Ring Signatures: Stronger Definitions, and Constructions Without Random Oracles

Adam Bender, Jonathan Katz*, and Ruggero Morselli**

Department of Computer Science, University of Maryland
{bender, jkatz, ruggero}@cs.umd.edu

Abstract. *Ring signatures*, first introduced by Rivest, Shamir, and Tauman, enable a user to sign a message so that a *ring* of possible signers (of which the user is a member) is identified, without revealing exactly *which member* of that ring actually generated the signature. In contrast to group signatures, ring signatures are completely "ad-hoc" and do not require any central authority or coordination among the various users (indeed, users do not even need to be aware of each other); furthermore, ring signature schemes grant users fine-grained control over the level of anonymity associated with any particular signature.

This paper has two main areas of focus. First, we examine previous definitions of security for ring signature schemes and suggest that most of these prior definitions are too weak, in the sense that they do not take into account certain realistic attacks. We propose new definitions of anonymity and unforgeability which address these threats, and then give separation results proving that our new notions are strictly stronger than previous ones. Next, we show two constructions of ring signature schemes in the standard model: one based on generic assumptions which satisfies our strongest definitions of security, and a second, more efficient scheme achieving weaker security guarantees and more limited functionality. These are the first constructions of ring signature schemes that do not rely on random oracles or ideal ciphers.

1   Introduction

Ring signatures enable a user to sign a message so that a "ring" of possible signers (of which the user is a member) is identified, without revealing exactly which member of that ring actually generated the signature. This notion was first formally introduced by Rivest, Shamir, and Tauman [20], and ring signatures — along with the related notion of ring/ad-hoc identification schemes — have been studied extensively since then [5, 19, 1, 23, 16, 11, 22, 18, 2]. Ring signatures are related, but incomparable, to the notion of group signatures [6]. On the one hand, group signatures have the additional feature that the anonymity of a signer can be revoked (i.e., the signer can be traced) by a designated group manager.

* This research was supported in part by NSF Trusted Computing Grants #0310499 and #0310751, NSF-ITR #0426683, and NSF CAREER award #0447075.
** Supported by NSF Trusted Computing Grant #0310499 and NSF-ITR #0426683.

S. Halevi and T. Rabin (Eds.): TCC 2006, LNCS 3876, pp. 60–79, 2006.

On the other hand, ring signatures allow greater flexibility: no centralized group manager or coordination among the various users is required (indeed, users may be unaware of each other at the time they generate their public keys), rings may be formed completely "on-the-fly" and in an ad-hoc manner, and users are given fine-grained control over the level of anonymity associated with any particular signature (via selection of an appropriate ring).

Ring signatures naturally lend themselves to a variety of applications which have been suggested already in previous work (see especially [20, 19, 11, 2]). The original motivation was to allow secrets to be leaked anonymously. Here, for example, a high-ranking government official can sign information with respect to the ring of *all* similarly high-ranking officials; the information can then be verified as coming from *someone* reputable without exposing the actual signer. Ring signatures can also be used to provide a member of a certain class of users access to a particular resource without explicitly identifying this member; note that there may be cases when third-party verifiability is required (e.g., to prove that the resource has been accessed) and so ring signatures, rather than ad-hoc identification schemes, are needed. Finally, we mention the application to designated-verifier signatures [17] especially in the context of e-mail. Here, ring signatures enable the sender of an e-mail to sign the message with respect to the ring containing the sender and the receiver; the receiver is then assured that the e-mail originated from the sender but cannot prove this to any third party. We remark that for this latter application it is sufficient to use a ring signature scheme which supports only rings of size two. See also [7] for another proposed application of ring signatures which support only rings of size two.

1.1 Our Contributions in Relation to Previous Work

This paper focuses on both definitions and constructions. We summarize our results in each of these areas, and relate them to prior work.

Definitions of security. Prior work on ring signature/identification schemes provides definitions of security that are either rather informal or seem (to us) unnaturally weak, in that they do not address what seem to be valid security concerns. One example is the failure to consider the possibility of *adversarially-chosen* public keys. Specifically, both the anonymity and unforgeability definitions in most prior work assume that honest users always sign with respect to rings consisting entirely of *honestly-generated* public keys; no security is provided if users sign with respect to a ring containing even one adversarially-generated public key. Clearly, however, a scheme which is not secure in the latter case is of limited use; this is especially true since rings are constructed in an ad-hoc fashion using keys of (possibly unknown) users which are not validated as being correctly constructed by any central authority. We formalize security against such attacks (as well as others), and show separation results proving that our definitions are strictly stronger than those considered in previous work. In addition to the new, strong definitions we present, the *hierarchy* of definitions we give is useful for characterizing the security of ring signature constructions.

Constructions. We show two constructions of ring signature schemes which are proven secure in the standard model. We stress that these are the *first* such constructions, as all previous constructions of which we are aware rely on random oracles/ideal ciphers.[1] It is worth remarking that ring identification schemes are somewhat easier to construct (using, e.g., techniques from [9]); ring signatures can then easily be derived from such schemes using the Fiat-Shamir methodology in the random oracle model [14]. This approach, however, is no longer viable (at least, based on our current understanding) when working in the standard model.

Our first construction is based on generic assumptions, and satisfies the strongest definitions of anonymity and unforgeability considered here. This construction is inspired by the generic construction of group signatures due to Bellare, et al. [3] and, indeed, the constructions share some similarities at a high level. However, a number of subtleties arise in our context that do not arise in the context of group signatures, and the construction given in [3] does not immediately lend itself to a ring signature scheme. Two issues in particular that we need to deal with are the fact that we have no central group manager to issue "certificates" as in [3], and that we additionally need to take into account the possibility of adversarially-generated public keys as discussed earlier (this is not a concern in [3] where there is only a single group public key published by a (semi-)trusted group manager).

Our second construction is more efficient than the first, but relies on specific number-theoretic assumptions. Furthermore, it provides more limited functionality and security guarantees than our first construction; most limiting is that it only supports rings of size two. Such a scheme is still useful for certain applications (as discussed earlier); furthermore, constructing an efficient 2-user ring signature scheme without random oracles seems difficult, as we still do not have the Fiat-Shamir methodology available in our toolbox. This second scheme is based on the (standard) signature scheme recently proposed by Waters [21] in the context of ID-based encryption; in fact, we reduce the security of our scheme directly to the security of his scheme.

2   Preliminaries

We use the standard definitions of public-key encryption schemes and semantic security, signature schemes and existential unforgeability under adaptive chosen-message attacks, and computational indistinguishability. In this paper we will assume public-key encryption schemes for which, with all but negligible probability over (pk, sk) generated at random using the specified key generation algorithm, $\mathsf{Dec}_{sk}(\mathsf{Enc}_{pk}(M)) = M$ holds with probability 1.

---

[1] Although Xu, Zhang, and Feng [22] claim a ring signature scheme in the standard model based on specific assumptions, their proof was later found to be flawed (personal communication from J. Xu, March 2005). Concurrently to our work, Chow, Liu and Yuen [8] show a ring signature scheme that they prove secure in the standard model (for rings of *constant* size) based on a new number-theoretic assumption.

We will also use the notion of a *ZAP*, which is a 2-round, public-coin, witness-indistinguishable proof system for any language in $\mathcal{NP}$ (the formal definition is given in Appendix A). ZAPs were introduced by Dwork and Naor [12], who show that ZAPs can be constructed based on trapdoor permutations. For notational purposes, we represent a ZAP by a triple $(\ell, \mathcal{P}, \mathcal{V})$ such that (1) the initial message r from the verifier has length $\ell(k)$ (where k is the security parameter); (2) the prover $\mathcal{P}$, on input the verifier-message r, statement x, and witness w, outputs $\pi \leftarrow \mathcal{P}_r(x, w)$; finally, (3) $\mathcal{V}_r(x, \pi)$ outputs 1 or 0, indicating acceptance or rejection of the proof.

3 Definitions

We begin by presenting the functional definition of a ring signature scheme. We refer to an ordered list $R = (PK_1, \ldots, PK_n)$ of public keys as a *ring*, and let $R[i] = PK_i$. We will also freely use set notation, and say, e.g., that $PK \in R$ if there exists an index i such that $R[i] = PK$. We will always assume, without loss of generality, that the keys in a ring are ordered lexicographically.

Definition 1 (Ring signature). *A ring signature scheme is a triple of* PPT *algorithms* (Gen, Sign, Vrfy) *that, respectively, generate keys for a user, sign a message, and verify the signature of a message. Formally:*

- Gen(1^k), *where k is a security parameter, outputs a public key PK and secret key SK.*
- $\mathsf{Sign}_{s,SK}(M, R)$ *outputs a signature σ on the message M with respect to the ring $R = (PK_1, \ldots, PK_n)$. We assume the following: (1) $(R[s], SK)$ is a valid key-pair output by* Gen; *(2) $|R| \geq 2$ (since a ring signature scheme is not intended[2] to serve as a standard signature scheme); and (3) each[3] public key in the ring is distinct.*
- $\mathsf{Vrfy}_R(M, \sigma)$ *verifies a purported signature σ on a message M with respect to the ring of public keys R.*

We require the following completeness condition to hold: for any integer k, any $\{(PK_i, SK_i)\}_{i=1}^n$ output by Gen(1^k), *any $s \in [n]$, and any M, we have* $\mathsf{Vrfy}_R(M, \mathsf{Sign}_{s,SK_s}(M, R)) = 1$ *where $R = (PK_1, \ldots, PK_n)$.*

A c-user ring signature scheme is a variant of the above that only supports rings of fixed size c (i.e., the Sign *and* Vrfy *algorithms only take as input rings R for which $|R| = c$, and correctness is only required to hold for such rings).*

To improve readability, we will generally omit the input "s" to the signing algorithm (and simply write $\sigma \leftarrow \mathsf{Sign}_{SK}(M, R)$), with the understanding that the signer can determine an index s for which SK is the secret key corresponding to public key $R[s]$. Strictly speaking, there may not be a unique such s when

---

[2] Furthermore, it is easy to modify any ring signature scheme to allow signatures with $|R| = 1$ by including a special key for just that purpose.

[3] This is without loss of generality, since the signer/verifier can simply take the sub-ring of distinct keys in R and correctness is unchanged.

R contains incorrectly-generated keys; in real-world usage of a ring signature scheme, though, a signer will certainly be able to identify their public key.

A ring signature scheme is used as follows: At various times, some collection of users runs the key generation algorithm Gen to generate public and secret keys. We stress that no coordination among these users is assumed or required. When a user with secret key SK wishes to generate an anonymous signature on a message M, he chooses a ring R of public keys which includes his own, computes $\sigma \leftarrow \text{Sign}_{SK}(M, R)$ and outputs (σ, R). (In such a case, we will refer to the holder of SK as the *signer* of the message and to the holders of the other public keys in R as the *non-signers*.) Anyone can now verify that this signature was generated by *someone* holding a key in R by running $\text{Vrfy}_R(M, \sigma)$.

We remark that although our functional definition of a ring signature scheme (cf. Def. 1) requires users to generate keys specifically for that purpose (in contrast to the requirements of [1, 2]), our first construction can be easily modified to work with any ring of users as long as they each have a public key for both encryption and signing (see Sect. 5).

As discussed in the Introduction, ring signatures must satisfy two independent notions of security: anonymity and unforgeability. There are various ways each of these notions can be defined (and various ways these notions have been defined in the literature); we present our definitions in Sections 3.1 and 3.2, and compare them in Sect. 4.

3.1   Definitions of Anonymity

The anonymity condition requires, informally, that an adversary not be able to tell which member of a ring generated a particular signature.[4] We begin with a basic definition of anonymity which is already stronger than that considered in most previous work in that we give the adversary access to a signing oracle (this results in a stronger definition even in the case of unconditional anonymity).

Definition 2 (Basic anonymity). *Given a ring signature scheme* (Gen, Sign, Vrfy), *a polynomial* $n(\cdot)$, *and a* PPT *adversary* $\mathcal{A}$, *consider the following game:*

1. *Key pairs* $\{(PK_i, SK_i)\}_{i=1}^{n(k)}$ *are generated using* Gen(1^k), *and the set of public keys* $S \stackrel{\text{def}}{=} \{PK_i\}_{i=1}^{n(k)}$ *is given to* $\mathcal{A}$.
2. $\mathcal{A}$ *is given access (throughout the entire game) to an oracle* Osign($\cdot, \cdot, \cdot$) *such that* Osign(s, M, R) *returns* $\text{Sign}_{SK_s}(M, R)$, *where we require* $R \subseteq S$ *and* $PK_s \in R$.
3. $\mathcal{A}$ *outputs a message* M, *distinct indices* i_0, i_1, *and a ring* $R \subseteq S$ *for which* $PK_{i_0}, PK_{i_1} \in R$. *A random bit* b *is chosen, and* $\mathcal{A}$ *is given the signature* $\sigma \leftarrow \text{Sign}_{SK_{i_b}}(M, R)$.
4. *The adversary outputs a bit* b', *and succeeds if* $b' = b$.

---

[4] All the anonymity definitions that follow can be phrased in either a *computational* or an *unconditional* sense (where, informally, in the former case anonymity holds for polynomial-time adversaries while in the latter case anonymity holds even for all-powerful adversaries). For simplicity, we only present the computational versions.

(Gen, Sign, Vrfy) *achieves* basic anonymity *if, for any* PPT $\mathcal{A}$ *and any polynomial* $n(\cdot)$, *the success probability of* $\mathcal{A}$ *in the above game is negligibly close to* $1/2$.

(Some previous papers consider a variant of the above in which the adversary is given a signature computed by a randomly-chosen member of R, and should be unable to guess the actual signer with probability better than $1/|R| + \mathsf{negl}(k)$. A hybrid argument shows that such a variant is equivalent to the above.)

Unfortunately, the above definition of basic anonymity leaves open the possibility of the following attack: (1) an adversary generates public keys in some arbitrary manner (which may possibly depend on the public keys of the honest users), and then (2) a legitimate signer generates a signature with respect to a ring containing some of these adversarially-generated public keys. The definition above offers no protection in this case! This attack, considered also in [19] (in a slightly different context) is quite realistic since, by their very nature, ring signatures are intended to be used in settings where there is not necessarily any central authority checking validity of public keys. This motivates the following, stronger definition:

Definition 3 (Anonymity w.r.t. adversarially-chosen keys). *Given a ring signature scheme* (Gen, Sign, Vrfy), *a polynomial* $n(\cdot)$, *and a* PPT *adversary* $\mathcal{A}$, *consider the following game:*

1. *As in Definition 2.*
2. *As in Definition 2, except that we no longer require* $R \subseteq S$.
3. *As in Definition 2, except that we no longer require* $R \subseteq S$.
4. *The adversary outputs a bit* b', *and succeeds if* $b' = b$.

(Gen, Sign, Vrfy) *achieves* anonymity w.r.t. adversarially-chosen keys *if for any* PPT $\mathcal{A}$ *and polynomial* $n(\cdot)$, *the success probability of* $\mathcal{A}$ *in the above game is negligibly close to* $1/2$.

The above definition only guarantees anonymity of a particular signature as long as there are at least two honest users in the ring. In some sense this is inherent, since if an honest signer U chooses a ring in which all *other* public keys (i.e., except for the public key of U) are owned by an adversary, then that adversary "knows" that U must be the signer (since the adversary did not generate the signature itself).

A weaker requirement one might consider when the signer U is the only honest user in the ring is that the adversary should be unable to *prove* to a third party that U generated the signature (we call this an *attribution attack*). Preventing such an attack in general seems to require the involvement of a trusted party (or at least a common random string), something we would like to avoid. We instead define a slightly weaker notion which, informally, can be viewed as offering some protection against attribution attacks as long as at least one other user U' in the ring generated her public key honestly. The honest users in the ring (other[5] than U), however, may later cooperate with the adversary by revealing

---

[5] The idea is that everyone in the ring is trying to "frame" U, but U is (naturally) refusing to divulge her secret key. Although this itself might arouse suspicion, the point is that it still cannot be proved — in court, say — that U was the signer.

their secret keys. (Actually, we even allow these users to reveal the *randomness*[6] used to generate their secret keys.) Note that this also ensures some measure of security in case secret keys are exposed or stolen.

In addition to the above, we consider also the stronger variant in which the secret keys of *all* honest users (i.e., including U) are exposed. This parallels (in fact, is stronger than) the anonymity definition given by Bellare, et al. in the context of group signatures [3]. For simplicity, we also protect against adversarially-chosen keys, although one could consider the weaker definition which does not.

Definition 4 (Anonymity against attribution attacks/full key exposure). *Given* (Gen, Sign, Vrfy), $n(\cdot)$, *and* $\mathcal{A}$ *as in Definition 3, consider the following game:*

1. *For* $i = 1$ *to* $n(k)$, *generate* $(PK_i, SK_i) \leftarrow$ Gen$(1^k; \omega_i)$ *for randomly-chosen* ω_i. *Give to* $\mathcal{A}$ *the set of public keys* $\{PK_i\}_{i=1}^{n(k)}$. *The adversary* $\mathcal{A}$ *is also given access to a signing oracle as in Definition 3.*
2. $\mathcal{A}$ *outputs a message* M, *distinct indices* i_0, i_1, *and a ring* R *for which* $PK_{i_0}, PK_{i_1} \in R$. *Adversary* $\mathcal{A}$ *is given* $\{\omega_i\}_{i \neq i_0}$. *Furthermore, a random bit* b *is chosen and* $\mathcal{A}$ *is given* $\sigma \leftarrow$ Sign$_{SK_{i_b}}(M, R)$.
3. *The adversary outputs a bit* b', *and succeeds if* $b' = b$.

(Gen, Sign, Vrfy) *achieves* anonymity against attribution attacks *if, for any* PPT $\mathcal{A}$ *and polynomial* $n(\cdot)$, *the success probability of* $\mathcal{A}$ *in the above game is negligibly close to* $1/2$. *If, in the second step,* $\mathcal{A}$ *is instead given* $\{\omega_i\}_{i=1}^{n(k)}$ *then we say* (Gen, Sign, Vrfy) *achieves* anonymity against full key exposure.

Linkability. Another desideratum of a ring signature scheme is that it be *unlinkable*; that is, it be infeasible to determine whether two signatures (possibly generated with respect to different rings) were generated by the same signer. As in [3], all our definitions imply (appropriate variants of) unlinkability.

3.2   Definitions of Unforgeability

The intuitive notion of unforgeability is, as usual, that an adversary should be unable to output (R, M, σ) such that Vrfy$_R(M, \sigma) = 1$ unless either (1) the adversary explicitly knows a secret key corresponding to one of the public keys in R, or (2) a user whose public key is in R explicitly signed M previously (with respect to the same ring R). Some subtleties arise, however, when defining what it means to allow the adversary a chosen-message attack on the scheme. Many previous works (e.g., [20]), assume a definition like the following:

Definition 5 (Unforgeability against fixed-ring attacks). *A ring signature scheme* (Gen, Sign, Vrfy) *is* unforgeable against fixed-ring attacks *if for any* PPT *adversary* $\mathcal{A}$ *and for any polynomial* $n(\cdot)$, *the probability that* $\mathcal{A}$ *succeeds in the following game is negligible:*

---

[6] This ensures security when erasure cannot be guaranteed, or when it cannot be guaranteed that all users will comply with the directive to erase their random coins.

1. *Key pairs $\{(PK_i, SK_i)\}_{i=1}^{n(k)}$ are generated using $\mathsf{Gen}(1^k)$, and the set of public keys $R \stackrel{\text{def}}{=} \{PK_i\}_{i=1}^{n(k)}$ is given to $\mathcal{A}$.*
2. *$\mathcal{A}$ is given access to a signing oracle $\mathsf{OSign}(\cdot, \cdot)$, where $\mathsf{OSign}(s, M)$ outputs $\mathsf{Sign}_{SK_s}(M, R)$.*
3. *$\mathcal{A}$ outputs (M^*, σ^*), and succeeds if $\mathsf{Vrfy}_R(M^*, \sigma^*) = 1$ and also $\mathcal{A}$ never made a query of the form $\mathsf{Osign}(\star, M^*)$.*

Note that not only is $\mathcal{A}$ restricted to making signing queries with respect to the *full* ring R, but its forgery is required to verify with respect to R as well. The following stronger, and more natural, definition was used in, e.g., [1]:

Definition 6 (Unforgeability against chosen-subring attacks). *A ring signature scheme* (Gen, Sign, Vrfy) *is* unforgeable against chosen-subring attacks *if for any* PPT *adversary $\mathcal{A}$ and for any polynomial $n(\cdot)$, the probability that $\mathcal{A}$ succeeds in the following game is negligible:*

1. *Key pairs $\{(PK_i, SK_i)\}_{i=1}^{n(k)}$ are generated using $\mathsf{Gen}(1^k)$, and the set of public keys $S \stackrel{\text{def}}{=} \{PK_i\}_{i=1}^{n(k)}$ is given to $\mathcal{A}$.*
2. *$\mathcal{A}$ is given access to a signing oracle $\mathsf{OSign}(\cdot, \cdot, \cdot)$, where $\mathsf{OSign}(s, M, R)$ outputs $\mathsf{Sign}_{SK_s}(M, R)$ and we require that $R \subseteq S$ and $PK_s \in R$.*
3. *$\mathcal{A}$ outputs (R^*, M^*, σ^*), and succeeds if $R^* \subseteq S$, $\mathsf{Vrfy}_{R^*}(M^*, \sigma^*) = 1$, and $\mathcal{A}$ never queried $(\star, M^*, R^*)$ to its signing oracle.*

While the above definition is an improvement, it still leaves open the possibility of an attack whereby honest users are "tricked" into generating signatures using rings containing adversarially-generated public keys. (Such an attack was also previously suggested by [19, 18].) The following definition takes this into account as well as (for completeness) an adversary who adaptively corrupts honest participants and obtains their secret keys. Since either of these attacks may be viewed as the outcome of corrupting an "insider," we use this terminology.[7]

Definition 7 (Unforgeability w.r.t. insider corruption). *A ring signature scheme* (Gen, Sign, Vrfy) *is* unforgeable w.r.t. insider corruption *if for any* PPT *adversary $\mathcal{A}$ and for any polynomial $n(\cdot)$, the probability that $\mathcal{A}$ succeeds in the following game is negligible:*

1. *Key pairs $\{(PK_i, SK_i)\}_{i=1}^{n(k)}$ are generated using $\mathsf{Gen}(1^k)$, and the set of public keys $S \stackrel{\text{def}}{=} \{PK_i\}_{i=1}^{n(k)}$ is given to $\mathcal{A}$.*
2. *$\mathcal{A}$ is given access to a signing oracle $\mathsf{OSign}(\cdot, \cdot, \cdot)$, where $\mathsf{OSign}(s, M, R)$ outputs $\mathsf{Sign}_{s, SK_s}(M, R)$ and we require that $PK_s \in R$.*
3. *$\mathcal{A}$ is also given access to a corrupt oracle $\mathsf{Corrupt}(\cdot)$, where $\mathsf{Corrupt}(i)$ outputs SK_i.*
4. *$\mathcal{A}$ outputs (R^*, M^*, σ^*), and succeeds if $\mathsf{Vrfy}_{R^*}(M^*, \sigma^*) = 1$, $\mathcal{A}$ never queried $(\star, M^*, R^*)$, and $R^* \subseteq S \setminus C$, where C is the set of corrupted users.*

---

[7] We are aware that, technically speaking, there are not really any "insiders" in the context of ring signatures.

We remark that Herranz [15] considers, albeit informally, a definition intermediate between our Definitions 6 and 7 in which corruptions of honest players are
allowed but adversarially-chosen public keys are not explicitly mentioned.

4   Separations Between the Security Definitions

In the previous section, we presented various definitions of anonymity and unforgeability. Here, we show that these definitions are in fact distinct, in the sense
that there exist (under certain assumptions) schemes satisfying a weaker definition but not a stronger one. First, we show separations for the definitions
of anonymity, considering in each case a scheme simultaneously satisfying the
strongest definition of unforgeability (all proofs appear in the full version [4]):

Claim 1. If there exists a scheme which achieves **basic** anonymity and is unforgeable w.r.t. insider corruption, then there exists a scheme which achieves
these same properties but which is **not** anonymous w.r.t. **adversarially-chosen
keys**.

Claim 2. If there exists a scheme which is anonymous w.r.t. **adversarially-
chosen keys** and is unforgeable w.r.t. insider corruption, then there exists
a scheme which achieves these same properties but which is **not** anonymous
against **attribution attacks**.

We also show separations for the definitions of unforgeability, considering now
schemes which simultaneously achieve the strongest definition of anonymity:

Claim 3. If there exists a scheme which is anonymous against full key exposure and unforgeable w.r.t. insider corruption, then there exists a scheme which
is anonymous against full key exposure and unforgeable against **fixed-ring
attacks**, but **not** unforgeable against **chosen-subring attacks**.
In contrast to the rest of the claims, the assumption in the above claim is not
minimal. We remark that the scheme of [16] serves as a *natural* example of
a scheme that is unforgeable against fixed-ring attacks, but which is **not** unforgeable against chosen-subring attacks (in the random oracle model); this was
subsequently fixed in [15]. We defer a detailed discussion to the full version [4].

Claim 4. If there exists a scheme which is anonymous against full key exposure and unforgeable against **chosen-subring attacks**, then there exists a
scheme achieving these same properties which is **not** unforgeable w.r.t. **insider
corruption**.

5   Ring Signatures Based on General Assumptions

We now describe our construction of a ring signature scheme that satisfies the
strongest of our proposed definitions, and is based on general assumptions. In
what follows, we let $(\mathsf{EGen}, \mathsf{Enc}, \mathsf{Dec})$ be a semantically-secure public-key encryption scheme, let $(\mathsf{Gen}', \mathsf{Sign}', \mathsf{Vrfy}')$ be a (standard) signature scheme, and let

$(\ell, \mathcal{P}, \mathcal{V})$ be a ZAP (for an $\mathcal{NP}$-language that will become clear once we describe the scheme). We denote by $C^* \leftarrow \mathsf{Enc}^*_{R_E}(m)$ the probabilistic algorithm that takes as input a set of encryption public keys $R_E = \{pk_{E,1}, \ldots, pk_{E,n}\}$ and a message m, and does the following: it first chooses random $s_1, \ldots, s_{n-1} \in \{0,1\}^{|m|}$ and then outputs:

$$C^* = \left(\mathsf{Enc}_{pk_{E,1}}(s_1), \mathsf{Enc}_{pk_{E,2}}(s_2), \cdots, \mathsf{Enc}_{pk_{E,n-1}}(s_{n-1}), \mathsf{Enc}_{pk_{E,n}}\left(m \oplus \bigoplus_{j=1}^{n-1} s_j\right) \right).$$

Note that, informally, encryption using Enc^* is semantically secure as long as at least one of the corresponding secret keys is unknown.

The idea of our construction is the following. Each user has an encryption key pair (pk_E, sk_E) and a standard signature key pair (pk_S, sk_S). To generate a ring signature with respect to a ring R of n users, the signer produces a standard signature σ' with her signing key. Next, the signer produces ciphertexts $C_1^*, \ldots, C_n^*$ using the Enc^* algorithm and the set R_E of all the *encryption* public keys in the ring; one of these ciphertexts will be an encryption of σ'. Finally, the signer produces a proof π, using the ZAP, that one of the ciphertexts is an encryption of a valid signature on the message with respect to the signature public key of one of the ring members.

Toward a formal description, let L denote the $\mathcal{NP}$ language:

$$\left\{ (pk_S, M, R_E, C^*) : \exists \sigma, \omega \text{ s.t. } C^* = \mathsf{Enc}^*_{R_E}(\sigma; \omega) \bigwedge \mathsf{Vrfy}'_{pk_S}(M, \sigma) = 1 \right\} ;$$

i.e., $(pk_S, M, R_E, C^*) \in L$ if C^* is an encryption (using $\mathsf{Enc}^*_{R_E}$) of a valid signature of M with respect to the public key pk_S. We now give the details of our construction, which is specified by the key-generation algorithm Gen, the ring signing algorithm Sign, and the ring verification algorithm Vrfy:

$\mathsf{Gen}(1^k)$:

1. Generate signing key pair $(pk_S, sk_S) \leftarrow \mathsf{Gen}'(1^k)$.
2. Generate encryption key pair $(pk_E, sk_E) \leftarrow \mathsf{Gen}(1^k)$ and erase sk_E.
3. Choose an initial ZAP message $r \leftarrow \{0,1\}^{\ell(k)}$.
4. Output the public key $PK = (pk_S, pk_E, r)$, and the secret key $SK = sk_S$.

$\mathsf{Sign}_{i^*, SK_{i^*}}(M, (PK_1, \ldots, PK_n))$:

1. Parse each PK_i as $(pk_{S,i}, pk_{E,i}, r_i)$, and parse SK_{i^*} as sk_{S,i^*}. Set $R_E := \{pk_{E,1}, \ldots, pk_{E,n}\}$.
2. Set $M^* := M \mid PK_1 \mid \cdots \mid PK_n$, where "$\mid$" denotes concatenation. Compute the signature $\sigma'_{i^*} \leftarrow \mathsf{Sign}'_{sk_{S,i^*}}(M^*)$.
3. Choose random coins $\omega_1, \ldots, \omega_n$ and: (1) compute $C_{i^*}^* = \mathsf{Enc}^*_{R_E}(\sigma'_{i^*}; \omega_{i^*})$ and (2) for $i \in \{1, \ldots, n\} \setminus \{i^*\}$, compute[8] $C_i^* = \mathsf{Enc}^*_{R_E}(0^{|\sigma'_{i^*}|}; \omega_i)$.

---

[8] We assume for simplicity that valid signatures w.r.t. the public keys $\{pk_{S,i}\}_{i \neq i^*}$ always have the same length as valid signatures w.r.t. pk_{S,i^*}. The construction can be adapted when this is not the case.

4. For $i \in [n]$, let x_i denote the statement: " $(pk_{S,i}, M^*, R_E, C_i^*) \in L$ ", and let $x := \bigvee_{i=1}^n x_i$. Compute the proof $\pi \leftarrow \mathcal{P}_{r_1}(x, (\sigma'_{i^*}, \omega_{i^*}))$.
5. The signature is $\sigma = (C_1^*, \ldots, C_n^*, \pi)$.

$\underline{\mathsf{Vrfy}_{PK_1, \ldots, PK_n}(M, \sigma)}$

1. Parse each PK_i as $(pk_{S,i}, pk_{E,i}, r_i)$. Set $M^* := M \mid PK_1 \mid \cdots \mid PK_n$ and $R_E := \{pk_{E,1}, \ldots, pk_{E,n}\}$. Parse σ as $(C_1^*, \ldots, C_n^*, \pi)$.
2. For $i \in [n]$, let x_i denote the statement " $(pk_{S,i}, M^*, R_E, C_i^*) \in L$ " and set $x := \bigvee_{i=1}^n x_i$.
3. Output $\mathcal{V}_{r_1}(x, \pi)$.

It is easy to see that the scheme above satisfies the functional definition of a ring signature scheme (recall that the $\{PK_i\}$ in a ring are always ordered lexicographically). We now prove that the scheme satisfies strong notions of anonymity and unforgeability:

Theorem 1. *If encryption scheme* (EGen, Enc, Dec) *is semantically secure, signature scheme* (Gen', Sign', Vrfy') *is existentially unforgeable under adaptive chosen-message attacks, and* $(\ell, \mathcal{P}, \mathcal{V})$ *is a ZAP for the language* $L' = \{(x_1, \ldots, x_n) : \exists i : x_i \in L\}$, *then the above ring signature scheme is (computationally) anonymous against attribution attacks, and unforgeable w.r.t. insider corruption.*

The proof is given in Appendix B.1.

Extension. The scheme above can also be used (with a few easy modifications) in a situation where some users in the ring have not generated a key pair according to Gen, as long as (1) every ring member has a public key both for encryption and for signing and (2) at least one of the members has included a sufficiently-long random string in his public key. Furthermore, the encryption (signature) public keys of different members of the ring may be associated with different encryption (signature) schemes. Thus, a *single* user who establishes a public key for a ring signature scheme suffices to provide anonymity for everyone. This also provides a way to include "oblivious" users in the signing ring [1, 2].

Achieving a stronger anonymity guarantee. The above scheme is not secure against full key exposure, and essential to our proof of anonymity is that the adversary not be given the random coins used to generate *all* (honest) ring signature keys.[9] (If the adversary gets all sets of random coins, it can decrypt ciphertexts encrypted using $\mathsf{Enc}^*_{R_E}$ for any ring of honest users R and thereby determine the true signer of a message.) It is possible to achieve anonymity against full key exposure using an enhanced form of encryption for which, informally, there exists an "oblivious" way to generate a public key without generating a corresponding secret key. This notion, introduced by Damgård and Nielsen [10],

---

[9] We remark that anonymity still holds if the adversary is given all *secret keys* (but not the randomness used to generate all secret keys). This is because the decryption key sk_E is erased, and not included in SK.

can be viewed as a generalization of *dense cryptosystems* in which the public key is required to be a uniformly distributed string (in particular, dense cryptosystems satisfy the definition below). We review the formal definition here.

Definition 8. *An* oblivious key generator *for the public-key encryption scheme* (EGen, Enc, Dec) *is a pair of* PPT *algorithms* (OblEGen, OblRand) *such that:*

- OblEGen, *on input* 1^k *and random coins* $\omega \in \{0,1\}^{n(k)}$, *outputs a key* pk;
- OblRand, *on input a key* pk, *outputs a string* ω;

and the following distribution ensembles are computationally indistinguishable:

$$\left\{ \omega \leftarrow \{0,1\}^{n(k)} : (\omega, \mathsf{OblEGen}(1^k; \omega)) \right\}$$

and

$$\left\{ (pk, sk) \leftarrow \mathsf{EGen}(1^k); \omega \leftarrow \mathsf{OblRand}(pk) : (\omega, pk) \right\}.$$

Note that if (EGen, Enc, Dec) is semantically secure, then (informally speaking) it is also semantically secure to encrypt messages using a public key pk generated by OblEGen, even if the adversary has the random coins used by OblEGen in generating pk. We remark for completeness that the El Gamal encryption scheme (over the group of quadratic residues modulo a prime) is an example of a scheme having an oblivious key generator.

Given the above, we adapt our construction in the natural way: specifically, the Gen algorithm is changed so that instead of generating pk_E using EGen (and then erasing the secret key sk_E and the random coins used), we now generate pk_E using OblEGen. Adapting the proof of Theorem 1, we can easily show:

Theorem 2. *Under the assumptions of Theorem 1 and assuming* (EGen, Enc, Dec) *has an oblivious key generator, the modified ring signature scheme described above is (computationally) anonymous against full key exposure, and unforgeable w.r.t. insider corruption.*

The proof is given in Appendix B.2.

6   An Efficient 2-User Ring Signature Scheme

In this section, we present a more efficient construction of a 2-user ring signature scheme based on specific assumptions. The scheme is based on the (standard) signature scheme constructed by Waters [21] which we briefly review now.

6.1   The Waters Scheme

Let $\mathbb{G}, \mathbb{G}_1$ be groups of prime order q such that there exists an efficiently computable bilinear map $\hat{e} : \mathbb{G} \times \mathbb{G} \to \mathbb{G}_1$. We assume that $q, \mathbb{G}, \mathbb{G}_1, \hat{e}$, and a generator $g \in \mathbb{G}$ are publicly known. The Waters signature scheme for messages of length n is defined as follows:

Key Generation. Choose $\alpha \leftarrow \mathbb{Z}_q$ and set $g_1 = g^\alpha$. Additionally choose random elements $h, u', u_1, \ldots, u_n \leftarrow \mathbb{G}$. The public key is $(g_1, h, u', u_1, \ldots, u_n)$ and the secret key is h^α.

Signing. To sign the n-bit message M, first compute $w = u' \cdot \prod_{i:M_i=1} u_i$. Then choose random $r \leftarrow \mathbb{Z}_q$ and output the signature $\sigma = (h^\alpha \cdot w^r, \ g^r)$.

Verification. To verify the signature (A, B) on message M with respect to public key $(g_1, h, u', u_1, \ldots, u_n)$, compute $w = u' \cdot \prod_{i:M_i=1} u_i$ and then check whether $\hat{e}(g_1, h) \cdot \hat{e}(B, w) \stackrel{?}{=} \hat{e}(A, g)$.

6.2   A 2-User Ring Signature Scheme

The main observation we make with regard to the above scheme is the following: element h is arbitrary, and only knowledge of h^α is needed to sign. So, we can dispense with including h in the public key altogether; instead, a user U with secret α and the value $g_1 = g^\alpha$ in his public key will use as his "h-value" the value $\bar{g}_1$ contained in the public key of a *second* user $\bar{U}$. This provides anonymity since $\bar{U}$ could *also* have computed the same value $(\bar{g}_1)^\alpha$ using the secret value $\bar{\alpha} = \log_g \bar{g}_1$ known to him (because $\bar{g}_1^\alpha = g_1^{\bar{\alpha}}$). We now proceed with the details.

Key Generation. Choose $\alpha \leftarrow \mathbb{Z}_q$ and set $g_1 = g^\alpha$. Additionally choose random elements $u', u_1, \ldots, u_n \leftarrow \mathbb{G}$. The public key is $(g_1, u', u_1, \ldots, u_n)$ and the secret key is α. (We again assume that $q, \mathbb{G}, \mathbb{G}_1, \hat{e},$ and g are system-wide parameters.)

Ring Signing. To sign message $M \in \{0,1\}^n$ with respect to the ring $R = \{PK, \overline{PK}\}$ using secret key α (where we assume without loss of generality that α is the secret corresponding to PK), proceed as follows: parse PK as $(g_1, u', u_1, \ldots, u_n)$ and $\overline{PK}$ as $(\bar{g}_1, \bar{u}', \bar{u}_1, \ldots, \bar{u}_n)$, and compute $w = u' \cdot \prod_{i:M_i=1} u_i$ and $\bar{w} = \bar{u}' \cdot \prod_{i:M_i=1} \bar{u}_i$. Then choose random $r \leftarrow \mathbb{Z}_q$ and output the signature

$$\sigma = (\bar{g}_1^\alpha \cdot (w\bar{w})^r, \ g^r) .$$

Ring Verification. To verify the signature (A, B) on message M with respect to the ring $R = \{PK, \overline{PK}\}$ (parsed as above), compute $w = u' \cdot \prod_{i:M_i=1} u_i$ and $\bar{w} = \bar{u}' \cdot \prod_{i:M_i=1} \bar{u}_i$ and then check whether $\hat{e}(g_1, \bar{g}_1) \cdot \hat{e}(B, (w\bar{w})) \stackrel{?}{=} \hat{e}(A, g)$.

It is not hard to see that correctness holds. We prove the following regarding the above scheme:

Theorem 3. *Assume the Waters signature scheme is existentially unforgeable[10] under adaptive chosen message attack. Then the 2-user ring signature scheme described above is unconditionally anonymous against full key exposure, and unforgeable against chosen-subring attacks.*

---

[10] This holds [21] under the computational Diffie-Hellman assumption in $\mathbb{G}$.

Proof. Unconditional anonymity against full key exposure follows easily from the observation made earlier: namely, that only the value $\bar{g}_1^\alpha = g_1^{\bar{\alpha}}$ (where $\bar{\alpha} \stackrel{\text{def}}{=} \log_g \bar{g}_1$) is needed to sign, and either of the two (honest) parties can compute this value.

We now prove that the scheme satisfies Definition 6. We do this by showing how an adversary $\mathcal{A}$ that forges a signature with respect to the ring signature scheme with non-negligible probability can be used to construct an adversary $\hat{\mathcal{A}}$ that forges a signature with respect to the Waters signature scheme (in the standard sense) with the same probability. For simplicity in the proof, we assume that $\mathcal{A}$ only ever sees the public keys of two users, requests all signatures to be signed with respect to the ring R containing these two users, and forges a signature with respect to that same ring R. By a hybrid argument, it can be shown that (for this scheme) this is equivalent to the more general case when $\mathcal{A}$ may see multiple public keys, request signatures with respect to various (different) 2-user subsets, and then output a forgery with respect to any 2-user subset of its choice.

Construct $\hat{\mathcal{A}}$ as follows: $\hat{\mathcal{A}}$ is given the public key $(\hat{g}_1, \hat{h}, \hat{u}', \hat{u}_1, \ldots, \hat{u}_n)$ of an instance of the Waters scheme. $\hat{\mathcal{A}}$ constructs two user public keys as follows: first, it sets $g_1 = \hat{g}_1$ and $\bar{g}_1 = \hat{h}$. Then, it chooses random $u', u_1, \ldots, u_n \leftarrow \mathbb{G}$ and sets $\bar{u}' = \hat{u}'/u'$ and $\bar{u}_i = \hat{u}_i/u_i$ for all i. It gives to $\mathcal{A}$ the public keys $(g_1, u', u_1, \ldots, u_n)$ and $(\bar{g}_1, \bar{u}', \bar{u}_1, \ldots, \bar{u}_n)$. Note that both public keys have the appropriate distribution. When $\mathcal{A}$ requests a ring signature on a message M with respect to the ring R containing these two public keys, $\hat{\mathcal{A}}$ requests a signature on M from its signing oracle, obtains in return a signature (A, B), and gives this signature to $\mathcal{A}$. Note that this is indeed a perfect simulation, since

$$\left(\hat{h}^{\log_g \hat{g}_1} \cdot \left(\hat{u}' \prod_{i:M_i=1} \hat{u}_i \right)^r, g^r \right) = \left(\bar{g}_1^{\log_g g_1} \cdot \left(u'\bar{u}' \prod_{i:M_i=1} u_i\bar{u}_i \right)^r, g^r \right),$$

which is an appropriately-distributed ring signature with respect to the public keys given to $\mathcal{A}$.

When $\mathcal{A}$ outputs a forgery (A^*, B^*) on a message M^*, this same forgery is output by $\hat{\mathcal{A}}$. Note that $\hat{\mathcal{A}}$ outputs a valid forgery whenever $\mathcal{A}$ does, since

$$\hat{e}(g_1, \bar{g}_1) \cdot \hat{e}\left(B^*, (u'\bar{u}' \textstyle\prod_{i:M_i^*=1} u_i\bar{u}_i) \right) = \hat{e}(A^*, g)$$

implies

$$\hat{e}(\hat{g}_1, \hat{h}) \cdot \hat{e}\left(B^*, (\hat{u}' \textstyle\prod_{i:M_i^*=1} \hat{u}_i) \right) = \hat{e}(A^*, g).$$

We conclude that $\hat{\mathcal{A}}$ outputs a forgery with the same probability as $\mathcal{A}$. Since, by assumption, the Waters scheme is secure, this completes the proof.

We remark that the security reduction in the above proof is tight.

An efficiency improvement. A (slightly) more efficient variant of the above scheme is also possible. Key generation is the same as before, except that an additional, random *identifier* $I \in \{0,1\}^k$ is also chosen and included in the public key. Let $<_{lex}$ denote lexicographic order. To sign message $M \in \{0,1\}^n$ with respect to the ring $R = \{PK, \overline{PK}\}$, first parse PK as $(I, g_1, u', u_1, \ldots, u_n)$ and $\overline{PK}$ as $(\bar{I}, \bar{g}_1, \bar{u}', \bar{u}_1, \ldots, \bar{u}_n)$. Choose random $r \leftarrow \mathbb{Z}_q$. If $I \leq_{lex} \bar{I}$, compute $w = u' \cdot \prod_{i:M_i=1} u_i$ and the signature

$$\sigma = (s \cdot w^r, \ g^r) \ ;$$

if $\bar{I} <_{lex} I$, compute $\bar{w} = \bar{u}' \cdot \prod_{i:M_i=1} \bar{u}_i$ and the signature

$$\sigma = (s \cdot \bar{w}^r, \ g^r) \ ,$$

where, in each case, $s = \bar{g}_1^\alpha = g_1^{\bar{\alpha}}$ is computed using whichever secret key is known to the signer. Verification is changed in the obvious way. A proof similar to the above shows that this scheme satisfies the same security properties as in Theorem 3.

References

1. M. Abe, M. Ohkubo, and K. Suzuki. 1-out-of-n signatures from a variety of keys. In *Advances in Cryptology — Asiacrypt 2002*.
2. B. Adida, S. Hohenberger, and R.L. Rivest. Ad-hoc-group signatures from hijacked keypairs. Available at `http://theory.lcs.mit.edu/~srhohen/papers/AHR.pdf`, 2005.
3. M. Bellare, D. Micciancio, and B. Warinschi. Foundations of group signatures: Formal definitions, simplified requirements, and a construction based on general assumptions. In *Advances in Cryptology — Eurocrypt 2003*.
4. A. Bender, J. Katz, and R. Morselli. Ring signatures: Stronger definitions, and constructions without random oracles. Cryptology ePrint Archive, 2005. `http://eprint.iacr.org/2005/304`.
5. E. Bresson, J. Stern, and M. Szydlo. Threshold ring signatures and applications to ad-hoc groups. In *Advances in Cryptology — Crypto 2002*.
6. D. Chaum and E. van Heyst. Group signatures. In *Advances in Cryptology — Eurocrypt '91*.
7. L. Chen, C. Kudla, and K.G. Patterson. Concurrent signatures. In *Advances in Cryptology — Eurocrypt 2004*.
8. S. S.M. Chow, J.K. Liu, and T. H. Yuen. Ring signature without random oracles. Cryptology ePrint Archive, 2005. `http://eprint.iacr.org/2005/317`.
9. R. Cramer, I. Damgård, and B. Schoenmakers. Proofs of partial knowledge and simplified design of witness hiding protocols. In *Advances in Cryptology — Crypto '94*.
10. I. Damgård and J.B. Nielsen. Improved non-committing encryption schemes based on a general complexity assumption. In *Advances in Cryptology — Crypto 2000*.
11. Y. Dodis, A. Kiayias, A. Nicolosi, and V. Shoup. Anonymous identification in ad-hoc groups. In *Advances in Cryptology — Eurocrypt 2002*.
12. C. Dwork and M. Naor. Zaps and their applications. In *Proc. 41st Annual Symposium on Foundations of Computer Science (FOCS)*. IEEE, 2000.
13. U. Feige, D. Lapidot, and A. Shamir. Multiple non-interactive zero knowledge proofs under general assumptions. *SIAM J. Computing*, 29(1):1–28, 1999.

14. A. Fiat and A. Shamir. How to prove yourself: Practical solutions to identification and signature problems. In *Advances in Cryptology — Crypto '86*.
15. J. Herranz. *Some digital signature schemes with collective signers*. PhD thesis, Universitat Politècnica de Catalunya, Barcelona, April 2005. Available at http://www.lix.polytechnique.fr/~herranz/thesis.htm.
16. J. Herranz and G. Sáez. Forking lemmas for ring signature schemes. In *Progress in Cryptology — Indocrypt 2003*.
17. M. Jakobsson, K. Sako, and R. Impagliazzo. Designated verifier proofs and their applications. In *Advances in Cryptology — Eurocrypt '96*.
18. J.K. Liu, V.K. Wei, and D.S. Wong. Linkable spontaneous anonymous group signatures for ad hoc groups. In *ACISP 2004*.
19. M. Naor. Deniable ring authentication. In *Advances in Cryptology — Crypto 2002*.
20. R.L. Rivest, A. Shamir, and Y. Tauman. How to leak a secret. In *Asiacrypt 2001*. Full version available at http://www.mit.edu/~tauman and to appear in *Essays in Theoretical Computer Science: in Memory of Shimon Even*.
21. B. Waters. Efficient identity-based encryption without random oracles. In *Advances in Cryptology — Eurocrypt 2005*.
22. J. Xu, Z. Zhang, and D. Feng. A ring signature scheme using bilinear pairings. In *Workshop on Information Security Applications (WISA)*, 2004.
23. F. Zhang and K. Kim. ID-based blind signature and ring signature from pairings. In *Advances in Cryptology — Asiacrypt 2002*.

A   ZAPs

Let L be an $\mathcal{NP}$ language with associated polynomial-time and polynomially-bounded *witness relation* $\mathcal{R}_L$ (i.e., such that $L \stackrel{\text{def}}{=} \{x \mid \exists w : (x, w) \in \mathcal{R}_L\}$). If $(x, w) \in \mathcal{R}_L$ we refer to x as the *statement* and w as the associated *witness* for x. We now recall the definition of a ZAP from [12]:

Definition 9 (ZAP). *A ZAP for an $\mathcal{NP}$ language L (with associated witness relation $\mathcal{R}_L$) is a triple $(\ell, \mathcal{P}, \mathcal{V})$, where $\ell(\cdot)$ is a polynomial, $\mathcal{P}$ is a PPT algorithm, and $\mathcal{V}$ is polynomial-time deterministic algorithm, and such that.*

Completeness. *For[11] any $(x, w) \in \mathcal{R}_L$ and any $r \in \{0, 1\}^{\ell(k)}$:*

$$\mathbf{Pr}\left[\pi \leftarrow \mathcal{P}_r(x, w) : \mathcal{V}_r(x, \pi) = 1\right] = 1 \ .$$

Adaptive soundness. *There exists a negligible function ε such that*

$$\mathbf{Pr}\left[r \leftarrow \{0, 1\}^{\ell(k)} : \exists (x, \pi) : \quad x \notin L \text{ and } \mathcal{V}_r(x, \pi) = 1\right] \leq \varepsilon(k) \ .$$

Witness indistinguishability. *(Informal) For any $x \in L$, any pair of witnesses w_0, w_1 for x, and any $r \in \{0, 1\}^{\ell(k)}$, the distributions $\{\mathcal{P}_r(x, w_0)\}$ and $\{\mathcal{P}_r(x, w_1)\}$ are computationally indistinguishable. (Note: more formally, we need to speak in terms of sequences $\{r_k \in \{0, 1\}^{\ell(k)}\}$, $\{x_k\}$, and $\{(w_{k,0}, w_{k,1})\}$ but we avoid doing so for simplicity of exposition.)*

---

[11] We remark that the definition in [12] allows for a negligible completeness error. However, their construction achieves perfect completeness when instantiated using the NIZK of [13].

A ZAP is used in the following way: The verifier generates a random first message $r \leftarrow \{0,1\}^{\ell(k)}$ and sends it to the prover $\mathcal{P}$. The prover, given r, a statement x, and associated witness w, sends $\pi \leftarrow \mathcal{P}_r(x, w)$ to the verifier. The verifier then runs $\mathcal{V}_r(x, \pi)$ and accepts iff the output is 1.

B   Proofs of Theorems 1 and 2

B.1   Proof of Theorem 1

We restate Theorem 1 for convenience:

If encryption scheme (EGen, Enc, Dec) is semantically secure, signature scheme (Gen′, Sign′, Vrfy′) is existentially unforgeable under adaptive chosen-message attacks, and $(\ell, \mathcal{P}, \mathcal{V})$ is a ZAP for L as described above, then the above ring signature scheme is (computationally) anonymous against attribution attacks, and unforgeable w.r.t. insider corruption.

Proof. We prove each of the desired security properties in turn.

Anonymity. For simplicity of exposition, we consider Definition 4 with $n = 2$; i.e., we assume only two users. By a straightforward hybrid argument, this implies the general case. Given a PPT adversary $\mathcal{A}$, we consider a sequence of experiments E_0, Hybrid_0, Hybrid_1, E_1 such that E_0 (resp., E_1) corresponds to the experiment of Definition 4 with $b = 0$ (resp., $b = 1$), and such that each experiment is computationally indistinguishable from the one before it. This implies that $\mathcal{A}$ has negligible advantage in distinguishing E_0 from E_1, as desired.

For convenience, we review experiment E_0. Here, two key pairs $(PK_0 = (pk_{S,0}, pk_{E,0}, r_0), SK_0)$ and $(PK_1 = (pk_{S,1}, pk_{E,1}, r_1), SK_1)$ are generated and $\mathcal{A}$ is given PK_0 and the randomness used to generate (PK_1, SK_1) (by hybrid argument, we can assume that $i_0 = 0$ and $i_1 = 1$). The adversary is also given access to a signing oracle (which can be used to obtain signatures computed using SK_0). $\mathcal{A}$ then outputs a message M along with a ring of public keys R containing both PK_0 and PK_1. Finally, $\mathcal{A}$ is given $\sigma \leftarrow \mathsf{Sign}_{SK_0}(M, R)$.

Experiment Hybrid_0 is the same as experiment E_0 except that we change how the signature σ is generated. In particular, step 3 of the ring signing algorithm is modified as follows: let R_E and M^* be as in the description of the ring signing algorithm given earlier. In step 3, instead of setting C_1^* to be an encryption of all zeros, we now compute $\sigma_1' \leftarrow \mathsf{Sign}_{sk_{S,1}}(M^*)$ and then set $C_1^* = \mathsf{Enc}_{R_E}^*(\sigma_1'; \omega_1)$. We stress that, as in E_0, the ciphertext C_0^* is still set to be an encryption of the signature σ_0', and the remaining ciphertexts are still encryptions of all zeros.

It is not hard to see that experiment Hybrid_0 is computationally indistinguishable from experiment E_0, assuming semantic security of the encryption scheme (EGen, Enc, Dec). This follows from the observations that (1) adversary $\mathcal{A}$ is *not* given the random coins used in generating PK_0 and so, in particular, it is not given the coins used to generate $pk_{E,0}$; (2) (informally) semantic security of encryption under $\mathsf{Enc}_{pk_{E,0}}$ implies semantic security of encryption using $\mathsf{Enc}_{R_E}^*$ as long as $pk_{E,0} \in R_E$ (a formal proof is straightforward); and, finally, (3) the coins ω_1 used in generating C_1^* are not used in the remainder of the ring signing algorithm.

Experiment Hybrid_1 is the same as Hybrid_0 except that we use a different witness when computing the proof π for the ZAP. In particular, instead of using witness (σ_0', ω_0) we use the witness (σ_1', ω_1). The remainder of the signing algorithm is unchanged.

It is relatively immediate that experiment Hybrid_1 is computationally indistinguishable from Hybrid_0, assuming witness indistinguishability of the ZAP. (We remark that the use of a ZAP, rather than non-interactive zero-knowledge, is essential here since the adversary may choose the "random string" component of all the adversarially-chosen public keys any way it likes.) In more detail, we can construct the following malicious verifier algorithm $\mathcal{V}^*$ using $\mathcal{A}$: verifier $\mathcal{V}^*$ generates (PK_0, SK_0) and (PK_1, SK_1) exactly as in experiments Hybrid_0 and Hybrid_1, and gives these keys and the appropriate associated random coins to $\mathcal{A}$. The signing queries of $\mathcal{A}$ can easily be answered by $\mathcal{V}^*$. When $\mathcal{A}$ makes its signing query, $\mathcal{V}^*$ computes the C_i^* exactly as in Hybrid_1 and then gives to the prover $\mathcal{P}$ the keys $\{pk_{S,i}\}_{i \in R}$, the message M^*, the set of keys R_E, and the ciphertexts $\{C_i^*\}_{i \in R}$; this defines the $\mathcal{NP}$-statement x exactly as in step 4 of the ring signing algorithm. In addition, $\mathcal{V}^*$ gives the two witnesses (σ_0', ω_0) and (σ_1', ω_1) to $\mathcal{P}$. Finally, $\mathcal{V}^*$ sends as its first message the "random string" component r of the lexicographically-first public key in R (note that this r is the random string that would be used to generate the proof π in step 4 of the ring signing algorithm). The prover responds with a proof $\pi \leftarrow \mathcal{P}_r(x, (\sigma_b', \omega_b))$ (for some $b \in \{0,1\}$), and then $\mathcal{V}^*$ outputs $(C_1^*, \ldots, C_n^*, \pi)$.

Note that if the prover uses the first witness provided to it by $\mathcal{V}^*$ then the output of $\mathcal{V}^*$ is distributed exactly according to Hybrid_0, while if the prover uses the second witness provided to it by $\mathcal{V}^*$ then the output of $\mathcal{V}^*$ is distributed exactly according to Hybrid_1. Witness indistinguishability of the ZAP thus implies computational indistinguishability of Hybrid_0 and Hybrid_1.

We may now notice that Hybrid_1 is computationally indistinguishable from E_1 by exactly the same argument used to show the indistinguishability of Hybrid_0 and E_0. This completes the proof.

Unforgeability. Assume there exists a PPT adversary $\mathcal{A}$ that breaks the above ring signature scheme (in the sense of Definition 7) with non-negligible probability. We construct an adversary $\mathcal{A}'$ that breaks the underlying signature scheme (Gen', Sign', Vrfy') (in the standard sense of existential unforgeability) with non-negligible probability.

$\mathcal{A}'$ receives as input a public key pk_S. Let $n = n(k)$ be a bound on the number of (honest user) public keys that $\mathcal{A}$ expects to be generated. $\mathcal{A}'$ runs $\mathcal{A}$ with input public keys $S = \{PK_1, \ldots, PK_n\}$, that $\mathcal{A}'$ generates as follows. $\mathcal{A}'$ chooses $i^* \leftarrow \{1, \ldots, n\}$ and sets $pk_{S,i^*} = pk_S$. The remainder of public key PK_{i^*} is generated exactly as prescribed by the Gen algorithm, with the exception that the decryption key sk_{E,i^*} that is generated is *not* erased. Public keys PK_i for $i \neq i^*$ are also generated exactly as prescribed by the Gen algorithm, again with the exception that the decryption keys $\{sk_{E,i}\}$ are not erased.

$\mathcal{A}'$ then proceeds to simulate the oracle queries of $\mathcal{A}$ in the natural way:

1. When $\mathcal{A}$ requests a signature on message M, with respect to ring R (which may possibly contain some public keys generated in an arbitrary manner by $\mathcal{A}$), to be signed by user $i \neq i^*$, then $\mathcal{A}'$ can easily generate the response to this query by running the Sign algorithm completely honestly;

2. When $\mathcal{A}$ requests a signature on message M, with respect to ring R (which, again, may possibly contain some public keys generated in an arbitrary manner by $\mathcal{A}$) to be signed by user i^*, then $\mathcal{A}'$ cannot directly respond to this query since it does not have sk_{S,i^*}. Instead, $\mathcal{A}'$ sets M^* appropriately, submits M^* to its signing oracle, and obtains in return a signature σ'_{i^*}. It then computes the remainder of the ring signature by following the rest of the Sign algorithm; note, in particular, that sk_{S,i^*} is not needed for this;

3. Any corruption query made by $\mathcal{A}$ for a user $i \neq i^*$ can be faithfully answered by $\mathcal{A}'$. On the other hand, if $\mathcal{A}$ ever makes a corruption query for i^*, then $\mathcal{A}'$ simply aborts.

At some point, $\mathcal{A}$ outputs a forgery $\bar{\sigma} = (\bar{C}_1^*, \ldots, \bar{C}_n^*, \bar{\pi})$ on a message $\bar{M}$ with respect to some ring of honest-user public keys $\bar{R} \subseteq S$. If $PK_{i^*} \notin \bar{R}$, then $\mathcal{A}'$ aborts. Otherwise, since $\mathcal{A}'$ knows all relevant decryption keys (recall that the ring $\bar{R}$ contains public keys of honest users only, and these keys were generated by $\mathcal{A}'$) it can decrypt $\bar{C}_{i^*}^*$ and obtain a candidate signature $\bar{\sigma}_{i^*}$. Finally, $\mathcal{A}'$ sets $\bar{M}^* = \bar{M} \mid \overline{PK}_1 \mid \cdots \mid \overline{PK}_{n'}$ (where $\bar{R} = \{\overline{PK}_i\}$) and outputs $(\bar{M}^*, \bar{\sigma}_{i^*})$. Note that (by requirement) $\mathcal{A}$ never requested a signature on message $\bar{M}$ with respect to the ring $\bar{R}$, and so $\mathcal{A}'$ never requested a signature on message $\bar{M}^*$ from its own oracle.

We claim that if $\mathcal{A}$ forges a signature with non-negligible probability $\varepsilon = \varepsilon(k)$, then $\mathcal{A}'$ forges a signature with probability at least $\varepsilon' = \varepsilon/n - \mathsf{negl}(k)$. To see this, note first that if $\mathcal{A}$ outputs a valid forgery then with all but negligible probability (by soundness of the ZAP) it holds that $(\overline{pk}_{S,i}, \bar{M}^*, \bar{R}_E, \bar{C}_i^*) \in L$ for some i (where $\overline{pk}_{S,i}$ and $\bar{R}_E$ are defined in the natural way based on the ring $\bar{R}$ and the public keys it contains). Conditioned on this, with probability $1/n$ it is the case that (1) $\mathcal{A}'$ did not abort and furthermore (2) $(\overline{pk}_{S,i^*}, \bar{M}^*, \bar{R}_E, \bar{C}_{i^*}^*) \in L$. When this occurs, then with all but negligible probability $\mathcal{A}'$ will recover (by decrypting as described above) a valid signature $\bar{\sigma}_{i^*}$ on the message $\bar{M}^*$ with respect to the given public key $\overline{pk}_{S,i^*} = pk_S$ (relying here on the fact that with all but negligible probability over choice of encryption public keys the encryption scheme Enc^* has zero decryption error). Security of $(\mathsf{Gen}', \mathsf{Sign}', \mathsf{Vrfy}')$ thus implies that ε must be negligible.

B.2 Proof of Theorem 2

We restate Theorem 2 for convenience:

Under the assumptions of Theorem 1 and assuming $(\mathsf{EGen}, \mathsf{Enc}, \mathsf{Dec})$ *has an oblivious key generator, the modified ring signature scheme described above is (computationally) anonymous against full key exposure, and unforgeable w.r.t. insider corruption.*

Proof. The proof of unforgeability follows immediately from Theorem 1 since, by Definition 8, the adversary cannot distinguish between the original scheme (in which the encryption key is generated using EGen) and the modified scheme (in which the encryption key is generated using OblEGen).

We now argue that the modified scheme achieves anonymity against full key exposure. First we note that the anonymity against attribution attacks claimed in Theorem 1 holds even when the adversary is given all random coins used to generate (PK_0, SK_0) *except* for those coins used to generate $pk_{E,0}$ (using EGen). Now, if there exists a PPT adversary $\mathcal{A}$ that breaks anonymity of the modified scheme in the sense of full key exposure, we can use it to construct a PPT adversary $\mathcal{A}'$ that breaks anonymity of the original scheme against attribution attacks. $\mathcal{A}'$ receives PK_0, the random coins $\omega_{S,1}, \omega_{E,1}$ used to generate (PK_1, SK_1), and the random coins $\omega_{S,0}$ used to generate $pk_{S,0}$ (i.e., $\mathcal{A}$ is *not* given the coins used to generate $pk_{E,0}$). Next, $\mathcal{A}'$ runs $\omega'_{E,0} \leftarrow \mathsf{OblRand}(pk_{E,0})$ and $\omega'_{E,1} \leftarrow \mathsf{OblRand}(pk_{E,1})$ and gives to $\mathcal{A}$ the public key PK_0 it received as well as the random coins $\omega_{S,0}, \omega'_{E,0}, \omega_{S,1}, \omega'_{E,1}$. The remainder of $\mathcal{A}$'s execution is simulated in the natural way by $\mathcal{A}'$.

Now, Definition 8 implies that the advantage of $\mathcal{A}$ in the above is negligibly close to the advantage of $\mathcal{A}$ in attacking the modified scheme in the sense of full key exposure. But the advantage of $\mathcal{A}$ in the above is exactly the advantage of $\mathcal{A}'$ in attacking the original scheme via key attribution attack. Since we have already proved that the original scheme is anonymous against attribution attacks (cf. Theorem 1), we see that the modified scheme is anonymous against full key exposure.

Efficient Blind and Partially Blind Signatures Without Random Oracles

Tatsuaki Okamoto

NTT Laboratories, Nippon Telegraph and Telephone Corporation,
1-1 Hikarino-oka, Yokosuka, 239-0847 Japan
okamoto.tatsuaki@lab.ntt.co.jp

Abstract. This paper proposes a new efficient signature scheme from bilinear maps that is secure in the standard model (i.e., without the random oracle model). Our signature scheme is more effective in many applications (e.g., blind signatures, group signatures, anonymous credentials etc.) than the existing secure signature schemes in the standard model such as the Boneh-Boyen [6], Camenisch-Lysyanskaya [10], Cramer-Shoup [15] and Waters [33] schemes (and their variants). The security proof of our scheme requires a slightly stronger assumption, the 2SDH assumption, than the SDH assumption used by Boneh-Boyen. As typical applications of our signature scheme, this paper presents efficient blind signatures and partially blind signatures that are secure in the standard model. Here, partially blind signatures are a generalization of blind signatures (i.e., blind signatures are a special case of partially blind signatures) and have many applications including electronic cash and voting. Our blind signature scheme is much more efficient than the existing secure blind signature schemes in the standard model such as the Camenisch-Koprowski-Warinsch [8] and Juels-Luby-Ostrovsky [22] schemes, and is also almost as efficient as the most efficient blind signature schemes whose security has been analyzed heuristically or in the random oracle model. Our partially blind signature scheme is the first one that is secure in the standard model and it is very efficient (almost as efficient as our blind signatures). We also present a blind signature scheme based on the Waters signature scheme.

1 Introduction

1.1 Background

Digital Signatures. The concept of digital signatures was invented by Diffie and Hellman [17], and their security was formalized by Goldwasser, Mical and Rivest [21]. A secure signature scheme exists if and only if a one-way function exists [26,32]. However, the general solution is far from yielding any practical applications.

Using the random oracle model, much more efficient secure signature schemes have been presented such as RSA-FDH, RSA-PSS, Fiat-Shamir and Schnorr signature schemes. However, the random oracle model cannot be realized in the

S. Halevi and T. Rabin (Eds.): TCC 2006, LNCS 3876, pp. 80–99, 2006.

standard (plain) model. In addition, signatures with hash functions (random oracles) are less suitable to several applications (e.g., group signatures).

Several efficient schemes that are secure in the standard model have recently been presented. There are two classes of such schemes, ones are based on the strong RSA assumption (i.e., based on the integer factoring (IF) problem), while the others are based on bilinear maps (i.e., based on the discrete logarithm (DL) problem). The Camenisch-Lysyanskaya [10], Cramer-Shoup [15], Fischlin [19] and Gennaro-Halevi-Rabin [20] schemes are based on the strong RSA assumption. The Boneh-Boyen [6], Camenisch-Lysyanskaya [10], and Waters [33] schemes are based on bilinear maps.

Digital signatures not only provide basic signing functionality but also are important building blocks for many applications such as blind signatures (for electronic voting and electronic cash), group signatures and credentials. In the light of these applications, the schemes based on bilinear maps (i.e., based on the discrete logarithm problem) are better than those based on the strong RSA assumption (i.e., based on the integer factoring problem), since we can often more easily construct efficient protocols based on the DL problem (because the order of a DL-based group can be published but the order of an IF-based multiplicative group cannot), and the data size is shorter with bilinear maps than with IF problems.

Among the bilinear-map-based schemes, the Boneh-Boyen scheme is not suitable to many applications such as blind signatures and credentials, since the signature forms $\sigma \leftarrow g^{1/(x+m+sy)}$, where (x, y) is the secret key, m is a message and (σ, s) is the signature, so it is hard to separate an operation (blinding, encryption etc.) with m from another operation that uses the secret key.

The Waters scheme is better than the Boneh-Boyen scheme, since a message operation, through the form $\prod_{i \in \mathcal{M}} u_i$, can be separated from another operation that uses the secret key. However, as shown in Section 9 the protocol of proving the knowledge of a message is not so efficient.

Blind Signatures. Since the concept of blind signatures was introduced by Chaum [13], it has been used in numerous applications, most prominently in electronic voting and electronic cash. Informally, blind signatures allow a user to obtain signatures from a signer on any document in such a manner that the signer learns nothing about the message that is being signed. The security of blind signatures was formalized by [22,28].

Even in the random oracle model, only a few secure blind signature schemes have been proposed [1,4,27,28,29,30]; [4] requires a non-standard strong assumption and [28,29,30] only allow a user to make a poly-logarithmically (not polynomially) bounded number of interactions with a signer, while [1,27] are secure for a polynomially number of interactions.

Only two secure blind signature schemes have been presented in the standard model [8,22]. However, the construction of [22] is based on a general two-party protocol and is thus extremely inefficient. The solution of [8] is much more efficient than that of [22], but it is still much less efficient than the secure blind signature schemes in the random oracle model [1,4,27,28,29,30]. For example, the protocol of [8] is much more complicated (where proofs of knowledge for at least 40 variables are required for a user) than that of [4,28,29], and requires many interactions between user and signer. Recently, a new blind signature scheme that is concurrently secure without random oracles has been presented [23], but it is not in the standard model but in the common reference string (CRS) model.

Partially Blind Signatures. One particular shortcoming of the concept of blind signatures is that, since the singer's view of the message to be signed is completely blocked, the signer has no control over the attributes except for those bound by the public key. For example, a shortcoming can be seen in a simple electronic cash system where a bank issues a blind signature as an electronic coin. Since the bank cannot set the value on any blindly issued coin, it has to use different public keys for different coin values. Hence the shops and customers must always carry a list of those public keys in their electronic wallet, which is typically a smart card whose memory is very limited. Some electronic voting schemes also face the same problem.

A *partially* blind signature scheme allows the signer to explicitly include common information in the blind signature under some agreement with the receiver. This concept is a generalization of blind signatures since the (normal) blind signatures are a special case of *partially* blind signatures where the common information is a null string.

The notion of partially blind signatures was introduced in [2], and the formal security definition and a secure partially blind signature scheme in the random oracle model were presented by [3]. However, no partially blind signature scheme secure in the standard model has been proposed.

1.2   Our Result

This paper proposes new digital signatures, blind signatures, and partially blind signatures that are secure in the standard model:

- (Digital signatures:)
  We propose a new efficient signature scheme secure in the standard model that is more suitable to many applications than the existing signature schemes secure in the standard model [6,10,15,33]. The security proof of our scheme requires a slightly stronger assumption, the 2SDH assumption, than the SDH assumption used by [6].
- (Blind signatures:)
  We propose a secure blind signature scheme in the standard model that is almost as efficient as the most efficient blind signature schemes whose security has been analyzed heuristically or in the random oracle model.
- (Partially blind signatures:)
  We propose the first secure partially blind signature scheme in the standard model. This scheme is almost as efficient as our blind signatures.

The proposed (partially) blind signature scheme is secure for polynomially many synchronized (or constant-depth concurrent) attacks, but not for general concurrent attacks. This paper presents an efficient way to convert our (partially) blind signature scheme in the standard model to a scheme secure for general concurrent attacks in the common reference string (CRS) model.

This paper also presents (partially) blind signatures from the Waters scheme that are secure in the standard model under the BDH assumption. The (partially) blind signatures are much less practical than the above-mentioned proposed scheme.

2   Preliminaries

2.1   Definition of Secure (Partially) Blind Signature Scheme

In this section we recall the definition of a secure *partially blind* signature scheme [3,8]. Note that this definition includes that of a secure *blind* signature scheme [22] as a special case where the piece of information shared by the signer and user, info, is a null string, $\perp$ (i.e., info $= \perp$).

Although our definition is based on [3,8], our *blindness* definition is slightly stronger than [3,8] as follows:

- Signer S^* can arbitrarily choose pk in ours, while pk must be honestly generated in [3,8].
- Even if only one of two users, U_0 or U_1, outputs a valid signature, S^* is allowed to obtain the valid signature and output the decision, b', in our definition, while only when both users, U_0 and U_1, output valid signatures, S^* is allowed to obtain them in [3,8].

Partially Blind Signature Scheme. In the scenario of issuing a partially blind signature, the signer and the user are assumed to agree on a piece of common information, denoted as info. In some applications, info may be decided by the signer, while in other applications it may just be sent from the user to the signer. Anyway, this negotiation is done outside of the signature scheme, and we want the signature scheme to be secure regardless of the process of agreement.

Definition 1. *(Partially Blind Signature Scheme) A Partially blind signature scheme is made up of four (interactive) algorithms (machines)* $(\mathcal{G}, \mathcal{S}, \mathcal{U}, \mathcal{V})$.

- $\mathcal{G}$ *is a probabilistic polynomial-time algorithm that takes security parameter* n *and outputs a public and secret key pair* (pk, sk).
- $\mathcal{S}$ *and* $\mathcal{U}$ *are a pair of probabilistic interactive Turing machines each of which has a public input tape, a private input tape, a private random tape, a private work tape, a private output tape, a public output tape, and input and output communication tapes. The random tape and the input tapes are read-only, and the output tapes are write-only. The private work tape is read-write. The public input tape of* $\mathcal{U}$ *contains* pk *generated by* $\mathcal{G}(1^n)$ *and* info. *The public input tape of* $\mathcal{S}$ *contains* info. *The private input tape of* $\mathcal{S}$ *contains* sk, *and that for* $\mathcal{U}$ *contains message* m. $\mathcal{S}$ *and* $\mathcal{U}$ *engage in the signature*

issuing protocol and stop in polynomial-time in n. When they stop, the public output tape of S contains either completed *or* not-completed*. Similarly, the private output tape of U contains either* $\perp$ *or* (m, σ).

– V *is a (probabilistic) polynomial-time algorithm that takes* $(pk, info, m, \sigma)$ *and outputs either* accept *or* reject.

Definition 2. *(Completeness) If S and U follow the signature issuing protocol with common input $(pk, info)$, then, with probability of at least $1 - 1/n^c$ for sufficiently large n and some constant c, S outputs* completed*, and U outputs (m, σ) that satisfies $V(pk, info, m, \sigma) =$ accept*. *The probability is taken over the coin flips of G, S and U.*

We say message-signature tuple $(info, m, \sigma)$ is *valid* with regard to pk if it leads V to accept.

Partial Blindness. To define the blindness property, let us introduce the following game among adversarial signer S^* and two honest users U_0 and U_1.

1. Adversary $S^*(1^n, info)$ outputs pk and (m_0, m_1).
2. Set up the input tapes of U_0, U_1 as follows:
   – Randomly select $b \in \{0, 1\}$ and put m_b and $m_{\bar{b}}$ on the private input tapes of U_0 and U_1, respectively ($\bar{b}$ denotes $1 - b$ hereafter).
   – Put $(info, pk)$ on the public input tapes of U_0 and U_1.
   – Randomly select the contents of the private random tapes.
3. Adversary S^* engages in the signature issuing protocol with U_0 and U_1.
4. If U_0 and U_1 output valid signatures $(info, m_b, \sigma_b)$ and $(info, m_{\bar{b}}, \sigma_{\bar{b}})$, respectively, then give those outputs to S^* in random order. If either U_0 or U_1 outputs a valid signature, $(info, m_b, \sigma_b)$ or $(info, m_{\bar{b}}, \sigma_{\bar{b}})$, then give this output to S^*. Give $\perp$ to S^* otherwise.
5. S^* outputs $b' \in \{0, 1\}$.

We define
$$\mathrm{Adv}_{\mathrm{PBS}}^{\mathrm{blind}} = 2 \cdot \Pr[b' = b] - 1,$$

where the probability is taken over the coin tosses made by S^*, U_0 and U_1.

Definition 3. *(Partial Blindness) Adversary S^* (t, ϵ)-breaks the blindness of a partially blind signature scheme if S^* runs in time at most t, and $\mathrm{Adv}_{\mathrm{PBS}}^{\mathrm{blind}}$ is at least ϵ. A partially blind signature scheme is (t, ϵ)-blind if no adversary S^* (t, ϵ)-breaks the blindness of the scheme.*

Remark. (Partially Perfect Blindness) As usual, one can go for a stronger notion of blindness depending on the power of the adversary and its success probability. A scheme provides partially *perfect* blindness if it is $(\infty, 0)$-blind.

Unforgeability. To define unforgeability, let us introduce the following game among adversarial user U^* and an honest signer S.

1. (pk, sk) is generated by $G(1^n)$, pk is put on the public input tapes of U^* and S, and sk is put on the private input tape of S.

2. For each run of the signature issuing protocol with $\mathcal{S}$, adversary $\mathcal{U}^*$ outputs info, which is put on the public input tape of $\mathcal{S}$. Then, $\mathcal{U}^*$ engages in the signature issuing protocol with $\mathcal{S}$ in a concurrent and interleaving way.
3. For each info, let ℓ_{info} be the number of executions of the signature issuing protocol where $\mathcal{S}$ outputs completed, given info on its input tape. (For info that has not appeared on the input tape of $\mathcal{S}$, define $\ell_{\mathsf{info}} = 0$.) Even when info $= \perp$, $\ell_\perp$ is also defined in the same manner.
4. $\mathcal{U}^*$ wins the game if $\mathcal{U}^*$ output ℓ valid signatures (info, m_1, σ_1), ..., (info, m_ℓ, σ_ℓ) for some info such that
   (a) $m_i \neq m_j$ for any pair (i, j) with $i \neq j$ ($i, j \in \{1, \ldots, \ell\}$).
   (b) $\ell > \ell_{\mathsf{info}}$.

We define $\mathbf{Adv}_{\mathrm{PBS}}^{\mathrm{unforge}}$ to be the probability that $\mathcal{U}^*$ wins the above game, taken over the coin tosses made by $\mathcal{U}^*$, $\mathcal{G}$ and $\mathcal{S}$.

Definition 4. *(Unforgeability) An adversary $\mathcal{U}^*$ (t, q_S, ϵ)-forges a partially blind signature scheme if $\mathcal{U}^*$ runs in time at most t, $\mathcal{U}^*$ executes at most q_S times the signature issuing protocol, and $\mathbf{Adv}_{\mathrm{PBS}}^{\mathrm{unforge}}$ is at least ϵ. A partially blind signature scheme is (t, q_S, ϵ)-unforgeable if no adversary $\mathcal{U}^*$ (t, q_S, ϵ)-forges the scheme.*

2.2 Bilinear Groups

This paper follows the notation regarding bilinear groups in [7,6]. Let $(\mathbb{G}_1, \mathbb{G}_2)$ be bilinear groups as follows:

1. $\mathbb{G}_1$ and $\mathbb{G}_2$ are two cyclic groups of prime order p, where possibly $\mathbb{G}_1 = \mathbb{G}_2$,
2. g_1 is a generator of $\mathbb{G}_1$ and g_2 is a generator of $\mathbb{G}_2$,
3. ψ is an isomorphism from $\mathbb{G}_2$ to $\mathbb{G}_1$, with $\psi(g_2) = g_1$,
4. e is a non-degenerate bilinear map $e : \mathbb{G}_1 \times \mathbb{G}_2 \rightarrow \mathbb{G}_T$, where $|\mathbb{G}_1| = |\mathbb{G}_2| = |\mathbb{G}_T| = p$, i.e.,
   (a) Bilinear: for all $u \in \mathbb{G}_1$, $v \in \mathbb{G}_2$ and $a, b \in \mathbb{Z}$, $e(u^a, v^b) = e(u, v)^{ab}$,
   (b) Non-degenerate: $e(g_1, g_2) \neq 1$ (i.e., $e(g_1, g_2)$ is a generator of $\mathbb{G}_T$),
5. e, ψ and the group action in $\mathbb{G}_1$, $\mathbb{G}_2$ and $\mathbb{G}_T$ can be computed efficiently.

3 Assumptions

Here we introduce a new assumption, the 2-variable strong Diffie-Hellman (2SDH) assumption on which the security of the proposed signature scheme is based.

q 2-Variable Strong Diffie-Hellman (q-2SDH) Problem. Let $(\mathbb{G}_1, \mathbb{G}_2)$ be bilinear groups shown in Section 2.2. The q-2SDH problem in $(\mathbb{G}_1, \mathbb{G}_2)$ is defined as follows: given a $(2q+6)$-tuple $(g_1, g_2, g_2^x, \ldots, g_2^{x^q}, g_2^y, g_2^{yx}, \ldots, g_2^{yx^q}, g_2^{\frac{y+b}{x+a}}, a, b)$ as input, output pair $(g_1^{\frac{1}{x+c}}, c)$ where $c \in \mathbb{Z}_p^*$. Algorithm $\mathcal{A}$ has advantage, $\mathbf{Adv}_{2SDH}(q)$, in solving q-2SDH in $(\mathbb{G}_1, \mathbb{G}_2)$ if

$$\mathbf{Adv}_{2SDH}(q) \leftarrow \Pr[\, \mathcal{A}(g_1, g_2, g_2^x, \ldots, g_2^{x^q}, g_2^y, g_2^{yx}, \ldots, g_2^{yx^q}, g_2^{\frac{y+b}{x+a}}, a, b) = (g_1^{\frac{1}{x+c}}, c) \,],$$

where the probability is taken over the random choices of $g_2 \in \mathbb{G}_2$, $x, y, a, b \in \mathbb{Z}_p^*$, and the coin tosses of $\mathcal{A}$.

Definition 5. *Adversary $\mathcal{A}$ (t, ϵ)-breaks the q-2SDH problem if $\mathcal{A}$ runs in time at most t and $\text{Adv}_{2SDH}(q)$ is at least ϵ. The (q, t, ϵ)-2SDH assumption holds if no adversary $\mathcal{A}$ (t, ϵ)-breaks the q-2SDH problem.*

Variant of q 2-Variable Strong Diffie-Hellman (q-2SDH$_S$) Problem.
The q-2SDH$_S$ problem in $(\mathbb{G}_1, \mathbb{G}_2)$ is defined as follows: given a $(3q + 4)$-tuple
$(g_1, g_2, g_2^x, g_2^y, g_2^{\frac{y+b_1}{x+a_1}}, \ldots, g_2^{\frac{y+b_q}{x+a_q}}, g_2^{a_1}, \ldots, g_2^{a_q}, b_1, \ldots, b_q)$ as input, output a pair
$(g_1^{\frac{y+d}{x+c}}, g_2^c, d)$ where $b_1, \ldots, b_q, d \in \mathbb{Z}_p^*$ and $d \notin \{b_1, \ldots, b_q\}$. Algorithm $\mathcal{A}$ has
advantage, $\text{Adv}_{2SDH_S}(q)$, in solving q-2SDH$_S$ in $(\mathbb{G}_1, \mathbb{G}_2)$ if

$$\text{Adv}_{2SDH_S}(q)$$
$$\leftarrow \Pr[\ \mathcal{A}(g_1, g_2, g_2^x, g_2^y, g_2^{\frac{y+b_1}{x+a_1}}, \ldots, g_2^{\frac{y+b_q}{x+a_q}}, g_2^{a_1}, \ldots, g_2^{a_q}, b_1, \ldots, b_q) = (g_1^{\frac{y+d}{x+c}}, g_2^c, d)],$$

where $b_1, \ldots, b_q, d \in \mathbb{Z}_p^*$ and $d \notin \{b_1, \ldots, b_q\}$, and the probability is taken over the random choices of $g_2 \in \mathbb{G}_2$, $x, y, a_1, b_1, \ldots, a_q, b_q \in \mathbb{Z}_p^*$, and the coin tosses of $\mathcal{A}$.

Definition 6. *Adversary $\mathcal{A}$ (t, ϵ)-breaks the q-2SDH$_S$ problem if $\mathcal{A}$ runs in time at most t and $\text{Adv}_{2SDH_S}(q)$ is at least ϵ. The (q, t, ϵ)-2SDH$_S$ assumption holds if no adversary $\mathcal{A}$ (t, ϵ)-breaks the q-2SDH$_S$ problem.*

Remark 1. We occasionally drop t and ϵ and refer to the q-2SDH (or q-2SDH$_S$) assumption rather than the (q, t, ϵ)-2SDH (or (q, t, ϵ)-2SDH$_S$) assumption. We also sometimes drop q- and $_S$ and refer to the 2SDH assumption rather than the q-2SDH or q-2SDH$_S$ assumption.

Remark 2. (Relation between the 2SDH and 2SDH$_S$ assumptions)

The 2SDH and 2SDH$_S$ assumptions are closely related in a manner similar to the equivalence of $(q-1)$-wDHA assumption and q-CAA assumption [25], where the q-wDHA problem is to output $g_1^{\frac{1}{x}}$, given a $(q + 2)$-tuple $(g_1, g_2, g_2^x, \ldots, g_2^{x^q})$ as input, and the q-CAA problem is to output pair $(g_1^{\frac{1}{x+c}}, c)$ where $c \in \mathbb{Z}_p^*$ and $c \notin \{a_1, \ldots, a_q\}$, given a $(2q + 3)$-tuple $(g_1, g_2, g_2^x, g_2^{\frac{1}{x+a_1}}, \ldots, g_2^{\frac{1}{x+a_q}}, a_1, \ldots, a_q)$ as input.

4   The Proposed Signature Scheme

This section presents the proposed secure signature scheme in the standard model under the 2SDH assumption.

Let $(\mathbb{G}_1, \mathbb{G}_2)$ be bilinear groups as shown in Section 2.2. Here, we assume that the message, m, to be signed is an element in $\mathbb{Z}_p^*$, but the domain can be extended to all of $\{0, 1\}^*$ by using a collision resistant hash function $H : \{0, 1\}^* \rightarrow \mathbb{Z}_p^*$, as mentioned in Section 3.5 in [6].

4.1 Signature Scheme

Key Generation. Randomly select generators $g_2, u_2, v_2 \in \mathbb{G}_2$ and set $g_1 \leftarrow \psi(g_2)$, $u_1 \leftarrow \psi(u_2)$, and $v_1 \leftarrow \psi(v_2)$. Randomly select $x \in \mathbb{Z}_p^*$ and compute $w_2 \leftarrow g_2^x \in \mathbb{G}_2$. The public and secret keys are:

Public key: g_1, g_2, w_2, u_2, v_2
Secret key: x

Signature Generation. Let $m \in \mathbb{Z}_p^*$ be the message to be signed. Signer $\mathcal{S}$ randomly selects r and s from $\mathbb{Z}_p^*$, and computes

$$\sigma \leftarrow (g_1^m u_1 v_1^s)^{1/(x+r)}.$$

Here $1/(x+r) \bmod p$ (and $m/(x+r) \bmod p$ and $s/(x+r) \bmod p$) are computed. In the unlikely event that $x+r \equiv 0 \mod p$, we try again with a different random r. (σ, r, s) is the signature of m.

Signature Verification. Given public-key $(g_1, g_2, w_2, u_2, v_2)$, message m, and signature (σ, r, s), check that $m, r, s \in \mathbb{Z}_p^*$, $\sigma \in \mathbb{G}_1$, $\sigma \neq 1$, and

$$e(\sigma, w_2 g_2^r) = e(g_1, g_2^m u_2 v_2^s).$$

If they hold, the verification result is `valid`; otherwise the result is `invalid`.

Remark. Here we assume that $g_1 = \psi(g_2)$ has been confirmed when the public-key is registered. Alternatively, $g_1 = \psi(g_2)$ can be confirmed in the signature verification procedure, or g_1 is not included in the public-key and $g_1 = \psi(g_2)$ is calculated in the signature verification process.

4.2 A Performance Improvement Technique (Precomputation)

By introducing additional secret key $y, z \in \mathbb{Z}_p^*$ such that $u_2 = g_2^y$ and $v_2 = g_2^z$, we can apply a precomputation technique for signature generation.

Before getting message m, signer $\mathcal{S}$ randomly selects r, δ from $\mathbb{Z}_p^*$, and computes $\sigma \leftarrow g_1^{\delta/(x+r)}$ as the precomputation of a signature. Given message m, $\mathcal{S}$ computes s such that $s \leftarrow (\delta - m - y)/z \bmod p$, where $1/z \bmod p$ can be also precomputed.

4.3 Security

Theorem 1. *If the $(q_S + 1, t', \epsilon')$-2SDH assumption holds in $(\mathbb{G}_1, \mathbb{G}_2)$, the proposed signature scheme is (t, q_S, ϵ)-strongly-existentially-unforgeable against adaptive chosen message attacks, provided that*

$$\epsilon \geq 3q_S \epsilon', \quad and \quad t \leq t' - \Theta(q_S^2 T),$$

where T is the maximum time for a single exponentiation in $\mathbb{G}_1$ and $\mathbb{G}_2$.

Proof. (Sketch) Assume $\mathcal{A}$ is an adversary that (t, q_S, ϵ)-forges the signature scheme. We will then construct algorithm $\mathcal{B}$ that breaks the $(q_S + 1)$-2SDH assumption with (t', ϵ'). Hereafter, we often use $q \leftarrow q_S + 1$ (as well as q_S).

An informal outline of our proof is as follows: First we classify the output (forgery) of $\mathcal{A}$ into three types (Types-1,2,3). We will then show that any type of output allows $\mathcal{B}$ to break the q-2SDH assumption. Type-1 forgery leads to breaking the q-SDH (to which q-2SDH is reducible) assumption in a manner similar to that in [6]. Type-2 forgery leads to breaking the q-2SDH assumption by producing $g_2^{\frac{1}{x+b}}$ from the q-2SDH problem including $g_2^{\frac{y+a}{x+b}}$. Type-3 forgery leads to breaking the discrete logarithm (to which q-2SDH is reducible).

First, we introduce three types of forgers, $\mathcal{A}$. Let $(g_1, g_2, w_2, u_2, v_2)$ be given to $\mathcal{A}$ as a public-key, and $z \leftarrow \log_{g_2} v_2 \in \mathbb{Z}_p^*$ (i.e., $v_2 = g_2^z$). Suppose $\mathcal{A}$ asks for signatures on messages $m_1, \ldots, m_{q_S} \in \mathbb{Z}_p^*$ and is given signatures (σ_i, r_i, s_i) for $i = 1, \ldots, q_S$ on these messages. The three types of forgers are as follows:

Type-1 forger outputs forged signature $(m*, \sigma^*, r^*, s^*)$ such that $r^* \notin \{r_1, r_2, \ldots, r_{q_S}\}$.

Type-2 forger outputs forged signature $(m^*, \sigma^*, r^*, s^*)$ such that $r^* \in \{r_1, r_2, \ldots, r_{q_S}\}$ (i.e., $r^* = r_k$ for some $k \in \{1, \ldots, q_S\}$) and $m^* + s^*z \not\equiv m_k + s_k z \pmod{p}$.

Type-3 forger outputs forged signature $(m*, \sigma^*, r^*, s^*)$ such that $r_1^* \in \{r_1, r_2, \ldots, r_{q_S}\}$ (i.e., $r^* = r_k$ for some $k \in \{1, \ldots, q_S\}$) and $m^* + s^*z \equiv m_k + s_k z \pmod{p}$. Note that in this case $s^* \neq s_k$, since $s^* = s_k$ implies $m^* = m_k$ and $\sigma^* = \sigma_k$.

Algorithm $\mathcal{B}$ is constructed as follows:

1. (Input:)
   $(g_1, A_0, A_1, \ldots, A_q, B_0, B_1, \ldots, B_q, C, a, b)$, where $A_i = g_2^{x^i}$, $B_i = g_2^{yx^i}$, and $C = g_2^{\frac{y+b}{x+a}}$ $(i = 0, 1, \ldots, q)$.

2. (Coin flip:)
   Algorithm $\mathcal{B}$ first picks a random value $c_{\text{type}} \in \{1, 2, 3\}$ that indicates its guess for the type of forger that $\mathcal{A}$ will emulate. The subsequent actions performed by $\mathcal{B}$ differ with $c_{\text{type}} \in \{1, 2, 3\}$ as follows:

3. (If $c_{\text{type}} = 1$;)
   In this case, q-SDH assumption is broken in a manner similar to that shown in [6].

4. (If $c_{\text{type}} = 2$;)

   (a) (Key setup)
       $\mathcal{B}$ randomly selects $z, r_i(\neq a)$ $(i = 1, \ldots, q - 1)$ from $\mathbb{Z}_p^*$. Let $f(X) \leftarrow \prod_{i=1}^{q-1}(X + r_i) \bmod p = \sum_{i=0}^{q-1} \beta_i X^i$. $\mathcal{B}$ can efficiently calculate $\beta_i \in \mathbb{Z}_p^*$ $(i = 0, \ldots, q - 1)$ from r_i $(i = 1, \ldots, q - 1)$.
       $\mathcal{B}$ computes

$$g_2' \leftarrow \prod_{i=0}^{q-1} A_i^{\beta_i} = g_2^{f(x)}, \quad w_2' \leftarrow \prod_{i=0}^{q-1} A_{i+1}^{\beta_i} = (g_2')^x,$$

$$u_2' \leftarrow \prod_{i=0}^{q-1} B_i^{\beta_i} = (g_2')^y, \quad v_2' \leftarrow (g_2')^z.$$

Let $g_1' \leftarrow \psi(g_2')$, $u_1' \leftarrow \psi(u_2')$ and $v_1' \leftarrow \psi(v_2')$.
$\mathcal{B}$ gives $(g_1', g_2', w_2', u_2', v_2')$ to $\mathcal{A}$ as a public-key of the signature scheme.

(b) (Simulation of signing oracle)
Upon receiving a query to the signing oracle, $\mathcal{B}$ simulates the reply to $\mathcal{A}$ as follows:
Let $f_i(X) \leftarrow f(X)/(X+r_i) \bmod p = \prod_{j=1, j \neq i}^{q-1}(X+r_i) \bmod p = \sum_{j=0}^{q-2} \gamma_j$
X^j. $\mathcal{B}$ can efficiently calculate $\gamma_j \in \mathbb{Z}_p^*$ $(j = 0, \dots, q-2)$ from r_l $(l \neq i \wedge l = 1, \dots, q-1)$.
First, $\mathcal{B}$ randomly selects $k \in \{1, 2, \dots, q-1\}$.
For each query $i \in \{1, 2, \dots, k-1, k+1, q-1\}$ (i.e., $i \neq k$) with message m_i from $\mathcal{A}$ to the signing oracle, $\mathcal{B}$ randomly selects $s_i \in \mathbb{Z}_p^*$, and computes

$$\sigma_i \leftarrow \left(\prod_{j=0}^{q-2} \psi(A_j)^{\gamma_j}\right)^{m_i + s_i z} \left(\prod_{j=0}^{q-2} \psi(B_j)^{\gamma_j}\right) = (g_1')^{(m_i + y + s_i z)/(x + r_i)}.$$

$\mathcal{B}$ returns (σ_i, r_i, s_i) to $\mathcal{A}$ as the reply to the query. Clearly this is a valid signature for public-key $(g_1', g_2', w_2', u_2', v_2')$.
For the k-th query with message m_k from $\mathcal{A}$ to the signing oracle, $\mathcal{B}$ computes $\omega_i, d \in \mathbb{Z}_p^*$ $(i = 1, \dots, q-2)$ such that $f(X) = c(X)(X+a) + d \bmod p$, $c(X) \leftarrow \sum_{i=0}^{q-2} \omega_i X^i$ and $d \in \mathbb{Z}_p^*$, and computes

$$\sigma_k \leftarrow \psi(C)^d \left(\prod_{i=0}^{q-2} \psi(A_i)^{\omega_i}\right)^b \prod_{i=0}^{q-2} \psi(B_i)^{\omega_i} = (g_1')^{(m_k + y + s_k z)/(x + r_k)},$$

$$s_k \leftarrow (b - m_k)/z \bmod p, \quad r_k \leftarrow a.$$

$\mathcal{B}$ returns (σ_k, r_k, s_k) to $\mathcal{A}$ as the reply to the query.

(c) (Output) When $\mathcal{A}$ outputs a (valid) forgery $(m^*, \sigma^*, r^*, s^*)$, $\mathcal{B}$ checks whether $r^* = a$ and $m^* + s^* z \not\equiv m_k + s_k z \pmod{p}$. If $r^* \neq a$ or $m^* + s^* z \equiv m_k + s_k z \pmod{p}$, then $\mathcal{B}$ outputs failure and aborts. Otherwise, $m^* + s^* z \not\equiv m_k + s_k z \pmod{p}$. Let $b^* \leftarrow m^* + s^* z \bmod p$. (Here $b = m_k + s_k z \bmod p$.) Since $b^* \neq b$, $\mathcal{B}$ can compute

$$\eta \leftarrow \left(\frac{(\sigma^*/\sigma_k)^{1/(b^* - b)}}{\prod_{i=0}^{q-2} \psi(A_i)^{\omega_i}}\right)^{1/d} = g_1^{1/(x+a)}.$$

$\mathcal{B}$ outputs (η, a).

5. (If $c_{\text{type}} = 3$;)

   (a) (Key setup)

   $\mathcal{B}$ randomly selects x', y' from $\mathbb{Z}_p^*$.

   $\mathcal{B}$ computes

   $$g_2' \leftarrow A_0 = g_2, \quad w_2' \leftarrow (g_2')^{x'}, \quad u_2' \leftarrow (g_2')^{y'}, \quad v_2' \leftarrow A_1 = g_2^x.$$

   Here we rename x as z' just for representation, so

   $$v_2' = (g_2')^{z'}.$$

   Let $g_1' \leftarrow g_1$.

   $\mathcal{B}$ gives $(g_1', g_2', w_2', u_2', v_2')$ to $\mathcal{A}$ as a public-key of the signature scheme.

   (b) (Simulation of signing oracle) Since $\mathcal{B}$ knows x', the simulation of the signing oracle exactly replicates the signing oracle.

   (c) (Output) When $\mathcal{A}$ outputs a (valid) forgery $(m^*, \sigma^*, r^*, s^*)$, $\mathcal{B}$ checks whether $r^* \in \{r_1, \ldots, r_{qs}\}$ (i.e., $r^* = r_k$ for some $k \in \{1, \ldots, qs\}$) and $s^* \neq s_k$. If $r^* \notin \{r_1, \ldots, r_{qs}\}$ or $s^* \neq s_k$, then $\mathcal{B}$ outputs `failure` and aborts. Otherwise, $\mathcal{B}$ computes

   $$z^* \leftarrow (m_k - m^*)/(s^* - s_k) \bmod p,$$

   and checks whether $A_1 = A_0^{z^*}$. If it holds, $z^* = z' = x$. $\mathcal{B}$ then randomly selects $c \in \mathbb{Z}_p^*$ and can compute $\eta \leftarrow g_1^{1/(z^*+c)} = g_1^{1/(x+c)}$.

   $\mathcal{B}$ outputs (η, c).

Since the value of c_{type} is independent from the type of forgery, $\mathcal{B}$ breaks the q-2SDH assumption with probability at least $\epsilon/(3q_S)$.   ⊣

5 Variant of the Proposed Signature Scheme

This section presents a slight variant of the proposed signature scheme presented in the previous section. This variant is used by our blind signatures.

5.1 Signature Scheme

The variant scheme is the same as the proposed signature scheme except for the signature generation and verification parts as follows: in this variant, the signature is

$$(\sigma \leftarrow (g_1^m u_1 v_1^s)^{1/(x+r)}, \alpha \leftarrow g_2^r, s),$$

while in the proposed signature scheme in Section 4, the signature is (σ, r, s).

The signature verification equation of this variant is

$$e(\sigma, w_2 \alpha) = e(g_1, g_2^m u_2 v_2^s),$$

while the proposed signature scheme in Section 4, the signature verification equation is $e(\sigma, w_2 g_2^r) = e(g_1, g_2^m u_2 v_2^s)$.

5.2 Security

Theorem 2. *If the (q_S, t', ϵ')-2SDH$_S$ assumption holds in $(\mathbb{G}_1, \mathbb{G}_2)$, the proposed signature scheme is (t, q_S, ϵ)-existentially-unforgeable against adaptive chosen message attacks, provided that*

$$\epsilon \geq 2\epsilon', \quad and \quad t \leq t' - O(q_S T),$$

where T is the maximum time for a single exponentiation in $\mathbb{G}_1$ and $\mathbb{G}_2$.

The proof is shown in the full paper version.

6 The Proposed (Partially) Blind Signature Scheme

This section shows the proposed *partially blind* signature scheme, which includes our *blind* signature scheme as a special case where $m_0 = 0$ or $h_2 = 1$.

6.1 Partially Blind Signature Scheme

Let $(\mathbb{G}_1, \mathbb{G}_2)$ be bilinear groups as shown in Section 2.2. Here, we also assume that the messages, m_0 and m_1, to be (partially blindly) signed are elements in $\mathbb{Z}_p^*$, but the domain can be extended to all of $\{0,1\}^*$ by using a collision resistant hash function $H : \{0,1\}^* \rightarrow \mathbb{Z}_p^*$, as mentioned in Section 3.5 in [6].

Key Generation. Randomly select generators $g_2, u_2, v_2, h_2 \in \mathbb{G}_2$ and set $g_1 \leftarrow \psi(g_2)$, $u_1 \leftarrow \psi(u_2)$, $v_1 \leftarrow \psi(v_2)$, and $h_1 \leftarrow \psi(h_2)$. Randomly select $x \in \mathbb{Z}_p^*$ and compute $w_2 \leftarrow g_2^x \in \mathbb{G}_2$. The public and secret keys are:

Public key: $g_1, g_2, w_2, u_2, v_2, h_2$
Secret key: x

Partially Blind Signature Generation.

1. Signer S and user U agree on common information m_0 (which is info in Section 2.1) in an predetermined way.
2. U randomly selects $s, t \in \mathbb{Z}_p^*$, computes

$$X \leftarrow h_1^{m_0 t} g_1^{m_1 t} u_1^t v_1^{st},$$

and sends X to S. Here, m_1 is the message to be blindly signed along with common information m_0. In addition, U proves to S that U knows $(t, m_1 t, t, st)$ for $X = (h_1^{m_0})^t g_1^{m_1 t} u_1^t v_1^{st}$ using the witness indistinguishable proof as follows:

   (a) U randomly selects a_1, a_2, a_3 from $\mathbb{Z}_p^*$, computes

$$W \leftarrow (h_1^{m_0})^{a_2} g_1^{a_1} u_1^{a_2} v_1^{a_3},$$

   and sends W to S.
   (b) S randomly selects $\eta \in \mathbb{Z}_p^*$ and sends η to U.

(c) U computes

$$b_1 \leftarrow a_1 + \eta m_1 t \bmod p, \quad b_2 \leftarrow a_2 + \eta t \bmod p, \quad b_3 \leftarrow a_3 + \eta s t \bmod p,$$

and sends (b_1, b_2, b_3) to S.

(d) S checks whether the following equation holds or not:

$$(h_1^{m_0})^{b_2} g_1^{b_1} u_1^{b_2} v_1^{b_3} = W X^{\eta}.$$

If it holds, S accepts. Otherwise, S rejects and aborts.

3. If S accepts the above protocol, S randomly selects $r \in \mathbb{Z}_p^*$. In the unlikely event that $x + r \equiv 0 \bmod p$, S tries again with a different random r. S also randomly selects $\ell \in \mathbb{Z}_p^*$, computes

$$Y \leftarrow (X v_1^{\ell})^{1/(x+r)} \quad \text{and} \quad R \leftarrow g_2^r,$$

and sends (Y, R, ℓ) to U.

Here, $Y = (X v_1^{\ell})^{1/(x+r)} = (h_1^{m_0} g_1^{m_1} u_1 v_1^{s+\ell/t})^{t/(x+r)}$.

4. U randomly selects $f \in \mathbb{Z}_p^*$, and computes

$$\tau = (ft)^{-1} \bmod p, \quad \sigma \leftarrow Y^{\tau}, \quad \alpha \leftarrow w_2^{f-1} R^f, \quad \beta \leftarrow s + \ell/t \bmod p.$$

Here, $\sigma = (h_1^{m_0} g_1^{m_1} u_1 v_1^{s+\ell/t})^{1/(fx+fr)} = (h_1^{m_0} g_1^{m_1} u_1 v_1^{s+\ell/t})^{1/(x+(f-1)x+fr)} = (h_1^{m_0} g_1^{m_1} u_1 v_1^{\beta})^{1/(x+\delta)}$, and $\alpha = w_2^{f-1} R^f = g_2^{(f-1)x+fr} = g_2^{\delta}$, where $\delta = (f-1)x + fr \bmod p$.

5. (σ, α, β) is the partially blind signature of (m_0, m_1), where m_0 is common information between S and U, and m_1 is blinded to S.

Signature Verification. Given public-key $(g_1, g_2, w_2, u_2, v_2, h_2)$, common information m_0, message m_1, and signature (σ, α, β), check that $m_0 \in \mathbb{Z}_p^*$, $m_1 \in \mathbb{Z}_p^*$, $\beta \in \mathbb{Z}_p$, $\sigma \neq 1$, $\sigma \in \mathbb{G}_1$, $\alpha \in \mathbb{G}_2$, and

$$e(\sigma, w_2 \alpha) = e(g_1, h_2^{m_0} g_2^{m_1} u_2 v_2^{\beta}).$$

6.2 Security

Theorem 3. *The proposed blind signature scheme ($m_0 = 0$ or $h_2 = 1$) is perfectly blind.*

Proof. Even if dishonest signer S^* outputs any public-key, $(g_2, w_2, u_2, v_2) \in (\mathbb{G}_2)^4$ and $g_1 = \psi(g_2)$, the view of S^*, $(X, W, \eta, b_1, b_2, b_3)$ as well as S's randomness, in the signature generation protocol is perfectly (information theoretically) independent from the value of (m, s, f), since $X = (g_1^m u_1 v_1^s)^t$ is perfectly independent from (m, s), the protocol is witness indistinguishable with respect to (m, s) against any dishonest S^*, and f is not used in the protocol with S^*.

Hence, the value of (m, δ, β) is perfectly independent from the view of S^*, where $\delta = (x+r)f - x \bmod p$ and $\beta \leftarrow s + \ell/t \bmod p$. Here, $\sigma = (g_1^m u_1 v_1^{\beta})^{1/(x+\delta)}$, $\alpha = g_2^{\delta}$, and (σ, α, β) is the (blind) signature of m. Therefore, the signature along with m, $(m, \sigma, \alpha, \beta)$, is also perfectly independent from the view of S^*, since σ and α are perfectly dependent on (m, δ, β). ⊣

Definition 7. *Let suppose a protocol between two parities, Alice and Bob. In a round of the protocol, Alice and Bob exchange messages, $a, b, c, \ldots, d$, where the first move is Alice (i.e., Alice sends a and Bob returns b etc.). We now consider q rounds of the protocol execution. Here $(a_i, b_i, c_i, \ldots, d_i)$ is the exchanged messages in the i-th round $(i = 1, \ldots, q)$. We say that a protocol between Alice and Bob is executed in a synchronized run of q rounds of the protocol, if the q rounds of the protocol consists of L sequential intervals and each interval, or the j-th interval $(j = 1, \ldots, L)$, consists of the parallel run of q_j $(q_j \in \{1, \ldots, q\}$ rounds of the protocol. $q = q_1 + \cdots + q_L$. Therefore, the first interval consists of: the first move from Alice is $(a_1, a_2, \ldots, a_{q_1})$, the second move from Bob is $(b_1, a_2, \ldots, b_{q_1})$, and so on. After completing the first interval, the second interval starts and consists of: the first move from Alice is $(a_{q_1+1}, a_{q_1+2}, \ldots, a_{q_1+q_2})$, the second move from Bob is $(b_{q_1+1}, b_{q_1+2}, \ldots, b_{q_1+q_2})$, and so on.*

Clearly the synchronized run is a generalization of the parallel and sequential runs.

Theorem 4. *If the (q_S, t', ϵ')-2SDH$_S$ assumption holds in $(\mathbb{G}_1, \mathbb{G}_2)$, the proposed blind signature scheme ($m_0 = 0$ or $h_2 = 1$) is (t, q_S, ϵ)-unforgeable against an L-interval synchronized run of adversaries, provided that*

$$\epsilon' \le \frac{1 - 1/(L+1)}{16} \cdot \epsilon, \quad and \quad t' \ge \frac{24L \log{(L+1)}}{\epsilon} \cdot (t + \Theta(T)) + \Theta(q_S T),$$

where T is the maximum time for a single exponentiation in $\mathbb{G}_1$ and $\mathbb{G}_2$.

Proof. (Sketch)

Assume $\mathcal{A}$ is an adversary that (t, q_S, ϵ)-forges the blind signature scheme. We will then construct an algorithm $\mathcal{B}$ that (t'', q_S, ϵ'')-forges the proposed signature scheme (basic signature scheme) presented in Section 5. This leads to an algorithm that breaks the 2SDH$_S$ assumption with $(q_S, t'' + O(q_S T), \epsilon''/2)$ by Theorem 2.

$\mathcal{B}$, given $(g_1, g_2, w_2, u_2, v_2)$ as a public key of the basic signature scheme, provides them to $\mathcal{A}$ as a public key for blind signatures.

$\mathcal{B}$ is allowed to access the signing oracle of the basic signature scheme q_S times. By using this signing oracle, $\mathcal{B}$ plays the role of an honest signer against $\mathcal{A}$ (dishonest user).

First, $\mathcal{A}$ requests $\mathcal{B}$ to sign X along with the witness indistinguishable (WI) protocol on witness $(mt \bmod p, t, st \bmod p)$ against $\mathcal{B}$'s random challenge $\eta \in \mathbb{Z}_p^*$. After completing the WI protocol, $\mathcal{B}$ resets $\mathcal{A}$ to the initial state of the WI protocol and runs the same procedure with the same commitment value of W and another random challenge $\eta' \in \mathbb{Z}_p^*$ $(\eta \ne \eta')$. If $\mathcal{B}$ succeeds in completing the WI protocol twice with different challenges η and η' such that

$$g_1^{b_1} u_1^{b_2} v_1^{b_3} = WX^{\eta}, \quad g_1^{b_1'} u_1^{b_2'} v_1^{b_3'} = WX^{\eta'}, \tag{1}$$

$\mathcal{B}$ can compute

$$m' \leftarrow (b_1 - b'_1)/(\eta - \eta') \bmod p,$$
$$t \leftarrow (b_2 - b'_2)/(\eta - \eta') \bmod p, \qquad (2)$$
$$s' \leftarrow (b_3 - b'_3)/(\eta - \eta') \bmod p,$$

such that

$$X = g_1^{m'} u_1^t v_1^{s'}.$$

$\mathcal{B}$ computes

$$m \leftarrow m'/t \bmod p, \quad s \leftarrow s'/t \bmod p. \qquad (3)$$

$\mathcal{B}$ then resumes the protocol just after the WI protocol, and sends m to the signing oracle. The signing oracle returns to $\mathcal{B}$ (σ, α, β) such that $(\sigma \leftarrow g_1^m u_1 v_1^\beta)^{1/(x+r)}$ and $\alpha = g_2^r$. $\mathcal{B}$ computes

$$Y \leftarrow \sigma^t, \quad \ell \leftarrow t(\beta - s) \bmod p, \qquad (4)$$

and returns $\mathcal{A}$ (Y, ℓ).

$\mathcal{B}$ repeats the above procedures (at the request of $\mathcal{A}$) q_S times. If all q_S rounds of the above procedures are completed, $\mathcal{A}$ finally outputs at least $q_S + 1$ valid signatures with distinct messages. From the pigeon-hole principle, among at least $q_S + 1$ distinct messages with valid signatures that $\mathcal{A}$ outputs, at least one message with valid signature is different from the q_S messages with valid signatures given by the signing oracle. This contradicts the q_S-unforgeability of the basic signature scheme.

The remaining problem in this strategy is how to execute all q_S rounds of the WI protocol twice with distinct challenges η and η' in a synchronized run with $\mathcal{A}$.

Claim. $\mathcal{B}$ can execute all q_S rounds of the WI protocol twice with distinct challenges η and η' in a synchronized run with $\mathcal{A}$ with probability at least $(1 - 1/(L + 1))\epsilon/8$ under the condition that $\mathcal{B}$ rewinds $\mathcal{A}$ with random challenges at most $24L \log (L + 1)/\epsilon$ times in total (or in L intervals).

Combining this result with Theorem 2 we obtain this theorem.   ⊣

Theorem 5. *The proposed partially blind signature scheme is perfectly blind.*

The proof is almost the same as that in Theorem 3.

Theorem 6. *If the (q_S, t', ϵ')-$2SDH_S$ assumption holds in $(\mathbb{G}_1, \mathbb{G}_2)$, the proposed partially blind signature scheme is (t, q_S, ϵ)-unforgeable against an L-interval synchronized run of adversaries, provided that*

$$\epsilon' \leq \frac{1 - 1/(L + 1)}{32} \cdot \epsilon, \quad and \quad t' \geq \frac{48L \log (L + 1)}{\epsilon} \cdot (t + \Theta(T)) + \Theta(q_S T),$$

where T is the maximum time for a single exponentiation in $\mathbb{G}_1$ and $\mathbb{G}_2$.

Remark. (Constant-depth concurrency) We can define a specific type of concurrent runs, *constant-depth concurrent* runs, in which, informally speaking, only a constant depth of purely inner rounds is allowed in all paths. Synchronized runs are a specific type of depth-1 concurrent runs. We can show that our blind signature scheme is still secure against a constant-depth concurrent run of adversaries under the same assumption and model. The result is presented in the full paper version.

6.3   Generalization

(m_0, m_1) with an additional key h_2 is generalized to $(m_0, \dots, m_l)$ with additional key $(h_{2,1}, \dots, h_{2,l})$. Arbitrary subset in $\{m_0, \dots, m_l\}$ can be blinded messages and the remaining be common messages.

7   Conversion to Fully Concurrent Security in the CRS Model

As mentioned above, the proposed (partially) blind signature scheme is secure against a synchronized run of adversaries (or more generally, a constant-depth concurrent run of adversaries). In this section, we show how to convert the proposed scheme to a scheme secure against a fully-concurrent run of adversaries. Our proposed blind signature scheme is secure in the *plain model* (without any setup assumptions), while the converted scheme is secure in the *common reference string (CRS) model*. The key idea is similar to [23], and uses the Paillier encryption for a simulator to extract blind messages with the help of the CRS model, and also uses a trapdoor commitment [16] to realize a concurrent zero-knowledge protocol. For simplicity of description, we will show a blind signature scheme, but it is straightforward to extend it to our partially blind signature scheme.

Key Generation. Randomly select generators $g_2, u_2, v_2 \in \mathbb{G}_2$ and set $g_1 \leftarrow \psi(g_2)$, $u_1 \leftarrow \psi(u_2)$, and $v_1 \leftarrow \psi(v_2)$. Randomly select $x \in \mathbb{Z}_p^*$, and compute $w_2 \leftarrow g_2^x \in \mathbb{G}_2$. In addition, randomly select secret and public keys of the Paillier encryption, two prime integers P and Q, and $(N = PQ, G)$, where $|N| = (6 + 3c_0)|p|$ (c_0 is a constant and $0 < c_0 < 1$). The public and secret keys, (pk, sk), of a trapdoor commitment, *commit*, [16] are also generated.

The public and secret keys and CRS are:

Public key: g_1, g_2, w_2, u_2, v_2
Secret key: x
CRS: N, G, pk
Trapdoor of CRS: P, Q, sk

Blind Signature Generation.

1. U checks whether $g_2, w_2, u_2, v_2 \in \mathbb{G}_2$ and $g_1 = \psi(g_2)$. If they hold, U proceeds the following signature generation protocol.

2. U randomly selects $s, t \in \mathbb{Z}_p^*$ and $A \in \mathbb{Z}_{N^2}$, computes

$$X \leftarrow g_1^{mt} u_1^t v_1^{st}, \quad D \leftarrow G^{(mt \bmod p) + t2^K + (st \bmod p)2^{2K}} A^N \bmod N^2,$$

and sends (X, D) to S. Here $K = (2 + c_0)|p|$, and $m \in \mathbb{Z}_p^*$ is the message to be blindly signed. In addition, U proves to S that U knows $(mt \bmod p, t, st \bmod p)$ for X as follows:

(a) U randomly selects a_1, a_2, a_3 from $\{0,1\}^{(2+c_1)|p|}$ (c_1 is a constant and $0 < c_1 < c_0 < 1$), $B \in \mathbb{Z}_{N^2}$ and r^* from the domain, computes

$$W \leftarrow g_1^{a_1} u_1^{a_2} v_1^{a_3}, \quad E \leftarrow G^{a_1 + a_2 2^K + a_3 2^{2K}} B^N \bmod N^2,$$

$$C \leftarrow commit(E, r^*, pk),$$

and sends (W, C) to S.

(b) S randomly selects $\eta \in \mathbb{Z}_p^*$ and sends η to U.

(c) U computes

$$b_1 \leftarrow a_1 + \eta(mt \bmod p), \quad b_2 \leftarrow a_2 + \eta t, \quad b_3 \leftarrow a_3 + \eta(st \bmod p),$$

$$F \leftarrow BA^\eta \bmod N^2,$$

and sends (b_1, b_2, b_3, F) as well as (E, r^*) to S.

(d) S checks whether the following equation holds or not:

$$|b_i| \leq (2 + c_1)|p| \quad (i = 1, 2, 3), \quad C = commit(E, r^*, pk).$$

$$g_1^{b_1} u_1^{b_2} v_1^{b_3} = WX^\eta, \quad G^{b_1 + b_2 2^K + b_3 2^{2K}} F^N \equiv ED^\eta \pmod{N^2}$$

If it holds, S accepts. Otherwise, S rejects and aborts.

3. The remaining procedure is the same as that of the original blind signature scheme.

Signature Verification. Same as that of the original blind signature scheme.

Security. The signature generation protocol is statistically WI except D, which is the Paillier encryption of a message. Since the Paillier encryption is semantically secure under the N-th residue assumption, this blind signature scheme satisfies blindness under this assumption.

If the WI protocol in the signature generation protocol is accepted by signer, simulator can extract (m, s, t) by decrypting D without rewinding $\mathcal{A}$ with high probability, by using the trapdoor of CRS (i.e., P, Q). That is, this scheme is unforgeable against any concurrent run of adversaries under the $2SDH_S$ assumption. (The proof is shown in the full paper version.)

8 Other Applications

We have shown the application of the proposed signature scheme to blind and partially blind signatures. The proposed signature scheme also supports other applications such as restrictive (partially) blind signatures, group signatures [24], verifiably encrypted signatures, anonymous credentials and chameleon hash signatures. (The full paper version presents restrictive (partially) blind signatures based on our (partially) blind signatures.)

9   (Partially) Blind Signatures from the Waters Scheme

9.1   The Proposed Blind Signature Scheme from the Waters Scheme

Key Generation. Let a symmetric bilinear group, $(\mathbb{G}_1, \mathbb{G}_1)$, be used in this scheme. Randomly select $\alpha \in \mathbb{Z}_p^*$. Randomly select generators $g, g_2, u', u_1, \ldots,$ $u_n \in \mathbb{G}_1$ and set $g_1 \leftarrow g^\alpha$.

Public key: $g, g_1, g_2, u', u_1, \ldots, u_n$
Secret key: g_2^α

Blind Signature Generation. Let m be the n-bit message to be signed, m_i the ith bit of m.

1. User U randomly selects $t \in \mathbb{Z}_p^*$, computes

$$X \leftarrow (u' \prod_{i=1}^{n} u_i^{m_i})^t,$$

and sends X to S. In addition, U proves to S that U knows $(t, m_1, \ldots, m_n)$ with $m_i \in \{0, 1\}$ for $X = (u' \prod_{i=1}^{n} u_i^{m_i})^t$ using the witness indistinguishable Σ protocols. For example,
   (a) U randomly selects $\delta_1, \ldots, \delta_n \in \mathbb{Z}_p^*$, computes $M_i = u_i^{m_i}(u')^{\delta_i}$ ($i = 1, \ldots, n$), and sends $(M_1, \ldots, M_n)$ to S.
   (b) U proves to S that U knows δ_i such that $M_i = (u')^{\delta_i}$ or $M_i = u_i(u')^{\delta_i}$ ($i = 1, \ldots, n$). Such an OR-proof can be efficiently realized by a Σ protocol [3].
   (c) U proves to S that U knows $(t, \beta, \gamma_1, \ldots, \gamma_n)$ such that $X = (\prod_{i=1}^{n} M_i)^t$ $(u')^\beta$, and $X = (u')^t \prod_{i=1}^{n} u_i^{\gamma_i}$, where $\beta \leftarrow t - t(\sum_{i=1}^{n} \delta_i) \bmod p$ and $\gamma_i \leftarrow tm_i$.
2. If S accepts the above protocol, S randomly selects $r \in \mathbb{Z}_p^*$, computes

$$Y_1 \leftarrow g_2^\alpha X^r, \quad Y_2 \leftarrow g^r,$$

and sends (Y_1, Y_2) to U.
3. U randomly selects $s \in \mathbb{Z}_p^*$, and computes

$$\sigma_1 \leftarrow Y_1(u' \prod_{i=1}^{n} u_i^{m_i})^s, \quad \sigma_2 \leftarrow Y_2^t g^s$$

4. $\sigma \leftarrow (\sigma_1, \sigma_2)$ is a blind signature.

Signature Verification. Given public-key $(g, g_1, g_2, u', u_1, \ldots, u_n)$, message $m \in \mathbb{Z}_p^*$, and signature $\sigma = (\sigma_1, \sigma_2)$, check

$$e(\sigma_1, g)/e(\sigma_2, u' \prod_{i=1}^{n} u_i^{m_i}) = e(g_1, g_2).$$

If it holds, the verification result is `valid`; otherwise the result is `invalid`.

Remark: If adversary $\mathcal{A}$ executes in a synchronized (or constant-depth concurrent) run with simulator $\mathcal{B}$ (as signer), $\mathcal{B}$ can effectively extract $(m_1, \ldots, m_n)$ and t from $\mathcal{A}$. $\mathcal{B}$ can then reduce the basic Waters signature scheme attack to the proposed blind signature scheme attack. It is straightforward to realize a partially blind signature scheme in a similar manner. The major problem in the efficiency of the signing process is in proving the knowledge of many $(O(n))$ variables in the WI Σ protocols.

Acknowledgements

The author would like to thank anonymous reviewers of TCC 2006 for their invaluable comments and suggestions.

References

1. Abe, M., A Secure Three-Move Blind Signature Scheme for Polynomially Many Signatures, Eurocrypt'01, LNCS 2045, pp.136-151, Springer-Verlag (2001).
2. Abe, M. and Fujisaki, E., How to Date Blind Signatures, Asiacrypt'96, LNCS 1163, pp.244-251, Springer-Verlag (1996).
3. Abe, M. and Okamoto, T., Provably Secure Partially Blind Signatures, Crypto'00, LNCS 1880, pp.271-286, Springer-Verlag (2000).
4. Bellare, M., Namprempre, C., Pointcheval, D. and Semanko, M., The power of RSA inversion oracles and the security of Chaum's RSA-based blind signature scheme, Financial Cryptography'01, LNCS, Springer-Verlag (2001).
5. Boldyreva, A., Threshold Signature, Multisignature and Blind Signature Schemes Based on the Gap-Diffie-Hellman-Group Signature Scheme, PKC'03, LNCS 2567, pp.31-46, Springer-Verlag (2003).
6. Boneh, D. and Boyen, X., Short Signatures Without Random Oracles, Crypto'04, LNCS, Springer-Verlag (2004).
7. Boneh, D., Lynn, B. and Shacham, H., Short Signatures from the Weil Pairing, Asiacrypt'01, LNCS, Springer-Verlag (2001).
8. Camenisch, J., Koprowski, M. and Warinschi, B., Efficient Blind Signatures without Random Oracles, Forth Conference on Security in Communication Networks - SCN '04, LNCS, Springer-Verlag (2004).
9. Camenisch, J. and Lysyanskaya, A., Efficient non-transferable anonymous multi-show credential system with optional anonymity revocation, Eurocrypt'01, LNCS 2045, pp. 93-118, Springer-Verlag (2001).
10. Camenisch, J. and Lysyanskaya, A., A signature scheme with efficient protocols, Security in communication networks, LNCS 2576, pp.268-289, Springer-Verlag (2002).

11. Camenisch, J. and Shoup, V., Practical verifiable encryption and decryption of discrete logarithms, Crypto'03, LNCS, pp. 126-144. Springer-Verlag (2003).
12. Camenisch, J. and Lysyanskaya, A., Signature Schemes and Anonymous Credentials from Bilinear Maps, Crypto'04, LNCS, Springer-Verlag (2004)
13. Chaum, D., Blind signatures for untraceable payments, Crypto'82, pp. 199-203. Plenum Press (1983).
14. Chow, S., Hui, L., Yiu, S. and Chow, K., Two Improved Partially Blind Signature Schemes from Bilinear Pairings, IACR Cryptology ePrint Archive, 2004/108 (2004).
15. Cramer, R. and Shoup, V., Signature schemes based on the strong RSA assumption, 6th ACM CCS, pp. 46-52. ACM press (1999).
16. Damgård, I., Efficient Concurrent Zero-Knowledge in the Auxiliary String Model, Eurocrypt'00, LNCS 1807, pp.418-430, Springer-Verlag (2000).
17. Diffie, W. and Hellma, M.E., New directions in cryptography, IEEE Trans. on Information Theory, IT-22(6), pp.644-654 (1976).
18. Fiat, A. and Shamir, A., How to prove yourself: Practical solution to identification and signature problems, Crypto'86, LNCS 263, Springer-Verlag (1987).
19. Fischlin, M., The Cramer-Shoup strong-RSA signature scheme revisited, PKC 2003, LNCS 2567, Springer-Verlag (2003).
20. Gennaro, R., Halevi, S. and Rabin, T., Secure hash-and-sign signatures without the random oracle, Eurocrypt'99, LNCS 1592, pp.123-139, Springer-Verlag (1999).
21. Goldwasser, S., Micali, S., and Rivest, R., A digital signature scheme secure against adaptive chosen-message attacks, SIAM Journal on Computing, 17, 2, pp.281-308 (1988).
22. Juels, A., Luby, M. and Ostrovsky, R., Security of blind digital signatures, Crypto'97, LNCS 1294, pp. 150-164, Springer-Verlag (1997).
23. Kiayias, A. and Zhou, H., Two-Round Concurrent Blind Signatures without Random Oracles, IACR Cryptology ePrint Archive, 2005/435 (2005)
24. Makita, T., Manabe, Y. and Okamoto, T., Short Group Signatures with Efficient Flexible Join, Manuscript (2005).
25. Mitsunari, S., Sakai, R. and Kasahara, M., A New Traitor Tracing, IEICE Trans. E-85-A, 2, pp. 481-484 (2002).
26. Naor, M. and Yung, M., Universal one-way hash functions and their cryptographic applications, 21st STOC, pp. 33-43, ACM (1989).
27. Pointcheval, D., Strengthened security for blind signatures, Eurocrypt'98, LNCS, pp.391-405, Springer-Verlag (1998).
28. Pointcheval, D. and Stern, J., Provably secure blind signature schemes, Asiacrypt'96, LNCS, Springer-Verlag (1996).
29. Pointcheval, D. and Stern, J., New blind signatures equivalent to factorization, ACM CCS, pp. 92-99. ACM Press (1997).
30. Pointcheval, D. and Stern, J., Security arguments for digital signatures and blind signatures, Journal of Cryptology, 13, 3, pp.361-396, Springer-Verlag (2000).
31. Schnorr, C.P., Security of Blind Discrete Log Signatures against Interactive Attacks, ICICS'01, LNCS 2229, pp.1-12, Springer-Verlag (2001).
32. Rompel, J., One-way functions are necessary and sufficient for secure signatures, STOC, pp.387-394, ACM (1990).
33. Waters, B., Efficient Identity-Based Encryption Without Random Oracles, Eurocrypt'05, LNCS 3494, pp. 114-127, Springer-Verlag (2005).
34. Zhang, F., Safavi-Naini, R. and Susilo, W, Efficient Verifiably Encrypted Signature and Partially Blind Signature from Bilinear Pairings, Indocrypt'03, LNCS 2904, pp. 191-204, Springer-Verlag (2003). Revised version available at http://www.uow.edu.au/susilo.

Key Exchange Using Passwords and Long Keys*

Vladimir Kolesnikov and Charles Rackoff

Dept. Comp. Sci., University of Toronto, Toronto, ON, M5S 3G4, Canada
{vlad, rackoff}@cs.utoronto.ca

Abstract. We propose a new model for key exchange (KE) based on a combination of different *types* of keys. In our setting, *servers* exchange keys with *clients*, who memorize short passwords and carry (stealable) storage cards containing long (cryptographic) keys. Our setting is a generalization of that of Halevi and Krawczyk [16] (HK), where clients have a password and the public key of the server.

We point out a subtle flaw in the protocols of HK and demonstrate a practical attack on them, resulting in a full password compromise. We give a definition of security of KE in our (and thus also in the HK) setting and discuss many related subtleties. We define and discuss protection against *denial of access* attacks, which is not possible in any of the previous KE models that use passwords. Finally, we give a very simple and efficient protocol satisfying all our requirements.

1 Introduction

We consider the goal of enabling multiple independent secure conversations between pairs of parties over an insecure network. The most convenient and natural way to achieve this is to perform a *key exchange* (KE), that is to provide the parties with matching randomly chosen keys that can be used for securing (only) a particular conversation. Of course, each player wants to communicate with a particular person, and even a powerful adversary *Adv* should not be able to match him up with a wrong partner. Therefore, players must possess some secret information with which they can authenticate themselves. The kind of information that is available to players determines the setting of KE. The simplest KE setting is when players have a shared random string. KE is more complicated in the public key setting, where parties have public/private key pairs with the public keys securely published. The most difficult setting is the pure password setting, where parties only have a short (presumably memorizable) shared password. We note that pure password KE protocols, at least in the standard model, are currently rather complicated and inefficient, due to the complexity of the setting.

1.1 Our Setting

Consider the client-server setting where both long keys and short keys (passwords) are used for KE. Assume that the server's (e.g. bank's) keys are securely stored. We

---

* The full version of this paper, containing a rigorous proof of security, appears in the Eprint archive [17].

S. Halevi and T. Rabin (Eds.): TCC 2006, LNCS 3876, pp. 100–119, 2006.
© Springer-Verlag Berlin Heidelberg 2006

take advantage of the inherent *logistical* differences in how keys are stored by the client (password in memory, long key on a storage card), to achieve more robust security than what is possible by using either type of key alone. Indeed, possession of long keys allows strong security guarantees against an online attacker. However, long keys can not be memorized, and thus must be stored, perhaps on a convenient plastic storage card. This is the vulnerability of this solution – the card may be (relatively) easily stolen by a physical attacker. On the other hand, passwords may be memorized, need not be stored, and thus can not be stolen. However, the protection against an online attacker one can hope to achieve with passwords is rather weak – passwords can always be guessed with relatively high probability. The only (somewhat satisfactory) protection against guessing attacks is recognizing them and refusing connection after a predetermined number of password failures[1].

Combining the benefits of both settings allows us to obtain a system, secure against both *types* of attack, and thus suitable for protection of sensitive information. This model is even more appealing due to its wide acceptance – it is natural for us to think of a card and a password, when we do, say, personal banking. More motivation is given in Sect. 3.

1.2   Our Contributions

We demonstrate a dangerous practical attack on the Halevi and Krawczyk (HK) [16] protocols, resulting in full compromise of any client's password (Sect. 2). The elegance, simplicity and practicality of the HK model and protocols resulted in their widespread practical use (e.g. their variants are being considered for parts of the IETF key exchange standard [12,10]). Therefore, the discovery of our attack may also have an important practical impact.

We propose and advocate the above *Combined Key* model of key exchange (ckKE). To the best of our knowledge, it has never been formally discussed. ckKE is a generalization of the HK model.

We give a formal definition of security of ckKE (Sect 3). Defining KE even in simpler settings has proven to be notoriously difficult, with a variety of (only seemingly!) innocuous decisions to be made. We discuss the subtleties of many of our choices, such as the necessity of tightness in the allowed success of the adversary, distinguishing the types of failures and reporting them, etc. Much of our discussion (e.g. on tightness of allowed success of the adversary *Adv*) also applies to and benefits pure password models.

We aim to make our definition as simple and natural as possible. For example, we require the server to explicitly indicate in its output whether a password failure occurred. We find this more intuitive than defining password guessing attack as an act of interference by the adversary (e.g. a successful impersonation!), as done in previous formalizations, such as [16,2]. Moreover, in previous formalizations, such as [16,2,5], the attacks are accounted by the environment; the server may not even "know" they occurred (e.g. in case of successful impersonation),

---

[1] We mention (but do not explicitly address) a variation of this defense against "too many" password guessing attacks. There the server limits the rate with which logins can be made, e.g. by exponentially increasing wait times between unsuccessful logins.

which makes attack recognition in practice less intuitive. We also find the game style of definitions (used in this paper) generally simpler and less prone to error than the simulation style (see discussion on the style of definition in Sect. 3.1 for more details).

We discuss unique security features available in ckKE "for free", such as the possibility of protection against the following *Denial of Access* (DoA) attack. *Adv*, attacking a player P, tries to connect to P's partner Q, using any password *pwd*. If *pwd* is correct, *Adv* wins; if not, *Adv* continues until he wins or Q refuses to connect to P. Then a legitimate P can no longer connect to Q. This easy to mount attack is unavoidable in any password-based setting (including HK) and is highly disruptive. We are not aware of the prevention of this attack being previously formalized. We formalize this attack and show how to prevent it in our model.

Finally, we give a very simple and efficient *two flow* KE protocol and prove its security (Sect. 4). An important feature of our protocol is that its flows are *independent* of each other, and thus can be sent in any order (or simultaneously), allowing for more flexibility and round efficiency.

1.3   Related Work

The problem of key exchange has deservedly received a vast amount of attention (e.g. [11,3,18,1,20,8,9]). The more complicated setting of pure password-based KE (pwKE) was first considered by Bellovin and Merritt [4]. Formal definitions (and protocols) in this setting were given by Bellare, Pointcheval and Rogaway [2], Boyko, Mackenzie and Patel [6], Goldreich and Lindell [13], and, recently, by Canetti et al. [7], as well as by many others.

Most relevant to our work is the problem of password-based KE in the asymmetric client-server setting, where the client has a password and the public key of the server. The question of resistance to off-line password-guessing attacks in this setting was first raised by Gong, et al. [14]. Later, Halevi and Krawczyk [16] formalized the notion of *one-way password authentication* in this setting and gave very simple and efficient protocols realizing it. They also extended their protocols to achieve key exchange with mutual authentication and perfect forward secrecy. The HK model is much simpler than the pure password model. The work of HK was the inspiration of our paper.

Further, Boyarsky [5] criticised the protocols of the earlier version [15] of [16] and suggested his own formalization of the same model. He showed several ways to amend a variant of protocols of [15] to satisfy his definition. We stress that he does not criticize protocols of the later version [16] we are considering.

Pinkas and Sander [19] consider heuristic approaches to securing password-only based authentication. They increase the cost of password-guessing and DoA attacks by using reverse Turing tests (RTT), that is, problems that are easy to solve for humans, but not for computers. We approach a different problem. In particular, RTT techniques can not increase security of a particular client against a determined attacker.

2   Attacking the Protocols of Halevi and Krawczyk [16]

Halevi and Krawczyk give four versions of their protocol (suitable for different tasks: password transmission, one-way authentication, and key exchange in two settings). Three of the four versions (with the exception of the Encrypted Password Transmission protocol) are (similarly) affected. We demonstrate our attack on their key exchange protocol.

The Halevi-Krawczyk protocol. Let S be a server with the public key pk_S, and p be the password shared between S and the client C. Let function $f(\cdot; \cdot)$ be *one-to-one on its components*, i.e. for every fixed strings p, x, functions $f(p; \cdot)$ and $f(\cdot; x)$ are one-to-one. Let $E = (Gen, Enc, Dec)$ be a CCA2 secure encryption scheme.

Construction 1. *(The Halevi-Krawczyk Mutual Authentication and Key Exchange Protocol (Π_{HK}))*

S		C
pick a nonce n	$n, pk_S \rightarrow$	*verify pk_S*
		pick random long key k
	$\leftarrow C, n, Enc_{pk_S}(k, f(p; C, S, k, n))$	
decrypt and verify		
$y := PRF_k(n, S, C)$	$y \rightarrow$	*check* $y = PRF_k(n, S, C)$
set $K = PRF_k(y)$		*set* $K = PRF_k(y)$

The "decrypt and verify" step outputs "FAIL" if the encryption is invalid or the received value of f does not match what S computes himself. The *nonces* must satisfy the *only* requirement that they never repeat.

Our Attack exploits the structure of f. We show that the conditions imposed on f are insufficient. The flaw of the proof of security of the protocol seems to be in the incorrect conclusion in Footnote 9 on p. 258 of [16]. We note that it is possible to make the proof (of security of one-way password authentication protocol) of Halevi and Krawczyk go through by additionally requiring that $f(\cdot; C, \cdot) \neq f(\cdot; C', \cdot)$ for any unequal client names C, C'.

For simplicity, we describe our attack on a specific instantiation of Π_{HK}. We stress that natural variants of our attack apply to many choices for f, and for nonce strategies, as well as for other parameter settings.

Let client names and passwords be 10 bits long, and nonces be 30 bits long. For a variable V, let v_i be the i-th bit of V. For example, $C = \langle c_1, c_2, ..., c_{10} \rangle$ is the name of the honest player, and $n = \langle n_1, n_2, ..., n_{30} \rangle$ is the nonce. Let the function be $f(p; C, S, k, n) = \langle c_1, ...c_9, c_{10} \oplus p_1, n_1, ...n_{21}, n_{22} \oplus p_2, ..., n_{30} \oplus p_{10}, S, k \rangle$. Finally, let nonces be chosen sequentially starting from 0. Note that this is a valid configuration of Π_{HK}.

The attack proceeds as follows. *Adv* creates an honest server S, an honest client C with any name $C = \langle c_1, c_2, ..., c_{10} \rangle$, and a bad client B with the name $B = \langle c_1, c_2, ..., c_{10} \oplus 1 \rangle$ and a randomly chosen password $p' = \langle p'_1, ..., p'_{10} \rangle$. Let p be C's password. Suppose for now that $p_1 \neq p'_1$, i.e. passwords of C and B differ in the high order bit. *Adv* observes one execution of KE between S and C. *Adv* records the encryption e sent by C and the nonce n (for concreteness, say $n = 00..00$, e.g. n is the first nonce). Now, B logs into S as himself, as follows. S sends the nonce $n' = n+1 = 00..01$, and B replies with $\langle B, n', e \rangle$. Now, if S doesn't fail, the password of C is computed as $pwd = \langle p'_1 \oplus 1, n_{22} \oplus n'_{22} \oplus p'_2, ..., n_{30} \oplus n'_{30} \oplus p'_{10} \rangle$ (since for $i = 22, ..., 30$, it must be that $n_{i-20} \oplus p_i = n'_{i-20} \oplus p'_i$). Also, if $p = pwd$, then S must accept, since $f(p'; B, S, k, n') = f(pwd, C, S, k, n)$. Thus, if S fails, pwd is eliminated from the possible passwords list.

B proceeds logging in as himself another $2^9 - 2$ times, eliminating different passwords one by one, until S accepts and that fact determines C's password. If S does not accept after B logged in $2^9 - 1$ times, B changes the first bit of his password with the server, and repeats the above entire attack (say, starting with a nonce ending with nine zeros), searching the other half-space. Finally, the two possible unchecked passwords can be verified by the same approach (and changing the password of B).

We stress that there were no attempts at impersonating C or S, and *all* failures are attributed to B. Neither C nor S know that C was attacked, thus C's account is never blocked. If B's account is blocked due to failures, B can claim mistyping and restore access. Moreover, there is no need to attack from only B's account; the attack can be easily distributed to try only a few passwords from each of many bad accounts. Again, it is easy to see that our attack is naturally generalizable to many practical instantiations of Π_{HK}.

On Boyarsky's [5] amendments of HK. The earlier version [15] of [16] had essentially the same protocol as [16], with the exception of the imposed requirements on the encryption scheme ([15] only required so-called *one-ciphertext verification attack* resistance, vs *ciphertext verification attack* resistance in [16]). Boyarsky [5] (independently from the revision resulting in the current version [16]) discovered the insufficiency of the weaker encryption. He gives his own formalization of the model and suggests three different amendments (see Sect. 5 of [5]) of the protocols of [15]. Boyarsky limits his consideration to the case where f is a concatenation function; thus our attack is not applicable to his protocols.

3   Key Exchange in the Combined Keys Model

Recall from the discussion in the Introduction that our setting (client carrying a plastic storage card and remembering a password) allows the advantage of robustness, that is graceful degradation of security in case one of the two types of keys is compromised. In particular, if the client's password is compromised, the security of KE should not suffer. On the other hand, if the card is compromised (e.g. copied), the remaining security should be that of the HK password model.

On resistance to server compromise. Halevi and Krawczyk briefly discuss resistance to insider attacks, i.e. attacks by rogue server employees who have access to some, but not all, private information stored on the server (see Sect. 3.3 in [16] for discussion of heuristic defense approaches). As another advantage of our setting, we mention that it allows stronger protection against server compromise. For example, public/private key pairs for each client C_i can be set up and used appropriately. Of course, an attacker who steals all the server data would now be able to successfully pose as the server. However, he can be prevented from posing as a client, as long as the client's private key remains secret. We note that such protection will require significant additional complexity of the definition and the protocol, and we leave it outside the scope of this paper. Therefore, as do Halevi and Krawczyk, in our exposition we assume that the server is secure, and his private information is never compromised.

On Denial of Access (DoA) attacks resistance. Recall that in the HK (and also in the pure password) setting, security critically depends on the ability of servers to suspend clients' accounts if there are "too many" password failures. At the same time, it is all too easy for Adv to cause them, making systems unusable by a trivial and easily mounted attack. In our combined key setting, it is natural to introduce protection against such DoA attacks. This can be done by requiring that polytime attackers can not cause password failures (and thus account suspension) without possession of long keys, stored on the card of the client. Of course, Adv may attempt attacks even without having the long keys, and furthermore, such attacks may be noticed by the servers. However, it is not hard to ensure that Adv does not learn anything from such attacks. This can be done, for example, by server first verifying possession of the long key (e.g. in form of a MAC), and immediately failing, if such verification failed. Then Adv does not learn anything about pwd, since it was not even used by the server. Therefore, such password guessing attacks are not a threat, and can be ignored. We formalize resistance to DoA attacks in our definition.

In our view, the main reason for using two types of keys is the two qualitatively different layers of protection against compromise. DoA resistance, although an important bonus, may not alone justify the cost of long key storage and management.

The reader may ask why one can't simply do two KE's in the two relevant models (one with parties sharing long keys, and the HK model) and combine the keys to obtain a KE protocol in our model. There are a number of issues to be addressed there. Firstly, a definition of security has to be given anyway – which is the bulk of our work. Secondly, natural ways of combining the two KE protocols (such as establishing a secure session using long keys, and sending the password over it) result in less efficient protocols.

3.1   Pre-definition Discussion

We start by briefly recalling the general setting for KE. There is a number of players (in our case, they are divided into two types – clients and servers) who have associated credentials, and pairs of whom may have shared common information. We think of a player as an *identity*, which may have many *instantiations*. Whenever a player P wishes to talk to another player Q, an instance of P is created with the required credentials passed. Thus an instance can be thought of as a participant of a particular conversation.

It is convenient to separate the notions of identity and instance for several reasons. Firstly, it is easier to talk about the independence of instances. Independence is highly desirable to avoid maintaining state and worry about communication and synchronization between instances. Secondly, a need often arises to have several channels of communication open between two or more parties simultaneously. Then the notion of instance makes it easier to implement and model concurrent executions of KE by a player.

We do not discuss how a player P knows that he wants to talk to a player Q. This may be done as part of previous (possibly insecure) communication, scheduled to happen at some predetermined time, or be requested by a higher level protocol. We give Adv the power to initiate conversations between players to model all possible scenarios.

Our goal is to enable a secure conversation, or *session*, between the instances of two players. Key exchange provides corresponding pairs of participants with matching keys that can be used for securing their communication. Of course, the keys of honest parties must appear random to the adversary Adv, and Adv must not be able to cause instances to match up in an inconsistent way[2].

To formalize the latter requirement, we need to define the notion of *partners* – instances who end up having a (n intended) conversation. We use session IDs (SID) to partner instances of players. There are several ways of using SID for this purpose, and we choose what we find to be the most natural – requiring each party that output a key to have an additional output *sid*. The other ways (e.g. requiring *sid* to be an input to parties, or requiring existence of a partnering function) seem to be less intuitive. We note that many natural protocols can be naturally modified to produce session ids. The *sid* output is not necessary in real protocols; it is only used for the purpose of defining and analyzing security of KE protocols.

Definition 1. *(KE Partners) Let P be a player. We denote by P_i the i-th instance of P. We write P_i^Q to emphasize that P_i intends to do KE with (some instance of) player Q. We say that an instance C_i^S of a client C and an instance S_j^C of a server S are* partners, *if they have output the same session id sid.*

Note that no two instances are partners when they are created; they may become partners once they've executed their KE protocols. We stress that P_i and P_i^Q refer to the *same* instance of P. We may omit the superscript in P_i^Q, when it is clear from the context.

Mutual authentication (MA) is an assurance that, if P_i^Q successfully completed and output a key, there must have been a Q_j^P "communicating" with him. We choose not to require it, because it can be achieved at the cost of two additional "key confirmation" flows (and refreshing the session key). Moreover, P_i^Q can never be sure that Q_j^P "is there" anyway, since Q_j^P may go offline at any time. Note, it is rather common and accepted to not require explicit mutual authentication for these reasons (e.g. [7]). Further, if we required MA, we must

---

[2] We note that Adv can cause confusion by mismatching instances of players and making them output *unrelated* keys. We don't regard this as a problem.

use a special $\perp$ output symbol to denote failure. In our definition we allow $\perp$, but don't insist on its use.

On the notions of attacks and failures. We first note that a special kind of failure – the *password failure* – must be introduced in our model to allow protection against DoA attacks. Intuitively, if *Adv*'s attack is such that the act of failure of the server may reveal some information about the client's password, then such failure is a password failure.

A natural approach to define adversary's ability to attack the system is by counting password checking *attempts*. However, it is less natural to define what an "attempt" is. Indeed, previous works on password-based key exchange (e.g. [16,2]) define "attempt" essentially as the act of *Adv*'s interference with the exchange of messages between two parties. However, it is less clear, for example, whether an act of *Adv* changing an insignificant bit of a message or an act of successful impersonation is such an attempt. Moreover, previously, the number of attempts was counted not by the server instances (they are not required to "know" whether a password guessing attack occurred), but by the environment.

An important feature of our definition is that servers themselves determine when, whether and what type of failure occurred. This explicates the notion of a *failed password attempt*, and ensures server's ability to identify a threat and react to it. Therefore, depending on the kind of failure, we allow servers to output either a *failure* symbol $\perp$, or a *password failure* symbol $P\perp$. We count password failures as $P\perp$'s reported by the servers, and clients accounts are suspended (to prevent further password guessing) based solely on that information and a predetermined threshold q. Therefore, a misidentification of an attack by the server is an omission of the protocol (opening a possibility of either password checking or DoA attacks), and we deem such protocols insecure.

We note that previous definitions, such as those of [16,2,5], can be similarly amended to ensure "explicit authentication" by additionally requiring that the server output $P\perp$ when he thinks a password attack has occurred. However, as discussed above, it seems to be cleaner to use the server's output as the only criterion for determining whether such an attack took place. Further, to ensure that the server does not misidentify the attacks, his output would need to be incorporated into the definitions, further complicating them.

The use of smart cards vs storage cards is briefly discussed in Sect. 4.

On the style of definition. As mentioned earlier, we prefer the game style of KE definitions in this paper. We find it easier to understand, since the game of the definition naturally corresponds to the actions and abilities of the adversary. We don't seem to need the complexity of simulation style definitions. An exception seems to be the very complex universally composable (UC) definitions, which can model very subtle issues such as password mistyping (see [7] and discussion in Sect 3.3). In addition to their complexity, UC-secure protocols currently are significantly less efficient than protocols in other frameworks. From another point of view, it is highly desirable to have different styles of definitions to discuss their relative strengths and, hopefully, prove equivalence in some settings.

On modelling the adversary. We consider a powerful Adv, who schedules events (such as creation of players and their instances) and controls all communications. This latter is modelled by the parties not sending messages to each other, but giving them to Adv for delivery. Adv is allowed to arbitrarily modify the messages (including dropping and injecting them) and schedule delivery. We allow Adv to create and arbitrarily initialize a polynomial number of accounts for corrupted clients. Note that in this model the actions of corrupt players need not be discussed separately from the actions of Adv, since Adv can simulate all their actions. For example, a message sent by a corrupted party can be viewed as a message injected by Adv.

Recall, Adv steals either the long key or the password of a client, and attacks one of the several security features of the protocol. We describe the (five) possible settings as games the attacker plays. (These games cover all cases – the cases that are not discussed explicitly are implicitly covered by stronger settings.)

Game KE_1 models the most complicated setting where Adv stole the long key of the client, and is attacking a server (that is trying to distinguish server's session key from random). This is the only game where Adv can benefit from guessing a password. Thus, in KE_1 Adv is allowed a limited number of P⊥'s.

Game KE_2 models the setting where Adv stole the long key and the password of the client, but is attacking a client.

Game KE_3 models the setting where Adv stole only the password of the client, and is attacking a server.

Game DOA models the inability of Adv to cause password failures without stealing the long key.

Game SID models the inability of Adv to cause two honest parties output different session keys, and is included for technical reasons (see discussion before the game's definition in Sect. 3.2 below).

One way to define security is to describe one adversary who, at some point in his attack, decides which of the five games above he really wants to play. However, since Adv's breaking abilities vary significantly among the games, defining allowed success of Adv in a "combined" game would be unnecessarily complicated. Therefore, we choose to describe five adversaries, each playing the corresponding game. We define the security of ckKE by inability of any of adversaries to win any of these games "too often". We note that it is possible to define the "combined" adversary model carefully, and to prove that any protocol that is secure with respect to the five adversaries would also be secure with respect to one "combined" adversary.

Liveness. Note that protocols may never terminate (e.g. when Adv cuts the communication channels). Instances may also output special failure symbols instead of (sid, key) pairs (e.g. when they detect Adv's interference). To ensure usability of KE protocols, we disallow these exceptional cases, unless Adv indeed attacks the system. Thus, we require that in the absence of an adversary, when processes communicate as intended, all sessions terminate, and intended partners output the same session id and key.

3.2 Formal Definition of Security of Key Exchange in the Combined Keys Model

Let n be a security parameter, and m be the number of bits in the password. In general, m can be a function of n; interesting cases are when m is constant or logarithmic in n. WLOG, say, the password domain is $D = \{0,1\}^m$. All players (*Adv*, clients and servers) are p.p.t. machines. Recall, the notion of partnering is defined in Def. 1.

We start by presenting the KE games. Recall, the first game models the setting where *Adv* obtained the long key of the client, is attacking a server, and is allowed a limited number of P⊥'s.

Game KE$_1$. *The adversary Adv starts by deterministically choosing the active attack threshold $q \in 1..|D|$ (based on the security parameter n) and creating an (honest) server S. Adv chooses S's name; then S's public and private keys are set up, and only the public key revealed to Adv. Adv then runs the parties by executing steps 1-5 multiple times, in any order:*

1. *Adv creates an honest client C. Adv is allowed to pick any unused name for the client; the client C is registered with S, and long key ℓ and password pwd are set up and associated with C. Only one honest client can be created. Adv is given the long key ℓ, but not pwd.*

2. *Adv creates a corrupt client B^i. Adv is allowed to initialize him in any way, choosing any unused name, long key and password for him.*

3. *Adv creates an instance C_i of the honest client C. C_i is given (secretly from Adv) as input: his name C, the partner server's name S, the public key of S, the long key and the password of C.*

4. *Adv creates an instance S_j of the honest server S. S_j is given (secretly from Adv) as input: his name S, the private key of S, the partner client's name (C or B^i) and that client's long key and password.*

5. *Adv delivers a message m to an honest party instance. That instance immediately responds with a reply (by giving it to Adv) and/or terminates and outputs the result (either a (sid,session key) pair or the failure symbol $\perp$) according to the protocol. The server instance can additionally output the password failure symbol P⊥. If the total number of P⊥ for the honest client is equal to the threshold q, Adv becomes restricted – he can not deliver messages to any instances S_j^C.*
   Adv learns the output, with the exception of its session key part. Additionally, at any time Adv may "open" any completed honest instance – then Adv is given the session key output by that instance.

Then Adv asks for a challenge on an instance S_j^C of the server S. S_j^C, who has been instantiated to talk to the honest client C, must have completed and not failed. The challenge is, equiprobably, either the key output by S_j^C or a random string of the same length. Adv must not have opened S_j^C or a partner of S_j^C, and is not allowed to do it in the future.

Then Adv continues to run the game as before (execute steps 2-5). Finally, Adv outputs a single bit b which denotes Adv's guess at whether the challenge *string was random. Adv wins if he makes a correct guess, and loses otherwise. Adv cannot "withdraw" from a challenge, and must produce his guess.*

Note the following technicality of KE_1. It is possible that Adv may find himself unable to complete the game. This may happen only when he had just caused the q-th P⊥ (and hence he is not allowed to deliver messages to servers) and he has no completed instances whom he is allowed to challenge. One way to handle this would be to require Adv flip a coin to determine whether he won or lost. We prefer to simply disallow, by this discussion, such behaviour of Adv, since the stalemate can be easily avoided by Adv having a "safety instance" complete before he risks the q-th P⊥.

In all other KE games (KE_2, KE_3, SID and DOA) below, it is possible (and natural) to require that the knowledge of pwd does not help Adv. We thus choose to reveal the password to Adv and remove restrictions on the number of P⊥'s (thus removing the definition of q). These games are presented by modifying KE_1. All of the above three modifications are included in all games below (and the last two are omitted in individual descriptions for conciseness).

Game KE_2 models the setting where Adv stole the long key and the password of the client, but is attacking a client.

Game KE_2. *This game is identical to KE_1, with the following additional exceptions.*

- *Adv is given pwd (in addition to ℓ) and must challenge an honest client instance C_i^S, who is talking to S.*

Game KE_3 models the setting where Adv stole only the password of the client, and is attacking a server.

Game KE_3. *This game is identical to KE_1, with the following additional exceptions.*

- *Adv is given pwd, but not the long key ℓ.*

Game SID enforces a non-triviality condition, preventing parties from improperly partnering up (e.g. by unnecessarily outputting the same session ids). Recall, Adv is not allowed to challenge parties whose partner has been opened, and we need to ensure that Adv is not unfairly restricted.

Game SID. *This game is identical to KE_1, with the following additional exceptions.*

- *Adv is given pwd (in addition to ℓ) and does not ask for (nor answers) the challenge.*
- *Adv wins if any two honest partners output different session keys.*

Finally, game DOA models resistance to the Denial of Access (DoA) attacks.

Game DOA. *This game is identical to KE_1, with the following additional exceptions.*

– *Adv is given pwd, but not the long key ℓ.*
– *Adv does not ask for (nor answers) the challenge.*
– *Adv wins if a server instance S_j^C outputs $\mathbb{PL}$.*

Definition 2. *(Secure Key Exchange in the Combined Keys Model.) We say that a key exchange protocol Π is* secure in the Combined Keys model, *if for every polytime adversaries Adv_1, Adv_2, Adv_3, Adv_{sid} and Adv_{doa} playing games KE_1, KE_2, KE_3, SID and DOA, their probabilities of winning (over the randomness used by the adversaries, all players and generation algorithms) is at most only negligibly (in n) better than:*

– $1/2 + \frac{q}{2|D|}$, *for KE_1,*
– $1/2$, *for KE_2 and KE_3,*
– 0, *for SID and DOA.*

KE definition for the HK setting. We note that Halevi and Krawczyk do not formally define the full notion of KE in their setting, but concentrate on the *one-way password authentication* of the client to the server. Because ckKE is a generalization of the HK setting and thanks to the modularity of our presentation, it is not hard to extract the KE definition for the HK setting from Def. 2. The only difference between our and the HK settings is that we additionally allow for the use of the long shared key ℓ. It turns out that it suffices to remove the games that do not allow *Adv* to know ℓ from Def. 2, to obtain a definition for the HK setting. (Of course, we also need to remove the uses of the long key ℓ from the remaining games.) Indeed, it is not hard to verify that the remaining games cover all possible attacks *Adv* can do in the HK setting. We explicate this definition below.

Definition 3. *(Secure Key Exchange in the HK Model.) We say that a key exchange protocol Π is* secure in the Halevi-Krawczyk, or *hybrid, model, if for every polytime adversaries Adv_1, Adv_2 and Adv_{sid} playing (amended as described above) games KE_1, KE_2 and SID, their probabilities of winning (over the randomness used by the adversaries, all players and generation algorithms) is at most only negligibly (in n) better than:*

– $1/2 + \frac{q}{2|D|}$, *for KE_1,*
– $1/2$, *for KE_2,*
– 0, *for SID.*

We note that although the pre- and post-definition discussion (of Sect. 3.1 and 3.3) discusses the ckKE setting, much of it applies to the HK setting as well.

3.3 Post-definition Discussion

On the sufficiency of only one honest server and one honest client. We note that definition of security is not strengthened by allowing *Adv* to create additional (good or bad) servers or good clients. The reason for this is that we assume independence in the initialization procedures of each pair of identities, and each instance is initialized only with information relevant to its partner. More detail follows.

Consider an adversary who wishes to attack a particular player – a client C or a server S. Suppose we allowed creation of additional good or bad servers. Note that initialization of a client C proceeds independently for servers S^1 and S^2, and, further, $C_{i_1}^{S^1}$ has no information about $C_{i_2}^{S^2}$, that is not known to Adv. Therefore, creating accounts for C with more than one server and instances of C talking to them does not help Adv, since it can be simulated by Adv. On the other hand, the ability to create many clients with a server is essential, since server instances talking to different clients do share common information among themselves – the secret key of the server. In fact, we exploit that in our attack on Π_{HK}. Only one honest client is sufficient, however, since additional honest clients can be played by Adv. We note that had we allowed clients to possess information common to two or more servers, we would have to allow Adv to create additional bad servers.

Addressing Boyarsky's criticism of the single-user case ([5]), we note that our definition allows Adv to determine whether two honest clients have the same password, causing at least an (expected) one P⊥ on each of the two clients. However, we don't see it as a problem, since, with high probability, clients' passwords differ. Therefore, determining a large clique of users with the same password would cause a large number of system-wide password failures and not cause bigger than expected "bang for the buck".

On the order of creation of good client and revealing the long key ℓ. Adv should first create the good client, and only then be allowed to see ℓ. This is the way the attack works in real life. Had we reversed the order, it would be easy to construct good protocols that would be defined insecure (e.g., a server leaks some information, if the client's name is the same as ℓ.)

On the allowed success of Adv in KE_1. Consider the success an adversary can always achieve (and therefore must be allowed in our definition). After q queries, Adv can guess the password with probability $q/|D|$, and if he fails to guess it, he can distinguish the key from random with probability $1/2$. Therefore, we should allow Adv's probability of success of at least $\frac{q}{|D|} + \frac{1}{2}\frac{|D|-q}{|D|} = \frac{1}{2}\frac{q+|D|}{|D|} = \frac{1}{2} + \frac{q}{2|D|}$.

On independence of the states of instances. In our model, there is no global information, and state is not preserved between executions of instances of players. Therefore, for example, it is not possible for an instance to know exactly how many P⊥'s occurred. Nevertheless, some communication and preservation of state can be achieved with the help of the adversary, as follows. The private key of S now additionally includes an n-bit MAC key k_M. Whenever S_j wants to publish a message m, he gives $(m, MAC_{k_M}(m))$ to Adv. The server's protocol has an optional field in one of the expected messages. S_j only accepts the properly MAC'ed messages in that field (this is essential, so that Adv cannot forge messages). We stress that communication may only happen if it is in the interest of Adv. Therefore, it can not be used to increase security of protocols, but mainly to uncover weaknesses of definitions (see example in the next topic).

On continuing the game after q P⊥'s. In the real world, at least ideally, after q P⊥'s, the server knows there is an attack on C, and will not accept new connections and will terminate all incomplete instances. How should we model this in our KE games? Although S may have cut communication with C, old sessions may still exist, and we need to ensure that they remain secure. That is why we allow the game to continue as before, but disallow sending messages to the server instances after q P⊥'s occurred.

Observe that once Adv got the challenge, "trying" another password may not help him much. Therefore, in particular, it is crucial to allow to challenge instances *after* q P⊥'s occurred.

It is not hard to design a concrete protocol demonstrating the necessity of our choice. Take a secure protocol Π. Modify it as follows to obtain Π'. Once a P⊥ of an honest client C occurred in the game (see above discussion on independence of states), in all future sessions with instances of C the all-zero session key is chosen with fixed small, but non-negligible probability (say $prob = \frac{1}{|D|^3}$). Clearly, this is a bad protocol, since after performing only one active attack, an attacker certainly breaks into one of the next few sessions. However, Π' would be deemed secure according to the definition, if Adv is not allowed to challenge after q P⊥'s (this is because Adv is allowed only one challenge, and he does not know which is the weak session. The expected advantage of Adv is less than what he gets from the q-th password try.)

On the necessity of tightness in defining the allowed success of Adv. Note that for every non-negligible slack allowed in Adv's success, there is a natural variant of Π' above, deemed secure by such definition. While one may be tempted to not be very careful in denying Adv "a few extra password tries", Π' has a much more dangerous vulnerability, which really should be prevented. We remark that in the password-only setting, if an indistinguishability of challenge based security definition does not require tightness, a simpler variant of Π', where players *always* output an all zero key with sufficiently small (yet non-negligible) probability, would be deemed secure.

On clients mistyping the passwords. How should we model the case when an honest client mistypes the password and causes P⊥? Consider the following protocol. Take a secure protocol, and modify it, so that S_j^C reveals ℓ once P⊥ occurred. It is easy to see that the new protocol remains secure in our definition, since we implicitly assume that C never mistypes the password. Indeed, in our definition, if a P⊥ occurred, it must have been caused by Adv. Since Adv cannot cause P⊥ without possession of ℓ, it is OK if S_j^C reveals ℓ. However, intuitively, we would not want to call such a protocol secure.

The only way to formally address the issue in our model is to allow C to mistype the password. A natural first idea is to allow Adv to instantiate clients with the password of his choice. However, it is not clear that this models real life – most often clients mistype their passwords to something related. Further, this would not address the protocol that reveals ℓ if the pwd is mistyped as $pwd + 1$, or, more generally, if the pwd is mistyped as a function of pwd.

A natural next idea is to instantiate clients with the password being $f(pwd)$, where the deterministic function f is specified by Adv. Only such an f that does not allow to check more than one password at a time may be allowed, and therefore strong restrictions on f are necessary. Indeed, setting $f(pwd) = 0$ on the first half of password domain D, and $f(pwd) = pwd$ on the second half, allows Adv to check half of password domain in one try. Restricting f to be a permutation does not work either, since applying such f allows to check whether pwd is a fixed point of f. Therefore functions f that have more than one or fewer than $|D| - 1$ fixed points are not allowed. At the same time, it is not hard to see that functions with $0, 1, |D|-1$ or $|D|$ fixed points do not allow Adv to check more than one password at a time when server is running a secure protocol, and thus may be allowed in our definition. Indeed, a function with 0 fixed points always causes S_j^C to P$\perp$; one with 1 fixed point fp always causes P$\perp$, unless $p = pwd$, and thus allows to check precisely one password; one with $|D|$ fixed points (identity) never causes P$\perp$; one with $|D| - 1$ fixed points always succeeds, unless pwd is the non-fixed point, and thus allows to check precisely one password.

At the same time, the most natural mistyping functions (e.g. confusing the order of digits) do not satisfy the requirements on f and do help the adversary (e.g. Adv can quickly test if the pin consists of the same decimal digits). More generally, Adv may infer a lot from simply observing a large volume of traffic, noting the patterns of honest clients mistyping their passwords, and matching them with expected patterns. However, it is not clear how to analyze this advantage, so we choose not to include password mistypes in our model at all, with the understanding that protocol designers take this discussion into account.

This subtlety also arises in KE in the pure password model, when passwords need not be chosen uniformly from D. Indeed, let $D_1 \subset D$ be all elements of D that end with a 0, and $pwd \in D$ is chosen uniformly from D_1. Then a protocol Π that reveals pwd iff pwd is mistyped only in the last digit, would be secure under a natural definition that does not allow mistyping. This is because pwd would not be revealed, unless Adv already had tried it. At the same time, such protocol Π should not be deemed secure. We note that the recent definition of password based KE in the complex Universal Composability model ([7]) addresses the issue of mistyping by allowing the environment both *choose* and *type* passwords.

On reporting failures to Adv immediately after failing. Consider a modification of Π_{HK}, where, upon a password failure, the server does not report it to Adv, but produces a random key and simulates successful completion of KE. This change would have prevented our attack of Sect 2. However, the achieved security would be illusory, since, in practice, it is hard to simulate successful completion well. Indeed, the fact of P$\perp$ must be somehow registered and used by S. This changes the state of S (in particular, the counter of active attacks is incremented). Since C can login after $q - 1$, but not after q P$\perp$'s, Adv is able to infer some information about S' outputs. To account for such "side channels", we require that players don't have private failure outputs (either $\perp$ or P$\perp$), and Adv is informed of failure as soon as it output. Note that this discussion relates to the *Additional discussion* in Sect. 2.1 of [7], where the authors argue that *Adv* need not know whether the passwords of two honest partners matched.

To further illustrate this point, suppose S_j at some point "knows" he is going to output $P\bot$, that is, S_j entered a state from which all execution paths lead to outputting $P\bot$, and *Adv* learned this fact. Suppose S_j does not terminate yet, but is waiting to receive another message. Then *Adv* can delay the delivery of the message indefinitely, S_j would never report $P\bot$, and we don't count it. In particular, adding an extra round of communication to a secure protocol Π, in which parties say whether they failed, makes Π insecure. This is consistent with our desire to force a server to correctly and timely report active attacks.

4   Our Protocol

Let n be a security parameter. To simplify discussion, we present our constructions with the domains and ranges of PRFG and MAC equal to $\{0,1\}^n$. Let $E = (Gen, Enc, Dec)$ be a CCA2 secure public key encryption scheme, $F : \{0,1\}^n \times \{0,1\}^n \mapsto \{0,1\}^n$ be a PRFG, and $MAC : \{0,1\}^n \times \{0,1\}^* \mapsto \{0,1\}^n$ be a message authentication code. Let N_C be the name of the client C, drawn from $\{0,1\}^n$. Shorter names can be used for efficiency, if desired.

Consider the following KE protocol Π, with two types of players, a server S and a client C who have secretly agreed on a password $pwd \in_R D$, a long secret key $\ell \in_R \{0,1\}^n$. Also, S has generated public/private key pair (pk_S, sk_S), and gave pk_S to C.

Construction 2. *(KE in the Combined Key Model (Π).)*

S^C	C^S
choose $r \in_R \{0,1\}^n$	choose $k \in_R \{0,1\}^n$,
	set $\alpha = Enc_{pk_S}(N_C, pwd, k)$
	$r \rightarrow \cdots \leftarrow \alpha, MAC_\ell(\alpha)$	
verify $MAC_\ell(\alpha)$ and N_C;	output
if fail, output $\bot$ and halt	$K = F_k(r), sid = (r, \alpha)$
verify pwd;	
if fail, output $P\bot$ and halt	
else output	
$K = F_k(r), sid = (r, \alpha)$	

WLOG, we assume that all protocol messages are formed properly (i.e. values are drawn from the appropriate domains, etc.). Then a client instance never fails, while a server instance may. Note that *Adv* may cause non-partnered parties to output unrelated keys. This is not a problem (see Sect. 3.1 and Footnote 2).

We stress that the two flows of the protocol are independent, and thus either of the parties can be the initiator. The DoA attacks are prevented if *Adv* does not have ℓ, even though, in particular, *Adv* is able to resend old messages of the client. The latter causes a server to output a random (from the point of view of *Adv*) session key, thus *Adv* is not able to take advantage of it. This also does not enable Adv to "reset" the fail counter in real executions (and thus try many passwords undetected), since the same effect can be achieved by Adv executing a KE between honest S_j^C and C_i^S, and then cutting the communication.

We treat the policies of account suspension and resetting of failure counters as external to our discussion, but stress that care should be taken in designing and implementing them. In particular, the client's explicit consent (communicated over a secure session) should be necessary for resetting the failed attempts counter, since otherwise Adv can be undetected when trying passwords between legitimate client logins. A natural scenario would be that the server asks the client whether he mistyped the password a certain number of times, and when client confirms, the fail counter is reset.

We further note that we can prevent Adv from resending C's old replies to S (e.g. if it is undesirable to have "hanging" sessions) by including r in the encryption of the client's reply and adding the corresponding verification step to S. We chose not to include it because it disallows the independence of flows of KE, and it is unclear whether hanging sessions are "worse" than hanging KE.

An alert reader will notice that smart cards may be gainfully used in place of client's storage cards. A smart card may hide the long key ℓ, only exposing the MAC'ing interface. An interesting setting is when Adv can "borrow" and return (but not copy) the card, obtaining only a period of ability to MAC strings of his choice. Our protocol will not benefit from such security improvements: C's messages are independent of S's, and thus Adv can MAC all the strings he might possibly need for an attack (e.g. strings containing all possible passwords) in one batch. Again, including r in the encryption of C's reply resolves this problem.

Π **is secure.** We first observe that for every Adv_{sid} and Adv_{doa} playing games SID and DOA, their probability of winning is negligible. Indeed, in our protocol, partners never output different keys (since the session key is determined by sid). As for Adv_{doa}, for a server to output $P\bot$, it is necessary to forge a MAC on an encryption not produced by any of the honest clients. This is only possible with negligible probability without the knowledge of the long key ℓ, assuming security of MAC.

Due to the lack of space, we formally consider the remaining games KE_i and adversaries in the full version [17], which appears in the Eprint archive. The structure of our proof is as follows. We start by reducing the KE adversaries to ones playing much simpler games. As a second step, we show that existence of new adversaries implies insecurity of either of the employed primitives. To give a flavor of the proof within the limited space, we include the most interesting intermediate game in Appendix A. Altogether, we've proven

Theorem 1. *The protocol Π of Constr. 2 is a secure key exchange protocol in the combined keys model.*

On generalizing Constr. 2. Consider creating a family of protocols parameterized by a function f similarly to the approach of Halevi and Krawczyk. The goal is to shorten the plaintext of the encryption α sent by C, which may improve the performance of the protocol. We note that we already reduce the amount of data under the CCA2-secure encryption – it is smaller than in any member of the HK families of KE protocols (but note that HK KE additionally achieve mutual authentication). We do not see how to further significantly increase efficiency by applying the HK idea to our protocols.

KE protocols for the HK setting. It is easy to see that removing the uses of the long key ℓ from the protocol of Constr. 2 casts it into the HK setting. The obtained protocol (explicated in Constr. 3 below) is a secure KE protocol in the HK setting, according to Def. 3. This conclusion immediately follows from the method of construction and Theorem 1.

Construction 3. *(KE in the HK setting.)*

S^C	C^S
choose $r \in_R \{0,1\}^n$	*choose* $k \in_R \{0,1\}^n$,
	set $\alpha = Enc_{pk_S}(N_C, pwd, k)$
	$r \rightarrow \cdots \leftarrow \alpha$
verify N_C;	*output*
if fail, output $\perp$ *and halt*	$K = F_k(r), sid = (r, \alpha)$
verify pwd;	
if fail, output $P\perp$ *and halt*	
else output	
$K = F_k(r), sid = (r, \alpha)$	

Acknowledgements. We thank Shai Halevi and the anonymous referees of TCC 2006 for very helpful comments on earlier versions of this work. We also thank Ian F. Blake for several stimulating discussions. The authors were in part supported by Natural Sciences and Engineering Research Council of Canada (NSERC) grants. The first author was also supported by Ontario Graduate Scholarship (OGS).

References

1. Mihir Bellare, Ran Canetti, and Hugo Krawczyk. A modular approach to the design and analysis of authentication and key exchange protocols (extended abstract). In *STOC '98: Proceedings of the thirtieth annual ACM symposium on Theory of computing*, pages 419–428, New York, NY, USA, 1998. ACM Press.
2. Mihir Bellare, David Pointcheval, and Phillip Rogaway. Authenticated key exchange secure against dictionary attacks. In *EUROCRYPT 2000*, pages 139–155, 2000.
3. Mihir Bellare and Phillip Rogaway. Entity authentication and key distribution. In *CRYPTO '93: Proceedings of the 13th annual international cryptology conference on Advances in cryptology*, pages 232–249, New York, NY, USA, 1994. Springer-Verlag New York, Inc.
4. Steven M. Bellovin and Michael Merritt. Encrypted key exchange: Password-based protocols secureagainst dictionary attacks. In *SP '92: Proceedings of the 1992 IEEE Symposium on Security and Privacy*, page 72, Washington, DC, USA, 1992. IEEE Computer Society.

5. Maurizio Kliban Boyarsky. Public-key cryptography and password protocols: the multi-user case. In *CCS '99: Proceedings of the 6th ACM conference on Computer and communications security*, pages 63–72, New York, NY, USA, 1999. ACM Press.

6. V. Boyko, P. MacKenzie, and S. Patel. Provably Secure Password-Authenticated Key Exchange Using Diffie-hellman. In B. Preneel, editor, *Proceedings EURO-CRYPT 2000*, pages 156–171, 2000.

7. Ran Canetti, Shai Halevi, Jonathan Katz, Yehuda Lindell, and Philip D. MacKenzie. Universally composable password-based key exchange. In *EUROCRYPT 2005*, pages 404–421, 2005.

8. Ran Canetti and Hugo Krawczyk. Analysis of key-exchange protocols and their use for building secure channels. In *EUROCRYPT '01: Proceedings of the International Conference on the Theory and Application of Cryptographic Techniques*, pages 453–474, London, UK, 2001. Springer-Verlag.

9. Ran Canetti and Hugo Krawczyk. Universally composable notions of key exchange and secure channels. In *EUROCRYPT '02: Proceedings of the International Conference on the Theory and Applications of Cryptographic Techniques*, pages 337–351, London, UK, 2002. Springer-Verlag.

10. T. Clancy. Eap password authenticated exchange, draft archive. http://www.cs.umd.edu/ clancy/eap-pax/, 2005.

11. Whitfield Diffie and Martin E. Hellman. New directions in cryptography. *IEEE Transactions on Information Theory*, IT-22(6):644–654, 1976.

12. Internet Engineering Task Force. Eap password authenticated exchange. http://www.ietf.org/internet-drafts/draft-clancy-eap-pax-03.txt, 2005.

13. Oded Goldreich and Yehuda Lindell. Session-key generation using human passwords only. In *CRYPTO '01: Proceedings of the 21st Annual International Cryptology Conference on Advances in Cryptology*, pages 408–432, London, UK, 2001. Springer-Verlag.

14. L. Gong, M. A. Lomas, R. M. Needham, and J. H. Saltzer. Protecting poorly chosen secrets from guessing attacks. *IEEE Journal on Selected Areas in Communications*, 11(5):648–656, 1993.

15. Shai Halevi and Hugo Krawczyk. Public-key cryptography and password protocols. In *CCS '98: Proceedings of the 5th ACM conference on Computer and communications security*, pages 122–131, New York, NY, USA, 1998. ACM Press.

16. Shai Halevi and Hugo Krawczyk. Public-key cryptography and password protocols. *ACM Trans. Inf. Syst. Secur.*, 2(3):230–268, 1999.

17. Vladimir Kolesnikov and Charles Rackoff. Key exchange using passwords and long keys. Manuscript, available from Eprint archive, http://eprint.iacr.org.

18. H. Krawczyk. Skeme: a versatile secure key exchange mechanism for internet. In *SNDSS '96: Proceedings of the 1996 Symposium on Network and Distributed System Security (SNDSS '96)*, page 114, Washington, DC, USA, 1996. IEEE Computer Society.

19. Benny Pinkas and Tomas Sander. Securing passwords against dictionary attacks. In *CCS '02: Proceedings of the 9th ACM conference on Computer and communications security*, pages 161–170, New York, NY, USA, 2002. ACM Press.

20. Victor Shoup. On formal models for secure key exchange. Technical Report RZ 3120 (#93166), 1999.

A   An Intermediate Game in the Proof of Theorem 1

We include the following game G_1 as an important intermediate step that gives the flavor of the proof of the most subtle case – the inability of Adv playing KE_1 win "too often". In the proof (included in the full version [17]) we show that Adv winning KE_1 implies an adversary $Dist_1$ winning G_1. We then show that $Dist_1$ winning "too often" implies insecurity of one of the employed primitives. Let n be a security parameter.

Game G_1. A maximum number of "password tries" q is deterministically (based on n) chosen by $Dist_1$ and fixed. The game initializes a CCA2 secure encryption scheme (by generating public and private keys pk_S and sk_S) and randomly chooses the password $pwd \in_R D$. Only the public key pk_S is given to $Dist_1$. $Dist_1$ queries the decryption oracle $O_D(e') = Dec_{sk_S}(e')$ to obtain decryptions of chosen strings. Then $Dist_1$ chooses a "client name" N_C. Then, for $i = 1, ..., u$, $Dist_1$ queries the encryption oracle O_E that produces random encryptions $e_i = Enc_{pk_S}(N_C, pwd, k_i)$, where $k_i \in_R \{0,1\}^n$ are chosen randomly and unknown to $Dist_1$. Here u is chosen by $Dist_1$. Then $Dist_1$ proceeds by executing Steps 1 - 2 multiple times, in any order:

1. $Dist_1$ queries the PRFG oracle $O_F(i,r) = F_{k_i}(r)$, where k_i was chosen (but not revealed) by O_E during it's i-th query. Here $r \in \{0,1\}^n$ and $i \in \{1..u\}$ are chosen by $Dist_1$.
2. $Dist_1$ queries the decryption oracle $O_D(e')$, where e' is chosen by $Dist_1$. He is not allowed to query O_D on any e_i obtained from O_E.

   Then $Dist_1$ chooses $i \in \{1, ..., u\}$ and $r_0 \in \{0,1\}^n$ and queries the challenge oracle $O_C(i, r_0)$. O_C produces a challenge as follows: it randomly chooses a bit b and a string $\rho \in_R \{0,1\}^n$. Then $O_C(i, r_0) = F_{k_i}(r_0)$ if $b = 0$, and $O_C(i, r_0) = \rho$ if $b = 1$. $Dist_1$ is not allowed to query $O_C(i, r_0)$, if he queried $O_F(i, r_0)$.

   Then, $Dist_1$ continues running Steps 1-2, with the exception that he is not allowed to query $O_F(i, r_0)$.

   Finally, $Dist_1$ generates a list of q password guesses $PL = \{p_1, ..., p_q\}$ and outputs a bit b'. $Dist_1$ wins if $pwd \in PL$ or if $b = b'$.

Mercurial Commitments: Minimal Assumptions and Efficient Constructions

Dario Catalano[1], Yevgeniy Dodis[2], and Ivan Visconti[3]

[1] CNRS-Ecole Normale Supérieure, Laboratoire d'Informatique, 45 Rue d'Ulm,
75230 Paris Cedex 05 - France
dario.catalano@ens.fr

[2] Department of Computer Science, New York University, 251 Mercer Street,
New York, NY 10012, USA
dodis@cs.nyu.edu

[3] Facoltà di Scienze Matematiche, Fisiche e Naturali, Università di Salerno,
via S. Allende n. 2, 84081 Baronissi (SA) - Italy
visconti@unisa.it

Abstract. (Non-interactive) *Trapdoor Mercurial Commitments* (TMCs) were introduced by Chase et al. [8] and form a key building block for constructing zero-knowledge sets (introduced by Micali, Rabin and Kilian [28]). TMCs are quite similar and certainly imply ordinary (non-interactive) *trapdoor commitments* (TCs). Unlike TCs, however, they allow for some additional freedom in the way the message is opened: informally, by allowing one to claim that "if this commitment can be opened at all, then it would open to this message". Prior to this work, it was not clear if this addition is critical or not, since all the constructions of TMCs presented in [8] and [28] used strictly stronger assumptions than TCs. We give an affirmative answer to this question, by providing simple constructions of TMCs from any trapdoor bit commitment scheme. Moreover, by plugging in various trapdoor bit commitment schemes, we get, in the trusted parameters (TP) model, *all* the efficient constructions from [28] and [8], as well as several immediate new (either generic or efficient) constructions. In particular, we get a construction of TMCs *from any one-way function* in the TP model, and, by using a special flavor of TCs, called *hybrid* TCs [6], even in the (weaker) shared random string (SRS) model.

Our results imply that (a) *mercurial commitments can be viewed as surprisingly simple variations of trapdoor commitments*; and (b) *the existence of non-interactive zero-knowledge sets is equivalent to the existence of collision-resistant hash functions*. Of independent interest, we also give a stronger and yet much simpler definition of mercurial commitments than that of [8], which is also met by our constructions in the TP model.

1 Introduction

Commitment schemes are important cryptographic primitives. They allow one party to commit to some value v so that v is kept secret from the rest of the world

S. Halevi and T. Rabin (Eds.): TCC 2006, LNCS 3876, pp. 120–144, 2006.
© Springer-Verlag Berlin Heidelberg 2006

(this is called *hiding*), and yet everybody knows that the value v is uniquely defined at the time v was committed (this is called *binding*). In particular, binding ensures that the party cannot announce the commitment first, and then decide later how to open it depending on the circumstances. In this sense, commitment schemes force the party to fully decide on what he is committing to.

At Eurocrypt 2005, Chase et al. [8] introduced an intriguing variant of commitments called *mercurial commitments*. The main difference comes from the fact that mercurial commitments allow for a small, and yet noticeable relaxation of the strict binding property of regular commitments. Namely, they allow for a two-stage opening protocol. In the *soft-open stage* the committer can claim that "if I committed to anything at all, then this value is m", while in the hard-opening stage he would indeed declare that "Yes, I really committed to the value m." In particular, any committed value c can either be both soft- and hard-opened only to one (correct!) message m, or can be soft-opened to arbitrary messages, but then it cannot be hard-opened at all! Moreover, the committer must decide before forming the commitment which one of the two cases suits him better: to commit to only one value, or not to commit to anything at all. Although this is seemingly not much better than regular commitments, the extra freedom of the committing party comes from the fact that by showing a soft-opening of his commitment to some value m, the receivers still cannot tell if m was really committed to by c, or if c was simply a "non-commitment" to anything (and the committer might be just going around and soft-opening c to arbitrary values m'). The receivers are sure, however, that it is impossible to hard-open c to any $m' \neq m$.

Chase et al. [8] distilled the above natural primitive to abstract away a relatively complicated (but efficient!) construction of *zero-knowledge sets* by Micali et al. [28]. Such ZK sets allow one to commit to some secret set S over some universe, and then to be able to non-interactively prove statements of the form $x \in S$ and $x \notin S$, and yet no other information (which cannot be deduced from the inclusions/exclusions above) about S is leaked — not even its size! With the abstraction of mercurial commitments, Chase et al. [8] obtained an elegant and easy-to-follow general "explanation" of the construction from [28]. Namely, they showed that the construction of [28] is an instance of a general construction of ZK sets from *any* mercurial commitment scheme and any collision-resistant hash function.

PLAIN VS. TRAPDOOR MERCURIAL COMMITMENTS. We remark that to match a very strong zero-knowledge definition of ZK sets from [28], Chase et al. [8] had to require that mercurial commitments satisfy the following "equivocation" property: there exists some trapdoor information msk (ordinarily not available to anybody) which enables one to completely destroy all the binding properties of mercurial commitments. Namely, using msk one can construct fake commitments, which look just like regular commitments and yet can be soft- or hard-opened to completely arbitrary values. (This is very similar to the notion of regular *trapdoor commitments* [4], where the knowledge of the corresponding trapdoor key can enable somebody to create fake regular commitments which can be opened to any message.) As already observed by [8], this strong equivocation property does not seem to be inherent for the "plain" primitive of mercurial commitments, but they chose to insist on this extra property since it was need for their main application. Since we believe that mercurial commitments are also interesting without equivocation, in our results we will distinguish between *plain* and *trapdoor* mercurial commitments. (Although our results described below will hold equally naturally for either case.) Indeed, we observe that one can define a weaker notion of ZK sets, which we informally call *indistinguishable sets*, which have the same functionality as ZK sets, but the privacy property is relaxed to only state that for any two sets and any sequence of inclusion/exclusion assertions which does not "separate" these sets, seeing the proofs of the corresponding assertions does not allow one to distinguish between these two sets. (This is somewhat similar to the distinction between witness indistinguishable [17] and ZK proofs [21].) And then it is easy to see that the same generic construction from [8] would give indistinguishable sets when applied to plain mercurial commitments. To summarize, we believe that both plain and trapdoor mercurial commitments are useful and deserve investigation.

MINIMAL ASSUMPTIONS FOR MERCURIAL COMMITMENTS. Having introduced a new cryptographic primitive, it is always very important to understand where it lies in the hierarchy of cryptographic assumptions. Towards this goal, [8] gave a general construction of (trapdoor) mercurial commitments in the plain model from any zero-knowledge proof system for NP. This construction, however, requires interaction. Then [8] showed a construction of non-interactive (trapdoor) mercurial commitments from non-interactive zero-knowledge proofs (NIZK) for NP, which are known, for example, to be implied by trapdoor permutations in the shared random string (SRS) model. However, this construction is mainly of theoretical interest, since it is very inefficient in practice. They also gave a more efficient (although bit-by-bit) construction of non-interactive (trapdoor) mercurial commitments from an even stronger assumption of claw-free permutations [22] in the trusted parameters (TP) model. On the other hand, [8] observed that (trapdoor) mercurial commitments are similar and trivially imply (trapdoor) regular commitments, although they pointed out some important differences as well. Thus, the following two questions were left open:

Question 1. *What minimal cryptographic and set-up assumptions are sufficient for non-interactive plain/trapdoor mercurial commitments?*

Question 2. *Can plain/trapdoor mercurial commitments be (efficiently) built from plain/trapdoor commitments?*

Our first result resolves these questions in a surprisingly simple fashion. We show a very simple and efficient construction of (bit) plain/trapdoor mercurial commitments from any bit plain/trapdoor regular commitment. The construction is a very simple generalization of the claw-free construction from [8], and since regular/trapdoor commitments are in principle equivalent to one-way functions in the SRS/TP model, we get

Theorem 1. *There exists a simple and efficient construction of (non-interactive) bit plain/trapdoor mercurial commitments from bit plain/trapdoor commitments. In particular, (non-interactive) plain/trapdoor mercurial commitments exist in the SRS/TP model if and only if one-way functions exist.*

The above result leaves open a question of basing (only trapdoor) mercurial commitments on one-way functions in the SRS model, which is a weaker set-up assumption than the TP model. Luckily, we observe that by using the same construction with a slight relaxation of trapdoor commitments, called *hybrid* trapdoor commitments, which were introduced by Catalano and Visconti [6] and shown to be equivalent to one-way functions *even in the SRS model*, we get

Theorem 2. *There exists a simple and efficient construction of (non-interactive) bit trapdoor mercurial commitments from (non-interactive) bit hybrid trapdoor commitments. In particular, non-interactive trapdoor mercurial commitments exist in the SRS model if and only if one-way functions exist.*

EFFICIENCY? Having resolved the question of feasibility, we can turn to the question of efficiency. Of course, we can plug in various efficient bit trapdoor commitment schemes to our previous construction, but this will only result in bit-by-bit constructions for long messages, which is pretty inefficient for practical use (e.g., for the ZK sets application). On the other hand, Chase et al. [8] gave two efficient constructions for long messages based on specific number-theoretic constructions (discrete log and factoring; the discrete log construction was implicit in [28]). Examining these constructions, one can see that there seems to be some kind of similarity between them, although it is not obvious exactly where this similarity comes from. Also, it is relatively hard to understand why each construction is really secure, without going into the details of the proof. Motivated by this, we ask

Question 3. *Is there an efficient and yet reasonably* **general** *construction of plain/trapdoor mercurial commitments, which would abstract and explain the efficient number-theoretic constructions from [8]?*

Our second result gives a surprisingly general answer to this question. Namely, we present a construction which directly transforms a plain/trapdoor bit commitment C into an efficient and (typically) multi-bit plain/mercurial commitment C'. Namely, we still base it on general plain/trapdoor commitment, just like in Theorem 1. However, a small catch is that we will need to assume an extra property from C (see Section 2.1 for a definition of Σ-protocol):

Theorem 3. *Assume C is a plain/trapdoor bit commitment which has an efficient Σ-protocol Π proving that one knows a witness d that a given (regular) commitment c can be opened to 0.[1] Then one one can construct an efficient plain/trapdoor mercurial commitment C' whose message space is equal to the challenge space of Π.*

---

[1] As explained in Section 5 proving this theorem, we will need a slight extra property (*) from such Σ-protocols, but it will always hold in any practical construction we are aware of. So we omit it from this statement.

Thus, to get message-efficient constructions, it will be important to design "challenge-efficient" Σ-protocols for our plain/trapdoor commitment schemes. While such Σ-protocol's Π in principle (see Theorem 5 below) can always be built from one-way functions, in general this will not outperform the simple construction in Theorem 1. However, the utility of this transformation comes from the fact that all number-theoretic (trapdoor) bit commitment schemes have very efficient Σ-protocols, and usually with rich challenge spaces. Plugging in various such commitment schemes with efficient protocols, we get many efficient constructions of mercurial commitments. In particular, both the discrete log and the factoring construction of [8] become special cases of our general transformation, when applied to an appropriate trapdoor commitment scheme! And several new constructions can be obtained as well (e.g., from RSA and Paillier [32] assumptions, as well as new discrete log and factoring constructions; see Section 5.1). More generally, we also believe that our construction is much easier to understand and sheds more light onto why the previous number-theoretic constructions where built in this particular way.

SIMPLER DEFINITION. As another small contribution, by *strengthening* the definition of trapdoor mercurial commitments as compared to the definition of [8], we considerably simplified the equivocation property of mercurial commitments. Since all our constructions (with the exception of Theorem 2) satisfy the stronger definition, and it results in easier and shorter proofs, we believe our strengthening is justified and could be of independent interest.

IMPLICATION TO ZK SETS. It is known from [8] and [31] that collision-resistant hash functions (CRHF) suffices for constructing interactive ZK sets in the plain model. Chase et al. [8] also made a simple observation that ZK sets imply the existence of CRHFs and therefore interactive ZK sets in the plain model and CRHFs are equivalent. Chase et al. in [8] also show that non-interactive indistinguishable/ZK sets can be constructed from any non-interactive plain/trapdoor mercurial commitment scheme and a collision-resistant hash function (CRHF). Using Theorem 1, Theorem 2, Theorem 3, and the fact that CRHFs imply both one-way functions (and, thus, plain/trapdoor/hybrid commitments in the SRS/TP/SRS models) and efficient *plain* commitment schemes (see [12] and [25]), we immediately obtain:

Theorem 4. *The existence of ZK (and, thus, indistinguishable) sets in the SRS model is equivalent to the existence of CRHFs. Moreover, ZK sets can be efficiently constructed from CRHFs and trapdoor bit commitment schemes, while indistinguishable sets can be efficiently constructed using CRHFs alone (by also building commitments out of them). The constructions become even more efficient if the commitment scheme in question has a challenge-efficient Σ-protocol needed for Theorem 3.*

2   Definitions

2.1 Σ-Protocols

Let $\mathcal{R} = \{(x, w)\}$ be some NP-relation (i.e., it is efficiently testable to see if $(x, w) \in \mathcal{R}$ and $|w| \leq \mathsf{poly}(|x|)$). We usually call x the input, and w — the witness (for x). Consider a three move protocol run between a PPT prover P, with input $(x, w) \in R$, and a PPT verifier V with input x, of the following form. P chooses a random string r_p, computes $a = \mathsf{Start}(x, w; r_p)$, and sends a to V. V then chooses a random string e (called "challenge") from some appropriate domain E (see below) and sends it to P. Finally, P responds with $z = \mathsf{Finish}(x, w, e; r_p)$. The verifier V then computes and returns a bit $b = \mathsf{Check}(x, a, e, z)$. We require that Start, Finish, and Check be polynomial-time algorithms, and that $|e| \leq \mathsf{poly}(|x|)$. Such a protocol (given by procedures $\mathsf{Start}, \mathsf{Finish}, \mathsf{Check}$) is called a Σ-*Protocol* for $\mathcal{R}$ if it satisfies the following properties, called completeness, special soundness, and special honest-verifier zero-knowledge:

- **Completeness:** If $(x, w) \in R$ then the verifier outputs $b = 1$ (with all but negligible probability).
- **Special Soundness:** There exists a PPT algorithm $\mathsf{Extract}$, called the (knowledge) extractor, such that it is computationally infeasible to produce an input tuple (x, a, e, z, e', z') such that $e \neq e'$ both lie in the proper "challenge" domain, $\mathsf{Check}(x, a, e, z) = \mathsf{Check}(x, a, e', z') = 1$, and yet $\mathsf{Extract}(x, a, e, z, e', z')$ fails to output a witness w such that $(x, w) \in R$. Intuitively, if some prover can correctly respond to two different challenges e and e' on the same first flow a, then the prover must "know" a correct witness w for x (in particular, x has a witness).
- **Special HVZK:** There exists a PPT algorithm Simul, called the simulator, such that for any $(x, w) \in R$ and for any fixed challenge e, the following two distributions are computationally indistinguishable. The first distribution (x, a, e, z) is obtained by running an honest prover P (with some fresh randomness r_p) against a verifier whose challenge is fixed to e. The second distribution (x, a, e, z) is obtained by computing the output $(a, z) \leftarrow \mathsf{Simul}(x, e)$ (with fresh randomness r_s). Intuitively, this says that for any a-priori fixed challenge e, it is possible to produce a protocol transcript computationally indistinguishable from an actual run with the prover (who knows w).

Since the standard zero-knowledge protocol for the Hamiltonian Cycle (see [16] and [23]) language is a (binary challenge) Σ-protocol, we get

Theorem 5 ([23],[16]). *Any* NP-*relation* $\mathcal{R}$ *has a (binary challenge)* Σ-*protocol if secure commitment schemes exist (in particular, in the SRS model if one-way functions exist).*

Of course, we will see and crucially exploit the fact that many natural specific languages have much more efficient Σ-protocols. We also notice that, aside from computational efficiency, a good quality measure for a given Σ-protocol is the size of its challenge space E (the larger the better). One reason for this dependency comes because the special soundness property easily implies that if a malicious prover does not "know" a valid witness w for x, then he can succeeds in fooling the verifier with probability at most (only negligibly better than) $1/|E|$. In our application, we will also see that the large size of E will also naturally translate to more efficient solutions, and we will therefore strive to use "challenge-efficient" Σ-protocols.

GENERALIZATIONS. First, we will allow $\mathcal{R}$ to depend on some honestly generated public parameter pk (known to everybody after generation); e.g. the standard discrete-log relation would be $\mathcal{R}_{p,g}(x, w) = 1$ if and only if $x = g^w \bmod p$, where the prime p and the generator g could be randomly generated. In this case the corresponding properties of the Σ-protocol should computationally hold over the choice of such parameters. However, for one of our applications we will require an even stronger technical property. Namely, we will say that a family of relations $\{\mathcal{R}_{pk}\}$ has a Σ-protocol which is *strongly hiding w.r.t. instance generation procedure* $\mathcal{P}$ if the special HVZK property holds even in the following experiment: $\mathcal{P}$ produces $(pk, x, w, \mathcal{I})$, where pk is the public key for $\mathcal{R}$, x is the input, w is the witness, and $\mathcal{I}$ is some side information available to attacker. Then we either give to the distinguisher a tuple $(\mathcal{I}, pk, x, a, e, z)$ obtained by having the prover run the real protocol with x and w, or where (a, z) is produced by the simulator $\mathsf{Simul}_{pk}(x, e)$. To put it differently, the side information $\mathcal{I}$ does not help the distinguisher to break the special HVZK property. We notice that, essentially all of the practical Σ-protocols known (including all the ones we will actually consider) will satisfy the statistical HVZK property, in which case they will be strongly-hiding w.r.t. any $\mathcal{P}$. Also, the generic protocol from Theorem 5 will also be strongly-hiding w.r.t. any efficient procedure $\mathcal{P}$ which only depends on the public parameters of the commitments used inside the protocol. This, once again, includes essentially all interesting procedures (including the specific one we will need later). In other words, for all practical purposes this extra property is just a technicality we need for the proof to go through.

As a second, orthogonal generalization, we can also consider "auxiliary-input" Σ-protocols, where in order to run the protocol, the prover P might need some extra information aux satisfying some property (which, presumably can be generated together with (x, w)), in addition to w. Notice, w alone is enough to allow for verification that $(x, w) \in \mathcal{R}$, so aux is only needed by the prover to fulfill his completeness requirement (in particular, the simulator does not need to know aux and special soundness and HVZK stay the same as before).

EFFICIENT Σ-PROTOCOLS. We briefly survey the following efficient Σ-protocols which we will use in the sequel. (The exact details will not be crucial for our purposes, so we will not present them here.) We notice that most of them will be unconditional: the security assumption behind the relation (such as discrete log) will be used later in the application; for example, in claiming that the hypothetical extraction of the witness contradicts the corresponding assumption.

The Schnorr Σ-protocol [34] allows one to unconditionally prove the knowledge of the discrete log in cyclic groups of prime order. A less known fact [19,9] is that (a slightly modified)[2] Schnorr protocol also works over the subgroup of quadratic residues Q_n over $\mathbb{Z}_n^*$, where n is the product of two safe primes. Interestingly, unlike in prime order groups, where the special soundness holds

---

[2] In particular, the prover works over the integers instead of over $\mathbb{Z}_{|Q_n|}$, since he does not know $|Q_n|$. Because of that the special HVZK guarantee is statistical here rather than perfect.

unconditionally, here it will hold computationally under the strong RSA assumption. In both of these cases the challenge space is exponential.

Very similar to Schnorr protocol, Gilliou-Quisquater (GQ) [24] protocol proves the knowledge of the e-th root over $\mathbb{Z}_n^*$ (i.e., solution to RSA), where $\gcd(e, \varphi(n)) = 1$ and n is the produce of two safe primes. Here, however, the challenge space should be smaller than the exponent e, so this protocol is challenge-efficient only if e is large (which is typically required when this protocol is used).

The Fiat-Shamir protocol is an unconditional binary-challenge Σ-protocol proving the knowledge of the square root over $\mathbb{Z}_n^*$, where n is the product of two primes. One way to make it challenge-efficient is to repeat it in parallel, but this is computationally inefficient. A better way is to use the elegant technique of Ong-Schnorr [30], at the expense of working over the set of quadratic residues Q_n, requiring n to be a Blum integer, and, more crucially, requiring an auxiliary witness to the prover. Namely, in order to make the challenge space to be of size 2^ℓ, the prover not only needs to know a square root of the input $x \in Q_n$, but also the 2^ℓ-root root $u \in Q_n$ of x (which is well defined when n is a Blum integer): see Lemma 3.1 in [1] explicitly stating the special soundness of this protocol. Of course, to run this protocol in practice one would first pick u and then set $w = u^{2^{\ell-1}} \bmod n$ (by repeated squaring) and $x = w^2 \bmod n$.

All the above mentioned protocols have statistical special HVZK, so they always satisfy strong-hiding. To summarize, natural relations arising from well established cryptographic assumptions have very computationally and challenge-efficient Σ-protocols.

2.2   Commitments and Trapdoor Commitments

COMMITMENTS. A (non-interactive) commitment scheme consists of four efficient algorithms: $\mathcal{C} = (\mathsf{Com\text{-}Gen}, \mathsf{Com}, \mathsf{Open}, \mathsf{Ver})$. The generation algorithm $\mathsf{Com\text{-}Gen}(1^k)$, where k is the security parameter, outputs a public commitment key pk (possibly empty, but usually consisting of public parameters for the commitment scheme). Given a message m from the associated message space $\mathcal{M}$ (e.g., $\{0,1\}^k$, although we will mainly concentrate on bit commitments), $\mathsf{Com}_{pk}(m; r)$ (computed using the public key pk and additional randomness r) produces a commitment string c for the message m. We will sometimes omit r and write $c \leftarrow \mathsf{Com}_{pk}(m)$. Similarly, the opening algorithm $\mathsf{Open}_{pk}(m; r)$ (which is supposed to be run using the same value r as the commitment algorithm) produces a decommitment value d for c. Finally, the verification algorithm $\mathsf{Ver}_{pk}(m, c, d)$ accepts (i.e., outputs 1) if it thinks the pair (c, d) is a valid commitment/decommitment pair for m. We require that for all $m \in \mathcal{M}$, $\mathsf{Ver}_{pk}(m, \mathsf{Com}_{pk}(m; r), \mathsf{Open}_{pk}(m; r)) = 1$ holds with all but negligible probability. We remark that without loss of generality we could have assumed that the opening algorithm simply outputs its randomness r as the decommitment, and the verification algorithm simply checks if $c = \mathsf{Com}_{pk}(m; r)$. However, we will find our more general notation more convenient for our purposes. When clear form the context, we will sometimes omit pk from our notation. Regular commitment schemes have two security properties:

Hiding. No PPT adversary (who knows pk) can distinguish the commitments to any two message of its choice: $\mathsf{Com}_{pk}(m_1) \approx \mathsf{Com}_{pk}(m_2)$. That is, $\mathsf{Com}_{pk}(m)$ reveals "no information" about m.

Binding. Having the knowledge of pk, it is computationally hard for the PPT adversary $\mathcal{A}$ to come up with c, m, d, m', d' such that (c, d) and (c, d') are valid commitment pairs for m and m', but $m \neq m'$ (such a tuple is said to cause a *collision*). That is, $\mathcal{A}$ cannot find a value c which it can open in two different ways.

Commitments are known to be theoretically equivalent to one-way functions [29],[26] (at least in the SRS model). However, efficient commitments can be built from collision-resistant hash functions [12],[25], and many number-theoretic assumptions (such as factoring, discrete log and RSA, and Paillier [32]; see below). In fact, most of these number-theoretic construct give a stronger kind of commitment — called trapdoor commitment — which we explain next.

TRAPDOOR COMMITMENTS. A (non-interactive) trapdoor commitment scheme consists of six efficient algorithms: $\mathcal{C} = (\mathsf{TrCom\text{-}Gen}, \mathsf{Com}, \mathsf{Open}, \mathsf{Ver}, \mathsf{Fake}, \mathsf{Equiv})$. The generation algorithm $\mathsf{TrCom\text{-}Gen}(1^k)$, where k is the security parameter, outputs a public commitment key pk and and a secret *trapdoor key sk*. Once pk is fixed, the meaning of Com, Open and Ver is exactly the same as for regular commitments. In particular, we will require that these algorithms satisfy the usual hiding and binding properties of the commitment schemes.

The trapdoor key sk is used in the algorithms Fake and Equiv to break the binding property of commitments. Namely, $\mathsf{Fake}_{sk}(;r)$ (which takes no input except for randomness r) produces "fake" commitment c, initially not associated to any message m. On other other hand, for any message m, $\mathsf{Equiv}_{sk}(m;r)$ (which is supposed to be run using the same value r as the fake commitment algorithm) produces a "fake decommitment" value d for $c = \mathsf{Fake}_{sk}(;r)$. In particular, we require that such fake (c, d) still satisfy the verification equation: for all $m \in \mathcal{M}$, $\mathsf{Ver}_{pk}(m, \mathsf{Fake}_{sk}(;r), \mathsf{Equiv}_{sk}(m;r)) = 1$ holds with all but negligible probability. Even stronger, we require that

Equivocation. for any $m \in \mathcal{M}$ (chosen by the adversary), a "true" commitment tuple $(m, \mathsf{Com}_{pk}(m;r), \mathsf{Open}_{pk}(m;r))$ should look computationally indistinguishable (over r) from the fake tuple $(m, \mathsf{Fake}_{sk}(;r), \mathsf{Equiv}_{sk}(m;r))$. More importantly, we require that these distributions should look indistinguishable even if the distinguisher knows not only the commitment key pk, but also *the trapdoor key sk* (we will explain the rational for this shortly)!

We notice that equivocation easily implies that trapdoor commitments satisfy the usual hiding property of commitments (since all commitments $\mathsf{Com}_{pk}(m)$ are indistinguishable from a single distribution $\mathsf{Fake}_{sk}()$): in fact, this indistinguishability holds even if the distinguisher knows sk! Thus, binding and equivocation are enough to argue the security of trapdoor commitment schemes.

We briefly give the rational of why we need such a strong equivocation property. This is done for the purposes of *composition*. Indeed, we would like to argue that given several "real" pairs (c, d), we can replace all of them by the corresponding "fake" pairs (c', d'), without anybody "noticing". However, the standard left-to-right hybrid argument requires us to be able to generate not only the "real left-pairs" (c, d), which we can do using pk, but also "fake right-pairs" (c', d'), and this we cannot do without the knowledge of sk. Requiring the indistinguishability to hold even with the knowledge of sk resolves this problem, and gives us all the natural composition properties.

CONSTRUCTIONS. There are many constructions of trapdoor commitments (and each of them also gives a regular commitment, of course). For example, efficient trapdoor commitments exist based on a variety of number-theoretic assumptions: factoring [27],[33], discrete log [3],[4]), RSA (combining [16],[24]), Paillier [5],[11]. In fact, some of these schemes (e.g., those based on discrete log and RSA) are special cases of a beautiful general construction by Feige and Shamir [16]. This construction *efficiently* transforms any Σ-protocol corresponding to a "hard" language in NP into a trapdoor commitment scheme. In particular, since we mentioned that all of NP has such Σ-protocols if one-way functions exists (see Theorem 5), and the latter also imply that some languages in NP are "hard" (at least, the the TP model), one can in principle construct a trapdoor commitment scheme *from any one-way function* in the TP model (see sec. 4.9.2.3 of [20]). We note that the message space for the resulting trapdoor commitment will be exactly the challenge space of the corresponding Σ-protocol, which, once again, demonstrates why we want to construct challenge-efficient Σ-protocols.[3] Quite interestingly, this construction *of* trapdoor commitments will be somewhat reminiscent to our main construction *from* trapdoor commitments (possessing a certain Σ-protocols; see Section 5), although this seems to be more of a coincidence.[4]

We also mention another, less general construction [27] of trapdoor commitments from claw-free permutation pairs [22]. This construction is only efficient for bit trapdoor commitments (which, once again, are sufficient for us). Looking at various known claw-free permutation constructions (e.g., see [14] for such a list), we immediately get efficient bit trapdoor commitment constructions from various assumptions, such as the already mentioned constructions from factoring [27], Paillier [11] and the bit-version of the discrete log construction of [3],[4]. In regards to discrete log, we finally mention the following "ad-hoc" construction of trapdoor bit commitments. The public key consists of two random generators g and $h = g^x$ of some prime order q cyclic group $\mathbb{G}$, where the discrete log is hard (here x is a random non-zero element of $\mathbb{Z}_q$), while the trapdoor key is x. To commit to 0, one computes g^{r_0} (for random non-zero $r_0 \in \mathbb{Z}_q$), while to commit

---

[3] Of course, since both of our generic mercurial commitment constructions only use *bit* commitments, even binary Σ-protocols for hard languages suffice for our purpose.

[4] Perhaps partially explained by the fact that mercurial commitment are trapdoor commitments with several very special properties (see Section 2.3). Correspondingly, in our main construction we will need "hard" languages also satisfying some special properties. Somehow remarkably, though, these extra properties have more or less led us to trapdoor commitments themselves! See Section 5.

to 1 one similarly computes h^{r_1}. The openings are r_0 and r_1, respectively. To break binding one needs to satisfy $g^{r_0} = h^{r_1}$, which means that one can compute $x = r_0 r_1^{-1} \bmod q$ (and this contradicts discrete log). On the other hand, if x is known, it is trivial to open a "fake" commitment h^{r_1} both to 1 (by simply presenting r_1) and to 0 (by presenting $r_1 x \bmod q$).

HYBRID TRAPDOOR COMMITMENTS. In [6] Catalano and Visconti presented the notion of *hybrid* trapdoor commitment schemes (in the context of constructing concurrent zero-knowledge proofs). Informally an hybrid trapdoor commitment scheme is a general commitment primitive that allows for two commitment parameters generation algorithms HGen and HTGen. If the commitment parameters are obtained as the output of HGen, then the resulting scheme is an unconditionally binding commitment scheme, while if the parameters are generated by HTGen, the produced scheme is actually a trapdoor commitment scheme. Moreover, no polynomially bounded adversary, taking as input only the (public) commitment parameters, should be able to tell the difference between parameters generated from HGen and parameters produced by HTGen. In [6], the authors show that 1) non-interactive hybrid trapdoor commitments can be constructed from any one-way function in the SRS model; and 2) efficient non-interactive hybrid trapdoor commitments can be constructed under standard number-theoretic assumptions in both the SRS and the TP models.

2.3   Mercurial Commitments

We now define mercurial commitments introduced by Chase et al. [8]. Our definition will be similar, but *stronger* than the definition from [8]. There are two reasons for making the change. First, all the efficient constructions in [8] and here will anyway satisfy the stronger definition. More importantly, by making our definition stronger we will also make it noticeably simpler (and shorter!) than the definition of [8]. More detailed comparison will be presented later.

PLAIN MERCURIAL COMMITMENTS. Such commitment schemes consist of seven efficient algorithms: $\mathcal{C} = (\mathsf{MCom\text{-}Gen}, \mathbb{H}\mathsf{Com}, \mathbb{H}\mathsf{Open}, \mathbb{H}\mathsf{Ver}, \mathbb{S}\mathsf{Com}, \mathbb{S}\mathsf{Open}, \mathbb{S}\mathsf{Ver})$. The first four algorithms $(\mathsf{MCom\text{-}Gen}, \mathbb{H}\mathsf{Com}, \mathbb{H}\mathsf{Open}, \mathbb{H}\mathsf{Ver})$ follow the syntax (and the functionality!) of regular commitment schemes (see Section 2.2). Namely, generation algorithm $\mathsf{MCom\text{-}Gen}(1^k)$, where k is the security parameter, outputs a public mercurial commitment key mpk. Given a message $M \in \mathcal{M}$, the *hard-commit* algorithm $\mathbb{H}\mathsf{Com}_{mpk}(M; R)$ produces a *hard-commitment* string C for M. We will sometimes write $C \leftarrow \mathbb{H}\mathsf{Com}_{mpk}(M)$. Similarly, the *hard-opening* algorithm $\mathbb{H}\mathsf{Open}_{mpk}(M; R)$ (which is supposed to be run using the same value R as the hard-commit algorithm) produces a *hard-decommitment* value π for C. Finally, the *hard-verification* algorithm $\mathbb{H}\mathsf{Ver}_{mpk}(M, C, \pi)$ accepts (i.e., outputs 1) if it thinks π proves that C is indeed a valid hard-commitment to M. We require that for all $M \in \mathcal{M}$, $\mathbb{H}\mathsf{Ver}_{mpk}(m, \mathbb{H}\mathsf{Com}_{mpk}(M; R), \mathbb{H}\mathsf{Open}_{mpk}(M; R)) = 1$ holds with all but negligible probability.

We now turn to the novel "soft algorithms". The *soft-commit* algorithm $\mathbb{S}\mathsf{Com}_{mpk}(; R)$ produces a *soft-commitment* string C (to no message in particular). We will sometimes write $C \leftarrow \mathsf{SCom}_{mpk}()$. The *soft-opening* algorithm $\mathsf{SOpen}_{mpk}(M, \mathsf{flag}; R)$, where $M \in \mathcal{M}$ and $\mathsf{flag} \in \{\mathbb{H}, \mathbb{S}\}$ now produces a *soft-decommitment* τ to M, which should say that "if the commitment produced using R can be hard-opened at all, then it would open to M". A bit more precisely, if $\mathsf{flag} = \mathbb{H}$, then τ is supposed to "correspond" to the hard-commitment $C = \mathbb{H}\mathsf{Com}_{mpk}(M; R)$, and if $\mathsf{flag} = \mathbb{S}$, then τ is a fake soft-decommitment "corresponding" to the soft-commitment $C = \mathsf{SCom}_{mpk}(; R)$. Either one of these cases is verified using the *soft-verification* algorithm $\mathsf{SVer}_{mpk}(M, C, \tau)$, which outputs 1 if it thinks that C could potentially be hard-opened to M in the future (which, intuitively, should be the case only when τ was produced from a hard-commitment). Specifically, we require that for all $M \in \mathcal{M}$, $\mathsf{SVer}_{mpk}(M, \mathbb{H}\mathsf{Com}_{mpk}(M; R), \mathsf{SOpen}_{mpk}(M, \mathbb{H}; R)) = 1$ holds with all but negligible probability, and similarly $\mathsf{SVer}_{mpk}(M, \mathsf{SCom}_{mpk}(; R), \mathsf{SOpen}_{mpk}(M, \mathbb{S}; R)) = 1$ holds with all but negligible probability.

We notice that in many cases (including all our constructions) the soft-decommitment τ to a hard-commitment C will consist of some proper part of the hard-decommitment π, and, correspondingly, the soft-verification algorithm will perform a proper subset of the tests performed by the hard-verification algorithm. For a lack of better name, we call such natural mercurial commitments *proper*.

SECURITY. The binding property of plain mercurial commitments consists of two requirements, stating that a valid hard- or soft-opening of C to some M implies that C can not be then hard-opened to any other message $M' \neq M$:

> **Mercurial Binding:** Having the knowledge of mpk, it is computationally hard for the PPT adversary $\mathcal{A}$ to come up with C, M, π, M', π' (resp. C, M, τ, M', π') such that π (respectively, τ) is a valid hard- (respectively soft-) decommitment of C to M and π' is a valid hard-decommitment of C to M', but $M \neq M'$ (such a tuple is said to cause a *hard* (respectively *soft*) *collision*). That is, $\mathcal{A}$ cannot find a value C which it can hard- or soft-open in one way and then hard-open in a different way.

We remark that for proper mercurial commitments it suffices to prove that no soft collisions can be found.

As for the analog of the hiding property, we require that not only hard-commitments to some M look indistinguishable from soft-commitments (to "nothing"), but this continues to hold even if they are both soft-opened to M (notice that by the mercurial binding property, the hard-commitment to M cannot be soft-opened to anything other than M).

- **Mercurial Hiding.** No PPT adversary (who knows mpk) can find $M \in \mathcal{M}$ for which it can distinguish a random "real" hard-commitment/soft-decommitment tuple $(M, \mathbb{H}\mathsf{Com}_{mpk}(M; R), \mathsf{SOpen}_{mpk}(M, \mathbb{H}; R))$ from a random "fake" soft-commitment/soft-decommitment tuple $(M, \mathsf{SCom}_{mpk}(; R), \mathsf{SOpen}_{pk}(M, \mathbb{S}; R))$.

(Trapdoor) Mercurial Commitments. Such commitment schemes consist of ten efficient algorithms: $\mathcal{C} = (\mathsf{TrMCom\text{-}Gen}, \mathbb{H}\mathsf{Com}, \mathbb{H}\mathsf{Open}, \mathbb{H}\mathsf{Ver}, \mathbb{S}\mathsf{Com}, \mathbb{S}\mathsf{Open},$ $\mathbb{S}\mathsf{Ver}, \mathsf{MFake}, \mathbb{H}\mathsf{Equiv}, \mathbb{S}\mathsf{Equiv})$. The generation algorithm $\mathsf{TrMCom\text{-}Gen}(1^k)$, where k is the security parameter, outputs a public mercurial commitment key mpk and and a secret mercurial *trapdoor key msk*. Once mpk is fixed, the meaning of $\mathbb{H}\mathsf{Com}$, $\mathbb{H}\mathsf{Open}$, $\mathbb{H}\mathsf{Ver}$, $\mathbb{S}\mathsf{Com}$, $\mathbb{S}\mathsf{Open}$ and $\mathbb{S}\mathsf{Ver}$ is exactly the same as for plain mercurial commitments. In particular, we will require that these algorithms satisfy the usual mercurial hiding and binding properties of the plain mercurial commitment schemes.

The trapdoor key msk is used in the algorithms MFake, $\mathbb{H}\mathsf{Equiv}$ and $\mathbb{S}\mathsf{Equiv}$ to break the binding property of commitments. The algorithm $\mathsf{MFake}_{msk}(; R)$ is somewhat similar in spirit to the soft-commitment algorithm $\mathbb{S}\mathsf{Com}_{mpk}$ and produces "fake" commitment C, initially not associated to any message M. The meaning of the other two algorithms $\mathbb{H}\mathsf{Equiv}_{msk}(M; R)$ and $\mathbb{S}\mathsf{Equiv}_{msk}(m; R)$ is also similar to that of the corresponding algorithms $\mathbb{H}\mathsf{Open}_{mpk}$, $\mathbb{S}\mathsf{Open}_{mpk}$, except they *always* operate on the fake commitments C not really associated to any message. Specifically, $\mathbb{H}\mathsf{Equiv}(M; R)$ produces a supposedly valid hard-opening π (called *hard-fake*) of the fake commitment $C = \mathsf{MFake}(; R)$ to M, while $\mathbb{S}\mathsf{Equiv}(M; R)$ produces a supposedly valid soft-opening τ (called *soft-fake*) of the fake commitment $C = \mathsf{MFake}(; R)$ to M. In particular, we require that for all $M \in \mathcal{M}$, $\mathbb{H}\mathsf{Ver}_{mpk}(M, \mathsf{MFake}_{mpk}(; R), \mathbb{H}\mathsf{Equiv}_{mpk}(M; R)) = 1$ holds with all but negligible probability, and similarly $\mathbb{S}\mathsf{Ver}_{mpk}(M, \mathsf{MFake}_{mpk}(; R),$ $\mathbb{S}\mathsf{Equiv}_{mpk}(M; R)) = 1$ holds with all but negligible probability. While the ability to soft-fake such bogus commitments is consistent with the previous ability of soft-opening, the ability to hard-fake them certainly contradicts the binding property that we had, and this is exactly the function of the trapdoor key msk!

Somewhat similar to the equivocation property of trapdoor commitments, we require that trapdoor mercurial commitments satisfy three related equivocation conditions. In each of them we say that no efficient distinguisher $\mathcal{A}$ can non-negligibly tell apart the corresponding "real" from the corresponding "ideal" game, *even if it is given the* <u>trapdoor key msk</u> *at the beginning of each real or ideal game*. In the following, the value R is always random.

- **HH Equivocation:** The real game consists of $\mathcal{A}$ choosing $M \in \mathcal{M}$ and getting back $(M, \mathbb{H}\mathsf{Com}_{mpk}(M; R), \mathbb{H}\mathsf{Open}_{mpk}(M; R))$; while the ideal game consists of $\mathcal{A}$ choosing $M \in \mathcal{M}$ and getting back $(M, \mathsf{MFake}_{msk}(; R),$ $\mathbb{H}\mathsf{Equiv}_{msk}(M; R))$.
- **HS Equivocation:** The real game consists of $\mathcal{A}$ choosing $M \in \mathcal{M}$ and getting back $(M, \mathbb{H}\mathsf{Com}_{mpk}(M; R), \mathbb{S}\mathsf{Open}_{mpk}(M, \mathbb{S}; R))$; while the ideal game consists of $\mathcal{A}$ choosing $M \in \mathcal{M}$ and getting back $(M,$ $\mathsf{MFake}_{msk}(; R), \mathbb{S}\mathsf{Equiv}_{msk}(M; R))$.
- **SS Equivocation:** The real game consists of $\mathcal{A}$ getting the value $C = \mathbb{S}\mathsf{Com}_{mpk}(; R)$, then choosing $M \in \mathcal{M}$, and finally getting $\mathbb{S}\mathsf{Open}_{mpk}(M,$ $\mathbb{S}; R)$; while ideal game consists of $\mathcal{A}$ getting the value $C = \mathsf{MFake}_{mpk}(; R)$, then choosing $M \in \mathcal{M}$, and finally getting $\mathbb{S}\mathsf{Equiv}_{mpk}(M; R)$.

Notice that similar-looking SH condition does not make sense in the real life (due to mercurial binding). Next, HS and SS Equivocations easily imply the Mercurial Hiding property, so it does not need to be checked. Also, for proper mercurial commitments it is easy to see that HH Equivocation implies HS Equivocation, so it is enough to check only HH and SS Equivocations.

RELATION TO THE ORIGINAL DEFINITION IN [8]. The main difference from [8] is in the equivocation property, which is considerably simpler to state and verify in our case. Moreover, it is also *stronger* than the definition of [8]. Essentially, the latter definition consists of playing an arbitrary composition of HH, HS and SS Equivocation games either in the real, or in the ideal world,[5] but where the distinguisher $\mathcal{A}$ is *not given the trapdoor key msk*. In this scenario the usual hybrid argument does not work (since $\mathcal{A}$ cannot simulate stuff in the ideal world by himself), so one cannot reduce the composed game to the one of the three atomic HH, SE or SS games. As a result, one has to build a full-fledged simulator, and formally argue that it fools the distinguisher. In contrast, in our scenario the hybrid argument easily works, so the security of our 3 atomic games easily implies the security of the composed game *even if the distinguisher knows msk*.

KNOWN CONSTRUCTIONS. Chase et al. [8] gave several elegant constructions of (trapdoor) mercurial commitments from the following assumptions:

- *One-way functions.* This construction works in the plain model but unfortunately is interactive. All next constructions are non-interactive.
- *Non-interactive zero-knowledge (NIZK) proofs for all of NP* [2], [15] and *unconditionally-binding commitment schemes.* However, this construction is mainly of theoretical interest, since all known NIZK constructions (especially for all of NP) are extremely inefficient. Interestingly, it also does *not* satisfy our stronger definition. However, in the sequel we will provide more general constructions (from one-way functions) which satisfy our stronger definition in the trusted parameters model and are still more efficient than this construction.[6]

---

[5] There is one other, more syntactic strengthening that we had to make in order to simplify the definition. Namely, in the more general definition of [8] one could have syntactically unrelated real and ideal experiments for generating mpk, so it did not make sense to give msk to $\mathcal{A}$ in the real game. In contrast, we insist that the public key generation even in the real world can be carried by generating both the public and the trapdoor key. While slightly more restrictive, since (1) all our efficient constructions in the trusted parameters model satisfy this restricted notion of key generation and (2) it considerably simplifies (and also *strengthens*) the definition, we feel it is very justified.

[6] We remark, however, that the NIZK is in the SRS model, while our OWF-based construction satisfying the stronger definition will be in the TP model. We can make SRS-based constructions by either building trapdoor commitments in the SRS model (which is known how to do from one-way *permutations* or specific number-theoretic assumptions), or by using our technique from Section 4 (while reverting to the weaker definition of [8]).

- *Claw-free permutations [22].* This construction give only bit mercurial commitment, and will be a special case of our first general construction from bit trapdoor commitments.
- *Discrete log.* This is a "distillation" of the original construction implicitly used in [28]; it supports long messages and is pretty efficient. It will be a special case of our second construction when used with the corresponding discrete-log based bit trapdoor commitment.
- *Factoring.* This is a new construction which supports long messages and is relatively efficient. It will be a special case of our second construction when used with the corresponding factoring-based bit trapdoor commitment.

IMPLICATIONS TO (TRAPDOOR) COMMITMENTS. It is simple to see that by "ignoring" all the "soft" algorithms of a secure plain/trapdoor commitment scheme, we immediately get a plain/trapdoor regular commitment scheme. (Concentrating, for example, on a slightly more complicated "trapdoor case", $\mathbb{H}$Com plays the role of Com, $\mathbb{H}$Open — of Open, $\mathbb{H}$Ver — of Ver, MFake — of Fake, and $\mathbb{H}$Equiv — of Equiv.) In the following, we show two simple constructions proving that the converse of this statement is true as well.

3   General Construction from (Trapdoor) Bit Commitments

As advocated in the introduction, we will first consider the construction of plain mercurial bit commitments from regular bit commitments, and then argue that the same construction extends to the trapdoor case as well.

BUILDING PLAIN MERCURIAL COMMITMENTS. Assume $\mathcal{C} = (\text{Com-Gen}, \text{Com}, \text{Open}, \text{Ver})$ is a regular bit commitment scheme. Define plain mercurial commitment $\mathcal{C}' = (\text{MCom-Gen}, \mathbb{H}\text{Com}, \mathbb{H}\text{Open}, \mathbb{H}\text{Ver}, \mathbb{S}\text{Com}, \mathbb{S}\text{Open}, \mathbb{S}\text{Ver})$ for a bit $b \in \{0, 1\}$ as follows (we set MCom-Gen = Com-Gen and let pk be the corresponding public key):

- $\mathbb{H}\text{Com}_{pk}(b; (r_0, r_1))$: output $(c_0, c_1) = (\text{Com}_{pk}(b; r_0), \text{Com}_{pk}(1 - b; r_1))$. Notice, commitment to 0 changes its place from left to right depending on b.
- $\mathbb{H}\text{Open}_{pk}(b; (r_0, r_1))$: output $(d_0, d_1) = (\text{Open}(b; r_0), \text{Open}(1 - b; r_1))$.
- $\mathbb{H}\text{Ver}_{pk}(b, (c_0, c_1), (d_0, d_1))$: accept if and only if $\text{Ver}_{pk}(b, c_0, d_0) = \text{Ver}_{pk}(1 - b, c_1, d_1) = 1$.
- $\mathbb{S}\text{Com}_{pk}(; (r_0, r_1))$: output $(c_0, c_1) = (\text{Com}_{pk}(0; r_0), \text{Com}_{pk}(0; r_1))$.
- $\mathbb{S}\text{Open}_{pk}(b, \text{flag}; (r_0, r_1))$: irrespective of flag $\in \{\mathbb{H}, \mathbb{S}\}$, output $d = \text{Open}(0; r_b)$.
- $\mathbb{S}\text{Ver}_{pk}(b, (c_0, c_1), d)$: accept if and only if $\text{Ver}_{pk}(0, c_b, d) = 1$.

The correctness of the scheme is obvious. Intuitively, mercurial commitment to $b = 0$ looks $(0, 1)$, to 1 — $(1, 0)$, and the fake — $(0, 0)$. Since the soft-opening of the hard commitment only opens the corresponding left or right 0, the fake commitment can indeed be soft-opened in both way, by honestly opening the appropriate left of right 0. On the other hand, seeing a hard-opening of some commitment $C = (c_0, c_1)$ (to some bit b) opens to 1 one of the two regular commitments, while the subsequent soft-opening of C to $(1-b)$ would then open this regular commitment to 0, which contradicts binding. Below, we formalize this is a straightforward manner.

Mercurial Binding. Since the mercurial commitment is proper, we only need to rule out soft collisions. For that, assume the attacker can find a soft collision. By symmetry, let us assume that 1 is the softly-opened message, and 0 is the hardly-opened one). So we denote this collision by $C = ((c_0, c_1), d_0, d_1, d_1')$ where $\mathsf{Ver}(0, c_0, d_0) = \mathsf{Ver}(1, c_1, d_1) = \mathsf{Ver}(0, c_1, d_1') = 1$. But then c_1 can be opened to both 0 and 1, a contradiction to the binding property of C.

Mercurial Hiding. Assume first $b = 0$. Then, the "real" hard-commitment/ soft-decommitment tuple $(\mathbb{H}\mathsf{Com}(0; (r_0, r_1)), \mathbb{S}\mathsf{Open}(0, \mathbb{H}; (r_0, r_1))$ looks like $(\mathsf{Com}(0; r_0), \mathsf{Com}(1; r_1), \mathsf{Open}(0; r_0))$, while the corresponding "fake" tuple $(\mathsf{Fake}(; (r_0, r_1)), \mathbb{S}\mathsf{Open}(0, \mathbb{S}; (r_0, r_1))$ looks like $(\mathsf{Com}(0; r_0), \mathsf{Com}(0; r_1), \mathsf{Open}(0; r_0))$. Clearly, such distribution are indistinguishable if $\mathsf{Com}(0)$ cannot be distinguished from $\mathsf{Com}(1)$, which follows from the hiding property of C. A similar argument holds for $b = 1$ as well.

TRAPDOOR CASE. The extension to the trapdoor case is simple as well. We now have additional algorithms Fake and Equiv for trapdoor commitments, and need to build the corresponding algorithms MFake, $\mathbb{H}$Equiv and $\mathbb{S}$Equiv for mercurial commitments.

- $\mathsf{MFake}_{sk}(; (r_0, r_1))$: output $(\mathsf{Fake}_{sk}(; r_0), \mathsf{Fake}_{sk}(; r_1))$.
- $\mathbb{H}\mathsf{Equiv}_{sk}(b; (r_0, r_1))$: output $(\mathsf{Equiv}_{sk}(b; r_0), \mathsf{Equiv}_{sk}(1 - b; r_1))$.
- $\mathbb{S}\mathsf{Equiv}_{sk}(b; (r_0, r_1))$: output $\mathsf{Equiv}_{sk}(0; r_b)$.

Correctness is obvious from definition. As for hiding, we only need to argue HH and SS Equivocations (since this is a proper mercurial commitment). Both are simple corollaries of the regular Equivocation properties of trapdoor commitments.

HH Equivocation. Let us assume $b = 0$, since $b = 1$ is symmetric. Then $(\mathbb{H}\mathsf{Com}(0; (r_0, r_1)), \mathbb{H}\mathsf{Open}(0; (r_0, r_1)))$ is equal to $D_{real} = (\mathsf{Com}(0; r_0), \mathsf{Com}(1; r_1), \mathsf{Open}(0; r_0), \mathsf{Open}(1; r_1))$, while $(\mathsf{MFake}(; (r_0, r_1)), \mathbb{H}\mathsf{Equiv}(0; (r_0, r_1)))$ is equal to $D_{ideal} = (\mathsf{Fake}(; r_0), \mathsf{Fake}(; r_1), \mathsf{Equiv}(0; r_0), \mathsf{Equiv}(1; r_1))$. Since r_0 and r_1 are independent, this amount to two independent applications of the regular Equivocation property to bits 0 and 1, respectively. Notice, though, already for this simple hybrid argument *we are using the fact that the attacker knows the trapdoor key* sk! To be precise, we must first consider a hybrid distribution $D_{hyb} = (\mathsf{Fake}(; r_0), \mathsf{Com}(1; r_1), \mathsf{Equiv}(0; r_0), \mathsf{Open}(1; r_1))$, and then show $D_{real} \approx D_{hyb}$ (here we only need *pk* to sample $(\mathsf{Com}(1; r_1), \mathsf{Open}(1; r_1))$) and $D_{hyb} \approx D_{ideal}$ (here we *need* sk to sample $(\mathsf{Fake}(; r_0), \mathsf{Equiv}(0; r_0))$).

SS Equivocation. In the real experiment, the attacker is first getting $(\mathsf{Com}(0; r_0), \mathsf{Com}(0; r_1))$, then he has to choose a bit b, after which he gets $\mathsf{Open}(0; r_b)$. In the ideal game, the attacker is getting $(\mathsf{Fake}(; r_0), \mathsf{Fake}(; r_1))$, then he has to choose a bit b, after which he gets $\mathsf{Equiv}(0; r_b)$. By symmetry, the choice of b does not matter here, so we can assume $b = 0$, so it suffices to argue $(\mathsf{Com}(0; r_0), \mathsf{Com}(0; r_1))$,

$\mathsf{Open}(0; r_0)) \approx (\mathsf{Fake}(; r_0), \mathsf{Fake}(; r_1), \mathsf{Equiv}(0; r_0))$. Once again, this follow by the hybrid argument, by considering an intermediate distribution $(\mathsf{Fake}(; r_0), \mathsf{Com}(; r_1), \mathsf{Equiv}(0; r_0))$ and using the fact that in the second hybrid the attacker can compute $(\mathsf{Fake}(; r_0), \mathsf{Open}(0; r_0))$.

COMPARISON TO [8]. The above construction is a very simple generalization of the one in [8], who used the following family of trapdoor bit commitments [27] obtained from any family of claw-free permutations [22] (f_0, f_1). Informally, recall that these are pairs of permutations where one cannot find a "claw" (r_0, r_1) satisfying $f_0(r_0) = f_1(r_1)$; also it is assumed that there exists a trapdoor f_0^{-1} allowing one to invert f_0 (in our application, we will not need a similar trapdoor for f_1). Now, to trapdoor commit to a bit b we can sample $f_b(r_b)$ (decommitment is r_b), while the knowledge of the trapdoor f_0^{-1} provides easy fake pairs: the fake commitment $c = f_1(r_1)$ (for random r_1) can be opened to 0 by giving $r_0 = f_0^{-1}(c)$), and to 1 — by giving r_1.

We remark, though, that the equivocality proof of our extension is indeed considerably shorter, — which is what it should be for such a simple construction! — than the corresponding proof [8]. Also, our construction implies mercurial commitments from other bit commitments which are not necessarily induced by claw-free permutations, such as the general construction of [16] from any Σ-protocol for a hard language, the factoring construction of [33], the Paillier construction of [5] or the ad hoc (g^{r_0}, h^{r_1})-construction mentioned in Section 2.2 (and, of course, the one-way function construction from Theorem 5).

4   Mercurial Commitments from One-Way Functions in the SRS Model

We now discuss a construction of a non-interactive bit trapdoor mercurial commitment scheme in the SRS model which requires only a non-interactive bit hybrid trapdoor commitment scheme as underlying building block. Since the latter can be constructed from one-way functions [6], we have that the same holds for the former. We stress that for this construction we use the original definition of mercurial commitments given in [8] where the shared random string used in the simulated game is computationally indistinguishable from the real random string. For lack of space we omit the original definitions of mercurial commitments given in [8] and of hybrid trapdoor commitments given in [6] (in particular the reader is referred to [6] for details about the constructions of hybrid trapdoor commitments).

OVERVIEW OF THE TECHNIQUE. In Section 3 we have shown a construction of mercurial commitments that can be based on the trapdoor commitment scheme proposed by Feige and Shamir [16]. This scheme needs, as common parameter, an Hamiltonian graph for which it is hard to compute an Hamiltonian cycle, but such that knowledge of a cycle allows one to equivocate. This can be realized in the trusted parameters model by generating the required instance of the Hamiltonian cycle language from the hardness of inverting a one-way function. In the

SRS model, this construction is known to work assuming that one-way *permutations* exist. In a nutshell, this is because a piece of the random string infers the computationally infeasible problem of inverting a one-way permutation (that, in turn, can be reduced to an instance of finding a cycle in an Hamiltonian graph), and in order for this to work one needs to make sure that a corresponding inverse actually exists.

In order to build a solution based on any one-way function in the SRS model, we start from the following observation. Since the possibility of computing commitments that can be equivocated is required by the simulator only (we stress that we are now using the original definition of mercurial commitments given in [8]), we could construct mercurial commitments by using regular (i.e., we do not require the equivocal property) commitments in the SRS model for the real game and trapdoor commitments in the trusted parameters model for the simulated game. In order for this idea to work, however, we need a scheme that can be used either as a trapdoor commitment in the trusted parameters model or as a regular one in the SRS model, but it is infeasible for the adversarial receiver to distinguish the two cases. In particular the trusted parameters of the simulated game must be computationally indistinguishable from a random string.

Commitment schemes realizing this requirement have been recently studied, defined and constructed by Catalano and Visconti [6] under the sole assumption that one-way functions exist.

A construction of non-interactive bit trapdoor mercurial commitments on top of a non-interactive bit hybrid trapdoor commitment scheme can be very easily obtained from the construction shown in Section 3. Indeed, it suffices to replace the algorithms of the regular non-interactive bit trapdoor commitment scheme with the corresponding ones of the hybrid one. The only additional step is that since the hybrid trapdoor commitment scheme has two algorithms for the generation of the commitment parameters, one is used in the real game and the other is used in the simulated game. In the SRS model, the former simply outputs a random string (i.e., the parties can deterministically extract the commitment parameters from a random string) while the latter outputs a random/pseudorandom string along with a trapdoor.

5   Efficient Construction from (Trapdoor) Bit Commitments with Σ-Protocols

The problem with the previous generic constructions is the fact that they only allows one to commit to one bit. Of course, we can always commit to many bits by following the "bit-by-bit" approach, but this is inefficient. Alternatively, we can try to utilize a multi-bit plain/trapdoor commitment scheme in the previous construction, but it is easy to see that the resulting length of the commitment will be linearly proportional to the number of *messages* that we want to commit to. This essentially means that setting this number to 2 — as we did in Section 3 — and doing the bit-by-bit composition is the best we can do if we try to extend the previous approach.

Instead, in this section we present our main construction which will directly transforms a plain/trapdoor bit commitment $\mathcal{C}$ into an efficient and (potentially) multi-bit plain/mercurial commitment $\mathcal{C}'$. However, we will need to assume an extra property from $\mathcal{C}$: there exists an efficient Σ-protocol Π proving that one knows a witness d that a given commitment c can be opened to 0. In this case, the message space of $\mathcal{C}'$ will be the challenge space of the corresponding Σ-protocol. Thus, if Π will be challenge-efficient, we would get a direct, large-message mercurial commitment $\mathcal{C}'$.

CONSTRUCTION. Let $\mathcal{C} = (\mathsf{Com\text{-}Gen}, \mathsf{Com}, \mathsf{Open}, \mathsf{Ver})$ be a regular bit commitment scheme which has a Σ-protocol $\Pi = (\mathsf{Start}, \mathsf{Finish}, \mathsf{Extract}, \mathsf{Simul})$ for the relation (family) $\mathcal{R}_{pk} = \{(c, d) \mid \mathsf{Ver}_{pk}(0, c, d) = 1\}$. Recall, this means that the verifier only gets a commitment c, and the prover also gets, as a witness, a valid opening d of c to 0. Also, assume $\mathcal{M}$ is the challenge space for Π.

We then define plain mercurial commitment $\mathcal{C}' = (\mathsf{MCom\text{-}Gen}, \mathbb{H}\mathsf{Com}, \mathbb{H}\mathsf{Open}, \mathbb{H}\mathsf{Ver}, \mathbb{S}\mathsf{Com}, \mathbb{S}\mathsf{Open}, \mathbb{S}\mathsf{Ver})$ for message space $\mathcal{M}$ as follows (we set $\mathsf{MCom\text{-}Gen} = \mathsf{Com\text{-}Gen}$ and let pk be the corresponding public key):

- $\mathbb{H}\mathsf{Com}_{pk}(m; (r_s, r_1))$: let $c_1 = \mathsf{Com}_{pk}(1; r_1)$ be a commitment to 1, and $(a_1, z_1) = \mathsf{Simul}_{pk}(c_1, m; r_s)$ be a fake first and last messages of Π which (here incorrectly) claim that c_1 is a commitment to 0 on challenge m. Output (c_1, a_1).
- $\mathbb{H}\mathsf{Open}_{pk}(m; (r_s, r_1))$: let $c_1 = \mathsf{Com}_{pk}(1; r_1)$ and $(a_1, z_1) = \mathsf{Simul}_{pk}(c_1, m; r_s)$ be as before. Set $d_1 = \mathsf{Open}_{pk}(1; r_1)$ and output (d_1, z_1).
- $\mathbb{H}\mathsf{Ver}_{pk}(m, (c_1, a_1), (d_1, z_1))$: accept if and only if $\mathsf{Ver}_{pk}(1, c_1, d_1) = 1$ (d_1 is correct decommitment to 1) and $\mathsf{Check}_{pk}(c_1, a_1, m, z_1) = 1$ (the fake transcript on challenge m that c_1 is a commitment to 0 looks good).
- $\mathbb{S}\mathsf{Com}_{pk}(; (r_p, r_0))$: let $c_0 = \mathsf{Com}_{pk}(0; r_0)$ be a commitment to 0, and $d_0 = \mathsf{Open}_{pk}(0; r_0)$ be the corresponding opening, and $a_0 = \mathsf{Start}_{pk}(c_0, d_0; r_p)$ be a real first messages of Π which (correctly!) claims that c_0 is a commitment to 0.[7] Output (c_0, a_0).
- $\mathbb{S}\mathsf{Open}_{pk}(m, \mathbb{H}; (r_s, r_1))$: let $c_1 = \mathsf{Com}_{pk}(1; r_1)$ and $(a_1, z_1) = \mathsf{Simul}_{pk}(c_1, m; r_s)$ be the fake transcript on challenge m that c_1 is a commitment to 0. Output z_1.
- $\mathbb{S}\mathsf{Open}_{pk}(m, \mathbb{S}; (r_p, r_0))$: let $c_0 = \mathsf{Com}_{pk}(0; r_0)$, $d_0 = \mathsf{Open}_{pk}(0; r_0)$, $a_0 = \mathsf{Start}_{pk}(c_0, d_0; r_p)$, and $z_0 = \mathsf{Finish}_{pk}(c_0, d_0, m; r_p)$ be the correct last flow to challenge m. Output z_0.
- $\mathbb{S}\mathsf{Ver}_{pk}(m, (c, a), z)$: accept if and only $\mathsf{Check}_{pk}(c_b, a, m, z) = 1$ (the transcript (a, m, z) stating that c is a commitment to 0 is correct).

Intuitively, the honest hard-committer is supposed to send a commitment c to 1, but *fake* the transcript that he in fact committed to 0. On the other hand, a lying soft-committer can simply send a commitment c to 0, and now

---

[7] Notice, here the prover actually knows the value r_0, and not just d_0. So for efficiency reasons we might consider auxiliary-input Σ-protocols where P's witness is actually r_0 itself. We will return to this point later.

can (*honestly!*) respond to any challenge/message m that he gets subsequently, which allows him to soft-open the first flow to any message m.[8] The binding security of this scheme comes from the fact that a hard-opening of c to 1, coupled with two soft-opening of the first flow a, must enable one to extract a legal witness, which is the hard-opening of c to 0, contradicting the binding of C. Similarly, the hiding property of C coupled with the zero-knowledge property of Σ-protocols imply that, without the hard-opening of c (which will tell if c is a commitment to 0 or 1), the real and fake behavior cannot be told apart. More formally,

Mercurial Binding. Since our commitment is proper, we only need to rule our soft collisions. This means that the attacker cannot find a commitment value (c, a), a decommitment d_1 proving that c is a commitment to 1, two messages $m \neq m'$, and two valid responses z and z' claiming that c is a commitment to 0. By the special soundness of the Σ-protocol, $\mathsf{Extract}(c, a, m, z, m', z')$ must be equal to a valid decommitment d_0 of c to 0. But then we found a way to open c to both 0 and 1 (via d_0 and d_1), contradicting the binding property of C.

Mercurial Hiding. Take any message/challenge m. Then, the "real" hard-commitment/soft-decommitment tuple for m looks like is given by three values $(c = \mathsf{Com}(1; r_1), (a, z) = \mathsf{Simul}(c, m; r_s))$. Since our commitment is hiding, and $\mathsf{Simul}(c, m)$ is publicly computable, we get that the above distribution is indistinguishable from $(c = \mathsf{Com}(0; r_0), (a, z) = \mathsf{Simul}(c, m; r_s))$. Now, since c has a proper witness $d_0 = \mathsf{Open}(0; r_0)$, the special HVZK property of Π states that the distribution on (a, z) looks indistinguishable than the one obtained by a running a real protocol on input c, witness d_0 and challenge m. But this means that the above distribution is indistinguishable from $(c = \mathsf{Com}(0; r_0), a = \mathsf{Start}(c, d_0; r_p), z = \mathsf{Finish}(c, d_0, m; r_p))$, which is exactly the triple corresponding to the "fake" soft-commitment/soft-decommitment procedures.

TRAPDOOR CASE. Recall, we now have additional algorithms Fake and Equiv for trapdoor commitments, and need to build the corresponding algorithms MFake, HEquiv and SEquiv for mercurial commitments. As a new technical property about the Σ-protocol, however, we will have to assume that $\Pi = \Pi_{pk}$ is strongly hiding w.r.t. a particular parameter generation procedure $\mathcal{P}$ (see Section 2.1). The parameter generation procedure we will need generates random keys $(pk, sk) \leftarrow \mathsf{Com\text{-}Gen}(1^k)$, picks a random r, computes $c = \mathsf{Fake}_{sk}(; r)$, $d_0 = \mathsf{Equiv}_{sk}(0; r)$, $d_1 = \mathsf{Equiv}_{sk}(1; r)$, and sets the side information to (sk, d_1), the input to be c, and the witness to be d_0. As explained in Section 2.1, this is more of a technicality which seems to be always satisfied in any non-pathological scenario arising in practice. We call this property (*), and can now describe the claimed extension.

---

[8] Is might appear peculiar that we require an honest party to cook-up a fake proof in order to succeed, while having a dishonest party perform such a proof correctly! Here, however, the primitive we build legally allows a dishonest party to look "slightly like an honest party". So the we force the honest party to do something slightly bad which might be "matched" by a good action of a dishonest party.

- $\mathsf{MFake}_{sk}(;(r_p,r))$: let $c{=}\mathsf{Fake}_{sk}(;r)$ be a fake commitment, $d_0 = \mathsf{Equiv}_{sk}(0;r)$ be its fake opening to 0, and $a_0 = \mathsf{Start}_{pk}(c,d_0;r_p)$ be a correct first flow of the Σ-protocol. Output (c,a_0).
- $\mathbb{H}\mathsf{Equiv}_{sk}(m;(r_p,r))$: let $c = \mathsf{Fake}_{sk}(;r)$, $d_0 = \mathsf{Equiv}_{sk}(0;r)$, and $a_0 = \mathsf{Start}_{pk}(c,d_0;r_p)$ be as before. Compute the fake opening $d_1 = \mathsf{Equiv}_{sk}(1;r)$ of c to 1, and the correct last message $z_0 = \mathsf{Finish}_{pk}(c,d_0,m;r_p)$. Output (d_1,z_0).
- $\mathbb{S}\mathsf{Equiv}_{sk}(m;(r_p,r))$: let $c = \mathsf{Fake}_{sk}(;r)$, $d_0 = \mathsf{Equiv}_{sk}(0;r)$, and $a_0 = \mathsf{Start}_{pk}(c,d_0;r_p)$ be as before. Compute the correct last message $z_0 = \mathsf{Finish}_{pk}(c,d_0,m;r_p)$ and output z_0.

Correctness is obvious from definition. As for hiding, we only need to argue HH and SS Equivocations (since this is a proper mercurial commitment).

HH Equivocation. Take any message m. Then $(\mathbb{H}\mathsf{Com}(m;(r_s,r_1)),\mathbb{H}\mathsf{Open}(m;(r_s,r_1)))$ is equal to $D_{real} = (c_1,d_1,a_1,z_1)$, where $c_1 = \mathsf{Com}(1;r_1)$, $d_1 = \mathsf{Open}(1;r_1)$, and $(a_1,z_1) = \mathsf{Simul}(c_1,m;r_s)$. Since $\mathsf{Simul}(c_1,m)$ is a public transformation, the Equivocality of $\mathcal{C}$ implies that the above distribution is indistinguishable from $(c = \mathsf{Fake}(;r), d_1 = \mathsf{Equiv}(1;r), a_1, z_1)$, where $(a_1,z_1) = \mathsf{Simul}(c,m;r_s)$. We are almost done, except we need to replace the above (a_1,z_1) by (a_0,z_0) obtained by running an honest execution of Π with witness $d_0 = \mathsf{Equiv}(0;r)$. This is almost exactly the HVZK property, except we formally need to use the strong hiding property (*) described above. Indeed, in addition to the input c and the public parameter pk, which are allowed in the usual HVZK property, here the distinguisher also knows two extra pieces of information: the trapdoor key sk (given to him at the beginning of the game) and the fake decommitment $d_1 = \mathsf{Equiv}(1;r)$. This is why we needed to to assume that this extra information does not violate the HVZK property.

SS Equivocation. In the real soft-commit/soft-open experiment, the distinguisher (who knows sk) is first getting $c_0 = \mathsf{Com}(0;r_0)$ and the correct first flow of the Σ-protocol showing that c_0 is indeed a commitment to 0 (using witness $d_0 = \mathsf{Open}(0;r_0)$). He then chooses a message m, and gets a correct third flow to message m. To put differently, he simply plays the role of (malicious) verifier in the honest run of the Σ-protocol on pair (c_0,d_0). Notice that the distinguisher's view can be perfectly simulated using some public probabilistic procedure $A_{sk}(c_0,d_0)$. Using the equivocation property of $\mathcal{C}$, the resulting distribution should be indistinguishable from $A_{sk}(c,d_0)$, where $c = \mathsf{Fake}(;r)$ and $d_0 = \mathsf{Equiv}(0;r)$. But, once again, it is easy to see that this view is exactly what the attacker gets in the ideal soft-commit/soft-open experiment.

GENERALIZATION. We already noticed in Footnote 7 that in the above definition of soft-commitment, the Prover actually knows the entire randomness r_0 and not just a witness $d_0 = \mathsf{Open}(0;r_0)$. This, of course, is of any value only in a very few schemes where $r_0 \neq d_0$. However, it will come up in one of our examples (see Section 5.1). To accommodate this extension, we can consider Σ-protocol's where the prover needs all of r_0 for the completeness of the protocol (special soundness is still only for d_0). For plain mercurial commitments, this is all we need to change. For the trapdoor variant, however, we will need an extra property from our trapdoor commitment scheme in regards to equivocation. Namely, in the fake commitment algorithm we need to be able to equivocate $c = \mathsf{Fake}(;r)$ to 0 by obtaining not only a good looking value d_0, but the entire randomness r_0. Once this is ensured, we can easily support auxiliary input Σ-protocols.

5.1   Examples

Below we briefly give several efficient instantiations of our construction, by applying it to several efficient trapdoor commitment schemes with challenge-efficient Σ-protocols. Our examples will cover *all* the previous efficient schemes, and several more, all as part of one general framework. For each scheme we will just briefly mention which trapdoor commitment and Σ-protocol to use, since the remaining details are obvious and not very illuminating.

DISCRETE LOG CONSTRUCTION FROM [8],[28]. We will consider the ad-hoc scheme from Section 2.2, where $\mathsf{Com}(0;r_0) = g^{r_0}$, $\mathsf{Com}(1;r_1) = h^{r_1}$, and the trapdoor $sk = \log_g h$ (here $r_0, r_1 \neq 0$). We need a Σ-protocol to prove the knowledge of $r_0 = \log_g(c)$, where c is the claimed commitment to 0. Of course, a natural thing to do is to take Schnorr protocol, but this will result in a slightly different (but equally efficient) scheme than what we are after. Instead, we will use a bit less esthetic but equally effective Σ-protocol. In the first flow the prover sends a value $T = g^t$ (for random t), he gets challenge m, and responds with $z = (t - m)/r_0 \bmod q$ (which is defined since $r_0 \neq 0$). The verifier checks if $g^m c^z = T$ (indeed, $m + r_0 z = t$, as needed). It is simple to see that this is indeed a Σ-protocol for the knowledge of the discrete log, and that by plugging it into our construction we get exactly the discrete log construction from [8,28].

We also remark what we could use a better known discrete-log commitment $\mathsf{Com}(0) = g^{r_0}$, $\mathsf{Com}(1) = hg^{r_1}$, coupled with either Schnorr Σ-protocol, or the one presented above. We will get yet another (equally efficient) solution.

FACTORING CONSTRUCTION FROM [8]. This scheme will use the generalization of our constriction to use auxiliary inputs, as explained earlier. Let us start with a well-known factoring-based trapdoor bit commitment from a corresponding claw-free permutation pair: the public parameter is a random square U, and $\mathsf{Com}(0;r_0) = r_0^2 \bmod n$, $\mathsf{Com}(1;r_1) = Ur_1^2 \bmod n$ (the trapdoor is the square root of U). Here we need a Σ-protocol for the knowledge of the square root. As we mentioned in Section 2.1, using Fiat-Shamir protocol [18] is not communication- or challenge-efficient. Instead, we use the auxiliary input Ong-Schnorr protocol [30]. For that one need to know 2^ℓ-th square root of $\mathsf{Com}(0)$, so we modify $\mathsf{Com}(0;r_0) = r_0^{2^\ell} \bmod n$ (but leave $\mathsf{Com}(1;r_1) = Ur_1^2 \bmod n$). We notice, that although the decommitment to 0 is "only" the square root $d_0 = r_0^{2^{\ell-1}}$, and not r_0 itself, the fake commitment should enable us to extract (using sk) the 2^ℓ-th root from c_0, and not just a mere square root. Of course, this is easy to achieve by defining $\mathsf{Fake}(;r) = r^{2^\ell}$, and "fully opening" it to 0 by giving r, and to 1 — by giving $r^{2^{\ell-1}}/\sqrt{U}$. With these changes we get *precisely* the factoring construction from [8]. We also notice that by using a different claw-free permutation $(r_0^2, 4r_1^2)$ [22] defined over the so called Williams integers, we can slightly simplify the scheme and set $U = 4$.

NEW RSA-BASED CONSTRUCTION. Here we could use the RSA-based trapdoor commitment $\mathsf{Com}(0; r_0) = r_0^e \bmod n$, $\mathsf{Com}(1; r_1) = y r_1^e \bmod n$, where y is a public parameter, whose e-th root is the trapdoor key. Here we simply need the Σ-protocol proof of knowledge of the e-th root, which is just the GQ protocol [24]. To have the protocol to be challenge-efficient, though, we will need to use a relatively large e.

ALTERNATIVE FACTORING CONSTRUCTION. We can use the following factoring-based commitment of [33] (slightly modified for easier Σ-protocols and specialized to bits). The public key is $n = p, q$, where $p = 2p' + 1, q = 2q' + 1$ are safe primes, and all the operations are performed in the subgroup Q_n of quadratic residues whose generator g is also part of parameters. Notice, $|Q_n| = p'q'$. Let C be a large enough constant (anything larger than n will do). Then $\mathsf{Com}(0; r_0) = g^{C+r_0} \bmod n$, $\mathsf{Com}(1; r_1) = g^{r_1} \bmod n$ (here r_0, r_1 are random from 0 to n (which is statistically close to $\varphi(n)$, which is the "true range" we are aiming for). The trapdoor is the value $|Q_n| = p'q'$. In this case the Σ-protocol we need to again the one of knowledge of discrete-log, but in the groups of unknown order. As mentioned before, such (computationally sound) protocol is given by [19,9].

PAILLIER-BASED SCHEME. Finally, we mention another trapdoor commitment based on the hardness of finding n-th roots over Z_{n^2} (where n is the the product of two safe primes, for simplicity), which is implicit in [11]. Here the public parameters will include a generator g in the subgroup S of n-th powers in $\mathbb{Z}_{n^2}^*$, and the n-th root u of g will be the trapdoor. Next, $\mathsf{Com}(0; r_0) = r_0^n \bmod n^2$, $\mathsf{Com}(1; r_1) = g r_1^n \bmod n^2$ (here $r_0, r_1 \in \mathbb{Z}_n^*$). This scheme is perfectly hiding and computationally binding assuming it is hard to take n-th root over $\mathbb{Z}_{n^2}$, and could be viewed as yet another claw-free based construction. The Σ-protocol for commitment to 0 is simply the Σ-protocol for knowing the n-th root. This protocol is very similar to the GQ protocol and is formally analyzed by [10].

6   Concluding Remarks

We believe that our results elucidate the notion of mercurial commitments, put them in their place on the map of cryptographic assumptions, and better explain the rational following the previous constructions of [28],[8]. We hope that mercurial commitments will find more interesting applications in the future.

This paper joins two independent papers that can be found at [7],[13].

Acknowledgments. The second author would like to thank Leonid Reyzin for several insightful conversations about mercurial commitments, and Tal Malkin for giving a talk inspiring this research. The work of the first and third authors has been supported in part by the European Commission through the IST Programme under Contract IST-2002-507932 ECRYPT. The work of the second author was supported in part through NSF Career Award CCR-0133806 and NSF grant CCR-0311095. The work of the third author is also supported in part through the FP6 program under contract FP6-1596 AEOLUS.

References

1. M. Abdalla and L. Reyzin. A new forward-secure digital signature scheme. In T. Okamoto, editor, *Advances in Cryptology—ASIACRYPT 2000*, volume 1976 of *Lecture Notes in Computer Science*, pages 116–129, Kyoto, Japan, 3–7 Dec. 2000. Springer-Verlag. Full version available from the Cryptology ePrint Archive, record 2000/002, http://eprint.iacr.org/.

2. Manuel Blum, Alfredo De Santis, Silvio Micali, and Guiseppe Persiano. Noninteractive zero-knowledge. *SIAM Journal of Computing*, 20(6), 1991.

3. Joan Boyar, S. A. Kurtz, Mark W. Krentel. A Discrete Logarithm Implementation of Perfect Zero-Knowledge Blobs. In *J. of Cryptology*, 2(2):63–76, 1990.

4. G. Brassard, D. Chaum, and C. Crépeau. Minimum disclosure proofs of knowledge. *Journal of Computer and System Sciences*, 37(2):156–189, Oct. 1988.

5. Dario Catalano, Rosario Gennaro, Nick Howgrave-Graham, Phong Q. Nguyen. Paillier's cryptosystem revisited. In *ACM Conference on Computer and Communications Security 2001*, pp. 206–214.

6. D. Catalano and I. Visconti. Hybrid Trapdoor Commitments and Their Applications. In *32nd International Colloquium on Automata, Languages, and Programming (ICALP 05)*, volume 3580 of *Lecture Notes in Computer Science*, pages 298–310. Springer-Verlag, 2005.

7. D. Catalano and I. Visconti. Non-Interactive Mercurial Commitments from One-Way Functions. Cryptology ePrint Archive, 2005.

8. Melissa Chase, Alexander Healy, Anna Lysysanskaya, Tal Malkin and Leonid Reyzin. Mercurial Commitments with Applications to Zero-Knowledge Sets. In *Proc. of EUROCRYPT*, pp. 422–439, 2005.

9. Ivan Damgård, Eiichiro Fujisaki. A Statistically-Hiding Integer Commitment Scheme Based on Groups with Hidden Order. In *ASIACRYPT 2002*, pp. 125–142.

10. Ivan Damgård, Mats Jurik. A Generalisation, a Simplification and Some Applications of Paillier's Probabilistic Public-Key System. *Public Key Cryptography 2001*, pp. 119–136.

11. I. Damgård and J. B. Nielsen. Perfect hiding and perfect binding universally composable commitment schemes with constant expansion factor. In M. Yung, editor, *Advances in Cryptology—CRYPTO 2002*, Lecture Notes in Computer Science. Springer-Verlag, 18–22 Aug. 2002.

12. I. B. Damgård, T. P. Pedersen, and B. Pfitzmann. On the existence of statistically hiding bit commitment schemes and fail-stop signatures. *Journal of Cryptology*, 10(3):163–194, Summer 1997.

13. Y. Dodis. Minimal Assumptions for Efficient Mercurial Commitments. Cryptology ePrint Archive, Report 2005/438.

14. Y. Dodis and L. Reyzin. On the power of claw-free permutations. In *Conference on Security in Communication Networks*, 2002.

15. U. Feige, D. Lapidot, and A. Shamir. Multiple noninteractive zero knowledge proofs under general assumptions. *SIAM J. Computing*, 29(1), 1999.

16. U. Feige and A. Shamir. Zero knowledge proofs of knowledge in two rounds. In G. Brassard, editor, *Advances in Cryptology—CRYPTO '89*, volume 435 of *Lecture Notes in Computer Science*, pages 526–545. Springer-Verlag, 1990, 20–24 Aug. 1989.

17. U. Feige and A. Shamir. Witness indistinguishability and witness hiding protocols. In *Proceedings of the Twenty Second Annual ACM Symposium on Theory of Computing*, pages 416–426, Baltimore, Maryland, 14–16 May 1990.

18. A. Fiat and A. Shamir. How to prove yourself: Practical solutions to identification and signature problems. In A. M. Odlyzko, editor, *Advances in Cryptology—CRYPTO '86*, volume 263 of *Lecture Notes in Computer Science*, pages 186–194. Springer-Verlag, 1987, 11–15 Aug. 1986.

19. E. Fujisaki and T. Okamoto. Statistical zero knowledge protocols to prove modular polynomial relations. In B. S. Kaliski Jr., editor, *Advances in Cryptology—CRYPTO '97*, volume 1294 of *Lecture Notes in Computer Science*, pages 16–30. Springer-Verlag, 17–21 Aug. 1997.

20. O. Goldreich. *Foundations of Cryptography: Basic Tools*. Cambridge University Press, 2001.

21. S. Goldwasser, S. Micali, and C. Rackoff. Knowledge complexity of interactive proofs. In *Proceedings of the Seventeenth Annual ACM Symposium on Theory of Computing*, pages 291–304, Providence, Rhode Island, 6–8 May 1985.

22. S. Goldwasser, S. Micali, and R. Rivest. A digital signature scheme secure against adaptive chosen-message attacks. *SIAM J. Computing*, 17(2), 1988.

23. O. Goldreich, S. Micali, and A. Wigderson. Proofs that yield nothing but their validity or all languages in NP have zero-knowledge proof systems. *Journal of the ACM*, 38(1):691–729, 1991.

24. L. C. Guillou and J.-J. Quisquater. A "paradoxical" identity-based signature scheme resulting from zero-knowledge. In S. Goldwasser, editor, *Advances in Cryptology—CRYPTO '88*, volume 403 of *Lecture Notes in Computer Science*, pages 216–231. Springer-Verlag, 1990, 21–25 Aug. 1988.

25. S. Halevi and S. Micali. Practical and provably-secure commitment schemes from collision-free hashing. In N. Koblitz, editor, *Advances in Cryptology—CRYPTO '96*, volume 1109 of *Lecture Notes in Computer Science*, pages 201–215. Springer-Verlag, 18–22 Aug. 1996.

26. J. Håstad, R. Impagliazzo, L. Levin, and M. Luby. A pseudorandom generator from any one-way function. *SIAM J. Computing*, 28(4), 1999.

27. H. Krawczyk and T. Rabin. Chameleon signatures. In *Network and Distributed System Security Symposium*, pages 143–154. The Internet Society, 2000.

28. Silvio Micali, Michael Rabin, and Joe Kilian. Zero-knowledge sets. In *Proc. 44th IEEE Symposium on Foundations of Computer Science (FOCS)*, 2003.

29. Moni Naor. Bit commitment using pseudorandomness. *Journal of Cryptology*, 4(2):51–158, 1991.

30. H. Ong and C. P. Schnorr. Fast signature generation with a Fiat Shamir-like scheme. In I. B. Damgård, editor, *Advances in Cryptology—EUROCRYPT 90*, volume 473 of *Lecture Notes in Computer Science*, pages 432–440. Springer-Verlag, 1991, 21–24 May 1990.

31. R. Ostrovsky, C. Rackoff, and A. Smith. Efficient consistency proofs for generalized queries on a committed database. In *31st International Colloquium on Automata, Languages, and Programming (ICALP 04)*, volume 3142 of *Lecture Notes in Computer Science*, pages 1041–1053. Springer-Verlag, 2004.

32. P. Paillier. Public-key cryptosystems based on composite degree residuosity classes. In J. Stern, editor, *Advances in Cryptology—EUROCRYPT '99*, volume 1592 of *Lecture Notes in Computer Science*. Springer-Verlag, 2–6 May 1999.

33. A. Shamir and Y. Tauman. Improved online/offline signature schemes. In J. Kilian, editor, *Advances in Cryptology—CRYPTO 2001*, volume 2139 of *Lecture Notes in Computer Science*, pages 355–367. Springer-Verlag, 19–23 Aug. 2001.

34. C.-P. Schnorr. Efficient signature generation by smart cards. *Journal of Cryptology*, 4(3):161–174, 1991.

Efficient Collision-Resistant Hashing from Worst-Case Assumptions on Cyclic Lattices[*]

Chris Peikert[1] and Alon Rosen[2]

[1] MIT Computer Science and AI Laboratory (CSAIL), Cambridge, MA
[2] DEAS, Harvard, Cambridge, MA

Abstract. The generalized knapsack function is defined as $f_{\mathbf{a}}(\mathbf{x}) = \sum_i a_i \cdot x_i$, where $\mathbf{a} = (a_1, \ldots, a_m)$ consists of m elements from some ring R, and $\mathbf{x} = (x_1, \ldots, x_m)$ consists of m coefficients from a specified subset $S \subseteq R$. Micciancio (FOCS 2002) proposed a specific choice of the ring R and subset S for which inverting this function (for random $\mathbf{a}, \mathbf{x}$) is at least as hard as solving certain worst-case problems on cyclic lattices.

We show that for a different choice of $S \subset R$, the generalized knapsack function is in fact *collision-resistant*, assuming it is infeasible to approximate the shortest vector in n-dimensional cyclic lattices up to factors $\tilde{O}(n)$. For slightly larger factors, we even get collision-resistance for *any* $m \geq 2$. This yields *very* efficient collision-resistant hash functions having key size and time complexity almost linear in the security parameter n. We also show that altering S is necessary, in the sense that Micciancio's original function is *not* collision-resistant (nor even universal one-way).

Our results exploit an intimate connection between the linear algebra of n-dimensional cyclic lattices and the ring $\mathbb{Z}[\alpha]/(\alpha^n - 1)$, and crucially depend on the factorization of $\alpha^n - 1$ into irreducible cyclotomic polynomials. We also establish a new bound on the discrete Gaussian distribution over general lattices, employing techniques introduced by Micciancio and Regev (FOCS 2004) and also used by Micciancio in his study of compact knapsacks.

1   Introduction

A function family $\{f_a\}_{a \in A}$ is said to be *collision-resistant* if given a uniformly chosen $a \in A$, it is infeasible to find elements $x_1 \neq x_2$ so that $f_a(x_1) = f_a(x_2)$. Collision-resistant hash functions are one of the most widely-employed cryptographic primitives. Their applications include integrity checking, user and message authentication, commitment protocols, and more.

Many of the applications of collision-resistant hashing tend to invoke the hash function only a small number of times. Thus, the efficiency of the function has a direct effect on the efficiency of the application that uses it. This is in contrast to primitives such as one-way functions, which typically must be invoked many times in their applications (at least when used in a black-box way) [9].

---

[*] Part of this work done while at MIT CSAIL.

S. Halevi and T. Rabin (Eds.): TCC 2006, LNCS 3876, pp. 145–166, 2006.
© Springer-Verlag Berlin Heidelberg 2006

Collision-resistance can be obtained from many well-studied complexity assumptions, but the resulting hash functions are not efficient enough for practical use. Instead, faster *heuristic* constructions such as MD5 and SHA-1 are often employed. Unfortunately, recent cryptanalytic analysis of many popular hash functions casts doubt on the heuristic approach [22,21]. This presents the theoretical community with a great opportunity and challenge: propose a *practical* hash function with *rigorous* security guarantees.

In this paper we present an *efficient* collision-resistant hash function whose security is based on a well-defined and plausible complexity assumption.

1.1 Generalized Knapsacks

Our constructions are based on a generalization of the well-known *knapsack* function. For a ring R, key $\mathbf{a} = (a_1, \ldots, a_m) \in R^m$, and input $\mathbf{x} = (x_1, \ldots, x_m)$, the generalized knapsack function is defined as

$$f_{\mathbf{a}}(\mathbf{x}) = \sum_{i=1}^{m} a_i \cdot x_i,$$

where each x_i is restricted to some large subset $S \subseteq R$. This generalization was proposed by Micciancio, who suggested a specific choice of the ring R and subset S for which inverting the function (for random $\mathbf{a}, \mathbf{x}$) is at least as hard as solving certain worst-case problems on *cyclic* lattices [14].

Knapsacks have a long and infamous history in cryptography; we refer the interested reader to Micciancio's account of various knapsack proposals and their cryptanalysis [14]. The bottom line is that even though many knapsack systems have been broken heuristically, there is still no *asymptotically-efficient* attack on the general function.

Micciancio's result might be viewed as an indication that knapsack functions (or at least, some version of them) are secure after all. In this paper, we continue Micciancio's line of study, and show that, for a different choice of $S \subset R$, the generalized knapsack function can enjoy even stronger cryptographic properties.

1.2 Lattices, Hardness, and Cryptography

Lattices are a great source of cryptographic hardness. First of all, lattices have been subject to hundreds of years of mathematical scrutiny, which lends support to conjectures on the computational hardness of problems related to lattices. Indeed, many lattice problems are NP-hard to approximate for small factors, e.g. the closest vector [20,4,7] and shortest vector problems [2,5,15,12].

Secondly, lattices admit *worst-case* to *average-case* reductions. In his groundbreaking result, Ajtai first constructed a one-way function [1], which was later observed to also be collision-resistant [10]. Public-key cryptosystems [11,3,18,19] soon followed, based on presumably stronger worst-case assumptions. As a bonus, these constructions tended to be asymptotically more efficient than those based on, e.g., modular exponentiation.

An interesting special case is presented by *cyclic* lattices. A lattice Λ is said to be cyclic if for any vector $\mathbf{x} \in \Lambda$, its *cyclic rotation* also belongs to Λ. The cyclic rotation of $\mathbf{x} = (x_0, \ldots, x_{n-1})^T \in \mathbb{R}^n$ is defined as $(x_{n-1}, x_0, \ldots, x_{n-2})^T$.

Micciancio's work [14] opened the door to the use of cyclic lattices as a new source of hardness assumptions, and motivates their study from a computational perspective. Currently no hardness results are known for problems on cyclic lattices (even in their exact versions), and the additional structure may indeed reduce the underlying hardness.

However, state-of-the-art lattice algorithms appear not to benefit from cyclicity, and it seems reasonable to conjecture that standard problems on cyclic lattices are intractable, at least for small approximation factors.

1.3 Our Results

Our main result is that certain instantiations of the generalized knapsack function are collision-resistant, assuming it is infeasible to approximate the shortest vector in cyclic lattices up to factors $\tilde{O}(n)$ almost linear in the dimension n.

Assuming hardness for slightly larger approximation factors $n^{1+\epsilon}$, our functions remain secure even when m is taken to be a *constant*. The functions have key size almost linear in the security parameter n and can be evaluated with m Fast Fourier Transform operations, making them potentially practical. To motivate our choice of knapsack function, we also show that Micciancio's original one-way function is not collision-resistant, nor even universal one-way.

In the course of proving our main results, we formulate special worst-case problems on cyclic lattices, and relate them to the more standard lattice problems. Most interestingly, we demonstrate that for cyclic lattices of *prime* dimension n, the short *independent* vectors problem SIVP reduces to (a slight variant of) the shortest vector problem SVP with only a factor of 2 loss in approximation factor. For general lattices, the best known reduction loses a $\sqrt{n}$ factor [16]; furthermore, that reduction performs manipulations on its input lattice that can destroy the cyclicity property. Hence our reduction can be seen as the first connection between SIVP and SVP on cyclic lattices.

Finally, in using the Gaussian techniques of [17], we also establish a new bound on the discrete Gaussian distribution over general lattices, which may be of independent interest.

1.4 Techniques and Ideas

The overarching theme of our paper is the tight relationship shared by cyclic lattices, the algebra of polynomials modulo $(\alpha^n - 1)$, and linear algebra in $\mathbb{R}^n$.

Cyclic lattices are closed under *cyclic convolution* with integer vectors. Furthermore, the lattice points naturally correspond to polynomials in $\mathbb{Z}[\alpha]/(\alpha^n - 1)$. Because convolution is equivalent to polynomial multiplication in $\mathbb{Z}[\alpha]/(\alpha^n - 1)$, this implies that integer cyclic lattices are isomorphic to *ideals* in $\mathbb{Z}[\alpha]/(\alpha^n - 1)$.

The divisors of $(\alpha^n - 1)$ in $\mathbb{Z}[\alpha]$ correspond to special *cyclotomic* linear subspaces of $\mathbb{R}^n$. These subspaces admit a natural partitioning into complementary pairs of orthogonal subspaces. Even more importantly, the subspaces are closed under *cyclic rotation* of vector coordinates, and under certain other conditions, these rotations are *linearly independent*. These facts imply a new connection between the SIVP and SVP problems in cyclic lattices.

The security of our knapsack function comes from using all this structure to impose an algebraic restriction on the function domain. Looking ahead to the security reduction, this restriction ensures that collisions in the function are very likely to yield "useful" and short lattice points in a desired subspace.

1.5 Comparison with Related Work

This work takes its inspiration from, and is most similar to, Micciancio's work on cyclic lattices [14]. However, while our knapsack function is very similar to Micciancio's, the reduction used to establish collision-resistance differs in many significant ways. First of all, Micciancio's function is proven to be one-way, while ours is collision-resistant. On the other hand, Micciancio relies on a presumably weaker worst-case assumption than we do. Our stronger assumption, combined with our algebraic view of cyclic lattices, makes our security reduction tighter and conceptually simpler.

Figure 1 gives a comparison of our work with other major results in worst-case to average-case reductions, in chronological order. Important considerations in these works include: provable security properties of the cryptographic function, efficiency of that function, class of lattice on which the function is based, type of worst-case problem that is assumed to be hard for that class of lattice, and its hardness of approximation factor. Our work compares very favorably in many of these considerations, at the cost of a qualitatively stronger assumption.

	Security	Efficiency	Lattice Class	Assumption	Approx. Factor
Ajtai [1]	CRHF	$O(n^2)$	General	SVP etc.	$\text{poly}(n)$
Cai, Nerurkar [6]	CRHF	$O(n^2)$	General	SVP etc.	$n^{4+\epsilon}$
Micciancio [14]	OWF	$\tilde{O}(n)$	Cyclic	GDD	$n^{1+\epsilon}$
Micciancio, Regev [17]	CRHF	$O(n^2)$	General	SVP etc.	$\tilde{O}(n)$
This work	CRHF	$\tilde{O}(n)$	Cyclic	SVP etc.	$\tilde{O}(n)$

Fig. 1. Comparison of results in lattice-based cryptographic functions with worst-case to average-case security reductions, to date. "Efficiency" means the key size and computation time, as a function of the lattice dimension n. "Security" denotes the function's main cryptographic property.

The actual worst-case assumption underlying our hash function is that SVP is hard on cyclic lattices for all sufficiently large *prime* dimensions n. Therefore, the discovery of an efficient algorithm for SVP on, say, all *even* dimensions would have no immediate effect on the security of our hash function. Conveniently, the *concrete* hardness of the cyclic lattice problems we study appears to be *greatest* when the dimension is prime! More specifically: problems in composite dimensions n seem to reduce to problems in the smaller prime (or prime-power) dimensions dividing n.

In an independent and concurrent work, Lyubashevsky and Micciancio [13] have obtained exceedingly similar results, but expressed in different mathematical language. In particular, by making many of the same algebraic insights, they construct collision-resistant hash functions with nearly identical parameters, based on a worst-case hardness assumption that can be shown to be equivalent to ours. They also present a more general algebraic framework for constructing hash functions, which can be related to problems in algebraic number theory. Due to its generality, their framework may have the potential to admit better constructions, though its current best application essentially matches ours.

2 Preliminaries

In this section we present basic definitions and results about statistical distance, hash functions, cyclic lattices, cyclotomic polynomials and Gaussian probability distributions. In many places we follow [17] almost verbatim.

For any real $a \geq 0$, $\lfloor a \rfloor$ denotes the largest integer not greater than a and $\lfloor a \rceil$ denotes the closest integer to a (i.e., $\lfloor a \rceil = \lfloor a + 1/2 \rfloor$). For any reals $a, b \geq 0$, $[a, b)$ denotes the set of all reals $a \leq r < b$. The uniform probability distribution over a set S is denoted $U(S)$. We let I denote $U([0, 1))$. A function $f(n)$ is said to be negligible (denoted $f(n) = n^{-\omega(1)}$) if for every $c > 0$ there exists an n_0 such that $|f(n)| < 1/n^c$ for all $n > n_0$.

The set of real numbers is denoted by $\mathbb{R}$, and the quotient ring of integers modulo a positive integer p is denoted by $\mathbb{Z}_p$. For a value $v \in \mathbb{Z}_p$, $|v|$ denotes the absolute value of the unique integer $r \in (-p/2, p/2]$ representing v's residue class. We use bold lower case letters (e.g., $\mathbf{x}$) to denote vectors and bold upper case letters (e.g., $\mathbf{A}$) to denote matrices. Vectors are represented as columns and we use $(\cdot)^T$ to denote matrix transposition. We adopt the convention that vector indices are *zero-based*, i.e. for $\mathbf{x} \in \mathbb{R}^n$ we write $\mathbf{x} = (x_0, \dots, x_{n-1})^T$. The ith coordinate of $\mathbf{x}$ is denoted x_i or $(\mathbf{x})_i$, depending on context. The *Euclidean norm* of a vector $\mathbf{x}$ (in either $\mathbb{R}^n$ or $\mathbb{Z}_p^n$) is the quantity $\|\mathbf{x}\| = (\sum_i |x_i|^2)^{1/2}$. The Euclidean norm of a matrix $\mathbf{S} = (\mathbf{s}_1, \dots, \mathbf{s}_t)$ is $\|\mathbf{S}\| = \max_i \|\mathbf{s}_i\|$. Other norms used in this paper (for vectors in either $\mathbb{R}^n$ or $\mathbb{Z}_p^n$) are the ℓ_1 norm $\|\mathbf{x}\|_1 = \sum_i |x_i|$ and the ℓ_∞ norm $\|\mathbf{x}\|_\infty = \max_i |x_i|$, which are similarly extended to matrices. These norms are related through the following inequalities, valid for any n-dimensional vector $\mathbf{x} \in \mathbb{R}^n$:

$$\|\mathbf{x}\| \leq \|\mathbf{x}\|_1 \leq \sqrt{n}\|\mathbf{x}\|$$
$$\|\mathbf{x}\|_\infty \leq \|\mathbf{x}\| \leq \sqrt{n}\|\mathbf{x}\|_\infty$$

We use standard definitions of statistical distance $\Delta(X, Y)$ between two random (discrete or continuous) variables X, Y. We also use the standard notions of one-wayness, universal one-wayness, and collision-resistance for function ensembles.

2.1   Lattices

A *lattice* in $\mathbb{R}^n$ is the set of all integer combinations

$$\Lambda = \left\{ \sum_{i=1}^{d} c_i \mathbf{b}_i \mid c_i \in \mathbb{Z} \text{ for } 1 \leq i \leq d \right\}$$

of d linearly independent vectors $\mathbf{b}_1, \ldots, \mathbf{b}_d \in \mathbb{R}^n$. We say that the lattice *spans* the d-dimensional subspace of $\mathbb{R}^n$ generated by $\mathbf{b}_1, \ldots, \mathbf{b}_d$. The set of vectors $\mathbf{b}_1, \ldots, \mathbf{b}_d$ is called a *basis* for the lattice, which can be written in matrix form as $\mathbf{B} = [\mathbf{b}_1 | \cdots | \mathbf{b}_d]$ with the basis vectors as columns. The lattice generated by $\mathbf{B}$ is denoted $\mathcal{L}(\mathbf{B})$. For any basis $\mathbf{B}$, we define the *fundamental parallelepiped* $\mathcal{P}(\mathbf{B}) = \{\mathbf{B} \cdot \mathbf{x} : \forall i, 0 \leq x_i < 1\}$.

The *minimum distance* $\lambda_1(\Lambda)$ of a lattice Λ is the length of the shortest nonzero lattice vector: $\lambda_1(\Lambda) = \min_{0 \neq \mathbf{x} \in \Lambda} \|\mathbf{x}\|$. More generally, the ith successive minimum $\lambda_i(\Lambda)$ is the smallest radius r such that the closed ball $\overline{\mathcal{B}}(r) = \{\mathbf{x} : \|\mathbf{x}\| \leq r\}$ contains i linearly independent lattice vectors.

Let H be a subspace of $\mathbb{R}^n$ and let Λ be a lattice that spans H. Then we define the *dual lattice* $\Lambda^* = \{\mathbf{x} \in H \mid \forall \mathbf{v} \in \Lambda, \langle \mathbf{x}, \mathbf{v} \rangle \in \mathbb{Z}\}$.

Cyclic lattices and convolution. For any $\mathbf{x} = (x_0, \ldots, x_{n-1})^T \in \mathbb{R}^n$, define the *rotation of* $\mathbf{x}$, denoted as $\mathrm{rot}(\mathbf{x})$, to be the vector $(x_{n-1}, x_0, \ldots, x_{n-2})^T$; similarly $\mathrm{rot}^i(\mathbf{x}) = \mathrm{rot}(\cdots \mathrm{rot}(\mathbf{x}) \cdots)$ is defined to be the rotation of $\mathbf{x}$, taken i times. A lattice Λ is *cyclic* if for all $\mathbf{x} \in \Lambda$, $\mathrm{rot}(\mathbf{x}) \in \Lambda$. For any integer $d \geq 1$, define the *rotation matrix* $\mathrm{Rot}^d(\mathbf{x})$ to be the matrix $[\mathbf{x} | \mathrm{rot}(\mathbf{x}) | \cdots | \mathrm{rot}^{d-1}(\mathbf{x})]$. ($\mathrm{Rot}^n(\mathbf{x})$ is known as the *circulant matrix* of $\mathbf{x}$.)

For any ring R, the (cyclic) convolution product of $\mathbf{x}, \mathbf{y} \in R^n$ is the vector $\mathbf{x} \otimes \mathbf{y} = \mathrm{Rot}^n(\mathbf{x}) \cdot \mathbf{y}$, with entries

$$(\mathbf{x} \otimes \mathbf{y})_k = \sum_{i+j=k \bmod n} x_i \cdot y_j.$$

Observe that in a cyclic lattice Λ, the convolution of any $\mathbf{x} \in \Lambda$ with any integer vector $\mathbf{y} \in \mathbb{Z}^n$ is also in the lattice: $\mathbf{x} \otimes \mathbf{y} \in \Lambda$. This is because all the columns of $\mathrm{Rot}^n(\mathbf{x})$ are in Λ, and any integer combination of points in Λ is also in Λ.

The convolution product is commutative, associative, and distributive over vector addition; also, it satisfies the following inequalities, valid for any n-dimensional vectors $\mathbf{x}, \mathbf{y} \in \mathbb{R}^n$:

$$\|\mathbf{x} \otimes \mathbf{y}\|_\infty \leq \|\mathbf{x}\| \cdot \|\mathbf{y}\|$$
$$\|\mathbf{x} \otimes \mathbf{y}\|_\infty \leq \|\mathbf{x}\|_1 \cdot \|\mathbf{y}\|_\infty$$

2.2   Polynomial Rings and Linear Algebra

Convolution and polynomial multiplication are intimately related. Specifically, for any ring R, we identify an element $(x_0, \ldots, x_{n-1}) = \mathbf{x} \in R^n$ with the polynomial $\mathbf{x}(\alpha) \in R[\alpha]/(\alpha^n - 1)$ defined as $\mathbf{x}(\alpha) = x_0 + x_1 \alpha + \ldots + x_{n-1}\alpha^{n-1}$.

Then it is easy to show that for any $\mathbf{x}, \mathbf{y} \in R^n$, $\mathbf{x} \otimes \mathbf{y}$ is identified with $\mathbf{x}(\alpha) \cdot \mathbf{y}(\alpha) \in R[\alpha]/(\alpha^n - 1)$. In words, convolution of two vectors is equivalent to taking the product of their polynomials modulo $\alpha^n - 1$. Throughout the paper, we will switch between vector and polynomial notation as is convenient.

In the following lemma, we relate the algebra of $\mathbb{R}[\alpha]/(\alpha^n - 1)$ to the linear algebra of $\mathbb{R}^n$.

Lemma 2.1. *Let* $\mathbf{a}, \mathbf{b} \in \mathbb{R}^n$ *with* $\mathbf{a}(\alpha) \cdot \mathbf{b}(\alpha) = 0 \bmod (\alpha^n - 1)$. *Then* $\langle \mathbf{a}, \mathbf{b} \rangle = 0$.

Proof. Let $\mathbf{F}$ be the $n \times n$ matrix with (zero-indexed) entries given by

$$(\mathbf{F})_{j,k} = \frac{e^{2\pi i j k / n}}{\sqrt{n}} = \frac{\omega^{jk}}{\sqrt{n}},$$

where ω is the principal nth root of unity ($\mathbf{F}$ is known as a *Fourier matrix*). It is well-known that $\mathbf{F}$ is a unitary matrix, so $\langle \mathbf{a}, \mathbf{b} \rangle = \langle \mathbf{Fa}, \mathbf{Fb} \rangle$. By definition, $(\mathbf{Fa})_i = \mathbf{a}(\omega^i)/\sqrt{n}$ and $(\mathbf{Fb})_i = \mathbf{b}(\omega^i)/\sqrt{n}$. Now because $\mathbf{a}(\alpha)\mathbf{b}(\alpha)$ is divisible by $\alpha^n - 1$, then $\mathbf{a}(\omega^i) \cdot \mathbf{b}(\omega^i) = 0$ (in $\mathbb{C}$) for every i. Therefore

$$\langle \mathbf{a}, \mathbf{b} \rangle = \langle \mathbf{Fa}, \mathbf{Fb} \rangle = \frac{1}{n} \sum_{i=1}^{n} \mathbf{a}(\omega^i) \mathbf{b}(\omega^i) = 0.$$

In the polynomial ring $\mathbb{Z}[\alpha]$, $(\alpha^n - 1)$ has a special structure: it uniquely factors into the product of *cyclotomic polynomials* (see e.g. [8] for a detailed treatment). For integer $k \geq 1$, the kth cyclotomic polynomial $\Phi_k(\alpha)$ is defined:

$$\Phi_k(\alpha) = \prod_{\substack{1 \leq c \leq k \\ (c,k)=1}} (\alpha - e^{2\pi i c / k}),$$

where (c, k) denotes the greatest common divisor of c and k. The cyclotomic polynomial $\Phi_k(\alpha)$ is irreducible in $\mathbb{Z}[\alpha]$, has integer coefficients, and has degree $\phi(k)$ (where ϕ denotes Euler's totient function). The factorization of $\alpha^n - 1$ in $\mathbb{Z}[\alpha]$ is: $\alpha^n - 1 = \prod_{\substack{k \mid n \\ k \geq 1}} \Phi_k(\alpha)$.

In the following lemmas, we establish connections between cyclotomic polynomials and the linear algebra of integer cyclic lattices:

Lemma 2.2. *Let* $\mathbf{c} \in \mathbb{Z}^n$, *and suppose* $\Phi(\alpha) \in \mathbb{Z}[\alpha]$ *divides* $(\alpha^n - 1)$ *and is coprime to* $\mathbf{c}(\alpha)$. *Then* $\mathbf{c}, \mathrm{rot}(\mathbf{c}), \ldots, \mathrm{rot}^{\deg(\Phi)-1}(\mathbf{c})$ *are linearly independent.*

Proof. Suppose that there exist $t_0, \ldots, t_{\deg(\Phi)-1} \in \mathbb{R}$ such that $\sum_{i=0}^{\deg(\Phi)-1} t_i \mathrm{rot}^i(\mathbf{c}) = 0$. Define $\mathbf{t} = (t_0, t_1, \cdots, t_{\deg(\overline{\Phi})-1}, 0, \cdots, 0)^T$, so $\mathbf{c} \otimes \mathbf{t} = 0$ (where the convolution is performed in $\mathbb{R}^n$). Therefore in $\mathbb{R}[\alpha]$, $(\alpha^n - 1)$ divides $\mathbf{c}(\alpha)\mathbf{t}(\alpha)$.

We recall two basic facts from field theory (see, e.g., [8, Proposition 9, Chapter 13]): first, $\Phi_k(\alpha)$ is the *minimal polynomial*[1] of any primitive kth root of unity,

---

[1] The minimal polynomial of an algebraic number ζ is the unique irreducible monic (i.e., with leading coefficient 1) polynomial $p(\alpha) \in \mathbb{Q}[\alpha]$ of minimum degree such that $p(\zeta) = 0$.

and has exactly the primitive kth roots of unity as its roots. Second, the minimal polynomial of any algebraic number ζ divides any polynomial $p(\alpha) \in \mathbb{Q}[\alpha]$ such that $p(\zeta) = 0$.

Now, because $\Phi(\alpha) \mid (\alpha^n - 1)$, $\Phi(\alpha)$ is a product of cyclotomic polynomials. Because $\Phi(\alpha)$ is coprime to $\mathbf{c}(\alpha)$ and $\mathbf{c}(\alpha) \in \mathbb{Z}[\alpha] \subset \mathbb{Q}[\alpha]$, none of the roots of $\Phi(\alpha)$ are roots of $\mathbf{c}(\alpha)$. Therefore all the roots of $\Phi(\alpha)$ must be roots of $\mathbf{t}(\alpha)$. Because $\deg(\mathbf{t}(\alpha)) < \deg(\overline{\Phi})$, we must have $\mathbf{t} = 0$. $\square$

Suppose $\Phi(\alpha) \in \mathbb{Z}[\alpha]$ divides $\alpha^n - 1$, i.e. $\Phi(\alpha)$ is a product of cyclotomic polynomials. We define the *cyclotomic subspace*

$$H_\Phi = \{\mathbf{x} \in \mathbb{R}^n : \Phi(\alpha) \text{ divides } \mathbf{x}(\alpha) \text{ in } \mathbb{R}[\alpha]\}.$$

Lemma 2.3. *H_Φ is closed under* rot*: that is, if $\mathbf{c} \in H_\Phi$, then* $\mathrm{rot}(\mathbf{c}) \in H_\Phi$.

Proof. Observe that the vector $\mathrm{rot}(\mathbf{c})$ is identified with the residue $\alpha \cdot \mathbf{c}(\alpha)$ mod $(\alpha^n - 1)$. Let $\alpha \cdot \mathbf{c}(\alpha) = Q(\alpha) \cdot (\alpha^n - 1) + R(\alpha)$, for $Q(\alpha), R(\alpha) \in \mathbb{R}[\alpha]$, where $\deg(R(\alpha)) < n$. Then because $\Phi(\alpha) \mid \alpha \cdot \mathbf{c}(\alpha)$ and $\Phi(\alpha) \mid Q(\alpha) \cdot (\alpha^n - 1)$, it must be that $\Phi(\alpha) \mid R(\alpha)$. Therefore $\Phi(\alpha)$ divides $\mathrm{rot}(\mathbf{c})(\alpha)$ in $\mathbb{R}[\alpha]$, as desired. $\square$

Lemma 2.4. *H_Φ is a linear subspace of $\mathbb{R}^n$ of dimension $n - \deg(\Phi)$.*

Proof. It is evident that H_Φ is closed under addition and scalar multiplication, so it is a linear subspace. To establish the dimension, define $\overline{\Phi}(\alpha) = (\alpha^n - 1)/\Phi(\alpha)$. By Lemma 2.1, because $\Phi(\alpha) \cdot \overline{\Phi}(\alpha) = 0$ mod $(\alpha^n - 1)$, H_Φ and $H_{\overline{\Phi}}$ are orthogonal subspaces. Therefore $\dim(H_\Phi) + \dim(H_{\overline{\Phi}}) \leq n$.

By Lemma 2.2, the vectors $\Phi, \mathrm{rot}(\Phi), \ldots, \mathrm{rot}^{\deg(\overline{\Phi})-1}(\Phi)$ are linearly independent. By Lemma 2.3, they all lie in H_Φ. Therefore $\dim(H_\Phi) \geq \deg(\overline{\Phi}) = n - \deg(\Phi)$. Symmetrically, $\dim(H_{\overline{\Phi}}) \geq n - \deg(\overline{\Phi})$. All three inequalities can be satisfied only with equality, hence $\dim(H_\Phi) = n - \deg(\Phi)$. $\square$

2.3   Gaussian Distributions

For any d-dimensional subspace H of $\mathbb{R}^n$, any $\mathbf{c} \in H$ and any $s > 0$, define

$$\rho_{H,s,\mathbf{c}}(\mathbf{x}) = \begin{cases} \exp(-\pi\|(\mathbf{x} - \mathbf{c})/s\|^2) & \text{if } \mathbf{x} \in H \\ 0 & \text{if } \mathbf{x} \notin H \end{cases}$$

to be the Gaussian function (over H) centered at $\mathbf{c}$, with radius s. By normalizing $\rho_{s,\mathbf{c}}$ by its total measure $\int_{\mathbf{x} \in H} \rho_{s,\mathbf{c}}(\mathbf{x})d\mathbf{x} = s^d$, we get a continuous distribution with density function

$$D_{H,s,\mathbf{c}}(\mathbf{x}) = \frac{\rho_{H,s,\mathbf{c}}(\mathbf{x})}{s^d}.$$

The center $\mathbf{c}$ is taken to be zero when not explicitly specified.

Given an orthonormal basis (consisting of d vectors in $\mathbb{R}^n$) for H, $D_{H,s,\mathbf{c}}$ can be written as the sum of d orthogonal 1-dimensional Gaussian distributions, each along one of the basis vectors. Therefore sampling from $D_{H,s,\mathbf{c}}$ can be efficiently approximated. For simplicity we will assume that our algorithms can work with infinite-precision real numbers and sample from Gaussians exactly.

The Fourier transform. For a d-dimensional subspace H of $\mathbb{R}^n$, the Fourier transform (over H) of a function $h : H \to \mathbb{C}$ is a function $\hat{h} : H \to \mathbb{C}$, defined as $\hat{h}(\mathbf{w}) = \int_{\mathbf{x} \in H} h(\mathbf{x}) e^{-2\pi i \langle \mathbf{x}, \mathbf{w} \rangle} \, d\mathbf{x}$. It follows directly from the definition that if, for all $\mathbf{x} \in H$, h satisfies $h(\mathbf{x}) \equiv g(\mathbf{x}+\mathbf{v})$ for some $\mathbf{v} \in H$ and some function $g : H \to \mathbb{R}$, then $\hat{h}(\mathbf{w}) = e^{2\pi i \langle \mathbf{v}, \mathbf{w} \rangle} \hat{g}(\mathbf{w})$. The Fourier transform of a Gaussian function (over H, centered at 0) is another Gaussian (also centered at 0); specifically, $\widehat{\rho_{H,s}} = s^d \cdot \rho_{H,1/s}$.

2.4   Gaussian Measures on Lattices

For any countable set S and any function f, define $f(S) = \sum_{x \in S} f(x)$. For a lattice $\Lambda \subset H$ that spans H and for any $\mathbf{x} \in \Lambda$, define

$$D_{\Lambda,s,\mathbf{c}}(\mathbf{x}) = \frac{D_{H,s,\mathbf{c}}(\mathbf{x})}{D_{H,s,\mathbf{c}}(\Lambda)}$$

to be the conditional probability of $\mathbf{x}$ sampled from $D_{H,s,\mathbf{c}}$, given $\mathbf{x} \in \Lambda$.

One fact connecting lattices and the Fourier transform is the Poisson summation formula:

Lemma 2.5. *Let H be a subspace of $\mathbb{R}^n$. For any lattice $\Lambda \subset H$ that spans H and any "well-behaved"[2] function f, $f(\Lambda) = \det(\Lambda^*)\hat{f}(\Lambda^*)$, where $\hat{f}$ is the Fourier transform (over H) of f.*

The smoothing parameter. Micciancio and Regev [17] defined a new lattice parameter related to Gaussian measures, called the *smoothing parameter.* The following is a generalization of their definition to lattices of possibly less than full rank:

Definition 2.1 (Smoothing parameter). *Let H be a subspace of $\mathbb{R}^n$. For a lattice $\Lambda \subset H$ that spans H and positive real $\epsilon > 0$, the smoothing parameter $\eta_\epsilon(\Lambda)$ is defined to be the smallest s such that $\rho_{H,1/s}(\Lambda^* \backslash \{0\}) \leq \epsilon$.*

The name "smoothing parameter" is justified by the following fact (stated formally in Lemma 2.6): if random noise chosen from a Gaussian distribution of radius $\eta_\epsilon(\Lambda)$ is added to a lattice Λ that spans H, the resulting distribution is almost uniform over H.

Lemma 2.6 ([17], Lemma 4.1, generalized to subspaces). *For any subspace H of $\mathbb{R}^n$, lattice $\mathcal{L}(\mathbf{B})$ that spans H, $\mathbf{c} \in H$, and $s \geq \eta_\epsilon(\mathcal{L}(\mathbf{B}))$, we have*

$$\Delta(D_{H,s,\mathbf{c}} \bmod \mathcal{P}(\mathbf{B}), U(\mathcal{P}(\mathbf{B}))) \leq \epsilon/2.$$

Micciancio and Regev also establish relationships between η_ϵ and other standard lattice parameters like λ_n. Here we generalize to lattices of possibly less than full rank:

---

[2] The precise condition is technical, but all functions we consider are well-behaved.

Lemma 2.7 ([17], Lemma 3.3, generalized to subspaces). *For any super-logarithmic function $f(n) = \omega(\log n)$, there exists a negligible function $\epsilon(n)$ such that: for any d-dimensional subspace H of $\mathbb{R}^n$ and lattice Λ that spans H, $\eta_\epsilon(\Lambda) \leq \sqrt{f(n)} \cdot \lambda_d(\Lambda)$.*

Finally, we will need to bound the norm of the convolution of two vectors, where one of the vectors is chosen from a discrete Gaussian distribution.

Lemma 2.8 ([14], Lemma 3.2, generalized to subspaces). *For any d-dimensional subspace H of $\mathbb{R}^n$, lattice Λ that spans H, positive reals $\epsilon \leq 1/3$, $s \geq 2\eta_\epsilon(\Lambda)$ and vectors $\mathbf{c}, \mathbf{x} \in H$,*

$$E_{\mathbf{v} \sim D_{\Lambda,s,\mathbf{c}}} \left[\| (\mathbf{v} - \mathbf{c}) \otimes \mathbf{x} \|^2 \right] \leq s^2 \cdot d \cdot \| \mathbf{x} \|^2.$$

2.5   A New Lemma on Gaussian Distributions over Lattices

In [17] it is shown that, for a full-rank lattice Λ and large enough s, $D_{\Lambda,s,\mathbf{c}}$ behaves very much like $D_{\mathbb{R}^n,s,\mathbf{c}}$, i.e. their moments are similar. In this work, we will need a different fact about $D_{\Lambda,s,\mathbf{c}}$, specifically, a bound on its maximum value over all points in Λ.

In order to prove such a bound, we need a lemma which is implicit in [17]:

Lemma 2.9 ([17]). *Let H be a d-dimensional subspace of $\mathbb{R}^n$, and Λ be a lattice that spans H. For any $s \geq \eta_\epsilon(\Lambda)$ and any $\mathbf{c} \in H$:*

$$s^d \det(\Lambda^*) \cdot (1 - \epsilon) \quad \leq \quad \rho_{H,s,\mathbf{c}}(\Lambda) \quad \leq \quad s^d \det(\Lambda^*) \cdot (1 + \epsilon).$$

Now we are ready to bound the maximum value of $D_{\Lambda,s,\mathbf{c}}(\cdot)$:

Lemma 2.10. *Let H be a d-dimensional subspace of $\mathbb{R}^n$ and let Λ be a lattice that spans H. For any $\epsilon > 0$, $s \geq 2 \cdot \eta_\epsilon(\Lambda)$, $\mathbf{y} \in \Lambda$, and $\mathbf{c} \in H$,*

$$D_{\Lambda,s,\mathbf{c}}(\mathbf{y}) \leq 2^{-d} \cdot \frac{1 + \epsilon}{1 - \epsilon}.$$

Proof. First, observe

$$D_{\Lambda,s,\mathbf{c}}(\mathbf{y}) = \frac{\rho_{H,s,\mathbf{c}}(\mathbf{y})}{\rho_{H,s,\mathbf{c}}(\Lambda)} \leq \frac{1}{s^d \det(\Lambda^*) \cdot (1 - \epsilon)},$$

because $\rho_{H,s,\mathbf{c}}(\mathbf{y}) \leq 1$ and by Lemma 2.9. Now we also have

$$1 \leq \rho_{H,s/2}(\Lambda) \leq (s/2)^d \det(\Lambda^*) \cdot (1 + \epsilon),$$

again by Lemma 2.9 and because $s/2 \geq \eta_\epsilon(\Lambda)$. Combining the inequalities, we get the result. $\square$

3   Worst-Case Problems on Cyclic Lattices

In this section we introduce a variety of worst-case computational problems on cyclic lattices, and exhibit some (worst-case to worst-case) reductions among them. We specify these problems in their *search* versions, rather than as decisional problems. Due to the algebraic nature of cyclic lattices and our hash function, we will find it useful to formulate problems that ask for short lattice vectors *within a specified cyclotomic subspace* of $\mathbb{R}^n$; as a group, we call these *cyclotomic* problems. After defining these problems, we show that certain cyclotomic problems are as hard as the more standard problems on cyclic lattices.

When formulating computational lattice problems it is customary to assume that the input basis contains integer entries (and we do so implicitly in all the problem definitions below). This restriction is without loss of generality, because rational entries can always be multiplied by their least common denominator, which just scales the lattice by some constant.

For generality, the problems below are parameterized by some arbitrary function ζ of the input lattice, and the quality of a solution is measured relative to ζ. Typically, ζ will be some appropriate lattice parameter, e.g. λ_1 or the lattice's smoothing parameter.

3.1   Definitions

Definition 3.1 (SubSIVP). *The* cyclotomic (generalized) short independent vectors problem, $\text{SUBSIVP}_\gamma^\zeta$, *given an n-dimensional full-rank cyclic lattice basis* $\mathbf{B}$ *and an integer polynomial* $\Phi(\alpha) \neq 0 \bmod (\alpha^n - 1)$ *that divides* $\alpha^n - 1$, *asks for a set of* $\dim(H_\Phi)$ *linearly independent (sub)lattice vectors* $\mathbf{S} \subset \mathcal{L}(\mathbf{B}) \cap H_\Phi$ *such that* $\|\mathbf{S}\| \leq \gamma(n) \cdot \zeta(\mathcal{L}(\mathbf{B}) \cap H_\Phi)$.

Definition 3.2 (SubSVP). *The* cyclotomic (generalized) short vector problem, $\text{SUBSVP}_\gamma^\zeta$, *given an n-dimensional full-rank cyclic lattice basis* $\mathbf{B}$ *and an integer polynomial* $\Phi(\alpha) \neq 0 \bmod (\alpha^n - 1)$ *that divides* $\alpha^n - 1$, *asks for a (sub)lattice vector* $\mathbf{c} \in \mathcal{L}(\mathbf{B}) \cap H_\Phi$ *such that* $\|\mathbf{c}\| \leq \gamma(n) \cdot \zeta(\mathcal{L}(\mathbf{B}) \cap H_\Phi)$.

Definition 3.3 (SubIncSVP). *The* cyclotomic incremental (generalized) short vector problem, $\text{SUBINCSVP}_\gamma^\zeta$, *given an n-dimensional full-rank cyclic lattice basis* $\mathbf{B}$, *an integer polynomial* $\Phi(\alpha) \neq 0 \bmod (\alpha^n - 1)$ *that divides to* $\alpha^n - 1$, *and a nonzero (sub)lattice vector* $\mathbf{c} \in \mathcal{L}(\mathbf{B}) \cap H_\Phi$ *such that* $\|\mathbf{c}\| > \gamma(n) \cdot \zeta(\mathcal{L}(\mathbf{B}) \cap H_\Phi)$, *asks for a nonzero (sub)lattice vector* $\|\mathbf{c}'\| \in \mathcal{L}(\mathbf{B}) \cap H_\Phi$ *such that* $\|\mathbf{c}'\| \leq \|\mathbf{c}\|/2$.

Note that Definitions 3.2 and 3.3 are slightly more general than the standard (incremental) shortest vector problems, because their approximation factors are relative to an arbitrary function ζ of the sublattice, rather than λ_1.

The standard well-studied lattice problems (on cyclic lattices) are simply special cases of the above problems. For example, the *shortest vector problem* SVP_γ is simply $\text{SUBSVP}_\gamma^\zeta$ with $\zeta = \lambda_1$ and $\Phi(\alpha) = 1$. The *generalized independent vectors problem* GIVP_γ^ζ, as described by Micciancio, is simply $\text{SUBSIVP}_\gamma^\zeta$ with $\Phi(\alpha) = 1$. The *shortest independent vectors problem* SIVP_γ is GIVP_γ^ζ with $\zeta = \lambda_n$.

3.2 Reductions Among Problems

In this section we give some standard (worst-case to worst-case) reductions among the the cyclotomic problems defined above, and the more standard lattice problems from the literature.

Micciancio coined the term *lattice-preserving* to describe a reduction from problem A to problem B which invokes its B-oracle only on the lattice specified in the instance of problem A. Following in this vein, we define a *sublattice-preserving* reduction between two *cyclotomic* problems to have the property that all calls to the B oracle are on the same cyclic lattice *and* cyclotomic subspace as specified in the problem A instance.

Proposition 3.1. *For any $\zeta, \gamma(n)$, there is a deterministic, polynomial-time sublattice-preserving reduction from $\mathrm{SUBSVP}_\gamma^\zeta$ to $\mathrm{SUBINCSVP}_\gamma^\zeta$.*

Proof. Given an instance $(\mathbf{B}, \Phi(\alpha))$ of $\mathrm{SUBSVP}_\gamma^\zeta$, we will use the following basic strategy: starting from some (possibly very long) nonzero $\mathbf{c} \in \mathcal{L}(\mathbf{B}) \cap H_\Phi$, iteratively reduce the length of $\mathbf{c}$ by invoking the oracle for $\mathrm{SUBINCSVP}_\gamma^\zeta$ on $(\mathbf{B}, \Phi(\alpha), \mathbf{c})$ until the oracle fails, which indicates that $\|\mathbf{c}\| \leq \gamma(n) \cdot \zeta(\mathcal{L}(\mathbf{B}) \cap H_\Phi)$.

It now suffices to show how to find such an initial $\mathbf{c}$ and bound its norm (and hence, the number of iterations). We claim that for some i, $\mathbf{c}(\alpha) = \mathbf{b}_i(\alpha)\Phi(\alpha) \bmod (\alpha^n - 1)$ is nonzero. For suppose not: then by Lemma 2.1, $\Phi \neq 0$ is orthogonal to $\mathbf{b}_i$ for every i, so the space spanned by $\mathbf{B}$ is not full-dimensional, which contradicts the assumption that $\mathbf{B}$ is full-rank.

Now, because $\Phi(\alpha)$ divides $\alpha^n - 1$, it is the product of cyclotomic factors of $\alpha^n - 1$. All such factors are computable in time $\mathrm{poly}(n)$, and there are at most n such factors, so any $\Phi(\alpha)$ has coefficients of length $\mathrm{poly}(n)$. This implies that $\|\mathbf{c}\| \leq 2^{\mathrm{poly}(n)}$, so the number of iterations in the reduction is $\mathrm{poly}(n)$. $\square$

The following lemma will help us reduce problems asking for *many linearly independent vectors* to problems asking for a *single vector*:

Lemma 3.1. *Let $\Phi(\alpha) \in \mathbb{Z}[\alpha]$ equal $(\alpha^n - 1)/\Phi_k(\alpha)$ for some $k \mid n$. Then for any cyclic lattice $\Lambda \subseteq \mathbb{Z}^n$ and any nonzero $\mathbf{c} \in \Lambda \cap H_\Phi$, vectors*

$$\mathbf{c}, \mathrm{rot}(\mathbf{c}), \ldots, \mathrm{rot}^{\deg(\Phi_k)-1}(\mathbf{c})$$

are linearly independent. As a consequence,

$$\lambda_1(\Lambda \cap H_\Phi) = \cdots = \lambda_{\dim(H_\Phi)}(\Lambda \cap H_\Phi).$$

Proof. Because $\mathbf{c} \neq 0$, $\mathbf{c}(\alpha) \in \mathbb{Z}[\alpha]$, and $\Phi(\alpha) \mid \mathbf{c}(\alpha)$, $\mathbf{c}(\alpha)$ is not divisible by $\Phi_k(\alpha)$. Then by Lemma 2.2, the rotations of $\mathbf{c}$ are linearly independent. Now let $\mathbf{c} \in \Lambda \cap H_\Phi$ be such that $\|\mathbf{c}\| = \lambda_1(\Lambda \cap H_\Phi)$. By Lemma 2.4, $\dim(H_\Phi) = \deg(\Phi_k)$. Because $\|\mathrm{rot}^i(\mathbf{c})\| = \|\mathbf{c}\|$ for any i, the result follows. $\square$

Corollary 3.1. *For any $\zeta, \gamma(n)$, there exists a deterministic, polynomial-time sublattice-preserving reduction from $\mathrm{SUBSIVP}_\gamma^\zeta$ instances $(\mathbf{B}, \Phi(\alpha))$ where $\Phi(\alpha) = (\alpha^n - 1)/\Phi_k(\alpha)$ for some $k \mid n$ to $\mathrm{SUBSVP}_\gamma^\zeta$, which makes exactly one oracle call.*

When the dimension n of a cyclic lattice is *prime*, $\alpha^n - 1$ factors as $\Phi_n(\alpha) \cdot \Phi_1(\alpha)$. In this case, there is a very tight connection between SIVP and SVP (in an appropriate subspace):

Proposition 3.2. *For any $\gamma(n)$, there is a deterministic, polynomial-time lattice-preserving reduction from* $\mathrm{SIVP}_{\max(n,2\gamma)}$ *on a cyclic lattice of prime dimension n to* $\mathrm{SUBSVP}_\gamma^{\lambda_1}$. *The reduction makes exactly one oracle call, on an instance for which* $\Phi(\alpha) = \Phi_1(\alpha) = \alpha - 1$.

Proof. The main idea behind the proof is as follows: first, we use the SUBSVP oracle to find a short vector in $\mathcal{L}(\mathbf{B}) \cap H_{\Phi_1}$, then rotate it to yield $n - 1$ linearly independent vectors. For the nth vector, we take the shortest vector in $\mathcal{L}(\mathbf{B}) \cap H_{\Phi_n}$, which can be found efficiently; furthermore, it is an n-approximation to the shortest vector in $\mathcal{L}(\mathbf{B}) \backslash H_{\Phi_1}$.

We now give the full proof. Given an integer lattice basis $\mathbf{B}$ of a cyclic lattice of prime dimension n, invoke the SUBSVP oracle on $(\mathbf{B}, \Phi_1(\alpha))$, yielding a lattice vector $\mathbf{c} \in \mathcal{L}(\mathbf{B}) \cap H_{\Phi_1}$ such that $\|\mathbf{c}\| \leq \gamma(n) \cdot \lambda_1(\mathcal{L}(\mathbf{B}) \cap H_{\Phi_1})$. Looking ahead, the rotations of $\mathbf{c}$ will provide $n - 1$ linearly independent vectors of length $\|\mathbf{c}\|$, however we will need one more vector (outside H_{Φ_1}) to solve SIVP.

Now let $s_i = \sum_{j=1}^n (\mathbf{b}_i)_j = \mathbf{b}_i(1)$ for $i = 1, \ldots n$. Because $\alpha - 1$ cannot divide every $\mathbf{b}_i(\alpha)$ (otherwise $\mathcal{L}(\mathbf{B}) \subset H_{\Phi_1}$, so $\mathcal{L}(\mathbf{B})$ would not be full-rank), some s_i must be non-zero. Let $g = \gcd(s_1, \ldots, s_n) \neq 0$, and let $\mathbf{g} = (g, g, \ldots, g)$. Output the vectors $\mathbf{S} = (\mathbf{c}, \mathrm{rot}(\mathbf{c}), \ldots, \mathrm{rot}^{n-2}(\mathbf{c}), \mathbf{g})$.

To prove correctness of the reduction, we first show that $\mathbf{g} \in \mathcal{L}(\mathbf{B})$. Note that for every i, $\mathbf{s}_i = \mathbf{b}_i \otimes (1,1,\ldots,1) = (s_i, s_i, \ldots, s_i) \in \mathcal{L}(\mathbf{B})$. By the extended Euclidean algorithm, $\mathbf{g}$ is an integer combination of the $\mathbf{s}_i$ vectors, hence $\mathbf{g} \in \mathcal{L}(\mathbf{B})$.

Claim. The vectors in $\mathbf{S}$ are linearly independent.

Proof. Because n is prime, $(\alpha^n - 1)/\Phi_1(\alpha) = \Phi_n(\alpha)$ is irreducible in $\mathbb{Z}[\alpha]$, so by Lemma 3.1 the $n - 1$ rotations of $\mathbf{c}$ in $\mathbf{S}$ are linearly independent. Further, $\mathbf{g} \notin H_{\Phi_1}$ while $\mathrm{rot}^i(\mathbf{c}) \in H_{\Phi_1}$ for every i (Lemma 2.3), so $\mathbf{S}$ consists of n linearly independent vectors from $\mathcal{L}(\mathbf{B})$.   □

We now analyze the approximation factor of the reduction. First, we bound $\lambda_n(\mathcal{L}(\mathbf{B}))$:

Claim.
$$\lambda_n(\mathcal{L}(\mathbf{B})) \geq \max\left(\frac{g}{\sqrt{n}}, \frac{\lambda_1(\mathcal{L}(\mathbf{B}) \cap H_{\Phi_1})}{2} \right).$$

Proof. Let $\mathbf{T}$ be some full-rank set of nonzero vectors in $\mathcal{L}(\mathbf{B})$ such that $\|\mathbf{T}\| = \lambda_n(\mathcal{L}(\mathbf{B}))$. Then $\mathbf{T}$ must contain some $\mathbf{u} \in \mathcal{L}(\mathbf{B}) \backslash H_{\Phi_1}$, because $\dim(H_{\Phi_1}) = n-1$. Let $\mathbf{u} = \sum_{i=1}^n a_i \mathbf{b}_i$ for integers $a_1, \ldots, a_n$. Because $\Phi_1(\alpha)$ does not divide $\mathbf{u}(\alpha)$, $\mathbf{u}(1) = \sum_{j=1}^n \mathbf{u}_j \neq 0$. Further, $\mathbf{u}(1) = \sum_{i=1}^n a_i \mathbf{b}_i(1)$, so g divides $\mathbf{u}(1)$. Therefore $\|\mathbf{u}\|_1 \geq |\mathbf{u}(1)| \geq g$, which implies $\lambda_n(\mathcal{L}(\mathbf{B})) = \|\mathbf{T}\| \geq \|\mathbf{u}\| \geq \|\mathbf{u}\|_1/\sqrt{n} \geq g/\sqrt{n}$.

Furthermore, $\mathbf{T}$ must contain some $\mathbf{v} \in \mathcal{L}(\mathbf{B})\backslash H_{\Phi_n}$, because $\dim(H_{\Phi_n}) = 1$. Now $\mathbf{v}' = \text{rot}(\mathbf{v}) - \mathbf{v}$ is identified with the polynomial $(\alpha - 1)\cdot\mathbf{v}(\alpha) \bmod (\alpha^n - 1)$, so $0 \neq \mathbf{v}' \in \mathcal{L}(\mathbf{B}) \cap H_{\Phi_1}$. Then by the triangle inequality we have

$$\lambda_1(\mathcal{L}(\mathbf{B}) \cap H_{\Phi_1}) \leq \|\mathbf{v}'\| \leq 2\|\mathbf{v}\| \leq 2\|\mathbf{T}\| = 2\lambda_n(\mathcal{L}(\mathbf{B})).$$

Now, $\|\mathbf{S}\| = \max(g\sqrt{n}, \gamma(n) \cdot \lambda_1(\mathcal{L}(\mathbf{B}) \cap H_{\Phi_1}))$. By taking both cases of $\|\mathbf{S}\|$ and invoking Claim 3.2 with each, we get

$$\frac{\|\mathbf{S}\|}{\lambda_n(\mathcal{L}(\mathbf{B}))} \leq \max(n, 2\gamma(n)).$$

We also have, for arbitrary (not necessarily prime) n, a reduction from SVP to SUBSVP:

Proposition 3.3. *For any $\gamma(n)$, there is a deterministic, polynomial-time lattice-preserving reduction from* $\text{SVP}_{\max(n,\gamma)}$ *to* $\text{SUBSVP}_{\gamma}^{\lambda_1}$*. The reduction calls the oracle exactly once, on an instance for which $\Phi(\alpha) = \Phi_1(\alpha) = \alpha - 1$.*

Proof. The reduction and proof of correctness are very similar to the one from the proof of Proposition 3.2: on input $\mathbf{B}$, call the SUBSVP oracle on $(\mathbf{B}, \Phi_1(\alpha))$, yielding a vector $\mathbf{c} \in \mathcal{L}(\mathbf{B}) \cap H_{\Phi_1}$ such that $\|\mathbf{c}\| \leq \gamma(n) \cdot \lambda_1(\mathcal{L}(\mathbf{B}) \cap H_{\Phi_1})$. Additionally, construct the vector $\mathbf{g}$ as above, and output the shorter of $\mathbf{c}$ and $\mathbf{g}$.

Using reasoning as above, we can show that $\lambda_1(\mathcal{L}(\mathbf{B})) \geq \min(g/\sqrt{n}, \lambda_1(\mathcal{L}(\mathbf{B})\cap H_{\Phi_1}))$. Then by considering both cases of $\lambda_1(\mathcal{L}(\mathbf{B}))$, we can show that

$$\frac{\min(\|\mathbf{g}\|, \|\mathbf{c}\|)}{\lambda_1(\mathcal{L}(\mathbf{B}))} \leq \max(n, \gamma(n)).$$

4   Generalized Compact Knapsacks

Definition 4.1 ([14], Definition 4.1). *For any ring R, subset $S \subset R$ and integer $m \geq 1$, the generalized knapsack function family $\mathcal{H}(R, S, m) = \{f_\mathbf{a} : S^m \rightarrow R\}_{\mathbf{a}\in R^m}$ is defined by*

$$f_\mathbf{a}(\mathbf{x}) = \sum_{i=1}^{m} x_i \cdot a_i.$$

In our knapsack function for security parameter n, R is the ring $R = (\mathbb{Z}_p^n, +, \otimes)$ of n-dimensional vectors over $\mathbb{Z}_p$, where $p = n^{O(1)}$ but *need not be prime*, with vector addition and convolution product $\otimes$.

This choice of ring admits very efficient implementations of the knapsack function: using a Fast Fourier Transform algorithm (which works for any n), convolution can be performed in $O(n \log n)$ operations in $\mathbb{Z}_p$, and addition of two vectors takes time $O(n \log p) = O(n \log n)$. Furthermore, by choosing a p such that $\mathbb{Z}_p$ has an element of multiplicative order n, we can compute the Fourier transform mod p using modular (rather than floating-point) arithmetic. The resulting time complexity of the function is $O(m \cdot n \cdot \text{poly}(\log n))$, with key size $O(m \cdot n \log n)$.

4.1   How to Find Collisions

Here we show how to find collisions in the compact knapsack function when $S = [0, D]^n$ for some $D = p^{\Theta(1)}$, for which Micciancio proved that the function was one-way (under suitable assumptions). Our attacks actually do more than just find arbitrary collisions; in fact, they find second preimages for many elements of the domain, thereby violating the definition of universal one-wayness as well. In the following we write $\mathbf{X} \in S^m \subset \mathbb{Z}_p^{n \times m}$ as an element of the domain, and $\mathbf{A} \in R^m = \mathbb{Z}_p^{n \times m}$ as a uniformly-chosen key.

First observe that $f_\mathbf{A}$ is linear: $f_\mathbf{A}(\mathbf{X}) + f_\mathbf{A}(\mathbf{X}') = f_\mathbf{A}(\mathbf{X} + \mathbf{X}')$. Therefore, for any fixed $\mathbf{X}'$ such that $\|\mathbf{X}'\|_\infty < D$ and a random key $\mathbf{A}$, to find a collision with $\mathbf{X}'$ it suffices to find a nonzero $\mathbf{X} \in S^m$ such that $f_\mathbf{A}(\mathbf{X}) = 0$ and $\|\mathbf{X}\|_\infty = 1$. In fact, our attack will be even stronger: we demonstrate a *fixed* $\mathbf{X} \neq 0$, *oblivious* to the key $\mathbf{A}$, for which $f_\mathbf{A}(\mathbf{X}) = 0$ with non-negligible probability (over the choice of $\mathbf{A}$).

We define $\mathbf{X}$ by its representation as an m-tuple of polynomials in the ring $\mathbb{Z}_p[\alpha]/(\alpha^n - 1)$. In this polynomial representation, $f_\mathbf{A}(\mathbf{X})$ corresponds to $\sum_{i=1}^m \mathbf{x}_i(\alpha) \cdot \mathbf{a}_i(\alpha) \bmod (\alpha^n - 1)$. For any small positive integer divisor q of n (including $q = 1$), we can define $\mathbf{X} = (\mathbf{x}_1, \ldots, \mathbf{x}_m)$ as follows: let

$$\mathbf{x}_1(\alpha) = \frac{\alpha^n - 1}{\alpha^q - 1} = \alpha^{n-q} + \alpha^{n-2q} + \cdots + 1,$$

and let $\mathbf{x}_j(\alpha) = 0$ for all $j \neq 1$. Then $\mathbf{X} \in S^m$, $\|\mathbf{X}\|_\infty = 1$, and $f_\mathbf{A}(\mathbf{X})$ corresponds to $\mathbf{a}_1(\alpha) \cdot \mathbf{x}_1(\alpha)$. Now suppose $\mathbf{a}_1(\alpha)$ is divisible by $\alpha^q - 1$, which happens with probability $1/p^q$ over the uniform choice of $\mathbf{A}$. Then $f_\mathbf{A}(\mathbf{X}) = 0$ because $(\alpha^n - 1)$ divides $\mathbf{a}_1(\alpha) \cdots \mathbf{x}_1(\alpha)$.

4.2   How to Achieve Collision-Resistance

The essential fact enabling the above attack is that $(\alpha^n - 1)$ is not *irreducible* in $\mathbb{Z}_p[\alpha]$, so $\mathbb{Z}_p[\alpha]/(\alpha^n - 1)$ is not an *integral domain*. That is, for many non-zero $\mathbf{a}(\alpha)$, it is easy to find non-zero $\mathbf{x}(\alpha)$ (having small coefficients) such that $\mathbf{a}(\alpha) \cdot \mathbf{x}(\alpha) = 0 \bmod (\alpha^n - 1)$. In particular, when we examine $\mathbf{a}(\alpha), \mathbf{x}(\alpha) \bmod (\alpha^n - 1)$ in their Chinese remainder representations, each of the components is zero for either $\mathbf{a}(\alpha)$ or $\mathbf{x}(\alpha)$ (or both).

To circumvent our particular attack, we can enforce an *algebraic constraint* on $\mathbf{X}$. Informally, we require every $\mathbf{x}_i(\alpha)$ to be divisible *over* $\mathbb{Z}[\alpha]$ by $\frac{\alpha^n - 1}{\Phi_k(\alpha)}$ for some fixed $k \mid n$. Then in the Chinese remainder representation, all but one component of $\mathbf{x}_i(\alpha)$ is zero, so the evaluation of $f_\mathbf{A}(\mathbf{X})$ is essentially performed mod $\Phi_k(\alpha)$.

Note that while $\Phi_k(\alpha)$ is irreducible over $\mathbb{Z}[\alpha]$, it *may* still be reducible over $\mathbb{Z}_p[\alpha]$. Therefore constraining $\mathbf{X}$ in the above way may *not* necessarily place the calculation of $f_\mathbf{A}(\mathbf{X})$ in an integral domain. Furthermore, the constraint is crafted specifically to prevent our attack, but not to prevent any other potential attacks on the function that may remain undiscovered. Nevertheless (and perhaps quite surprisingly), it proves to be exactly what is needed to attain collision-resistance, as our security reduction will demonstrate.

Formally, we consider the generalized compact knapsack function where the set $S = S_{D,\Phi} \subset \mathbb{Z}_p^n$ for some bound D on the max-norm of $\mathbf{X}$ (recall that $\|\mathbf{x}\|_\infty \in [0, p/2]$ for any $\mathbf{x} \in \mathbb{Z}_p^n$), and $\Phi(\alpha) = \frac{\alpha^n - 1}{\Phi_k(\alpha)}$ for some $k \mid n$. For a value $v \in \mathbb{Z}_p$, define $v_\mathbb{Z}$ to be the unique integer in the range $(-p/2, p/2]$ representing v as a residue, and for a vector $\mathbf{x} \in \mathbb{Z}_p^n$ define the vector $\mathbf{x}_\mathbb{Z} \in \mathbb{Z}^n$ similarly. Now we define $S_{D,\Phi}$ as:

$$S_{D,\Phi} = \{\mathbf{x} \in \mathbb{Z}_p^n : \|\mathbf{x}\|_\infty \leq D \text{ and } \Phi(\alpha) \text{ divides } \mathbf{x}_\mathbb{Z}(\alpha) \text{ in } \mathbb{Z}[\alpha]\}. \qquad (1)$$

4.3   How to Get a (Useful) Hash Function

In order to verify that our knapsack is a *hash* function, we must compare the size of the domain $S_{D,\Phi}^m$ to the size of the function's range. In addition, practical usage requires efficient one-to-one encodings of *bit strings* into elements of the domain, and of range elements back to bit strings.

Both tasks are most easily done when n is prime and $\Phi(\alpha) = \alpha - 1$. Given a string $w \in \{0,1\}^\ell$, where $\ell = m \cdot (n-1) \cdot \lfloor \log D \rfloor$, encode w in the following way: first, break w into m chunks representing vectors $\mathbf{w}_i \in [0, D-1]^{n-1}$ for $i = 1, \ldots, m$. For each i, and for $j = 0, \ldots, n-2$, let $(\mathbf{x}_i)_j = \pm(\mathbf{w}_i)_j$, where the signs are iteratively chosen to satisfy the invariant that every partial sum $\sum_{k=0}^{j}(x_i)_k \in [-D, D]$. Finally, for every i let $(\mathbf{x}_i)_{n-1} = -\sum_{j=0}^{n-2}(x_i)_j \in [-D, D]$, so that $\mathbf{x}_i(1) = \sum_{j=0}^{n-1}(\mathbf{x}_i)_j = 0$, hence $\alpha - 1$ divides $\mathbf{x}_i(\alpha)$ and $\|\mathbf{x}_i\|_\infty \leq D$.

To encode the output, first notice that $\alpha - 1$ divides $\mathbf{y}(\alpha)$, where $\mathbf{y} = f_\mathbf{A}(\mathbf{X})$. Therefore it is sufficient to write $(\mathbf{y})_j$ in binary for $j = 0, \ldots, n-2$. This can be done using $(n-1) \cdot \lceil \log p \rceil$ bits. Therefore, the function shrinks its input by a factor of $\frac{m \lfloor \log D \rfloor}{\lceil \log p \rceil}$, which for appropriate choices of parameters is larger than 1.

5   The Main Reduction

Due to the reductions among worst-case problems on cyclic lattices explored in Section 3.2, the security of our hash function can be established by reducing the worst-case problem $\text{SUBINCSVP}_\gamma^{\eta_\epsilon}$ to finding collisions in $\mathcal{H}(\mathbb{Z}_p^n, S_{D,\Phi}, m)$. Because collision-resistance is meaningful even for functions that do not shrink their input, we exhibit a general reduction in Theorem 5.1, then consider special cases of hash functions in the corollaries that follow.

Theorem 5.1. *For any polynomially-bounded functions $D(n)$, $m(n)$, $p(n)$ and negligible function $\epsilon(n)$ such that $p(n) \geq 8n^{2.5} \cdot m(n)D(n)$ and $\gamma(n) \geq 16n \cdot m(n)D(n)$, there is a probabilistic polynomial-time reduction from $\text{SUBINCSVP}_\gamma^{\eta_\epsilon}$ instances $(\mathbf{B}, \Phi(\alpha), \mathbf{c})$ where $\frac{\alpha^n - 1}{\Phi(\alpha)} = \Phi_k(\alpha)$ for some $k \mid n$ to finding collisions in $\mathcal{H}(\mathbb{Z}_{p(n)}^n, S_{D(n),\Phi}, m(n))$.*

Roadmap to the proof. First we describe a reduction that, given a collision-finding oracle $\mathcal{F}$, attempts to solve SUBINCSVP. The remainder of the proof is a series of claims that establish the correctness of the reduction. Claim 5 shows that the reduction feeds $\mathcal{F}$ a properly-distributed input. Claim 5 establishes that the reduction's output vector is in the proper sublattice. Claims 5 and 5 show that, with good likelihood, the output is both nonzero and significantly shorter than the input lattice vector (respectively).

Proof. Assume that $\mathcal{F}$ finds collisions in the specified hash family, for infinitely many n and $\Phi(\alpha)$, with probability at least $1/q(n)$ for some polynomial $q(\cdot)$. For shorthand, we will abbreviate $H = H_\Phi$ and let $d = \dim(H)$ throughout the proof. We assume wlog that $d \geq 3$, because efficient algorithms are known for SVP when $d = 1, 2$ (we omit details).

Our reduction proceeds as follows: on input $(\mathbf{B}, \mathbf{c})$ where $\mathbf{c} \in \mathcal{L}(\mathbf{B}) \cap H$,

1. For $i = 1$ to m,
    - Generate uniform $\mathbf{v}_i \in \mathcal{L}(\mathbf{B}) \cap H \cap \mathcal{P}(\mathrm{Rot}^d(\mathbf{c}))$. (See [16] for algorithms.)
    - Generate noise $\mathbf{y}_i \in H$ according to $D_{H,s}$ for $s = 2\|\mathbf{c}\|/\gamma(n)$. Let $\mathbf{y}_i' = \mathbf{y}_i \bmod \mathbf{B}$.
    - Choose $\mathbf{b}_i$ (as described below) so that $\mathrm{Rot}^n(\mathbf{c}) \cdot \mathbf{b} = \mathbf{v}_i + \mathbf{y}_i'$, and let $\mathbf{a}_i = \lfloor \mathbf{b}_i \cdot p \rceil$.
      Choosing $\mathbf{b}_i$ is done by breaking it into two parts: $\mathbf{b}_i^1 = ((\mathbf{b}_i)_0, \ldots, (\mathbf{b}_i)_{d-1})^T$, and $\mathbf{b}_i^2 = ((\mathbf{b}_i)_d, \ldots, (\mathbf{b}_i)_{n-1})^T$. First, pick $\mathbf{b}_i^2$ according to $I^{n-d} = U([0,1))^{n-d}$. Then solve for $\mathbf{b}_i^1$ as follows: let $\mathbf{G} \in \mathbb{R}^{d \times n}$ be such that $\mathbf{G} \cdot \mathrm{Rot}^d(\mathbf{c}) = \mathbf{I}_d$, the $d \times d$ identity matrix. (Such a $\mathbf{G}$ exists because $\mathrm{Rot}^d(\mathbf{c})$ has column rank d, and it can be found via Gaussian elimination.) Then $\mathbf{b}_i^1 = \mathbf{G} \cdot (\mathbf{v}_i + \mathbf{y}_i' - \mathbf{w}_i)$, where $\mathbf{w}_i = \mathrm{Rot}^n(\mathbf{c}) \cdot (0, \ldots, 0, (\mathbf{b}_i)_d, \ldots, (\mathbf{b}_i)_{n-1})^T$.
2. Give $\mathbf{A} = (\mathbf{a}_1 \bmod p, \ldots, \mathbf{a}_m \bmod p)$ to the collision-finding oracle $\mathcal{F}$. Get a collision $\mathbf{X} \neq \mathbf{X}'$ such that $\|\mathbf{X}\|_\infty, \|\mathbf{X}'\|_\infty \leq D$, and $\Phi(\alpha)$ divides every $\mathbf{x}_i(\alpha), \mathbf{x}_i'(\alpha)$. Let $\mathbf{Z} = \mathbf{X} - \mathbf{X}'$, and note that $\|\mathbf{Z}\|_\infty \leq 2D$ and $\Phi(\alpha)$ divides every $\mathbf{z}_i(\alpha)$.
3. Output the vector

$$\mathbf{c}' = \sum_{i=1}^m (\mathbf{v}_i + \mathbf{y}_i' - \mathbf{y}_i) \otimes \mathbf{z}_i - \mathbf{c} \otimes \frac{\sum_{i=1}^m \mathbf{a}_i \otimes \mathbf{z}_i}{p} \tag{2}$$

$$= \sum_{i=1}^m (\mathbf{v}_i + \mathbf{y}_i' - \mathbf{y}_i - \frac{\mathbf{c} \otimes \mathbf{a}_i}{p}) \otimes \mathbf{z}_i. \tag{3}$$

The following claim follows from Lemma 2.6 and straightforward manipulations of statistical distance:

Claim. The probability that $\mathcal{F}$ outputs a valid collision is non-negligible:

$$\Pr[(\mathbf{X}, \mathbf{X}') \text{ is a valid collision}] \geq 1/q(n) - m(n) \cdot \epsilon(n)/2.$$

Proof. It suffices to bound the statistical distance $\Delta(\mathbf{A}, U(\mathbb{Z}_p^{nm}))$ by $m\epsilon/2$. Each $\mathbf{a}_i$ is independently generated, so by the triangle inequality, $\Delta(\mathbf{A}, U(\mathbb{Z}_p^{nm})) \leq m \cdot \Delta(\mathbf{a}_i \bmod p, U(\mathbb{Z}_p^n))$. Now $\mathbf{a}_i \bmod p = \lfloor (\mathbf{b}_i \bmod 1) \cdot p \rceil$, so $\Delta(\mathbf{a}_i \bmod p, U(\mathbb{Z}_p^n)) \leq \Delta(\mathbf{b}_i \bmod 1, I^n)$.

Let $\mathbf{b}_i^1 = ((\mathbf{b}_i)_0, \ldots, (\mathbf{b}_i)_{d-1})^T$, and $\mathbf{b}_i^2 = ((\mathbf{b}_i)_d, \ldots, (\mathbf{b}_i)_{n-1})^T$. By construction, $\mathbf{b}_i^2$ is uniform over $[0,1)^{n-d}$. Additionally, we have

$$\mathbf{b}_i^1 = \mathbf{G} \cdot (\mathbf{v}_i + \mathbf{y}_i' - \mathbf{w}_i) = \mathbf{G} \cdot (\mathbf{v}_i + \mathbf{y}_i') - \mathbf{G} \cdot \mathbf{w}_i, \tag{4}$$

where $\mathbf{w}_i$ is a function of $\mathbf{b}_i^2$. Notice that $\mathbf{y}_i'$ is distributed according to $D_{H,s} \bmod \mathcal{P}(\mathbf{B})$, so by Lemma 2.6,

$$\Delta(\mathbf{y}_i', U(\mathcal{P}(\mathbf{B}))) \leq \epsilon/2.$$

Because $\mathbf{v}_i$ is uniform over $\mathcal{L}(\mathbf{B}) \cap H \cap \mathcal{P}(\mathrm{Rot}^d(\mathbf{c}))$, we get

$$\Delta(\mathbf{v}_i + \mathbf{y}_i' \bmod \mathrm{Rot}^d(\mathbf{c}), U(\mathcal{P}(\mathrm{Rot}^d(\mathbf{c})))) \leq \epsilon/2,$$

which by definition of $\mathbf{G}$ implies

$$\Delta(\mathbf{G} \cdot (\mathbf{v}_i + \mathbf{y}_i') \bmod 1, I^d) \leq \epsilon/2.$$

By Equation (4), we have that conditioned on *any* value $\mathbf{v} \in [0,1)^{n-d}$,

$$\Delta(\{\mathbf{b}_i^1 \bmod 1 \mid \mathbf{b}_i^2 = \mathbf{v}\}, I^d) \leq \epsilon/2.$$

Using standard manipulations of statistical distance, we conclude that $\Delta(\mathbf{b}_i \bmod 1, I^n) \leq \epsilon/2$, as desired. $\qquad\square$

Claim. If $\mathcal{F}$ outputs a valid collision, $\mathbf{c}' \in \mathcal{L}(\mathbf{B}) \cap H$.

Proof. First observe that $\mathcal{L}(\mathbf{B}) \cap H$ is a sublattice of $\mathcal{L}(\mathbf{B})$. We now examine the terms in Equation (2). By construction, $\mathbf{v}_i + \mathbf{y}_i' - \mathbf{y}_i \in \mathcal{L}(\mathbf{B}) \cap H$, and $\mathbf{z}_i \in \mathbb{Z}^n$, so the first summation is in $\mathcal{L}(\mathbf{B}) \cap H$. Next, $f_{\mathbf{A}}(\mathbf{Z}) = \sum_i \mathbf{a}_i \otimes \mathbf{z}_i = 0 \bmod p$ by the assumption that $\mathcal{F}$ outputs a valid collision, so $\frac{\sum_i \mathbf{a}_i \otimes \mathbf{z}_i}{p} \in \mathbb{Z}^n$. Since $\mathbf{c} \in \mathcal{L}(\mathbf{B}) \cap H$, the second term of Equation (2) is also in $\mathcal{L}(\mathbf{B}) \cap H$. $\qquad\square$

Claim. Conditioned on $\mathcal{F}$ outputting a collision, $\Pr[\mathbf{c}' \neq 0] \geq 3/4$.

Proof. The main idea: because $\mathbf{c}' \in H$, $\mathbf{c}' = 0$ iff $\varPhi_k(\alpha)$ divides $\mathbf{c}'(\alpha)$. Because $\varPhi_k(\alpha)$ is irreducible, we can show that $\mathbf{c}'(\alpha) = 0 \bmod \varPhi_k(\alpha)$ only when a sample from $D_{\mathcal{L}(\mathbf{B}) \cap H, s, -\mathbf{y}_1'}$ hits a certain target lattice point exactly. By Lemma 2.10, the probability of this event is small.

Throughout the proof we implicitly condition all probabilities on the event that $\mathcal{F}$ outputs a collision. Because $\varPhi(\alpha)$ divides $\mathbf{c}'(\alpha)$ and $\varPhi(\alpha) \cdot \varPhi_k(\alpha) = (\alpha^n - 1)$, by Equation (3) we get

$$\mathbf{c}' = 0 \iff \sum_{i=1}^m \left(\mathbf{v}_i(\alpha) + \mathbf{y}_i'(\alpha) - \mathbf{y}_i(\alpha) + \frac{\mathbf{c}(\alpha)\mathbf{a}_i(\alpha)}{p} \right) \cdot \mathbf{z}_i(\alpha) = 0 \bmod \varPhi_k(\alpha).$$

Since $\mathbf{Z} \neq 0$, there exists i such that $\mathbf{z}_i \neq 0$; assume without loss of generality that $i = 1$. Then let $\mathbf{h}(\alpha) = \sum_{i>1} (\mathbf{v}_i(\alpha) + \mathbf{y}_i'(\alpha) - \mathbf{y}_i(\alpha) + \frac{\mathbf{c}(\alpha) \cdot \mathbf{a}_i(\alpha)}{p}) \cdot \mathbf{z}_i(\alpha)$ and rearrange terms, yielding

$$\left(\mathbf{v}_1(\alpha) + \mathbf{y}_1'(\alpha) - \mathbf{y}_1(\alpha) + \frac{\mathbf{c}(\alpha) \cdot \mathbf{a}_1(\alpha)}{p} \right) \cdot \mathbf{z}_1(\alpha) = -\mathbf{h}(\alpha) \bmod \varPhi_k(\alpha). \tag{5}$$

Now because $\mathbf{z}_1 \neq 0$ and $\Phi(\alpha)$ divides $\mathbf{z}_1(\alpha)$, it must be that $\mathbf{z}_1(\alpha) \neq 0 \bmod \Phi_k(\alpha)$. Since $\mathbb{Z}[\alpha]/\Phi_k(\alpha)$ is an integral domain, there exists at most one element $\mathbf{w}(\alpha) \in \mathbb{Z}[\alpha]/\Phi_k(\alpha)$ such that $\mathbf{w}(\alpha) \cdot \mathbf{z}_1(\alpha) = -\mathbf{h}(\alpha) \bmod \Phi_k(\alpha)$. If no such $\mathbf{w}(\alpha)$ exists, then $\mathbf{c}' \neq 0$ always, and we're done. If such a $\mathbf{w}(\alpha)$ exists, then $\mathbf{c}' = 0$ only when the multiplicand of $\mathbf{z}_1(\alpha)$ in Equation (5) equals $\mathbf{w}(\alpha)$. Then $\mathbf{c}' = 0$ only if:

$$(\mathbf{y}_1' - \mathbf{y}_1)(\alpha) = \mathbf{w}(\alpha) - \frac{\mathbf{c}(\alpha) \cdot \mathbf{a}_1(\alpha)}{p} - \mathbf{v}_1(\alpha) \bmod \Phi_k(\alpha).$$

Now, $\mathbf{y}_1$ is independent of $\mathbf{v}_1$ and the coins of $\mathcal{F}$. Furthermore, conditioned on $\mathbf{y}_1'$, $\mathbf{y}_1$ is independent of $\mathbf{h}, \mathbf{z}_1$, and $\mathbf{a}_1$, because these variables depend only on $\mathbf{y}_1'$ and other independent coins. Therefore by averaging over these variables, it suffices to bound

$$M = \max_{\mathbf{h}'(\alpha)} \Pr\left[(\mathbf{y}_1' - \mathbf{y}_1)(\alpha) = \mathbf{h}'(\alpha) \bmod \Phi_k(\alpha) \mid \mathbf{y}_1'\right].$$

Because $\Phi(\alpha)$ divides $(\mathbf{y}_1' - \mathbf{y}_1)(\alpha)$,

$$M = \max_{\mathbf{h}'(\alpha)} \Pr\left[(\mathbf{y}_1' - \mathbf{y}_1)(\alpha) = \mathbf{h}'(\alpha) \bmod (\alpha^n - 1) \mid \mathbf{y}_1'\right].$$

Now given $\mathbf{y}_1'$, $\mathbf{y}_1 - \mathbf{y}_1'$ is distributed according to $D_{\mathcal{L}(\mathbf{B}) \cap H_\Phi, s, -\mathbf{y}_1'}$ because $\mathbf{y}_1 - \mathbf{y}_1' \mathcal{L}(\mathbf{B}) \cap H_\Phi$. By Lemma 2.10 and because $d \geq 3$,

$$M \leq 2^{-d} \cdot \frac{1+\epsilon}{1-\epsilon} \leq 1/4$$

for sufficiently large n. $\qquad\square$

Claim. Conditioned on $\mathcal{F}$ outputting a collision, $\Pr\left[\|\mathbf{c}'\| \leq \frac{\|\mathbf{c}\|}{2}\right] \geq 1/2$.

Proof. Throughout the proof we implicitly condition all probabilities on the event that $\mathcal{F}$ outputs a collision. First, it is sufficient to establish the bound $E[\|\mathbf{c}'\|] \leq \frac{\|\mathbf{c}\|}{4}$, because by Markov's inequality, this implies $\Pr\left[\|\mathbf{c}'\| > \frac{\|\mathbf{c}\|}{2}\right] \leq 1/2$. Now by Equation (2) and the triangle inequality,

$$\|\mathbf{c}'\| \leq \sum_{i=1}^{m} \left\|(\mathbf{v}_i + \mathbf{y}_i' - \frac{\mathbf{c} \otimes \mathbf{a}_i}{p}) \otimes \mathbf{z}_i\right\| + \sum_{i=1}^{m} \|\mathbf{y}_i \otimes \mathbf{z}_i\|. \qquad (6)$$

Now using the fact that $\mathrm{Rot}^n(\mathbf{c}) \cdot \mathbf{b}_i = \mathbf{v}_i + \mathbf{y}_i'$, we get

$$\mathbf{v}_i + \mathbf{y}_i' - \frac{\mathbf{c} \otimes \mathbf{a}_i}{p} = \frac{\mathrm{Rot}^n(\mathbf{c}) \cdot \mathbf{b}_i \cdot p - \mathrm{Rot}^n(\mathbf{c}) \cdot \mathbf{a}_i}{p} = \frac{\mathrm{Rot}^n(\mathbf{c})(\mathbf{b}_i \cdot p - \mathbf{a}_i)}{p}.$$

Since $\|\mathbf{b}_i \cdot p - \mathbf{a}_i\|_\infty \leq 1/2$, we get

$$\left\|\mathbf{v}_i + \mathbf{y}_i' - \frac{\mathbf{c} \otimes \mathbf{a}_i}{p}\right\|_\infty \leq \frac{n\|\mathbf{c}\|}{2p}.$$

Now we use the fact that $\|\mathbf{z}_i\|_1 \le 2n \cdot D$, yielding

$$\left\| (\mathbf{v}_i + \mathbf{y}'_i - \frac{\mathbf{c} \otimes \mathbf{a}_i}{p}) \otimes \mathbf{z}_i \right\|_\infty \le \left\| \mathbf{v}_i + \mathbf{y}'_i - \frac{\mathbf{c} \otimes \mathbf{a}_i}{p} \right\|_\infty \cdot \|\mathbf{z}_i\|_1 \le \frac{n^2 \|\mathbf{c}\| D}{p}.$$

Finally, using the fact that $\|\mathbf{w}\| \le \sqrt{n}\|\mathbf{w}\|_\infty$ for any n-dimensional vector $\mathbf{w}$ and summing over $i = 1, \ldots, m$, we get that the first summation in Equation (6) is at most $\frac{mn^{2.5}\|\mathbf{c}\| D}{p}$.

Next we analyze the second term of Equation (6). Conditioned on $\mathbf{y}'_i$, the distribution of $\mathbf{y}_i - \mathbf{y}'_i \in \mathcal{L}(\mathbf{B}) \cap H$ is $D_{\mathcal{L}(\mathbf{B}) \cap H, s, -\mathbf{y}'_i}$, and is independent of $\mathbf{A}, \mathbf{Z}$, and the coins of $\mathcal{F}$. Recall that $s = 2\|\mathbf{c}\|/\gamma(n) > 2\eta_\epsilon(\mathcal{L}(\mathbf{B}) \cap H)$, by assumption on the input to SUBINCSVP. Also recall that $\mathbf{y}_i$ is chosen according to $D_{H,s}$, and that $\mathbf{z}_i \in H$. So by Lemma 2.8,

$$E\left[\|\mathbf{y}_i \otimes \mathbf{z}_i\|^2 \mid \mathbf{y}'_i\right] = E_{(\mathbf{y}_i - \mathbf{y}'_i) \leftarrow D_{\mathcal{L}(\mathbf{B}) \cap H, s, -\mathbf{y}'_i}} \left[\|((\mathbf{y}_i - \mathbf{y}'_i) - (-\mathbf{y}'_i)) \otimes \mathbf{z}_i\|^2\right]$$
$$\le s^2 \|\mathbf{z}_i\|^2 \cdot d$$
$$\le s^2 n^2 D^2.$$

Because $\mathrm{Var}[X] = E[X^2] - E[X]^2 \ge 0$ for any random variable X, it must be that $E\left[\|\mathbf{y}_i \otimes \mathbf{z}_i\| \mid \mathbf{y}'_i\right] \le n \cdot s \cdot D$. Adding up and averaging over all $\mathbf{y}'_i$, we get

$$\sum_{i=1}^m E\left[\|\mathbf{y}_i \otimes \mathbf{z}_i\|\right] \le m \cdot n \cdot s \cdot D = \frac{2m \cdot n \cdot \|\mathbf{c}\| \cdot D}{\gamma(n)}.$$

Combining everything, we get:

$$E[\|\mathbf{c}'\|] \le \frac{m \cdot n^{2.5} \cdot \|\mathbf{c}\| \cdot D}{p} + \frac{2m \cdot n \cdot \|\mathbf{c}\| \cdot D}{\gamma(n)}$$
$$= \|\mathbf{c}\| \cdot \left(\frac{m \cdot n^{2.5} \cdot D}{p} + \frac{2m \cdot n \cdot D}{\gamma(n)} \right).$$

Using the hypotheses $p \ge 8mn^{2.5}D$ and $\gamma(n) \ge 16mnD$, we get $E[\|\mathbf{c}'\|] \le \|\mathbf{c}\|/4$, as desired. $\square$

Then by the two claims and the union bound, we get that (conditioned on $\mathcal{F}$ producing a collision) the probability that $\mathbf{c}'$ is a solution to the SUBINCSVP instance is at least $1/4$. By Claim 5, the reduction solves SUBINCSVP in the worst case with non-negligible probability, which can be amplified to high probability by standard repetition techniques. This completes the proof. $\square$

Putting it all together. Using the relationship between η_ϵ and λ_{n-1}, restricting n to be prime, and setting the knapsack parameters appropriately, we get collision-resistant hash functions:

Corollary 5.1. *For any $m(n) = \Theta(\log n)$, there exist $D(n) = \Theta(1)$ and $p(n) = n^{2.5 + \Theta(1)}$ such that: $\mathcal{H}(\mathbb{Z}_{p(n)}^n, S_{D(n), \Phi_1(\alpha)}, m(n))$ is a hash function ensemble for which finding collisions for infinitely many prime n is at least as hard as solving SVP_γ with high probability in the worst case for infinitely many prime n within a factor $\gamma(n) = n \cdot \mathrm{poly}(\log n)$.*

Proof. We can choose $D(n)$ and $p(n)$ such that $\frac{m(n)\log D(n)}{\log p(n)} = \Theta(1)$ is greater than 1 (yielding a hash function) and satisfying the hypothesis of Theorem 5.1. Because n is prime, $(\alpha^n - 1)/\Phi_n(\alpha) = \Phi_1(\alpha)$, so by Theorem 5.1 and Lemma 3.1 we have an algorithm for $\text{SUBSVP}^{\eta_\epsilon(n)}_{\Theta(n\log n)}$ in H_{Φ_1}. By Lemma 2.7, this is an algorithm for $\text{SUBSVP}^{\lambda_{n-1}}_{n\cdot\text{poly}(\log n)}$ in H_{Φ_1}. Again because n is prime, by Lemma 3.1 we have $\lambda_{n-1} = \lambda_1$ on $\mathcal{L}(\mathbf{B}) \cap H_{\Phi_1}$, so (finally) by Proposition 3.3 we get an algorithm for $\text{SVP}_{n\cdot\text{poly}(\log n)}$. $\qquad\square$

Corollary 5.2. *For any constant $\delta > 0$, there exist $D(n) = n^{\Theta(1)}$, $p(n) = n^{2.5+\Theta(1)}$, and $m(n) = \Theta(1)$ such that: $\mathcal{H}(\mathbb{Z}^n_{p(n)}, S_{D(n),\Phi_1(\alpha)}, m(n))$ is a hash function ensemble for which finding collisions for infinitely many prime n is at least as hard as solving SVP_γ with high probability in the worst case for infinitely many prime n within a factor $\gamma(n) = n^{1+\delta}$.*

Proof. We can choose $D(n) = \Theta(n^{\delta/2})$ and a large enough $m(n) = \Theta(1)$ so that $\frac{m(n)\log D(n)}{\log p(n)} > 1$. The chain of reductions is the same as in the proof of Corollary 5.1, yielding an SVP algorithm with approximation factor $n \cdot m(n) \cdot D(n) \cdot \text{poly}(\log n) \le n^{1+\delta}$. $\qquad\square$

Acknowledgements

We thank the anonymous reviewers for their helpful and thorough comments, and especially for a simplified proof of Lemma 2.10.

References

1. M. Ajtai. Generating hard instances of lattice problems (extended abstract). In *Proc. 28th Annual ACM Symposium on Theory of Computing (STOC 1996)*, pages 99–108, 1996.
2. M. Ajtai. The shortest vector problem in L_2 is NP-hard for randomized reductions (extended abstract). In *Proc. 30th Annual ACM Symposium on Theory of Computing (STOC 1998)*, pages 10–19, 1998.
3. M. Ajtai and C. Dwork. A public-key cryptosystem with worst-case/average-case equivalence. In *Proc. 29th Annual ACM Symposium on Theory of Computing (STOC 1997)*, pages 284–293, 1997.
4. S. Arora, L. Babai, J. Stern, and Z. Sweedyk. The hardness of approximate optima in lattices, codes, and systems of linear equations. *J. Computer and System Sciences*, 54(2):317–331, 1997.
5. J.-Y. Cai and A. Nerurkar. Approximating the SVP to within a factor $(1+1/\dim^\epsilon)$ is NP-hard under randomized reductions. *Jounal of Computer and System Sciences*, 59(2):221–239, 1999.
6. J.-Y. Cai and A. P. Nerurkar. An improved worst-case to average-case connection for lattice problems. In *Proc. 38th Annual Symposium on Foundations of Computer Science (FOCS 1997)*, page 468, 1997.
7. I. Dinur, G. Kindler, and S. Safra. Approximating-CVP to within almost-polynomial factors is NP-hard. In *Proc. 39th Annual Symposium on Foundations of Computer Science (FOCS 1998)*, pages 99–111. IEEE Computer Society, 1998.

8. D. S. Dummit and R. M. Foote. *Abstract Algebra*. Prentice Hall, Upper Saddle River, NJ, USA, second edition, 1999.
9. R. Genarro, Y. Gertner, J. Katz, and L. Trevisan. Bounds on the efficiency of generic cryptographic constructions. *SIAM J. Computing*, 35(1):217–246, 2005.
10. O. Goldreich, S. Goldwasser, and S. Halevi. Collision-free hashing from lattice problems. Electronic Colloquium on Computational Complexity (ECCC) Report TR96-042, 1996.
11. O. Goldreich, S. Goldwasser, and S. Halevi. Public-key cryptosystems from lattice reduction problems. In *Proc. 17th Annual Conference on Advances in Cryptology (CRYPTO 1997)*, pages 112–131. Springer-Verlag, 1997.
12. S. Khot. Hardness of approximating the shortest vector problem in lattices. In *Proc. 45th Symposium on Foundations of Computer Science (FOCS 2004)*, pages 126–135. IEEE Computer Society, 2004.
13. V. Lyubashevsky and D. Micciancio. Generalized compact knapsacks are collision resistant. Electronic Colloquium on Computational Complexity (ECCC) Report TR05-142, 2005.
14. D. Micciancio. Generalized compact knapsaks, cyclic lattices, and efficient one-way functions from worst-case complexity assumptions. In *Proc. 43rd Annual Symposium on Foundations of Computer Science (FOCS 2002)*.
15. D. Micciancio. The shortest vector problem is NP-hard to approximate to within some constant. *SIAM J. Computing*, 30(6):2008–2035, Mar. 2001.
16. D. Micciancio and S. Goldwasser. *Complexity of Lattice Problems: a cryptographic perspective*, volume 671 of *The Kluwer International Series in Engineering and Computer Science*. Kluwer Academic Publishers, Boston, Massachusetts, 2002.
17. D. Micciancio and O. Regev. Worst-case to average-case reductions based on Gaussian measure. pages 371–381.
18. O. Regev. New lattice-based cryptographic constructions. *J. ACM*, 51(6):899–942, 2004.
19. O. Regev. On lattices, learning with errors, random linear codes, and cryptography. In *Proc. 37th Annual ACM Symposium on Theory of Computing (STOC 2005)*, pages 84–93, 2005.
20. P. van Emde Boas. Another NP-complete problem and the complexity of computing short vectors in a lattice. Technical Report 81-04, University of Amsterdam, 1981.
21. X. Wang, Y. L. Yin, and H. Yu. Finding collisions in the full SHA-1. In *CRYPTO*, 2005.
22. X. Wang and H. Yu. How to break MD5 and other hash functions. In *EURO-CRYPT*, pages 19–35, 2005.

On Error Correction in the Exponent

Chris Peikert

MIT CSAIL, 32 Vassar St, Cambridge, MA, 02139
cpeikert@theory.csail.mit.edu

Abstract. Given a corrupted word $\mathbf{w} = (w_1, \ldots, w_n)$ from a Reed-Solomon code of distance d, there are many ways to efficiently find and correct its errors. But what if we are instead given $(g^{w_1}, \ldots, g^{w_n})$ where g generates some large cyclic group — can the errors still be corrected efficiently? This problem is called *error correction in the exponent*, and though it arises naturally in many areas of cryptography, it has received little attention.

We first show that *unique decoding* and *list decoding* in the exponent are no harder than the computational Diffie-Hellman (CDH) problem in the same group. The remainder of our results are negative:

- Under mild assumptions on the parameters, we show that *bounded-distance decoding* in the exponent, under $e = d - k^{1-\epsilon}$ errors for any $\epsilon > 0$, is as hard as the discrete logarithm problem in the same group.
- For *generic* algorithms (as defined by Shoup, Eurocrypt 1997) that treat the group as a "black-box," we show lower bounds for decoding that exactly match known algorithms.

Our generic lower bounds also extend to decisional variants of the decoding problem, and to groups in which the decisional Diffie-Hellman (DDH) problem is easy. This suggests that hardness of decoding in the exponent is a qualitatively new assumption that lies "between" the DDH and CDH assumptions.

1 Introduction

Reed-Solomon codes and cryptography. The Reed-Solomon (RS) family of error-correcting codes [19] has proven incredibly useful throughout several areas of theoretical computer science and in many real-world applications. They are very simple to define: for any field F_q of size q, any *message length* k and *code length* n such that $k \leq n \leq q$, and any *evaluation set* of n distinct points $\mathcal{E} = \{\alpha_1, \ldots, \alpha_n\} \subseteq F_q$, the *Reed-Solomon (RS) code* $\mathbb{RS}_q(\mathcal{E}, k)$ is the set of all codewords $(p(\alpha_1), \ldots, p(\alpha_n))$, where $p(x) \in F_q[x]$, $\deg(p) < k$.

In addition to their elegant definition and many beautiful combinatorial properties, Reed-Solomon codes also admit efficient algorithms for correcting errors. The algorithm of Berlekamp and Welch [1] corrects up to $d/2 = (n - k + 1)/2$ errors in any codeword $\mathbf{w} \in \mathbb{RS}_q(\mathcal{E}, k)$, while the *list-decoding* algorithm of Guruswami and Sudan [12] (building on groundbreaking work by Sudan [23]) can find all codewords within Hamming distance $n - \sqrt{nk}$ of a given word.

S. Halevi and T. Rabin (Eds.): TCC 2006, LNCS 3876, pp. 167–183, 2006.

Reed-Solomon codes also play a fundamental role in modern cryptography, but are often known by a different name: *Shamir* (or *polynomial*) *secret-sharing* [21]. McEliece and Sarwate first observed [15] that sharing a secret using Shamir's scheme is equivalent to encoding the secret under an RS code: a random low-degree polynomial p is chosen so that $p(\alpha_0)$ is the value of the secret, and the shares are the evaluations of p at many other distinct points $\alpha_1, \ldots, \alpha_n$. Moreover, reconstructing the secret when players withhold or mis-report their shares is equivalent to decoding a codeword that has been corrupted with erasures or errors (respectively).

Placing shares in the exponent. Many cryptographic schemes rely on the presumed hardness of computing discrete logs in some cyclic group G of prime order q generated by an element g. In constructing threshold versions of such schemes, distributing trust over many players often involves distributing the secret key via polynomial secret sharing/RS encoding (where the alphabet is the field $\mathbb{Z}_q$). To perform the cryptographic task, typically the players must collectively compute some value of the form g^w, where w depends on the secret key and must remain secret. For example, to decrypt an ElGamal [10] ciphertext $(c, d) = (g^r, m \cdot y^r)$ where $y = g^x$ and x is the secret key, the players must collectively compute the value $c^x = g^{xr}$ without revealing their individual shares of x.

The basic protocol for computing g^w usually works as follows:

1. Player i uses its share of the secret key to compute g^{w_i}, where $w_i = p(\alpha_i)$ is a share of the secret value $w = p(\alpha_0)$ under a polynomial p of degree less than k.
2. The players broadcast their respective values of g^{w_i}, for $i = 1, \ldots, n$.
3. The broadcast values are (efficiently) *"interpolated in the exponent"* [1] to recover g^w.

   Specifically, for any $S \subset \{1, \ldots, n\}$ such that $|S| = k$, given the values $g^{w_i} = g^{p(\alpha_i)}$ for $i \in S$, each player locally computes

$$g^w = g^{p(\alpha_0)} = g^{\sum_{i \in S} \lambda_i^S p(\alpha_i)} = \prod_{i \in S} (g^{w_i})^{\lambda_i^S}$$

using appropriate Lagrange coefficients λ_i^S:

$$\lambda_i^S = \prod_{j \in S, j \neq i} \frac{\alpha_j - \alpha_0}{\alpha_j - \alpha_i} \mod q.$$

In Step 3 above, notice that any subset S of size k suffices, and that the values from players outside S are unused in the interpolation formula. Therefore interpolation in the exponent is robust against a "halting" adversary — i.e., one that may refuse to broadcast some shares, but always correctly reports the values of those shares it does broadcast.

---

[1] Using the language of coding theory, we might call this "erasure-decoding in the exponent."

Introducing errors in the exponent. A *malicious* adversary, on the other hand, may lie about its shares. This introduces *errors* in the exponent, instead of erasures. Without a way to separate correct shares from incorrect shares, the interpolation formula may produce different results depending on which shares are used.

This motivates the natural question of whether it is possible to efficiently *correct errors "in the exponent."* More specifically: if a vector $\mathbf{x} = (x_1, \ldots, x_n)$ differs from some RS codeword $\mathbf{w} = (w_1, \ldots, w_n)$ in at most e positions, then given $g^{\mathbf{x}} = (g^{x_1}, \ldots, g^{x_n})$, is it possible to efficiently recover $g^{\mathbf{w}} = (g^{w_1}, \ldots, g^{w_n})$?

The goal of this paper is to investigate the computational complexity of error correction in the exponent, and to relate it to well-known computational problems in cyclic groups (such as discrete log and Diffie-Hellman).

Relationships among parameters. Error correction in the exponent involves several different parameters, and its complexity depends upon the relationships among these parameters. For analyzing asymptotic behavior, all these parameters (q, n, k, e) will be seen as functions of a single security parameter ℓ. We will focus our attention on those parameter values which are most common in cryptographic settings:

- Complexity of algorithms will always be measured relative to the security parameter ℓ. An *efficient* algorithm is one that runs in time polynomial in the security parameter. A function is said to be *negligible* if it asymptotically decreases faster than the inverse of any fixed polynomial in ℓ; otherwise it is said to be *non-negligible*.
- The alphabet size q is exponential in ℓ; that is, $q = 2^{O(\ell)}$.
- The codeword length n (which often corresponds to the number of players in a protocol) may be an arbitrary polynomial in ℓ. Therefore n is some polynomial in $\log q$.
- The message length k (which often corresponds to the number of "curious" — i.e., semi-honest — players) is at most n.
- The number of errors e (which often corresponds to the number of malicious players) is at most n.

In protocols, often it is assumed that either $e = 0$ (corresponding to an honest-but-curious adversary) or $e = k - 1$ (corresponding to a fully-malicious adversary). In order to understand the problem more generally, we will consider e and k independently.

1.1   Applications

While error correction in the exponent is a very interesting problem in its own right, it is also heavily motivated by existing work.

In the positive direction, an error correction algorithm would be highly desirable, because it would lead to improvements in robustness (i.e., correctness in the presence of cheating players) and efficiency of many multiparty cryptographic protocols. Currently, these protocols often require either expensive zero-knowledge proofs of correct operation, or more efficient tools like verifiable secret sharing. In either case, these steps cost extra rounds of communication and computation, which could be avoided by instead having the parties perform local error correction (with the side-effect of also identifying cheating parties).

There are many concrete cryptographic systems in the literature which would benefit from error correction in the exponent, including (but not limited to): threshold DSS key generation and signature protocols [11], threshold ElGamal protocols [17], protocols for multiplication of shared secrets in the exponent [18], distributed pseudorandom generators, functions, and verifiable random functions [16,9,6], traitor-tracing schemes [2], and others. The last example of a traitor-tracing scheme is interesting because, unlike the others, it is not a threshold cryptographic protocol. This indicates that error correction in the exponent may have relevance in many other areas of cryptography as well.

On the other hand, if in our study of this problem we discover that it appears to be hard, then it can be used as a basis for new assumptions that may provide a foundation for new kinds of cryptographic schemes, or improved constructions of existing primitives.

1.2   Our Results

We consider the problem of correcting errors in the exponent for the (family of) codes $\mathbb{RS}_q(\mathcal{E}, k)$, defined over the field $\mathbb{Z}_q$ for prime q.

First we observe that *unique decoding* and *list decoding* in the exponent, when the number of errors e does not exceed the classical error bounds for those problems, is no harder than the computational Diffie-Hellman (CDH) problem [8] in the same group. The remainder of our results are negative:

- Under mild assumptions on the parameters, we show that *bounded-distance decoding* in the exponent under $e = d - k^{1-\epsilon}$ errors is as hard as the discrete logarithm problem in the same group, for any constant $\epsilon > 0$.
- For *generic* algorithms (as defined by Shoup [22]) that only perform "black-box" group operations, we show lower bounds for decoding that exactly match known algorithms.

Our generic lower bounds also extend to decisional variants of the decoding problem, and to groups in which the decisional Diffie-Hellman (DDH) problem is easy. This suggests that hardness of decoding in the exponent is a qualitatively new assumption that lies "between" the DDH and CDH assumptions.

Taken together, our positive and negative results may also hint at new connections between other popular problems on cyclic groups (e.g., discrete log and Diffie-Hellman), which may be illuminated by further study of error correction in the exponent.

1.3   Related Work

We are aware of only one work which directly addresses error correction in the exponent: Canetti and Goldwasser [4] gave a simple, efficient decoding algorithm which works when $e+1 = k = O(\sqrt{n})$. (See Proposition 2.1 for a generalization.)

This provides an inexpensive way to achieve mild robustness in their threshold version of the Cramer-Shoup cryptosystem [7].

A few recent works have investigated the hardness of various "plain" (i.e., not in the exponent) decoding tasks for Reed-Solomon codes. Cheng and Wan [5], somewhat surprisingly, showed that (under an appropriate number of errors) certain list- and bounded-distance decoding problems are as hard as computing discrete logs. However, their setting differs from ours in many important ways: in their work, q is necessarily small (polynomial in n), and list-decoding is related to the discrete log problem in the field $\mathbb{F}_{q^h}$ for a somewhat large h. In contrast, we are concerned mainly with unique decoding and bounded-distance decoding as they relate to computational problems in groups of order q, where q is exponentially large in n.

Guruswami and Vardy [13] resolved a long-standing open problem, showing that maximum-likelihood decoding (i.e., finding the nearest codeword) of Reed-Solomon codes is NP-hard. More specifically, they showed that it is hard to distinguish whether a word is at distance $n - k$ or $n - k - 1$ from a Reed-Solomon code. Of course, the problem remains NP-hard when placed "in the exponent." However, their results are also incomparable to ours: they show a stronger form of hardness, but only in the worst case, for a very large number of errors, and for a carefully-crafted evaluation set $\mathcal{E}$. In contrast, we show weaker forms of hardness, but in the average case, under many fewer errors, and for any $\mathcal{E}$.

We again stress that both of the above works [5, 13] are concerned with the hardness of plain decoding (not in the exponent).

Notation. We denote a vector $\mathbf{x}$ in boldface and its value at index i by x_i. For two vectors $\mathbf{x}, \mathbf{y}$ of the same length, define $\Delta(\mathbf{x}, \mathbf{y})$ to be the Hamming distance between $\mathbf{x}$ and $\mathbf{y}$, i.e. the number of indices i for which $x_i \neq y_i$. Define $\mathrm{wt}(\mathbf{x}) = \Delta(0, \mathbf{x})$. For a code $\mathbb{C}$ and a vector $\mathbf{x}$, define $\Delta(\mathbf{x}, \mathbb{C}) = \min_{\mathbf{y} \in \mathbb{C}} \Delta(\mathbf{x}, \mathbf{y})$. Denote $\{1, \ldots, n\}$ by $[n]$.

2   Initial Observations and Upper Bounds

Unique decoding with a Diffie-Hellman oracle. Clearly, unique decoding in the exponent under $e < (n - k + 1)/2$ errors is no harder than the discrete log problem: given $(g^{x_1}, \ldots, g^{x_n})$, taking discrete logs yields $(x_1, \ldots, x_n)$, which can be corrected using the standard algorithms [1]. However, this approach is actually overkill: it is, in fact, enough to have an oracle for the (computational) Diffie-Hellman problem in G. The main ingredient of the Berlekamp-Welch algorithm is simply a linear system, which can be solved in the exponent if we have a way to perform multiplication and inversion mod q (in the exponent). Multiplication is immediately provided by the Diffie-Hellman oracle: on g^a, g^b, the oracle gives us g^{ab}. Inversion can be implemented as follows: on input g^a, compute $g^{a^{-1} \bmod q} = g^{a^{q-1} \bmod q}$ by repeated squaring in the exponent. (Note that this approach requires that q be known.)

Unique decoding by enumeration. Another approach to unique decoding (under $e < (n - k + 1)/2$ errors) is to merely enumerate over all subsets of size k of received shares. For each subset K, interpolate the shares (in the exponent) to each point in $\mathcal{E}$, counting the number of points in $\mathcal{E}$ for which the interpolated value disagrees with the received share. It is easy to show that when the number of disagreements is at most e, the shares in K are all correct, and the entire codeword can be recovered from them. Unfortunately, this approach takes time $\binom{n}{k}$, which in general is not polynomial in the security parameter.

A similar, but more efficient randomized approach was given in [4] for the case $e + 1 = k = O(\sqrt{n})$. Here we generalize it to arbitrary e, k:

Proposition 2.1. *For any $e, k < n$ such that $e < (n - k + 1)/2$, there is an algorithm for unique decoding in the exponent which performs $O\left(nk(\log q) \cdot \binom{n}{k}/\binom{n-e}{k}\right)$ group operations and succeeds with all but negligible (in n) probability. When $ek = O(n \log n)$, the algorithm performs $\mathrm{poly}(n) \cdot O(\log q)$ group operations.*

Proof. The algorithm works exactly the same as the enumeration algorithm, except with an independent, random subset K for each iteration, for some suitable number of attempts.

Correctness of the algorithm immediately follows from the distance property of $\mathrm{RS}_q(\mathcal{E}, k)$. We now analyze the runtime: each iteration requires $O(nk \log q)$ group operations, using repeated squaring to exponentiate each share to its appropriate Lagrange coefficient. An iteration succeeds if and only if all k of the chosen shares are correct, and the probability of this event is:

$$\frac{\binom{n-e}{k}}{\binom{n}{k}} = \frac{(n-e)!(n-k)!}{(n)!(n-e-k)!}.$$

There are two ways to bound this quantity from below: we can write $\frac{(n-e)!}{n!} \geq n^{-e}$ and $\frac{(n-k)!}{(n-e-k)!} \geq (n - e - k)^e$, *or* we can write $\frac{(n-k)!}{n!} \geq n^{-k}$ and $\frac{(n-e)!}{(n-e-k)!} \geq (n - e - k)^k$. Taking the best of the two options, we get a bound of:

$$\left(1 - \frac{e + k}{n}\right)^{\min(e,k)} = \exp(-O(\min(e, k)(e + k)/n)) = \exp(-O(ek/n)),$$

which is $1/\mathrm{poly}(n)$. Therefore the algorithm can be made to run in $\mathrm{poly}(n)$ time and succeed with high probability. $\square$

Taking the best of all the above approaches, we see that the complexity of unique decoding in the exponent is upper-bounded by the complexity of the CDH problem and by $nk \cdot (\log q) \cdot \binom{n}{k}/\binom{n-e}{k}$.

List decoding. When the number of errors is larger than the unique decoding radius (i.e., half the distance of the code), the technique of *list decoding* can still be used to recover *all* codewords within a given radius of the received word. For example, the list decoding algorithm of Guruswami and Sudan [12] for Reed-Solomon codes can recover all codewords within a radius of $n - \sqrt{nk}$ (which is always at least as large as the unique decoding radius $(n - k + 1)/2$).

The list decoding algorithm of [12] performs operations which are much more sophisticated than those of the Berlekamp-Welch unique decoding algorithm [1]. (For example, the list decoding algorithm needs to compute polynomial GCDs, perform Hensel liftings, and factor univariate polynomials.) However, it turns out that all these operations can still be performed "in the exponent" with the aid of a CDH oracle. Therefore, correcting significantly more errors (i.e., $n - \sqrt{nk}$) in the exponent also reduces to the CDH problem. (We thank abhi shelat for his assistance with these observations.)

The remainder of this paper will be devoted to establishing hardness results and lower bounds.

3   Bounded-Distance Decoding in the Exponent

In this section, we show that *bounded-distance decoding* (a relaxation of unique decoding) in the exponent, under a large number of errors, is as hard as the discrete log problem. We define the following code for a generator g of a cyclic group G of order q: $\mathbb{C}_q(\mathcal{E}, k, g) = \{(g^{w_1}, \ldots, g^{w_n}) \; : \; \mathbf{w} \in \mathbb{RS}_q(\mathcal{E}, k)\}$. Note that this code's alphabet is the group G. The Hamming distance Δ is defined over G^n as it is for any other alphabet.

> **Problem:** Bounded-distance decoding of $\mathbb{C}_q(\mathcal{E}, k, g)$ under e errors. We denote this problem by BDDE-RS$_{q, \mathcal{E}, k, e}$.
> **Instance:** A generator g of G, and $\mathbf{x}$ such that $\Delta(\mathbf{x}, \mathbb{C}_q(\mathcal{E}, k, g)) \leq e$.
> **Output:** Any codeword $\mathbf{p} \in \mathbb{C}_q(\mathcal{E}, k, g)$ such that $\Delta(\mathbf{p}, \mathbf{x}) \leq e$.

We will relate BDDE-RS to the problem of finding a non-trivial *representation* of the identity element relative to a random base, as proposed by Brands [3]:

> **Problem:** Finding a nontrivial representation of the identity element $1 \in G$, with respect to a uniform base of n elements. We denote this problem FIND-REP.
> **Instance:** A base $(x_1, \ldots, x_n) \in G^n$, chosen uniformly.
> **Output:** Any nontrivial $(a_1, \ldots, a_n) \in \mathbb{Z}_q^n$ such that $\prod_{i=1}^n x_i^{a_i} = 1$.

Brands showed that solving FIND-REP in G is as hard as computing discrete logs in G. For completeness, we briefly recall the result and its proof.

Proposition 3.1 ([3], Proposition 3). *If there exists an efficient randomized algorithm to solve FIND-REP in G with non-negligible probability, then there exists an efficient randomized algorithm which, on input $(g, y = g^z)$ for any generator $g \in G$ and uniform $z \in \mathbb{Z}_q$, outputs z with overwhelming probability.*

Proof. Suppose algorithm $\mathcal{B}$ solves FIND-REP in G with non-negligible probability. We construct the following algorithm to solve the discrete log problem in G: on input (g, y) where $\log_g y$ is desired, choose $(r_1, \ldots, r_n)$ and $(s_1, \ldots, s_n)$ from $\mathbb{Z}_q^n$ uniformly and independently, and let $x_i = g^{r_i} y^{s_i}$. Run $\mathcal{B}$ on $(x_1, \ldots, x_n)$, receiving correct output $(a_1, \ldots, a_n)$ with non-negligible probability. If $\sum s_i a_i \neq 0 \bmod q$, output $-\frac{\sum r_i a_i}{\sum s_i a_i} \bmod q$.

The analysis is straightforward: first observe that the constructed $(x_1, \ldots, x_n)$ is uniform over G^n, because g is a generator of prime order. Furthermore, the s_i are independent of x_i, so they are independent of $\mathcal{B}$'s output. Therefore if $(a_1, \ldots, a_n)$ is nontrivial, $\Pr[\sum s_i a_i = 0 \bmod q] = 1/q$, which is negligible. Now suppose $z = \log_g y$. Then $1 = \prod x_i^{a_i} = \prod g^{a_i(r_i + z s_i)}$, which implies $\sum a_i(r_i + z s_i) = 0 \bmod q$. Solving for z, we see that the algorithm's output is correct.

Finally, because the discrete log problem is random self-reducible, an efficient algorithm that solves discrete log with non-negligible probability can be converted into one which succeeds with overwhelming probability.   □

3.1   Our Reduction

Our reduction from FIND-REP to BDDE-RS relies chiefly on the following technical lemma, which bounds the probability that a *random* word in G^n (i.e., an instance of FIND-REP) is very far from an RS codeword (in the exponent). This lemma may be of independent interest, and any improvements to it will automatically reduce the error bound in our discrete log reduction.

Lemma 3.1. *For any positive integer $c \leq n - k$, and any code $\mathbb{C}_q(\mathcal{E}, k, g)$,*

$$\Pr_{\mathbf{x}}\left[\Delta(\mathbf{x}, \mathbb{C}_q(\mathcal{E}, k, g)) > n - k - c\right] \leq \frac{q^c \cdot n^{2c}}{\binom{n}{k+c}},$$

where the probability is taken over the uniform choice of $\mathbf{x}$ from G^n.

Proof. It is apparent that $\Delta(\mathbf{x}, \mathbb{C}_q(\mathcal{E}, k, g)) \leq n - k - c$ if (and only if) there exists some set of indices $S \subseteq [n]$, $|S| = k + c$, satisfying the following condition, which we call the "low-degree" condition for the set S:

The points $\{(\alpha_i, \log_g x_i)\}_{i \in S}$ lie on a polynomial of degree $< k$.

Define $\mathcal{S} = \{S \subseteq [n] : |S| = k + c\}$. For every $S \in \mathcal{S}$, define X_S to be the 0-1 random variable indicating whether S satisfies the low degree condition, taken over the random choice of $\mathbf{x}$. Let $X = \sum_{S \in \mathcal{S}} X_S$.

Now for all $S \in \mathcal{S}$, $\Pr_{\mathbf{x}}[X_S = 1] = q^{-c}$, because any k points of $\{(\alpha_i, \log_g x_i)\}_{i \in S}$ define a unique polynomial of degree at most k, and the remaining c points independently lie on that polynomial each with probability $1/q$. Then by linearity of expectation, $E[X] = \binom{n}{k+c}/q^c$. Now by Chebyshev's inequality,

$$\begin{aligned}
\Pr[\Delta(\mathbf{x}, \mathbb{C}_q(\mathcal{E}, k, g)) > n - k - c] &= \Pr[X = 0] \\
&\leq \Pr[|X - E[X]| \geq E[X]] \\
&\leq \frac{\sigma_X^2}{E[X]^2},
\end{aligned}$$

where σ_Z^2 denotes the variance of a random variable Z.

It remains to analyze $\sigma_X^2 = E[X^2] - E[X]^2$. The central observation is that for a large fraction of $S, S' \in \mathcal{S}$, X_S and $X_{S'}$ are independent, hence they contribute little to the variance. In particular, if $|S \cap S'| \leq k$, then $E[X_S|X_{S'} = 1] = E[X_S]$, i.e. X_S and $X_{S'}$ are independent and $E[X_S X_{S'}] = E[X_S]E[X_{S'}]$.

For all other distinct pairs S, S' such that $|S \cap S'| > k$, $E[X_S X_{S'}] \leq E[X_S] \leq 1/q^c$. The number of such pairs can be bounded (from above) as follows: we have $\binom{n}{k+c}$ choices for S, then $\binom{k+c}{k+1}$ choices of some $k + 1$ elements of S to include in S', then $\binom{n-k-1}{c-1}$ remaining arbitrary values to complete the choice of S'. So the total number of pairs is at most $\binom{n}{k+c}\binom{k+c}{k+1}\binom{n-k-1}{c-1}$.

Putting these observations together, we obtain the following bound on σ_X^2:

$$
\begin{aligned}
\sigma_X^2 \;&=\; \sum_{S \in \mathcal{S}} \left(E[X_S^2] - E[X_S]^2\right) + \sum_{\substack{S, S' \in \mathcal{S} \\ S \neq S'}} \left(E[X_S X_{S'}] - E[X_S]E[X_{S'}]\right) \\
&\leq\; \sum_{S \in \mathcal{S}} E[X_S] + \sum_{\substack{S, S' \in \mathcal{S} \\ S \neq S'}} \left(E[X_S X_{S'}] - E[X_S]E[X_{S'}]\right) \\
&\leq\; E[X] + \sum_{\substack{S, S' \in \mathcal{S} \\ |S \cap S'| > k}} E[X_S X_{S'}] \;\leq\; E[X]\left[1 + \binom{k+c}{c+1}\binom{n-k-1}{c-1}\right].
\end{aligned}
$$

Since $k + c \leq n$, we may apply the (very loose) bound of $\binom{n}{y} \leq n^y$ to the two binomial coefficients to get $\sigma_X^2 \leq E[X] \cdot n^{2c}$, and the claim follows. $\qquad\square$

Theorem 3.1. *For any n, k, c and q such that $\binom{n}{k+c} \geq 2q^c n^{2c}$, if an efficient randomized algorithm exists to solve BDDE-RS$_{q,\varepsilon,k,n-k-c}$ with non-negligible probability (over a uniform instance and the randomness of the algorithm), then an efficient randomized algorithm exists to solve the discrete log problem in G.*

The following corollary gives concrete relationships among $n, k, q,$ and decoding radius for which the theorem applies.

Corollary 3.1. *For any constant $\epsilon > 0$, $\delta \in (0, 1]$, and any $q = 2^{O(\ell)}$ exponential in the security parameter ℓ, for any polynomial $n(\ell) = \omega(\ell^{1/\delta\epsilon})$, any $k = \Omega(n^\delta)$, $k \leq (1 - \Omega(1)) \cdot n$ and any $c \leq k^{1-\epsilon}$, the discrete log problem in cyclic groups of order q reduces to BDDE-RS$_{q,\varepsilon,k,n-k-c}$.*

Example 3.1. For $k = n/2$ and $c = k^{0.99}$, we certainly have $k \leq (1-\Omega(1)) \cdot n$ and $k = \Omega(n^1)$. Then a poly-time algorithm for bounded-distance decoding in the exponent for RS words of length $n = \ell^{100}$ under $n/2 - k^{0.99}$ errors would imply a poly-time algorithm for discrete log in groups of size about $q = 2^\ell$. In contrast, the unique decoding radius of this code is $n/4 = n/2 - k/2$, and the list decoding radius is $n - \sqrt{nk} \approx n/2 - k \cdot 0.414$; both are close to the bounded-distance radius above. Because RS codes can efficiently be uniquely- and list-decoded in the exponent using an oracle for the Diffie-Hellman problem, the error radius of our reduction comes tantalizingly close to providing a reduction from the discrete log problem to the Diffie-Hellman problem. (We thank abhi shelat for this interpretation of the result.)

Proof (of Corollary 3.1). Because $\binom{n}{k+c} \geq (\frac{n}{k+c})^{k+c}$ and $q^c \geq 2n^{2c}$ for sufficiently large ℓ, then by Theorem 3.1, it suffices to establish that for $n = \omega(\ell^{1/\delta\epsilon})$ and sufficiently large ℓ,

$$\left(\frac{n}{k+c}\right)^{k+c} \geq q^{2c} \quad \Longleftrightarrow \quad (k+c) \log \frac{n}{k+c} \geq 2c \log q.$$

We will establish the second inequality by bounding the left side from below by $\Omega(k)$, and bounding the right side from above by $o(k)$, which suffices.

First we analyze the left term: because $c = k^{1-\epsilon}$,

$$\lim_{\ell \to \infty} \frac{n}{k+c} = \frac{n}{k} \geq 1 + \Omega(1),$$

so $\log \frac{n}{k+c} = \Omega(1)$. Therefore the left term is $\Omega(k)$.

On the right, we have $2c \log q = c \cdot O(\ell)$. Because $c \leq k^{1-\epsilon}$ and $n = \omega(\ell^{1/\delta\epsilon}) \Longleftrightarrow \ell = o(n^{\delta\epsilon})$, the right side is $k^{1-\epsilon} \cdot o(n^{\delta\epsilon})$. Finally $n^\delta = O(k)$, so we get $k^{1-\epsilon} \cdot o(k^\epsilon) = o(k)$, as desired. $\qquad\square$

Proof (of Theorem 3.1). Suppose that algorithm $\mathcal{D}$ solves BDDE-RS$_{q,\mathcal{E},k,n-k-c}$ with non-negligible probability. By Proposition 3.1, it will suffice to construct an algorithm $\mathcal{A}$ that solves FIND-REP in G with non-negligible probability.

$\mathcal{A}$ works as follows: on input $\mathbf{x} = (x_1, \ldots, x_n)$, where $\mathbf{x}$ is uniform over G^n, immediately run $\mathcal{D}(g, \mathbf{x})$. By Lemma 3.1, $(g, \mathbf{x})$ is an instance of BDDE-RS$_{q,\mathcal{E},k,n-k-c}$ with probability at least $1/2$. Then conditioned on this event, the instance is uniform, and with non-negligible probability $\mathcal{D}$ outputs some $\mathbf{p} = (p_1, \ldots, p_n)$ where $\Delta(\mathbf{p}, \mathbf{x}) \leq n - k - c$. Take any $k + 1$ indices $E \subseteq [n]$ such that $x_i = p_i$ for $i \in E$. Then any k of the $\mathbf{x}_i$ linearly interpolate (in the exponent) to the remaining x_i. That is, we can compute non-trivial Lagrange coefficients λ_i for all $i \in E$ such that $\prod_{i \in E} x_i^{\lambda_i} = 1$. Let $\lambda_i = 0$ for all $i \notin E$, and output $(\lambda_1, \ldots, \lambda_n)$, which is a solution to FIND-REP. $\qquad\square$

4   Generic Algorithms for Noisy Polynomial Interpolation

Generic algorithms. Shoup proposed the *generic algorithms* framework [22] for computational problems in groups. Informally, a generic algorithm only performs group operations in a black-box manner; it does not use any particular property of the *representation* of group elements.

Formally, we consider a group G, an arbitrary set $S \subset \{0,1\}^*$ with $|S| \geq |G|$, and a random injective *encoding function* $\sigma : G \to S$. We are only concerned with cyclic groups G of prime order q, independent of their representation. Such group are all isomorphic to $\mathbb{Z}_q$ under addition, so we will assume without loss of generality that $G = \mathbb{Z}_q$ under group operation $+$.

A generic algorithm $\mathcal{A}$ has access to an *encoding list* $(\sigma(x_1), \ldots, \sigma(x_t))$ of elements $x_1, \ldots, x_t \in \mathbb{Z}_q$. $\mathcal{A}$ can make unit-time queries of the form $x_i \pm x_j$ to a *group oracle* by specifying the operation and the indices i, j into the encoding list; the answer $\sigma(x_{t+1})$, where $x_{t+1} = x_i \pm x_j$, is appended to the list. The query complexity of a generic algorithm is the number of elements in its encoding list (including any provided as input) when it terminates.

The probability space of an execution of $\mathcal{A}$ consists of the random choice of input, the random function σ, and the coins of $\mathcal{A}$. If we bound the success probability of $\mathcal{A}$ over this space, then it follows that for *some* encoding function σ, the same bound applies when the probability is taken only over the input and $\mathcal{A}$'s coins. Therefore any algorithm which uses the group in a "black-box" manner is subject to the bound.

We remark that most general-purpose algorithms for discrete log and related problems are indeed generic. One exception is the index calculus method, which requires a notion of "smoothness" in the group G. Thus far, index calculus methods have not been successfully applied to groups over the kinds of elliptic curves that are typically used in cryptography.

Schwartz's lemma. A key tool in the analysis of generic algorithms is *Schwartz's Lemma*, which bounds the probability that a multivariate nonzero polynomial, defined over a finite field, is zero at a random point.

Lemma 4.1 ([20]). *For any nonzero polynomial* $f \in \mathbb{F}_q[X_1, \ldots, X_t]$ *of total degree* d,
$$\Pr[f(x_1, \ldots, x_t) = 0] \leq d/q,$$
where the probability is taken over a uniform choice of $(x_1, \ldots, x_t) \in \mathbb{F}_q^t$.

Noisy polynomial interpolation. We now consider a problem which we call "noisy polynomial interpolation," which is closely related to decoding for Reed-Solomon codes. (See Remark 4.1 below for details on this relationship.) This is exactly the problem which tends to appear in many multiparty cryptographic protocols.

> **Problem:** Generic noisy polynomial interpolation at a fixed point $\alpha_0 \notin \mathcal{E}$ under $e < (n - k + 1)/2$ errors. We denote this problem by $\mathsf{GNPI}_{q,\mathcal{E},\alpha_0,k,e}$.
> **Instance:** An initial encoding list $(\sigma(P(\alpha_1) + e_1), \ldots, \sigma(P(\alpha_n) + e_n), \sigma(1))$ for a random $P(x) \in \mathbb{Z}_q[x]$, $\deg(P) < k$, and a random $\mathbf{e} \in \mathbb{Z}_q^n$ such that $\mathrm{wt}(\mathbf{e}) = e$.
> **Output:** $\sigma(P(\alpha_0))$.

Remark 4.1. GNPI is potentially a *strictly easier* problem than full decoding: it could be the case that interpolating a noisy polynomial at some specific, rare point α_0 is easier than recovering the entire codeword (i.e., interpolating at all points $\alpha_1, \ldots, \alpha_n$). Conversely, recovering the entire codeword would permit generic Lagrange interpolation of the polynomial at *any* point α_0. Therefore, the bound for GNPI provided by Theorem 4.1 is potentially stronger than one which might be provided for the full-decoding task.

Theorem 4.1. *A generic algorithm for* $\mathsf{GNPI}_{q,\mathcal{E},\alpha_0,k,e}$ *making* m *queries succeeds with probability at most* $(m + 1)^2 \left(1/q + \binom{n-k}{e} / \binom{n}{e}\right)$.

Corollary 4.1. *If $ek = \omega(n \log n)$, then no efficient generic algorithm solves $\text{GNPI}_{q,\mathcal{E},\alpha_0,k,e}$, except with probability negligible in the security parameter. In particular, the algorithm of Canetti and Goldwasser [4] (described in Section 2) is optimal.*

Proof (of Corollary 4.1). First, $\binom{n-k}{e}/\binom{n}{e} \leq \left(\frac{n-k}{n}\right)^e = (1 - k/n)^e = \exp(-\Omega(ek /n))$, which is negligible in n, and hence in the security parameter. Since $1/q$ is negligible as well, the total success probability is negligible. $\qquad\square$

Proof (of Theorem 4.1). We can write the real interaction between a generic algorithm $\mathcal{A}$ and its oracle as a game, which proceeds as follows: let $P_0, \ldots, P_{k-1}$ and $E_1, \ldots, E_n$ be indeterminants. First, the game chooses $\mathbf{p} = (p_0, \ldots, p_{k-1}) \leftarrow \mathbb{Z}_q^k$ and $\mathbf{e} \in \mathbb{Z}_q^n$ uniformly, such that $\text{wt}(\mathbf{e}) = e$. While interacting with $\mathcal{A}$, the game will maintain a list of linear polynomials $F_1, \ldots, F_t \in \mathbb{Z}_q[P_0, \ldots, P_{k-1}, E_1, \ldots, E_n]$. Concurrently, $\mathcal{A}$ will have an encoding list $(\sigma(x_1), \ldots, \sigma(x_t))$ where $x_j = F_j(\mathbf{p}, \mathbf{e})$. Furthermore, the game defines an "output polynomial" F_0, which corresponds to the correct output.

Initially, $t = n + 1$, $F_j = E_j + \sum_{i=0}^{k-1} P_i \alpha_j^i$ for $j \in [n]$, and $F_{n+1} = 1$. The output polynomial is $F_0 = \sum_{i=0}^{k-1} P_i \alpha_0^i$.

Whenever $\mathcal{A}$ makes a query for $x_i \pm x_j$, the game computes $F_{t+1} = F_i \pm F_j, x_{t+1} = F_{t+1}(\mathbf{p}, \mathbf{e}), \sigma_{t+1} = \sigma(x_{t+1})$, and appends σ_{t+1} to $\mathcal{A}$'s encoding list. When $\mathcal{A}$ terminates, we may assume that it always outputs some σ_j it received from the oracle (otherwise $\mathcal{A}$ only succeeds with probability at most $\frac{1}{q-m}$). Then $\mathcal{A}$ succeeds iff $\sigma_j = \sigma(F_0(\mathbf{p}, \mathbf{e}))$.

The ideal game. We now consider an "ideal game" between $\mathcal{A}$ and a different oracle, in which each *distinct polynomial* F_j is mapped to a *distinct*, random σ_j, independent of the value $F_j(\mathbf{p}, \mathbf{e})$. More formally, the game proceeds as follows: initially, $(\sigma_1, \ldots, \sigma_{n+1})$ is just a list of distinct random elements of S corresponding to polynomials $F_1, \ldots, F_{n+1}$ defined above. Whenever $\mathcal{A}$ asks for $x_i \pm x_j$ as its $(t+1)$st query, the game computes $F_{t+1} = F_i \pm F_j$. If $F_{t+1} = F_\ell$ for any $\ell \leq t$, the game sets $\sigma_{t+1} = \sigma_\ell$, otherwise it chooses σ_{t+1} to be a random element of $S - \{\sigma_1, \ldots, \sigma_t\}$. Finally, when $\mathcal{A}$ terminates, the game chooses a random value σ_0 from $S - \{\sigma_1, \ldots, \sigma_m\}$, corresponding to F_0. $\mathcal{A}$ succeeds in this game if it outputs σ_0; since $\mathcal{A}$ only produces output from $\{\sigma_1, \ldots, \sigma_m\}$, the success probability in the ideal game is zero.

It is easy to see that $\mathcal{A}$'s success probability in the real game is identical to its success probability in the ideal game, *conditioned* on a "failure event" $\mathcal{F}$ not occurring. The event $\mathcal{F}$ is that $F_i(\mathbf{p}, \mathbf{e}) = F_{i'}(\mathbf{p}, \mathbf{e})$ for some $F_i \neq F_{i'}$, where $i, i' \in \{0, \ldots, m\}$, and the probability is taken over $\mathbf{p}, \mathbf{e}$.

Analysis of the games. We now analyze $\Pr[\mathcal{F}]$: for any $F_i \neq F_{i'}$, consider $F = (F_i - F_{i'}) \in \mathbb{Z}_q[P_0, \ldots, P_{k-1}, E_1, \ldots, E_n]$. Suppose that in $\mathbf{e}$, the values e_j for indices $j \in M = \{m_1, \ldots, m_e\}$ are chosen uniformly, while the others are zero. Then we can consider a polynomial F' in the indeterminants $P_0, \ldots, P_{k-1}$ and $E_{m_1}, \ldots, E_{m_e}$, where F' is simply F with zero substituted for each $E_j, j \notin M$.

Let $\mathbf{e}' = (\mathbf{e}_{m_1}, \ldots, \mathbf{e}_{m_e})$. We are then interested in $Pr_{\mathbf{p},\mathbf{e}'}[F'(\mathbf{p}, \mathbf{e}') = 0]$. There are two cases: if F' is nontrivial, then this probability is $1/q$ by Lemma 4.1, because $\mathbf{p}$ and $\mathbf{e}'$ are chosen uniformly. Therefore it remains to bound $Pr_{\mathbf{p},\mathbf{e}}[F' = 0]$.

In order to have $F' = 0$, the constant term and all the coefficients of P_ℓ must be zero in F', and hence also in F. By its construction, F is a nontrivial linear combination of $F_0, \ldots, F_n$, and $F_{n+1} = 1$: i.e., there exist $\mathbf{c} = (c_0, \ldots, c_n) \in \mathbb{Z}_q^{n+1}$ and $d \in \mathbb{Z}_q$ such that

$$F = d + \sum_{j=0}^{n} c_j F_j = d + \sum_{j=1}^{n} c_j E_j + \sum_{\ell=0}^{k-1} P_\ell \cdot \sum_{j=0}^{n} c_j \alpha_j^\ell.$$

Therefore we have $d = 0$ and $A\mathbf{c} = 0$, where A is a Vandermonde matrix with $A_{\ell+1,j+1} = \alpha_j^\ell$ for $j = 0, \ldots, n$ and $\ell = 0, \ldots, k-1$. Because any k columns of A are linearly independent and F is nontrivial, we have $\mathrm{wt}(\mathbf{c}) \geq k+1$. In order for $F' = 0$, it must be that $c_j = 0$ for every $j \in M$. Because the set M is chosen independently of $\mathbf{c}$, the probability of this event is at most $\binom{n-k}{e}/\binom{n}{e}$. Finally, by a union bound over all pairs $F_i \neq F_{i'}$, we obtain the result. $\qquad\square$

4.1   Relation to the DDH Problem

In this section, we show evidence that the noisy polynomial interpolation problem in G is not as easy as the Decisional Diffie-Hellman (DDH) problem in G. Specifically, for the GNPI problem, we show lower bounds for generic algorithms that are augmented with a DDH oracle.

Such lower bounds imply that, even in groups in which the DDH problem is easy, noisy polynomial interpolation may still be hard. Such a scenario is not just idle speculation: there are reasonable instances of so-called "gap Diffie-Hellman" groups [14], in which the DDH problem is *known* to be easy, but the *computational* Diffie-Hellman problem is believed to be hard. Recalling from Section 2 that GNPI is no harder than the CDH problem, this suggests that GNPI may be a problem of intermediate hardness, located strictly between the (easy) DDH problem and the (assumed hard) CDH problem.

Augmented generic algorithms. We augment a generic algorithm $\mathcal{A}$ with a DDH oracle as follows: at any time, $\mathcal{A}$ can submit to the DDH oracle a triple (a, b, z) of indices into its encoding list. The oracle replies whether $x_a \cdot x_b = x_z \bmod q$.

Theorem 4.2. *A generic algorithm for $GNPI_{q,\mathcal{E},\alpha_0,k,e}$, augmented with a DDH oracle, making m_G queries to its group oracle and m_D queries to its DDH oracle succeeds with probability at most $\left((m_G + 1)^2 + 2m_D\right)\left(1/q + \binom{n-k}{e}/\binom{n}{e}\right)$.*

Corollary 4.2. *If $ek = \omega(n \log n)$, no efficient generic algorithm augmented with a DDH oracle solves $GNPI_{q,\mathcal{E},\alpha_0,k,e}$, except with probability negligible in the security parameter.*

Proof (Sketch of Theorem 4.2). As in the proof of Theorem 4.1, we consider "real" and "ideal" games, and bound the probability of a failure event.

Both games proceed much in the same way: they maintain a list of polynomials F_i and answer queries to the group oracle as before. The games answer DDH queries (a, b, z) in the following way:

- In the real game, respond "yes" if $F_a(\mathbf{p}, \mathbf{e}) \cdot F_b(\mathbf{p}, \mathbf{e}) = F_z(\mathbf{p}, \mathbf{e})$, where the multiplication is done in $\mathbb{Z}_q$.
- In the ideal game, respond "yes" if $F_a \cdot F_b = F_z$, where the multiplication is of formal polynomials in $\mathbb{Z}_q[P_0, \dots, P_{k-1}, E_1, \dots, E_n]$. (Because every F_i is linear, the ideal game will only respond "yes" when at least one of F_a, F_b is a constant.)

The failure event $\mathcal{F}$ is the union of the old failure event (from the proof of Theorem 4.1) with the event that, for some query (a, b, z) to the DDH oracle, $F_a(\mathbf{p}, \mathbf{e}) \cdot F_b(\mathbf{p}, \mathbf{e}) - F_z(\mathbf{p}, \mathbf{e}) = 0$ when $F_a \cdot F_b - F_z \neq 0$.

As before, suppose $M = \{m_1, \dots, m_e\}$ is the set of indices such that $\{e_j\}_{j \in M}$ are chosen uniformly, while the others are zero, and let $\mathbf{e}' = (e_{m_1}, \dots, e_{m_e})$. For a particular query (a, b, z) such that $F = F_a \cdot F_b - F_z \neq 0$, consider the polynomial $F' \in \mathbb{Z}_q[P_0, \dots, P_{k-1}, E_{m_1}, \dots, E_{m_e}]$ which is defined to be F with zero substituted for all $E_j, j \notin M$. Define F_a', F_b', F_z' similarly, so $F' = F_a' F_b' - F_z'$. Certainly the total degree of F' is at most 2. If $F' \neq 0$, then by Lemma 4.1, $\Pr[F'(\mathbf{p}, \mathbf{e}') = 0] \leq 2/q$.

It remains to bound $\Pr_\mathbf{e}[F' = 0 \mid F \neq 0]$. In order to have $F \neq 0$ and $F' = 0$, we consider two mutually exclusive cases: (1) F_a or F_b (or both) is a constant polynomial, or (2) F_a, F_b are both non-constant polynomials, i.e. of positive degree.

In case (1), F is nonzero, linear, and is a linear combination of $F_1, \dots, F_{n+1}$. As argued in the proof of Theorem 4.1, $\Pr[F' = 0 \mid F \neq 0] \leq \binom{n-k}{e}/\binom{n}{e}$.

For case (2), we first introduce some notation: for a polynomial H and a monomial Z, define $\text{coeff}_Z(H)$ to be the coefficient of Z in H. We claim that for either $i = a$ or $i = b$, F_i' is a constant polynomial. Suppose not: then there exist two indeterminants X, Y such that $\text{coeff}_X(F_a') \neq 0$ and $\text{coeff}_Y(F_b') \neq 0$. If $X = Y$, we see that $\text{coeff}_{X^2}(F') \neq 0$, a contradiction. If $X \neq Y$, we have

$$\text{coeff}_{XY}(F') = \text{coeff}_X(F_a')\text{coeff}_Y(F_b') + \text{coeff}_X(F_b')\text{coeff}_Y(F_a') = 0.$$

Then $\text{coeff}_X(F_b) \neq 0$, which implies that $\text{coeff}_{X^2}(F') \neq 0$, a contradiction.

Using reasoning as in the proof of Theorem 4.1, we see that

$$\Pr[F_a' \text{ or } F_b' \text{ is constant} \mid F_a, F_b \text{ are non-constant}] \leq 2\binom{n-k}{e}/\binom{n}{e}.$$

Taking a union bound over all queries to the DDH oracle, we get the claimed result.   □

4.2   Decisional Variants

Certain *decisional* versions of the noisy polynomial interpolation problem are also hard for generic algorithms. Here, in addition to the noisy points of the polynomial, the algorithm is given the correct value $P(\alpha_0)$ and a truly random value (in random order), and simply must decide which is which. We denote this problem by $\mathsf{DGNPI}_{q,\mathcal{E},\alpha_0,k,e}$. The hardness of DGNPI implies that $P(\alpha_0)$ "looks random," given the noisy values of the polynomial.

> **Problem:** Decisional generic noisy polynomial interpolation at a fixed point $\alpha_0 \notin \mathcal{E}$ under $e < (n-k+1)/2$ errors. We denote this problem by $\mathsf{DGNPI}_{q,\mathcal{E},\alpha_0,k,e}$.
>
> **Instance:** Encoding list $(\sigma(P(\alpha_1)+e_1), \ldots, \sigma(P(\alpha_n)+e_n), \sigma(1), \sigma(z_0),$ $\sigma(z_1))$ for a random $P(x) \in \mathbb{Z}_q[x]$, $\deg(P) < k$, a random $\mathbf{e} \in \mathbb{Z}_q^n$ such that $\mathrm{wt}(\mathbf{e}) = e$, and a random bit b where $z_b = P(\alpha_0)$ and z_{1-b} is random.
>
> **Output:** The bit b.

Theorem 4.3. *A generic algorithm for* $\mathsf{DGNPI}_{q,\mathcal{E},\alpha_0,k,e}$ *making m queries succeeds with probability at most* $\frac{1}{2} + 2m^2 \left(1/q + \binom{n-k}{e}/\binom{n}{e} \right)$.

Proof (Sketch). The proof is very similar to the proof of Theorem 4.1. We again imagine a game which maintains a list of polynomials F_i in the indeterminants $P_0, \ldots, P_{k-1}, E_1, \ldots, E_n$, and two new indeterminants Z_0, Z_1. In the ideal game, the two input polynomials corresponding to z_0 and z_1 are just Z_0 and Z_1, respectively. In the ideal game, every distinct polynomial is mapped to a different string, and the algorithm succeeds with probability $1/2$ because its view is independent of b. The failure event is that for some $F_i \neq F_{i'}$, either $F(\mathbf{p}, \mathbf{c}, \sum_{j=0}^{k-1} p_j \alpha_0^j, z) = 0$ or $F(\mathbf{p}, \mathbf{e}, z, \sum_{j=0}^{k-1} p_j \alpha_0^j) = 0$ where $F = F_i - F_{i'}$ and z is chosen at random. From here, the analysis proceeds as in Theorem 4.1. $\square$

In fact, we can extend the definition of DGNPI instances to include the value of the polynomial P at *several* distinct points $\beta_0, \ldots, \beta_r \notin \mathcal{E}$, instead of just at α_0. These evaluations "look random" to generic algorithms, with a distinguishing advantage bounded by $2m^2 \left(1/q + \binom{n-(k-r)}{e}/\binom{n}{e} \right)$. Also, as in Section 4.1, we can prove that DGNPI is hard for generic algorithms that are augmented with a DDH oracle. We defer the details to the full version.

5   Conclusions and Open Problems

We have shown evidence that error correction (of Reed-Solomon codes) in the exponent is hard, and that its hardness seems to be qualitatively different than that of the Diffie-Hellman problems. We can think of several related open problems, including:

- Find some other family of codes which admits an efficient (preferably generic) algorithm for decoding in the exponent, and which can be used as the basis of a secret-sharing scheme — or show that the two goals are mutually incompatible.

- Demonstrate a non-generic decoding algorithm for a specific class of cyclic groups with performance better than the generic bounds (perhaps using ideas from index calculus methods).
- Provide new constructions of standard (or new) cryptographic primitives, assuming error correction in the exponent is hard. Such constructions would be useful both as a hedge against possible attacks on other (stronger) assumptions, and for any unique functionality properties they may have.
- Show new connections between the discrete log and Diffie-Hellman problems, using the fact that decoding is often easy with a CDH oracle.

In addition, the general idea of correcting errors in "partially hidden" data (i.e., data that has been obscured by some one-way function) seems ripe with interesting problems.

Acknowledgements

The author gratefully thanks Shafi Goldwasser, Ran Canetti, Alon Rosen, Adam Smith, Tal Rabin, and abhi shelat for helpful comments and discussions, and the anonymous reviewers for their valuable and constructive suggestions.

References

1. E. Berlekamp and L. Welch. Error correction of algebraic block codes. US Patent Number 4,633,470, 1986.
2. D. Boneh and M. K. Franklin. An efficient public key traitor tracing scheme. In *CRYPTO '99: Proceedings of the 19th Annual International Cryptology Conference on Advances in Cryptology*, pages 338–353, London, UK, 1999. Springer-Verlag.
3. S. Brands. Untraceable off-line cash in wallet with observers. In *CRYPTO '93: Proceedings of the 13th annual international cryptology conference on Advances in cryptology*, pages 302–318, New York, NY, USA, 1994. Springer-Verlag New York, Inc.
4. R. Canetti and S. Goldwasser. An efficient threshold public key cryptosystem secure against chosen ciphertext attack. In *Advances in Cryptology — EUROCRYPT '99*, volume 1592, pages 90–106. Springer-Verlag, 1999.
5. Q. Cheng and D. Wan. On the list and bounded distance decodability of the Reed-Solomon codes. In *Proc. FOCS 2004*, pages 335–341. IEEE Computer Society, 2004.
6. R. Cramer and I. Damgård. Secret-key zero-knowlegde and non-interactive verifiable exponentiation. In *1st TCC*, pages 223–237, 2004.
7. R. Cramer and V. Shoup. A practical public key cryptosystem provably secure against adaptive chosen ciphertext attack. In *Advances in Cryptology — CRYPTO'98*, 1998.
8. W. Diffie and M. E. Hellman. New directions in cryptography. *IEEE Transactions on Information Theory*, IT-22(6):644–654, 1976.
9. Y. Dodis. Efficient construction of (distributed) verifiable random functions. In *6th PKC*, pages 1–17, 2003.
10. T. E. Gamal. A public-key cryptosystem and a signature scheme based on discrete logarithms. *IEEE Transactions on Information Theory*, 31:469–472, 1985.

11. R. Gennaro, S. Jarecki, H. Krawczyk, and T. Rabin. Robust threshold dss signatures. In *Advances in Cryptology — Eurocrypt '96*, pages 354–371, 1996.
12. V. Guruswami and M. Sudan. Improved decoding of reed-solomon and algebraic-geometric codes. In *IEEE Symposium on Foundations of Computer Science*, pages 28–39, 1998.
13. V. Guruswami and A. Vardy. Maximum-likelihood decoding of Reed-Solomon codes is NP-hard. In *SODA*, 2005.
14. A. Joux and K. Nguyen. Separating decision Diffie-Hellman from computational Diffie-Hellman in cryptographic groups. *J. Cryptology*, 16(4):239–247, 2003.
15. R. J. McEliece and D. V. Sarwate. On sharing secrets and Reed-Solomon codes. *Comm. ACM*, 24(9):583–584, 1981.
16. M. Naor, B. Pinkas, and O. Reingold. Distributed pseudo-random functions and kdcs. In *Advances in Cryptology — Eurocrypt '99*, pages 327–346, 1999.
17. C. Park and K. Kurosawa. New ElGamal type threshold digital signature scheme. *IEICE Trans. Fundamentals*, E79-A(1):86–93, January 1996.
18. M. D. Raimondo and R. Gennaro. Secure multiplication of shared secrets in the exponent. Cryptology ePrint Archive, Report 2003/057, 2003.
19. I. S. Reed and G. Solomon. Polynomial codes over certain finite fields. *J. SIAM*, 8(2):300–304, June 1960.
20. J. T. Schwartz. Fast probabilistic algorithms for verification of polynomial identities. *J. ACM*, 27(4):701–717, 1980.
21. A. Shamir. How to share a secret. *Comm. ACM*, 22(11):612–613, 1979.
22. V. Shoup. Lower bounds for discrete logarithms and related problems. In *Proc. Eurocrypt '97*, pages 256–266, 1997.
23. M. Sudan. Decoding of Reed-Solomon codes beyond the error-correction bound. *Journal of Complexity*, 13(1):180–193, 1997.

On the Relation Between the Ideal Cipher and the Random Oracle Models

Yevgeniy Dodis* and Prashant Puniya

Courant Institute of Mathematical Sciences,
New York University
{dodis, puniya}@cs.nyu.edu

Abstract. The Random Oracle Model and the Ideal Cipher Model are two of the most popular idealized models in cryptography. It is a fundamentally important practical and theoretical problem to compare the relative strengths of these models and to see how they relate to each other. Recently, Coron et al. [8] proved that one can securely instantiate a random oracle in the ideal cipher model. In this paper, we investigate if it is possible to instantiate an ideal block cipher in the random oracle model, which is a considerably more challenging question. We conjecture that the *Luby-Rackoff construction* [19] with a sufficient number of rounds should suffice to show this implication. This does not follow from the famous Luby-Rackoff result [19] showing that 4 rounds are enough to turn a pseudorandom function into a pseudorandom permutation, since the results of the intermediate rounds are known to everybody. As a partial step toward resolving this conjecture, we show that random oracles imply ideal ciphers in the *honest-but-curious model*, where all the participants are assumed to follow the protocol, but keep all their intermediate results. Namely, we show that the *Luby-Rackoff construction* with a superlogarithmic number of rounds can be used to instantiate the ideal block cipher in any honest-but-curious cryptosystem, and result in a similar honest-but-curious cryptosystem in the random oracle model. We also show that securely instantiating the ideal cipher using the Luby Rackoff construction with upto a logarithmic number of rounds is equivalent in the honest-but-curious and malicious models.

1   Introduction

Designing provably secure as well as efficient cryptographic protocols is never an easy task. When one tries to achieve provable security without making any assumptions, it often comes at the expense of simplicity and efficiency of the design. On the other hand, practical and efficient schemes are often based on heuristics that cannot be justified with a formal proof. In the late 1990s, this problem was addressed and several ideas were proposed to strike a balance between these two conflicting requirements.

RANDOM ORACLE MODEL. One of these was the formalization of the well known *Random Oracle Model* (ROM) by Bellare and Rogaway [3]. In this model, we

---

* Supported in part by NSF career award CCR-0133806 and NSF grant CCR-0311095.

S. Halevi and T. Rabin (Eds.): TCC 2006, LNCS 3876, pp. 184–206, 2006.
© Springer-Verlag Berlin Heidelberg 2006

assume the existence of a publicly accessible ideal random function and prove protocol security based on this assumption. As was shown by a huge body of literature (for a small set of representative examples, see [3,6,4,15,24,25]), the ROM often allows one to design very simple, intuitive and efficient protocols for many tasks, while simultaneously providing a seemingly convincing security guarantee for such practical constructions. Of course, in practice an ideal random function is instantiated by a concrete, "heuristically-secure" hash function, such as one of the *SHA* functions. The hope of the security of such a substitution comes from the optimistic belief that, — although no security proof is currently found with the heuristic hash function, — the only way such composition can fail is if some unexpected inter-dependency between a protocol and the *code* of a concrete hash function is found. For practical protocols and real-life, "messy" hash functions, it seems unlikely that such unexpected inter-dependency should be found, at least not without directly attacking a carefully-designed heuristic hash function, which is also considered unlikely. On the other side, in theory such security proofs in the ROM have came under scrutiny, after a series of results showed artificial schemes that are provably secure in the ROM, but are uninstantiable in the standard model [10,22,16,11,2]. Still, none of these results directly attack any of the widely used cryptographic schemes, such as OAEP [6] or PSS [4], that rely on secure hash functions. In particular, all the practical applications of the random oracle methodology still appear to be "plausibly secure". Additionally, in some cases the protocols in the ROM came before and *influenced* the first (often slower) solutions in the standard model, and in some other cases the ROM solutions are still the *only* known solutions. To summarize, the random oracle model remains a useful and popular tool in the protocol design.

IDEAL CIPHER MODEL. Another example of such an ideal assumption model is the *Ideal Cipher Model* (ICM) (also known as the "Shannon Model"). In this model, we assume the existence of a publicly accessible Ideal Block Cipher. This is a block cipher, with a k bit key and a n bit input, that is chosen uniformly from all block ciphers of this form. All parties in the ICM can make both forward (encryption) or inverse (decryption) queries to the ideal block cipher. One proves the security of a cryptosystem under this assumption, and then instantiates the ideal block cipher with a practical block cipher construction, such as AES. Although the ICM is not as popular as the random oracle model, there are still several examples of schemes where this model has been used [5,13,14,17,18].

Several questions have been raised regarding security in the ideal cipher model. Existing block cipher constructions, such as DES, AES etc. are vulnerable to related key attacks and have distinguishing patterns that are unlikely to occur in a random permutation. Hence it may not be entirely secure to use these constructions to instantiate the ideal block cipher. As in the case of the random oracle model, uninstantiable schemes that are secure in the ideal cipher model have also been presented (see [1]). But, all these problems withstanding, the ideal cipher model does provide security against generic attacks that do not exploit weaknesses of the underlying block cipher.

COMPARING THE MODELS. From a theoretical viewpoint, it is interesting to compare different ideal assumption models (such as ROM and ICM). That is, compare two ideal assumption models to see which one provides a better security guarantee. There was no satisfactory definition that captured this idea until recently. In TCC 2004, Maurer et al [20] proposed an extension of the classical notion of indistinguishability, called *indifferentiability*. Based on this notion of indifferentiability, Coron et al [8] gave the definition of an "indifferentiable construction" of one ideal primitive (F) using another (G). If a construction satisfies this definition, then any application that is provably secure in the former ideal model (F) remains provably secure in the latter model (G) as well, when instantiated using this construction.

It is an interesting question to analyze the relationship between the *Random Oracle Model* and the *Ideal Cipher Model* using this notion of indifferentiability. It had been believed for quite some time that it should be possible to instantiate a *random oracle* in the *ideal cipher model*. This is because an ideal block cipher seems to be a much stronger primitive than a random oracle, as it seems plausible that one can construct "unstructured" functions from permutations. In [8], a formal proof of this conjecture was given. The authors analyzed the Merkle-Damgård construction [12,21] for extending the domain of a random function in the indifferentiability scenario. The Merkle-Damgård construction is the basis of almost all practical hash functions, such as SHA or MD5. It was shown in [8] that, although the plain Merkle-Damgård construction does *not* work in extending the domain of a random oracle in the indifferentiability model, several slight (and easily implementable) modifications of this construction formally satisfy the indifferentiability requirement. In fact, they also extended these constructions to the *ideal cipher model* and showed that, by using the *Davies Meyer hash function* [26] in place of a "fixed-size" random oracle, any of these modified constructions still satisfy the indifferentiability definition. This result, in turn, implies that *a random oracle can be securely instantiated in the ideal cipher model*.

What about the other direction of this question? *Can one securely instantiate an ideal cipher in the random oracle model?* This direction seems much more difficult to tackle. Actually, it is widely believed that a positive answer holds in this direction too [9]. In fact, it is conjectured that, with a sufficient number of rounds, the *Luby-Rackoff (LR) construction* [19] (with independent random oracles, indexed by the ideal cipher key and the round number, as round functions) is a secure construction of an ideal block cipher in the random oracle model.[1] In spite of this, there has not been much progress in getting a formal proof of this conjecture.

OUR MAIN RESULT. In this paper, we take a step toward finding such a proof. Namely, we will show that the Luby-Rackoff construction works in the *honest-but-curious* model, where all the participants are assumed to follow the pre-

---

[1] We notice that the famous Luby-Rackoff result [19], showing that 4 rounds are enough to turn a pseudorandom function into a pseudorandom permutation, is not applicable here, since it crucially relies on the secrecy of the intermediate round values, while in our setting such intermediate round values are public.

scribed protocols, but keep all the intermediate results (such as the intermediate round values in the LR construction). Namely, we show that the LR-construction (with a superlogarithmic number of rounds in the security parameter) can be used to instantiate the ideal block cipher in *any* honest-but-curious cryptosystem, and result in a similar honest-but-curious cryptosystem in the random oracle model. While weaker than a result in the malicious model, we stress that the conclusion works for *any application in the honest-but-curious model*, even a "maliciously chosen one". In essence, we are using the honest-but-curious aspect only in assuming that the participants will not use the random oracle for purposes other than honestly evaluating the LR construction on *adversarially chosen* points. We now describe our results in more detail.

1.1   Our Results in More Detail

We will start by recalling the definition of *indifferentiability* of a construction of an ideal primitive. This is the same definition as described in [8]. We will then describe what it means to implement an ideal primitive G using an ideal primitive F in the "honest-but-curious model". We will present a restricted version of the definition of general indifferentiability that captures this notion, which we will call *indifferentiability in the honest-but-curious model*. This definition is weaker than general indifferentiability, but is considerably stronger than the classical notion of indistinguishability (see below). We will also describe special types of constructions, which we call *transparent constructions*, for which this restricted definition is equivalent to general indifferentiability.

Once we have a suitable definition, we will describe the *random permutation model* where we assume the existence of a publicly accessible random permutation π (and its inverse π^{-1}). Note that this can be thought of as a very special case of the *ideal block cipher*, where the key space has a single element. We will show that if we can find an indifferentiable construction of a *random permutation* from a random oracle, it can be easily extended to get an indifferentiable construction of an *ideal block cipher* from a random oracle. This is simply done by prepending the key to the block cipher to the input of the random oracle. Thus, it is (necessary and) sufficient to study constructions of a single random permutation from a random oracle.

We will then describe a construction of a random permutation from a random oracle: namely, the *LR-construction* described above, where we derive the round functions from the *random oracle* (indexed by the round number). We conjecture that the LR-construction is indifferentiable from a *random permutation*, with a sufficient number of rounds. As we said, though, we will not be able to prove this result in general. Our main result, however, will prove this implication in the honest-but-curious model, *as long as the number of rounds is super-logarithmic in the security parameter λ*. The proof of this theorem is quite non-trivial, and will essentially show that any distinguisher needs to make an exponential number of queries to have a non-negligible chance of telling apart this construction from a true random permutation in the honest-but-curious indifferentiability scenario.

We conjecture that our result is sub-optimal in a sense that the LR construction seems to be secure even with a "large enough" *constant* number of rounds (see below on what large enough could be), and even in the malicious model. However, we show its "optimality" in the following sense: we prove that for upto a logarithmic number of rounds the LR-construction is a *transparent construction*. Thus, short of resolving our conjecture in the malicious model, any improvement in the number of rounds even in the *honest-but-curious* model will right away imply the same result in the *malicious* model as well. From a positive spin, for upto logarithmic number of rounds one can *without loss of generality* concentrate on the honest-but-curious model (although we have no indication if such proof will be any simpler). From a negative side, we show that for superlogarithmic number of rounds the LR-construction is *provably not transparent*, which means that our positive result in the honest-but-curious model does not trivially imply the same result in the malicious model.

Finally, we mention that for any less than 6 rounds, the LR-construction is not an indifferentiable construction of a random permutation. (The same will also hold in the honest-but-curious model since for less than 6 rounds the LR-construction is a transparent construction.) Aside from showing that at least 6 rounds are needed, this result can be seen as a separation between indifferentiability, even in the honest-but-curious model, and the classical notion of indistinguishability. This is because, in [19], Luby and Rackoff proved that for ≥ 4 rounds this construction is *indistinguishable* from a random permutation. To put it differently, even in the context of the LR-construction the ability to observe "intermediate results" gives a noticeable edge to the adversary, partially explaining why the indifferentiability result seems to be much harder to get.

2 Definitions

In this section, we introduce the main notations and definitions that we will use henceforth. An *ideal primitive* is an algorithmic entity that receives a query from one of the parties and responds to the querying party immediately, and which implements some functionality in an ideal fashion. The ideal primitives we will consider in this paper are random oracles and ideal ciphers. A random oracle is an ideal implementation of a function that assigns a uniformly random value (chosen from a prespecified range) to each input. An ideal cipher is an ideal implementation of a block cipher $E : \{0,1\}^\kappa \times \{0,1\}^n \to \{0,1\}^n$. Each key $k \in \{0,1\}^\kappa$ to the block cipher E defines a random permutation $E_k = E(k, \cdot)$ on $\{0,1\}^n$. The ideal cipher E accepts both forward queries (E) as well as inverse queries (E^{-1}) ($(0,k,m)$ or $(1,k,c)$ resp.).

2.1 Preliminaries

Let us first establish some basic notation that we will be using. We denote the set of all functions $\{0,1\}^n \to \{0,1\}^n$ by F_n and the set of all permutations on $\{0,1\}^n$ by P_n (clearly, $P_n \in F_n$). For a bit string x, $x|_L$ and $x|_R$ denote the left and right halves of x, respectively. $\oplus$ denotes bit by bit XOR of two bit strings.

Definition 1 (Feistel Permutation). *Given a function $f \in F_n$, the Feistel permutation Ψ_f is a permutation in P_{2n} that outputs $x|_R \parallel x|_L \oplus f(x|_R)$ where $x|_L$ and $x|_R$ are the left and right halves of the 2n bit input x, respectively.*

It is easy to see that Ψ_f is a permutation in P_{2n} for any function $f \in F_n$. In fact, it is really easy to invert the Feistel permutation as well. Indeed, $\Psi_f^{-1}(S \parallel T) = (f(S) \oplus T) \parallel S$.

Luby and Rackoff [19] define *pseudorandom permutation ensembles* (PPE) to be distributions of permutations that are indistinguishable from the uniform distribution for any efficient distinguisher. When the distinguisher has access to both the forward and inverse permutation, it is called a *strong pseudorandom permutation ensemble* (SPPE). It was proven in [19], that a 3 (4 resp.) round application of the Feistel permutation, with independent round functions in each round is a PPE (SPPE resp.)

2.2 Indifferentiability and the Honest-But-Curious Model

We will use the notion of *indifferentiability* introduced by Maurer et al [20] to define a secure implementation of an ideal primitive. The ideal primitive that we will attempt to implement is an *ideal cipher*. In [8], the notion of indifferentiability was used to define the security of hash functions (as random oracles). Thus the treatment in [8] is suitable for our problem as well. We now briefly recall the main definitions involved here:

Definition 2. *A Turing machine $C_{\mathcal{G}}$ with oracle access to an ideal primitive $\mathcal{F}$ is said to be (t_D, t_S, q, ϵ) indifferentiable from an ideal primitive $\mathcal{G}$ if there exists a simulator S, such that for any distinguisher D it holds that $|Pr[D^{C_{\mathcal{G}}, \mathcal{F}} = 1] - Pr[D^{\mathcal{G}, S} = 1]| < \epsilon$. The simulator has oracle access to the ideal primitive $\mathcal{G}$ and runs in time t_S. The distinguisher runs in time at most t_D and makes at most q queries.*

It is shown in [20] that if $C_{\mathcal{G}}^{\mathcal{F}}$ is indifferentiable from $\mathcal{G}$, the $C_{\mathcal{G}}^{\mathcal{F}}$ can replace $\mathcal{G}$ in any cryptosystem, and the resulting cryptosystem will be at least as secure in the $\mathcal{F}$ model as in the $\mathcal{G}$ model. See [8] for more details.

The above definition works for any malicious adversary. We will now present a relaxed version of this notion that we will refer to as *indifferentiability in the honest-but-curious model* for reasons that will be clear soon. In the new definition, the distinguisher effectively has active access to only one oracle. To illustrate this, in the $\mathcal{F}$ model the distinguisher can only query the $\mathcal{G}$ construction $C_{\mathcal{G}}^{\mathcal{F}}$, and not the $\mathcal{F}$ oracle. In addition, it also has access to the queries made by the construction $C_{\mathcal{G}}$ to $\mathcal{F}$, which we will denote as the communication transcript $\mathcal{T}_{C_{\mathcal{G}} \leftrightarrow \mathcal{F}}$. Thus the role of the simulator S in the $\mathcal{G}$ model changes from trying to simulate $\mathcal{F}$ in the general indifferentiability (defn. 2), to trying to simulate the communication transcript $\mathcal{T}_{C_{\mathcal{G}} \leftrightarrow \mathcal{F}}$ in $\mathcal{G}$ model. When the distinguisher has access to $C_{\mathcal{G}}$ and $\mathcal{F}$, its queries can be divided into two types. Those for which it does not observe the queries of $C_{\mathcal{G}}$, and those for which it does. In the $\mathcal{G}$ mode, the former queries are sent directly to the $\mathcal{G}$ oracle and the responses of $\mathcal{G}$ are sent back. While the latter queries are made through the simulator S, which forwards

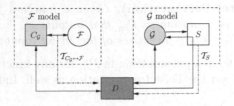

Fig. 1. Indifferentiability in honest-but-curious model: The distinguisher D either interacts with $C_{\mathcal{G}}$ and gets the transcript $\mathcal{T}_{C_{\mathcal{G}} \leftrightarrow \mathcal{F}}$ or it interacts with $\mathcal{G}$ and gets the simulated transcript $\mathcal{T}_S$

the same query to the $\mathcal{G}$ oracle. But apart from sending back $\mathcal{G}$'s response to the distinguisher, it also sends a simulated communication transcript $\mathcal{T}_S$. These two views of the distinguisher are depicted in figure 1.

Definition 3. *A Turing machine $C_{\mathcal{G}}$ (with oracle access to $\mathcal{F}$) is said to be (t_D, t_S, q, ϵ) indifferentiable from an ideal primitive $\mathcal{G}$ in the honest-but-curious model if there exists a simulator S such that for any distinguisher D it holds that:*

$$\left| Pr\left[D^{C_{\mathcal{G}}, \mathcal{T}_{C_{\mathcal{G}} \leftrightarrow \mathcal{F}}} = 1 \right] - Pr\left[D^{\mathcal{G}, \mathcal{T}_S} = 1 \right] \right| < \epsilon$$

The simulator S simulates the transcript $\mathcal{T}_S$ for queries made by the distinguisher to it and runs in time t_S. The distinguisher D runs in time at most t_D and makes at most q queries to its oracle. The distinguishing advantage ϵ is a negligible function of the security parameter λ. If t_S and q are both polynomial in λ then the construction $C_{\mathcal{G}}$ is said to be polynomially indifferentiable from $\mathcal{G}$ in the honest-but-curious model.

Note that the simulator S does not make any extra queries to $\mathcal{G}$ apart from forwarding the queries made by the distinguisher D. This fact is crucial since we want the property that the distinguisher should not learn anything from observing the internal functioning of $C_{\mathcal{G}}$ (i.e. queries made to $\mathcal{F}$), that it cannot learn from an ideal $\mathcal{G}$ itself.

Consider a construction $C_{\mathcal{G}}$ that is (polynomially) indifferentiable from $\mathcal{G}$ in the honest-but-curious model. Our new definition guarantees that any cryptosystem $\mathcal{P}$, possibly involving honest-but-curious parties, that uses the construction $C_{\mathcal{G}}$ in the $\mathcal{F}$ model behaves in exactly the same way as it does in the $\mathcal{G}$ model. This fact is formally stated in the following lemma.

Lemma 1. *If a construction $C_{\mathcal{G}}$ using $\mathcal{F}$ is indifferentiable from $\mathcal{G}$ in the honest-but-curious model, as stated in definition 3, then any cryptographic protocol $\mathcal{P}$ (involving honest-but-curious parties possibly) using $C_{\mathcal{G}}$ in the $\mathcal{F}$ model behaves exactly the same way as in the $\mathcal{G}$ model.*

Proof: [also see figure 2] Say there exists a protocol $\mathcal{P} = (\mathcal{P}_{hon}, \mathcal{P}_{cur})$ that behaves differently when using $C_{\mathcal{G}}$ in $\mathcal{F}$ model. $\mathcal{P}_{hon}$ represents the conventional honest parties of the protocol, and $\mathcal{P}_{cur}$ represents the curious ones. We claim that the curious parties $\mathcal{P}_{cur}$ do not gain any extra information when using the

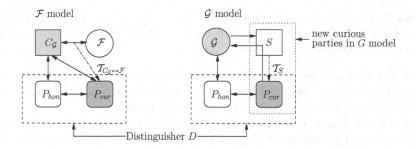

Fig. 2. An idea of the proof of lemma 1. The conventional honest parties P_{hon} along with the curious ones P_{cur} can be seen together as a distinguisher D.

construction $C_{\mathcal{G}}$. We will prove this by simulating the view of all parties in $\mathcal{P}$ in the $\mathcal{F}$ model, in the $\mathcal{G}$ model as well. But this is exactly what definition 3 guarantees. We simply replace the construction $C_{\mathcal{G}}$ with $\mathcal{G}$. And we use the simulator S guaranteed by our definition to simulate the transcript $\mathcal{T}_{C_{\mathcal{G}} \leftrightarrow \mathcal{F}}$ for the curious parties $\mathcal{P}_{cur}$. Thus the queries made by the curious parties $\mathcal{P}_{cur}$ are directed through the simulator S, which along with the response of $\mathcal{G}$ adds a fake transcript $\mathcal{T}_S$ for the curious parties. The conventional honest parties $\mathcal{P}_{hon}$ are given direct access to the ideal primitive $\mathcal{G}$. And the indistinguishability of the two scenarios $(C_{\mathcal{G}}, \mathcal{T}_{C_{\mathcal{G}} \leftrightarrow \mathcal{F}})$ and $(\mathcal{G}, \mathcal{T}_S)$ implies that the views of all parties in the protocol remains the same.   □

We note here that the notion of "indifferentiability of $C_{\mathcal{G}}$ from $\mathcal{G}$ in the honest but curious model" is at least as strong as (in fact, as we shall see later, strictly stronger than) the notion of "indistinguishability of $C_{\mathcal{G}}$ and $\mathcal{G}$". Clearly, a distinguisher in the indistinguishability scenario will work in the former scenario (def. 3) simply by ignoring the transcripts $\mathcal{T}_{C_{\mathcal{G}} \leftrightarrow \mathcal{F}}$ (or $\mathcal{T}_S$).

2.3   Transparent Constructions

Even though general indifferentiability (definition 2) seems to be much stronger than indifferentiability in the honest-but-curious model (definition 3), we now show that for certain types of constructions these two definitions are, in fact, equivalent.

Definition 4 (Transparent Constructions). *A construction $C_{\mathcal{G}}$ of $\mathcal{G}$ (using oracle access to $\mathcal{F}$) is a (t_E, q_E) transparent construction if there exists a Turing machine E (called an "extracting algorithm") such that for any $x \in dom(\mathcal{F})$ it is the case that $E^{C_{\mathcal{G}}^{\mathcal{F}}, \mathcal{T}_{C_{\mathcal{G}} \leftrightarrow \mathcal{F}}}(x) = \mathcal{F}(x)$. Here $\mathcal{T}_{C_{\mathcal{G}} \leftrightarrow \mathcal{F}}$ denotes the transcript of all the communication between $C_{\mathcal{G}}$ and $\mathcal{F}$. E runs in time t_E and makes at most q_E queries to $C_{\mathcal{G}}^{\mathcal{F}}$ for any input x, while $dom(\mathcal{F})$ represents the domain of $\mathcal{F}$. And $|x|$, t_D and q_E are polynomial in the security parameter λ.*

Thus a transparent construction $C_{\mathcal{G}}^{\mathcal{F}}$ is such that it is possible to efficiently compute $\mathcal{F}(x)$ at any input x by making a polynomial number of queries to $C_{\mathcal{G}}$ and observing the communication between $C_{\mathcal{G}}$ and its oracle $\mathcal{F}$.

Lemma 2. If a transparent construction $C_{\mathcal{G}}$ (using $\mathcal{F}$) is (polynomially) in-differentiable from $\mathcal{G}$ in the honest-but-curious model (defn. 3) then it is also (polynomially) indifferentiable from $\mathcal{G}$ (defn. 2).

Proof: Say that a construction $C_{\mathcal{G}}$ is indifferentiable from ideal primitive $\mathcal{G}$ in the honest-but-curious model. Then we have a simulator S_{hon} that successfully fakes the transcript $\mathcal{T}_{C_{\mathcal{G}} \leftrightarrow \mathcal{F}}$ (with $\mathcal{T}_{S_{hon}}$) in the $\mathcal{G}$ model.

First, we will design a simulator S_{mal} for general indifferentiability using the simulator S_{hon}. The simulator S_{mal} needs to simulate the ideal primitive $\mathcal{F}$ in $\mathcal{G}$ model. On getting a query $x \in dom(\mathcal{F})$, S_{mal} uses the extracting algorithm E (for $C_{\mathcal{G}}$) to compute $\mathcal{F}(x)$. The extracting algorithm needs oracle access to the construction $C_{\mathcal{G}}$ and the communication transcript $\mathcal{T}_{C_{\mathcal{G}} \leftrightarrow \mathcal{F}}$. The simulator S_{mal} replaces the construction $C_{\mathcal{G}}$ with the ideal G oracle, which it has access to. And it uses the "honest-but-curious" simulator S_{hon} to produce a fake transcript for E. By definition 3 the extracting algorithm E has no way to tell that it has oracle access to $(\mathcal{G}, \mathcal{T}_{S_{hon}})$ instead of $(C_{\mathcal{G}}, \mathcal{T}_{C_{\mathcal{G}} \leftrightarrow \mathcal{F}})$. This simulator conversion is illustrated in figure 3a.

Now we will show that the simulator S_{mal} designed above actually works. To the contrary, say there is a distinguisher D_{mal} with non-negligible advantage in the general indifferentiability game. Then we will design a distinguisher D_{hon} for the honest-but-curious indifferentiability scenario. D_{hon} simply runs the "malicious" distinguisher D_{mal} and uses the extracting algorithm E to simulate the $\mathcal{F}$ oracle for D_{mal}. Note that it is easy for D_{hon} to run the extracting algorithm E, which needs the exact same oracles that D_{hon} has access to. The new distinguisher is illustrated in figure 3b.

Say $C_{\mathcal{G}}$ is a (t_E, q_E) transparent construction. Then if the simulator S_{hon} runs in time $t_{S_{hon}}$ for every query, then S_{mal} runs in time $\mathcal{O}(t_{S_{hon}} \cdot q_E + t_E)$. And if D_{mal} makes $q_{D_{mal}}$ queries and runs in time $t_{D_{mal}}$ then D_{hon} makes at most $\mathcal{O}(q_{D_{mal}} \cdot q_E)$ queries and runs in time $\mathcal{O}(t_{D_{mal}} \cdot t_E)$. $\square$

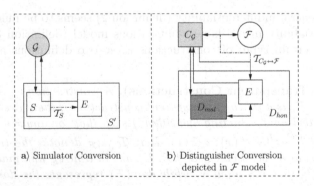

a) Simulator Conversion   b) Distinguisher Conversion
depicted in $\mathcal{F}$ model

Fig. 3. a. Conversion of the simulator S in honest-but-curious model to simulator S' in general indifferentiability. **b.** Conversion of the malicious distinguisher D_{mal} into an honest-but-curious distinguisher D_{cur}.

This theorem essentially implies that if one is able to find a transparent construction $C_{\mathcal{G}}$ for an ideal primitive $\mathcal{G}$ and prove its indifferentiability in the honest-but-curious model. This will also imply the general indifferentiability of the construction $C_{\mathcal{G}}$.

3   The Luby-Rackoff Construction

In this section, we will give a construction of an ideal cipher $E : \{0,1\}^{\kappa} \times \{0,1\}^{2n} \to \{0,1\}^{2n}$ from a random oracle $H : \{0,1\}^* \to \{0,1\}^n$. Note that it suffices to give a construction C_{π} of a single random permutation $\pi : \{0,1\}^{2n} \to \{0,1\}^{2n}$ using H. Similar to the ideal cipher oracle, the random permutation oracle π accepts both forward and inverse queries, but it has a key space of cardinality 1. On input $(0, x)$ the oracle outputs $y = \pi(x)$ and on input $(1, y)$ it outputs x such that $\pi(x) = y$. A construction for the ideal cipher E can be easily derived from this random permutation construction by prepending the key of the ideal cipher to every query C_{π} makes to H.

We will now concentrate on getting an indifferentiable construction of a random permutation from a random oracle, and all our results can be carried over to the ideal cipher model using the technique suggested above.

THE RANDOM PERMUTATION CONSTRUCTION. We first note that the constructions in [19,23] etc. are not necessarily indifferentiable from a random permutation, since all these results are proven in the classical indistinguishability model. Here we will give an indifferentiable construction of *random permutation* (RP) from the *random oracle* (RO) $H : \{0,1\}^* \to \{0,1\}^n$. Similar to [19,23], our construction is based on multiple rounds of the Feistel permutation. However, our proofs will be in the indifferentiability model. We first formally define a "k round LR-construction".

Definition 5 (k round LR-construction). *Given functions $h_i \in F_n : i = 1 \ldots k$, the k round LR-construction $\Psi_{h_1,\ldots,h_k}$ is essentially the composition of k rounds of Feistel permutation, $\Psi_{h_k} \circ \Psi_{h_{k-1}} \circ \ldots \circ \Psi_{h_1}$.*

We will basically use a k round LR-construction (with sufficiently large k) to get a random permutation $\pi : \{0,1\}^{2n} \to \{0,1\}^{2n}$. We will use independent random functions h_i for each round of the k round LR-construction $\Psi_{h_1,\ldots,h_k}$. Note that it is easy to get these independent random functions $h_i \in F_n$ from the random oracle H. These can be simply defined as $h_i(x) = H(\langle i \rangle \parallel x)$ for $i = 1 \ldots k$. Here $\langle i \rangle$ represents the $\log(k)$-bit binary representation of i. The k round LR construction with round functions derived in this fashion is denoted as $C_{\pi,k}$. We conjecture that for sufficient number of rounds k this is an indifferentiable construction of RP from RO.

Conjecture 1. *For a sufficient number of rounds k, the k round construction $C_{\pi,k}$ (using a random oracle $H : \{0,1\}^* \to \{0,1\}^n$) is an indifferentiable construction of a random permutation $\pi : \{0,1\}^{2n} \to \{0,1\}^{2n}$.*

Even though we believe this conjecture to hold, we have been unable to prove it formally. However, we will formally show that the k round LR construction is indifferentiable from a random permutation in the honest-but-curious scenario with a sufficient number of rounds k.

3.1 Transparency for $O(\log \lambda)$ Rounds

The question now is how many rounds should suffice to prove indifferentiability in the *honest-but-curious model?* We first show that for upto a logarithmic (in security parameter λ) number of rounds proving indifferentiability of the LR-construction in the honest-but-curious model is no simpler than proving its indifferentiability in general. Recall from section 2 that a *transparent construction* is one for which indifferentiability in the honest-but-curious model implies its indifferentiability in the general model. We prove that for upto a logarithmic (in λ) number of rounds the LR-construction is a transparent construction.

Theorem 2. *The k round LR-construction $\mathcal{C}_{\pi,k}$ is a (t_E, q_E) transparent construction of the random permutation π from random oracle H for number of rounds $k = \mathcal{O}(\log(\lambda))$. The running time t_E and number of queries q_E are both polynomial in the security parameter λ.*

Proof: Consider the k round LR-construction $\mathcal{C}_{\pi,k}$ for number of rounds $k = \mathcal{O}(\log(\lambda))$. We will describe an extracting algorithm E that when given access to $(\mathcal{C}_{\pi,k}, \mathcal{T}_{\mathcal{C}_{\pi,k} \leftrightarrow H})$ can extract the values of $H(\langle i \rangle \parallel x)$ for any $x \in \{0,1\}^n$ and $i = 1 \ldots k$. Note that such an algorithm E will suffice for our purpose. This is because the random oracle output at any other input is never used by the construction $\mathcal{C}_{\pi,k}$. Thus we will assume that E gets inputs of the form $(\langle i \rangle \parallel x)$, and it outputs the value $H(\langle i \rangle \parallel x)$ (or $h_i(x)$). We will describe this algorithm E in an inductive fashion.

- **Input** $(\langle 1 \rangle \parallel x)$: On this input, E chooses an arbitrary n bit string, R_0. It then assigns $R_1 = x$ and makes the query $\mathcal{C}_{\pi,k}(0, R_0 \parallel R_1)$. This is a forward RP query. In response, it gets the transcript $\mathcal{T}_{\mathcal{C}_{\pi,k} \leftrightarrow H}$, which includes the value $h_1(R_1)$.
- **Input** $(\langle i \rangle \parallel x)$, $i \geq 2$: Such round function values are computed recursively.
  - Choose arbitrary $R_0, R_1 \in \{0,1\}^n$, and query $\mathcal{C}_{\pi,k}(0, R_0 \parallel R_1)$. This will give us a random round value R_{i-1}^1 and corresponding round function value $h_{i-1}(R_{i-1}^1)$.
  - Compute $R_{i-2}^1 = h_{i-1}(R_{i-1}^1) \oplus x$. Recursively invoke $E(\langle i-2 \rangle \parallel R_{i-2}^1)$ to get $h_{i-2}(R_{i-2}^1)$.
  - Compute $R_{i-3}^1 = h_{i-2}(R_{i-2}^1) \oplus R_{i-1}^1$. Recursively invoke $E(\langle i-3 \rangle \parallel R_{i-3}^1)$ to get $h_{i-3}(R_{i-3}^1)$.
  - Continue in this fashion to get $(R_{i-4}^1, h_{i-4}(R_{i-4}^1))$, $\ldots$, $(R_1^1, h_1(R_1^1))$.
  - Compute $R_0^1 = h_1(R_1^1) \oplus R_2^1$. Now query $\mathcal{C}_{\pi,k}(0, R_0^1 \parallel R_1^1)$. This will give us the round function values $(R_i^1, h_i(R_i^1))$. But $R_i^1 = h_{i-1}(R_{i-1}^1) \oplus R_{i-2}^1 = x$. Thus we have $h_i(x)$.

For a query ($\langle i \rangle \parallel x$), let the worst case running time of E be $t_E(i)$ and number of queries be $q_E(i)$. From the above algorithm, we can deduce that $t_E(i) = t_E(i-2) + t_E(i-3) + \ldots + t_E(1) + \mathcal{O}(1)$ and $q_E(i) = q_E(i-2) + q_E(i-3) + \ldots + q_E(1) + \mathcal{O}(1)$. Now one can verify that $t_E(i)$ and $q_E(i)$ are both approximately equal to the ith Fibonacci number. And hence in the worst case $t_E = q_E = \mathcal{O}\left(\phi^k\right)$, where $\phi = \frac{\sqrt{5}+1}{2}$. And thus when $k = \mathcal{O}(\log(\lambda))$, both t_E and q_E are polynomial in the security parameter λ. Hence $C_{\pi,k}$ is a transparent construction when $k = \mathcal{O}(\log(\lambda))$. $\qquad\qquad\square$

Thus one can hope to prove indifferentiability of the LR-construction for $\mathcal{O}(\log(\lambda))$ rounds in the honest-but-curious model, and it will imply the general indifferentiability of the construction. However, there is no indication to suggest that this task might be any easier than the general result.

3.2 Main Result: Equivalence for $\omega(\log \lambda)$ Rounds

On the positive side, we prove the indifferentiability of the LR-construction in the honest-but-curious model for a super-logarithmic number of rounds.

Theorem 3. *The k round construction $C_{\pi,k}$ is $\left(t_D,\ t_S,\ q,\ \mathcal{O}\left((q \cdot k)^4 \cdot 2^{-n}\right)\right)$ indifferentiable from a random permutation $\pi : \{0,1\}^{2n} \to \{0,1\}^{2n}$ (with security parameter λ) in the honest-but-curious model for $k = \omega\left(\log(\lambda)\right)$ rounds. t_S, n and q are all polynomial in λ.*

Proof Intuition: The proof of this theorem consists of two parts. First, we will describe the simulator S that fakes the communication between $C_{\pi,k}$ and H, in the random permutation model. The input to the simulator is either of the form $(0, x)$ (forward π query) or $(1, y)$ (inverse π query), where $x, y \in \{0,1\}^{2n}$. In the random oracle model, if the input $(0, x)$ is given to the construction $C_{\pi,k}$ in the random oracle model, then $C_{\pi,k}$ makes queries to the random oracle H and computes the values $R_1 \ldots R_k$ where $R_0 = x|_L$, $R_1 = x|_R$ and $R_i = h_{i-1}(R_{i-1}) \oplus R_{i-2}$ for $i \in \{2, \ldots k+1\}$. Inverse queries $(1, y)$ are handled in a similar fashion, albeit in reverse starting from $R_k = y|_L$ and $R_{k+1} = y|_R$ and computing $R_i = h_{i+1}(R_{i+1}) \oplus R_{i+2}$ for $i \in \{k-1 \ldots 0\}$.

In the random permutation model, the simulator performs essentially the same computation except that it simulates the round functions h_i itself. It maintains a table T_{h_i} for each of the round functions h_i, in which it stores all previously generated round function values. Consider a forward query $(0, x)$, thus $R_0 = x|_L$ and $R_1 = x|_R$. The simulator S generates a fake transcript for this query as follows:

1. First, it forwards this query $(0, x)$ to the random permutation π and gets $y = \pi(x)$. Thus, in our representation of the LR-construction $R_k = y|_L$ and $R_{k+1} = y|_R$.
2. Next, it checks to see if $h_k(R_k)$ is already defined. If so then it checks the tables $T_{h_{k-1}}, T_{h_{k-2}}, \ldots$ and so on to see if there exists a chain of defined values of the form $[R_{i-1} = h_i(R_i) \oplus R_{i+1}]_{i=k \ldots bot}$, where $bot \in \{1, k\}$. If $bot = 1$ then the

entire chain is already defined, so it checks to see if the $(R_{bot-1} \parallel R_{bot}) = x$. If so, S returns this sequence of values as the transcript to the distinguisher, otherwise the simulator *exits with failure* since there is no way to define the round function values consistent with π.

3. If $bot > 1$ then it checks to see if similarly there exists a chain of defined round function values going down from $R_0 = x|_L, R_1 = x|_R$. That is, a sequence of round values $[R_{i+1} = h_i(R_i) \oplus R_{i-1}]_{i=1...top}$, where $top \in \{1, k\}$. It then checks to see if $top \geq bot - 2$. If so then it *exits with failure* since it cannot be consistent with both π and its previous responses.

4. If everything goes well until now, then the simulator S starts defining the missing round function values between top and bot. It defines the function values $h_{top+1}(R_{top+1}) \ldots h_{bot-2}(R_{bot-2})$ at random. It joins the top and bottom chains by defining $h_{bot-1}(R_{bot-1}) = R_{bot} \oplus R_{bot-2}$ and $h_{bot}(R_{bot}) = R_{bot+1} \oplus R_{bot-1}$.

5. After completing the entire chain in this fashion, S sends it to D.

Thus the simulator S simply tries to define all intermediate round function values randomly. However, it first scans to see if part of the chain of round function values is already defined. It does so both starting from top and bottom, and defines the undefined values in the middle at random but making sure that it joins the two partial chain. If it so happens that the two chains are so long that there are no undefined round values left in the middle, then it realizes that it cannot be consistent with both these chains simultaneously and exits with failure.

The next task is to prove the indistinguishability of the random oracle model, with the LR-construction $\mathcal{C}_{\pi,k}$ and the transcript of its communication with the RO H, and the random permutation model, with the random permutation π and the fake transcript generated by the simulator S described above. Our proof consists of a hybrid argument that starts in the random permutation model and through a series of indistinguishable hybrid models it ends up in the random oracle model. The most non-trivial part of the proof consists of the combinatorial lemma 3, which involves counting the number of queries needed by D to induce an inconsistency in the responses of S. This number is shown to be exponential in the number of rounds k, and hence super-polynomial in the security parameter λ when $k = \omega(\log \lambda)$. The formal proof is given below. $\square$

A formal proof of the fact that the simulator described above works is given in appendix A.

3.3   Non-transparency for $\omega(log\lambda)$ Rounds

One can deduce from theorem 3 that if the LR-construction with $\omega(\log \lambda)$ rounds is a transparent construction, then it will imply the general indifferentiability of this construction too. Unfortunately, we show that for number of rounds $\omega(\log(\lambda))$ the LR-construction is not a transparent construction.

Theorem 4. *The k round LR-construction $\mathcal{C}_{\pi,k}$ is not a transparent construction of the random permutation π for number of rounds $k = \omega(\log(\lambda))$.*

Proof: Say that we are given an extracting algorithm E that given oracle access to $\mathcal{C}_{\pi,k}$ along with the transcript of the communication between $\mathcal{C}_{\pi,k}$ and the RO H, is supposed to compute H on any input. We will give a query x for which E cannot find $H(x)$ with non-negligible probability.

In the proof of theorem 3, we used a hybrid argument to prove the indistinguishability of $(\mathcal{C}_{\pi,k}, \mathcal{T}_{\mathcal{C}_{\pi,k} \leftrightarrow H})$ from $(\pi, \mathcal{T}_S)$. Recall the hybrid scenario in figure 5b, where we had the simulator S_1 that avoids XOR of any 3 of previously defined round (function) values, and the relaying algorithm $\mathcal{M}_1$ that uses the simulator S_1 to respond to the random permutation queries made by the simulator. By our hybrid argument in the proof of theorem 3, we can see that the random oracle scenario $(\mathcal{C}_{\pi,k}, \mathcal{T}_{\mathcal{C}_{\pi,k} \leftrightarrow H})$ is also indistinguishable from this hybrid scenario $(\mathcal{M}_1, \mathcal{T}_{S_1})$.

Coming back to our current proof, if we give the extracting algorithm E access to $(\mathcal{M}_1, \mathcal{T}_{S_1})$, then it should be able to compute the output of any of the round functions simulated by S_1 on any input, just as it does in the random oracle model. If this is not the case, then we can use the extracting algorithm E to design a distinguisher that can tell apart the random oracle model from this hybrid model with high probability. Let us denote the round functions simulated by S_1 as $h_1 \ldots h_k$ and the corresponding round values as $R_0, \ldots, R_{k+1}$.

We will ask the extracting algorithm E to compute $h_{\frac{k}{2}}(x)$. Say E finds out $h_{\frac{k}{2}}(x)$ in query number m, which can be assumed to be a forward query without loss of generality. Denote the round values in query number m as $R_0^{(m)}, \ldots, R_{k+1}^{(m)}$. We can deduce that $R_{(k/2)}^{(m)} = x$ since it is in this query that E finds the values $h_{\frac{k}{2}}(x)$. Now if the round value $R_{(k/2)-1}^{(m)}$ is a new round value then $h_{\frac{k}{2}-1}(R_{(k/2)-1}^{(m)})$ would have been assigned a random value and $h_{\frac{k}{2}-1}(R_{(k/2)-1}^{(m)}) \oplus R_{(k/2)-2}^{(m)}$ would have been equal to x with only a negligible probability. So it must have been the case that $R_{(k/2)-1}^{(m)}$ was defined in some query prior to query number m. We can make similar deductions to show that all the round values $R_0^{(m)}, \ldots, R_{(k/2)-2}^{(m)}$ were also defined in queries previous to the mth query.

After this the proof of the theorem follows in pretty much the same way as the combinatorial lemma 3. We show that the extracting algorithm must have already made a $\phi^{\frac{k}{4}}$ (for $\phi = \frac{\sqrt{5}+1}{2}$) queries prior to the mth query. For a super-logarithmic number of rounds k, this is super-polynomial in the security parameter λ. $\square$

3.4   Negative Results for Constant Rounds

Finally, we mention that one does need to use sufficient number of rounds of the Feistel permutation in the construction, to have any hope of proving it indifferentiable. Coron [7] showed that for less than 6 rounds the LR-construction is not indifferentiable from a random permutation.

Theorem 5 ([7]). *Let* $C_{\pi,k}$ *be the* k *round LR-construction of a random permutation* π*, with number of rounds* $k < 6$*. Then there is an efficient distinguisher* D *such that for any simulator* S*,* D *can distinguish the oracle pair* $(C_{\pi,k}, H)$ *and* (π, S) *with non-negligible probability.*

It is easy to see that the construction $(C_{\pi,k}, H)$ cannot work for $k < 4$, since in this case it does not even satisfy the classical indistinguishability definition [19]. Coron [7] gave attacks on 4 and 5 round LR-constructions in the indifferentiability scenario. We give an attack on the 4 round LR construction in appendix B for illustration.

This theorem also implies that indifferentiability (even in the honest-but-curious model) is strictly stronger than classical indistinguishability. This is because the LR-construction with 4 rounds or more is known to satisfy the latter [19]. Thus we can derive the following corollary from theorem 5.

Corollary 1. *A 4 round LR-construction is indistinguishable , but not indifferentiable, from a random permutation (even in the honest-but-curious model).*

4   Conclusions and Future Work

In this paper, we have shown that the Luby-Rackoff construction with a super-logarithmic number of rounds can be used to instantiate the ideal block cipher in any honest-but-curious cryptosystem. We have also proved that improving this result to upto a logarithmic number of rounds will imply that this construction is indifferentiable from the ideal cipher in general. The main question that still remains unanswered is whether the Luby-Rackoff construction is indifferentiable from the ideal cipher in general.

Acknowledgements. We would like to thank Jean-Sébastien Coron and Joel Spencer for useful discussions.

References

1. J. Black, *The Ideal-Cipher Model, Revisited: An Uninstantiable Blockcipher-Based Hash Function*, eprint 2005/210, (2005).
2. M. Bellare, A. Boldyreva and A. Palacio. *An Uninstantiable Random-Oracle-Model Scheme for a Hybrid-Encryption Problem*, To appear in *Proccedings of Eurocrypt* (2004).
3. M. Bellare, and P. Rogaway, *Random oracles are practical: A paradigm for designing efficient protocols*, In *Proceedings of the 1st ACM Conference on Computer and Communications Security* (1993), 62 -73.
4. M. Bellare and P. Rogaway, *The exact security of digital signatures - How to sign with RSA and Rabin*. Proceedings of Eurocrypt'96, LNCS vol. 1070, Springer-Verlag, 1996, pp. 399-416.
5. J. Black, P. Rogaway, T. Shrimpton, *Black-Box Analysis of the Block-Cipher-Based Hash-Function Constructions from PGV*, in Advances in Cryptology - CRYPTO 2002, California, USA.

6. M. Bellare and P. Rogaway, *Optimal Asymmetric Encryption*, Proceedings of Eurocrypt'94, LNCS vol. 950, Springer-Verlag, 1994, pp. 92–111.

7. J.-S. Coron, *personal communication*.

8. J.-S. Coron, Y. Dodis, C. Malinaud and P. Puniya, *Merkle-Dåmgard Revisited: How to Construct a Hash Function*, In *Advances in Cryptology - Crypto 2005 Proceedings* (2005), 430 -448.

9. J.-S. Coron, A. Joux, and D. Pointcheval, *Equivalence Between the Random Oracle Model and the Random Cipher Model*, *Dagstuhl Seminar 02391: Cryptography*, (2002).

10. R. Canetti, O. Goldreich, and S. Halevi, *The random oracle methodology, revisited*, In *Proceedings of the 30th ACM Symposium on the Theory of Computing* (1998) , ACM Press, pp. 209 -218.

11. R. Canetti, O. Goldreich, and S. Halevi, *On the random-oracle methodology as applied to length-restricted signature schemes*, In *First Theory of Cryptography Conference* (2004).

12. I. Dåmgard, *A Design Principle for Hash Functions*, In Crypto '89, pages 416-427, 1989. LNCS No. 435.

13. A. Desai, *The security of all-or-nothing encryption: Protecting against exhaustive key search*, In *Advances in Cryptology - Crypto'00* (2000), LNCS vol. 1880, Springer-Verlag.

14. S. Even, and Y. Mansour, *A construction of a cipher from a single pseudorandom permutation*, In *Advances in Cryptology - ASIACRYPT'91* (1992), LNCS vol. 739, Springer-Verlag, pp. 210 -224.

15. A. Fiat, and A. Shamir, *How to prove yourself: Practical solutions to identification and signature problems*, In *Advances in Cryptology - Crypto'86* (1986), Lecture Notes in Computer Science, Springer-Verlag, pp. 186 -194.

16. S. Goldwasser and Y. Tauman. *On the (In)security of the Fiat-Shamir Paradigm*, In *Proceedings of the 44th Annual IEEE Symposium on Foundations of Computer Science* (2003), 102-114.

17. E. Jaulmes, A. Joux, and F. Valette, *On the security of randomized CBC-MAC beyond the birthday paradox limit: A new construction*, In *Fast Software Encryption (FSE 2002)* (2002), vol. 2365 of Lecture Notes in Computer Science, Springer-Verlag, pp. 237 -251.

18. J. Kilian, and P. Rogaway, *How to protect DES against exhaustive key search (An analysis of DESX)*, *Journal of Cryptology 14*, 1 (2001), 17 -35.

19. M. Luby and C. Rackoff, *How to construct pseudo-random permutations from pseudo-random functions*, SIAM J. Comput., Vol. 17, No. 2, April 1988.

20. U. Maurer, R. Renner, and C. Holenstein, *Indifferentiability, Impossibility Results on Reductions, and Applications to the Random Oracle Methodology*, Theory of Cryptography - TCC 2004, Lecture Notes in Computer Science, Springer-Verlag, vol. 2951, pp. 21-39, Feb 2004.

21. R. Merkle, *One way hash functions and DES*, in *Advances in Cryptology, Proc. Crypto'89, LNCS 435*, G. Brassard, Ed., Springer-Verlag, 1990, pp. 428-446.

22. J. B. Nielsen. *Separating Random Oracle Proofs from Complexity Theoretic Proofs: The Non-Committing Encryption Case*, In *Advances in Cryptology - Crypto 2002 Proceedings* (2002), 111-126

23. M. Naor and O. Reingold, *On the construction of pseudo-random permutations: Luby-Rackoff revisited*, J. of Cryptology, vol 12, 1999, pp. 29-66.

24. D. Pointcheval, and J. Stern, *Security proofs for signature schemes*, In *Advances in Cryptology - Eurocrypt 1996 proceedings*, 387 -398.

25. C.-P. Schnorr, *Efficient signature generation by smart cards*, In *Journal of Cryptology 4*, 3 (1991), 161 -174.
26. R. Winternitz, *A secure one-way hash function built from DES*, in Proceedings of the IEEE Symposium on Information Security and Privacy, pages 88-90. IEEE Press, 1984.

A   Formal Proof of Indifferentiability

Now we will prove that when the simulator S described in theorem 3 above is used in the indifferentiability game, then any distinguisher D that makes at most q queries to its oracles has only a negligible distinguishing advantage. Here q and n (the output length of H) are both polynomial functions of the security parameter λ, while the number of rounds in the LR construction is $k = \omega(\log(\lambda))$. As we mentioned, our proof proceeds via a hybrid argument.

Hiding the random permutation π. Let us start in the random permutation scenario. Here the distinguisher has oracle access to π and the simulator S. Our first modification is to prevent D from directly accessing π, by replacing it with a simple relaying algorithm $\mathcal{M}$ that acts as an interface to π. When $\mathcal{M}$ gets a query from the distinguisher, it simply relays this query to the random permutation π and sends back the response of π. In this new scenario, the distinguisher has oracle access to $\mathcal{M}^\pi$ and S^π (see figure 5a). Since we have made no real change from the point of view of the distinguisher, we have $Pr[D^{(\pi, \mathcal{T}_{S^\pi})} = 1] = Pr[D^{(\mathcal{M}^\pi, \mathcal{T}_{S^\pi})} = 1]$.

Bounding out the "bad events". Now we will modify the simulator S, so that it never outputs certain types of collisions that will affect our analysis later. Recall that the simulator S needs to define the round function values $h_1(R_1) \ldots h_k(R_k)$ in order to generate the transcript $\mathcal{T}_S$ for every query made to it. And S tries to assign random values to $h_i(R_i)$ for any new R_i.

Now we introduce a slightly modified simulator S_1 that is essentially the same as S except that it chooses round function values more carefully. Let us first fix a little notation. We will number the queries made to the simulator in the order they are made, query number 1 followed by 2 and so on. And for the mth query made to the simulator, we will label its round values as $R_0^{(m)}, R_1^{(m)}, \ldots, R_k^{(m)}, R_{k+1}^{(m)}$.

Assume for now that query number m is a forward query. When assigning a new round function value $h_i(R_i^{(m)})$ in this query, the distinguisher makes sure that $h_i(R_i^{(m)})$ is cannot be represented as an *XOR of upto three previously defined values*. This includes all values $R_j^{(\ell)}$ or $h(R_j^{(\ell)})$ for $\ell < m$, $j \in \{0, k+1\}$ and all values $R_j^{(m)}$ or $h_j(R_j^{(m)})$ for $j < i$ (and $j > i$ for an inverse query m). More formally, S_1 assigns a value $h_i(R_i^{(m)})$ for the mth query (a forward query) that does not satisfy the following equality for values $x_1, x_2, x_3 \in \{R_{j_1}^{(\ell)}, h_{j_1}(R_{j_1}^{(\ell)}), R_{j_2}^{(m)}, h_{j_2}(R_{j_2}^{(m)}) \mid \ell < m \ , \ j_1 \leq k+1, \ j_2 < i\}$

$$h_i(R_i^{(m)}) = x_1 \ or \ (x_1 \oplus x_2) \ or \ (x_1 \oplus x_2 \oplus x_3)$$

The distinguisher cannot tell if it has oracle access to $(\mathcal{M}, S)$ or $(\mathcal{M}, S_1)$ unless the old simulator S outputs a round function value that satisfies one of the above equalities. Let us denote this event by B_1. Hence for any distinguisher D making q queries,

$$\left| Pr\left[D^{(\mathcal{M}^\pi, \mathcal{T}_{S^\pi})} = 1 \right] - Pr\left[D^{(\mathcal{M}^\pi, \mathcal{T}_{S_1^\pi})} = 1 \right] \right| \leq Pr\left[B_1 \right]$$

We can bound the probability of B_1 occurring by noticing that for randomly assigned round function values, $Pr\left[B_1 \right] = \mathcal{O}\left(\frac{(q \cdot k)^4}{2^n} \right)$. This can be derived by using the birthday paradox to bound the probability that any XOR of upto 4 round (or round function) values is 0^n.

Transferring Control to the Simulator. Next we will modify the relaying algorithm $\mathcal{M}$ so that it does not simply act as a channel between the distinguisher and π. The new relaying algorithm, which we will call $\mathcal{M}_1$, responds to the π queries by making the same queries to the simulator S_1 and computing $\pi(x)$ (or $\pi^{-1}(y)$) from the responses of S_1 (see figure 5b).

To illustrate this point, say $\mathcal{M}_1$ gets a query $(0, x)$ from the distinguisher D (that is, a forward query to π). Then $\mathcal{M}_1$ forwards this query to S_1, which in turn gets $y = \pi(x)$ from the random permutation and constructs a fake transcript $\mathcal{T}_{S_1}(0, x)$ (or round values $R_0 = x|_L, R_1 = x|_R, \ldots, R_{k+1}$). If all goes well this transcript is consistent with π. The simulator sends this transcript $\mathcal{T}_{S_1}(0, x)$ to $\mathcal{M}_1$, which can compute $\pi(x)$ from $\mathcal{T}_{S_1}$ and send it to the distinguisher D with this value. Inverse queries $1, y)$ are handled in a similar fashion.

From the view of D, everything in this scenario is same as in the previous one unless the simulator S_1 exits with failure on some query made by $\mathcal{M}_1$. This happens if and only if S_1 fails to be consistent with the random permutation π on some query. We claim that if the number of queries q made by the distinguisher D is polynomial in the security parameter λ then the simulator S_1 is always consistent with π.

Lemma 3. For a polynomial number of queries q made to the simulator S_1, the responses of the simulator are always consistent with the random permutation π.

Proof: Say query number m is the first time S_1 is inconsistent with π. Without loss of generality assume this to be a forward query $(0, x)$, with $\pi(x) = y$. Thus $R_0^{(m)} = x|_L, R_1^{(m)} = x|_R$ and $R_k^{(m)} = y|_L, R_{k+1}^{(m)} = y|_R$. Since S_1 is inconsistent on this query, there exist partial round value chains, $R_0^{(m)}, R_1^{(m)} \ldots R_{top}^{(m)}, R_{top+1}^{(m)}$ and $R_{bot-1}^{(m)}, R_{bot}^{(m)} \ldots R_k^{(m)}, R_{k+1}^{(m)}$ with $top \geq bot - 2$. But in this case either $(top \geq \frac{k}{2})$ or $(bot \leq \frac{k}{2} + 1)$. That is, at least one of these two partial chains consists of more than $\frac{k}{2}$ defined round function values. Without loss of generality, assume that the $top \geq \frac{k}{2}$. Thus all round function values $h_1(R_1^{(m)}) \ldots h_{top}(R_{top}^{(m)})$ were defined before query number m was made. We will look at the queries where each of these round function values was defined for the first time. For any round value $R_i^{(j)}$, we denote by $first(R_i^{(j)})$ the query number where the round function value $h_i(R_i^{(j)})$ was first defined. Thus if $R_i^{(j)}$ is a new round value that appeared in query number j itself, then $first(R_i^{(j)}) = j$ otherwise $first(R_i^{(j)}) < j$. We can thus say that for $i = 1 \ldots top$, it is the case that $first(R_i^{(m)}) < m$.

Now consider any three consecutive round values $R_{i-1}^{(m)}$, $R_i^{(m)}$ and $R_{i+1}^{(m)}$ for $i \in \{2 \ldots top - 1\}$. Let $first(R_{i-1}^{(m)}) = \ell_{i-1}$, $first(R_i^{(m)}) = \ell_i$ and $first(R_{i+1}^{(m)}) = \ell_{i+1}$ ($\ell_{i-1}, \ell_i, \ell_{i+1} < m$). We wish to analyze the order of the queries ℓ_{i-1}, ℓ_i and ℓ_{i+1}. First, note that $\ell_{i-1} \neq \ell_i$ and $\ell_i \neq \ell_{i+1}$. Either case would imply that the ℓ_i^{th} query is the same as the mth query, and the inconsistency should have occurred there itself. Let us now look at the possible orders between ℓ_{i-1}, ℓ_i and ℓ_{i+1}.

1. $(\ell_i > \ell_{i-1} \geq \ell_{i+1})$ or $(\ell_i > \ell_{i+1} > \ell_{i-1})$. That is, query number ℓ_i occurs after ℓ_{i-1} and ℓ_{i+1}. We know that $h_i(R_i^{(m)}) = R_{i-1}^{(m)} \oplus R_{i+1}^{(m)}$ and hence $h_i(R_i^{(\ell_i)}) = R_{i-1}^{(\ell_{i-1})} \oplus R_{i+1}^{(\ell_{i+1})}$. But the round values $R_{i-1}^{(\ell_{i-1})}$ and $R_{i+1}^{(\ell_{i+1})}$ already exist when $h_i(R_i^{(\ell_i)})$ was defined for the first time in the ℓ_i^{th} query. And since the simulator S_1 avoids such an XOR collision, this order is *impossible*.
2. $(\ell_{i-1} > \ell_i > \ell_{i+1})$ or $(\ell_{i-1} < \ell_i < \ell_{i+1})$. These strictly increasing/decreasing orderings are *possible*.
3. $(\ell_i < \ell_{i-1} < \ell_{i+1})$ or $(\ell_i < \ell_{i+1} < \ell_{i-1})$. Here the ℓ_i^{th} query comes before both the ℓ_{i-1}^{th} and ℓ_{i+1}^{th} queries. These orders are *possible*.
4. $(\ell_i < \ell_{i-1} = \ell_{i+1})$. This is the same as above, except that the $\ell_{i-1} = \ell_{i+1}$. In this case, a short calculation gives that $h_i(R_i^{(\ell_{i-1})}) = h_i(R_i^{(\ell_i)})$, where $R_i^{(\ell_{i-1})} \neq R_i^{(\ell_i)}$. And since $R_i^{(\ell_{i-1})}$ exists before $h_i(R_i^{(\ell_i)})$ is defined, this order is *impossible*.

Thus we know that the possible orderings of the queries for any three consecutive round values are the configurations 2 and 3. Now we can apply the same to all the queries $first(R_1^{(m)}) = \ell_1$, $first(R_2^{(m)}) = \ell_2, \ldots, first(R_{top}^{(m)}) = \ell_{top}$, considering each triple of consecutive round values separately and then combining of these orderings together. Using this, we obtain that there is a $j \in \{1, k\}$ such that $(\ell_1 > \ell_2 > \ldots > \ell_j)$ and $(\ell_j < \ell_{j+1} < \ldots < \ell_{top})$. That is, the query numbers $\ell_1 \ldots \ell_{top}$ are strictly decreasing until some ℓ_j and strictly increasing after that. One can verify that any other configuration will involve one of the "impossible" triple orderings 1 or 4.

Now we will look for more structure in these queries. If $j \geq \frac{top}{2}$, then we will analyze the decreasing sequence of queries $\ell_1 \ldots \ell_j$, otherwise we will analyze the increasing sequence of queries $\ell_j \ldots \ell_{top}$. Without loss of generality, assume that $j \geq \frac{top}{2}$; the case $j < \frac{top}{2}$ is symmetrical. Since we earlier derived that $top \geq \frac{k}{2}$, we can also deduce that $j \geq \frac{k}{4}$.

Now we will show that these queries and others that led to the inconsistency in the mth query form a *Fibonacci tree* of depth j (which we know is $\geq \frac{k}{4}$). Each node of the Fibonacci tree corresponds to a different query, with m^{th} query at the root of the tree. This would imply that m is at least as large as the number of nodes in a *Fibonacci tree* of depth $\frac{k}{4}$. But since we know that $k = \omega(\log(\lambda))$ it also holds that m is superpolynomial in the security parameter λ. In turn, this implies that the simulator S_1 is always consistent with the random permutation for any polynomial number of queries.

The queries from $\ell_1 \ldots \ell_j$ form the first level of the *Fibonacci tree* which we will describe. To see this structure more explicitly, we will now move from the mth query to these first level queries. Consider any three consecutive queries in this ordering, $\ell_i, \ell_{i+1}, \ell_{i+2}$ (recall $\ell_i > \ell_{i+1} > \ell_{i+2}$). Let us look at the ℓ_i^{th} query. This query could be a forward or inverse query. For now we assume that it is a forward query. As it will turn out, if this is an inverse query then the *Fibonacci tree* of queries would be even larger, and so will the number of queries needed. We know that $R_i^{(\ell_i)}(= R_i^{(m)})$ is a new round value in this query. Consider the round function value $h_{i-1}(R_{i-1}^{(\ell_i)})$. Since $R_i^{(\ell_i)} = R_i^{(m)}$, we can deduce that

$$h_{i-1}\left(R_{i-1}^{(\ell_i)}\right) = R_{i-2}^{(\ell_i)} \oplus h_{i+1}\left(R_{i+1}^{(\ell_{i+1})}\right) \oplus R_{i+2}^{(\ell_{i+2})}$$

Note that the ℓ_{i+1}^{th} and ℓ_{i+2}^{th} queries were made before query number ℓ_i. And since the ℓ_i^{th} query is a forward one, $R_{i-2}^{(\ell_i)}$ is defined before $R_{i-1}^{(\ell_i)}$. Now if $R_{i-1}^{(\ell_i)}$ is a new round value then the simulator S_1 would have avoided the above XOR representation. Thus $h_{i-1}(R_{i-1}^{(\ell_i)})$ was already defined before the ℓ_i^{th} query. Using similar analysis, one can also deduce that the round function values $h_1(R_1^{(\ell_i)}) \ldots h_{i-2}(R_{i-2}^{(\ell_i)})$ also had to be defined prior to the ℓ_i^{th} query.

Let $first(R_1^{(\ell_i)}) = b_1, \ldots, first(R_{i-1}^{(\ell_i)}) = b_{i-1}$. Consider the queries b_{i-1} and b_{i-2}. Let us see in what order these queries could have occurred. We know that queries b_{i-1} and b_{i-2} were both made before the ℓ_i^{th} query. We also know that the ℓ_{i+1}^{th} query was also made before ℓ_i^{th} query. First note that $b_{i-1} \neq b_{i-2}$, since otherwise the b_{i-1}^{th} query would the same as query number ℓ_i, which is not possible since $R_i^{(\ell_i)}$ is a new round values in the ℓ_i^{th} query.

1. $b_{i-2} < b_{i-1} \leq \ell_{i+1}$ or $b_{i-1} < b_{i-2} \leq \ell_{i+1}$. A short calculation in this case gives $h_{i+1}(R_{i+1}^{(\ell_{i+1})}) = R_{i+2}^{(\ell_{i+2})} \oplus R_{i-2}^{(b_{i-2})} \oplus h_{i-1}(R_{i-1}^{(b_{i-1})})$. Since all 3 of these round (function) values existed before $h_{i+1}(R_{i+1}^{(\ell_{i+1})})$ was defined, the simulator S_1 would have avoided their XOR. Hence these orderings are *impossible*.

2. $b_{i-2} \leq \ell_{i+1} < b_{i-1}$ or $\ell_{i+1} < b_{i-2} < b_{i-1}$. These orderings are *impossible* since here we can make a similar argument for $h_{i-1}(R_{i-1}^{(b_{i-1})})$.

3. $b_{i-1} \leq \ell_{i+1} < b_{i-2}$. This ordering is *possible*.

4. $\ell_{i+1} < b_{i-1} < b_{i-2}$. This ordering is also *possible*.

We note here a couple of things about these possible orderings before we move on. First, query b_{i-1} could have only been made before query b_{i-2}. Secondly, query b_{i-2} could not have been made before the query ℓ_{i+1}. Now starting with this ordering defined between queries b_{i-1} and b_{i-2}, we can deduce the order in which queries $b_1 \ldots b_{i-3}$ could have been made. The analysis of this will be pretty much the same as that for $\ell_1 \ldots \ell_{top}$, with one major difference. Here the only possible order amongst $b_1 \ldots b_{i-1}$ we will get will be a descending order

$b_1 > b_2 > \ldots > b_{i-1}$. That is query b_1 was made before b_2 which was made before b_3 and so on. This happens because we were able to establish a strict order between b_{i-2} and b_{i-1}, which was not the case for ℓ_{top-1} and ℓ_{top}. Thus the $i-1$ queries, $b_1 \ldots b_{i-1}$, had to be made in strict decreasing order. This fact turns out to be really crucial since we do not lose half of the queries at this level of the "Query tree", as we did in the case of query number m.

Thus for each of the queries ℓ_i at the first level, we have at least $i-2$ queries that lie strictly in between ℓ_i and ℓ_{i+1}. Note that the same counting method can be extended to the b_i queries to show that there are $i-2$ queries strictly in between b_i and b_{i+1}, and so on. This query structure takes the shape of a *Fibonacci tree*. Since queries at any level lie strictly in between two consecutive parent level queries, it turns out that each of the queries in the tree is, in fact, different! An example this query structure is shown in figure 4.

Let $T(i)$ represents the number of queries in a "query tree" starting with $R_i^{(m)}$ (thus $T(1) = 1$). From the structure of the "query tree", we can compute that $T(i) = T(i-1) + T(i-2)$. But this is exactly the expression for the ith *Fibonacci number*. We will not recompute this expression here and just state that $T(i) = \mathcal{O}\left(\phi^i\right)$ where $\phi = \frac{\sqrt{5}+1}{2}$. And thus if an inconsistency occurs in query number m, then $m = \mathcal{O}\left(T\left(\frac{k}{4}\right)\right) = \mathcal{O}\left(\phi^{\frac{k}{4}}\right)$, which is superpolynomial in the security parameter λ, if $k = \omega(\log(\lambda))$. ◻

Thus for any distinguisher D that makes q queries ($q = poly(\lambda)$), it is the case that $Pr[D^{(\mathcal{M}^\pi, \mathcal{T}_{S_1^\pi})}] = Pr[D^{(\mathcal{M}^{\mathcal{T}_{S_1^\pi}}, \mathcal{T}_{S_1^\pi})}]$.

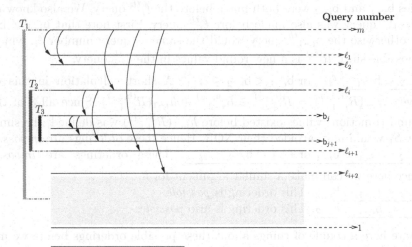

Fig. 4. An example of a "Fibonacci tree" formed by queries (showing three levels)

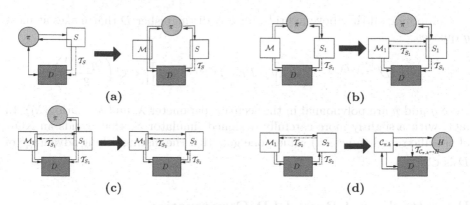

Fig. 5. Overall Game Structure

Removing the Random Permutation π. Until now, all queries are forced to be consistent with π. Now we will modify the simulator S_1 and get closer to the actual random oracle scenario. The new simulator, which we shall denote by S_2, does not attempt to output transcripts consistent with π. As before it implements the k round LR-construction with randomly assigned internal round functions. But now it also implements the last (or first) couple of round functions h_{k-1}, h_k (or h_2, h_1) with randomly chosen values (see figure 5c).

To illustrate this, when the new simulator S_2 gets a forward query $(0, x)$. It computes $R_0 = x|_L$, $R_1 = x|_R$ and assigns random values to $h_1(R_1), \ldots, h_k(R_k)$. It then sends the round values $R_0, \ldots, R_{k+1}$ as the transcript for the query $(0, x)$. Inverse queries are handled in a symmetrical fashion. The relaying algorithm, $\mathcal{M}_1$, as before uses these transcripts to compute its responses to D's queries.

Note that the distinguisher cannot tell this scenario apart from the previous scenario, unless

- the new simulator S_2 violates the XOR constraint satisfied by S_1. We call this event B_3.
- the old simulator S_1 exits with failure. We call this event B_4.

Lemma 3 implies that the event B_4 does not happen for any distinguisher D that makes a polynomial number of queries. Thus for any distinguisher D making at most a polynomial number of queries q,

$$\left| Pr\left[D^{(\mathcal{M}^{T_{S_1^\pi}}, T_{S_1^\pi})} \right] - Pr\left[D^{(\mathcal{M}^{T_{S_2}}, T_{S_2})} \right] \right| \leq Pr\left[B_3 \right] = \mathcal{O}\left(\frac{(q.k)^4}{2^n} \right)$$

Onto the Random Oracle Model. Note that the previous scenario is essentially the same as the random oracle scenario, since all round function values chosen by S_2 are random. Therefore for any distinguisher D (figure 5d), we have
$$Pr[D^{(\mathcal{M}^{T_{S_2}}, T_{S_2})}] = Pr[D^{(C_{\pi,k}^H, T_{C_{\pi,k} \leftrightarrow H})} = 1].$$

Combining all the above hybrids, for any distinguisher D that makes at most q queries,

$$\left| Pr\left[D^{(C^H_{\pi,k}, \mathcal{T}_{C_{\pi,k} \leftrightarrow H})} = 1 \right] - Pr\left[(D^{\pi, \mathcal{T}_{S^\pi}}) = 1 \right] \right| < \mathcal{O}\left(\frac{(q \cdot k)^4}{2^n} \right)$$

Here q and n are polynomial in the security parameter λ, and $k = \omega(\log(\lambda))$. In fact, with a slightly more carefully designed simulator S_1 that avoids an XOR of specific round (function) values, one gets that the distinguishing advantage of D is $\mathcal{O}\left(\frac{q^4}{2^n} \right)$

B   Attack on 4 Round LR-Construction

We will represent the round values of the construction $C_{\pi,4}$ as $R_0, R_1 \ldots R_4, R_5$, such that $C_{\pi,4}(R_0 \| R_1) = (R_4 \| R_5)$. And the round functions will be denoted as $h_1, \ldots, h_4$. Now consider any simulator S for which we get the two scenarios: $(C_{\pi,4}, H)$ and (π, S). We will design a distinguisher D that distinguishes these two with high probability for any simulator S.

The distinguisher D essentially forces the simulator to satisfy a constraint that holds with very low probability for an RP π. On the other hand, it always holds for the LR-construction $C_{\pi,4}$. The algorithm of D is as follows:

1. Choose 3 arbitrary n bit strings, R_2, R_2', R_3.
2. Query the random oracle H to get $h_2(R_2)$, $h_2(R_2')$ and $h_3(R_3)$, in this order.
3. Compute $R_1 = h_2(R_2) \oplus R_3$ and $R_1' = h_2(R_2') \oplus R_3$.
4. Query the random oracle to get $h_1(R_1)$ and $h_1(R_1')$. Compute $R_0 = h_1(R_1) \oplus R_2$ and $R_0' = h_1(R_1') \oplus R_2$.
5. Query the random permutation on $R_0 \| R_1$ and $R_0' \| R_1'$ to get the values $R_4 \| R_5$ and $R_4' \| R_5'$, respectively.
6. Check if $R_4 \oplus R_4' = R_2 \oplus R_2'$. If so, then output 1 else output 0

Note that the values R_2 and R_2' were queried upon before R_3. Hence the round values R_1 and R_1' are completely arbitrary round values controlled by the distinguisher. The distinguisher D always outputs 1 when given access to the construction $C_{\pi,4}$. But when given access to the random permutation, the simulator S will need to find $h_1(R_1)$ and $h_1(R_1')$ that satisfy the constraint:

$$\pi((h_1(R_1) \oplus R_2) \| R_1)|_L \oplus \pi((h_1(R_1') \oplus R_2') \| R_1')|_L = R_2 \oplus R_2'$$

In this equation R_1, R_1', R_2 and R_2' are all effectively chosen by the distinguisher. Hence no efficient simulator can find two round function values $h_1(R_1)$ and $h_1(R_1')$ that satisfy the above constraint with non-negligible probability for a random permutation π.

Intrusion-Resilience Via the Bounded-Storage Model*

Stefan Dziembowski**

Institute of Informatics,
Warsaw University, Poland
and
Institute for Informatics and Telematics
CNR Pisa, Italy

Abstract. We introduce a new method of achieving intrusion-resilience in the cryptographic protocols. More precisely we show how to preserve security of such protocols, even if a malicious program (e.g. a virus) was installed on a computer of an honest user (and it was later removed). The security of our protocols relies on the assumption that the amount of data that the adversary can transfer from the infected machine is limited (however, we allow the adversary to perform any efficient computation on user's private data, before deciding on what to transfer). We focus on two cryptographic tasks, namely: session-key generation and entity authentication. Our method is based on the results from the Bounded-Storage Model.

1 Introduction

In the contemporary Internet environment, computers are often exposed to attacks of malicious programs, which can monitor the machines and steal the secret data. This type of software can be secretly attached to seemingly harmless programs, or can be installed by worms or viruses. In order to protect against these threats a user is usually advised to use virus and spyware removal tools. These tools need to be frequently updated (as the new viruses spread out very quickly). Nevertheless, for an average PC user it is quite inevitable that his computer is from time to time infected by a malicious process (which is later removed by an appropriate tool).

This phenomenon can be particularly damaging if the user runs some cryptographic programs on his machine. This is because in most of cryptographic tasks (encryption, authentication) the user needs to posses (and store somewhere) a secret key s. If the user does not store s outside of the machine (e.g. on a trusted hardware that will later participate in the protocol), then it seems that there

---

* This is an extended version of a report [Dzi05] that appeared on the eprint archive.
** Partially supported by the EU ECRYPT grant IST-2002-507932 and by the Polish KBN grant 4 T11C 042 25. Part of this work was carried out during the tenure of an ERCIM fellowship. Another part of this work was done when the author was employed at the Institute of Mathematics of the Polish Academy of Sciences.

is little that can be done to preserve the security, as the malicious process can always steal s (and then impersonate the honest user, or decrypt his private communication). If the protocol is based on a password memorized by the user then the virus can wait until the password is typed and then record the key-strokes.

In this paper we propose a method for constructing *intrusion-resilient cryptographic protocols*, i.e. such protocols that remain secure even after the adversary gained access to the victim's machine (and later lost this access). The security of our protocols is based on a novel assumption that the amount of data that the adversary is allowed to transfer from the victim's machine is limited (however, we allow the adversary to perform any efficient computation on user's private data, before deciding on what to transfer). In the security proofs we make use of the theory of the Bounded Storage Model (see Section 3).

1.1   Previous Work

Intrusion-resilience was introduced in [IR02] (see also [DFK+03]) and can be viewed as a combination of forward and backward security.[1] A cryptosystem is *forward-secure* if an exposure of a secret key at some particular time t does not affect the security of the sessions of the protocol that ended *before t*. It was studied in context of key-exchange (see e.g. [DvOW92, Kra96]), digital signatures (this research was initiated by Ross Anderson, see [And02]) and public-key encryption [CHK03]. A cryptosystem is *backward-secure* if the exposure of a secret key at time t does not affect the security of the sessions of the protocol that started *after t*. So far this was achieved by distributing the secret key among a group of participants (e.g. in [IR02] this group consist of two players: a *signer* and a *home base*). One has to make an assumption that the entire group is never compromised by the adversary at the same time.

Cryptosystems that remain secure even in case of a partial leakage of the secret key were already studied in the area of *Exposure-Resilient Cryptography* (see e.g. [Dod00]). The differences from our model are as follows: (1) they consider only the leakage of *individual* bits of the secret keys and (2) the keys in their protocols are short.

Our model can be viewed as a generalization of the model of Kelsey and Schneier [KS99]. In their model the adversary is allowed to access individual bits of the secret key (this is justified by an assumption that the access to the memory is slow). In this model they show a simple authentication protocol (the secret key is a long random string of bits; in order to verify the authenticity of the client the server asks for the values of some randomly chosen positions of the secret key). In Sect. 5.2 we show that this protocol is also secure in our model.

Independently[2] (but earlier) a similar model was introduced by Dagon et al. [DLL05]. They propose a system (called *VAST*) for securely storing secret data on devices that can be subject to an intrusion e.g. by a virus. They assume that

---

[1] There seems to be some confusion in the literature about the terminology. What is called *forward security* in [And02] is called *backward security* in [IR02, DFK+04]. In this paper we use the terminology of [IR02].

[2] We became aware of this work after submitting our paper to TCC.

such data is encrypted by a weak (human-memorized) password (let T denote the resulting ciphertext and let π be the password). To prevent the adversary from downloading T and cracking the password (i.e. performing a dictionary attack on π) on his own machine, they design their protocols so that T is too large to be fully downloaded. In order for this to make sense they need to assume that the computing power of the virus is limited (so the virus cannot perform the password-cracking on the victim's machine). This is in contrast to our model, where we can grant the virus a right to perform an arbitrary (polynomial-time) computation on the victim's data. Another difference is that they assume that the adversary does not have a full access to the victim's machine. In particular when the user is interacting with VAST the virus should not have access to the keyboard. This is because when the user enters the password π to the machine the virus can learn π by recording the key-strokes.

1.2   Our Contribution

We propose a new method for constructing intrusion-resilient protocols for the session-key generation and entity authentication (the main novelty of our approach is the new method of achieving backward-security; the forward-security is achieved in a fairly standard way). The assumption that we make is that the secret key is of huge size (e.g. K is of size 5 GB). More precisely, we will grant the adversary the power to break into the honest user's machine and take full control over it. We will assume that the adversary is able to perform arbitrary (efficient) computation on victim's data. Clearly, during the period of the break-in one cannot hope for much security, since the adversary has a complete knowledge about the behavior of one of the honest users (and hence she can e.g. impersonate the user or steal the session key). So the intrusion-resilience is the maximum what we can hope for. We achieve it by assuming that the amount of data that the adversary can retrieve is much smaller than K (say it is 0.5 GB). This assumption may be quite practical as in many situations transmitting unnoticeably 0.5 GB of data is hard. Observe that if the secret key is of size 1 KB then the virus can e.g. post it on some Usenet group, so that the author of the virus can download it anonymously. Clearly this is much harder if the secret is huge.

Another motivation is that protocols that are secure in our model have a high level of resiliency against side-channel analysis [KSWH00]. Recall that the side-channel attacks allow the adversary to obtain some information about the users' secrets by observing the behavior of the implementation of the protocol. In practice the full protection against such attacks is hard, and we can only hope for minimizing the amount of leaked information. The assumptions that we make in our model guarantee that even if some information about the secrets is leaked, the protocols are still secure.

Our method is based on the theory of the Bounded-Storage Model (see Sect. 3). In the BSM one constructs protocols secure under the assumption that the amount of data that the adversary can store is smaller that the amount of data that can be broadcasted (e.g. by a satellite). The fact that he theory of the BSM has applications here may seem surprising at the first sight, as in some sense the assumptions in the BSM are opposite to ours. However, as it turns out, these models show similarities and in fact theorems that were proven in the BSM are useful for us.

Our exposition is rather informal, as we mostly aim at introducing the model and showing its power, not at providing ready to use practical solutions for concrete problems. For the same reason we do not provide numerical examples and we do not give comparisons between security levels of different schemes presented in the paper. Nevertheless, we believe that the protocols provided here (or their variants) may find practical applications.

Finally, let us note that our results are proven in the random oracle model (see Sect. 2.4).

1.3   The Contribution of [CDD+05]

The entity authentication protocol that we present in our paper was independently constructed and analyzed by Cash et al. [CDD+05]. Moreover, they improve our results by constructing a session-key generation protocol without the random oracle assumption. They also provide some concrete numerical examples of the parameter values that can be used in practical implementations.

2   Preliminaries

2.1   Probability Theory

The *min-entropy* of a probability distribution P_X is defined as

$$H_\infty(X) := \min_{x \in \mathcal{X}}(-\log_2(P_X(x))).$$

If X is a random variable and A is an event then P_X is the distribution of X and $P_{X|A}$ is a conditional distribution of X given A. In this case we define $H_\infty(X) := H_\infty(P_X)$ and $H_\infty(X \mid A) := H_\infty(P_{X|A})$. For more on min-entropy and its relation to the standard Shannon entropy see e.g. [Cac97].

Let the *statistical distance* between random variables X and X' distributed over the same set $\mathcal{X}$ be defined as

$$\delta(X, X') := \frac{1}{2} \sum_{x \in \mathcal{X}} |X(x) - X'(x)|$$

We will also say that X is $\delta(X, X')$-*far from* X'. If U is a random variable with uniform distribution over $\mathcal{X}$ then define $d(X) := \delta(X, U)$. The above notation extends in a natural way to probability distributions.

2.2   Message Authentication Codes

We will use the following (simplified) security definition of the Message Authentication Codes (*MACs*). For a more complete definition the reader may consult e.g. [Gol04]. *MAC* is an algorithm which takes as an input a security parameter 1^k, a random secret key $S \in \{0,1\}^{\lambda(k)}$ (where λ is some polynomial) and a message $M \in \{0,1\}^*$. It outputs an *authentication tag* $MAC_S(M,1^k))$ (we will sometimes drop 1^k). It is *secure against an adaptive chosen-message attack* if any probabilistic polynomial time (PPT) adversary (taking as input 1^k) has negligible[3] (in k) probability of producing a valid pair $(M, MAC_S(M,1^k))$, after seeing an arbitrary number of pairs

$$(M_1, MAC_S(M_1,1^k)), (M_2, MAC_S(M_2,1^k)) \ldots$$

(where $M \notin \{M_1, M_2, \ldots\}$), even when $M_1, M_2, \ldots$ were adaptively chosen by the adversary.

2.3 Public-Key Encryption

A *public-key encryption scheme* is a triple $(G, encr, decr)$, where G is a PPT *key-generation algorithm* taking as input 1^k and returning as output a (private-key, public-key) pair (E, D), $encr$ is an polynomial-time algorithm taking as input 1^k, a message $M \in \{0,1\}^*$ and a public key E and returning a *ciphertext* $C = encr_E(M)$, and $decr$ is an algorithm taking as input a private key D a ciphertext C and returning a message $M' = decr_D(C)$. We require that always $M = decr_D(encr_E(M))$. Let $\mathcal{E}$ be a polynomial time adversary which is given 1^k and E. Her goal is to win the following game. She produces two messages M_0 and M_1 (of the same length). Then, she is given a ciphertext $C = encr_S(M_r)$, where $r \in \{0,1\}$ is random. She has to guess r. We say that $(G, encr, decr)$ is *semantically secure* [GM84] if any polynomial time adversary has chances at most negligibly (in k) better that 0.5. More on the definitions of secure public-key encryption can be found e.g. in [Gol04].

2.4 Random Oracle Model

We prove the security of our protocol in the *Random Oracle Model* [BR93]. More precisely, we will model a hash function $H : \{0,1\}^i \to \{0,1\}^j$ as a *random oracle*, i.e. a black box containing a random function $h : \{0,1\}^i \to \{0,1\}^j$. We assume that every party (including the adversary) has access to this oracle, i.e. can ask it for the value of h on any (chosen by her) arguments.

3   Bounded Storage Model

We will use the results from the Bounded-Storage Model, introduced by Maurer in [Mau92]. So far, this model was studied in the context of *information-theoretically secure* encryption [ADR02, DM04b, Lu04, Vad04, Din05], key-agreement [CM97, DM04a], oblivious transfer [CCM98, Din01, DHRS04] and time-stamping [MSTS04]. In this model one assumes that a random t-bit string R (called a *randomizer*) is either temporarily available to the public (e.g. the signal of a deep

---

[3] A function $f : \mathcal{N} \to \mathcal{R}$ is *negligible (in k)* if for every $c \geq 1$ there exists k_0 such that for every $k \geq k_0$ we have $|f(k)| \leq k^{-c}$.

space radio source) or broadcast by one of the legitimate parties. We assume that the memory s of the adversary is smaller than t and therefore she can store only partial information about R. It has been shown in [ADR02, DM04b, Lu04, Vad04] that under this assumption the legitimate parties, Alice and Bob, sharing a short secret key Y initially, can generate a very long n-bit one-time pad X with $n \gg |Y|$ about which the adversary has essentially no information.

More formally, Alice and Bob share a short secret *initial key* Y, selected uniformly at random from a key space $\mathcal{Y}$, and they wish to generate a much longer n-bit *expanded key* X (i.e. $n \gg \log_2 |\mathcal{Y}|$). In a first phase, a t-bit random string R is available to all parties, i.e., the randomizer space is $\mathcal{R} = \{0,1\}^t$. Alice and Bob apply a known *key-expansion function*

$$f : \mathcal{R} \times \mathcal{Y} \to \{0,1\}^n$$

to compute the expanded key as $X = f(R, Y)$. Of course, the function f must be efficiently computable and based on only a very small portion of the bits of R, so that Alice and Bob need not read the entire string R.

The adversary Eve $\mathcal{E}$ can store arbitrary s bits of information about R, i.e., she can apply an arbitrary storage function

$$h : \mathcal{R} \to \mathcal{U}$$

for some $\mathcal{U}$ with the only restriction that $|\mathcal{U}| \leq 2^s$. The memory size during the evaluation of h does not need to be bounded. The value stored by Eve is $U = h(R)$. After storing U, Eve looses the ability to access R. All she knows about R is U. In order to prove as strong a result as possible, one assumes that Eve can now even learn Y, although in a practical system one would of course keep Y secret.

A key-expansion function f is secure in the bounded-storage model if, with overwhelming probability[4], Eve, knowing U and Y, has essentially no information about X. To be more precise, let us introduce a security parameter k which is an additional input of f and of Eve. Let us assume that the length of the randomizer, the size of Eve's memory and the length of the output of f are functions of k, i.e. $t = \tau(k)$, $s = \sigma(k)$, and $n = \nu(k)$ (with $\nu(k) \geq k$). Also, assume that the set of the initial keys is always equal to $\{0,1\}^{\mu(k)}$, for some function μ. We say that function f is (σ, τ, ν, μ)-*secure in the bounded-storage model* if for any Eve (with memory at most $\sigma(k)$) the statistical distance of the conditional probability distribution $P_{X|U=u,Y=y}$ from uniform distribution over the $\nu(k)$-bit strings is negligible, with overwhelming probability over values u and y. Above we assumed that the adversary and the function f are deterministic, but note that we would not loose any security by allowing them to be randomized.[5]

---

[4] Formally, a sequence of probabilities $p_0, p_1, \ldots$ is *overwhelming* if the function $f(k) = 1 - p_k$ is negligible.

[5] Formally we could do it by allowing $\mathcal{E}$ and f to take extra random inputs $r_\mathcal{E}$ and r_f, resp. This does not give any extra power to the adversary, for the following reasons: (1) the input r_f is obsolete since if $\mathcal{E}$ is randomized then having r_f clearly does not change anything as $\mathcal{E}$ can simply choose r_f herself and encode it into the description of f; (2) the input $r_\mathcal{E}$ is obsolete since a computationally unbounded $\mathcal{E}$ can always (for any value of k) find the optimal $r_\mathcal{E}$.

Several key expansion functions were proven secure in the past couple of years (see for example [ADR02, DM04b, Lu04, Vad04]). In the next section we present an example of such a function, taken from [DM04b]. We have chosen the function of [DM04b] because we believe that it is the simplest one. The reader familiar with the BSM literature can safely skip the next section.

3.1   The Scheme of [DM04b]

The randomizer $R \in \mathcal{R} = \{0,1\}^t$ is interpreted as being arranged in a matrix with m rows, denoted $R(1), \ldots, R(m)$, for some $m \geq 1$ called the *height* of the randomizer. Each row consists of $l + n - 1$ bits, for some $l \geq 1$ called the *width* of the randomizer. Hence $t = m(l+n-1)$ and R can be viewed as an $m \times (l+n-1)$ matrix (see Fig. 1). The initial key $Y = (Y_1, \ldots, Y_m) \in \mathcal{Y} = \{1, \ldots, l\}^m$ selects one starting point within each row, and the expanded key $X = (X_1, \ldots, X_n)$ is the component-wise XOR of the m blocks of length n beginning at these starting points Y_i, i.e.,

$$X = f(R, Y),$$

where $f : \mathcal{R} \times \mathcal{Y} \to \{0,1\}^n$ is defined as follows. For $r \in \mathcal{R}$ and $Y = (Y_1, \ldots, Y_m) \in \mathcal{Y}$,

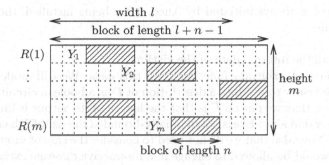

Fig. 1. Illustration of the scheme for deriving an expanded n-bit key $X = (X_1, \ldots, X_n)$, to be used as a one-time pad, from a short secret initial key $Y = (Y_1, \ldots, Y_m)$. The randomizer R is interpreted as a $m \times (l + n - 1)$ matrix with rows $R(1), \ldots, R(m)$ of length $l+n-1$. The expanded key X is the component-wise XOR of m blocks of length n, one selected from each row, where Y_i is the starting point of the ith block within the ith row $R(i)$.

$$f(R, Y) := \left(\bigoplus_{i=1}^{m} R(i, Y_i), \ldots, \bigoplus_{i=1}^{m} R(i, Y_i + n - 1) \right), \tag{1}$$

where $R(i,j)$ denotes the jth bit in the ith row of R. This is illustrated in Fig. 1.

The above function f was proven secure in [DM04b], assuming that memory of the adversary has a size that is a constant fraction $c < 1$ of the randomizer. For the practically looking parameters this constant should be around 8%, i.e. $\sigma(k) := \tau(k) \cdot 0.08$. See [DM04b] for details.

4 Intrusion-Resilient Session-Key Generation

By *session-key generation* we mean a protocol that allows two parties (that share a long-term symmetric key) to agree securely on a session key even in presence of a malicious adversary that can obstruct their communication. Below, we describe what we mean by *intrusion-resilient* session-key generation.

4.1 An Informal Description of the Model

First, let us fix the basic terminology. The honest users Alice A and Bob B will be attacked by a (polynomially bounded) *adversary* Eve $\mathcal{E}$. The adversary is allowed (1) to eavesdrop and to store the entire communication between Alice and Bob (2) to fabricate messages or to prevent them from arriving and (3) to (periodically) install malicious programs on the honest user's machines (see below). Such a program will be called a *virus*. We assume that the honest users share a long-term secret key K generated randomly. The time is divided into sessions $T_1, T_2, \ldots$ (the number of sessions will be bounded). At the beginning of the session the users are allowed to get some fresh random input. At the end of each session T_i the users output a new *session key* κ_i. (In practice, once κ_i is generated, the users will utilize κ_i for secure communication.) For simplicity assume that each execution of the protocol is always initiated by Alice. After being installed, the virus can do the following.

1. Read all the internal data of the victim.
2. Compute an arbitrary function Γ on this data. We will model it by asking the adversary to produce a description of Γ as a boolean circuit. The only restriction that we put on Γ is that the length of its output is limited (observe however that since Eve is polynomially-bounded the size of Γ has to be polynomial). Note also that we do not need to consider the case of *interactive* viruses (that would be allowed to engage in a interactive massage exchange with the adversary), since the circuit may contain the description of the entire state of the adversary.
3. Send the result of the computation back to the adversary.

Note, that we assume that the adversary is not allowed to modify the programs running on the users' machines. Informally speaking the goal of the adversary is to successfully *break some test session T_{test}* (of her choice), by achieving one of the following goals:

1. learn κ_{test},
2. convince at least one of the players to accept some κ'_{test} about which the adversary has some significant information, or
3. make A and B agree on different keys.

Clearly, if the adversary installs a virus on one of the users' machines in session T_{test} then she can instruct the virus to retrieve κ_{test} (since in a usual scenario the session key κ_i is short[6]). Therefore, we are interested only in the adversary breaking those sessions T_{test} during which no virus was installed (neither on the machine of A nor on the one of B).

Traditionally when considering forward security (see e.g. [Kra96]) one allows the adversary to learn all the session keys except of the challenge key κ_{test}. In our model this ability of $\mathcal{E}$ comes from the fact that the adversary can compromise all sessions except of T_{test} (we will actually allow the adversary to „compromise a session" that has already ended some time ago). Finally, let us remark that in this model we assume that the players can reliably erase their data (in particular, after the session T_i the players would erase κ_i). Actually, we will assume that the only data that is not erased between the sessions is the secret key K.

4.2   A More Formal Description of the Model

We are now going to define the model more formally. Our definitions are inspired by the definitions of the security of key-exchange protocols (esp. [CK01]). For the sake of simplicity we assume that the protocol is executed just between two fixed parties, and concurrent execution of the sessions is not allowed, i.e. the users simply execute one session after another. Giving a complete definition (e.g. in the style of [CK01]) remains an open task.

The *session-key generation scheme* is a tuple $(A, B, \alpha, \beta, \gamma, \delta, \chi)$, where $\alpha, \beta, \gamma, \delta, \chi$ are some polynomials and A and B are interactive Turing machines, taking as input a security parameter 1^k and a secret key $K \in \{0,1\}^{\alpha(k)}$. The adversary $\mathcal{E}$ is a PPT Turing Machine taking as input 1^k. The execution is divided into the sessions $T_1, T_2, \ldots, T_{\chi(k)}$. The execution of each T_i looks as follows:

1. The machines A and B receive uniformly (and independently) chosen random inputs $r_A \in \{0,1\}^{\beta(k)}$ and $r_B \in \{0,1\}^{\beta(k)}$ (respectively).
2. Machines start exchanging messages. The adversary can eavesdrop the messages. She can also prevent some of the messages from arriving to the destination and fabricate new messages. At the beginning A sends a unique message *start* to B (so the adversary knows that a new session started).
3. At the end of the session the machines (privately) output an agreed key $\kappa_i \in \{0,1\}^{\delta(k)}$. If the traffic was not disturbed by the adversary then they have to output the same value.
4. Now the adversary may choose to *compromise the session* T_i (each session T_i may be compromised at most once in the entire execution of the protocol). In this case the following happens.
    (a) Eve produces a description of a boolean circuit C (which models the virus) computing a function $\Gamma : \{0,1\}^w \to \{0,1\}^{\gamma(k)}$ (w is an arbitrary value). Clearly we will always have $\gamma(k)\chi(k) < \alpha(k)$, since otherwise Eve could retrieve the entire secret key K. The size of C is arbitrary (however, it has to be polynomial in the security parameter, as the adversary is polynomially-bounded).

---

[6] Even if one would develop a scheme in which κ_i is too large to be retrieved, the adversary could simply tell the virus to steal the data that is encrypted with κ_i.

Note that we assume a uniform bound $\gamma(k)$ on the amount of bits that the adversary is allowed to steal in each compromised session. More generally, one could give a bound on the *total* number of bits retrieved by the adversary in all compromised sessions.

(b) Eve learns the value of $\Gamma(r_A, r_B, K)$.

Observe that the function Γ ,,has a complete view" of the internal states of the parties during the session. Thus in particular the value of $\Gamma(r_A, r_B, K)$ may include the encoding of κ_i (if this is the wish of the adversary). Also note that our model is actually stronger than what we need in practice (as we assume that Γ has simultaneous access to both A and B, without restricting the amount of data that she needs to transfer between the parties, to perform the computation).

5. The adversary may decide to compromise a session (in the same way as in Point 4) even long time after the session T_i is finished (one can imagine that the descriptions of the states of A and B at the end of T_i are deposited somewhere and the adversary may decide to access them at any later time). This may seem an artificial strengthening of the model. However, in fact it simplifies things, as it allows us to model the fact that κ_i may become known to the adversary at some later point. Alternatively, we could introduce a special type of session-key-queries [CK01] that the adversary may ask to learn κ_i after the end of T_i.

Let $\mathcal{C}$ be the set of all compromised sessions. Clearly, the adversary wins if for some session $T_i \notin \mathcal{C}$ users A and B outputted different keys. If this is not the case then at the end of the execution the adversary decides that some $T_{test} \notin \mathcal{C}$ will be her *test-session*. In this case her task will be to distinguish κ_{test} from a truly random key of the same length. Of course we need to require that at least one of A and B actually outputted some key κ_{test} (by blocking the message flow the adversary can clearly prevent the parties from reaching any agreement). The distinguishing game is as follows:

1. Let $r \in \{0, 1\}$ be random.
2. If $r = 0$ then pass κ_{test} to the adversary. Otherwise generate a random $\kappa' \in \{0, 1\}^{\delta(k)}$ and pass it to the adversary. The adversary outputs some $r' \in \{0, 1\}$. We say that she *won the distinguishing game* if $r = r'$.

Definition 1. *We say that a key generation scheme* $(A, B, \alpha, \beta, \gamma, \delta, \chi)$ *as above is* intrusion-resilient *if for any PPT* $\mathcal{E}$

1. *the chances that in some session* $T_i \notin \mathcal{C}$ *machines A and B outputted different keys are negligible (in k), and*
2. *the chances that $\mathcal{E}$ wins the distinguishing game, are at most negligibly (in k) greater than* $1/2$.

4.3   The Protocol for Intrusion-Resilient Session-Key Generation

Preliminaries. Let f be (σ, τ, ν, μ)-secure in the BSM. Let MAC be a message authentication scheme secure against adaptive chosen message attack. Assume that for a security parameter 1^k the length the secret key of MAC is $\lambda(k)$. Let $H : \{0,1\}^{\nu(k)} \rightarrow \{0,1\}^{\lambda(k)}$ be a hash function (modeled as a random oracle). Let $(G, encr, decr)$ be a semantically secure public-key encryption scheme. In order to achieve forward-security we will use the public-key encryption in a standard way (see e.g. [DvOW92, Kra96]): Alice will (1) generate an ephemeral (public key, private key) pair[7] and (2) send the public key (in an authenticated way) to Bob, Bob will generate the session key κ and send it (encrypted with Alice's public key) back to Alice (who can later decrypt κ).[8] Afterwards, the ephemeral keys are erased.

The Protocol. Fix some value of the security parameter k. Let $\mathcal{R} = \{0,1\}^{\tau(k)}$ and let $\mathcal{Y} = \{0,1\}^{\mu(k)}$. Assume that Alice and Bob share a random secret key $K = (R_A, R_B) \in \mathcal{R}^2$ and hence $\alpha(k) := 2 \cdot \tau(k)$. In each session T_i the players execute the following protocol.

1. Alice generates a random $Y_A \in \mathcal{Y}$ and sends it to Bob.
2. Bob generates a random $Y_B \in \mathcal{Y}$ and sends it to Alice.
3. Both parties calculate $S := f(R_A, Y_A) \oplus f(R_B, Y_B)$ and $S' := H(S)$.
4. Alice generates a public key E and sends $(E, MAC_{S'}(\text{A}{:}E))$ to Bob.
5. Bob verifies the correctness of the authentication tag. If it is correct then he generates a random κ_i and sends $(encr_E(\kappa_i), MAC_{S'}(encr_E(\text{B}{:}\kappa_i)))$ to Alice. He outputs κ_i.
6. Alice verifies the correctness of the authentication tag. If it is correct then she decrypts κ_i and outputs it.
7. The players erase all their internal data (including κ_i and random inputs), except of the long-term key K.

The role of labels „A:" and „B:" is to prevent the adversary from bouncing the message sent by Alice in Step 4 back to her in Step 5.

The Bound on the Amount of Retrieved Data. An important parameter that needs to be fixed is the amount of data that the virus can retrieve in each session, i.e. the value of $\gamma(k)$. If the adversary compromises some sessions than at any point of the execution of the scheme, then she knows the value of some function $\tilde{h}$ of K. We can think about $\tilde{h}$ as changing dynamically after each session. After execution of i sessions the length of the output of $\tilde{h}$ is at most the sum of

- $i \cdot \gamma(k)$ (since she could have compromised at most i sessions so far), and
- $i \cdot \lambda(k)$ (since she could have learned i keys of the MAC scheme[9])

Since the maximal number of sessions is $\chi(k)$ we know that the output of $\tilde{h}$ is of a length at most

$$\chi(k) \cdot (\gamma(k) + \lambda(k)).$$

---

[7] *Ephemeral key* is a key that is generated just for some particular session (and it is erased later).
[8] In [DvOW92, Kra96] it is actually done by exchanging Diffie-Hellman ephemeral keys, i.e. doing authenticated Diffie-Hellman key agreement.
[9] We have to add it because the definition of the security of MAC does not imply the secrecy of all the bits of the key.

Therefore if we want this value to be at most $\sigma(k)$ we have to set

$$\gamma(k) := \sigma(k)/\chi(k) - \lambda(k). \tag{2}$$

This ensures that the information that Eve has about K is at most $\sigma(k)$ bits.

4.4 The Security of the Protocol

We prove the following.

Theorem 1. *The protocol in Sect. 4.3 is intrusion resilient.*

Proof (sketch). Fix some uncompromised session T_i. Let us first consider the case when the adversary wants to break it by disrupting (by stealing and substituting messages) the communication. Let S_A and S_B be the values of S computed by A and B (resp.) in Step 3. If the execution of the protocol was not disturbed by the adversary then we have $S_A = S_B$. By the security of f in the BSM, the adversary has almost no information about the values S_A and S_B (i.e. their distribution is negligibly far from uniform from her point of view). Note that this holds even if she was disrupting the communication between the parties. The only thing that the adversary could possibly do is to force S_A and S_B to be such that they are not equal, but they are not independent either. For example by modifying the message Y_A (sent in Step 1) she could make $S_A \oplus S_B$ to be equal to some value $S_\oplus$ chosen by her.[10]

This is why, before using S, we hash it (in Step 3): $S' := H(S)$. Let $S'_A := H(S_A)$ and let $S'_B := H(S_B)$. Clearly the chances of $\mathcal{E}$ of guessing S_A or S_B are negligible. This is because the distributions of S_A and S_B are negligibly far from a uniform distribution over $\{0,1\}^{\nu(k)}$ and we assumed that $\nu(k) \geq k$. Therefore (since we model the hash function as a random oracle) we can assume that (except with negligible probability) from the point of view of $\mathcal{E}$ the distributions of the values S'_A and S'_B are entirely uniform. Moreover, one of the following has to hold (except with negligible probability):

1. $S'_A = S'_B$, or
2. S'_A and S'_B are independent.

Assume that the first case holds. Then, the adversary is not able to fabricate messages in Steps 4 and 5, without breaking the MAC. The security of κ_i follows now from the security of the encryption scheme (if the adversary could distinguish κ_i from a random key, then she could clearly break the semantic security of $(G, encr, decr)$).

---

[10] Consider for example the scheme from Sect. 3.1. Write $Y_A = (Y_1, \ldots, Y_m)$. Suppose the adversary stored the first row $(R_A(1))$ of R_A (she should have enough memory to do it) and she modified $Y_A = (Y_1, \ldots, Y_m)$ (sent is Step 1) only on the first component (Y_1). Let Y'_A be the result of this modification. Clearly almost always $f_A(R_A, Y_A) \neq f_A(R_A, Y'_A)$; however, $f_A(R_A, Y_A) \oplus f_A(R_A, Y'_A)$ (and hence $S_A \oplus S_B$) is known to the adversary.

In the second case, the parties easily discover that the adversary was interfering with their communication. This is because if the adversary wants to prevent them from discovering this, then she needs to create (in Steps 4 and 5) valid pairs (message, MAC), without having any information about the secret keys. Again, she cannot do it without breaking the MAC.

Now suppose that the adversary wants to distinguish κ_i from a random key, after the session is completed. If she compromises some future session T_j then she can of course recover the key S' used in session T_i (if she stored Y_A and Y_B from T_i). However, now it is too late (as the key S' is used only for authentication). Therefore, the security of κ_i again follows from the semantic security of the encryption scheme.   □

4.5   An Alternative Protocol

In this section we show another variant of the protocol from Sect. 4.3. The main difference is that instead of using a BSM-secure key derivation function f, we will use a function $\tilde{f} : \mathcal{R} \times \mathcal{Y} \rightarrow \{0,1\}^k$ that is not BSM-secure, but still works for our purposes. Again, let k be a security parameter and suppose that the randomizer R is a random element from $\mathcal{R} = \{0,1\}^{\tau(k)}$. Let $\mathcal{Y} := \{(Y_1, \ldots, Y_k) \in \{1, \ldots, \tau(k)\}^k \mid Y_1 < \cdots < Y_k\}$. Thus $\mathcal{Y}$ can be viewed as a set of all k-element subsets of $\{1, \ldots, \tau(k)\}$. First, define

$$\varphi((R_1, \ldots, R_{\tau(k)}), (Y_1, \ldots, Y_k)) := (R_{Y_1}, \ldots, R_{Y_k}).$$

Let H be a hash function. We set

$$\tilde{f}(R, Y) := H(\varphi(R, Y)).$$

In other words: we just pick random positions of the secret key, concatenate them and hash the result. Of course usually $\tilde{f}$ is not secure in the BSM as the hash functions belong to the complexity-theoretic world. However, if we model H as a random oracle, then the value of $\tilde{f}(R, Y)$ is random from the point of view of the adversary, unless she managed to guess the value of $\varphi(R, Y)$. So, if we want to use $\tilde{f}$ instead of f in the protocol from Sect. 4.3, then we have to show that the probability of any adversary of guessing $\varphi(R, Y)$ correctly, is negligible (for the appropriate choice of the parameters), even when the adversary is given $h(R)$ and Y (for some $h : \{0,1\}^{\tau(k)} \rightarrow \{0,1\}^{\sigma(k)}$ chosen by her). If we model the adversary's guess as a function g we can formalize this requirement as follows.

Lemma 1. *Suppose* $\sigma(k) = (1 - \delta)\tau(k) - k$, *for an arbitrary* $\delta > 0$. *For arbitrary functions* $h : \{0,1\}^{\tau(k)} \rightarrow \{0,1\}^{\sigma(k)}$ *and* $g : \{0,1\}^{\sigma(k)} \rightarrow \{0,1\}^k$ *we have that*

$$P(\varphi(R, Y) = g(h(R), Y)) \tag{3}$$

is negligible.

For the proof we need two other lemmas. The first lemma (proven in [CM97], see Lemma 3) is quite simple. It roughly states that the knowledge of s bits of a random string R reduces its min-entropy by around s, with a high probability.

Lemma 2 ([CM97]). *Let R be a random variable uniformly distributed over $\{0,1\}^t$. Let $h : \{0,1\}^t \to \{0,1\}^s$ be an arbitrary function. Then, with probability at least $1 - 2^k$ the variable $h(R)$ takes a value u such that*

$$H_\infty(R \mid h(R) = u) \geq t - s - k.$$

The second lemma (proven in [NZ96], see Lemma 11) is more complicated. Informally speaking, it states that if $R \in \{0,1\}^t$ is a random string with min-entropy $\delta \cdot t$ and $Y \in \mathcal{Y}$ is chosen uniformly at random, then $\varphi(R, Y) \in \{0,1\}^k$ has (with high probability) a min-entropy close to $\delta'k$, where δ' is some constant.

Lemma 3 ([NZ96]). *Let P_R be a probability distribution over $\{0,1\}^t$ with min-entropy δt. Suppose R is chosen according to P_R. Then, with probability at least $1 - \epsilon$ (over the choice of $y = Y$) the distribution of $P_{\varphi(R,y)}$ is ϵ-far from some distribution $P_{X'}$ whose min-entropy is $\delta'k$ where $\delta' := c\delta/\log(\delta^{-1})$ and $\epsilon := \max(2^{-ck}, 2^{-c\delta'l})$ for some constant c.*

Actually, the lemma that is proven in [NZ96] is stronger, as it does not require Y to be entirely uniform (see [NZ96] for details). We are now ready for the proof of Lemma 1.

Proof (of Lemma 1). To simplify the notation we set $s := \sigma(k)$ and $t := \tau(k)$. First, observe that by Lemma 2 we have that (except with a negligible probability 2^{-k}) the variable $h(R)$ takes a value u such that

$$H_\infty(R \mid h(R) = u) \geq t - s - k = \delta t. \tag{4}$$

So, suppose that such u was selected. We are now going to apply Lemma 3. Thus set $\delta' = c\delta/(\log \delta^{-1})$ and $\epsilon = \max\left(2^{-ck}, 2^{-c\delta'k}\right)$ (where c is some constant). Observe that δ' is constant and ϵ is negligible. Therefore (by Lemma 3) we know that with overwhelming probability Y took a value y such that the conditional distribution of

$$P_{\varphi(R,Y) \mid h(R)=u, Y=y} \tag{5}$$

is at most ϵ-far from a distribution $P_{X'}$ with min-entropy $\delta' t$. Assume that this indeed happened. If we want to maximize (3) we have to let g choose an element with the maximal probability according to the distribution $P_{\varphi(R,Y) \mid h(R)=u, Y=y}$. Clearly this probability is at most $2^{-H_\infty(P_{X'})} + \epsilon = 2^{-\delta' t} + \epsilon$ which is negligible in k.

5   Intrusion-Resilient Entity Authentication

In this section we informally describe a practical intrusion-resilient method for entity authentication. In order to achieve such entity authentication one could of course use the scheme from Sect. 4; however, this is an overkill and for practical applications a much simpler method suffices. The idea is as follows. We will construct an *intrusion-resilient* scheme that allows a user U to authenticate to a server S. We will consider *only intrusions into* U. This corresponds to a practical situation in which the computers of the users are usually much more vulnerable for the attacks then the computer of the server.

Assume that the parties have already established a channel C between S and U that is authentic only from the point of view of the user, i.e. U knows that (1) whatever comes through this channel is sent by U and (2) whatever is sent through it can be read only by U. Now, the user wants to authenticate to the server. This is a typical scenario on the Internet, where C is established e.g. using SSL (and the server authenticates with a certificate). In practice usually U authenticates to S by sending his password over C. This method is clearly not intrusion-resilient because once a virus enters the machine of U he can retrieve the password (or record the key-strokes if the password is memorized by a human).

In this section we propose an authentication method that is intrusion-resilient (in the same sense as the protocols in the previous sections). Again, we will use the assumption that the secret key K of the user is too large to be fully downloaded. We allow the virus to perform arbitrary computations[11] of the victim's machine.

5.1   Our Protocol

Let f be a function that is (σ, τ, ν, μ)-secure in the BSM. Fix some security parameter k. The secret key K is simply the randomizer $R \in \{0,1\}^{\tau(k)}$. The key is stored both on the user's machine and on the server. The protocol is as follows (all the communication is done via the channel C).

1. The server selects a random $Y \in \{0,1\}^{\mu(k)}$ and sends it to Bob.
2. Bob replies with $f(R,Y)$.
3. Alice verifies the correctness of Bob's reply.

Now assume that the adversary retrieved at most $\sigma(k)$ bits of R. More precisely, assume that the adversary knows a value $h(R)$, where h is a function with the range $\{0,1\}^{\sigma(k)}$. It is easy to see that (by the security of f) she has negligible chances of being able to reply correctly to the challenge Y. Observe that if the adversary replies (in Step 2) with some value X, and Alice rejects this answer, than the adversary learns exactly one bit of information about R (namely that $f(R,T) \neq X$), which should be added to the total number of „retrieved" bits (if one want to achieve the security against multiple impersonation attempts).

Note that since we assume that the server is secure (i.e. there are no intrusions to him) hence one could generate K pseudo-randomly and just store the seed on the server. For example: the seed s could be a key to the block-cipher B and one could set $K := (B_s(1), B_s(2), \ldots, B_s(j))$, for some appropriate parameter j (this method allows for a quick access to any part of K).

---

[11] The computational power of the virus does not need to be limited in this case.

5.2   The Protocol of [KS99]

In this section we note that in the protocol from Sect. 5.1 one can use a simpler function f than the functions secure in the BSM. Namely, the server can simply ask (in Step 1) for the values of k random positions on K. Formally, the challenge in Step 1 is a random k-element subset of the set $\{1, \ldots, \tau(k)\}$. The function f in Step 2 is replaced with φ (where φ was defined in Sect. 4.5). This is exactly the protocol of [KS99] (however in that paper it was analyzed in a weaker model where the adversary is allowed to access only the individual bits of the secret key). The security of this protocol follows from the analysis in Sect. 4.5.

6   Discussion

The main drawback of our protocols is that during the intrusion the virus can impersonate the user (and the user may not even be aware that something wrong is happening). As a partial remedy we suggest that the user could be required to split the private key into 2 halves K_1 and K_2, and to store each of them on a separate DVD disc. In this case the authentication process would require physical action of replacing one DVD with another (assuming that there is only one DVD drive in the machine). Note that this method does not work if we assume that the adversary is able to store large amounts of data on user's hard-disc (as in this case she can make a local copy of the DVDs containing the key).

7   Open Problems

It remains an open problem to examine which variant of the protocols described above is the best for practical applications. We did not provide a comparison between the protocols based on the BSM key-expansion function and the protocols based on the function φ (Sect. 4.5 and 5.2), as such comparison should depend on the concrete parameters that one wants to optimize (the size of the communicated data, computing time, level of security). For some choice of these parameters (long computing time, high level of security) it may be even practical to use protocols that perform computations on the entire randomizer. For example in the protocol in Sect. 4.5 one could use function $\tilde{f}$ that simply hashes the entire randomizer R concatenated with Y (i.e. set $\tilde{f}(R, Y) = H(R \cdot Y)$).

Another open problem is to implement other cryptographic tasks (as asymmetric encryption and signature schemes) in our model.

Acknowledgments

We would like to thank Krzysztof Pietrzak and Bartosz Przydatek for helpful discussions, and the anonymous referees for their comments.

References

[ADR02]   Y. Aumann, Y. Z. Ding, and M. O. Rabin. Everlasting security in the bounded storage model. *IEEE Transactions on Information Theory*, 48(6):1668–1680, 2002.

[And02]   R. Anderson. Two remarks on public key cryptology. Technical report, University of Cambridge, Computer Laboratory, 2002.

[BR93]   M. Bellare and P. Rogaway. Random oracles are practical: A paradigm for designing efficient protocols. In *ACM Conference on Computer and Communications Security*, pages 62–73, 1993.

[Cac97]   Christian Cachin. *Entropy Measures and Unconditional Security in Cryptography*. PhD thesis, ETH Zurich, 1997. Reprint as vol. 1 of *ETH Series in Information Security and Cryptography*, ISBN 3-89649-185-7, Hartung-Gorre Verlag, Konstanz, 1997.

[CCM98]   C. Cachin, C. Crepeau, and J. Marcil. Oblivious transfer with a memory-bounded receiver. In *39th Annual Symposium on Foundations of Computer Science*, pages 493–502, 1998.

[CDD+05]   D. Cash, Y. Z. Ding, Y. Dodis, W. Lee, R. Lipton, and S. Walfish. Intrusion-resilient authentication and key agreement in the limited communication model. Manuscript, 2005.

[CHK03]   R. Canetti, S. Halevi, and J. Katz. A forward-secure public-key encryption scheme. In *Advances in Cryptology - EUROCRYPT 2003, International Conference on the Theory and Applications of Cryptographic Techniques, Warsaw, Poland, May 4-8, 2003, Proceedings*, volume 2656 of *Lecture Notes in Computer Science*, pages 255–271, 2003.

[CK01]   R. Canetti and H. Krawczyk. Analysis of key-exchange protocols and their use for building secure channels. In *Advances in Cryptology - EUROCRYPT 2001, International Conference on the Theory and Application of Cryptographic Techniques, Innsbruck, Austria, May 6-10, 2001, Proceeding*, volume 2045 of *Lecture Notes in Computer Science*, pages 453–474, 2001.

[CM97]   C. Cachin and U. Maurer. Unconditional security against memory-bounded adversaries. In Burton S. Kaliski Jr., editor, *CRYPTO*, volume 1294 of *Lecture Notes in Computer Science*, pages 292–306. Springer, 1997.

[DFK+03]   Y. Dodis, M. K. Franklin, J. Katz, A. Miyaji, and M. Yung. Intrusion-resilient public-key encryption. In *Topics in Cryptology - CT-RSA 2003, The Cryptographers' Track at the RSA Conference 2003, San Francisco, CA, USA, April 13-17, 2003, Proceedings*, volume 2612 of *Lecture Notes in Computer Science*, pages 19–32, 2003.

[DFK+04]   Y. Dodis, M. K. Franklin, J. Katz, A. Miyaji, and M. Yung. A generic construction for intrusion-resilient public-key encryption. In Tatsuaki Okamoto, editor, *CT-RSA*, volume 2964 of *Lecture Notes in Computer Science*, pages 81–98. Springer, 2004.

[DHRS04]   Y. Z. Ding, D. Harnik, A. Rosen, and R. Shaltiel. Constant-round oblivious transfer in the bounded storage model. In M. Naor, editor, *TCC*, volume 2951 of *Lecture Notes in Computer Science*, pages 446–472. Springer, 2004.

[Din01]   Y. Z. Ding. Oblivious transfer in the bounded storage model. In Joe Kilian, editor, *CRYPTO*, volume 2139 of *Lecture Notes in Computer Science*, pages 155–170. Springer, 2001.

[Din05]   Y. Z. Ding. Error correction in the bounded storage model. In J. Kilian, editor, *TCC*, volume 3378 of *Lecture Notes in Computer Science*, pages 578–599. Springer, 2005.

[DLL05]   D. Dagon, W. Lee, and R. J. Lipton. Protecting secret data from insider attacks. In *Financial Cryptography and Data Security, 9th International Conference, FC 2005, Roseau, The Commonwealth of Dominica, February 28 - March 3, 2005,*, pages 16–30, 2005.

[DM04a]   S. Dziembowski and U. Maurer. On generating the initial key in the bounded-storage model. In Jan Camenisch and Christian Cachin, editors, *Advances in Cryptology — EUROCRYPT '04*, volume 3027 of *Lecture Notes in Computer Science*, pages 126–137. Springer-Verlag, May 2004.

[DM04b]   S. Dziembowski and U. Maurer. Optimal randomizer efficiency in the bounded-storage model. *Journal of Cryptology*, 17(1):5–26, January 2004.

[Dod00]   Y. Dodis. *Exposure-Resilient Cryptography*. PhD thesis, Massachussetts Institute of Technology, August 2000.

[DvOW92]   W. Diffie, P. C. van Oorschot, and M. J. Wiener. Authentication and authenticated key exchanges. *Designs, Codes and Cryptography*, 2(2):107–125, 1992.

[Dzi05]   S. Dziembowski. Intrusion-resilience via the bounded-storage model. Cryptology ePrint Archive,Report 2005/179, 2005. http://eprint.iacr.org/.

[GM84]   S. Goldwasser and S. Micali. Probabilistic encryption. *Journal of Computer and System Sciences*, 28(2):270–299, 1984.

[Gol04]   O. Goldreich. *Foundations of Cryptography: Volume 2, Basic Applications*. Cambridge University Press, New York, NY, USA, 2004.

[IR02]   G. Itkis and L. Reyzin. Sibir: Signer-base intrusion-resilient signatures. In *Advances in Cryptology - CRYPTO 2002, 22nd Annual International Cryptology Conference, Santa Barbara, California, USA, August 18-22, 2002, Proceedings*, volume 2442 of *Lecture Notes in Computer Science*, pages 499–514, 2002.

[Kra96]   H. Krawczyk. A versatile secure key-exchange mechanism for the internet. In *Proceedings of the 1996 Symposium on Network and Distributed System Security (SNDSS '96)*, pages 114–127. IEEE Computer Society, 1996.

[KS99]   J. Kelsey and B. Schneier. Authenticating secure tokens using slow memory access. In *USENIX Workshop on Smart Card Technology*, pages 101–106. USENIX Press, 1999.

[KSWH00]   J. Kelsey, B. Schneier, D. Wagner, and C. Hall. Side channel cryptanalysis of product ciphers. *Journal of Computer Security*, 8(2/3), 2000.

[Lu04]   C.-J. Lu. Encryption against storage-bounded adversaries from on-line strong extractors. *Journal of Cryptology*, 17(1):27–42, January 2004.

[Mau92]   U. Maurer. Conditionally-perfect secrecy and a provably-secure randomized cipher. *Journal of Cryptology*, 5(1):53–66, 1992.

[MSTS04]   T. Moran, R. Shaltiel, and A. Ta-Shma. Non-interactive timestamping in the bounded storage model. In *Advances in Cryptology - CRYPTO 2004, 24th Annual International Cryptology Conference, Santa Barbara, California, USA, August 15-19, 2004, Proceedings*, volume 3152 of *Lecture Notes in Computer Science*, pages 460–476, 2004.

[NZ96]   N. Nisan and D. Zuckerman. Randomness is linear in space. *Journal of Computer and System Sciences*, 52(1):43–52, 1996.

[Vad04]   S. P. Vadhan. Constructing locally computable extractors and cryptosystems in the bounded-storage model. *Journal of Cryptology*, 17(1):43–77, January 2004.

Perfectly Secure Password Protocols in the Bounded Retrieval Model

Giovanni Di Crescenzo[1], Richard Lipton[2], and Shabsi Walfish[3]

[1] Telcordia Technologies, Piscataway, NJ, USA
`giovanni@research.telcordia.com`
[2] Georgia Institute of Technology, Atlanta, GA, USA
`rjl@cc.gatech.edu`
[3] New York University, New York, NY, USA
`walfish@cs.nyu.edu`

Abstract. We introduce a formal model, which we call the *Bounded Retrieval Model*, for the design and analysis of cryptographic protocols remaining secure against intruders that can retrieve a limited amount of parties' private memory. The underlying model assumption on the intruders' behavior is supported by real-life physical and logical considerations, such as the inherent superiority of a party's local data bus over a *remote* intruder's bandwidth-limited channel, or the detectability of voluminous resource access by any *local* intruder. More specifically, we assume a fixed upper bound on the amount of a party's storage retrieved by the adversary. Our model could be considered a non-trivial variation of the well-studied Bounded Storage Model, which postulates a bound on the amount of storage available to an adversary attacking a given system.

In this model we study perhaps the simplest among cryptographic tasks: user authentication via a password protocol. Specifically, we study the problem of constructing efficient password protocols that remain secure against offline dictionary attacks even when a large (but bounded) part of the storage of the server responsible for password verification is retrieved by an intruder through a remote or local connection. We show password protocols having satisfactory performance on both *efficiency* (in terms of the server's running time) and *provable security* (making the offline dictionary attack not significantly stronger than the online attack). We also study the tradeoffs between efficiency, quantitative and qualitative security in these protocols. All our schemes achieve *perfect security* (security against computationally-unbounded adversaries). Our main schemes achieve the interesting efficiency property of the server's lookup complexity being much smaller than the adversary's retrieval bound.

1 Introduction

Partially motivated by the recent press attention to intrusions from both external attackers and insiders into databases containing highly sensitive information

S. Halevi and T. Rabin (Eds.): TCC 2006, LNCS 3876, pp. 225–244, 2006.

(e.g., [27, 28]), we initiate a rigorous study of cryptographic protocols in the presence of intruders, under a novel and reasonable assumption on their power. This leads us to define a new formal model which we call the Bounded Retrieval Model since we assume a bound on the amount of a party's stored data that can be retrieved by the adversary. In practice, this bound would be due to both physical and logical considerations, as we now explain. With respect to internal attackers, this bound may result from the capabilities of a simple Intrusion Detection System (IDS), which can easily monitor any large and repeated access to the party's stored data. With respect to external attackers, this bound is further minimized as a consequence of the inherent gap between the (smaller) availability of bandwidth due to physical limits and the (larger) availability of storage memory: an attacker needing a large amount of time to retrieve large amounts of sensitive data will most likely be unable to maintain an unauthorized connection for enough time without being detected.

Our model could be considered a non-trivial variation of the well-studied Bounded Storage Model, introduced in [13] (see, e.g., [12, 14] and references therein for further studies of several cryptographic tasks, such as key-agreement, encryption, oblivious transfer, time-stamping, etc). This model postulates a fixed upper bound on the storage capacity (but no bound at all on the computational power) of the adversary attacking a cryptographic protocol. Thus, with respect to the standard model used in the security analysis of most cryptographic primitives, where the adversary is assumed to have a polynomial upper bound on both storage and computational power, this model achieves much higher security at the expense of a stronger assumption on the adversary's *storage* capability. Analogously, our model also avoids upper bounds on the computational power of the adversaries at the expense of a stronger assumption on the adversary's *retrieval* capability, which we argued before as being supported by reasonable considerations.

In this paper we use this model to analyze possibly the simplest cryptographic task: entity authentication via password verification, which we will briefly call a 'password protocol' in the rest of the paper.

Password Protocols. Despite their often noticed weaknesses, password protocols remain the most widely used method for authenticating computer users. In traditional UNIX-like password schemes, the server stores some one-way function of users' passwords in a single password database. In order to verify a login attempt, the server simply computes the same one-way function on a putative password supplied by the user attempting to login, and compares it to the stored value in the database. If the values match, the user is allowed to log in. An adversary trying to impersonate an authorized user can always try an *"online attack"* by entering different passwords in correspondence to the user's login name. However, if the user's password is chosen with enough entropy or randomness, each attempt is extremely unlikely to succeed, and modern servers are programmed to close the authentication session after just a few unsuccessful attempts. Unfortunately, the password database itself is typically small, and can be quickly and easily retrieved by any attacker capable of minimally compromising the security of the server. Although the password database does not directly contain any of the user's passwords, it opens up the possibility of an *"offline dictionary attack"* to the adversary. In such an attack, the adversary can utilize the information contained in *any single record* of the password database by attempting to apply the appropriate one-way function to every word in a dictionary in the hopes that it will match the content of that record (due to the users' tendency to supply dictionary words for their passwords). Although too large to be efficiently searched by a human, dictionaries are typically small enough so that they are efficiently searchable by a computer, thus making this offline dictionary attack quite feasible.

In this paper we explore the following simple but intriguing question: can we design a scheme so that an adversary is required to access *many records* when trying to carry out these attacks? Storage is a static, cheap resource, that is very available today (such that a server can easily provide it in huge quantities). External bandwidth, a time-dependent resource, is certainly much less available than storage. Furthermore, the bandwidth available to a remote or local attacker may be easily controlled by physical means (or even by monitoring traffic at the server's interface to the outside world). By using this gap between server's storage capacity and the adversary's ability to retrieve stored data, we show how to realize a significant server's security advantage over the adversary, thus making the off-line dictionary attack just slightly more powerful than the (practically unsuccessful) online attack.

Analysis in the Bounded Retrieval Model. Intuitively, we propose to construct password database files that are so large that either (1) they cannot be retrieved in their entirety by a local or remote intruder in any reasonable time (due to access or bandwidth limitations), or (2) any such huge retrieval operation is easily detected. Note that (2) can be obtained using very simple and efficient intrusion detection mechanisms (see, e.g., [1] for a survey and [5] for a theoretical model of intrusion detection). Specifically, a huge retrieval would be considered an anomalous event, thus triggering actions such as closing the adversary's access port or preventing access to the storage area from any insider. Furthermore, note that there are several typical scenarios where (1) can be true. The simplest is clearly that of an adversary with relatively limited bandwidth. In fact, this limitation is already present in existing networks, as even the high bandwidth connections commonly available today may require minutes to transfer modest data amounts such as 1 gigabyte. (See Appendix A for detailed numerical examples.) As another typical scenario, assume the server is distributing the password database in several locations and that the adversary is either unaware of the position of some of them, or cannot physically access some of them.

Formally, we place a bound on the quantity of information from the server's storage area that is retrieved by the adversary during an attack. Analogously to [13], security in our model can be information theoretic in nature, which is quite desirable due to the brute-force nature of off-line dictionary attacks. We consider two main classes of attacks in this model: (1) static retrievals, modeling the case where the adversary must pre-select the data he wishes to collect prior to the beginning of the actual data retrieval phase; and (2) adaptive retrievals, modeling an adversary selecting each single location to be retrieved based on the content of all previously retrieved locations. In both cases the total information retrieved by the adversary is bounded by a fixed parameter. Each of these two classes of attacks models a real world scenario. For example, static intrusions model any situation where the adversary may receive blocks of data chosen independently of their contents, such as data recovered from a damaged and discarded hard disk. Adaptive intrusions model the most general scenario, where the adversary may have arbitrary access to data blocks of its choice, such as retrieving data interactively from an insecure network file server (for example, via FTP). Although it would seem that an adversary, if possible, would always choose to perform an adaptive retrieval attack, one should keep in mind that adaptivity also requires the adversary to expend time to examine the information that is being retrieved and therefore may actually provide the adversary with less information than in the case of a static retrieval. We note that in this model access to the entire data file by the adversary is not ruled out, as it can be easily prevented using intrusion detection techniques. On the other hand, we caution the reader that the model does not include the case in which the server is totally compromised by an attacker, where the latter would actually be able to directly observe the passwords received from users during their login attempts anyway.

Our results. In designing password protocols in the bounded retrieval model, we pay attention to various parameters for security (e.g., the adversary's advantage over the online attack success probability, and the adversary's retrieval strategy) and efficiency (e.g., the server's lookup complexity). This allows us to appropriately set target goals for both. To that purpose, it is useful to keep in mind the following two important issues about parameters:

Adversary's advantage vs. online attack success probability. Although it is certainly desirable to have a password scheme with 0 or exponentially small adversary's advantage probability, in practice it is essentially just as desirable to have a password scheme with the adversary's advantage comparable to the online attack success probability (as the overall attacker's success probability is the sum of the two values).

Server's running time vs. adversary's retrieval bound. Although intuitively it would seem easier to design provably secure password schemes where at each user registration or verification the server reads more locations than the adversary is ever allowed, this severely restricts the efficiency of the scheme and its practical applicability.

Summarizing, the combination of efficiency and security properties we desire requires the adversary's advantage to be provably comparable to the online attack success probability, and the server's running time (as measured by the number of data blocks read) to be significantly smaller than the adversary's.

Towards this goal, our first result is a lower bound on the advantage of the adversary, which, among other things, relates the advantage to the adaptivity of the server's lookup strategy. We then start by exploring what schemes can be constructed using well-known cryptographic tools such as secret sharing schemes and all-or-nothing transforms. These result in schemes $\mathcal{P}_1$ and $\mathcal{P}_2$, having the smallest possible adversary's advantage (0 and exponentially small, respectively) in the strongest adversarial model (adaptive attacks), but requiring the server's running time to be larger than the adversary's retrieval bound.

Our main protocols, denoted as $\mathcal{P}_3$ and $\mathcal{P}_4$, achieve high efficiency in that the server's lookup complexity is much smaller than the adversary's retrieval limit. Protocol $\mathcal{P}_3$ is based on dispersers and pairwise-independent hash functions, and guarantees both security against adaptive adversaries and that the adversary's advantage is not significantly larger than the online attack success probability. Because of our previous lower bound, this protocol achieves an *optimal* bound on the adversary's advantage (up to a constant) for typical values of the adversary's retrieval bound (e.g. whenever the adversary's retrieval bound is a constant fraction of the storage). Protocol $\mathcal{P}_4$ is based on t-wise independent hash functions and strong extractors and achieves security against static retrieval attacks, ensuring exponentially-small adversary's advantage without any computational assumption. Protocols $\mathcal{P}_3$ and $\mathcal{P}_4$ can be combined, resulting in a single scheme that simultaneously enjoys both of their desirable security properties.

A more detailed account of our protocols' properties is in Figure 1. We note that none of our protocols is proved secure by assuming the existence of random oracles (we thus removed this assumption from one protocol in [7]).

Protocol name	Adversary's advantage	Adversary's strategy	Server's complexity	Server's strategy	Storage constraints
$\mathcal{P}_1$	0	adaptive	$l > q$	non-adaptive	$n \geq 2td$
$\mathcal{P}_2$	$O(2^{-\lambda})$	adaptive	$l > q$	non-adaptive	$n \geq 2d$
$\mathcal{P}_3$	$O(m^3/(m-q)^2 l 2^d)$	adaptive	$l < q$	non-adaptive	$n \geq 2d + 1$
$\mathcal{P}_4$	$(2t + 4) \cdot 2^{-\lambda}$	static	$l < q$	adaptive	$n \geq O(\lambda) + 2d$
none	$2^{-\lambda}$	static	$l < q$	non-adaptive	

Fig. 1. Any two protocols in the above table are incomparable, in the sense that each one is better in some features than the other. Specifically, protocol $\mathcal{P}_1$, based on secret sharing, is of interest as it achieves 0 adversary's advantage. Protocol $\mathcal{P}_2$, based on all-or-nothing transforms, is of interest as it achieves exponentially small adversary's advantage while improving storage constraints. Protocol $\mathcal{P}_3$, based on dispersers and pairwise-independent hash functions, is of interest as it achieves security against adaptive adversaries and server's lookup complexity l smaller than the adversary's retrieval bound q. Protocol $\mathcal{P}_4$, based on strong extractors and t-wise independent hash functions, is of interest as it achieves the efficiency property of $\mathcal{P}_3$ as well as exponentially small advantage against static adversaries. At the end of the paper we also discuss a protocol that combines features from protocols $\mathcal{P}_3$ and $\mathcal{P}_4$. The last line in the table points out that achieving exponentially small advantage against adaptive or static adversaries is impossible when the server's lookup strategy is non-adaptive, due to a lower bound in Section 3. Formal definitions, including parameters and performance measures used in the table, are given in Section 2.

Related work. The Bounded Retrieval Model is a novel variation of the Bounded Storage Model of [13], and furthermore in some of our solutions we use strong extractors, which are a common tool for protocols in the Bounded Storage Model. However, we point out that solutions and analysis for our password protocol problems have to address quite non-trivial obstacles, even given such tools. A bounded-retrieval notion similar to ours was also used implicitly in [10], in the context of smart cards with slow memory access. Whereas [10] studied the problem of token based authentication in a bounded retrieval context, we consider the problem of password based authentication.

The importance of securing the server's password file has been well-known for many years, and is discussed in detail, for instance, in [20, 8]. Various aspects of password protocols have been studied in the security literature. One important area is that of securing password protocols where the communication goes over an insecure network (e.g., see [9] for schemes based on public-key encryption and [2, 3, 18, 26] for heuristic schemes not using public keys). While this aspect is orthogonal to the server compromise security considered in our work, we stress that many of the cited results can be modularly combined with results in this paper to obtain network password protocols secure against bounded retrieval attacks. Other work on password-related protocols includes well-studied areas like password- authenticated key exchange, that are even farther from the scope of this work.

2   Model and Formal Definitions

We start by presenting the scenario for password protocols. We discuss the entities involved and the assumed connectivity among them, the phases, the (sub)protocols, and finally the requirements that a password protocol has to satisfy to be declared secure in the bounded retrieval model.

Entities, connectivity, resources. An arbitrary system (or network) containing a number of resources can be accessed locally or remotely through a password protocol controlled by a *server* S. The *users*, denoted as $U_1, U_2, \ldots, U_t$ for some integer t, are any entities that need resources in the system, and thus may require access to it. Although potentially all users are connected to each other as well as to the server through some communication link, for practical purposes, we are interested in password protocols where each user only interacts with the server, and not necessarily at the same time. For simplicity, we will assume that the communication link between each user and the server is private or not subject to attacks, although we note that the model in which this link is also subject to adversarial attacks is of orthogonal focus and can be separately studied (but will not be studied in this paper). The server's storage area contains a *password file*, that we denote as F, with m locations, each containing a record of n bits. We denote as $F[i]$ the content of the i-th location of F and, as $F[L]$ the set $\{F[i] : i \in L\}$.

Subprotocols and Phases. A password protocol can be divided into four main algorithms or subprotocols: a *setup* algorithm, a *password sampling* algorithm, a *registration* algorithm and an *identification* subprotocol.

A setup algorithm, that we denote as SET, is only run by the server. On input a security parameter λ in unary, algorithm SET returns an m-location password file F, for some $m = \text{poly}(\lambda)$ in time at most polynomial in λ.

A password sampling algorithm, that we denote as SAMPLE, is run by users to select their passwords. We will only consider the algorithm SAMPLE that, on input parameter 1^d, returns a uniformly chosen string from $\{0,1\}^d$. (In each of our constructions, by properly using tools such as extractors, we can modularly reduce to this case more general cases such as that of users choosing passwords from a smaller dictionary of strings with known min-entropy.) We will think of the *password length* d as a constant, this being much smaller than the security parameter λ.

The registration algorithm, denoted as REG, is a possibly probabilistic polynomial time (in n) algorithm that takes as input a user's login name log_i, her password pw_i, and the password file F, and returns an output F for S. Here, log_i denotes a login name somehow generated by S or by U_i (we won't deal with the details on how this happens but just assume that each user has a distinct login name that is, for simplicity of notation, d-bit long), and the output F is an updated version of the password file.

During an identification subprotocol, a user U_i sends both log_i and pw_i to S, which runs a deterministic polynomial time (in λ) algorithm VER on input log_i, pw_i, F, in addition to the various parameters and all login names, and returns *accept* (briefly, 1) or *reject* (briefly, 0), according to whether the user has been positively identified or not.

We will denote a *password protocol* as a quadruple of probabilistic algorithms $\mathcal{P} = (\text{SET}, \text{SAMPLE}, \text{REG}, \text{VER})$, and we will assume, for simplicity, that an execution of $\mathcal{P}$ can be divided into three phases: first, an *initialization phase*, where the server runs the setup algorithm; then, a *registration phase*, where each among the t users $U_1, \ldots, U_t$ chooses a password using SAMPLE and runs subprotocol REG with server S; finally, an *identification phase*: at any time, any among $U_1, \ldots, U_t$ can run the identification subprotocol with S. We denote as $Param$ the list of parameters (represented in unary) associated with $\mathcal{P}$, that can be any subset among: the *password length* d, the *number of users* t, the *number of locations* m, the *location size* n, the *security parameter* λ, which have been defined above, and the *lookup complexity* l, and the *retrieval bound* q, which will be defined later.

Correctness requirement. A basic requirement we expect from a password protocol is that, at any time, a server positively identifies previously registered users.

Definition 1. *Let* $\mathcal{P} = (\text{SET}, \text{SAMPLE}, \text{REG}, \text{VER})$ *be a password protocol with parameters* $Param = (d, t, m, n, l, q)$. *The* correctness requirement *for* $\mathcal{P}$ *is as follows: for each* $j \in \{1, \ldots, t\}$, *and any login-name* log_j, *it holds that*

$$\Pr\left[\begin{array}{c} F \leftarrow \text{SET}(1^n); \{pw_i \leftarrow \text{SAMPLE}(1^d); F \leftarrow \text{REG}(log_i, pw_i, F)\}_{i=1}^t : \\ \text{VER}(Param, \{log_i\}_{i=1}^t, log_j, pw_j, F) = 1 \end{array} \right] = 1.$$

Bounded Retrieval security requirement. All our results consider an adversary A that is *not* time-bounded. This is not only an interesting byproduct of our results but also an especially desired requirement in our model, as we want to withstand adversaries who can run dictionary attacks. Moreover, the adversary is given knowledge of all users' login names, and is allowed to retrieve up to q entries from the server S's password file F. We consider two levels of adaptivity (that is, dependency on the content of F) that the adversary can use in choosing the q entries from F. In practical applications, the adaptivity level plays an important role, as adaptivity may slow down the retrieval rate for the adversary. Specifically, we will restrict the adversarial attack to one of the following two types:

1. *Static Retrieval:* First, the adversary must select a set of at most q locations $L = \{l_1, \ldots, l_q\}$, without observing any of the data in the password file F. Then the adversary is given the contents $F[l_1], \ldots, F[l_q]$ of the selected locations. Finally, the adversary returns a pair (log_j, pw'_j), trying to guess the password of user U_j.
2. *Adaptive Retrieval:* As before, except each of the q locations can be selected by the adversary after seeing the contents of the previously selected ones.

Formally, for $x \in \{$ static, adaptive $\}$, we say that a *bounded retrieval attack* of type x is successful if the experiment $E_x^{\mathcal{P},A}$ returns 1, where

1. if $x = $ *static* then $E_x^{\mathcal{P},A} = E_{static}^{\mathcal{P},A}$
2. if $x = $ *adaptive* and $E_x^{\mathcal{P},A} = E_{adaptive}^{\mathcal{P},A}$,

and, for all parameters $Param = (d, t, m, n, l, q)$ described in unary and all login names $\{log_1, \ldots, log_t\}$, the experiments are defined as follows (here, the notation $y \leftarrow Alg(x_1, x_2, \ldots)$ denotes the process of running the (possibly probabilistic) algorithm Alg on input $x_1, x_2, \ldots$ and the necessary random coins, and obtaining y as output):

$E_{static}^{\mathcal{P},A}(Param, \{log_i\}_{i=1}^t)$
1. $F \leftarrow \text{SET}(1^n)$
2. for $i = 1, \ldots, t$,
    $pw_i \leftarrow \text{SAMPLE}(1^d)$
    $F \leftarrow \text{REG}(log_i, pw_i, F)$
3. $p \leftarrow (Param, \{log_i\}_{i=1}^t)$
4. $\{l_1, \ldots, l_q\} \leftarrow A(p)$
5. $(log', pw') \leftarrow A(p, \{l_i, F[l_i]\}_{i=1}^q)$
6. if $\text{VER}(p, log', pw', F) = 1$ then
    return: 1
    else **return:** 0.

$E_{adaptive}^{\mathcal{P},A}(Param, \{log_i\}_{i=1}^t)$
1. $F \leftarrow \text{SET}(1^n)$
2. for $i = 1, \ldots, t$,
    $pw_i \leftarrow \text{SAMPLE}(1^d)$
    $F \leftarrow \text{REG}(log_i, pw_i, F)$
3. $i \leftarrow 0; p \leftarrow (Param, \{log_i\}_{i=1}^t)$
4. repeat
    $i \leftarrow i + 1$
    $l_i \leftarrow A(p, \{l_j, F[l_j]\}_{j=1}^{i-1})$
    until $i = q$
5. $(log', pw') \leftarrow A(p, \{l_i, F[l_i]\}_{i=1}^q)$
6. if $\text{VER}(p, log', pw', F) = 1$ then
    return: 1 else **return:** 0.

We are now ready to define the security requirement for password protocols in the bounded retrieval model.

Definition 2. *Let* $\mathcal{P} = (\text{SET}, \text{SAMPLE}, \text{REG}, \text{VER})$ *be a password protocol with parameters* (d, t, m, n, l, q). *For* $x \in \{$ *static, adaptive* $\}$, *we say that* $\mathcal{P}$ *is* ϵ-*secure against a bounded retrieval attack of type* x *if for any algorithm* A, *all login names* $\{log_1, \ldots, log_t\}$, *and any* $j = 1, \ldots, t$, *it holds that*

$$\Pr \left[b \leftarrow E_x^{\mathcal{P},A}(Param, \{log_i\}_{i=1}^t) \, : \, b = 1 \wedge log' = log_j \right] \leq \frac{1}{2^d} + \epsilon.$$

Remarks. In the above definitions we only have addressed the most basic and practically relevant variant of a number of definitions that one could come up with. For instance, one could strengthen the security requirement by defining an adversary to be successful even if it obtains any nonzero information about the joint values of all passwords, rather than just being able to successfully login, as defined above. (This requirement seems stronger than what's desired in practice.)

Performance Metrics. In addition to the above different adversarial models, when designing password protocols secure under bounded retrieval attacks, we also consider various performance metrics, which we will now discuss in detail. In the rest of the paper we will present a lower bound on the *adversary's advantage*, denoted as ϵ, and protocols that exhibit tradeoffs between all these metrics, in the effort of balancing their security and efficiency.

Time, lookup strategy, storage complexity. An obviously important metric is the *time complexity* of algorithms SET, REG and VER; in particular, we will pay attention to the (possibly parallel) time complexity of VER, as it is run more frequently in applications. Additionally, we will pay special attention to the *lookup strategy* of algorithm VER, and specifically, to whether it is *adaptive* or *non-adaptive*; that is, based on location content or not. Also related to time complexity is the *storage complexity*; that is, the amount of storage used by the server during the initialization phase. Although storage is today an easily available resource, we will ensure that even a large increase in the storage complexity does not make the time complexity impractical.

Lookup complexity. Additionally, we will pay special attention to the *lookup complexity* of algorithms REG and VER, which we denote as l, and defined as the maximum number of locations from F that is read or written by either algorithm REG during its execution on an input log_i, pw_i, F or algorithm VER, when run on an input log, pw, F, in addition to all parameters and login-names. We will assume, without loss of generality, that this number is the same for all inputs to REG and VER. (We note that all algorithms REG, VER can be simply modified so that this holds).

Adversary's breaking advantage. Our model is of information-theoretic nature, as we will consider security against adversaries that are not time-bounded. Therefore, we will be interested in constructions that achieve *adversary's advantage* ϵ either $= 0$ or exponentially small (in the security parameter λ). Additionally, given that an on-line attack is always available in practice to an adversary, we will be interested in constructions that achieve $\epsilon = O(2^{-d})$, where d is the length of a password. (Note that 2^{-d} may not be exponentially small in the security parameter.)

Lookup complexity vs. Retrieval Bound. Given lookup complexity l and retrieval bound q for the adversary, it is of interest to achieve constructions that have the smallest possible value for l and the highest possible for q, in combination with satisfactory performance on the above metrics.

3   A Lower Bound on the Adversary's Advantage

We present a lower bound on the security of password protocols having lookup complexity smaller than the adversary's retrieval bound. This will be used to prove the protocol in Section 5 optimal up to a multiplicative constant.

Some definitions. Let $\mathcal{P} = (\text{SET}, \text{SAMPLE}, \text{REG}, \text{VER})$ be a password protocol and let l denote the lookup complexity of the verification subprotocol VER. We now define t distributions LocD_j, for $j = 1, \dots, t$, where each LocD_j is the distribution of the locations in F accessed by the algorithm VER on fixed input $(Param, \{log_i\}_{i=1}^t, log_j, pw_j, F)$ generated as in experiment $E_{static}^{\mathcal{P},A}$. Formally, we first define algorithm LVER as the algorithm that, given an input $(Param, \{log_i\}_{i=1}^t, log_j, pw_j, F)$, returns the set L of locations from F accessed during an execution of algorithm VER on the same input. Then we can define, for $j = 1, \dots, t$, the distribution LocD_j as

$$\{\text{run steps } 1, 2 \text{ of } E_{static}^{\mathcal{P},A} \,; L \leftarrow \text{LVER}(Param, \{log_i\}_{i=1}^t, log_j, pw_j, F) \,:\, L\};$$

that is, the distribution of locations read by the server during a login by U_j. Note that both VER and LVER are deterministic algorithms, and therefore the actual probability space for distribution LocD_j is given by the randomness contained in the public file F obtained during the execution of experiment $E_{static}^{\mathcal{P},A}$; and, specifically, by how the locations accessed by VER change, if at all, as an effect of such randomness. For instance, in the case of a non-adaptive lookup strategy, by definition, L can be a single value and therefore the distribution LocD_j trivializes to having a single value in its support. Recall that for a distribution D over support X, the *collision probability* $cp(D)$ is defined as $\sum_{x \in X}(\Pr[x' \leftarrow D : x' = x])^2$; where we note that if a distribution has a single value in its support, then its collision probability is 1.

Lower Bound Statement and Discussion. Informally, the following lower bound formalizes the intuition that if the servers' lookup complexity is smaller than the adversary's retrieval bound then the larger the amount of adaptivity in the server's lookup strategy, the harder is the adversary's job in finding a password. More formally:

Theorem 1. *Let* $\mathcal{P} = (\text{SET}, \text{SAMPLE}, \text{REG}, \text{VER})$ *be a password protocol with parameters* (d, n, t, l, m, q), *and assume that* $\mathcal{P}$ *is* ϵ-*secure against a bounded retrieval attack of static type. If* $l \leq q$ *then it holds that*

$$\epsilon \geq \max_{j \in \{1, \dots, t\}} \left(\left\lfloor \frac{q}{l} \right\rfloor \cdot \frac{1}{2^d} \cdot cp(\text{LocD}_j) \right),$$

where $cp(\text{LocD}_j)$ *is the collision probability of distribution* LocD_j *defined above.*

We note that in the case of non-adaptive lookup strategy from VER, distribution LocD_j returns a single value, its collision probability is equal to 1, and the bound in the above theorem becomes $\epsilon \geq \lfloor q/l \rfloor \cdot 2^{-d}$, under the hypothesis $l \leq q$. As in practice, work in the order of 2^d may be efficiently performed, we derive that non-adaptive strategies for VER can only result in password protocols ϵ-secure for values of ϵ that are *not* smaller than the on-line attack success probability (e.g., $\epsilon = \Omega(2^{-d})$).

The formal proof of Theorem 1 follows by showing an adversary that can run some modified version of the server's algorithm and always finds a password that would be accepted by the server with probability equal to the lower bound on ϵ in the statement of the theorem. Specifically, the adversary creates a new password file F' identically and independently distributed from the real one; then it starts an off-line dictionary attack by trying several passwords from the dictionary, as follows. For each password, it runs the server's registration and verification algorithms on input F' to determine the set of locations read or written by the server; then, it queries the same set of locations from the real file F, and runs the verification algorithm to see if that password would be accepted by the server. Details of the proof appear in the full version of the paper.

4   Strongly-Secure Constructions with Large Lookup Complexity

The purpose of this section is to present two very basic constructions of password protocols secure against bounded retrieval attacks, and show that they achieve very strong security at the expense of requiring an inefficient lookup strategy from the server. Specifically, these constructions achieve essentially the best possible security properties: the adversary's advantage can be 0 in one construction and exponentially small in the other one. Furthermore, these values are achieved against an adaptive adversary. The server's lookup strategy in these constructions is also non-adaptive. On the other hand, in both constructions the server's lookup complexity is larger than the adversary's retrieval bound. In fact, in one of the two constructions the server has to access the entire password file in order to verify a user's identity. (Constructions in the next sections will lower the server's lookup complexity and at the same time obtain desirable security properties.) Formally, we obtain the following:

Theorem 2. *For $i = 1, 2$ there exist protocols $\mathcal{P}_i = (\mathrm{SET}_i, \mathrm{SAMPLE}_i, \mathrm{REG}_i, \mathrm{VER}_i)$ with parameters $(n, t, d; m_i, q_i, l_i)$, that are ϵ_i-secure against a bounded retrieval attack of adaptive type, and such that*
1. *$\epsilon_1 = 0$, $m_1 \geq l_1 \geq q_1 + 1$ and $n \geq 2td$.*
2. *$\epsilon_2 = O(2^{-\lambda})$, $m_2 = l_2 \geq q_2 + \min(\lambda, o(q_2))$, and $n \geq 2d$.*

Note that in both constructions $l_i \geq q_i$. For practical applications, the fact that the server's lookup complexity is large constrains the size of the password file so that it is not very large (or otherwise the identification phase would not be efficient). As a consequence, the adversary's retrieval bound cannot be large either, which restricts the applicability of these schemes to settings where the adversary has a small retrieval rate (e.g., if the adversary has a slow connection). The two schemes satisfying Theorem 2 are based on secret sharing schemes for threshold access structures, as in [22] (using polynomial interpolation), and on adaptively-secure all-or-nothing transforms, as in [6] (using adaptively-secure exposure-resilient functions). Very informally, in the first scheme, the entire password file contains the shares of a threshold scheme, where the secret is the concatenation of all login names and passwords, and the threshold is set as strictly larger than the adversary's retrieval bound. Analogously, in the second scheme, the password file can be seen as an all-or-nothing transform of the concatenation of all login names and passwords. We provide a formal description of these schemes in the full version of this paper.

5   A Secure Construction with Small Lookup Complexity

The constructions in Section 4 showed how to achieve strong security (in terms of both the adversary's advantage and the attack type) and non-adaptive server lookup at the expense of a large lookup complexity. In this section we start exploring what security we can achieve if we target constructions with low lookup complexity, while still maintaining non-adaptive lookup. The lower bound of Section 3 implies that the best security that can be obtained under this setting is comparable to the security against on-line attack. In the rest of the section we give a construction that achieves this security level and is therefore essentially optimal (up to lower-order multiplicative factors) for this setting. We present a password protocol secure against bounded retrieval attacks, which we also call SCS, since the server's storage algorithm in this protocol is based on three basic actions: Select, Combine and Store. Specifically, on an input consisting of a login and a password, the server carefully selects several locations from the password files, combines their content according to some function, and stores the result of this function as a tag that can be associated with this password. We instantiate the 'select' action of the SCS scheme by using dispersers, and the 'combine' action using a pairwise-independent hash function. Our construction has server's lookup complexity lower than the adversary's retrieval bound, and, moreover, the following properties: adversary's advantage comparable with the security against on-line attack; non-adaptive server's lookup strategy; constant parallel time complexity; and security against adaptive adversaries. Formally, we obtain the following:

Theorem 3. *There exists a password protocol* $\mathcal{P}_3 = (\text{SET}, \text{SAMPLE}, \text{REG}, \text{VER})$ *with parameters* $Param = (n, t, d, m, q, l)$, *that is* ϵ-*secure against a bounded retrieval attack of adaptive type, and such that, for any* t, d, m, q, *it holds that* $\epsilon = \frac{m^3}{l \cdot (m-q)^2} \cdot \frac{1}{2^d}$, *for* $n \geq 2d + 1$ *and* $l = 2^b$, *where*
$b = \log^2(d) \cdot \text{poly}(\log \log d) + (\log d) \cdot (\log(m/(m-q)))$.

Note that the value of ϵ in the theorem matches (up to a constant) the bound from Theorem 1 in the typical case $q = cm$, for $0 < c < 1$. We also note that the constant factor c here can be made arbitrarily close to 1. We now prove Theorem 3.

A first tool: t-wise independent hash families. Informally, t-wise independence requires that for any fixed set of t elements, a uniformly selected function from the hash family will map those elements to t *uniformly distributed* and *independent* outputs. A formal definition of t-wise independent hash functions follows.

Definition 3 (t-wise Independent Hash Function). *A family $\mathcal{H}$ of functions $h_w : \{0,1\}^a \rightarrow \{0,1\}^b$ is t-wise independent if, for any distinct elements $x_1, \ldots, x_t \in \{0,1\}^a$, and any $r_1, \ldots, r_t \in \{0,1\}^b$, we have that*

$$\Pr_w[h_w(x_1) = r_1, \ldots, h_w(x_t) = r_t] = (2^{-b})^t$$

A commonly used t-wise independent hash function is defined, when $c = a = b$, by simply evaluating a $t-1$ degree polynomial over $GF(2^c)$. Specifically, define the following family $\mathcal{H}$, where $x, w_1, \ldots, w_t$ are viewed as elements of $GF(2^c)$, the field over which the computation is to be performed:

$$\mathcal{H} = \{h_{w_1, \ldots, w_t} \mid h_{w_1, \ldots, w_t}(x) = \sum_{j=1}^{t} w_j x^{j-1}\}$$

In our constructions we will use this construction of t-wise independent hash families both in the case $a > b$ (in this section, when $t = 2$) and in the case $a < b$ (in the next section, for larger values of t) where, in both cases, we set $c = \max(a, b)$ and we use trivial padding or truncation operations to satisfy length consistencies. We note that a function from this family can be indexed by exactly t strings of c bits each.

A second tool: Extractors and dispersers. Extractors and dispersers were first introduced in [17] and [24], respectively, and have received a significant amount of attention in several areas of computer science, mostly in the derandomization literature, but also in other areas including combinatorics, network theory and security. Both extractors and dispersers are often defined as bipartite graphs, while in this paper it will be easier to use their functional definition, which we now recall.

The *statistical distance* between two distributions D_1, D_2 over the same space S is defined as $sd(D_1, D_2) = \frac{1}{2} \sum_{x \in S} |\Pr[x \leftarrow D_1] - \Pr[x \leftarrow D_2]|$. We say that distributions D_1, D_2 are δ-*close* if it holds that $sd(D_1, D_2) \leq \delta$. We say that a distribution D is δ-*close to uniform* if it holds that $sd(D, U) \leq \delta$, where U denotes the uniform distribution over the same space S. The *min-entropy* of a distribution D over space S is defined as $H_\infty(D) = \min_x \{-\log_2(\Pr[x \leftarrow D])\}$.

A function Ext: $\{0,1\}^a \times \{0,1\}^b \rightarrow \{0,1\}^c$ is called a (k, δ)-*extractor* if for any distribution D on $\{0,1\}^a$ with min-entropy at least k, the distribution $N(D)$ is δ-close to uniform, where $N(D) = \{x \leftarrow D; e \leftarrow \{0,1\}^b; y \leftarrow \text{Ext}(x, e) : y\}$.

A function Disp: $\{0,1\}^a \times \{0,1\}^b \rightarrow \{0,1\}^c$ is called a (k, δ)-*disperser* if for any $A \subseteq \{0,1\}^a$ such that $|A| \geq 2^k$, it holds that $|N(A)| \geq (1 - \delta)2^c$, where $N(A) = \{z \mid z = \text{Disp}(x, y), x \in A, y \in \{0,1\}^b\}$.

We refer to [16, 23] for surveys of applications, constructions and related results for extractors and dispersers. (We use the formal definition of dispersers that appears in [16]; other papers such as [23] use a slightly different definition.)

Construction of protocol $\mathcal{P}_3$. The protocol $\mathcal{P}_3 = (\text{SAMPLE}, \text{SET}, \text{REG}, \text{VER})$ uses a polynomial-time computable function $\text{SELECT} : \{0, 1\}^d \times \{0, 1\}^d \to [m]^l$, that we later instantiate using extractors, and a family $\mathcal{H}$ of pairwise-independent hash functions $h_w : \{0, 1\}^{nl+2d} \to \{0, 1\}^n$ (selection of which parameterizes the REG and VER algorithms). We first describe algorithms SET, REG, and VER, and then one instantiation of the function SELECT.

Algorithm SET. Formally, algorithm SET, on input parameters d, t, n, l, m, q in unary, returns an m-location password file F, which can be parsed as $F = X \circ T$ with $|X| = m_x$ and $|T| = t$ (and thus $m = m_x + t$). X is initialized as an array of values $X[1], \ldots, X[m_x]$ uniformly chosen from $\{0, 1\}^n$, and T is initialized as an empty array of t locations.

Algorithm REG. The registration algorithm maps a login and a password to a subset of locations in the set of locations containing random elements, combines their content by computing a tag as their sum, and stores the tag. Formally, on input log_i, pw_i, F, algorithm REG runs the following steps:

1. compute $(loc_1, \ldots, loc_l) = \text{SELECT}(log_i, pw_i)$;
2. compute $tag_i = h_w(log_i|pw_i|X[loc_1]| \cdots |X[loc_l])$,
3. store tag_i into T by setting $T[i] = tag_i$.

Algorithm VER. The verification algorithm recomputes the tag corresponding to the input login and password and checks that it is equal to the tag stored during the registration phase. Formally, on input $Param, \{log_i\}_{i=1}^t, log', pw', F$, where $F = X|T$, algorithm VER runs the following steps:

1. compute $(loc'_1, \ldots, loc'_l) = \text{SELECT}(log', pw')$;
2. let $j \in \{1, \ldots, t\}$ be such that $log_j = log'$;
3. if there exists no such j then return: 0 and halt;
4. verify that $T[j] = h_w(log'|pw'|X[loc'_1]| \cdots |X[loc'_l])$;
5. if so, return: 1; else return: 0.

Instantiation of function SELECT. For our construction we only need to apply dispersers, but since (k, δ)-extractors are also (k, δ)-dispersers (this can be seen by setting D equal to the uniform distribution over subset A), and given that extractors have been much more studied in the literature, we will apply (a certain kind of) extractors. In particular, we are interested in extractors that firstly maximize the parameter c, denoting the extractor output, so that it is as close as possible to the sum of the min-entropy of the source and the number of real random bits used. Secondly, it is of interest to minimize the value of parameter b for that to happen. This choice criterion is based on that of minimizing the adversary's advantage first, and then, further minimizing the server's sequential running time. We note that in this scheme the parallel running time is constant with respect to the lookup complexity l, regardless of which extractor we choose.

A recent survey [23] summarizes most known results about extractors, and we can plug in some of the results in Table 1, pp. 11 of [23] to obtain a function SELECT with satisfactory performance. Bearing in mind the aforementioned criterion, we will use the following fact (obtained from Corollary 6.15 of [21]):

Fact 4. [21] *For any $0 \leq \alpha < a$ and $\delta > \exp(-\alpha/(\log^* \alpha)^{O(\log^* \alpha)})$, there exists an explicit (k, δ)-extractor Ext: $\{0,1\}^a \times \{0,1\}^b \rightarrow \{0,1\}^c$ such that $k = a - \alpha$, $b = O(\log^2(a) \cdot \text{poly}(\log \log a) + (\log a) \cdot (\log(1/\delta)))$ and $c = k + b - 2\log(1/\delta) - O(1)$.*

Informally, we can instantiate SELECT as the function returning all outputs of the above extractor, when given the password as a first input and all possible l values as a second input. (In the graph-based formulation, these would be all neighbors of the node associated with the password). More formally, for any parameters n, t, d, m, q, where $m = m_x + t$, we can instantiate SELECT as follows. For $log \in \{0,1\}^d$ and $pw \in \{0,1\}^d$, we define SELECT(log, pw) = $(loc_1, \ldots, loc_l)$, where $loc_j = \text{Ext}((log|pw), j)$, for $j = 1, \ldots, l$; algorithm Ext: $\{0,1\}^a \times \{0,1\}^b \rightarrow \{0,1\}^c$ is the (k, δ)-extractor guaranteed from Fact 4, where $\alpha = d$, $a = 2d$; $b = O(\log^2(2d) \cdot \text{poly}(\log \log 2d) + (\log 2d) \cdot (\log(m/(m - q))))$; $c = \log m$; $l = 2^b$; $\delta = 1 - q/m$; and $k = \log m - l - 2\log(m/(m - q)) - O(1)$.

Proving the security of the SCS protocol. Proving the security property of $\mathcal{P}_3$ makes crucial use of the properties of dispersers and of pairwise-independent hash functions. The main intuition is that if the adversary queries q locations from F, possibly using an adaptive querying strategy, even if he tries to run an off-line password attack, he will be able to test only a very small number of passwords. More specifically, we observe the following facts, using the properties of pairwise-independent hash functions: (1) the probability that the server accepts a false password, is very small. Then we observe that the content of the locations queried by the adversary define t partitions of the set of passwords into two sets: the set of passwords that are mapped to locations queried by the adversary and its complement. Furthermore, (2) the size of the first set is small, (i.e., $O(m^3/l(m - q)^2)$), and (3) any password in the second set does not give a significant advantage to the adversary in being successful; where (2) uses the properties of the disperser from Fact 4 and (3) uses the definition of pairwise-independent hash functions. The security property of $\mathcal{P}_3$ follows by combining the three mentioned facts.

6   Strong Security with Small, Adaptive Lookup Complexity

In the previous section we showed that it is possible to construct password protocols secure against bounded retrieval attacks by adaptive adversaries, and simultaneously have low lookup complexity. The adversary's advantage in the previous construction is not significantly larger than the on-line attack success probability, and essentially meets the lower bound in Section 3. In this section we investigate the possibility of achieving even smaller adversary's advantage (say, exponentially small) while maintaining an efficient lookup complexity. Since the server's lookup strategy will be adaptive, the lower bound in Section 3 does not apply. The scheme remains incomparable to the scheme in previous section though, as it is only secure against static adversaries. We call our new scheme HE, for Hashing and Extraction, according to the strategy used by the server's registration algorithm. Formally, we obtain the following:

Theorem 5. There exists a protocol $\mathcal{P}_4 = (\text{SET}_4, \text{SAMPLE}_4, \text{REG}_4, \text{VER}_4)$ with parameters $Param = (n, t, d, m, q, l)$, that is ϵ-secure against a bounded retrieval attack of static type, and such that, for any t, d, q, m, it holds that $\epsilon = (2t + 3) \cdot 2^{-\lambda}$, $m > q + t \geq [\, l = t + O(d + \lambda) \,]$ and $n = O(\lambda) + 2d$.

We stress that in $\mathcal{P}_4$ the server uses an adaptive lookup strategy and therefore the exponentially small upper bound on ϵ does not contradict the lower bound of Theorem 1. We now sketch the proof of Theorem 5.

Tools used by our HE protocol. The construction uses two tools: t-wise independent hash functions (see Definition 3), where t is the number of users, and locally computable and strong extractors.

Locally-computable and strong extractors. We recall two additional properties that extractors (defined in Section 5) may satisfy. Intuitively, the definition of strong extractors requires that the extractor's output remains statistically close to random even when conditioned on the value of the random seed; furthermore, the definition of locally computable extractors requires that the extractor reads only a small subset of the bits contained in the (large) input distribution that the entropy is to be extracted from (this is for efficiency reasons only). The formal definitions follow.

A function Ext: $\{0,1\}^a \times \{0,1\}^b \to \{0,1\}^c$ is called a *strong (k, δ)-extractor* if for any distribution D on $\{0,1\}^a$ with min-entropy at least k, the distribution $U(b) \times N(D, U(b))$ is δ-close to distribution $U(b) \times U(c)$, where $N(D, U(b))$ is defined as $\{x \leftarrow D; e \leftarrow \{0,1\}^b; y \leftarrow \text{Ext}(x, e) : y\}$, and, for any z, $U(z)$ denotes the uniform distribution over $\{0,1\}^z$.

An extractor Ext: $\{0,1\}^a \times \{0,1\}^b \to \{0,1\}^c$ is ℓ-*locally-computable* if for any $R \in \{0,1\}^b$, the value of R uniquely determines the bit locations in $x \in \{0,1\}^a$ used while computing $\text{Ext}(x, R)$ and the number of such locations is at most ℓ.

We will use the following strong and locally-computable extractor, guaranteed from Theorem 8.5 in [25]:

Fact 6 ([25]). *Let ρ, σ be arbitrary constants > 0. For every $a \in \mathbb{N}$, $\delta > \exp(-a/2^{O(\log^* a)})$, $c \leq (1 - \sigma)a\rho$, there is an explicit ℓ-locally computable and strong (k, δ) extractor* Ext : $\{0,1\}^a \times \{0,1\}^b \to \{0,1\}^c$ *such that:*

1. $k = a\rho$
2. $b = \log a + O(\log c + \log(1/\delta))$
3. $\ell = (1 + \sigma)kc/n + \log(1/\delta)$

Construction of protocol $\mathcal{P}_4$. We assume, for simplicity, that the algorithm SAMPLE just uniformly and independently selects a password from $\{0,1\}^d$.

Algorithm SET. Let the data block size $n = 2d + \lambda$, where λ is the security parameter. SET initializes a t-location array W with a uniformly chosen t-wise independent hash function $h_w : \{0,1\}^d \to \{0,1\}^n$. SET then initializes X as an array of m_x locations containing uniformly and independently chosen values in $\{0,1\}^n$. Additionally, SET initializes an empty array T with t empty locations of n bits each, and then sets $F = T \circ W \circ X$. Observe that the total number of data blocks in F is $m = 2t + m_x$.

Algorithm REG. The registration algorithm first hashes the login name and the password to a random value using the t-wise independent hash function specified by the W component of F, to produce a seed value R. (Note that this step makes the server's lookup strategy adaptive, as the computation of R depends on the contents of W, and subsequent lookup operations in X will depend on R.) The extractor is applied to the X component of F using the previously computed seed R in order to produce a (nearly) uniform random output. The resulting output R' may then be used as the tag associated with this password, and is stored in the T component of F. Formally, on input i, log_i, pw_i, F, where $F = T \circ W \circ X$, algorithm REG does the following:

1. set $w = (W[1], \ldots, W[t])$ and $R = h_w(log_i \mid pw_i)$.
2. set $R' = \text{Ext}(X, R)$;
3. store $tag_i = R'$ in location $T[i]$.

Algorithm VER. The VER algorithm is essentially identical to the REG algorithm, only after computing tag_i, rather than storing it in the $T[i]$ location in F, the value is compared with the previously stored value in $T[i]$, and the result of the comparison is output. The total number of lookups performed by VER is $l = t + \ell + 1$. Formally, VER, on input $Param, \{log_i\}_{i=1}^t, log', pw', F = T \circ W \circ X$, does the following:

1. set $w = (W[1], \ldots, W[t])$ and $R = h_w(log' \mid pw')$.
2. let $j \in \{1, \ldots, t\}$ be such that $log_j = log'$;
3. if there exists no such j then return: 0 and halt;
4. set $R' = \text{Ext}(X, R)$;
5. if $T[j] = R'$ then return: 1 else return: 0

Proof that $\mathcal{P}_4$ satisfies Theorem 5. Proving the security property of $\mathcal{P}_4$ makes crucial use of the properties of strong extractors and of t-wise independent hash functions, as follows. We use the properties of strong extractors to show that the first tag is statistically indistinguishable from a uniformly distributed tag, even conditioned on the value of all other tags and on the value of the hash function used to generate the seed for the extractor. In proving that the conditioning on the value of all other tags does not affect the statistical indistinguishability, we use the indistinguishability of the extractor's output from random, even conditioned over the result of a bounded-output function over the extractor's input. In proving that the conditioning on the seed does not affect the statistical indistinguishability, we use the extractor's 'strong' property. Then we replace the first tag with a random tag and repeat the analogous argument over the second tag, etc. (Note that independence of the R values used to compute each tag is guaranteed for up to t users by the t-wise independent hash function.) Finally, we compute an upper bound on the adversary's advantage when all tags are random by computing an upper bound on collisions on the t-wise independent hash function and on the extractor used. A formal proof is available in the full version of the paper.

An extension: Combining protocols P_3 and P_4. Recall that protocol P_3 is secure against adaptive adversaries but allows the adversary to achieve a non-negligible advantage (which is optimal in the setting of adaptive intrusions). Furthermore, protocol P_4 only allows the adversary to achieve at most negligible advantage, but is only secure against static adversaries. We would like to achieve the "best of both worlds" with a single scheme that limits the adversary to a negligible advantage in case of static intrusions, but remains secure even under an adaptive attack.

Fortunately, such a solution is indeed possible. We simply modify the construction of P_3, replacing the input pw_i with the tag_i computed as in protocol P_4. That is, the the final tag_i values computed using P_3 will now be based on "password" inputs taken from the tag_i values computed via P_4 using the user's actual password. It can be shown that the resulting scheme achieves security comparable to that of P_4 under static intrusion attacks, and comparable to that of P_3 under adaptive intrusion attacks.

Acknowledgment. The first author thanks Rajesh Talpade for interesting discussions on intrusion detection. This material is based upon work supported by the United States Air Force under Contract FA8750-04-C-0249. Any opinion, findings, and conclusions or recommendations expressed in this material are those of author(s) and do not necessarily reflect the view of the United States Air Force.

References

1. S. Axelsson. Research in Intrusion-Detection systems: A Survey, in *Technical Report 98-17*, Dept. of Comp. Eng., Chalmers, Univ. of Technology, Goteborg, Sweden, 1998, http://citeseer.ist.psu.edu/axelsson98research.html.
2. S. Bellovin and M. Merrit. Encrypted Key Exchange, in *Proc. of the 1992 Internet Society Network and Distributed System Security Symposium.*
3. S. Bellovin and M. Merrit. Augmented Encrypted Key Exchange, in *Proc. of the 1st ACM Conference on Computer and Communication Security*, pp. 224-250
4. G. R. Blakley. Safeguarding cryptographic keys. In *Proc. of the National Computer Conference*, v.48, pp. 242–268, 1979.
5. G. Di Crescenzo, A. Ghosh, and R. Talpade. Towards a Theory of Intrusion Detection. In *Proc. of European Symposium on Research in computer Security* (ESORICS 2005), vol. 3679 of LNCS, pp. 267-286, Springer-Verlag.
6. Y. Dodis, A. Sahai, A. Smith. On Perfect and Adaptive Security in Exposure-Resilient Cryptography. In *Proc. of EUROCRYPT 2001*, vol. 2045 of LNCS, pp. 301-324. Springer-Verlag.

7. Password Protocols provably secure in the Bounded Retrieval Model, first public version of this work, unpublished draft, April 2005.
8. D.C. Feldmeier and P.R. Karn. UNIX Password Security - Ten Years Later, in *Proceedings of Crypto'89*, LNCS, no. 435, Springer-Verlag, pp. 44-63
9. S. Halevi and H. Krawczyk. Public-key Cryptography and Password Protocols. In *Proc. of the 5th annual ACM conference on Computer and Communications Security*, pp. 122-131, 1998
10. John Kelsey and Bruce Schneier. Authenticating Secure Tokens Using Slow Memory Access. *USENIX Workshop on Smart Card Technology*, USENIX Press, pp. 101-106, 1999.
11. Chi-Jen Lu. Encryption against storage-bounded adversaries from on-line strong extractors. *Journal of Cryptology*, vol. 17, no. 1, pp. 27-42, 2004.
12. Stefan Dziembowski and Ueli Maurer. Optimal Randomizer Efficiency in the Bounded-Storage Model. In *Journal of Cryptology*, vol. 17, no. 1, pp. 5-26.
13. Ueli Maurer. Conditionally-Perfect Secrecy and a Provably-Secure Randomized Cipher. *Journal of Cryptology*, vol. 5, no. 1, pp. 53-66, 1992.
14. Tal Moran, Ronen Shaltiel, Amnon Ta-Shma. Non-interactive Timestamping in the Bounded Storage Model. In *Proc. of CRYPTO 2004*, vol. 3152 of LNCS, pp. 460-476. Springer-Verlag.
15. R. Morris and K. Thompson. Password Security: A Case History, in *Communications of the ACM*, Vol. 22, no. 11, 1979, pp. 594-597.
16. N. Nisan and A. Ta-Shma. Extracting Randomness: A Survey and New Constructions, in *Journal of Computer and System Sciences*, February 1999, vol. 58, no. 1, pp. 148-173(26)
17. N. Nisan and D. Zuckerman. More Deterministic Simulation in Logspace, in *Proc. of ACM STOC 93*.
18. S. Patel. Number theoretic attacks on secure password schemes, in *Proc. of the 1997 IEEE Symposium on Security and Privacy*.
19. Benny Pinkas and Tomas Sander. Securing Passwords Against Dictionary Attacks. *ACM CCS-9: Computer and Communications Security*, Nov., 2002.
20. N. Provos and D. Mazieres, A Future-Adaptable Password Scheme, In *Proceedings of the Annual USENIX Technical Conference*, 1999.
21. O. Reingold, S. Vadhan, and A. Wigderson. Entropy Waves, The Zig-Zag Graph Product, and New Constant-Degree Expanders and Extractors. in *Electronic Colloquium on Computational Complexity TR01-018*; other versions in *Proceedings of FOCS 2000* and *Annals of Mathematics*, vol. 155, pp. 157-187, 2002.
22. A. Shamir. How to Share a Secret. *Communications of the ACM*, Volume 22 , Issue 11 (November 1979)
23. R. Shaltiel. Recent developments in Explicit Constructions of Extractors. *Bulletin of the EATCS*, 77:67-95, 2002.
24. M. Sipser. Expanders, Randomness and Time vs. Space, in *Journal of Computer and System Sciences*, vol. 36, 1988.
25. S. P. Vadhan. On constructing locally computable extractors and cryptosystems in the bounded storage model. *Journal of Cryptology*, vol. 17, no. 1, pp. 43-77, 2004.
26. T. Wu, *The secure remote password protocol*, in Proc. of the 1998 Internet Society Network and Distributed System Security Symposium
27. http://money.cnn.com/2003/02/18/technology/creditcards/
28. http://www.detnews.com/2005/technology/0506/18/tech-219662.htm

A   Numerical Examples

Some typical parameters for instantiating protocol $\mathcal{P}_4$ might be as follows. Set $O(\lambda) = 176$ for a dictionary of size $\approx 2^d$, where, say $d = 40$ (this yields a dictionary of approximately 1 trillion words). We have that $n = 2d + O(\lambda) = 80 + 176 = 256 = 2^8$ (assuming a small constant under the O notation). This requires that data be read from storage in chunks not less than 48 bytes in size.

For a system with $t \approx 2^{12} = 4096$ maximum users, we can achieve the following parameters. Letting $m = 2t + \hat{m} = 2^{13} + 2^{35} \approx 2^{35}$, we obtain a total storage requirement of $mn = 2^8 2^{35} = 2^{43}$ bits, or approximately 1 TB (terabyte). It should be noted that 1 terabyte of storage can currently be purchased at very reasonable cost (under \$1000). Given storage of this size, we can safely set $\beta = 0.99$, allowing the adversary to retrieve up to 99 percent of the storage, which is about 990 MB (megabytes) of data. If we limit the server to an outgoing bandwidth of 8192 bits/sec = 1024 bytes/sec, it will take the adversary over 30 years to download that much data. With an outgoing bandwidth of 1024 bytes/sec, the server can process approximately 32 logins/sec. The lookup complexity will be $l = t + O(d + \lambda) \approx 2^{12} + C(40 + 176) \approx 2^{13}$, which is about 8000 blocks of 256-bits each, per login (a total of less than half a megabyte of data).

Polylogarithmic Private Approximations and Efficient Matching

Piotr Indyk[1] and David Woodruff[1,2]

[1] MIT CSAIL
{indyk, dpwood}@mit.edu
[2] Tsinghua University

Abstract. In [12] a *private approximation* of a function f is defined to be another function F that approximates f in the usual sense, but does not reveal any information about x other than what can be deduced from $f(x)$. We give the first two-party private approximation of the l_2 distance with polylogarithmic communication. This, in particular, resolves the main open question of [12].

We then look at the *private near neighbor* problem in which Alice has a query point in $\{0,1\}^d$ and Bob a set of n points in $\{0,1\}^d$, and Alice should privately learn the point closest to her query. We improve upon existing protocols, resolving open questions of [13, 10]. Then, we relax the problem by defining the *private approximate near neighbor problem*, which requires introducing a notion of secure computation of approximations for functions that return sets of points rather than values. For this problem we give several protocols with sublinear communication.

Keywords: private approximations, secure multiparty computation, nearest neighbor, communication complexity.

1 Introduction

Recent years witnessed the explosive growth of the amount of available data. Large data sets, such as transaction data, the web and web access logs, or network traffic data, are in abundance. Much of the data is stored or made accessible in a distributed fashion. This neccessitates the development of efficient protocols that compute or approximate functions over such data (e.g. see [2]).

At the same time, the availability of this data has raised significant privacy concerns. It became apparent that one needs cryptographic techniques in order to control data access and prevent potential misuse. In principle, this task can be achieved using the general results of secure function evaluation (SFE) [33, 18]. However, in most cases the resulting private protocols are much less efficient than their non-private counterparts[1]. Moreover, SFE applies only to algorithms that compute functions *exactly*, while for most massive data sets problems, only

---

[1] A rare exception is the result of [29], who show how to obtain private and communication-efficient versions of non-private protocols, as long as the communication cost is logarithmic.

S. Halevi and T. Rabin (Eds.): TCC 2006, LNCS 3876, pp. 245–264, 2006.

efficient *approximation* algorithms are known or are possible. Indeed, while it is true that SFE can be used to privately implement any efficient algorithm, it is of little use applying it to an approximation algorithm when the approximation leaks more information about the input than the solution itself.

In a pioneering paper [12], the authors introduced a framework for secure computation of approximations. They also proposed an $\tilde{O}(\sqrt{n})$-communication[2] two-party protocol for approximating the Hamming distance between two binary vectors. This improves over the linear complexity of computing the distance exactly via SFE, but still does not achieve the polylogarithmic efficiency of a non-private protocol of [25]. Improving the aforementioned bound was one of the main problems left open in [12].

In this paper we provide several new results for secure computation of approximations. Our first result is an $\tilde{O}(1)$-communication protocol for approximating the Euclidean (ℓ_2) distance between two vectors. This, in particular, solves the open problem of [12]. Since distance computation is a basic geometric primitive, we believe that our result could lead to other algorithms for secure approximations. Indeed, in [1] the authors show how to approximate the ℓ_2 distance using small space and/or short amount of communication, initiating a rich body of work on streaming algorithms.

In the second part of the paper, we look at secure computation of a *near neighbor* for a query point q (held by Alice) among n data points P (held by Bob) in $\{0,1\}^d$. We improve upon known results [10, 13] for this problem under various distance metrics, including ℓ_2, set difference, and Hamming distance over arbitrary alphabets. Our techniques also result in better communication for the *all-near neighbors* problem, where Alice holds n different query points, resolving an open question of [13], and yield a binary inner product protocol with communication $d + O(k)$ in the common random string model.

Complexity	Problem	Prior work	SFE
$O(n+d)$	Near neighbor under l_2, Hamming over $\{0,1\}^d$, Set difference	[10]	$\tilde{O}(nd)$
$\tilde{O}(dU+n)$	Near neighbor under distances $f(a,b) = \sum_{i=1}^{d} f_i(a_i, b_i),\ a_i, b_i \in [U]$	[10]	$\tilde{O}(nd \log U)$
$\lceil \log d \rceil d + O(k)$	Hamming distance	[14]	$O(kd)$
$\tilde{O}(nd^2 + n^2)$	All-near neighbors	[13]	$\tilde{O}(n^2 d)$

However, all of our protocols for the near neighbor problem have the drawback of needing $\Omega(n)$ bits of communication, though the dependence on d is often optimal. Thus, we focus on what we term the *approximate near neighbor problem*. For this we introduce a new definition of secure computation of approximations for functions that return points (or sets of points) rather than values.

---

[2] We write $f = \tilde{O}(g)$ if $f(n,k) = O\left(g(n,k) \log^{O(1)}(n)\mathrm{poly}(k)\right)$, where k is a security parameter.

Approximate privacy. Let $P_t(q)$ be the set of points in P within distance t from q. In the c-*approximate near neighbor* problem, the protocol is required to report a point in $P_{cr}(q)$, as long as $P_r(q)$ is nonempty. We say that a protocol solving this problem is c'-*private* (or just *private* if $c' = c$) if Bob learns nothing, while Alice learns nothing except what can be deduced from the set $P_{c'r}(q)$. In our paper we always set $c' = c$.

We believe this to be a natural definition of privacy in the context of the approximate near neighbor problem. First, observe that if we insist that Alice learns only the set P_r (as opposed to P_{cr}), then the problem degenerates to the *exact* near neighbor problem. Indeed, even though the definition of correctness allows the protocol to output a point $p \in P_{cr} - P_r$, in general Alice cannot simulate this protocol given only the set P_r. Thus, in order to make use of the flexibility provided by the approximate definition of the problem, it seems necessary to relax the definition of privacy as well.

Second, the above relaxation of privacy appears natural in the context of applications of near neighbor algorithms. In most situations, the distance function is only a heuristic approximation of the dis-similarity between objects, and there is no clear rationale for a sharp barrier between objects that can or cannot be revealed (still, it is important that the information leak is limited). Our model formalizes this intuition, and our algorithmic results shows that it is possible to exploit the model to obtain more efficient algorithms.

Specifically, within this framework, we give a c-approximate near neighbor protocol with communication $\tilde{O}(n^{1/2} + d)$ for any constant $c > 1$. The protocol is based on dimensionality reduction technique of [25]. We show how the dependence on d can be made polylogarithmic if Alice just wants a coordinate of a point in P_{cr}. We also give a protocol based on locality-sensitive hashing (LSH) [23], with communication $\tilde{O}(n^{1/2+1/(2c)} + d)$, but significantly less work (though still polynomial).

Finally, proceeding along the lines of [20], we say the protocol *leaks b bits of information* if it can be simulated given b extra bits which may depend arbitrarily on the input. With this definition, we give a protocol with $\tilde{O}(n^{1/3} + d)$ communication leaking only k bits, where k is a security parameter.

General vs specific solutions. As described above, this paper offers solutions to *specific* computational problems. In principle, a general "compiler-like" approach (as in [33, 18]) would be preferable. However, it appears unlikely that a compiler approach can be developed in the context of *approximate* problems. Indeed, there is no general method that, for a given problem, generates an efficient approximation algorithm (even ignoring the privacy issue). This implies that a compiler would have to start from a particular approximation to a given function. Unfortunately, as mentioned earlier, such approximation itself can leak too much information.

This argument leads us to believe that, in context of approximate algorithms, designing efficient private solutions to specific problems is the only possible approach.

2 Preliminaries

Background on homomorphic encryption, oblivious transfer (OT), and secure function evaluation (SFE) can be found in appendix A. We write $negl(k, n)$ to denote an arbitrary negligible function of k, n, that is a function which shrinks faster than any inverse polynomial in n, k.

We assume both parties are computationally bounded and semi-honest, meaning they follow the protocol but may keep message histories in an attempt to learn more than is prescribed. In [18, 7, 29], it is shown how to transform a semi-honest protocol into a protocol secure in the malicious model. Further, [29] does this at a communication blowup of at most a small factor of $poly(k)$. Therefore, we assume parties are semi-honest in the remainder of the paper.

We briefly review the semi-honest model, referring the reader to [17, 26] for more details. Let $f : \{0,1\}^* \times \{0,1\}^* \rightarrow \{0,1\}^* \times \{0,1\}^*$ be a function, the first element denoted $f_1(x_1, x_2)$ and the second $f_2(x_1, x_2)$. Let π be a two-party protocol for computing f. The views of players P_1 and P_2 during an execution of $\pi(x_1, x_2)$, denoted $\text{View}_1^\pi(x_1, x_2)$ and $\text{View}_2^\pi(x_1, x_2)$ respectively, are:

$$\text{View}_1^\pi(x_1, x_2) = (x_1, r_1, m_{1,1}, \ldots, m_{1,t}), \text{View}_2^\pi(x_1, x_2) = (x_2, r_2, m_{2,1}, \ldots, m_{2,t}),$$

where r_i is the random input and $m_{i,j}$ the messages received by player i respectively. The outputs of P_1 and P_2 during an execution of $\pi(x_1, x_2)$ are denoted $\text{output}_1^\pi(x_1, x_2)$ and $\text{output}_2^\pi(x_1, x_2)$. We define $\text{output}^\pi(x_1, x_2)$ to be $(\text{output}_1^\pi(x_1, x_2), \text{output}_2^\pi(x_1, x_2))$. We say that π privately computes a function f if there exist PPT algorithms S_1, S_2 for which for $i \in \{1, 2\}$ we have the following indistinguishability

$$\{S_i(x_i, f_i(x_1, x_2)), f(x_1, x_2)\} \overset{c}{\equiv} \{\text{View}_i^\pi(x_1, x_2), \text{output}^\pi(x_1, x_2)\}.$$

This simplifies to $\{S_i(x_i, f_i(x_1, x_2))\} \overset{c}{\equiv} \{\text{View}_i^\pi(x_1, x_2)\}$ if either $f_1(x_1 x_2) = f_2(x_1, x_2)$ or if $f(x_1, x_2)$ is deterministic or equals a specific value with probability $1 - negl(k, n)$, for k a security parameter.

We need a standard composition theorem [17] concerning private subprotocols. An *oracle-aided protocol* (see [26]) is a protocol augmented with a pair of oracle tapes for each party and oracle-call steps. In an oracle-call step parties write to their oracle tape and the oracle responds to the requesting parties. An oracle-aided protocol uses the *oracle-functionality* $f = (f_1, f_2)$ if the oracle responds to query x, y with $(f_1(x, y), f_2(x, y))$, where f_1, f_2 denote first and second party's output respectively. An oracle-aided protocol *privately reduces* g to f if it privately computes g when using oracle-functionality f.

Theorem 1. *[17] If a function g is privately reducible to a function f, then the protocol g' derived from g by replacing oracle calls to f with a protocol for privately computing f, privately computes g.*

We now define the *functional privacy* of an approximation as in [12]. For our approximation protocols we will have $f_1(x, y) = f_2(x, y) = f(x, y)$.

Definition 1. *Let $f(x, y)$ be a function, and let $\hat{f}(x, y)$ be a randomized function. Then $\hat{f}(x, y)$ is functionally private for f if there is an efficient simulator S s.t. for every x, y, we have $\hat{f}(x, y) \overset{c}{\equiv} S(f(x, y))$.*

A *private approximation* of f privately computes a randomized function $\hat{f}$ that is functionally private for f.

Finally, we need the notion of a protocol for securely evaluating a circuit *with ROM*. In this setting, the ith party has a table $R_i \in (\{0, 1\}^r)^s$ defined by his inputs. The circuit, in addition to the usual gates, is equipped with *lookup gates* which on inputs (i, j), output $R_i[j]$.

Theorem 2. *[29] If C is a circuit with ROM, then it can be securely computed with $\tilde{O}(|C|T(r, s))$ communication, where $T(r, s)$ is the communication of 1-out-of-s OT on words of size r.*

3 Private ℓ_2 Approximation

Here we give a private approximation of the ℓ_2 distance. Alice is given a vector $a \in [M]^n$, and Bob a vector $b \in [M]^n$. Note that $\|a - b\|^2 \leq T_{max} \overset{def}{=} nM^2$. In addition, parameters ϵ, δ and k are specified. For simplicity, we assume that $k = \Omega(\log(nM))$. The goal is for both parties to compute an estimate E such that $|E - \|x\|^2| \leq \epsilon\|x\|^2$ with probability at least $1 - \delta$, for $x \overset{def}{=} a - b$. Further, we want E to be a private approximation of $\|x\|$, as defined in section 2. As discussed there, wlog we assume the parties are semi-honest. We set the parameter $B = \Theta(k)$; this notation means $B = ck$ for a large enough constant c independent from $k, n, M, \delta, \epsilon$. In our protocol we make the following cryptographic assumptions.

1. There exists a PRG G stretching polylog(n) bits to n bits secure against poly(n)-sized circuits.
2. There exists an OT scheme for communicating 1 of n bits with communication polylog(n).

At the end of the section we discuss the necessity and plausibility of these assumptions. Our protocol relies on the following fact and corollary.

Fact 3. *[27] Let A be a random $n \times n$ orthonormal matrix (i.e., A is picked from a distribution defined by the Haar measure). Then there is $c > 0$ such that for any $x \in \Re^n$, any $i = 1, \ldots, n$, and any $t > 1$,*

$$\Pr[|(Ax)_i| \geq \frac{\|x\|}{\sqrt{n}}t] \leq e^{-ct^2}.$$

Corollary 1. *Suppose we sample A as in Fact 3 but instead generate our randomness from G, rounding its entries to the nearest multiple of $2^{-\Theta(B)}$. Then,*

$\forall x \in [M]^n$,

$$\Pr[(1 - 2^{-B})\|x\|^2 \leq \|Ax\|^2 \leq \|x\|^2 \text{ and } \forall_i (Ax)_i^2 < \frac{\|x\|^2}{n}B] > 1 - \text{neg}(k, n)$$

Proof. If there were an infinite sequence of $x \in [M]^n$ for which this did not hold, a circuit with x hardwired would contradict the pseudorandomness of G.

Protocol Overview: Before describing our protocol, it is instructive to look at some natural approaches and why they fail. We start with the easier case of approximating the Hamming distance, and suppose the parties share a common random string. Consider the following non-private protocol of [25] discussed in [12]: Alice and Bob agree upon a random $O(\log n) \times n$ binary matrix R where the ith row consists of n i.i.d. Bernoulli(β^i) entries, where β is a constant depending on ϵ. Alice and Bob exchange Ra, Rb, and compute $R(a - b) = Rx$. Then $\|x\|$ can be approximated by observing that $\Pr[(Ra)_i = (Rb)_i] \approx 1/2$ if $\|x\| \gg \beta^{-i}$, and $\Pr[(Ra)_i = (Rb)_i] \approx 1$ if $\|x\| \ll \beta^{-i}$. Let the output be E. The communication is $O(\log n)$, but it is not private since both parties learn Rx. Indeed, as mentioned in [12], if $a = 0$ and $b = e_i$, then Rx equals the ith column of R, which cannot be simulated without knowing i.

However, given only $\|x\|$, it is possible to simulate E. Therefore, as pointed out in [12], one natural approach to try to achieve privacy is to run an SFE with inputs Ra, Rb, and output E. But this also fails, since knowing E *together with the randomness R* may reveal additional information about the inputs. If E is a deterministic function of Ra, Rb, and if $a = 0$ and $b = e_i$, Alice may be able to find i from a and R.

In [12], two private protocols which each have $\Omega(n)$ communication for a worst-case choice of inputs, were cleverly combined to overcome these problems and to achieve $\tilde{O}(\sqrt{n})$ communication. The first protocol, **High-Distance Estimator**, works when $\|x\| > \sqrt{n}$. The idea is for the parties to obliviously sample random coordinates of x, and use these to estimate $\|x\|$. Since the sampling is oblivious, the views depend only on $\|x\|$, and since it is random, the estimate is good provided we take $\tilde{O}(\sqrt{n})$ samples.

The second protocol, **Low-Distance Estimator**, works when $\|x\| \leq \sqrt{n}$. Roughly, the idea is for the parties to perfectly hash their vectors into $O(\sqrt{n})$ buckets so that at most one coordinate j for which $a_j \neq b_j$ lies in any given bucket. The parties then run an SFE with their buckets as input, which can compute $\|x\|$ exactly by counting the number of buckets which differ.

Our protocol breaks this $O(\sqrt{n})$ communication barrier as follows. First, Alice and Bob agree upon a random *orthonormal* matrix A in $\mathbb{R}^{n \times n}$, and compute Aa and Ab. The point of this step is to uniformly spread the mass of the difference vector x over the n coordinates, as per Fact 3, while preserving the length. Since we plan to sample random coordinates of Ax to estimate $\|x\|$, it is crucial to spread out the mass of $\|x\|$, as otherwise we could not for instance, distinguish $x = 0$ from $x = e_i$. The matrix multiplication can be seen as an analogue to the perfect hashing in **Low-Distance Estimator**, and the coordinate sampling as an analogue to that in **High-Distance Estimator**.

To estimate $\|x\|$ from the samples, we need to be careful of a few things. First, the parties should not learn the sampled values $(Ax)_j$, since these can reveal too much information. Indeed, if $a = 0$, then $(Ax)_j = (Ab)_j$, which is not private.

To this end, the parties run a secure circuit with ROM (see section 2) Aa and Ab, which privately obtains the samples.

Second, we need the circuit's output distribution E to depend only on $\|x\|$. It is not enough for $\mathbf{E}[E] = \|x\|^2$, since a polynomial number of samples from E may reveal non-simulatable information about x based on E's higher moments. To this end, the circuit uses the $(Ax)_j$ to independently generate r.v.s z_j from a Bernoulli distribution with success probability depending only on $\|x\|$. Hence, z_j depends only on $\|x\|$.

Third, we need to ensure that the z_j contain enough information to approximate $\|x\|$. We do this by maintaining a loop variable T which at any point in time is guaranteed to be an upper bound on $\|x\|^2$ with overwhelming probability. Using Corollary 1, for all j it holds that $q \stackrel{\text{def}}{=} n(Ax)_j^2/(TB) \leq 1$ for a parameter B, so we can generate the z_j from a Bernoulli(q) distribution. Since T is halved in each iteration, for some iteration $\mathbf{E}[\sum_j z_j]$ will be large enough to ensure that E is tightly concentrated.

We now describe the protocol in detail. Set $\ell = \Theta(B)(1/\epsilon^2 \log(nM) \log(1/\delta) + k)$. In the following, if $q > 1$, then the distribution Bernoulli(q) means Bernoulli(1).

ℓ_2-Approx (a, b):

1. Alice, Bob exchange a seed of G and generate A as in Corollary 1
2. Set $T = T_{max}$
3. Repeat:
   (a) {Assertion: $\|x\|^2 \leq T$ }
   (b) A secure circuit with ROM Aa, Ab computes the following
      - Generate random $i_1, \ldots, i_\ell$ and compute $(Ax)_{i_1}^2, \ldots (Ax)_{i_\ell}^2$
      - Generate $\{z_j\}_{j \in [\ell]}$ from i.i.d. Bernoulli$\left(\frac{n(Ax)_{i_j}^2}{TB}\right)$ distributions
   (c) $T = T/2$
4. Until $\sum_i z_i \geq \frac{\ell}{4B}$ or $T < 1$
5. Output $E = \frac{2TB}{\ell} \sum_i z_i$ as an estimate of $\|x\|^2$

Note that the protocol can be implemented in $O(1)$ rounds by parallelizing the secure circuit invocations.

Lemma 1. *The probability that assertion 3a holds in every iteration of step 3 is $1 - \text{neg}(k, n)$. Moreover, if $\|x\|^2 \neq 0$, then when the algorithm exits, with probability $1 - \text{neg}(k, n)$ it holds that $\mathbf{E}[\sum_j z_j] = \Theta\left(\ell/B\right)$.*

Proof. By Corollary 1, $\Pr_A[(1 - 2^{-B})\|x\|^2 \leq \|Ax\|^2 \leq \|x\|^2$ and $\forall_i (Ax)_i^2 < \frac{\|x\|^2}{n} B] = 1 - \text{neg}(k, n)$, so we may condition on this event occurring. If $\|x\|^2 = 0$, then $Ax = 0$, and thus $\Pr[E = 0] = 1$.

Otherwise, $\|x\|^2 \geq 1$. Consider the smallest j for which $T_{max}/2^j < \|x\|^2$. We show for $T = T_{max}/2^{j-1} \geq \|x\|^2 \geq 1$ that $\Pr[\sum_j z_j < \ell/(4B)] = \text{neg}(k, n)$. The assertion holds at the beginning of the jth iteration by our choice of T. Thus,

$n(Ax)_i^2 \leq TB$ for all $i \in [n]$ by the properties of A. So for all j, $\Pr[z_j = 1] = \frac{\|Ax\|^2}{TB} \geq (1 - 2^{-B})/(2B)$, and thus $\mathbf{E}[\sum_j z_j] \geq \ell/(3B)$. By a Chernoff bound, $\Pr[\sum_j z_j < \ell/(4B)] = \text{neg}(k, n)$, so if ever $T = T_{max}/2^{j-1}$, then this is the last iteration with overwhelming probability.

Note that the second part of the lemma follows from standard Chernoff bounds. Indeed, if $\|x\|^2 \neq 0$, then we have shown with overwhelming probability that in some iteration, $T \geq 1$ and $\sum_i z_i \geq \ell/4B$, so we may condition on the event that the algorithm exits in such an iteration. But for a certain constant in the big-Oh notation, one can show (by Chernoff and union bounds) that the probability $\sum_i z_i \geq \ell/4B$ when $\mathbf{E}[\sum_j z_j] = O(\ell/B)$ is negligible. On the other hand, once $\mathbf{E}[\sum_j z_j] \geq \ell/3B$, we have shown that $\sum_i z_i \geq \ell/4B$ with overwhelming probability. Thus we exit with $\mathbf{E}[\sum_j z_j] = \Theta(\ell/B)$.

Correctness. We claim $\Pr[|E - \|x\|^2| \leq \epsilon] \geq 1 - \delta$. Since $Ax = 0$ if $\|x\|^2 = 0$, we have that $E = 0$ in this case, and the claim is immediate. So suppose $\|x\|^2 \neq 0$. By Lemma 1, when the algorithm exits, with probability $1 - \text{neg}(k, n)$, $\mathbf{E}[\sum_i z_i] = \Theta(\ell/B)$, so we assume this event occurs. By a Chernoff and a union bound over iterations, we may assume that whenever $\mathbf{E}[\sum_i z_i] = \Theta(\ell/B)$,

$$\Pr\left[\left|\sum_i z_i - E\left[\sum_i z_i\right]\right| \geq \frac{\epsilon}{2} E\left[\sum_i z_i\right]\right] \leq e^{-\Theta(\epsilon^2 \frac{\ell}{B})} < \frac{\delta}{2}.$$

Thus, this holds when the algorithm exits. By Lemma 1, assertion 3a holds, so that $\ell(1 - 2^{-B})\|x\|^2 \leq TB \cdot \mathbf{E}[\sum_i z_i] \leq \ell \|x\|^2$. Setting $E = \frac{2TB}{\ell} \sum_i z_i$ (recall that T is halved in step 3c) then shows that $\Pr[|E - \|x\|^2| \geq \epsilon\|x\|^2] \leq \delta$.

Privacy. We replace the secure circuit with ROM in step 3b of ℓ_2-Approx with an oracle (see section 2). We construct a single simulator Sim, which given $\Delta \overset{\text{def}}{=} \|x\|^2$, satisfies $\text{Sim}(\Delta) \overset{c}{\equiv} \text{View}_A^\pi(a, b)$ and $\text{Sim}(\Delta) \overset{c}{\equiv} \text{View}_B^\pi(a, b)$, where $\text{View}_A^\pi(a, b)$, $\text{View}_B^\pi(a, b)$ are Alice, Bob's real views respectively. This, in particular, implies functional privacy. It will follow that ℓ_2-Approx is a private approximation of Δ.

Sim (Δ):

1. Generate a random seed of G
2. Set $T = T_{max}$
3. Repeat:
   (a) Generate $\{z_j\}_{j \in [\ell]}$ from i.i.d. Bernoulli($\frac{\Delta}{TB}$) distributions
   (b) $T = T/2$
4. Until $\sum_i z_i \geq \frac{\ell}{4B}$ or $T < 1$
5. Output $E = \frac{2TB}{l} \sum_i z_i$

With probability $1 - \text{neg}(k, n)$, the matrix A satisfies the property in Corollary 1, so we assume this event occurs. In each iteration, the random variables z_j are independent in both the simulation and the protocol. Further, the probabilities that $z_j = 1$ in the simulated and real views differ only by a multiplicative factor of $(1 - 2^{-B})$ as long as $T \geq \Delta$. But the probability that, in either view, we encounter $T < \Delta$ is $\mathrm{neg}(k, n)$.

Complexity. Given our cryptographic assumptions, we use $\tilde{O}(1)$ communication and $O(1)$ rounds.

Remark 1. Our cryptographic assumptions are fairly standard, and similar to the ones in [12]. There the authors make the weaker assumptions that PRGs stretching n^γ bits to n bits and OT with n^γ communication exist for any constant γ. In fact, the latter implies the former [21, 15]. If we were to instead use these assumptions, our communication would be $O(n^\gamma)$, still greatly improving upon the $O(n^{1/2+\gamma})$ communication of [12]. A candidate OT scheme satisfying our assumptions can be based on the Φ-Hiding Assumption [6], and can be derived by applying the PIR to OT transformation of [30] to the scheme in that paper.

Remark 2. For the special case of Hamming distance, we have an alternative protocol based on the following idea. Roughly, both parties apply the perfect hashing of the Low-Distance Estimator protocol of [12] for a logarithmic number of levels j, where the jth level contains $\tilde{O}(2^j)$ buckets. To overcome the $\tilde{O}(\sqrt{n})$ barrier of [12], instead of exchanging the buckets, the set of buckets is randomly and obliviously sampled. From the samples, an estimate of $\Delta(a, b)$ is output. For some j, $2^j \approx \Delta(a, b)$, so the estimate will be tightly concentrated, and for reasons similar to ℓ_2-Approx, will be simulatable. We omit the details, but note that two advantages of this alternative protocol are that the time complexity will be $\tilde{O}(n)$ instead of $\tilde{O}(n^2)$, and that we don't need the PRG G, as we may use k-wise independence for the hashing.

4   Private Near Neighbor and c-Approximate Near Neighbor Problems

We consider the case in which Alice has a point q, and Bob a set of n points P.

4.1   Private Near Neighbor Problem

Suppose for some integer U, Alice has $q \in [U]^d$, Bob has $P = p_1, \ldots, p_n \in [U]^d$, and Alice should learn $\min_i f(q, p_i)$, where f is some distance function. In [10] protocols for ℓ_1, ℓ_2, Hamming distance over U-ary alphabets, set difference, and arbitrary distance functions $f(a, b) = \sum_{i=1}^d f_i(a_i, b_i)$ were proposed, using an untrusted third party. We improve the communication of these protocols and remove the third party using homomorphic encryption to implement polynomial evaluation as in [13], and various hashing tricks.

In [13], the authors consider the private all-near neighbors problem in which Alice has n queries $q_1, \ldots, q_n \in [U]^d$ and wants all p_i for which $\Delta(p_i, q_j) \leq t < d$ for some j and parameter t. Our techniques improve the $\tilde{O}(n^2 d)$ communication of a generic SFE and the $\tilde{O}(n\binom{d}{t})$ communication of [13] for this problem to

$\tilde{O}(nd^2 + n^2)$. Finally, in the common random string model we achieve $\lceil \log d \rceil + O(k)$ communication for the (exact) Hamming distance, and an inner product protocol with $d + O(k)$ communication.

For the details of our schemes, see the full version of our paper [24]. We do not focus on them here since they still suffer from an $\Omega(n)$ communication cost. We instead focus on how to privately approximate these problems.

4.2 Private c-Approximate Near Neighbor Problem

Suppose $q \in \{0,1\}^d$ and $p_i \in \{0,1\}^d$ for all i. Let $P_t = \{p \in P \mid \Delta(p,q) \leq t\}$, and $c > 1$ be a constant.

Definition 2. *A c-approximate NN protocol is correct if when $P_r \neq \emptyset$, Alice outputs a point $f(q,P) \in P_{cr}$ with probability $1 - 2^{-\Omega(k)}$. It is private if in the computational sense, Bob learns nothing, while Alice learns nothing except what follows from P_{cr}. Formally, Alice's privacy is implied by an efficient simulator Sim for which $\langle q, P, f(q,P) \rangle \overset{c}{\equiv} \langle q, P, Sim(1^n, P_{cr}, q) \rangle$ for $\text{poly}(d,n,k)$-time machines.*

Following [20], we say the protocol *leaks b bits of information* if there is a deterministic "hint" function $h : \{0,1\}^{(n+1)d} \to \{0,1\}^b$ such that the distributions $\langle q, P, f(q,P) \rangle$ and $\langle q, P, Sim(1^n, P_{cr}, q, h(P,q)) \rangle$ are indistinguishable. As motivated in section 1, we believe these to be natural extensions of private approximations in [12, 20] from values to sets of values.

We give a private c-approximate NN protocol with communication $\tilde{O}(\sqrt{n}+d)$ and a c-approximate NN protocol with communication $\tilde{O}(n^{1/3}+d)$ which leaks k bits of information. Both protocols are based on dimensionality reduction in the hypercube [25]. There it is shown that for an $O(\log n) \times d$ matrix A with entries i.i.d. Bernoulli($1/d$), there is an $\tau = \tau(r, cr)$ such that for all $p, q \in \{0,1\}^d$, the following event holds with probability at least $1 - 1/\text{poly}(n)$

If $\Delta(p,q) \leq r$, then $\Delta(Ap, Aq) \leq \tau$, and if $\Delta(p,q) \geq cr$, then $\Delta(Ap, Aq) > \tau$.

Here, arithmetic occurs in $\mathbb{Z}_2$. We use this idea in the following helper protocol DimReduce(τ, B, q, P). Let A be a random matrix as described above. Let $S = \{p \in P \mid \Delta(Ap, Aq) \leq \tau\}$. If $|S| > B$, replace S with the lexicographically first B elements of S. DimReduce outputs random shares of S.

DimReduce (τ, B, q, P):

1. Bob performs the following computation
   - Generate a matrix A as above, and initialize L to an empty list.
   - For each $v \in \{0,1\}^{O(\log n)}$, let $L(v)$ be the first B p_i for which $\Delta(Ap_i, v) \leq \tau$.
2. A secure circuit with ROM L and input (q, A) executes:
   - Compute Aq.
   - Lookup Aq in L to obtain S. If $|S| < B$, pad S so that all S have the same length.
   - Output random shares (S^1, S^2) of S so that $S = S^1 \oplus S^2$.

It is an easy exercise to show the correctness and privacy of DimReduce.

Remark 3. As stated, the communication is $\tilde{O}(dB)$. The dependence on d can be improved to $\tilde{O}(d + B)$ using homomorphic encryption. Roughly, Alice sends $E(q_1), \ldots, E(q_d)$ to Bob, who sets $L(v)$ to be the first B different $E(\Delta(p_i, q))$ for which $\Delta(Ap_i, v) \leq \tau$. Note that $E(\Delta(p_i, q))$ is efficiently computable, and has size $\tilde{O}(1) \ll d$.

It will be useful to define the following event $\mathcal{H}(r_1, r_2, P)$ with $r_1 < r_2$. Suppose we run DimReduce independently k times with matrices A_i. Then $\mathcal{H}(r_1, r_2, P)$ is the event that at least $k/2$ different i satisfy

$$\forall p \in P_{r_1}, \ \Delta(A_i p, A_i q) \leq \tau(r_1, r_2) \text{ and } \forall p \in P \setminus P_{r_2}, \ \Delta(A_i p, A_i q) > \tau(r_1, r_2).$$

The next lemma follows from the properties of the A_i and Chernoff bounds:

Lemma 2. $\Pr[\mathcal{H}(r_1, r_2, P)] = 1 - 2^{-\Omega(k)}$.

4.3 c-Approximate NN Protocol

Protocol Overview: Our protocol is based on the following intuition. When $|P_{cr}|$ is large, a simple solution is to run a secure function evaluation with Alice's point q as input, together with a random sample P' of roughly a $k/|P_{cr}|$ fraction of Bob's points P. The circuit returns a random point of $P' \cap P_{cr}$, which is non-empty with overwhelming probability. The communication is $\tilde{O}(n/|P_{cr}|)$.

On the other hand, when $|P_{cr}|$ is small, if for k independent trials Alice and Bob run DimReduce$(\tau(r, cr), |P_{cr}|, q, P)$, then with overwhelming probability $P_r \subseteq \cup_i S_i$, where S_i denotes the (randomly shared) output in the ith execution. A secure function evaluation can then take in the random shares of the S_i and output a random point of P_r. The communication of this scheme is $\tilde{O}(|P_{cr}|)$.

Our protocol combines these two protocols to achieve $\tilde{O}(\sqrt{n})$ communication, by sampling roughly an $n^{-1/2}$ fraction of Bob's points in the first protocol, and by invoking DimReduce with parameter $B = \tilde{O}(\sqrt{n})$ in the second protocol. This approach is similar in spirit to the "high distance / low distance" approach used to privately approximate the Hamming distance in [12].

---

c-**Approx** (q, P):

1. Set $B = \tilde{O}(\sqrt{n})$.
2. Independently run DimReduce$(\tau(r, cr), B, q, P)$ k times, generating shares (S_i^1, S_i^2).
3. Bob finds a random subset P' of P of size B.
4. On inputs q, S_i^1, S_i^2, P', a secure circuit executes:
   - Compute $S_i = S_i^1 \oplus S_i^2$ for all i.
   - Let $f(q, P)$ be a random point from $P_{cr} \cap P' \neq \emptyset$ if it is non-empty,
   - Else let $f(q, P)$ be a random point from $P_r \cap \cup_i S_i$ if it is non-empty, otherwise set $f(q, P) = \emptyset$.
   - Output $(f(q, P), \text{null})$.

Using the ideas in Remark 3, the communication is $\tilde{O}(d+B)$, since the SFE has size $\tilde{O}(B)$. Let $\mathcal{F}$ be the event that $P' \cap P_{cr} \neq \emptyset$, and put $\mathcal{H} = \mathcal{H}(r, cr, P)$.

Correctness. Suppose P_r is nonempty. The probability s of correctness is just the probability we don't output $\emptyset$. Thus $s \geq \Pr[\mathcal{F}] + \Pr[\neg\mathcal{F}] \Pr[f(q, P) \neq \emptyset \mid \neg\mathcal{F}]$.

Case $|P_{cr}| \geq \sqrt{n}$: For sufficiently large B, we have $s \geq \Pr[\mathcal{F}] = 1 - 2^{-\Omega(k)}$.

Case $|P_{cr}| < \sqrt{n}$: It suffices to show $\Pr[f(q, P) \neq \emptyset \mid \neg\mathcal{F}] = 1 - 2^{-\Omega(k)}$. But this probability is at least $\Pr[f(q, P) \neq \emptyset \mid \mathcal{H}, \neg\mathcal{F}] \Pr[\mathcal{H}]$, and if $\mathcal{H}$ occurs, then $f(q, P) \neq \emptyset$. By Lemma 2, $\Pr[\mathcal{H}] = 1 - 2^{-\Omega(k)}$.

Privacy. Note that Bob gets no output, so Alice's privacy follows from the composition of of DimReduce and the secure circuit protocol of step 5. Similarly, if we can construct a simulator Sim with inputs $1^n, P_{cr}, q$ so that the distributions $\langle q, P, f(q, P) \rangle$ and $\langle q, P, Sim(1^n, P_{cr}, q) \rangle$ are statistically close, Bob's privacy will follow by that of DimReduce and the secure circuit protocol of step 5.

Sim $(1^n, P_{cr}, q)$:

1. Set $B = \tilde{O}(n^{1/2})$.
2. With probability $1 - \binom{n-|P_{cr}|}{B}\binom{n}{B}^{-1}$, output a random element of P_{cr},
3. Else output a random element of P_r.

Let X denote the output of $Sim(1^n, P_{cr}, q)$. It suffices to show that for each $p \in P$, $|\Pr[f(q, P) = p] - \Pr[X = p]| = 2^{-\Omega(k)}$, since this also implies $|\Pr[f(q, P) = \emptyset] - \Pr[X = \emptyset]| = 2^{-\Omega(k)}$. We have

$$\Pr[f(q, P) = p] = \Pr[f(q, P) = p, \mathcal{F}] + \Pr[f(q, P) = p, \neg\mathcal{F}]$$
$$= \Pr[f(q, P) = p, \mathcal{F}] + \Pr[f(q, P) = p, \neg\mathcal{F} \mid \mathcal{H}] \pm 2^{-\Omega(k)}$$
$$= \Pr[\mathcal{F}]|P_{cr}|^{-1} + \Pr[\neg\mathcal{F}] \Pr[f(q, P) = p \mid \mathcal{H}, \neg\mathcal{F}] \pm 2^{-\Omega(k)},$$

where we have used Lemma 2. Since $\Pr[\mathcal{F}] = 1 - \binom{n-|P_{cr}|}{B}\binom{n}{B}^{-1}$, we have

$$|\Pr[f(q, P) = p] - \Pr[X = p]| \leq$$
$$\Pr[\neg\mathcal{F}]\left|\Pr[f(q, P) = p \mid \mathcal{H}, \neg\mathcal{F}] - \delta(p \in P_r)|P_r|^{-1}\right| + 2^{-\Omega(k)}.$$

If $|P_{cr}| \geq \sqrt{n}$, then $\Pr[\neg\mathcal{F}] = 2^{-\Omega(k)}$. If $|P_{cr}| < \sqrt{n}$, then $\Pr[f(q, P) = p \mid \mathcal{H}, \neg\mathcal{F}] = \delta(p \in P_r)|P_r|^{-1}$.

Reducing the dependence on d: The way the current problem is stated, there is an $\Omega(d)$ lower bound. We now sketch how, if Alice just wants to learn some coordinate of an element of P_{cr}, this dependence can be made polylogarithmic. The idea is to perform an approximation to the Hamming distance instead of using the $E(\Delta(p_i, q))$ in the current protocol (see, e.g., DimReduce, and the following remark). The approximation we use is that given in [25], namely, the parties will agree upon random matrices A_i for some subset of i in $[n]$, and from the $A_i p_i$ and $A_i q$ will determine $(1 \pm \epsilon)$ approximations to the $\Delta(p_i, q)$ with probability $1 - 2^{-k}$. We don't need private approximations since the parties will not learn these values, but rather, they will input the $A_i p_i, A_i q$ into a secure circuit which makes decisions based on these approximations.

More precisely, Bob samples B of his vectors p_i, and in parallel agrees upon B matrices A_i and feeds the $A_i p_i$ into a secure circuit. Alice feeds in the $A_i q$. Let $c \geq 1 + 8\epsilon$. The circuit looks for an approximation of at most $r(1 + 6\epsilon)$. If such a value exists, the circuit gives Alice the corresponding index. Observe that if $|P_{r(1+4\epsilon)}| > \sqrt{n}$, then with probability $1 - 2^{-k}$ an index is returned to an element in P_{cr}, and that this distribution is simulatable. So assume $|P_{r(1+4\epsilon)}| \leq \sqrt{n}$.

The parties proceed by running a variant of $\mathsf{DimReduce}(\tau(r, r(1+4\epsilon)), B, q, P)$, with the important difference being that the output no longer consists of shares of the $E(\Delta(p_i, q))$. Instead, for each entry $L(v)$, Bob pretends he is running the approximation of [25] with Alice's point q. That is, the parties agree on B different matrices A_i and Bob computes $A_i p$ for each $p \in L(v)$. A secure circuit obtains these products, and computes the approximations. It outputs an index to a random element with approximation at most $r(1 + 2\epsilon)$. If P_r is nonempty, such an index will exist with probability $1 - 2^{-k}$. Also, the probability that an index to an element outside of $P_{r(1+4\epsilon)}$ is returned is less than 2^{-k}, and so the distribution of the index returned is simulatable.

Finally, given the index of some element in P_{cr}, the parties perform OT and Alice obtains the desired coordinate, The communication is $\tilde{O}(\sqrt{n} + \mathrm{polylog}(d))$.

Locality-sensitive hashing (LSH): We also have a similar protocol based on LSH, which only achieves $\tilde{O}(n^{1/2+1/(2c)} + d)$ communication, but has much smaller time complexity (though still polynomial). More precisely, the work of the LSH scheme is $n^{O(1)}$, whereas the work of c-**Approx** is $n^{O(1/(c-1)^2)}$, which is polynomial only for constant c. See Appendix B for the details.

4.4 c-Approximate NN Protocol Leaking k Bits

Protocol Overview: We consider three balls $P_r \subseteq P_{br} \subseteq P_{cr}$, where $c - b, b - 1 \in \Theta(1)$. We start by trying to use dimensionality reduction to separate P_r from $P \setminus P_{br}$, and to output a random point of P_r. If this fails, we try to sample and output a random point of P_{cr}. If this also fails, then it will likely hold that $n^{1/3} \leq |P_{br}| \leq |P_{cr}| \leq n^{2/3}$. We then sample down the pointset P by a factor of $n^{-1/3}$, obtaining $\tilde{P}$ with survivors $\tilde{P}_{br}, \tilde{P}_{cr}$ of P_{br}, P_{cr} respectively. It will now likely hold that we can use dimensionality reduction to separate $\tilde{P}_{br}$ from $\tilde{P} \setminus \tilde{P}_{cr}$ to obtain and output a random point of $\tilde{P}_{br}$. The hint function will encode the probability, to the nearest multiple of 2^{-k}, that the first dimensionality reduction fails, which may be a non-negligible function of $P \setminus P_{cr}$. This hint will be enough to simulate the entire protocol.

The protocol can be implemented in polynomial time with communication $\tilde{O}(B + d) = \tilde{O}(n^{1/3} + d)$.

---

c-**ApproxWithHelp** (q, P):

1. Set $B = \tilde{O}(n^{1/3})$.
2. Independently run $\mathsf{DimReduce}(\tau(r, br), B, q, P)$ k times, generating shares (S_i^1, S_i^2).
3. Bob finds random subsets P', $\tilde{P}$ of P of respective sizes B and $n^{2/3}$.
4. Independently run $\mathsf{DimReduce}(\tau(br, cr), B, q, \tilde{P})$ k times, generating shares $(\tilde{S}_i^1, \tilde{S}_i^2)$.
5. On inputs $q, S_i^1, S_i^2, P', \tilde{S}_i^1, \tilde{S}_i^2$, a secure circuit executes:
   - Compute $S_i = S_i^1 \oplus S_i^2$ and $\tilde{S}_i = \tilde{S}_i^1 \oplus \tilde{S}_i^2$ for all i.
   - If for most i, $|S_i| < B$, let $f(q, P)$ be a random point in $P_r \cap \cup_i S_i$, or set it to $\emptyset$ if it is empty.
   - Else if $P_{cr} \cap P' \neq \emptyset$, let $f(q, P)$ be a random point in $P_{cr} \cap P'$.
   - Else let $f(q, P)$ be a random point in $P_{br} \cap \cup_i \tilde{S}_i$ if it is non-empty, otherwise set $f(q, P) = \emptyset$.
   - Output $(f(q, P), \mathsf{null})$.

To prove correctness and privacy, we introduce some notation. Let $\mathcal{E}_1$ be the event that the majority of the $|S_i|$ are less than B, and $\mathcal{E}_2$ the event that $P_r \subseteq \cup_i S_i$. Let $\mathcal{F}$ be the event that $P' \cap P_{cr} \neq \emptyset$. Let $\mathcal{G}_1$ be the event that $1 \leq \tilde{P}_{br} \leq \tilde{P}_{cr} \leq B$ and $\mathcal{G}_2$ the event that $\tilde{P}_{br} \subseteq \cup_i \tilde{S}_i$. Finally, let $\mathcal{H}_1 = \mathcal{H}(r, br, P)$ and $\mathcal{H}_2 = \mathcal{H}(br, cr, \tilde{P})$. Note that $\Pr[\mathcal{H}_1], \Pr[\mathcal{H}_2]$ are $1 - 2^{-\Omega(k)}$ by Lemma 2. We need two lemmas:

Lemma 3. $\Pr[\mathcal{E}_2 \mid \mathcal{E}_1] = 1 - 2^{-\Omega(k)}$.

Proof. If $\mathcal{H}_1$ and $\mathcal{E}_1$ occur, then there is an i for which $P_r \subseteq S_i$, so $\mathcal{E}_2$ occurs.

Lemma 4. $\Pr[\mathcal{G}_2 \mid \mathcal{G}_1] = 1 - 2^{-\Omega(k)}$.

Proof. If $\mathcal{H}_2$ and $\mathcal{E}_2$ occur, then the majority of the $\tilde{S}_i$ contain $\tilde{P}_{br}$, so $\mathcal{G}_2$ occurs.

Correctness. We may assume $P_r \neq \emptyset$. The probability s of correctness is just the probability the algorithm doesn't return $\emptyset$. Since $\mathcal{F}, \mathcal{E}_1$, and $\mathcal{G}_1$ are independent,

$$s \geq \Pr[\mathcal{E}_1] \Pr[\mathcal{E}_2 \mid \mathcal{E}_1] + \Pr[\neg \mathcal{E}_1](\Pr[\mathcal{F}] + \Pr[\neg \mathcal{F}] \Pr[\mathcal{G}_1] \Pr[\mathcal{G}_2 \mid \mathcal{G}_1]).$$

Case $|P_{br}| < B$: $\mathcal{H}_1$ implies $\mathcal{E}_1$ since $|P_{br}| < B$, and using Lemma 3, $s \geq \Pr[\mathcal{E}_1] \Pr[\mathcal{E}_2 \mid \mathcal{E}_1] = 1 - 2^{-\Omega(k)}$.

Case $|P_{br}| \geq B$: Since $\Pr[\mathcal{E}_2 \mid \mathcal{E}_1] = 1 - 2^{-\Omega(k)}$ by Lemma 3, we just need to show that $\Pr[\mathcal{F}] + \Pr[\neg \mathcal{F}] \Pr[\mathcal{G}_1] \Pr[\mathcal{G}_2 \mid \mathcal{G}_1] = 1 - 2^{-\Omega(k)}$. If $|P_{cr}| > n^{2/3}$, it suffices to show $\Pr[\mathcal{F}] = 1 - 2^{-\Omega(k)}$. This holds for large enough $B = \tilde{O}(n^{1/3})$. Otherwise, if $|P_{cr}| \leq n^{2/3}$, then it suffices to show $\Pr[\mathcal{G}_1] \Pr[\mathcal{G}_2 \mid \mathcal{G}_1] = 1 - 2^{-\Omega(k)}$. By assumption, $B \leq |P_{br}| \leq |P_{cr}| \leq n^{2/3}$. Therefore, for large enough B, $\Pr[\mathcal{G}_1] = 1 - 2^{-\Omega(k)}$, and thus by Lemma 4, $\Pr[\mathcal{G}_1]\Pr[\mathcal{G}_2 \mid \mathcal{G}_1] = 1 - 2^{-\Omega(k)}$.

Privacy. Note that Bob gets no output, so Alice's privacy follows from the composition of DimReduce and the secure circuit protocol of step 5. Similarly, if we can construct a simulator Sim with inputs $1^n, P_{cr}, q, h(P_{cr}, q)$ so that the distributions $\langle q, P, f(q, P)\rangle$ and $\langle q, P, Sim(1^n, P_{cr}, q, h(P_{cr}, q))\rangle$ are statistically close, Bob's privacy will follow by that of DimReduce and the secure circuit of step 5.

We define the hint function $h(P_{cr}, q)$ to output the nearest multiple of 2^{-k} to $\Pr[\mathcal{E}_1]$. In the analysis we may assume that Sim knows $\Pr[\mathcal{E}_1]$ exactly, since its output distribution in this case will be statistically close to its real output distribution.

$\underline{\textbf{Sim} \ (1^n, P_{cr}, q, \Pr[\mathcal{E}_1])}$:

1. Set $B = \tilde{O}(n^{1/3})$.
2. With probabiity $\Pr[\mathcal{E}_1]$, output a random element of P_r, or $\emptyset$ if $P_r = \emptyset$.
3. Else with probability $1 - \binom{n-|P_{cr}|}{B}\binom{n}{B}^{-1}$, output a random element of P_{cr},
4. Else output a random element of P_{br}.

Let X denote the output of $Sim(1^n, P_{cr}, q, \Pr[\mathcal{E}_1])$. It suffices to show that for each $p \in P$,

$$|\Pr[f(q, P) = p] - \Pr[X = p]| = 2^{-\Omega(k)},$$

since then we have $|\Pr[f(q, P) = \emptyset] - \Pr[X = \emptyset]| = 2^{-\Omega(k)}$. Using the independence of $\mathcal{F}, \mathcal{E}_1, \mathcal{G}_1$, and Lemmas 3, 4, we bound $\Pr[f(q, P) = p]$ as follows

$$\Pr[f(q, P) = p] = \Pr[\mathcal{E}_1, f(q, P) = p] + \Pr[\neg\mathcal{E}_1, f(q, P) = p]$$
$$= \Pr[\mathcal{E}_1]\Pr[f(q, P) = p \mid \mathcal{E}_2\mathcal{E}_1] + 2^{-\Omega(k)} + \Pr[\neg\mathcal{E}_1]\Pr[\mathcal{F}]\Pr[f(q, P) = p \mid F, \neg\mathcal{E}_1]$$
$$+ \Pr[\neg\mathcal{E}_1]\Pr[\neg\mathcal{F}]\Pr[f(q, P) = p \mid \neg F, \neg\mathcal{E}_1]$$
$$= \Pr[\mathcal{E}_1]|P_r|^{-1}\delta(p \in P_r) \pm 2^{-\Omega(k)} + \Pr[\neg\mathcal{E}_1]\Pr[\mathcal{F}]|P_{cr}|^{-1}$$
$$+ \Pr[\neg\mathcal{E}_1]\Pr[\neg\mathcal{F}]\Pr[\mathcal{G}_1]\Pr[f(q, P) = p \mid \mathcal{G}_1\mathcal{G}_2\neg\mathcal{F}\neg\mathcal{E}_1] \pm 2^{-\Omega(k)}$$
$$+ \Pr[\neg\mathcal{E}_1]\Pr[\neg\mathcal{F}]\Pr[\neg\mathcal{G}_1]\Pr[f(q, P) = p \mid \neg\mathcal{G}_1\neg\mathcal{F}\neg\mathcal{E}_1]$$
$$= \Pr[\mathcal{E}_1]|P_r|^{-1}\delta(p \in P_r) + \Pr[\neg\mathcal{E}_1]\Pr[\mathcal{F}]|P_{cr}|^{-1}$$
$$+ \Pr[\neg\mathcal{E}_1]\Pr[\neg\mathcal{F}]\Pr[\mathcal{G}_1]|P_{br}|^{-1}\delta(p \in P_{br})$$
$$+ \Pr[\neg\mathcal{E}_1]\Pr[\neg\mathcal{F}]\Pr[\neg\mathcal{G}_1]\Pr[f(q, P) = p \mid \neg\mathcal{E}_1\neg\mathcal{F}\neg\mathcal{G}_1] \pm 2^{-\Omega(k)}.$$

On the other hand, since $\Pr[\mathcal{F}] = 1 - \binom{n-|P_{cr}|}{B}\binom{n}{B}^{-1}$, then $\Pr[X = p]$ is

$$\Pr[\mathcal{E}_1]|P_r|^{-1}\delta(p \in P_r) + \Pr[\neg\mathcal{E}_1]\Pr[\mathcal{F}]|P_{cr}|^{-1} + \Pr[\neg\mathcal{E}_1]\Pr[\neg\mathcal{F}]|P_{br}|^{-1}\delta(p \in P_{br}),$$

so that,

$$| \Pr[f(q, P) = p] - \Pr[X = p]| \le$$
$$\Pr[\neg \mathcal{E}_1] \Pr[\neg \mathcal{F}] \Pr[\neg \mathcal{G}_1] \Pr[f(q, P) = p \mid \neg \mathcal{E}_1 \neg \mathcal{F} \neg \mathcal{G}_1] + 2^{-\Omega(k)}.$$

If $|P_{br}| < B$, $\Pr[\neg \mathcal{E}_1] = 2^{-\Omega(k)}$. If $|P_{cr}| \ge n^{2/3}$, $\Pr[\neg \mathcal{F}] = 2^{-\Omega(k)}$. Otherwise $B \le |P_{br}| \le |P_{cr}| \le n^{2/3}$, and as shown for correctness, $\Pr[\neg \mathcal{G}_1] = 2^{-\Omega(k)}$, which shows that $| \Pr[f(q, P) = p] - \Pr[X = p]| = 2^{-\Omega(k)}$.

Acknowledgments

The second author would like to thank Andrew Yao for support and hospitality while visiting Tsinghua University.

References

[1] N. Alon, Y. Matias, and M. Szegedy. *The space complexity of approximating the frequency moments*, STOC, 1996.

[2] K. Bharat and A. Broder. *Estimating the relative size and overlap of public web search engines*, WWW, 1998.

[3] A. Beimel, Y. Ishai, T. Malkin. *Reducing the servers computation in private information retrieval: PIR with preprocessing*, CRYPTO, 2000.

[4] J. D. C. Benaloh, *Verifiable secret-ballot elections*. PhD thesis, Yale University, 1987.

[5] C. Cachin, J. Camenisch, J. Kilian and J. Müller. *One-round secure computation and secure autonomous mobile agents*, ICALP, 2000.

[6] C. Cachin, S. Micali and M. Stadler. *Computationally private information retrieval with polylogarithmic communication*, Eurocrypt, 1999.

[7] R. Canetti, Y. Lindell, R. Ostrovsky, and A. Sahai. *Universally composable two-party computation*, STOC, 2002.

[8] B. Chor, N. Gilboa and M. Naor, *Private information retrieval by keywords*, Technical Report CS0917, Department of Computer Science, Technion, 1997.

[9] B. Chor, O. Goldreich, E. Kushilevitz and M. Sudan. *Private information retrieval*, FOCS, 1995.

[10] W. Du and M. J. Attalah. *Protocols for secure remote database access with approximate matching*, CCS - Workshop on Security and Privacy in E-commerce, 2000.

[11] S. Even, O. Goldreich and A. Lempel. *A randomized protocol for signing contracts*, Communications of the ACM, 1985.

[12] J. Feigenbaum, Y. Ishai, T. Malkin, K. Nissim, M. Strauss, and R. Wright. *Secure multiparty computation of approximations*, ICALP 2001.

[13] M. Freedman, K. Nissim and B. Pinkas. *Efficient private matching and set intersection*, Eurocrypt, 2004.

[14] B. Goethals, S. Laur, H. Lipmaa, and T. Mielikainen. *On secure scalar product computation for privacy-preserving data mining*, ICISC, 2004.

[15] J. Hastad, R. Impagliazzo, L. A. Levin, and M. Luby. *Construction of a pseudo-random generator from any one-way function*, Technical Report TR-91-068, International Computer Science Institute, 1991.

[16] Y. Gertner, Y. Ishai, E. Kushilevitz and T. Malkin. *Protecting data privacy in private information retrieval schemes*, STOC, 1998.

[17] O. Goldreich. *Secure multi-party computation*, 1998. Available at http://philby.ucsd.edu/

[18] O. Goldreich, S. Micali, and A. Wigderson. *How to play any mental game*, STOC, 1987.

[19] S. Goldwasser and S. Micali. *Probabilistic encryption*, JCSS, 1984.

[20] S. Halevi, R. Krauthgamer, E. Kushilevitz, and K. Nissim. *Private approximation of NP-hard functions*, STOC 2001.

[21] R. Impagliazzo and M. Luby. *One-way functions are essential for complexity-based cryptography*, FOCS, 1989.

[22] P. Indyk. *High-dimensional computational geometry*. PhD Thesis, Stanford University, 2000.

[23] P. Indyk and R. Motwani. *Approximate nearest neighbors: towards removing the curse of dimensionality*, STOC, 1998.

[24] P. Indyk and D. Woodruff. *Polylogarithmic private approximations and efficient matching*, ECCC, Technical Report TR05-117, 2005.

[25] E. Kushilevitz, R. Ostrovsky and Y. Rabani. *Efficient search for approximate nearest neighbor in high dimensional spaces*, STOC, 1998.

[26] Y. Lindell and B. Pinkas. *Privacy preserving data mining*, Crypto, 2000.

[27] V.D. Milman and G. Schechtman. Asymptotic Theory of Finite Dimensional Normed Spaces. Lecture Notes in Mathematics, **1200**, Springer Verlag, 1986.

[28] D. Naccache and J. Stern. *A new public key cryptosystem*, Eurocrypt, 1997.

[29] M. Naor and K. Nissim. *Communication complexity and secure function evaluation*, STOC, 2001.

[30] M. Naor and B. Pinkas. *Oblivious transfer and polynomial evaluation*, STOC, 1999.

[31] P. Paillier. *Public-key cryptosystems based on composite degree residuosity classes*, Eucrocrypt, 1999.

[32] M. Rabin. *How to exchange secrets by oblivious transfer*. Technical report TR81, Aiken Computation Lab, 1981.

[33] A. C. Yao. *Protocols for secure computations*, FOCS, 1982.

A   Cryptographic Tools

Homomorphic Encryption. An encryption scheme, $E : (G_1, +) \to (G_2, \cdot)$ is homomorphic if for all $a, b \in G_1$, $E(a + b) = E(a) \cdot E(b)$. For more background on this primitive see, for example, [19, 28]. We will make use of the Paillier homomorphic encryption scheme [31].

Oblivious Transfer and SPIR. Oblivious transfer is equivalent to the notion of symmetrically-private information retrieval (SPIR), where the latter usually refers to communication-efficient implementations of the former. SPIR was introduced in [16]. With each invocation of a SPIR protocol a user learns exactly one bit of a binary database while giving the server no information about which bit was learned. We rely on single-server SPIR schemes in our protocols. Such schemes necessarily offer computational, rather than unconditional, security [9]. Applying the transformation of [30] to the PIR scheme of [6] give SPIR constructions with $\tilde{O}(n)$ server work and $\tilde{O}(1)$ communication.

One issue is that in some of our schemes, we actually perform OT on *records* rather than on bits. It is a simple matter to convert a binary OT scheme into an OT scheme on records by running r invocations of the binary scheme in parallel, where r is the record size. This gives us a 1-round, $\tilde{O}(r)$ communication, $\tilde{O}(nr)$ server work OT protocol on records of size r. The dependence on r can be improved using techniques of [8].

Secure Function Evaluation. In [18, 33] it is shown how two parties holdings inputs x and y can privately evaluate any circuit C with communication $O(k(|C| + |x| + |y|))$, where k is a security parameter. In [5] it is shown how to do this in one round for the semi-honest case we consider. The time complexity is the same as the communication. We use such protocols as black boxes in our protocols.

B   Private c-Approximate NN Based on Locality Sensitive Hashing

We give an alternative private c-approximate NN protocol, with slightly more communication than that in section 4.2, but less work (though still polynomial). It is based on locality sensitive hashing (LSH) [23]. The fact we need is that there is a family of functions $\mathcal{G} : \{0,1\}^d \to \{0,1\}^{\tilde{O}(1)}$ such that each $g \in \mathcal{G}$ has description size $\tilde{O}(1)$, and $\mathcal{G}$ is such that for all $p, q \in \{0,1\}^d$,

$$\Pr_{g \in \mathcal{G}}[g(p) = g(q)] = \Theta\left(n^{-\Delta(p,q)/cr}\right)$$

Recall that Alice has a point $q \in \{0,1\}^d$ and Bob has n points $P \subseteq \{0,1\}^d$. For correctness, Alice should learn a point of P_{cr} provided $P_r \neq \emptyset$. For privacy, her view should be simulatable given only P_{cr}.

Our protocol is similar to that in section 4.2. When $|P_{cr}|$ is large, one can run a secure function evaluation with Alice's point q as input, together with a random sample P' of roughly a $k/|P_{cr}|$ fraction of Bob's points P. The circuit returns a random point of $P' \cap P_{cr}$ which is non-empty with probabiity $1 - 2^{-\Omega(k)}$. The communication is $\tilde{O}(n/|P_{cr}|)$.

On the other hand, when $|P_{cr}|$ is small, if Alice and Bob exchange functions g_i independently $\tilde{O}(n^{1/c})$ times, then with overwhelming probability $P_r \subseteq \cup_i S_i$, where S_i denotes the subset of Bob's points p with $g_i(p) = g_i(q)$. Using a secure ciruit with ROM, we can obtain these sets S_i, and output a random point of P_r. The communication is $\tilde{O}(n^{1/c}|P_{cr}|)$.

Our protocol balances these approaches to achieve $\tilde{O}(n^{1/2+1/(2c)})$ communication.

There are a few technicalities dodged by this intuition. First, even though the parties exchange $\tilde{O}(n^{1/c})$ different g_i, and can thus guarantee that each p is in some S_i with probability $1 - 2^{-\Omega(k)}$, it may be that whenever $p \in S_i$, many points from $P \setminus P_{cr}$ also land in S_i, so that S_i is very large. Even though we only expect $|P \setminus P_{cr}|O(1/n) = O(1)$ points from $P \setminus P_{cr}$ in S_i, since $\Pr[p \in S_i] = \Theta(n^{-1/c})$

is small, p may only be in S_i when S_i is large. Because the size of the S_i affects the communication of our protocol, we cannot always afford for the ROM to receive the whole S_i (sometimes we will truncate it). However, in the analysis, we show that the average S_i is small, and this will be enough to get by with low communication.

Second, we need to extend the notion of a lookup gate given in section 2. Instead of just mapping inputs (i, j) to output $R_i[j]$, the jth entry in the ith party's ROM, we also allow j to be a key, so that the output is the record in R_i keyed by j. This can be done efficiently using [8], and Theorem 2 is unchanged, assuming the length of the keys is $\tilde{O}(1)$.

LSH (q, P):

1. Set $B = \tilde{O}(n^{1/2+1/(2c)})$ and $C = \tilde{O}(n^{1/c})$.
2. Bob finds a random subset P' of P of size B .
3. For $i = 1$ to k,
    (a) Alice and Bob agree upon C random $g_{i,j} \in \mathcal{G}$.
    (b) Bob creates a ROM L with entries $L(v)$ containing p s.t. $g(p) = v$.
    (c) A secure circuit with ROM L on input $(q, \{g_{i,j}\})$ executes:
        − Compute $v_{i,j} = g_{i,j}(q)$ for each j.
        − Lookup the $L(v_{i,j})$ one by one for the different $v_{i,j}$ until the communication exceeds dB. If it is less, pad it to dB.
        − Output shares S_i^1, S_i^2 so that $S_i^1 \oplus S_i^2$ is the set of sets $L(v_j)$.
4. A secure circuit on inputs P', S_i^1, S_i^2 executes:
    − Compute the set $S_i = S_i^1 \oplus S_i^2 = \cup_j L(v_j)$ for all i.
    − Let $f(q, P)$ be random in $P_{cr} \cap P'$ if it is non-empty.
    − Else let $f(q, P)$ be random in $P_r \cap \cup_i S_i$ if it is non-empty, else set $f(q, P) = \emptyset$.
    − Output $(f(q, P), \text{null})$.

The communication is $\tilde{O}(dB)$. By using homomorphic encryption, one can reduce the dependence on d, as per remark 3. Let $\mathcal{E}$ be the event that $P_r \subseteq \cup_i S_i$, and let $\mathcal{F}$ be the event that $P_{cr} \cap P'$ is non-empty.

Correctness. Suppose $P_r \neq \emptyset$. The probability s of correctness is just the probability we don't output $\emptyset$. Thus $s \geq \Pr[\mathcal{F}] + \Pr[\neg\mathcal{F}] \Pr[f(q, P) \neq \emptyset \mid \neg\mathcal{F}]$.

Case $|P_{cr}| \geq n^{1/2-1/(2c)}$: For sufficiently large B, we have $s \geq \Pr[\mathcal{F}] = 1 - 2^{-\Omega(k)}$.

Case $|P_{cr}| < n^{1/2-1/(2c)}$: It is enough to show $\Pr[f(q, P) \neq \emptyset \mid \neg\mathcal{F}] = 1 - 2^{-\Omega(k)}$. Fix i. Put $Y = \sum_j |L(v_{i,j})|$, where $|L(v_{i,j})|$ denotes the number of points in $L(v_{i,j})$. The expected number of points in $P \setminus P_{cr}$ that are in $L(v_{i,j})$ is at most $n \cdot O(1/n) = O(1)$. Since $|P_{cr}| < n^{1/2-1/(2c)}$, $\mathbf{E}[L(v_{i,j})] < n^{1/2-1/(2c)} + O(1)$. Thus $\mathbf{E}[Y] \leq B/3$ for large enough B, so $\Pr[Y > B] \leq 1/3$ by Markov's inequality. Thus, with probability $1 - 2^{-\Omega(k)}$, for at least half of the i, S_i is not truncated in step 3c. Moreover, for large enough B, any i, and any $p \in P_r$,

$\Pr[p \in S_i] = 1 - 2^{-\Omega(k)}$ for large enough C. By a few union bounds then, $\Pr[P_r \subseteq \cup_i S_i] = \Pr[\mathcal{E}] = 1 - 2^{-\Omega(k)}$. Thus,

$$\Pr[f(q, P) \neq \emptyset \mid \neg\mathcal{F}] \geq \Pr[f(q, P) \neq \emptyset, \; \mathcal{E} \mid \neg\mathcal{F}]$$
$$= \Pr[f(q, P) \neq \emptyset \mid \mathcal{E}, \; \neg\mathcal{F}] \Pr[\mathcal{E}]$$
$$\geq 1 - 2^{-\Omega(k)}.$$

Privacy. Note that Bob gets no output, so Alice's privacy follows from that of the secure circuit protocol. We construct a simulator $Sim(1^n, P_{cr}, q)$ so that the distributions $\langle q, P, f(q, P) \rangle$ and $\langle q, P, Sim(1^n, P_{cr}, q) \rangle$ are statistically close. Bob's privacy then follows by the composition with the secure circuit protocol.

Sim $(1^n, P_{cr}, q)$:

1. Set $B = \tilde{O}(n^{1/2+1/(2c)})$.
2. With probabiity $1 - \binom{n-|P_{cr}|}{B}\binom{n}{B}^{-1}$, output a random element of P_{cr}.
3. Else output a random element of P_r.

Let X denote the output of $Sim(1^n, P_{cr}, q)$. It suffices to show that for each $p \in P$, $|\Pr[f(q, P) = p] - \Pr[X = p]| = 2^{-\Omega(k)}$, since this also implies $|\Pr[f(q, P) = \emptyset] - \Pr[X = \emptyset]| = 2^{-\Omega(k)}$. We have

$$\Pr[f(q, P) = p] = \Pr[f(q, P) = p, \mathcal{F}] + \Pr[f(q, P) = p, \neg\mathcal{F}]$$
$$= \Pr[\mathcal{F}] |P_{cr}|^{-1} + \Pr[f(q, P) = p, \neg\mathcal{F}]$$

Note that $\Pr[\mathcal{F}] = 1 - \binom{n-|P_{cr}|}{B}\binom{n}{B}^{-1}$. Therefore,

$$|\Pr[f(q, P) = p] - \Pr[X = p]| = \Pr[\neg\mathcal{F}] |\Pr[f(q, P) = p \mid \neg\mathcal{F}] - \delta(p \in P_r)|P_r|^{-1}|.$$

If $|P_{cr}| \geq n^{1/2-1/(2c)}$, this is $2^{-\Omega(k)}$, since then $\Pr[\neg\mathcal{F}] = 2^{-\Omega(k)}$. Otherwise, $|P_{cr}| < n^{1/2-1/(2c)}$, and as shown in the proof of correctness, we have $\Pr[\mathcal{E}] = \Pr[P_r \subseteq \cup_i S_i] = 1 - 2^{-\Omega(k)}$. Thus

$$\Pr[f(q, P) = p \mid \neg\mathcal{F}]$$
$$= \Pr[f(q, P) = p \mid \mathcal{E}, \; \neg\mathcal{F}] \Pr[\mathcal{E}] \pm 2^{-\Omega(k)}$$
$$= \delta(p \in P_r)|P_r|^{-1} \pm 2^{-\Omega(k)}$$

which completes the proof.

Calibrating Noise to Sensitivity in Private Data Analysis

Cynthia Dwork[1], Frank McSherry[1], Kobbi Nissim[2], and Adam Smith[3,*]

[1] Microsoft Research, Silicon Valley
{dwork, mcsherry}@microsoft.com
[2] Ben-Gurion University
kobbi@cs.bgu.ac.il
[3] Weizmann Institute of Science
adam.smith@weizmann.ac.il

Abstract. We continue a line of research initiated in [10, 11] on privacy-preserving statistical databases. Consider a trusted server that holds a database of sensitive information. Given a query function f mapping databases to reals, the so-called *true answer* is the result of applying f to the database. To protect privacy, the true answer is perturbed by the addition of random noise generated according to a carefully chosen distribution, and this response, the true answer plus noise, is returned to the user.

Previous work focused on the case of noisy sums, in which $f = \sum_i g(x_i)$, where x_i denotes the ith row of the database and g maps database rows to $[0, 1]$. We extend the study to general functions f, proving that privacy can be preserved by calibrating the standard deviation of the noise according to the *sensitivity* of the function f. Roughly speaking, this is the amount that any single argument to f can change its output. The new analysis shows that for several particular applications substantially less noise is needed than was previously understood to be the case.

The first step is a very clean characterization of privacy in terms of indistinguishability of transcripts. Additionally, we obtain separation results showing the increased value of interactive sanitization mechanisms over non-interactive.

1 Introduction

We continue a line of research initiated in [10, 11] on privacy in *statistical* databases. A statistic is a quantity computed from a sample. Intuitively, if the database is a representative sample of an underlying population, the goal of a privacy-preserving statistical database is to enable the user to learn properties of the population as a whole while protecting the privacy of the individual contributors.

We assume the database is held by a trusted server. On input a query function f mapping databases to reals, the so-called *true answer* is the result of applying f

---

* Supported by the Louis L. and Anita M. Perlman Postdoctoral Fellowship.

S. Halevi and T. Rabin (Eds.): TCC 2006, LNCS 3876, pp. 265–284, 2006.

to the database. To protect privacy, the true answer is perturbed by the addition of random noise generated according to a carefully chosen distribution, and this response, the true answer plus noise, is returned to the user.

Previous work focused on the case of noisy sums, in which $f = \sum_i g(x_i)$, where x_i denotes the ith row of the database and g maps database rows to $[0, 1]$. The power of the noisy sums primitive has been amply demonstrated in [6], in which it is shown how to carry out many standard datamining and learning tasks using few noisy sum queries.

In this paper we consider general functions f mapping the database to vectors of reals. We prove that privacy can be preserved by calibrating the standard deviation of the noise according to the *sensitivity* of the function f. This is the maximum amount, over the domain of f, that any single argument to f, that is, any single row in the database, can change the output.

We begin by defining a new notion of privacy leakage, ϵ-indistinguishability. An interaction between a user and a privacy mechanism results in a *transcript*. For now it is sufficient to think of transcripts corresponding to a single query function and response, but the notion is completely general and our results will handle longer transcripts.

Roughly speaking, a privacy mechanism is ϵ-indistinguishable if for all transcripts t and for all databases $\mathbf{x}$ and $\mathbf{x}'$ differing in a single row, the probability of obtaining transcript t when the database is $\mathbf{x}$ is within a $(1 + \epsilon)$ multiplicative factor of the probability of obtaining transcript t when the database is $\mathbf{x}'$. More precisely, we require the absolute value of the logarithm of the ratios to be bounded by ϵ. In our work, ϵ is a parameter chosen by *policy*.

We then formally define the sensitivity $S(f)$ of a function f. This is a quantity *inherent* in f; it is not chosen by policy. Note that $S(f)$ is independent of the actual database.

We show a simple method of adding noise that ensures ϵ-indistinguishability of transcripts; the noise depends only on ϵ and $S(f)$, and is independent of the database and hence of its size. Specifically, to obtain ϵ-indistinguishability it suffices to add noise according to the following distribution: $Pr[y] \propto e^{-\epsilon|y|/S(f)}$.

The extension to privacy-preserving approximations to "holistic" functions f that operate on the entire database broadens the scope of private data analysis beyond the orignal motivation of a purely statistical, or "sample population" context. Now we can view the database as an object that is *itself* of intrinsic interest and that we wish to analyze in a privacy-preserving fashion. For example, the database may describe a concrete interconnection network – not a sample subnetwork – and we wish to learn certain properties of the network without releasing information about individual edges, nodes, or subnetworks. The technology developed herein therefore extends the scope of the line of research, beyond privacy-preserving statistical databases to privacy-preserving analysis of data.

1.1   Additional Contributions

Definitions of Privacy. Definition of privacy requires care. In addition to our indistinguishability-based definition mentioned above we also consider notions based on semantic security and simulation and prove equivalences among these. A simple hybrid argument shows that utility requires non-negligible information leakage, hence all our definitions differ from their original cryptographic counterparts in that we accommodate non-negligible leakage. In particular, the standard measure of statistical difference is not a sufficiently good metric in our setting, and needs to be replaced with a more delicate one.

In previous work [10, 11, 6], the definitions were based on semantic security but the proofs were based on indistinguishability, so our move to ϵ-indistinguishability is a simplification. Also, semantic security was proved against *informed* adversaries. That is, an adversary with knowledge of the entire database except a single row, say, row i, could not glean any additional information about row i beyond what it knew before interaction with the privacy mechanism. This is fine; it says that without the database, seeing that X smokes does not necessarily increase our gambling odds that X will develop heart disease, but if the database teaches the correlation between smoking and heart disease improving our guessing odds should not be considered a violation of privacy. However, the new formulation immediately gives indistinguishability against an adversary with any amount of prior knowledge, and the above explanation is no longer necessary.

Examples of Sensitivity-Based Analysis. To illustrate our approach, we analyze the sensitivity of specific data analysis functions, including histograms, contingency tables, and covariance matrices, all of which have very high-dimensional output, and show that their sensitivities are independent of the dimension. Previous privacy-preserving approximations to these quantities used noise proportional to the dimension; the new analysis permits noise of size $O(1)$. We also give two general classes of functions which have low sensitivity: functions which estimate distance from a set (e.g minimum cut size in a network) and functions which can be approximated from a random sample.

Limits on Non-Interactive Mechanisms. There are two natural models of data sanitization: interactive and non-interactive. In the non-interactive setting, the data collector—a trusted entity—publishes a "sanitized" version of the collected data; the literature uses terms such as "anonymization" and "de-identification". Traditionally, sanitization employed some perturbation and data modification techniques, and may also have included some accompanying synopses and statistics. In the interactive setting, the data collector provides a mechanism with which users may pose queries about the data, and get (possibly noisy) answers.

The first of these seems quite difficult (see [12, 7, 8]), possibly due to the difficulty of supplying utility that has not yet been specified at the time the sanitization is carried out. In contrast, powerful results for the interactive approach have been obtained ([11, 6] and the present paper). We show that for any non-interactive mechanism San satisfying our definition of privacy, there exist low-sensitivity functions $f(\mathbf{x})$ which cannot be approximated at all based on San($\mathbf{x}$), unless the database is very large: If each database entry consists of d bits, then the database must have $2^{\Omega(d)}$ entries in order to answer all low-sensitivity queries—even to answer queries from a restricted class called *sum queries*. In other words, a non-interactive mechanism must be tailored to suit certain functions to the exclusion of others. This is not true in the interactive setting, since one can answer the query f with little noise regardless of n^1.

The separation results are significant given that the data-mining literature has focused almost exclusively on non-interactive mechanisms, specifically, randomized response (see Related Work below) and that statisticians have traditionally operated on "tables" and have expressed to us a strong preference for non-interactive "noisy tables" over an interactive mechanism.

1.2 Related Work

The literature in statistics and computer science on disseminating statistical data while preserving privacy is extensive; we discuss only directly relevant work here. See, e.g., [5] for pointers to the broader literature.

PRIVACY FROM PERTURBATION. The venerable idea of achieving privacy by adding noise is both natural and appealing. An excellent and detailed exposition of the many variants of this approach explored in the context of statistical disclosure control until 1989, many of which are still important elements of the toolkit for data privacy today, may be found in the survey of Adam and Wortmann [1]. The "classical" antecedent closest in spirit to our approach is the work of Denning [9].

Perturbation techniques are classified into two basic categories: (i) *Input perturbation techniques*, where the underlying data are randomly modified, and answers to questions are computed using the modified data; and (ii) *Output perturbation*, where (correct) answers to queries are computed exactly from the real data, but noisy versions of these are reported. Both techniques suffer from certain inherent limitations (see below); it seems that these limitations caused a decline in interest within the computer science community in designing perturbation techniques for achieving privacy^2.

The work of Agrawal and Srikant [3] rekindled this interest; their principal contribution was an algorithm that, given an input-perturbed database, learns the original input distribution. Subsequent work studied the applicability and limitations of perturbation techniques, and privacy definitions have started to evolve, as we next describe.

DEFINITIONAL WORK. Several privacy definitions have been put forward since [3]. Their definition measured privacy in terms of the noise magnitude added to a value. This was shown to be problematic, as the definition ignored what an

---

[1] It is also not true if one employs weaker definitions of security; the connection between the definitions and the separation between models of interaction is subtle and, in our view, surprising. See Section 4.

[2] The same is not true of the statistics community; see, for example, the work of Roque [14].

adversary knowing the underlying probability distribution might infer about the data [2]. Evfimievsky et al. [12] noted, however, that such an *average* measure allows for infrequent but noticeable privacy breaches, and suggested measuring privacy in terms of the *worst-case* change in an adversary's a-priori to a-posteriori beliefs. Their definition is a special case of Definition 1 for input perturbation protocols of a limited form. A similar, more general definition was suggested in [10, 11, 6]. This was modeled after semantic security of encryptions.

Our basic definition of privacy, ϵ-indistinguishability, requires that a change in one database entry induce a small change in the distribution on the view of the adversary, under a specific, "worst-case" measure of distance. It is the same as in [12], adapted to general interactive protocols. An equivalent, semantic-security-flavored formulation is a special case of the definition from [10, 11, 6]; those definitions allowed a large loss of privacy to occur with negligible probability.

We note that *k-anonymity* [15] and the similarly motivated notion of protection against *isolation* [7, 8]) have also been in the eye of privacy research. The former is a syntactic characterization of (input-perturbed) databases that does not immediately capture semantic notions of privacy; the latter definition is a geometric interpretation of protection against being brought to the attention of others. The techniques described herein yield protection against isolation.

SUM QUERIES. A cryptographic perspective on perturbation was initiated by Dinur and Nissim [10]. They studied the amount of noise needed to maintain privacy in databases where a query returns (approximately) the number of 1's in any given subset of the entries. They showed that if queries are not restricted, the amount of noise added to each answer must be very high – linear (in n, the size of the database) for the case of a computationally unbounded adversary, and $\Omega(\sqrt{n})$ for a polynomially (in n) bounded adversary. Otherwise, the adversary can reconstruct the database almost exactly, producing a database that errs on, say, 0.01% of the entries. In contrast, jointly with Dwork, they initiated a sequence of work [10, 11, 6] which showed that limiting the users to a sublinear (in n) number of queries ("SuLQ") allows one to release useful global information while satisfying a strong definition of privacy. For example, it was shown that the computationally powerful noisy sum queries discussed above, that is, $\sum_{i=1}^{n} g(i, x_i)$, where g maps rows to values in $[0, 1]$, can be safely answered by adding $o(\sqrt{n})$ noise (from a gaussian, binomial, or Laplace distribution)— a level well below the sampling error one would expect in the database initially.

2   Definitions

We model the adversary as a probabilistic interactive Turing machine with an advice tape. Given a database access protocol San, an adversary $\mathcal{A}$, and a particular database $\mathbf{x}$, let the random variable $\mathcal{T}_{\mathsf{San},\mathcal{A}}(\mathbf{x})$ denote the transcript. The randomness in $\mathcal{T}_{\mathsf{San},\mathcal{A}}(\mathbf{x})$ comes from the coins of San and of $\mathcal{A}$. Note that for non-interactive schemes, there is no dependence on the adversary $\mathcal{A}$. We will drop either or both of the subscripts San and $\mathcal{A}$ when the context is clear.

We model the database as a vector of n entries from some domain D. We typically consider domains D of the form $\{0,1\}^d$ or $\mathbb{R}^d$. The Hamming distance $d_H(\cdot, \cdot)$ over D^n is the number of entries in which two databases differ.

Our basic definition of privacy requires that close databases correspond to close distributions on the transcript. Specifically, for every transcript, the probabilities of it being produced with the two possible databases are close. We abuse notation somewhat and use $\Pr[A = a]$ to denote probability density for both continuous and discrete random variables.

Definition 1. *A mechanism is ϵ-indistinguishable if for all pairs* $\mathbf{x}, \mathbf{x}' \in D^n$ *which differ in only one entry, for all adversaries $\mathcal{A}$, and for all transcripts t:*

$$\left| \ln\left(\frac{\Pr[\mathcal{T}_\mathcal{A}(\mathbf{x}) = t]}{\Pr[\mathcal{T}_\mathcal{A}(\mathbf{x}') = t]} \right) \right| \le \epsilon. \tag{1}$$

We sometimes call ϵ the *leakage*. When ϵ is small, $\ln(1+\epsilon) \approx \epsilon$, and so the definition is roughly equivalent to requiring that for all transcripts t, $\frac{\Pr[\mathcal{T}_\mathcal{A}(\mathbf{x})=t]}{\Pr[\mathcal{T}_\mathcal{A}(\mathbf{x}')=t]} \in 1 \pm \epsilon$.

The definition is unusual for cryptography, in that in most cryptographic settings it is sufficient to require that distributions be *statistically close* (i.e. have small total variation distance) or that they be *computationally indistinguishable*. However, the requirement of Definition 1 is much more stringent than statistical closeness: one can have a pair of distributions whose statistical difference is arbitrarily small, yet where the ratio in Eqn. 1 is infinite (by having a point where one distribution assigns probability zero and the other, non-zero). We chose the more stringent notion because (a) it is achievable at very little cost, and (b) more standard distance measures do not yield meaningful guarantees in our context, since, as we will see, the leakage must be non-negligible. As with statistical closeness, Definition 1 also has more "semantic" formulations; these are discussed in Appendix A.

As we will next show, it is possible to release quite a lot of "global" information about the database while satisfying Definition 1. We first define the *Laplace* distribution, $\text{Lap}(\lambda)$. This distribution has density function $h(y) \propto \exp\left(-|y|/\lambda\right)$, mean 0, and standard deviation λ.

Example 1 (Noisy Sum). Suppose $\mathbf{x} \in \{0,1\}^n$, and the user wants to learn $f(\mathbf{x}) = \sum_i x_i$, the total number of 1's in the database. Consider adding noise to $f(\mathbf{x})$ according to a Laplace distribution:

$$\mathcal{T}(x_1, \ldots, x_n) = \sum_i x_i + Y, \quad \text{where } Y \sim \text{Lap}(1/\epsilon).$$

This mechanism is ϵ-indistinguishable. To see why, note that for any real numbers y, y' we have $\frac{h(y)}{h(y')} \le e^{\epsilon|y-y'|}$. For any two databases $\mathbf{x}$ and $\mathbf{x}'$ which differ in a single entry, the sums $f(\mathbf{x})$ and $f(\mathbf{x}')$ differs by one. Thus, for $t \in \mathbb{R}$, the ratio $\frac{\Pr(\mathcal{T}(\mathbf{x})=t)}{\Pr(\mathcal{T}(\mathbf{x}')=t)} = \frac{h(t-f(\mathbf{x}))}{h(t-f(\mathbf{x}'))}$ is at most $e^{\epsilon|f(\mathbf{x})-f(\mathbf{x}')|} \le e^\epsilon$, as desired.

NON-NEGLIGIBLE LEAKAGE AND THE CHOICE OF DISTANCE MEASURE. In the example above it is clear that even to get a constant-factor approximation to $f(\mathbf{x})$, we must have $\epsilon = \Omega(1/n)$, quite large by cryptographic standards where the usual requirement is for the leakage to drop faster than any polynomial in the lengths of the inputs. However, non-negligible or leakage is *inherent* for statistical utility: If the distance ϵ between the distributions induced by close databases is $o(1/n)$, then the distance between the distributions induced by *any* two databases is $o(1)$ and *no* statistic about the database can be usefully approximated.

Average-case distance measures such as statistical difference do not yield meaningful guarantees when $\epsilon = \Omega(1/n)$.

Example 2. Consider the candidate sanitization

$$\mathcal{T}(x_1, ..., x_n) = (i, x_i) \quad \text{where } i \in_R \{1, ..., n\}.$$

If $\mathbf{x}$ and $\mathbf{x}'$ differ in a single position, the statistical difference between $\mathcal{T}(\mathbf{x})$ and $\mathcal{T}(\mathbf{x}')$ is $1/n$, and yet it is clear that every transcript reveals private information about some individual.

Indeed, Definition 1 is not satisfied in this example, since if $\mathbf{x}$ and $\mathbf{x}'$ differ, say, in the ith coordinate, then the transcript (i, x_i) has probability zero when the database is $\mathbf{x}'$.

3   Sensitivity and Privacy

We now formally define sensitivity of functions, described informally in the Introduction. We will prove that choosing noise according to $\mathrm{Lap}(S(f)/\epsilon)$ ensures ϵ-indistinguishability when the query function f has sensitivity $S(f)$. We extend the analysis to vector-valued functions f, and even to adaptively chosen series of query functions. Intuitively, if ϵ is a "privacy budget" then this analysis explains how the budget is spent by a sequence of queries.

Definition 2 (L_1 Sensitivity). *The L_1 sensitivity of a function $f : D^n \to \mathbb{R}^d$ is the smallest number $S(f)$ such that for all $\mathbf{x}, \mathbf{x}' \in D^n$ which differ in a single entry,*

$$\|f(\mathbf{x}) - f(\mathbf{x}')\|_1 \leq S(f) .$$

Sensitivity is a Lipschitz condition on f: if $\mathsf{d}_H(\cdot, \cdot)$ is the Hamming metric on D^n, then for all pairs of databases $\mathbf{x}, \mathbf{x}' \in D^n$: $\frac{\|f(\mathbf{x}) - f(\mathbf{x}')\|_1}{\mathsf{d}_H(\mathbf{x}, \mathbf{x}')} \leq S(f)$. One can define sensitivity with respect to any metric on the output space; see Section 3.3.

Example 3 (Sums and Histograms). Consider the sum functionality above: if $D = \{0, 1\}$ and $f(\mathbf{x}) = \sum_{i=1}^{n} x_i$ (viewed as an real number), then the sensitivity of f with respect to the usual metric on $\mathbb{R}$ is $S_{L_1}(f) = 1$.

Now consider an arbitrary domain D which has been partitioned into d disjoint bins $B_1, ..., B_d$. The function $f : D^n \to \mathbb{Z}^d$ which computes the number of database points which fall into each bin is called a *histogram* for $B_1, ..., B_m$. Changing one point in the database can change at most two of these counts — one bin loses a point, another bin gains one. The L_1 sensitivity of f is thus 2, independent of d.

3.1   Calibrating Noise According to $S(f)$

Recall that if the noise Y is drawn from the Laplace distribution, then $h(y)/h(y')$ is at most $e^{|y-y'|/\lambda}$. A similar phenomenon holds in higher dimension. If Y is a vector of d independent Laplace variables, the density function at y is proportional to $\exp(-\|y\|_1/\lambda)$. A simple but important consequence is that the random variables $z + Y$ and $z' + Y$ are close in the sense of Definition 1: for all $t \in \mathbb{R}^d$,

$$\frac{\Pr(z + Y = t)}{\Pr(z' + Y = t)} \in \exp(\pm\frac{\|z - z'\|_1}{\lambda}).$$

Thus, to release a (perturbed) value $f(\mathbf{x})$ while satisfying privacy, it suffices to add Laplace noise with standard deviation $S(f)/\epsilon$ in each coordinate.

Proposition 1 (Non-interactive Output Perturbation). *For all $f : D^n \to \mathbb{R}^d$, the following mechanism is ϵ-indistinguishable:*
$$\mathsf{San}_f(\mathbf{x}) = f(\mathbf{x}) + (Y_1, ..., Y_d) \quad \text{where the } Y_i \text{ are drawn i.i.d. from } \mathrm{Lap}(S(f)/\epsilon)$$

The proposition is actually a special case of the privacy of a more general, possibly adaptive, interactive process.

Before continuing with our discussion, we will need to clarify some of the notation to highlight subtleties raised by adaptivity. Specifically, adaptivity complicates the nature of the "query function", which is no longer a predetermined function, but rather a strategy for producing queries based on answers given thus far. For example, an adaptive histogram query might ask to refine those regions with a substantial number of respondents, and we would expect the set of such selected regions to depend on the random noise incorporated into the initial responses.

Recalling our notation, a transcript $t = [Q_1, a_1, Q_2, a_2 ..., Q_d, a_d]$ is a sequence of questions and answers. For notational simplicity, we will assume that Q_i is a well defined function of $a_1, ..., a_{i-1}$, and that we can therefore truncate our transcripts to be only a vector $t = [a_1, a_2, ..., a_d]^3$. For any transcript t, we will let $f_t : D \to R^d$ be the function whose ith coordinate reflects the query Q_i, which we assume to be determined entirely by the first $i - 1$ components of t. As we now see, we can bound the privacy of an adaptive series of questions using the largest diameter among the functions f_t.

Consider a trusted server, holding $\mathbf{x}$, which receives an adaptive sequence of queries $f_1, f_2, f_3, ..., f_d$, where each $f_i : D^n \to \mathbb{R}$. For each query, the server San either (a) refuses to answer, or (b) answers $f_i(\mathbf{x}) + \mathrm{Lap}(\lambda)$. The server can limit the queries by refusing to answer when $S(f_t)$ is above a certain threshold. Note

---

[3] Although as written the choice of query is deterministic, this can be relaxed by adding coins to the transcript.

that the decision whether or not to respond is based on $S(f_t)$, which can be computed by the user, and hence is not disclosive.

Theorem 1. *For an arbitrary adversary $\mathcal{A}$, let $f_t(\mathbf{x}) : D^n \to \mathbb{R}^d$ be its query function as parameterized by a transcript t. If $\lambda = \max_t S(f_t)/\epsilon$, the mechanism above is ϵ-indistinguishable.*

Proof. Using the law of conditional probability, and writing t_i for the indices of t,

$$\frac{\Pr[\mathrm{San}_f(\mathbf{x}) = t]}{\Pr[\mathrm{San}_f(\mathbf{x}') = t]} = \prod_i \frac{\Pr[\mathrm{San}_f(\mathbf{x})_i = t_i | t_1, \ldots, t_{i-1}]}{\Pr[\mathrm{San}_f(\mathbf{x}')_i = t_i | t_1, \ldots, t_{i-1}]}$$

For each term in the product, fixing the first $i - 1$ coordinates of t fixes the values of $f_t(\mathbf{x})_i$ and $f_t(\mathbf{x}')_i$. As such, the conditional distributions are simple laplacians, and we can bound each term and their product as

$$\prod_i \frac{\Pr[\mathrm{San}_f(\mathbf{x})_i = t_i | t_1, \ldots, t_{i-1}]}{\Pr[\mathrm{San}_f(\mathbf{x}')_i = t_i | t_1, \ldots, t_{i-1}]} \leq \prod_i \exp(|f_t(\mathbf{x})_i - f_t(\mathbf{x}')_i|/\lambda)$$

$$= \exp(\|f_t(\mathbf{x}) - f_t(\mathbf{x}')\|_1/\lambda)$$

We complete the proof using the bound $S(f_t) \leq \lambda\epsilon$, for all t.

3.2   Specific Insensitive Functions

We describe specific functionalities which have low sensitivity, and which consequently can be released with little added noise using the protocols of the previous section.

HISTOGRAMS AND DISJOINT ANALYSES. There are many types of analyses that first partition the input space into disjoint regions, before proceeding to analyze each region separately. One very simple example of such an analysis is a histogram, which simply counts the number of elements that fall into each region. Imagining that D is subdivided into d disjoint regions, and that $f : D^n \to \mathbb{Z}^d$ is the function that counts the number of elements in each region, we saw in Example 3 that $S(f) = 2$. Notice that the output dimension, d, does not play a role in the sensitivity, and hence in the noise needed in an ϵ-indistinguishable implementation of a histogram. Comparing this with what one gets by applying the framework of [6] we note a significant improvement in the noise. Regarding each bin value as a query, the noise added in the original framework to each coordinate is $O(\sqrt{d}/\epsilon)$, and hence the total L_1 error is an $O(\sqrt{d})$ factor larger than in our scheme. This factor is especially significant in applications where bins outnumber the data points (which is often the case with contingency tables).

Clearly any analysis that can be run on a full data set can be run on a subset, and we can generalize the above observation in the following manner. Letting D be partitioned into d disjoint regions, let $f : D^n \to \mathbb{R}^d$ be a function whose output coordinates $f(x)_i$ depend only on those elements in the ith region. We can bound $S(f) \leq 2\max_i S(f_i)$. Again, and importantly, the value of d does not appear in this bound.

LINEAR ALGEBRAIC FUNCTIONS. One very common analysis is measuring the mean and covariance of attributes of the data. If $v : D \to \mathbb{R}^d$ is some function mapping rows in the database to column vectors in $\mathbb{R}^d$, the mean vector μ and covariance matrix C are defined as

$$\mu = \mu(x_1, \ldots, x_n) = \underset{i}{\text{avg}}\, v(x_i)$$

$$\text{and} \quad C = C(x_1, \ldots, x_n) = \underset{i}{\text{avg}}\, v(x_i)v(x_i)^T - \mu\mu^T .$$

These two objects have dimension d and d^2, respectively, and a crude bound on their sensitivity would incorporate these terms. However, if we are given an upper bound $\gamma = \max_x \|v(x)\|_1$, then we can incorporate it into our sensitivity bounds.

Specifically, the mean is simply a sum, and an arbitrary change to a single term results in a vector $\mu + \delta$ where $\|\delta\|_1 \le 2\gamma/n$. The covariance matrix is more complicated, but is also a sum at heart. Using the L_1 norm on matrices as one might apply the L_1 to a d^2 dimensional vector, we see that an arbitrary change to a single x_i can change the $\mu\mu^T$ term by at most

$$\mu\mu^T - (\mu + \delta)(\mu + \delta)^T = \mu\delta^T + \delta\mu^T + \delta\delta^T \tag{2}$$

$$= \mu\delta^T + \delta(\mu + \delta)^T \tag{3}$$

The first and second terms each have L_1 norm at most $2\gamma^2/n$. An arbitrary change to x_i can alter a $v(x_i)v(x_i)^T$ term by at most $4\gamma^2$. Hence a total $L1$ change of $8\gamma^2/n$.

Again, we witness an improvement in the noise magnitude when compared to applying the framework of [6]. As computing C amounts to performing d^2 queries, we get $L1$ noise that is $O(d)$ factor larger than with the current analysis.

DISTANCE FROM A PROPERTY. The functionalities discussed until now had a simple representation as sums of vectors, and the sensitivity was then (at most) twice the maximum L_1 norm of one of these vectors. However, one can bound the sensitivity of much more complex functions.

Given a set $S \subseteq D^n$, the distance $f_S(\mathbf{x})$ between a particular database $\mathbf{x}$ and S is the Hamming distance (in D^n) between $\mathbf{x}$ and the nearest point $\mathbf{x}'$ in S. For any set S, $f_S(\mathbf{x})$ has sensitivity 1. We can safely release $f_S(\mathbf{x}) + Y$ where $Y \sim \text{Lap}(1/\epsilon)$.

As an example, we could imagine social network described as a database of links between pairs of individuals. We might like to measure how "robust" the network is: how many social links would have to change (either added or removed) for the graph to become disconnected, non-expansive, or poorly clustered? Each of these counts, which change by at most one when a single edge is altered, can be released with only a small amount of noise added.

Suppose that $n = \binom{m}{2}$, $D = [0, 1]$, and we interpret the entries of the database as giving the weights of edges in a graph with m vertices (that is, the "individuals" here are the edges). Then the weight of the minimum edge-cut in the graph, which is the distance from the nearest disconnected graph, is a 1-sensitive function. It is easily computable, and so one can safely release this information about a network (approximately) without violating the privacy of the component edges.

Other interesting graph functionalities also have low sensitivity. For example, if $D = [0, 1]$, the weight of the minimum spanning tree is 1-sensitive.

FUNCTIONS WITH LOW SAMPLE COMPLEXITY. Any function f which can be accurately approximated by an algorithm which looks only at a small fraction of the database has low sensitivity, and so the value can be released safely with relatively little noise. In particular, functions which can be approximated based on a random sample of the data points fit this criterion.

Lemma 1. *Let* $f : D^n \to \mathbb{R}^d$. *Suppose there is a randomized algorithm A such that for all inputs* $\mathbf{x}$, *(1) for all i, the probability that A reads x_i is at most α and (2)* $\|A(\mathbf{x}) - f(\mathbf{x})\|_1 \leq \sigma$ *with probability at least* $\beta = \frac{1+\alpha}{2}$. *Then* $S(f) \leq 2\sigma$.

The lemma translates a property of f related to ease of computation into a combinatorial property related to privacy. It captures many of the low-sensitivity functions described in the preceding sections, although the bounds on sensitivity given by the lemma are often quite loose.

Proof. For any particular entry $i \in \{1, ..., n\}$, denote by $A(\mathbf{x})\big|_{-i}$ the distribution on the outputs of A conditioned on the event that A does not read position i. By the definition of conditional probability, we get that for all $\mathbf{x}$ the probability that $A(\mathbf{x})\big|_{-i}$ is within distance σ of $f(\mathbf{x})$ is strictly greater than $(\beta - \alpha)/(1 - \alpha) \geq \frac{1}{2}$. Pick any $\mathbf{x}, \mathbf{x}'$ which only differ in the ith position. By the union bound, there exists some point p in the support of $A(\mathbf{x})\big|_{-i}$ which is within distance σ of *both* $f(\mathbf{x})$ and $f(\mathbf{x}')$, and hence $\|f(\mathbf{x}) - f(\mathbf{x}')\|_1 \leq \|f(\mathbf{x}) - p\|_1 + \|p - f(\mathbf{x}')\|_1 \leq 2\sigma$. ∎

One might hope for a converse to Lemma 1, but it does not hold. Not all functions with low sensitivity can be approximated by an algorithm with low sample complexity. For example, let $D = GF(2^{\lceil \log n \rceil})$ and let $f(\mathbf{x})$ denote the Hamming distance between $\mathbf{x}$ and the nearest codeword in a Reed-Solomon code of dimension $k = n(1 - o(1))$. One cannot learn anything about $f(\mathbf{x})$ using fewer than k queries, and yet f has sensitivity 1 [4].

3.3   Sensitivity in General Metric Spaces

The intuition that *insensitive* functions of a database can be released privately is not specific to the L_1 distance. Indeed, it seems that if changing one entry in $\mathbf{x}$ induces a small change in $f(\mathbf{x})$ — under any measure of distance on $f(\mathbf{x})$ — then we should be able to release $f(\mathbf{x})$ privately with relatively little noise. We formalize this intuition for (almost) any metric $\mathsf{d}_\mathcal{M}$ on the output $f(\mathbf{x})$. We will use symmetry, i.e. $\mathsf{d}_\mathcal{M}(x, y) = \mathsf{d}_\mathcal{M}(y, x)$, and the triangle inequality: $\mathsf{d}_\mathcal{M}(x, y) \leq \mathsf{d}_\mathcal{M}(x, z) + \mathsf{d}_\mathcal{M}(z, y)$.

Definition 3. *Let $\mathcal{M}$ be a metric space with a distance function* $\mathsf{d}_\mathcal{M}(\cdot, \cdot)$. *The sensitivity $S_\mathcal{M}(f)$ of a function $f : D^n \to \mathcal{M}$ is the amount that the function value varies when a single entry of the input is changed.*

$$S_\mathcal{M}(f) \stackrel{def}{=} \sup_{\mathbf{x}, \mathbf{x}':\, \mathsf{d}_H(\mathbf{x}, \mathbf{x}')=1} \mathsf{d}_\mathcal{M}(f(\mathbf{x}), f(\mathbf{x}'))$$

Given a point $z \in \mathcal{M}$, (and a measure on $\mathcal{M}$) we can attempt to define a probability density function

$$h_{z,\epsilon}(y) \propto \exp\left(\frac{\epsilon \cdot \mathsf{d}_{\mathcal{M}}(y,z)}{2 \cdot S_{\mathcal{M}}(f)}\right).$$

There may not always exist such a density function, since the right-hand expression may not integrate to a finite quantity. However, if it is finite then the distribution given by $h_{z,\epsilon}()$ is well-defined.

To reveal an approximate version of $f(\mathbf{x})$ with sensitivity S, one can sample a value according to $h_{f(\mathbf{x}),\epsilon/S}()$.

$$\Pr[\mathcal{T}(\mathbf{x}) = y] = \frac{\exp\left(\frac{\epsilon}{2S_{\mathcal{M}}(f)} \cdot \mathsf{d}_{\mathcal{M}}(y, f(\mathbf{x}))\right)}{\int_{y \in \mathcal{M}} \exp\left(\frac{\epsilon}{2S_{\mathcal{M}}(f)} \cdot \mathsf{d}_{\mathcal{M}}(y, f(\mathbf{x}))\right) dy}. \tag{4}$$

Theorem 2. *In a metric space where $h_{f(\mathbf{x}),\epsilon}()$ is well-defined, adding noise to $f(\mathbf{x})$ as in Eqn. 4 yields an ϵ-indistinguishable scheme.*

Proof. Let $\mathbf{x}$ and $\mathbf{x}'$ be two databases differing in one entry. The distance $\mathsf{d}_{\mathcal{M}}(f(\mathbf{x}), f(\mathbf{x}'))$ is at most $S(f)$. For any y, the ratio $\frac{\exp(\mathsf{d}_{\mathcal{M}}(y,f(\mathbf{x})))}{\exp(\mathsf{d}_{\mathcal{M}}(y,f(\mathbf{x}')))}$ is thus at most $e^{S(f)}$, by the triangle inequality. Similarly, the ratio $\frac{\exp(\frac{\epsilon}{2S(f)} \cdot \mathsf{d}_{\mathcal{M}}(y,f(\mathbf{x})))}{\exp(\frac{\epsilon}{2S(f)} \cdot \mathsf{d}_{\mathcal{M}}(y,f(\mathbf{x}')))}$ is at most $e^{\epsilon/2}$. Finally, the normalization constant $\int_{y \in \mathcal{M}} \exp\left(\frac{\epsilon \cdot \mathsf{d}_{\mathcal{M}}(y,f(\mathbf{x}))}{2S(f)}\right) dy$ also differs by a factor of at most $e^{\epsilon/2}$ between $\mathbf{x}$ and $\mathbf{x}'$, since at all points in the space the integrand differs by at most $e^{\epsilon/2}$. The total ratio $h_{f(\mathbf{x}),\epsilon}(y) / h_{f(\mathbf{x}'),\epsilon}(y)$ differs by at most $e^{\epsilon/2} \cdot e^{\epsilon/2} = e^{\epsilon}$, as desired.

Remark 1. One can get rid of the factor of 2 in the definition of $h_{z,\epsilon}()$ in cases where the normalization factor does not depend on z. This introduces slightly less noise.

As a simple example, consider a function whose output lies in the Hamming cube $\{0,1\}^d$. By Theorem 2, one can release $f(\mathbf{x})$ safely by flipping each bit of the output $f(\mathbf{x})$ independently with probability roughly $\frac{1}{2} - \frac{\epsilon}{2S(f)}$.

4   Separating Interactive Mechanisms from Non-interactive Ones

In this section, we show a strong separation between interactive and non-interactive database access mechanisms. Consider the interactive setting of [10, 11, 6], that answers queries of the form $f_g(\mathbf{x}) = \sum_{i=1}^{n} g(i, x_i)$ where $g : [n] \times D \rightarrow [0,1]$. As the sensitivity of any f_g is 1, an interactive access mechanism can answer any such query with accuracy about $1/\epsilon$. This gives a good approximation to $f(\mathbf{x})$ as long as ϵ is larger than $1/n$.

Suppose the domain D is $\{0,1\}^d$. We show below that for any non-interactive, ϵ-indistinguishable mechanism San, there are many functions f_g which cannot be answered by $\mathcal{T}_{\mathsf{San}}$ unless the database consists of at least $2^{\Omega(d)}$ points. For these queries, it is not possible to distinguish the sanitization of a database in which *all* of the n entries satisfy $g(i, x_i) = 0$ from a database in which all of the entries satisfy $g(i, x_i) = 1$. We will consider Boolean functions $g_{\mathbf{r}}$ of a specific form. Given n non-zero binary strings $\mathbf{r} = (r_1, r_2, ..., r_n)$, $r_i \in \{0,1\}^d$, we define $g_{\mathbf{r}}(i, x)$ to be the inner product, modulo 2, of r_i and x, that is $g_{\mathbf{r}}(i, x) = \bigoplus_j x^{(j)} r_i^{(j)}$, denoted $r_i \odot x$. In the following we will usually drop the subscript $\mathbf{r}$ and write g for $g_{\mathbf{r}}$.

Theorem 3 (Non-interactive Schemes Require Large Databases). *Suppose that* San *is an ϵ-indistinguishablenon-interactive mechanism with domain $D = \{0,1\}^d$. For at least 2/3 of the functions of the form $f_g(\mathbf{x}) = \sum_i g(i, x_i)$, the following two distributions have statistical difference $O(n^{4/3}\epsilon^{2/3}2^{-d/3})$:*

$$Distribution \ 0: \ \mathcal{T}_{\mathsf{San}}(\mathbf{x}) \ where \ \mathbf{x} \in_R \{\mathbf{x} \in D^n : f_g(\mathbf{x}) = 0\}$$
$$Distribution \ 1: \ \mathcal{T}_{\mathsf{San}}(\mathbf{x}) \ where \ \mathbf{x} \in_R \{\mathbf{x} \in D^n : f_g(\mathbf{x}) = n\}$$

In particular, if $n = o(\frac{2^{d/4}}{\sqrt{\epsilon}})$, for most functions $g(i, x) = r_i \odot x$, it is impossible to learn the relative frequency of database items satisfying the predicates $g(i, x_i)$. We prove Theorem 3 below. First, a few remarks:

1. The order of the quantifiers is important: for any particular $f_g()$, it is easy to design a non-interactive scheme which answers that query accurately. However, no single non-interactive scheme can answer most queries of this form, unless $n \in \exp(d)$.
2. The strong notion of ϵ-indistinguishability in Definition 1 is essential to Theorem 3. For example, consider the candidate sanitization which outputs m pairs (i, x_i) chosen at random from the database. When $m = \theta(1)$ this is essentially Example 2; it fails to satisfy Definition 1 but yields $O(1/n)$-close distributions $O(1/n)$ on neighboring databases. However, it does permit estimating f_g with accuracy about $n/\sqrt{m}$ (the order of quantifiers is again important: for any particular query, the sample will be good with high probability). Thus, even for constant m, this is better than what is possible for any ϵ-indistinguishable scheme with $n = 2^{o(d)}$.

4.1   A Stronger Separation for Randomized Response Schemes

"Randomized response" refers to a special class of non-interactive schemes, in which each user's data is perturbed individually, and then the perturbed values are published. That is, there exists a randomization operator $Z : D \to \{0,1\}^*$ such that

$$\mathcal{T}_{\mathsf{San}}(x_1, ..., x_n) = Z(x_1), ..., Z(x_n).$$

This approach means that no central server need ever see the users' private data: each user i computes $Z(x_i)$ and releases only that.

We can strengthen Theorem 3 for randomized response schemes. We can consider functions f_g where the *same* predicate $g : D \to \{0,1\}$ is applied to all the entries in $\mathbf{x}$. I.e. $f(\mathbf{x}) = \sum_i g(x_i)$ (e.g. "how many people in the database have blue eyes?"). For most vectors r, the parity check $g_r(x) = r \odot x$ will be difficult to learn from $Z(x)$, and so $f(\mathbf{x})$ will be difficult to learn from $T_{\mathsf{San}}(\mathbf{x})$ unless n is very large.

Proposition 2 (Randomized Response). *Suppose that* San *is a ϵ-indistinguishable randomized response mechanism. For at least $2/3$ of the values $r \in \{0,1\}^d \setminus \{0^d\}$, the following two distributions have statistical difference $O(n\epsilon^{2/3} 2^{-d/3})$:*

Distribution 0: $T_{\mathsf{San}}(\mathbf{x})$ where each $x_i \in_R \{x \in \{0,1\}^d : r \odot x = 0\}$

Distribution 1: $T_{\mathsf{San}}(\mathbf{x})$ where each $x_i \in_R \{x \in \{0,1\}^d : r \odot x = 1\}$

In particular, if $n = o(2^{d/3}/\epsilon^{2/3})$, no user can learn the relative frequency of database items satisfying the predicate $g_r(x) = r \odot x$, for most values r.

4.2   Proving the Separation Results

The two proofs have the same structure: a hybrid argument with a chain of length $2n$, in which the bound on statistical distance at each step in the chain is given by Lemma 2 below. Adjacent elements in the chain will differ according to the domain from which one of the entries in the database is chosen, and the elements in the chain are the probability distributions of the sanitizations when the database is chosen according to the given n-tuple of distributions.

For any r, partition the domain D into two sets: $D_r = \{x \in \{0,1\}^d : r \odot x = 0\}$, and $\bar{D}_r = D \setminus D_r = \{x \in \{0,1\}^d : r \odot x = 1\}$. We abuse notation and let D_r also stand for a random vector chosen uniformly from that set (similarly for D and $\bar{D}_r$).

The intuition for the key step is as follows. Given a randomized map $Z : D \to \{0,1\}^*$, the quantity $\Pr[Z(D_r) = z]$ is with high probability an estimate for $\Pr[Z(D) = z]$. That is because when r is chosen at random, D_r consists of 0^d, along with $2^{d-1}-1$ points chosen pairwise independently in $\{0,1\}^d$. This allows us to show that the variance of the estimator $\Pr[Z(D_r) = z]$ is very small, as long as Z satisfies a strong indistinguishability condition implied by ϵ-indistinguishability. As a result, the distribution $Z(D_r)$ will be very close to $Z(D)$.

Lemma 2. *Let $Z : D \to \{0,1\}^*$ be a randomized map such that for all pairs $x, x' \in D$, and all outputs z, $\frac{\Pr[Z(x)=z]}{\Pr[Z(x')=z]} \in \exp(\pm\epsilon)$. For all $\alpha > 0$: with probability at least $1 - \alpha$ over $r \in \{0,1\}^d \setminus \{0^d\}$,*

$$\mathbf{SD}\left(Z(D_r), \ Z(D)\right) \leq O\left(\frac{\epsilon^2}{\alpha \cdot 2^d}\right)^{1/3}.$$

The same statement holds for $\bar{D}_r$.

The lemma is proved below, in Section 4.3. We first use it to prove the two separation results.

Proof (Proof of Theorem 3). "Distribution 0" in the statement is $T_{\mathsf{San}}(D_{r_1}, ..., D_{r_n})$. We show that with high probability over the choice of the r_i's, this is close the transcript distribution induced by a uniform input, i.e. $T(D, ..., D)$. We proceed by a hybrid argument, adding one constraint at a time. For each i, we want to show

$$T_{\mathsf{San}}(D_{r_1}, ..., D_{r_i},\quad D\quad , D, ..., D) \quad \text{is close to}$$
$$T_{\mathsf{San}}(D_{r_1}, ..., D_{r_i},\quad D_{r_{i+1}}\, , D, ..., D).$$

Suppose that we have chosen $r_1, ..., r_i$ already. For any $x \in \{0,1\}^d$, consider the randomized map where the $(i+1)$-th coordinate is fixed to x:

$$Z(x) = T_{\mathsf{San}}(D_{r_1}, ..., D_{r_i},\ x\ , D, ..., D) \tag{5}$$

Note that $Z(D)$ is equal to the i-th step in the hybrid, and $Z(D_{r_{i+1}})$ is equal to the $(i+1)$-st step.

The ϵ-indistinguishabilityof San implies that $Z()$ satisfies $\frac{\Pr[Z(x)=z]}{\Pr[Z(x')=z]} \in \exp(+\epsilon)$. Applying Lemma 2 shows that with probability at least $1 - \frac{1}{6n}$ over r_{i+1}, $Z(D_{r_i})$ is within statistical difference σ of $Z(D)$, where $\sigma = O(\sqrt[3]{n}\epsilon^2 2^{-d})$. That is, adding the i-th constraint on the inputs changes the output distribution by at most σ. By a union bound, all the steps in the hybrid have size at most σ with probability at least $\frac{5}{6}$. In that case, the total distance is $n\sigma$.

We can apply exactly the same reasoning to a hybrid starting with Distribution 1, and ending with $T(D, ..., D)$. Again, with probability at least $\frac{5}{6}$, the total distance is $n\sigma$. With probability at least $2/3$, both chains of hybrids accumulate statistical difference bounded by $n\sigma$, and the distance between Distributions 0 and 1 is at most $2n\sigma = O(n^{4/3}\epsilon^{2/3}2^{-d/3})$.

Proof (Proof of Proposition 2). If T_{San} is a randomized response scheme, then there is a randomized map $Z()$ from D to $\{0,1\}^*$, such that $T_{\mathsf{San}}(x_1, ..., x_n) = Z(x_1), ..., Z(x_n)$. If T_{San} is ϵ-indistinguishable, then for all pairs $x, x' \in D$, and for all outputs z, $\frac{\Pr[Z(x)=z]}{\Pr[Z(x')=z]} \in \exp(\pm\epsilon)$.

It is sufficient to show that with probability at least $2/3$ over a random choice r, $r \neq 0^d$, the distributions $Z(D_r)$ and $Z(\bar{D}_r)$ are within statistical difference $O(\epsilon^{2/3}2^{-d/3})$. This follows by applying Lemma 2 with $\alpha = 1/3$. By a hybrid argument, the difference between Distributions 0 and 1 above is then $O(n\epsilon^{2/3}2^{-d/3})$.

4.3 Proving that Random Subsets Approximate the Output Distribution

Proof (Proof of Lemma 2). Let $p(z|x)$ denote the probability that $Z(x) = z$. If x is chosen uniformly in $\{0,1\}^d$, then the probability of outcome z is $p(z) = \frac{1}{2^d} \sum_x p(z|x)$.

For symmetry, we will pick not only the string r but an offset bit b, and look at the set $D_{r,b} = \{x \in \{0,1\}^d : r \odot x = b\}$. This simplifies the calculations somewhat.

One can think of $\Pr[Z(D_{r,b}) = z]$ as estimating $p(z)$ by pairwise-independently sampling $2^d/2$ values from the set D and only averaging over that subset. Since, by the assumption on Z, the values $p(z|x)$ all lie in an interval of width about $\epsilon \cdot p(z)$ around $p(z)$, this estimator will have small standard deviation. We will use this to bound the statistical difference.

Let $\hat{p}(z) = \Pr[Z(D_{r,b}) = z]$, where the probability is taken over the coin flips of Z and the choice of $x \in D_{r,b}$. For a fixed z, $\hat{p}(z)$ is a random variable depending on the choice of r, b, and $\mathbb{E}_{r,b}[\hat{p}(z)] = p(z)$.

Claim 1. $\mathrm{Var}_{r,b}[\hat{p}(z)] \leq \dfrac{2 \cdot \tilde{\epsilon}^2 \cdot p(z)^2}{2^d}$, where $\tilde{\epsilon} = e^\epsilon - 1$.

The proof of Claim 1 appears below. We now complete the proof of Lemma 2. We say that a value z is δ-good for a pair (r,b) if $\hat{p}(z) - p(z) \leq \delta \cdot p(z)$. By the Chebyshev bound, for all z,

$$\Pr_{r,b}[z \text{ is not } \delta\text{-good for } (r,b)] \leq \frac{\mathrm{Var}\,[\hat{p}(z)]}{\delta^2 p(z)^2} \leq \frac{2\tilde{\epsilon}^2}{\delta^2 2^d}.$$

If we take the distribution on z given by $p(z)$, then with probability at least $1 - \alpha$ over pairs (r,b), the fraction of z's (under $p(\cdot)$) which are good is at least $1 - \frac{2\tilde{\epsilon}^2}{\alpha\delta^2 2^d}$.

Finally, if a $1 - \gamma$ fraction of the z's are δ-good for a particular pair (r,b), then the statistical difference between the distribution $\hat{p}(z)$ and $p(z)$ is at most $2(\gamma + \delta)$. Setting $\delta = \sqrt[3]{\frac{2\alpha\tilde{\epsilon}^2}{2^d}}$, we get a total statistical difference of at most 4δ. Since $\tilde{\epsilon} < 2\epsilon$ for $\epsilon \leq 1$, the total distance between $\hat{p}(\cdot)$ and $p(\cdot)$ is at most $4\sqrt[3]{12\epsilon^2 2^{-d}}$, for at least a $1-\alpha$ fraction of the pairs (r,b). The bit b is unimportant here since it only switches D_r and its complement $\bar{D}_r$. The distance between $Z(D_r)$ and $Z(D)$ is exactly the same as the distance between $Z(\bar{D}_r)$ and $Z(D)$, since $Z(D)$ is the mid-point between the two. Thus, the statement holds even over pairs of the form $(r,0)$. This proves Lemma 2.

Proof (Proof of Claim 1). Let p^* be the minimum over x of $p(z|x)$. Let $q_x = p(z|x) - p^*$ and $\bar{q} = p(z) - p^*$. The variance of $\hat{p}(z)$ is the same as the variance of $\hat{p}(z) - p^*$. We can write $\hat{p}(z) - p^*$ as $\frac{2}{2^d} \sum_x q_x \chi_0(x)$, where $\chi_0(x)$ is 1 if $x \in D_{r,b}$. The expectation of $\hat{p}(z) - p^*$ is $\bar{q}$, which we can write $\frac{1}{2^d} \sum_x q_x$.

$$\mathrm{Var}_{r,b}[\hat{p}(z)] = \mathbb{E}_{r,b}\left[\left(\tfrac{2}{2^d}\sum_x q_x\chi_0(x) - \tfrac{1}{2^d}\sum_x q_x\right)^2\right] = \mathbb{E}_{r,b}\left[\left(\tfrac{1}{2^d}\sum_x q_x(2\chi_0(x) - 1)\right)^2\right] \tag{6}$$

Now $(2\chi_0(x) - 1) = (-1)^{r\odot x\oplus b}$. This has expectation 0. Moreover, for $x \neq y$, the expectation of $(2\chi_0(x) - 1)(2\chi_0(y) - 1)$ is exactly $1/2^d$ (if we chose r with no restriction it would be 0, but we have the restriction that $r \neq 0^d$). Expanding the square in Eqn. 6,

$$\mathsf{Var}_{r,b}\left[\hat{p}(z)\right] = \frac{1}{2^{2d}} \sum_x q_x^2 + \frac{1}{2^{3d}} \sum_{x \neq y} q_x q_y$$

$$= \frac{1 - \frac{1}{2^d}}{2^{2d}} \sum_x q_x^2 + \frac{1}{2^d} \left(\frac{1}{2^d} \sum_x q_x \right)^2 \leq \frac{1}{2^d} \left(\max_x q_x^2 + \bar{q}^2 \right).$$

By the indistinguishability condition, both $(\max_x q_x)$ and $\bar{q}$ are at most $(e^\epsilon - 1)p^* \leq \tilde{\epsilon} \cdot p(z)$. Plugging this into the last equation proves Claim 1.

References

[1] N. R. Adam and J. C. Wortmann. Security-control methods for statistical databases: a comparative study. *ACM Computing Surveys*, 25(4), December 1989.

[2] Dakshi Agrawal and Charu C. Aggarwal. On the design and quantification of privacy preserving data mining algorithms. In *Proceedings of the Twentieth ACM SIGACT-SIGMOD-SIGART Symposium on Principles of Database Systems*. ACM, 2001.

[3] Rakesh Agrawal and Ramakrishnan Srikant. Privacy-preserving data mining. In Weidong Chen, Jeffrey F. Naughton, and Philip A. Bernstein, editors, *SIGMOD Conference*, pages 439–450. ACM, 2000.

[4] Eli Ben-Sasson, Prahladh Harsha, and Sofya Raskhodnikova. Some 3cnf properties are hard to test. In *STOC*, pages 345–354. ACM, 2003.

[5] Web page for the Bertinoro CS-Statistics workshop on privacy and confidentiality. Available from http://www.stat.cmu.edu/~hwainer, July 2005.

[6] Avrim Blum, Cynthia Dwork, Frank McSherry, and Kobbi Nissim. Practical privacy: The sulq framework. In *PODS*, 2005.

[7] Shuchi Chawla, Cynthia Dwork, Frank McSherry, Adam Smith, and Hoeteck Wee. Toward privacy in public databases. In *Theory of Cryptography Conference (TCC)*, pages 363–385, 2005.

[8] Shuchi Chawla, Cynthia Dwork, Frank McSherry, and Kunal Talwar. On the utility of privacy-preserving histograms. In *21st Conference on Uncertainty in Artificial Intelligence (UAI)*, 2005.

[9] Dorothy E. Denning. Secure statistical databases with random sample queries. *ACM Transactions on Database Systems*, 5(3):291–315, September 1980.

[10] Irit Dinur and Kobbi Nissim. Revealing information while preserving privacy. In *Proceedings of the Twenty-Second ACM SIGACT-SIGMOD-SIGART Symposium on Principles of Database Systems*, pages 202–210, 2003.

[11] Cynthia Dwork and Kobbi Nissim. Privacy-preserving datamining on vertically partitioned databases. In Matthew K. Franklin, editor, *CRYPTO*, volume 3152 of *Lecture Notes in Computer Science*, pages 528–544. Springer, 2004.

[12] Alexandre V. Evfimievski, Johannes Gehrke, and Ramakrishnan Srikant. Limiting privacy breaches in privacy preserving data mining. In *Proceedings of the Twenty-Second ACM SIGACT-SIGMOD-SIGART Symposium on Principles of Database Systems*, pages 211–222, 2003.

[13] Shafi Goldwasser and Silvio Micali. Probabilistic encryption. *Journal of Computer and System Sciences*, 28(2):270–299, April 1984.

[14] Gina Roque. *Masking microdata with mixtures of normal distributions*. University of California, Riverside, 2000. Doctoral Dissertation.

[15] Latanya Sweeney. *k*-anonymity: A model for protecting privacy. *International Journal on Uncertainty, Fuzziness and Knowledge-based Systems*, 10(5):557–570, 2002.

Appendix

A   "Semantically" Flavored Implications of Definition 1

Definition 1 equates privacy with the inability to distinguish two close databases. Indistinguishability is a convenient notion to work with (as is indistinguishability of encryptions [13]); however, it does not directly say what an adversary may do and learn. In this section we present some "semantically" flavored definitions of privacy, and their equivalence to Definition 1.

Because of the need to have some utility conveyed by the database, it is not possible to get as strong a notion of security as we can, say, with encryption. We discuss two definitions which we consider meaningful, suggestively named *simulatability* and *semantic security*. The natural intuition is that if the adversary learns very little about x_i for all i, then privacy is satisfied. Recall the discussion of smoking and heart disease, from the Introduction. What is actually shown is that the adversary cannot learn much more about any x_i than she could learn from knowing almost all the data points except x_i.

Extending terminology from Blum et al. [6], we say an adversary is *informed* if she knows some set of $n-k$ database entries before interacting with the mechanism, and tries to learn about the remaining ones. The parameter k measures her remaining uncertainty.

Definition 4. *A mechanism* San *is* (k, ϵ)*-simulatable if for every adversary* $\mathcal{A}$, *and for every set* $I \subseteq [n]$ *of size* $n - k$, *there exists an* informed *adversary* $\mathcal{A}'$ *such that for any* $\mathbf{x} \in D^n$:

$$\left| \ln\left(\frac{\Pr[\ \mathcal{T}_{\mathsf{San}, \mathcal{A}}(\mathbf{x}) = t\ ]}{\Pr[\ \mathcal{A}'(\mathbf{x}|_I) = t\ ]} \right) \right| \leq \epsilon$$

where $\mathbf{x}|_I$ *denotes the restriction of* $\mathbf{x}$ *to the index set* I.

For convenience in stating implications among definitions, we extend the definition of indistinguishability (Definition 1) to pairs of databases at Hamming distance k:

Definition 5. *A mechanism is* (k, ϵ)*-indistinguishable if for all pairs* $\mathbf{x}, \mathbf{x}'$ *which differ in at most* k *entries, for all adversaries* $\mathcal{A}$ *and for all transcripts* t, $\left| \ln\left(\frac{\Pr[\mathcal{T}_\mathcal{A}(\mathbf{x}) = t]}{\Pr[\mathcal{T}_\mathcal{A}(\mathbf{x}') = t]} \right) \right| \leq \epsilon$.

Any $(1, \frac{\epsilon}{k})$-indistinguishable mechanism is also (k, ϵ)-indistinguishable. To see why, consider a chain of at most k databases connecting $\mathbf{x}$ and $\mathbf{x}'$, where only one entry changes at each step. The probabilities change by a factor of $\exp(\pm\epsilon/k)$ at each step, so $\frac{\Pr[\mathcal{T}_\mathcal{A}(\mathbf{x})=t]}{\Pr[\mathcal{T}_\mathcal{A}(\mathbf{x}')=t]} \in \exp(\pm\epsilon/k)^k = \exp(\pm\epsilon)$.

Claim 2.

1. *A (k, ϵ)-indistinguishable mechanism is (k, ϵ)-simulatable.*
2. *A (k, ϵ)-simulatable mechanism is $(k, 2\epsilon)$-indistinguishable.*

Proof. (1) A mechanism that is (k, ϵ)-indistinguishable is (k, ϵ)-simulatable. The simulator fills in the missing entries of $\mathbf{x}$ with default values to obtain $\mathbf{x}'$ which differs from $\mathbf{x}$ in at most k entries, then simulates an interaction between $\mathsf{San}(\mathbf{x}')$ and $\mathcal{A}$.

(2) A mechanism that is (k, ϵ)-simulatable is $(k, 2\epsilon)$-indistinguishable. Suppose that $\mathbf{x}', \mathbf{x}''$ agree in a set I of $n - k$ positions. Definition 4 says that for all $\mathcal{A}$ and all subsets I of $n - k$ indices, there exists an $\mathcal{A}'$ that, seeing only the rows indexed by I, can relatively accurately simulate the distribution of transcripts induced when $\mathcal{A}$ interacts with the full database. Since $\mathbf{x}'|_I = \mathbf{x}''|_I$ the behavior of $\mathcal{A}'$ is close to both that of the privacy mechanism interacting with $\mathcal{A}$ on $\mathbf{x}'$ and $\mathcal{A}$ on $\mathbf{x}''$:

$$\left| \ln\left(\frac{\Pr[\mathcal{T}_\mathcal{A}(\mathbf{x}') = t]}{\Pr[\mathcal{T}_\mathcal{A}(\mathbf{x}'') = t]}\right) \right| \le \left| \ln\left(\frac{\Pr[\mathcal{T}_\mathcal{A}(\mathbf{x}') = t]}{\Pr[\mathcal{A}'(\mathbf{x}'|_I) = t]}\right) \right| + \left| \ln\left(\frac{\Pr[\mathcal{A}'(\mathbf{x}''|_I) = t]}{\Pr[\mathcal{T}_\mathcal{A}(\mathbf{x}'') = t]}\right) \right| \le 2\epsilon.$$

$$\tag{7}$$

Simulatability states that for any i, little more is learned about individual i by an adversary interacting with the access mechanism than what she might learn from studying the rest of the world.

Simulatability still leaves implicit what, exactly, the adversary can compute about the database. *Semantic security* captures a more computationally-flavored meaning of privacy. Given an informed adversary, who knows $\mathbf{x}|_I$, we say a $\mathbf{x}' \in D^n$ is *consistent* if it agrees with the adversary's knowledge; i.e. $\mathbf{x}'|_I = \mathbf{x}|_I$. A *consistent probability distribution $\mathcal{D}$* is a probability distribution over consistent databases.

Definition 6. *A mechanism is (k, ϵ)-semantically secure if every interaction with an informed adversary results in a bounded change in the a-posteriori probability distribution. That is, for all informed adversaries $\mathcal{A}$, for all consistent distributions $\mathcal{D}$, for all transcripts t, and for all predicates $f : D^n \to \{0, 1\}$:*

$$\left| \ln\left(\frac{\Pr[f(\mathbf{x}') = 1]}{\Pr[f(\mathbf{x}') = 1 | \mathcal{T}_\mathcal{A}(\mathbf{x}') = t]}\right) \right| \le \epsilon. \tag{8}$$

The probabilities are taken over the coins of $\mathcal{A}, \mathsf{San}$ and choices of consistent $\mathbf{x}'$ according to $\mathcal{D}$.

Claim 3. *A mechanism is (k, ϵ)-indistinguishable iff it is (k, ϵ)-semantically-secure.*

Proof. (1) Let San be a (k, ϵ)-indistinguishable mechanism, and assume San is not (k, ϵ)-semantically-secure. Using Bayes' rule, we get that for some f and t:

$$\ln(\frac{\Pr[f(\mathbf{x}) = 1]}{\Pr[f(\mathbf{x}) = 1 | \mathcal{T}_\mathcal{A}(\mathbf{x}) = t]}) = \ln(\frac{\Pr[\mathcal{T}_\mathcal{A}(\mathbf{x}) = t]}{\Pr[\mathcal{T}_\mathcal{A}(\mathbf{x}) = t | f(\mathbf{x}) = 1]}) > \epsilon. \quad (9)$$

Pick a consistent $\mathbf{x}_0$ that maximizes $\Pr[\mathcal{T}(\mathbf{x}_0) = t]$ subject to $f(\mathbf{x}_0) = 0$. Clearly, $\Pr[\mathcal{T}(\mathbf{x}_0) = t] \geq \Pr[\mathcal{T}_\mathcal{A}(\mathbf{x}) = t]$. Similarly, pick a consistent $\mathbf{x}_1 \in D$ that minimizes $\Pr[\mathcal{T}(\mathbf{x}_1) = t]$ subject to $f(\mathbf{x}_1) = 1$. We get that

$$\ln(\frac{\Pr[\mathcal{T}_\mathcal{A}(\mathbf{x}_0) = t]}{\Pr[\mathcal{T}_\mathcal{A}(\mathbf{x}_1) = t]}) > \epsilon. \quad (10)$$

Noting that $d_H(\mathbf{x}_1, \mathbf{x}_2) \leq k$ we get a contradiction to the mechanism being (k, ϵ)-indistinguishable.
(2) Let San be a (k, ϵ)-semantically-secure mechanism, and assume San is not (k, ϵ')-indistinguishable. That is, there exist $\mathbf{x}_0, \mathbf{x}_1$ such that $d_H(\mathbf{x}_0, \mathbf{x}_1) \leq k$ and a possible transcript t such that

$$\left|\ln(\frac{\Pr[\mathcal{T}_\mathcal{A}(\mathbf{x}_1) = t]}{\Pr[\mathcal{T}_\mathcal{A}(\mathbf{x}_0) = t]})\right| > \epsilon. \quad (11)$$

Wlog, assume $\Pr[\mathcal{T}_\mathcal{A}(\mathbf{x}_0) = t] > \Pr[\mathcal{T}_\mathcal{A}(\mathbf{x}_1) = t]$, and that $\mathbf{x}_0, \mathbf{x}_1$ agree on their first $K = n - k$ coordinates. Let $\mathcal{A}$ be an informed adversary that knows these entries, and $\mathcal{D}$ be a consistent distribution that assigns probability α to $\mathbf{x}_0$ and $1 - \alpha$ to $\mathbf{x}_1$. Finally, take f to be any predicate such that $f(\mathbf{x}_b) = b$. We get that

$$\Pr[f(\mathbf{x}') = 1 | \mathcal{T}_\mathcal{A}(x') = t] = \frac{\Pr[\mathcal{T}_\mathcal{A}(\mathbf{x}_1) = t] \cdot \Pr[f(\mathbf{x}') = 1]}{\alpha \cdot \Pr[\mathcal{T}_\mathcal{A}(\mathbf{x}_0) = t] + (1 - \alpha) \cdot \Pr[\mathcal{T}_\mathcal{A}(\mathbf{x}_1) = t]}, \quad (12)$$

and hence

$$\ln(\frac{\Pr[f(\mathbf{x}') = 1]}{\Pr[f(\mathbf{x}') = 1 | \mathcal{T}_\mathcal{A}(\mathbf{x}') = t]}) = \ln(1 - \alpha + \alpha\frac{Pr[\mathcal{T}_\mathcal{A}(\mathbf{x}_0) = t]}{\Pr[\mathcal{T}_\mathcal{A}(\mathbf{x}_1) = t]}) > \ln(1 - \alpha + \alpha e^\epsilon). \quad (13)$$

Taking $\alpha \to 1$ yields the claim.

Unconditionally Secure Constant-Rounds Multi-party Computation for Equality, Comparison, Bits and Exponentiation

Ivan Damgård[1], Matthias Fitzi[1,*], Eike Kiltz[2,**],
Jesper Buus Nielsen[1,***], and Tomas Toft[1,†]

[1] University of Aarhus,
Department of Computer Science,
DK-8200 Aarhus N, Denmark
[2] CWI Amsterdam,
The Netherlands

Abstract. We show that if a set of players hold shares of a value $a \in \mathbb{F}_p$ for some prime p (where the set of shares is written $[a]_p$), it is possible to compute, in constant rounds and with unconditional security, sharings of the bits of a, i.e., compute sharings $[a_0]_p, \ldots, [a_{\ell-1}]_p$ such that $\ell = \lceil \log_2 p \rceil$, $a_0, \ldots, a_{\ell-1} \in \{0,1\}$ and $a = \sum_{i=0}^{\ell-1} a_i 2^i$. Our protocol is secure against active adversaries and works for any linear secret sharing scheme with a multiplication protocol. The complexity of our protocol is $\mathcal{O}(\ell \log \ell)$ invocations of the multiplication protocol for the underlying secret sharing scheme, carried out in $\mathcal{O}(1)$ rounds.

This result immediately implies solutions to other long-standing open problems such as constant-rounds and unconditionally secure protocols for deciding whether a shared number is zero, comparing shared numbers, raising a shared number to a shared exponent and reducing a shared number modulo a shared modulus.

1 Introduction

Assume that n parties have shared values $a_1, \ldots, a_\ell$ from some field $\mathbb{F}$ using some linear secret sharing scheme, such as Shamir's. Let $f : \mathbb{F}^\ell \to \mathbb{F}^m$. By

---

* Supported by SECOQC, Secure Communication based on Quantum Cryptography, under the Information Societies Technology Programme of the European Commission, IST-2003-506813.

** The paper was written while the author was a visitor at University of California, San Diego, supported by a DAAD postdoc fellowship.

*** Supported by FICS, Foundations In Cryptology and Security, centre of the Danish National Science Research Council and ECRYPT, European Network of Excellence in Cryptology, under the Information Societies Technology Programme of the European Commission, IST-2002-507932.

† Supported by SCET, Secure Computing, Economy, and Trust, Alexandra Instituttet A/S.

S. Halevi and T. Rabin (Eds.): TCC 2006, LNCS 3876, pp. 285–304, 2006.
© Springer-Verlag Berlin Heidelberg 2006

computing f with unconditional security on the sharings we mean that the parties run among themselves a protocol using a network with perfectly secure point-to-point channels. The protocol results in the parties obtaining sharings of $(b_1, \ldots, b_m) = f(a_1, \ldots, a_\ell)$, while leaking no information on the values $a_1, \ldots, a_\ell$ or $b_1, \ldots, b_m$. The question *which functions can be computed with unconditional security on sharings, using a constant rounds protocol* is a long-standing open problem [BB89].

However, a number of functions are known to have unconditionally secure, constant-rounds protocols. The most general class with known solutions are functions with a constant-depth arithmetic circuit (counting unbounded fan-in addition and unbounded fan-in multiplication as one gate towards the depth).

The only non-trivial part needed in these solutions is unbounded fan-in multiplication $b = \prod_{i=1}^\ell a_i$. This can be done in constant rounds using the techniques by Bar-Ilan and Beaver [BB89], assuming a single multiplication can be done in constant rounds, which is indeed the case for standard linear (verifiable) secret-sharing schemes.

However, a number of functions do not have small constant-depth arithmetic solutions. Consider, e.g., the function $\overset{?}{<}: \mathbb{F}_p \times \mathbb{F}_p \to \mathbb{F}_p$, where $(a \overset{?}{<} b) \in \{0, 1\}$ and $(a \overset{?}{<} b) = 1$ iff $a < b$ (where a and b are considered as residues $a, b \in \{0, 1, \ldots, p-1\}$). This function has a huge number of zeros and is not constant zero. Therefore we cannot hope for an efficient arithmetic solution to computing $\overset{?}{<}$ (the function can of course be expressed as a polynomial over the field, and thus a constant-depth circuit, but the circuit would have a number of gates proportional to the size of the field).

On the other hand a number of results are known where if the inputs are given in a particular form, then any function which can be expressed by a binary Boolean circuit with g gates and depth d, can be computed unconditionally securely in constant rounds, by evaluating a constant-depth arithmetic circuit with $O(2^d g)$ gates (see e.g [BB89, IK00, IK02]).

If, in particular, the input a is delivered as bitwise sharings $[a_0]_p, \ldots, [a_{\ell-1}]_p$ and $b = f(a)$ can be computed using a binary Boolean circuit with depth d and g gates, then sharings of the bits of $b = f(a)$ can be computed with complexity[1] $O(2^d g)$, unconditionally secure in constant rounds. This can e.g. be done using Yao's circuit scrambling technique with an unconditionally secure encryption scheme — an observation first made by [IK02]. This would e.g. allow to compute the function $\overset{?}{<}: (\mathbb{F}_p)^\ell \times (\mathbb{F}_p)^\ell \to \mathbb{F}_p, ((a_0, \ldots, a_{\ell-1}), (b_0, \ldots, b_{\ell-1})) \mapsto \sum_{i=0}^{\ell-1} a_i 2^i \overset{?}{<} \sum_{i=0}^{\ell-1} b_i 2^i$ unconditionally securely in constant rounds.

So, different representations of the inputs allow different classes of functions to be computed unconditionally securely in constant rounds — at least with the

---

[1] For the rest of the paper we measure the complexity of protocols by the maximal number of invocations of the multiplication protocol, which is typically the dominating term in the communication complexity. The exact communication complexity then depends on the communication complexity of the multiplication protocol used.

current knowledge of the area. It would therefore be very useful to be able to change representations efficiently. Previously it was not known how to do this. For instance, this was the reason why the protocols of Cramer and Damgård [CD01] for linear algebra in constant rounds could not handle fields with large characteristic without assuming that the input was shared bitwise to begin with, which limits the applicability of those protocols. In this paper, we therefore investigate the problem of changing between sharings modulo a prime p and bitwise sharings.

1.1 Our Results

Given a prime p, let $\ell = \lceil \log_2 p \rceil$. We will show how to compute, unconditionally secure and in constant rounds, $[a_0]_p, \ldots, [a_{\ell-1}]_p$ from $[a]_p$ such that $a_0, \ldots, a_{\ell-1} \in \{0,1\} \subseteq \mathbb{Z}_p$ and such that $a = \sum_{i=0}^{\ell-1} a_i 2^i$. The complexity is bounded by $\mathcal{O}(1)$ rounds and $\mathcal{O}(\ell \log_2 \ell)$ invocations of the multiplication protocol.

The only assumptions we need about the underlying secret sharing scheme are the following: 1) the secret sharing scheme is linear (i.e., given sharings $[a]_p$ and $[b]_p$ and public constants $c, d \in \mathbb{Z}_p$, the parties can securely compute a sharing $[ac + bd \bmod p]_p$ without interaction) and 2) there exists a constant-round multiplication protocol for the secret sharing scheme (i.e., given sharings $[a]_p$ and $[b]_p$, the parties can securely compute a sharing $[ab \bmod p]_p$ by interacting). If the multiplication protocol (and the secret sharing scheme) is secure against active adversaries, our protocols will be actively secure too. Likewise, if secret sharing scheme and multiplication protocol are adaptively secure, our protocols inherit this property. The assumption on multiplication implies that the adversary structure must be $Q2$ which, in the standard threshold case, means that we need honest majority.

This result immediately implies efficient constant-rounds protocols for some interesting problems. In particular, we can also compute, in constant rounds, outputs from the following functions in shared form:

- The equality function asking whether a shared input value is zero or not. This function was exactly what was missing in [CD01] in order to handle fields with large characteristics.
- The less-than comparison function of two numbers from $\mathbb{F}_p$, when considered residues in $\{0, 1, \ldots, p-1\}$.
- Modulo reduction, performing a discrete modulo reduction (with respect to a public/shared modulus).
- Discrete Exponentiation (with respect to a public/shared exponent and modulus).

We note that, while unconditional security is typically defined by requiring that the information leaked by the protocol is exponentially small in some security parameter κ, our protocols obtain a slightly stronger notion, which has also been considered in the literature. In particular, our protocols are perfectly secure except with probability $O(2^{-\kappa})$ — i.e. with probability $1 - O(2^{-\kappa})$ no information is leaked at all. Furthermore, the parties will be able to detect when a run of the protocol is in progress which would leak information if completed, and have the power to abort such a run. This yields a *perfectly secure protocol*, except that with probability $O(2^{-\kappa})$ it might terminate with some abort symbol $\perp$.[2]

1.2   Related Work

There has been a considerable amount of previous work on unconditionally secure constant-rounds multi-party computation with honest majority (c.f. [BB89] and [FKN94, IK97, CD01, Bea00, IK00, IK02]). As mentioned, this work has shown that some functions can indeed be computed in constant rounds with unconditional security, but this has been limited to restricted classes of functions, such as NC_1 or non-deterministic log-space.

In [ACS02] Algesheimer, Camenisch and Shoup also present a protocol for securely computing the bit-decomposition $[a]_p \mapsto ([a_0]_p, \ldots, [a_{\ell-1}]_p)$. It however only provides correctness and privacy when a is guaranteed to be noticeably smaller than p. Furthermore, it is only passively secure and is not constant rounds.

1.3   Organization

In Section 2 we give some technical preliminaries. In Section 3 we give the high-level protocol for bit decomposition, assuming a number of results from subsequent sections, in particular that it is possible to add bitwise-shared numbers and compare bitwise-shared number within certain complexities. In Section 4 we show how to generate the sharing of a uniformly random bit. In Section 5 we give the protocol for comparing two bitwise-shared numbers and in Section 6 we give the protocol for adding two bitwise-shared numbers. Finally, in Section 7 we mention a couple of applications of the new bit-decomposition protocol.

2   Preliminaries

In this section we introduce some notation and some known techniques.

We assume that n parties are connected by perfectly secure channels in a synchronous network. Let $\mathbb{F}_p$ denote the finite field with p elements where p is

---

[2] Choosing between unconditional (but imperfect) termination, correctness or privacy, we find that settling for imperfect termination but perfect correctness (on termination) and perfect privacy is the better choice. Simply because the other unconditional notions can be obtained from such a solution. To get perfect termination and perfect correctness but only unconditional privacy: when the protocol aborts, reconstruct the inputs and compute the results. This yields a protocol which is perfect except that it leaks information with small probability. To get perfect termination, perfect privacy but only unconditional correctness: when the protocol aborts, simply return with some dummy guess at the results. This yields a protocol which is perfect except that it is incorrect with small probability. Finally, to get a perfectly secure protocol: rerun the protocol when it aborts. This gives a perfectly secure protocol. It, however, only runs in *expected* constant rounds.

a prime, and let $\ell = \lceil \log_2 p \rceil$. We will assume throughout that $p > 2^\kappa$, and so whenever one of our protocols abort with a probability that is $O(1/p)$, this will be considered negligible and will be ignored. If one needs to execute our (sub)protocol(s) with a given (small) prime p, one can always execute in parallel a sufficiently large number of instances to make the failure probability small enough.

By $[a]_p$ we denote a secret sharing of $a \in \mathbb{F}_p$ over $\mathbb{F}_p$. We assume that the secret-sharing scheme allows to compute a sharing $[a + b \bmod p]_p$ from $[a]_p$ and $[b]_p$ without communication, and that it allows to compute $[ab \bmod p]_p$ from $a \in \mathbb{F}_p$ and $[b]_p$ without communication; We write

$$[a + b \bmod p]_p \leftarrow [a]_p + [b]_p$$

and

$$[ab \bmod p]_p \leftarrow a[b]_p$$

for these operations. The secret-sharing scheme should of course also allow to take a sharing $[c]_p$ and reveal the value $c \in \mathbb{F}_p$ to all parties; We write

$$c \leftarrow \text{REVEAL}([c]_p) .$$

We also assume that the secret sharing scheme allows to compute a sharing $[ab \bmod p]_p$ from $[a]_p$ and $[b]_p$ with unconditional security. We denote the multiplication protocol by MULT, and write

$$[ab \bmod p]_p \leftarrow \text{MULT}([a]_p, [b]_p) .$$

Sometimes we will also write

$$[b \bmod p]_p \leftarrow \text{MULT}([a_1]_p, \ldots, [a_l]_p) ,$$

to avoid writing $b_2 \leftarrow \text{MULT}([a_1]_p, [a_2]_p)$, $b_3 \leftarrow \text{MULT}([b_2]_p, [a_3]_p)$, $\ldots$, $b \leftarrow \text{MULT}([b_{l-1}]_p, [a_l]_p)$. This costs $l - 1$ rounds and $l - 1$ invocations of MULT.

We will express the protocols' round complexities as the number of sequential rounds of MULT invocations — and their communication complexities as the overall number of MULT invocations. I.e., if we first run a copies of MULT in parallel and then run b copies of MULT in parallel, then we say that we have round complexity 2 and communication complexity $a + b$. Note that standard linear (verifiable) secret-sharing schemes have efficient constant-rounds protocols for multiplication.

For our protocols to be actively secure, the secret sharing scheme and the multiplication protocol should be actively secure. This in particular means that the adversary structure must be $Q2$. By the adversary structure we mean the set $\mathcal{A}$ of subsets $C \subset \{1, \ldots, n\}$ which the adversary might corrupt; It is $Q2$ if it holds for all $C \in \mathcal{A}$ that $\{1, \ldots, n\} \setminus C \notin \mathcal{A}$.

All our protocols can be proven secure in the UC model [Can01]. In the UC model our protocols can be expressed in a hybrid model with an ideal functionality F allowing the parties to privately load values in $\mathbb{F}_p$ into F and allowing the parties to add, multiply and output loaded and/or computed values. For an approach to formulate such an ideal functionality, see e.g., the Arithmetic Black-Box (ABB) from [DN03]. It can then be shown that an information theoretic VSS with a multiplication protocol implements this ideal functionality (as an example the VSS schemes from [CDM00] will do). The full version of this paper will contain more details on how our protocols can be proven secure in the UC model.

2.1   Some Known Techniques

The following known techniques will be of importance later on.

Random Elements. The parties can share a uniformly random, unknown field element. We write

$$[a]_p \leftarrow \text{RAN}_p() .$$

This is done by letting each party P_i deal a sharing $[a_i]_p$ of a uniformly random $a_i \in \mathbb{F}_p$. Then the parties compute the sharing $[a]_p = \sum_{i=1}^{n} [a_i]_p$. The communication complexity of this is given by n dealings, which we assume is upper bounded by the complexity of one invocation of the multiplication protocol.

If passive security is considered, this is trivially secure. If active security is considered and some party refuses to contribute with a dealing, the sum is just taken over the contributing parties. This means that the sum is at least taken over a_i for $i \in H$, where $H = \{1, \ldots, n\} \setminus C$ for some $C \in \mathcal{A}$. Since $\mathcal{A}$ is Q2 it follows that $H \notin \mathcal{A}$. So, at least one honest party will contribute to the sum, implying randomness and privacy of the sum.

Random Invertible Elements. Using [BB89] the parties can share a uniformly random, unknown, invertible field element along with a sharing of its inverse. We write

$$([a]_p, [a^{-1}]_p) \leftarrow \text{RAN}_p^*() ,$$

and it proceeds as follows: $[a]_p \leftarrow \text{RAN}_p()$ and $[b]_p \leftarrow \text{RAN}_p()$. $[c]_p = \text{MULT}([a]_p, [b]_p)$. $c \leftarrow \text{REVEAL}([c]_p)$. If $c \notin \mathbb{F}_p^*$, then abort. Otherwise, proceed as follows: $[a^{-1} \bmod p]_p \leftarrow (c^{-1} \bmod p)[a]_p$. Output $([a]_p, [a^{-1}]_p)$.

The correctness is straightforward. As for privacy, if $c \in \mathbb{F}^*$, then (a, b) is a uniformly random element from $\mathbb{F}^* \times \mathbb{F}^*$ for which $ab \bmod p = c$, and thus a is a uniformly random element in $\mathbb{F}_p^*$. If $c \notin \mathbb{F}^*$, then the algorithm aborts. This happens with probability less than $2/p$. The complexity is (at most) 2 rounds and 3 invocations of MULT.

Unbounded Fan-In Multiplication. Using the technique from [BB89] it is possible to do unbounded fan-in multiplication in constant rounds. For the special case where we compute all "prefix products" $\prod_{i=1}^{m} a_i$ ($m = 1, \ldots, \ell$), we write

$$([a_1]_p, \ldots, [(a_1 a_2 \cdots a_\ell) \bmod p]_p) \leftarrow \text{MULT}^*([a_1]_p, \ldots, [a_\ell]_p) .$$

In the following, we only need the case where we have inputs $[a_1]_p, \ldots, [a_\ell]_p$, where $a_i \in \mathbb{F}_p^*$. For $1 \leq i_0 \leq i_1 \leq \ell$, let $a_{i_0, i_1} = \left(\prod_{i=i_0}^{i_1} a_i \right) \bmod p$. We are often only interested in computing $a_{1,\ell}$, but the method allows to compute any other a_{i_0,i_1} at the cost of one extra multiplication. For the complexity analysis, let A denote the number of a_{i_0,i_1}'s which we want to compute.

First run $\mathrm{RAN}_p^* \; \ell+1$ times in parallel, to generate $[b_0 \in_R \mathbb{F}^*]_p, [b_1 \in_R \mathbb{F}^*]_p, \ldots, [b_\ell \in_R \mathbb{F}^*]_p$, along with $[b_0^{-1}]_p, [b_1^{-1}]_p, \ldots, [b_\ell^{-1}]_p$, using 2 rounds and $3(\ell+1)$ invocations of MULT. For simplicity we will use the estimate of 3ℓ invocations.

Then for $i = 1, \ldots, \ell$ compute and reveal $[d_i]_p = \mathrm{MULT}([b_{i-1}]_p, [a_i]_p, [b_i^{-1}]_p)$, using 2 rounds and 2ℓ invocations of MULT.

Now we have that $d_{i_0,i_1} = \prod_{i=i_0}^{i_1} d_i = b_{i_0-1}(\prod_{i=i_0}^{i_1} a_i)b_{i_1}^{-1} = b_{i_0-1}a_{i_0,i_1}b_{i_1}^{-1}$ (mod p), so we can compute $[a_{i_0,i_1}]_p = d_{i_0,i_1}\mathrm{MULT}([b_{i_0-1}^{-1}]_p, [b_{i_1}]_p)$, using 1 round and A invocations of MULT.

The overall complexity is 5 rounds and $5\ell + A$ invocations of MULT.

3 Bit-Decomposition

Let p be a prime $p \in [2^{\ell-1}, 2^\ell]$. We look at the bit-decomposition function $\mathrm{BITS} : \mathbb{F}_p \rightarrow (\mathbb{F}_p)^\ell, a \mapsto (a_0, \ldots, a_{\ell-1})$ given by $a_0, \ldots, a_{\ell-1} \in \{0,1\} \subseteq \mathbb{F}_p$ and $a = \sum_{i=0}^{\ell-1} a_i 2^i$, where $a \in \mathbb{F}_p$ is considered a residue $a \in \{0, 1, \ldots, p-1\}$. We denote a run of this protocol by

$$([a_0]_p, \ldots, [a_{\ell-1}]_p) \leftarrow \mathrm{BITS}([a]_p) .$$

The protocol for bit decomposition makes use of various sub-protocols which in turn draw on further sub-protocols. The dependency between the building blocks can be seen in Fig. 1. We now describe the highest level sub-protocols:

- Random solved BITS. This protocol has no inputs, and has outputs

$$([b_0]_p, \ldots, [b_{\ell-1}]_p, [b]_p) \leftarrow \mathrm{SOLVED\text{-}BITS}() ,$$

  where b is a uniformly random element $b \in \mathbb{F}_p$ and $(b_0, \ldots, b_{\ell-1}) = \mathrm{BITS}(b)$. As shown in the next subsection, this can be done using 21 rounds and 96ℓ invocations of the multiplication protocol.
- Bitwise sum. Let $[x]_\mathrm{B} = [x_0]_p, \ldots, [x_{l-1}]_p$ denote a bitwise sharing of an integer x. We use

$$[z]_\mathrm{B} \leftarrow \mathrm{BIT\text{-}ADD}([x]_\mathrm{B}, [y]_\mathrm{B})$$

  to denote the computation of a bitwise sharing $[z]_\mathrm{B} = [z_0]_p, \ldots, [z_l]_p$ of $x+y$ from bitwise sharings, $[x]_\mathrm{B} = [x_0]_p, \ldots, [x_{l-1}]_p$ and $[y]_\mathrm{B} = [y_0]_p, \ldots, [y_{l-1}]_p$, of integers x and y. The length l need not be the length ℓ of the prime p. In Section 6 it is shown how to implement BIT-ADD unconditionally securely in constant rounds. When $x, y \in \{0, \ldots, 2^l - 1\}$ the complexity is 37 rounds and $55l \log_2 l$ invocations of the multiplication protocol.
- Bitwise less-than. Finally we use

$$[x \overset{?}{<} y]_p \leftarrow \mathrm{BIT\text{-}LT}([x]_\mathrm{B}, [y]_\mathrm{B})$$

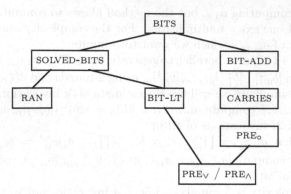

Fig. 1. *Protocol hierarchy*

to denote the computation of a sharing of the bit $(x \stackrel{?}{<} y) \in \{0, 1\}$, where $(x \stackrel{?}{<} y) = 1$ iff $x < y$, starting from bitwise sharings, $[x]_B = [x_0]_p, \ldots, [x_{l-1}]_p$ and $[y]_B = [y_0]_p, \ldots, [y_{l-1}]_p$, of integers x and y; Again l need not be ℓ. In Section 5 it is shown how to implement BIT-LT unconditionally securely in constant rounds. The complexity is 19 rounds and $22l$ invocations of the multiplication protocol.

We sometimes run the above protocols on non-shared inputs. If e.g. x is an integer known by all parties, then we let

$$[z]_B \leftarrow \text{BIT-ADD}(x, [y]_B) ,$$

mean the following: first compute the bitwise representation $(x_0, x_1, \ldots, x_{l-1})$ of x, then let $[x]_B = ([x_0]_p, \ldots, [x_{l-1}]_p)$ be some dummy bitwise sharing of x, and then run $[z]_B \leftarrow \text{BIT-ADD}([x]_B, [y]_B)$.

The bit decomposition of $[a]_p$ now proceeds as follows.

Protocol $[a]_B \leftarrow \textbf{BITS}([a]_p)$

1. The input is $[a]_p$, where $a \in \mathbb{F}_p$.
2. $([b_0]_p, \ldots, [b_{\ell-1}]_p, [b]_p) \leftarrow \text{SOLVED-BITS}()$.
3. $[a - b]_p \leftarrow [a]_p - [b]_p$.
4. $c \leftarrow \text{REVEAL}([a - b]_p)$, where $c \in \mathbb{F}_p$.
5. $[d]_B \leftarrow \text{BIT-ADD}(c, [b]_B)$, where $[d]_B = ([d_0]_p, \ldots, [d_\ell]_p)$.
6. $[q]_p \leftarrow \text{BIT-LT}(p, [d]_B)$.
7. $(f_0, \ldots, f_{\ell-1}) = \text{BITS}(2^\ell - p)$, the bitwise representation of the positive integer $2^\ell - p$.
8. For $i = 0, \ldots, \ell - 1$ in parallel: $[g_i]_p = f_i[q]_p$.
9. $[g]_B = ([g_0]_p, \ldots, [g_{\ell-1}]_p)$.
10. $[h]_B \leftarrow \text{BIT-ADD}([d]_B, [g]_B)$, where $[h]_B = ([h_0]_p, \ldots, [h_{\ell+1}]_p)$.
11. $[a]_B = ([h_0]_p, \ldots, [h_{\ell-1}]_p)$.
12. Output $[a]_B$.

As for the privacy, notice that assuming that the sub-protocols leak no information, the only place where information is potentially leaked is in Step 4, where c is leaked. Since b is assumed to be a uniformly random, unknown value from $\mathbb{F}_p$, independent of a, it however follows that c is uniformly random in $\mathbb{F}_p$ and leaks no information about a.

As for the correctness, notice that $c = a - b \bmod p$ and $d = c + b$ (in the integers). Therefore $d = a + qp$ for some $q \in \{0, 1\}$. Since $a \in \{0, 1, \ldots, p-1\}$ it follows that $q = 1$ iff $p < d$. A sharing of this q is computed in Step 6. Then let $f = 2^\ell - p$, let $g \in \mathbb{Z}$ be the integer which is bitwise shared in Step 9, and let $h \in \mathbb{Z}$ be the integer which is bitwise shared in Step 10. Clearly, $g = qf = q2^\ell - qp$ (in the integers). Therefore $h = d + g = (a + qp) + (q2^\ell - qp) = a + q2^\ell$. In Step 11 we then compute a as $h \bmod 2^\ell$ by dropping the two most significant bits of h.

As for the complexity, we generated one solved BITS, had two applications of BIT-ADD and one application of BIT-LT. This yields a total complexity of 114 rounds and $110\ell \log_2 \ell + 118\ell$ invocations.

3.1   Generating Random Solved BITS

We now describe the protocol SOLVED-BITS. As a sub-protocol we use a protocol RAN$_2$ for generating uniformly random shared bits. This protocol has no inputs, and outputs a sharing $[a]_p$, where $a \in \{0, 1\} \subseteq \mathbb{F}_p$ is uniformly random. We write

$$[a]_p \leftarrow \text{RAN}_2() .$$

In Section 4, we show how to implement RAN$_2$ in 2 rounds and 2 invocations of the multiplication protocol.

The generation of a random input/output pair for BITS proceeds as follows.

Protocol $([b]_B, [b]_p) \leftarrow$ **SOLVED-DITS**()

1. For $i = 0, \ldots, \ell - 1$ in parallel: $[b_i]_p \leftarrow \text{RAN}_2()$.
2. $[b]_B = ([b_0]_p, \ldots, [b_{\ell-1}]_p)$.
3. $[c]_p \leftarrow \text{BIT-LT}([b]_B, p)$.
4. $c \leftarrow \text{REVEAL}([c]_p)$.
5. If $c = 0$, then abort. Otherwise proceed as below.
6. $[b]_p \leftarrow \sum_{i=0}^{\ell-1} 2^i [b_i]_p$.
7. Output $([b]_B, [b]_p)$.

As for the correctness, notice that $[b]_B$ is by construction the bit-wise sharing of $[b]_p$. Furthermore, b is uniformly random from $\{0, 1, \ldots, 2^\ell - 1\}$. So under the condition that SOLVED-BITS does not abort, b is uniformly random from $\{0, 1, \ldots, p-1\}$, as desired.

As for the privacy, when SOLVED-BITS does not abort, the only information leaked is that $b < p$. This is however an *a priory* fact by the output requirement on SOLVED-BITS.

Let us examine the probability that SOLVED-BITS aborts. In case one is able to control the choice of the prime p, an optimal choice would be to let p be a Mersenne prime $p = 2^\ell - 1$ for some $\ell > \kappa$. In that case the probability that $b \geq p$ is less than $2^{-\kappa}$. Although the Mersenne primes soon become sparse, this would at least work for small values of ℓ. At the time of writing $p = 2^{25964951} - 1$ is the largest p for which we know that this works [NWKo]. Other primes close to powers of two work almost as nicely.

In the worst-case, where we have no control over p, our only guarantee is that $p \in [2^{\ell-1}, 2^\ell]$ for some ℓ. In that case the probability that $b \leq p$ when $b \in_R \{0, 1, \ldots, 2^\ell - 1\}$ can be as large as $1/2$. Using a Chernoff bound it can be seen that if one generates $n = 12\kappa$ candidates, then the probability that less than $n/4$ of them satisfy $b < p$ is upper bounded by $2^{-\kappa}$.

As for the complexity, one run of the basic SOLVED-BITS requires ℓ calls of RAN$_2$ and one call of BIT-LT, neglecting the cost of the one call to REVEAL. This gives a complexity of 21 rounds and 24ℓ invocations of the multiplication protocol. If the basic protocol has to be repeated in parallel to get a lower abort probability, the round complexity is still 21, and the amortized communication complexity goes up to 96ℓ.

4   Random Bits

We now describe a protocol RAN$_2$ for securely generating a sharing of a uniformly random bit. The protocol has no inputs, and the output is a sharing $[a]_p$ of a uniformly random $a \in \{0, 1\} \subseteq \mathbb{F}_p$. We assume that $p > 2$ such that $\mathbb{F}_p$ does not have characteristic 2.

First some notation. Let $\mathbb{F}_p^*$ be the set of non-zero elements of $\mathbb{F}_p$ and let $Q_p \subset \mathbb{F}_p^*$ be the subset of squares. For $a \in Q_p$, let $\sqrt{a}$ be the unique $b \in \{1, \ldots, (p-1)/2\}$ where $b^2 \bmod p = a$. We define $S : \mathbb{F}_p^* \to \mathbb{F}_p$ by $S(x) = 1$ if $0 < x < p/2$ and $S(x) = -1$ if $p/2 < x < p$. Note that it holds for all $x \in \mathbb{F}_p^*$ that $x = S(x)\sqrt{x^2} \bmod p$. Clearly, if $a \in_R \mathbb{F}_p^*$ is a uniformly random non-zero element, then $S(a)$ is uniformly random in $\{1, -1\}$ and, furthermore, $S(a) = a(\sqrt{a^2})^{-1}$. This suggests the following protocol.

Protocol $[d]_p \leftarrow$ RAN$_2$()

1. $[a]_p \leftarrow$ RAN$_p$().
2. $[a^2 \bmod p]_p = $ MULT($[a]_p, [a]_p$).
3. $a^2 \bmod p \leftarrow$ REVEAL($[a^2 \bmod p]_p$).
4. If $a^2 \bmod p = 0$, then abort. Otherwise, proceed as below.
5. $b = \sqrt{a^2} \bmod p$.
6. $[c]_p \leftarrow (b^{-1} \bmod p)[a]_p$.
7. $[d]_p \leftarrow 2^{-1}([c]_p + 1)$.
8. Output $[d]_p$.

As for correctness, notice that when RAN$_2$ does not abort, then $c = S(a)$, where c is the value shared in Step 6. Therefore c is uniformly random in $\{1, -1\}$. It then easily follows that d is uniformly random in $\{0, 1\}$.

As for privacy, note that when the protocol does not abort, then a is uniformly random from $\mathbb{F}_p^*$, and we are essentially using $S(a)$ as output. The only information leaked about a is $a^2 \bmod p$, which is independent of $S(a)$ when a is uniformly random in $\mathbb{F}_p^*$.

If $a = 0$, then the protocol aborts. This happens with probability $1/p$.

The complexity of generating $[a]_p$ is bounded by the complexity of one multiplication. Then one multiplication is needed to compute $[a^2 \bmod p]_p$. The rest is essentially for free. This gives a complexity of 2 rounds and 2 invocations.

5   Bitwise Less-Than

We show how to compare two bitwise-shared numbers in constant rounds. We first present two sub-protocols.

5.1   Symmetric Functions

Assume that we have inputs $[a_1]_p, \ldots, [a_\ell]_p$, where $a_1, \ldots, a_\ell \in \{0, 1\} \subseteq \mathbb{F}_p$, and want to compute a symmetric Boolean function f on these. We also need to assume that $\mathbb{F}_p$ has characteristic larger than $\ell + 1$, which here just means that we need that $\ell < p - 1$.

A symmetric Boolean function only depends on the number of 1's in its input, it can therefore be written as $f(x_1, \ldots, x_\ell) = \phi(1 + \sum_{i=1}^{\ell} x_i)$ for some function $\phi : \{1, 2, \ldots, \ell + 1\} \to \{0, 1\}$. By Lagrange interpolation, we can construct a polynomial with coefficients $\alpha_0, \ldots, \alpha_\ell$ such that $\phi(X) = \sum_{i=0}^{\ell} \alpha_i X^i \bmod p$ for $X \in \{1, 2, \ldots, \ell + 1\}$. This allows a particularly efficient secure computation, as follows.

Protocol $[f(a_1, \ldots, a_\ell) \bmod p]_p \leftarrow f([a_1]_p, \ldots, [a_\ell]_p)$

1. $[a]_p \leftarrow 1 + \sum_{i=1}^{\ell} [a_i]_p$.
2. $([a]_p, [a^2 \bmod p]_p, \ldots, [a^{\ell+1} \bmod p]_p) \leftarrow \text{MULT}^*([a]_p, \ldots, [a]_p)$.
3. $[f(a) \bmod p]_p \leftarrow \sum_{i=0}^{\ell} \alpha_i [a^i \bmod p]_p$.

In Step 2 we have that $a \in \mathbb{F}_p^*$, so we can apply the protocol MULT* securely. The protocol is clearly private and correct. The complexity is 5 rounds and 6ℓ invocations of MULT.

5.2   Prefix-Or

Assume that we have inputs $[a_1]_p, \ldots, [a_\ell]_p$, where $a_1, \ldots, a_\ell \in \{0, 1\} \subseteq \mathbb{F}_p$, and want to compute the prefix-or $[b_1]_p, \ldots, [b_\ell]_p$, where $b_i = \vee_{j=1}^{i} a_j$.

To obtain complexity linear in ℓ, we use the method by Chandra, Fortune and Lipton [CFL83a]. For notational convenience, assume that $\ell = \lambda^2$ for an integer λ. We will split a into λ blocks of λ bits each. For this purpose we rename each bit a_k as $a_{i,j}$ where $k = \lambda(i-1) + j$, and $i, j = 1, \ldots, \lambda$. Thus, $a = (a_{1,1}, a_{1,2}, \ldots, a_{1,\lambda}, a_{2,1}, \ldots, a_{2,\lambda}, \ldots, a_{\lambda,\lambda})$, and for $i = 1, \ldots, \lambda$, we call $a_{i,1}, \ldots, a_{i,\lambda}$ a block of a. The desired output will be split in blocks using the same notation. Note that we can compute an Or with unbounded fan-in, $[x]_p \leftarrow \vee_{j=1}^{\lambda}[x_j]_p$, using Section 5.1, as this is a symmetric function.

Protocol $([b_1]_p, \ldots, [b_\ell]_p) \leftarrow \text{PRE}_\vee([a_1]_p, \ldots, [a_\ell]_p)$

1. For $i = 1, \ldots, \lambda$, in parallel: $[x_i]_p = \vee_{j=1}^{\lambda}[a_{i,j}]_p$.
2. For $i = 1, \ldots, \lambda$, in parallel: $[y_i]_p = \vee_{k=1}^{i}[x_k]_p$.
3. $[f_1]_p = [x_1]_p$.
4. For $i = 2, \ldots, \lambda$, let $[f_i]_p = [y_i]_p - [y_{i-1}]_p$.
5. For $i = 1, \ldots, \lambda, j = 1, \ldots, \lambda$, in parallel: $[g_{i,j}]_p = \text{MULT}([f_i]_p, [a_{i,j}]_p)$.
6. For $j = 1, \ldots, \lambda$: $[c_j]_p = \sum_{i=1}^{\lambda}[g_{i,j}]_p$.
7. For $j = 1, \ldots, \lambda$, in parallel: $[b_{\cdot,j}]_p = \vee_{k=1}^{j}[c_k]_p$.
8. For $i = 1, \ldots, \lambda, j = 1, \ldots, \lambda$, in parallel: $[s_{i,j}]_p = \text{MULT}([f_i]_p, [b_{\cdot,j}]_p)$.
9. For $i = 1, \ldots, \lambda, j = 1, \ldots, \lambda$: $[b_{i,j}]_p \leftarrow [s_{i,j}]_p + [y_i]_p - [f_i]_p$.

The privacy follows from the fact that we only call private sub-protocols. As for the correctness, the variables have the following interpretation. We have that $x_i = 1$ iff the the i'th block contains a 1. Therefore $y_i = 1$ iff there is a 1 in one of the i first blocks, and $f_i = 1$ iff the i'th block is the first block to contain a 1. Hence the sequence of f_i values has form $f = (0, \ldots, 0, 1, 0, \ldots, 0)$, and we let i_0 be the position of the single 1-bit. Now, for $i < i_0$, the i'th block of the output should be all 0's. For $i > i_0$, the i'th block of the output should be all 1's. Finally, the i_0'th block of the output should the prefix-or of the i_0'th input block. The block $c = (c_1, \ldots, c_\lambda)$ is formed by taking the "inner product" of f and a and therefore, by the special form of f, equals the i_0'th block of a. The values $(b_{\cdot,1}, \ldots, b_{\cdot,\lambda})$ are the prefix-or bits of c. This means that the bits $s_{i,j}$ form an all-0 vector, except that the i_0'th block equals c. It now follows directly from the form of the $s_{i,j}$'s, f_i's and y_i's that the output bits $b_{i,j}$ get the correct value in the final step.

The protocol uses 3λ invocations of the protocol for symmetric functions, in three rounds and on problems of size λ. This gives a complexity of 15 rounds and 18ℓ invocations. Besides this there are two rounds of ℓ multiplications each, giving a total complexity of 17 rounds and 20ℓ invocations.

5.3   Bitwise Less-Than

We now describe the protocol BIT-LT. Note that given sharings of two bits $[a]_p$ and $[b]_p$ we can compute their Xor in one round by first computing $[d]_p \leftarrow [a_i]_p - [b_i]_p$ and then computing $[e]_p \leftarrow \text{MULT}([d]_p, [d]_p)$. Below we write this as $[e]_p \leftarrow \text{XOR}([a]_p, [b]_p)$.

Protocol $[c]_p \leftarrow$ **BIT-LT**$([a]_B, [b]_B)$

1. For $i = 0, \ldots, \ell - 1$: $[e_i]_p \leftarrow$ XOR$([a_i]_p, [b_i]_p)$.
2. $([f_{\ell-1}]_p, \ldots, [f_0]_p) =$ PRE$_\vee([e_{\ell-1}]_p, \ldots, [e_0]_p)$.
3. $[g_{\ell-1}]_p = [f_{\ell-1}]_p$.
4. For $i = 0, \ldots, \ell - 2$: $[g_i]_p \leftarrow [f_i]_p - [f_{i+1}]_p$.
5. For $i = 0, \ldots, \ell - 1$: $[h_i]_p \leftarrow$ MULT$([g_i]_p, [b_i]_p)$.
6. $[h]_p \leftarrow \sum_{i=0}^{\ell-1} [h_i]_p$.
7. Output $[h]_p$.

Privacy follows from the fact that we only call private sub-protocols. As for the correctness, assume that $a \neq b$, and let i_0 denote the largest index i, where $a_i \neq b_i$. Then $a < b$ iff $b_{i_0} = 1$. Note that i_0 is the largest i for which $f_i = 1$, and thus $g_i = 1$ iff $i = i_0$. Therefore $h = b_{i_0}$. In the special case $a = b$, clearly $h = 0$, as it should be.

The protocol uses one invocation of PRE$_\vee$ on an instance of size ℓ, costing 17 rounds and 20ℓ invocations of MULT. Then there are two rounds more, each of ℓ invocations of MULT, giving a total of 19 rounds and 22ℓ invocations of MULT.

6 Bitwise Sum

We show how to add two bitwise-shared numbers in constant rounds. We first present a sub-protocol.

6.1 Generic Prefix Computations

Assume that we have some alphabet $\Sigma \subseteq \{0,1\}^n$ and bitwise-shared inputs $[a_1]_B, \ldots, [a_\ell]_B$, where $a_i \in \Sigma$. That is, $[a_i]_B = [a_{i,1}]_p, \ldots, [a_{i,n}]_p$ consists of n sharings of bits, and $(a_{i,1}, \ldots, a_{i,n}) \in \Sigma$. Assume furthermore that an associative binary operator $\circ : \Sigma \times \Sigma \to \Sigma$ is given and that we want to compute sharings

$$([b_1]_B, \ldots, [b_\ell]_B) = \text{PRE}_\circ([a_1]_B, \ldots, [a_\ell]_B) ,$$

where $b_i = \circ_{j=1}^i a_j$. Assume that it is possible to securely compute a sharing $[b_\ell] = \circ_{j=1}^\ell [a_j]$ with complexity R rounds and $C(\ell)$ invocations of MULT. For short, we will refer to $\circ_{j=1}^\ell [a_j]$ as the "sum" of $a_1, \ldots, a_\ell$. We assume for notational convenience that $\ell = 2^k$ for some k.

We use the method by Chandra, Fortune and Lipton [CFL83b]. For each $i = 1, \ldots, k$ we will split the sequence $a_1, \ldots, a_\ell$ into consecutive blocks of size 2^i items each. We let $b_{i,j}$ be the "sum" of the j'th such block, i.e., $b_{i,j} = \circ_{m=j\cdot2^i+1}^{j\cdot2^i+2^i} a_m$. There are $\ell - 1$ of the "sums" $b_{i,j}$, namely one of length $\ell = 2^k$, two of length 2^{k-1}, up to $\ell/2$ of length two. The complexity for computing all of them in parallel is thus R rounds and $\sum_{i=1}^k 2^i C(\ell \cdot 2^{-i})$ invocations of MULT.

It is easy to see that each of the ℓ values b_i can be computed as a "sum" of at most k of the $b_{i,j}$'s. Doing this in parallel for all b_i's costs another R rounds and at most $\ell C(k)$ invocations. Therefore the total complexity is upper bounded by $2R$ rounds and $\sum_{i=1}^{\log_2 \ell} 2^i C(\ell \cdot 2^{-i}) + \ell C(\log_2 \ell) \leq \log_2 \ell \cdot C(\ell) + \ell C(\log_2 \ell)$ invocations of MULT.

6.2 Bitwise Sum

We now describe the protocol $[d]_B \leftarrow$ BIT-ADD$([a]_B, [b]_B)$.

For $i = 1, \ldots, \ell$, define the carry $c_i \in \{0, 1\}$ by $c_i = 1$ iff $\sum_{j=0}^{i-1} 2^j (a_j + b_j) > 2^i$. It is straightforward to verify that given a bitwise sharing of the carries we can compute a bitwise sharing of the sum as follows.

Protocol $[d]_B \leftarrow$ BIT-ADD$([a]_B, [b]_B)$

1. $([c_1]_p, \ldots, [c_\ell]_p) \leftarrow$ CARRIES$([a]_B, [b]_B)$.
2. $[d_0]_p = [a_0]_p + [b_0]_p - 2[c_1]_p$.
3. $[d_\ell]_p = [c_\ell]_p$.
4. For $i = 1, \ldots, \ell - 1$: $[d_i]_p = [a_i]_p + [b_i]_p + [c_i]_p - 2[c_{i+1}]_p$.
5. Output $[d]_B = ([d_0]_p, \ldots, [d_\ell]_p)$.

Evidently, the complexities of this protocol are the same as those of subprotocol CARRIES as presented below. We therefore get 37 rounds and $55\ell \log_2 \ell$ invocations of MULT.

6.3 Computing the Carry Bits

In order to compute the carries, we use the well-known *carry set/propagate/kill* algorithm. Let $\Sigma = \{S, P, K\}$. The algorithm uses an operator $\circ : \Sigma \times \Sigma \to \Sigma$, defined by $x \circ S = S$ for all $x \in \Sigma$, $x \circ K = K$ for all $x \in \Sigma$, and $x \circ P = x$ for all $x \in \Sigma$. This is the carry-propagation operator, and it can be verified to be associative[3].

For two bitwise-represented numbers $a = (a_0, \ldots, a_{\ell-1})$ and $b = (b_0, \ldots, b_{\ell-1})$, for $i = 0, \ldots, \ell - 1$, let $e_i = S$ iff a carry is set at position i (i.e., $a_i + b_i = 2$); $e_i = P$ iff a carry would be propagated at position i (i.e. $a_i + b_i = 1$); and $e_i = K$ iff a carry would be killed at position i, (i.e. $a_i + b_i = 0$). It is straightforward to verify that $c_i = 1$ (the i'th carry bit is set) if and only if $e_0 \circ \cdots \circ e_{i-1} = S$.

We represent S, P, and K with bit vectors $(1, 0, 0), (0, 1, 0)$ and $(0, 0, 1) \in \{0, 1\}^3$. The values of the e_i's in this representation can be easily computed from the a_i's and b_i's as shown below. Hence, given a protocol for unbounded fan-in computation of the carry-propagation operator $\circ$ on this representation, we compute carries as follows.

---

[3] Note that this definition is changed from the standard one to be consistent with the fact that we write numbers with the least significant bit first.

Protocol $[c]_B \leftarrow$ CARRIES$([a]_B, [b]_B)$

1. For $i = 0, \ldots, \ell - 1$, in parallel: $[s_i]_p = $ MULT$([a_i]_p, [b_i]_p)$.
2. For $i = 0, \ldots, \ell - 1$: $[p_i]_p = [a_i]_p + [b_i]_p - 2[s_i]_p$, $[k_i]_p = 1 - [s_i]_p - [p_i]_p$ and set $[e_i]_B = ([s_i]_p, [p_i]_p, [k_i]_p)$, i.e., interpret the sharings $[s_i]_p, [p_i]_p, [k_i]_p$ as a bit-wise sharing of a 3-bit string $e_i \in \Sigma$.
3. $([f_0]_B, \ldots, [f_{\ell-1}]_B) \leftarrow $ PRE$_\circ([e_0]_B, \ldots, [e_{\ell-1}]_B)$.
4. For $i = 0, \ldots, \ell - 1$, set $([s_i]_p, [p_i]_p, [k_i]_p) = [f_i]_B$, i.e., each $[f_i]_B$ consists of shares of 3 bits which we now name s_i, p_i and k_i.
5. Output $[c]_B = ([s_0]_p, [s_1]_p, \ldots, [s_{\ell-1}]_p)$.

The privacy follows from only using private sub-protocols, and correctness follows readily from the above arguments.

Section 6.1 describes how to compute PRE$_\circ([e_0]_B, \ldots, [e_{\ell-1}]_B)$ assuming a protocol for computing the $\circ$-operator with unbounded fan-in. The next section shows how to do this unbounded fan-in with complexity 18 rounds and 27ℓ invocations of MULT. This and the analysis of the protocol from Section 6.1 shows that we can compute all $f_0, \ldots, f_{\ell-1}$ with complexity 36 rounds and $54\ell \log_2 \ell$ invocations. Besides this, the CARRIES protocol has only one round containing a total of ℓ invocations of MULT, giving a total complexity of 37 rounds and $55\ell \log_2 \ell$ invocations of MULT.

6.4 Unbounded Fan-In Carry Propagation

We describe a protocol for computing $\circ_{i=1}^{\ell} e_i$, where we again represent e_i as (s_i, p_i, k_i). The protocol uses an unbounded fan-in And in Step 1 and a prefix-And in Step 2. These protocols are defined equivalently to a unbounded fan-in Or and prefix-Or, and implemented in the same complexity using DeMorgan's Rule.

Protocol $([a]_p, [b]_p, [c]_p) \leftarrow \circ_{i=1}^{\ell}([s_i]_p, [p_i]_p, [k_i]_p)$

1. $[b]_p \leftarrow \wedge_{i=1}^{\ell}[p_i]_p$.
2. $([q_\ell]_p, \ldots, [q_1]_p) \leftarrow $ PRE$_\wedge([p_\ell]_p, \ldots, [p_1]_p)$.
3. $[c_\ell] = [k_\ell]$.
4. For $i = 1, \ldots, \ell - 1$, in parallel: $[c_i]_p \leftarrow [k_i] \wedge [q_{i+1}]_p$.
5. $[c]_p = \sum_{i=1}^{\ell}[c_i]_p$.
6. $[a]_p \leftarrow 1 - [b]_p - [c]_p$.

As for correctness, it should be clear that $b = 1$ (a propagate) iff $p_i = 1$ for $i = 1, \ldots, \ell$, making b correct. Furthermore, we have that $c = 1$ (a kill) iff there exists some i such that $k_i = 1$ and $p_{i+1} = 1, \ldots, p_\ell = 1$. I.e. $c = \vee_{i=1}^{\ell}(k_i \wedge q_{i+1})$. Since k_i and p_i are never 1 simultaneously it can be seen that at most one of the expressions $k_i \wedge q_{i+1}$ equals one. This implies that $c = \sum_{i=1}^{\ell}(k_i \wedge q_{i+1})$. By our representation it follows that $a = 1 - b - c$.

Since we can compute $[b]_p$ and $[c]_p$ in parallel, the overall complexity for an unbounded fan-in carry propagation can be verified to be 18 rounds and 27ℓ invocations of MULT.

7   Applications

In this section we mention some secure multi-party protocols for specific tasks that use our new constant-rounds protocol for computing shares of the bit decomposition as an atomic sub-protocol. All application protocols are unconditionally secure constant-rounds protocols. We want to stress that even though the number of invocations of the underlying multiplication protocol is always polynomial in $\ell = \lceil \log_2 p \rceil$ and the number of rounds is constant we did not put much effort in optimizing the running time and round complexity.

For the remaining part of this section let p be a prime, $p \in [2^{\ell-1}, 2^\ell]$.

7.1   Comparison and Equality

In this subsection we look at the equality function $\overset{?}{=} : \mathbb{F}_p \to \mathbb{F}_p$, where $(x\overset{?}{=}y) \in \{0,1\}$ and $(x\overset{?}{=}y) = 1$ iff $x = y$, and the comparison function $\overset{?}{<} : \mathbb{F}_p \times \mathbb{F}_p \to \mathbb{F}_p$, where $(x\overset{?}{<}y) \in \{0,1\}$ and $(x\overset{?}{<}y) = 1$ iff $x < y$.

For equality, assume the shares $[x]_p, [y]_p$ are given and we want to compute shares $[x\overset{?}{=}y]_p$. Setting $z = x - y \in \mathbb{F}_p$ the problem clearly reduces to computing $[z\overset{?}{=}0]_p$. The latter one can be done by first computing shares of the bits $[z]_B = [z_0]_p, \ldots, [z_{\ell-1}]_p$ and then $[z\overset{?}{=}0]_p = \wedge_{i=0}^{\ell-1}[z_i]_p$, which can be computed in constant round using Section 5.1, as it is a symmetric function.

For the comparison function we are given shares $[x]_p, [y]_p$ and want to computes shares $[x\overset{?}{<}y]_p$. This can be done by first computing shares of the bits $[x]_B = [x_0]_p, \ldots, [x_{\ell-1}]_p$ and $[y]_B = [y_0]_p, \ldots, [y_{\ell-1}]_p$. Now shares of the comparison function can be computed using BIT-LT from Section 5.

7.2   Private Exponentiation

The exponentiation function $\exp : \mathbb{F}_p \times \mathbb{Z}_p \to \mathbb{F}_p$ is given by $\exp(x, a) = (x^a \bmod p) \in \mathbb{F}_p$.

PUBLIC EXPONENT a. We first deal with the case where the exponent a is publicly known and the value x is shared, i.e., given $[x]_p$ and a we want to compute $[x^a]_p$. Assume there exists a protocol that outputs random shares $[r]_p$ of a random non-zero value $r \in \mathbb{F}_p^*$ together with shares of its ath power $[r^a]_p$. We will show later how to implement such a protocol in constant rounds.

Assuming such a protocol exists, a protocol to securely compute the exponentiation function is straightforward (using the Bar-Ilan and Beaver [BB89] inversion trick): First the parties run the protocol to get shares of $[r]_p$ and $[r^a]_p$ for a random $r \in \mathbb{F}_p^*$. Then they compute $[xr]_p = \text{MULT}([x]_p, [r]_p)$, open $[xr]_p$ to get $xr \in \mathbb{F}_p$, and every player individually computes $y = (xr)^a = x^a r^a \in \mathbb{F}_p$. Now $[x^a]_p$ is obtained by computing $[x^a]_p = y[r^{-a}]_p$ where $[r^{-a}]_p = [(r^a)^{-1}]_p$ is obtained from $[r^a]_p$ using the Bar-Ilan and Beaver inversion protocol.

It is easy to see that this protocol is private as long as $x \neq 0$. We handle the case $x = 0$ as an "exception" using our protocol for evaluating the equality function from Section 7.1. The idea is to substitute x by $\tilde{x} = x + (x \overset{?}{=} 0)$. Note that this always assures $\tilde{x} \neq 0$. Then shares $[x^a]_p$ can be computed as

$$[x^a]_p = [\tilde{x}^a]_p - [(x \overset{?}{=} 0)]_p .$$

We note that this "exception trick" may also be used in some other places (like in the inversion protocol) to handle special shared inputs that may lead to information leakage. Any protocol that initially leaks information for m different shared input values can now be updated to a protocol providing perfect privacy by (roughly) the cost of additional m (parallel) executions of the equality protocol.

It remains to provide the protocol that, given a public a, outputs shares $[r]_p$ together with shares of its ath power $[r^a]_p$ for a random non-zero value $r \in \mathbb{F}_p^*$. In the honest-but-curious model this is simply done by letting each player j locally select a random non-zero value $r_j \in \mathbb{F}_p^*$ together with its ath power $r_j^a \in \mathbb{F}_p^*$. Each value r_j is shared among the players. Now define r as the product of all r_j such that r^a also equals to the product of all r_j^a. Shares of both products $r = \prod_{j=1}^n r_j$ and $r^a = \prod_{j=1}^n r_j^a$ can be computed using the unbounded fan-in multiplication protocol, MULT*. We now show how to make this protocol robust against active adversaries using a "cut-and-choose" technique: In addition to $[r_i]_p$, $[r_i^a]_p$, user i generates random sharings $[s_i]_p$, $[s_i^a]_p$. The players jointly form a random bit b. Then they compute and open (s_i, s_i^a) or $(s_i r_i, s_i^a r_i^a)$, according to the value of b and verify that the first number is non-zero and that the second number is the first raised to the public a. This can be repeated in parallel an appropriate number of times.

SHARED EXPONENT a. Now we consider the case where the exponent a is also given as a share, i.e., the users are given $[x]_p$ and $[a]_p$ and want to compute $[x^a]_p$. We show how this case can be reduced to the previous one.

First run the bit decomposition protocol to obtain shares of the bits $[a]_B = [a_0]_p, \ldots [a_{\ell-1}]_p$ of the exponent a such that $a = \sum_{i=0}^{\ell-1} 2^i a_i$ with $a_i \in \{0, 1\}$. Then, using unbounded fan-in multiplication, shares $[x^a]_p$ may now be obtained via the equation

$$x^a = x^{\sum_{i=0}^{\ell-1} 2^i a_i} = \prod_{i=0}^{\ell-1} x^{2^i a_i} = \prod_{i=0}^{\ell-1} (a_i x^{2^i} + 1 - a_i) \in \mathbb{F}_p \tag{1}$$

where the shares $[x^{(2^i)}]_p$ can be computed (in parallel for $1 \leq i \leq l-1$) with the exponentiation protocol above. If x is non-zero the protocol MULT* can be used for the unbounded fan-in multiplication, and the "exception trick" can be used use to deal with the case were x can be zero.

7.3 Modulo Reduction

Let $m \in [2, p-1]$ be a public integer. In this subsection we look at the "modulo m" function, $\mathrm{mod}_m : \mathbb{F}_p \to \mathbb{F}_p, x \mapsto x \bmod m \in \{0, \ldots, m-1\}$, where $x \in \mathbb{F}_p$ is considered a residue $x \in \{0, 1, \ldots, p-1\}$. We show how to privately compute the modulo m function in constant rounds, i.e., the players are given $[x]_p$ and want to compute $[\mathrm{mod}_m(x)]_p$ for a public integer m.

The players first compute shares $[x]_B = [x_0]_p, \ldots, [x_{\ell-1}]_p$ of the bits of x, i.e. $x = \sum_{i=0}^{\ell-1} x_i 2^i$. Note that if m is a power of 2, i.e., $m = 2^a$, then $[x \bmod m]_p$ can be computed using the equation $x \bmod 2^a = \sum_{i=0}^{a-1} 2^i x_i$. Otherwise define $y = \sum_{i=0}^{\ell-1} x_i (2^i \bmod m) \in \mathbb{Z}$. Then clearly $x \bmod m = (\sum_{i=0}^{\ell-1} x_i (2^i \bmod m)) \bmod m = y - tm$ for some integer t in the range $t \in [0, \ell-1]$. Define $y^{(i)} = y - im \in \mathbb{Z}$. The shares $[y^{(i)}]_B$ can be computed in parallel for all $i \in [0, \ell-1]$ using the bitwise sum protocol from Section 6. The value $x \bmod m$ is now the unique $y^{(t)}$ such that $0 \leq y^{(t)} < m$. Shares of such an $[x \bmod m]_B = [y^{(t)}]_B$ can be found using (ℓ parallel applications of) the comparison function and one conversion to shares of $[x \bmod m]_p$.

7.4 Private Modulo Reduction

The players are given shares $[x]_p$ and shares $[m]_p$ of an integer m of known bit-size $\ell_0 \ll \ell$. The problem is to compute shares $[x \bmod m]_p$. There already exists an efficient protocol to approximate $[x \bmod m]_p$ due to [ACS02] but it does not run in a constant number of rounds. In this section we note that combining the techniques of this paper with the results from [ACS02] and [KLM05] (the latter one approximates the fractional part of $1/m$ by a Taylor polynomial), we get an efficient constant-rounds protocol to compute an approximation of $[x \bmod m]_p = [x - \lfloor \frac{x}{m} \rfloor \cdot m]_p$. Shares of the exact value of $x \bmod m$ may then be obtained by running an appropriate number of comparison protocols to make sure that result lies in the interval $[0, m-1]$. With the results from Section 7.2 this enables us to build a constant-rounds protocol that privately computes shares $[x^a \bmod m]_p$, where all three inputs, x, a, and m are given as shares (together with the bit-size ℓ_0 of m). Here we only consider the case of prime m. First compute shares $[x \bmod m]_p$ and $[a \bmod m - 1]_p$. The prime p has to be large enough (of bit-size $\ell > \ell_0^2$) such that in Eqn. (1) no wrap-around modulo p appears: after computing $[x^a]_p$, the modulo reduction protocol is used again to compute shares $[x^a \bmod m]_p$.

7.5 Unrestricted Conversion to Additive Shares over the Integers

Informally, additive shares over the integers are $(n-1)$-out-of-n shares where each party P_j holds a random share $x_j \in [-2^\rho A, 2^\rho A]$ (where ρ is some security parameter). The secret x is then defined as $x = \sum_{j=1}^{n} x_j \in [-A, A]$ over the integers. We use $[x]_\mathbb{Z}$ to denote additive shares over the integers. See [ACS02] for a formal definition of additive shares and for applications.

Let p be a prime. We want to note that we now can give a constant-rounds protocol that converts shares $[x]_p$ to shares $[x]_\mathbb{Z}$. Prior to our work, by a result from [ACS02], this could only be done in constant rounds when x is guaranteed to be considerably smaller than the modulus p. As the protocol in [ACS02] our protocol is only passively secure.

First compute shares $[x]_B = [x_1]_p, \ldots, [x_{\ell-1}]_p$ of the bits of x. Then (in parallel) convert the shares $[x_j]_p$ to shares $[x_j]_{\mathbb{Z}}$ over the integers using the technique from [ACS02] (note that this can be carried out since the shares of the bits are now "small enough" compared to the modulus p). Finally, the integer shares of x can be computed without interaction via $[x]_{\mathbb{Z}} = \sum_{i=0}^{\ell-1} 2^i [x_i]_{\mathbb{Z}}$.

Acknowledgments

The authors would like to thank the anonymous referees for many useful suggestions, which helped improve the presentation considerably.

References

[ACS02]   Joy Algesheimer, Jan Camenisch, and Victor Shoup. Efficient computation modulo a shared secret with application to the generation of shared safe-prime products. In Moti Yung, editor, *Advances in Cryptology – CRYPTO 2002*, volume 2442 of *Lecture Notes in Computer Science*, pages 417–432, Santa Barbara, CA, USA, August 18–22, 2002. Springer-Verlag, Berlin, Germany.

[BB89]   Judit Bar-Ilan and Donald Beaver. Non-cryptographic fault-tolerant computing in constant number of rounds of interaction. In *Proc. ACM PODC'89*, pages 201–209, 1989.

[Bea00]   Donald Beaver. Minimal latency secure function evaluation. In Bart Preneel, editor, *Advances in Cryptology – EUROCRYPT 2000*, volume 1807 of *Lecture Notes in Computer Science*, pages 335–350, Bruges, Belgium, May 14–18, 2000. Springer-Verlag, Berlin, Germany.

[Can01]   Ran Canetti. Universally composable security: A new paradigm for cryptographic protocols. In *42nd Annual Symposium on Foundations of Computer Science*, pages 136–145, Las Vegas, Nevada, 14–17 October 2001. IEEE.

[CD01]   Ronald Cramer and Ivan Damgård. Secure distributed linear algebra in a constant number of rounds. In J. Kilian, editor, *Advances in Cryptology - Crypto 2001*, pages 119–136, Berlin, 2001. Springer-Verlag. Lecture Notes in Computer Science Volume 2139.

[CDM00]   Ronald Cramer, Ivan Damgård, and Ueli Maurer. General secure multiparty computation from any linear secret-sharing scheme. In Bart Preneel, editor, *Advances in Cryptology - EuroCrypt 2000*, pages 316–334, Berlin, 2000. Springer-Verlag. Lecture Notes in Computer Science Volume 1807.

[CFL83a] Ashok K. Chandra, Steven Fortune, and Richard J. Lipton. Lower bounds for constant depth circuits for prefix problems. In *Proceedings of ICALP 1983*, pages 109–117. Springer-Verlag, 1983. Lecture Notes in Computer Science Volume 154.

[CFL83b] Ashok K. Chandra, Steven Fortune, and Richard J. Lipton. Unbounded fan-in circuits and associative functions. In *15th Annual ACM Symposium on Theory of Computing*, pages 52–60, Boston, Massachusetts, USA, April 25–27, 1983. ACM Press.

[DN03]   Ivan Damgård and Jesper B. Nielsen. Universally composable efficient mul-
           tiparty computation from threshold homomorphic encryption. In D. Boneh,
           editor, *Advances in Cryptology - Crypto 2003*, Berlin, 2003. Springer-Verlag.
           Lecture Notes in Computer Science.

[FKN94]   Uri Feige, Joe Kilian, and Moni Naor. A minimal model for secure compu-
           tation. In *Proc. ACM STOC*, pages 554–563, 1994.

[IK97]   Yuval Ishai and Eyal Kushilevitz. Private simultaneous messages protocols
           with applications. In *Proc. 5th Israel Symposium on Theoretical Comp. Sc.
           ISTCS*, pages 174–183, 1997.

[IK00]   Yuval Ishai and Eyal Kushilevitz. Randomizing polynomials: A new repre-
           sentation with applications to round-efficient secure computation. In *41st
           Annual Symposium on Foundations of Computer Science*, pages 294–304,
           Las Vegas, Nevada, USA, November 12–14, 2000. IEEE Computer Society
           Press.

[IK02]   Yuval Ishai and Eyal Kushilevitz. Perfect constant-round secure computa-
           tion via perfect randomizing polynomials. In *Proceedings of ICALP 2002*,
           pages 244–256, Berlin, 2002. Springer-Verlag. Lecture Notes in Computer
           Science Volume 2380.

[KLM05]   Eike Kiltz, Gregor Leander, and John Malone-Lee. Secure computation of
           the mean and related statistics. In *TCC 2005: 2nd Theory of Cryptography
           Conference*, volume 3378 of *Lecture Notes in Computer Science*, pages 283–
           302, Cambridge, MA, USA, February 10–12, 2005. Springer-Verlag, Berlin,
           Germany.

[NWKo]   Martin Nowak, Georg Woltman, Scott Kurowski, and others. Mersenne.org
           project discovers new largest known prime number $2^{25,964,951} - 1$. Press
           release.

Efficient Multi-party Computation
with Dispute Control*

Zuzana Beerliová-Trubíniová and Martin Hirt

ETH Zurich, Department of Computer Science, CH-8092 Zurich
{bzuzana, hirt}@inf.ethz.ch

Abstract. Secure multi-party computation (MPC) allows a set of n players to securely compute an agreed function of their inputs, even when up to t players are under the control of an (active or passive) adversary. In the information-theoretic model MPC is possible if and only if $t < n/2$ (where active security with $t \geq n/3$ requires a trusted key setup).

Known passive MPC protocols require a communication of $\mathcal{O}(n^2)$ field elements per multiplication. Recently, the same communication complexity was achieved for active security with $t < n/3$. It remained an open question whether $\mathcal{O}(n^2)$ complexity is achievable for $n/3 \leq t < n/2$.

We answer this question in the affirmative by presenting an active MPC protocol that provides optimal ($t < n/2$) security and communicates only $\mathcal{O}(n^2)$ field elements per multiplication. Additionally the protocol broadcasts $\mathcal{O}(n^3)$ field elements *overall*, for the whole computation.

The communication complexity of the new protocol is to be compared with the most efficient previously known protocol for the same model, which requires *broadcasting* $\Omega(n^5)$ field elements per multiplication. This substantial reduction in communication is mainly achieved by applying a new technique called *dispute control*: During the course of the protocol, the players keep track of disputes that arise among them, and the ongoing computation is adjusted such that known disputes cannot arise again. Dispute control is inspired by the player-elimination framework. However, player elimination is not suited for models with $t \geq n/3$.

1 Introduction

1.1 Background

Secure multi-party computation (MPC) enables a set of n players to securely evaluate an agreed function of their inputs even when t of the players are corrupted by a central adversary. A *passive adversary* can read the internal state of the corrupted players, trying to obtain information about the honest players' inputs. An *active adversary* can additionally make the corrupted players deviate from the protocol, trying to falsify the outcome of the computation.

---

* This work was partially supported by the Zurich Information Security Center. It represents the views of the authors.

The MPC problem dates back to Yao [Yao82]. The first generic solutions presented in [GMW87, CDG87, GHY87] were based on cryptographic intractability assumptions. Later, MPC protocols with information-theoretic security were developed [BGW88, CCD88, RB89, Bea91b], which is the focus of this work.

Information-theoretic security against a passive or active adversary is possible if and only if $t < n/2$. The protocols with active security require broadcast channels, which can be simulated from scratch for $t < n/3$ [PSL80, BGP92, CW92], and can be simulated when a trusted key setup is available for $t < n$ [DS82, PW92].[1]

The communication complexity of MPC is measured in bits sent by honest parties. The function to be computed is represented as an arithmetic circuit over a finite field (with additions and multiplications). The classical MPC protocol with passive security (for $t < n/2$) requires a communication of $\mathcal{O}(n^2)$ field elements per multiplication [BGW88]. Recently, the same communication complexity was achieved for active security, including the costs for simulating the broadcast channels [HM01]; however, this protocol is only suitable for $t < n/3$. The most efficient actively secure MPC protocol for $t < n/2$ requires *broadcasting* $\Omega(n^5)$ field elements per multiplication [CDD+99], and each of these broadcasts must be simulated with an expensive broadcast protocol [PW92].

1.2   Contributions

In this work, we show that information-theoretic MPC with adaptive active security for $t < n/2$ is achievable with sending $\mathcal{O}(n^2)$ field elements per multiplication, and broadcasting $\mathcal{O}(n^3)$ field elements *overall*, for the whole computation. This improves on previous protocols which require *broadcasting* $\Omega(n^5)$ field elements per multiplication [CDD+99].

This result is of particular theoretical interest, as it shows that for all t for which information-theoretic MPC is possible, i.e., $t < n/2$, (adaptive) active security is achievable at essentially the same costs as passive security. This extends the result of [HM01], where only the range $t < n/3$ could be solved. The achieved communication complexity might well be optimal, as even in the passive model it seems unavoidable that for each multiplication gate, every player sends a value to every other player.

The following table summarizes the communication complexities of known and new MPC protocols, where κ denotes the security parameter (i.e., the bit-length of a field element), $\mathcal{BC}(\cdot)$ the number of broadcasted bits, and c_M the number of multiplication gates in the circuit. For simplicity, we assume that the function takes n inputs and gives n outputs.

Thresh.	Adv.	Communication	References
$t < n/2$	passive	$\mathcal{O}\left(c_M n^2 + n^2\right)\kappa$	[BGW88]
$t < n/3$	active	$\mathcal{O}\left(c_M n^2 + n^4\right)\kappa + \mathcal{O}\left(n^3\right)\mathcal{BC}(\kappa)$	[HM01]
$t < n/2$	active	$\mathcal{O}\left(c_M n^5 + n^4\right)\kappa + \mathcal{O}\left(c_M n^5 + n^4\right)\mathcal{BC}(\kappa)$	[CDD+99]
$t < n/2$	active	$\mathcal{O}\left(c_M n^2 + n^5\kappa\right)\kappa + \mathcal{O}\left(n^3\right)\mathcal{BC}(\kappa)$	this paper

---

[1] Even cryptographically secure broadcast and MPC require a trusted key setup for $t \geq n/3$.

Technically, the new protocol improves the approach of [CDD+99], which requires $\Omega(n^5)$ broadcasts per multiplication. We introduce a new concept, so-called *dispute control*, that allows to substantially reduce the communication complexity. The goal of dispute control is to reduce the frequency of faults that the adversary can provoke by identifying a pair of disputing players (at least one of them corrupted) whenever a fault is observed and preventing this pair from getting into dispute ever again. Hence, the number of faults that can occur during the whole protocol is limited to $t(t + 1)$. This technique is inspired by the player-elimination framework [HMP00], and shares many advantages with it. However, player elimination is not to be suited for models with $t \geq n/3$. Furthermore, player elimination is not applicable in the input stage, which results in our protocol being more efficient than the protocol in [HM01] when the number of inputs is large ($n^2\kappa$ bits per input in our protocol versus $n^4\kappa$ bits in [HM01]).

2   Protocol Overview

2.1   Model

We consider a set $\mathcal{P}$ of n players, $\mathcal{P} = \{P_1, \ldots, P_n\}$, which are connected with a complete network of secure synchronous channels. Furthermore, we assume the availability of broadcast channels. These can be simulated when a trusted setup is available [PW92]. The adversary corrupts up to t players for any fixed t with $t < n/2$, and makes them deviate from the protocol in any desired manner. The adversary is computationally unbounded, active, adaptive and rushing. The security of our protocols is information-theoretic with a negligible error probability of $2^{-\mathcal{O}(\kappa)}$ for some security parameter κ.

For the ease of presentation, we always assume that the messages sent through the channels are from the right domain — if a player receives a message which is not in the right domain (e.g., no message at all), he replaces it with an arbitrary message from the specified domain.

The function to be computed is specified as an arithmetic circuit over a finite field $\mathcal{F} = GF(2^\kappa)$, with input, addition, multiplication, random, and output gates. We denote the number of gates of each type by c_I, c_A, c_M, c_R, and c_O.

2.2   Dispute Control

In the active model, the adversary can provoke inconsistencies among the honest players, who therefore regularly have to check their views and, in case of inconsistencies, invoke some fault-recovery procedure. These checks tend to be very expensive (they require invocations to a Byzantine agreement primitive), and must be performed even when no player deviates from the protocol.

The goal of *dispute control* is to reduce the frequency of faults by publicly identifying (localizing) a pair of disputing players (at least one of them corrupted) whenever a fault is observed and preventing this pair from getting into dispute ever again. Hence, the number of faults that can occur during the whole protocol is limited to $t(t + 1)$.

The localized disputes are filed in a publicly known *dispute set* $\Delta \subseteq \mathcal{P} \times \mathcal{P}$, a set of unordered pairs of players that are in dispute with each other. A pair $\{P_i, P_j\} \in \Delta$ means that there is a dispute between P_i and P_j, hence either P_i or P_j (or both) are corrupted. Note that from the point of view of P_i, the players $\{P_j \mid \{P_i, P_j\} \in \Delta\}$ are corrupted, and P_i doesn't care for them; in particular, he won't send or receive any private messages from them. As no honest player can be in dispute with more than t players, we automatically include the pairs $\{P_i, P_j\}$ for every $P_j \in \mathcal{P}$ once P_i is involved in more than t disputes. Furthermore, we define the set $\mathcal{X}$ to be the set of players who are undoubtedly detected to be corrupted, i.e., those players who are in dispute with more than t other players.

Once dispute control is in place, we can take advantage of the fact that the number of faults during the protocol is limited and reduce the number of expensive consistency checks: We divide the protocol into n^2 *segments*, run each segment without any consistency checks and only at the end of the segment check all operations of the segment in a single verification step. If the verification fails, a new dispute is localized, and the segment is repeated. At most $t(t+1)$ segments can fail, and the total number of segment evaluations (including repetitions) is at most $n^2 + t(t+1)$, hence the overhead for repeating failed segments is only a factor of 2. Formally the evaluation of each segment proceeds as follows:

1. **Private (dispute-aware) computation.** The effective protocol is computed very efficiently but non-robustly. This computation is adjusted to prevent faults due to disputes that are already registered in the dispute set Δ. In particular, players in dispute do not communicate with each other privately.
2. **Fault detection.** The players jointly find out whether or not a fault has occurred. This step typically requires each player to broadcast one bit indicating whether or not he observed an inconsistency within the current segment. If no fault is reported, then the computation of the segment is completed, and the next segment is evaluated. If at least one fault is reported, we say that the segment has failed, and the following step is performed.
3. **Fault localization and dispute control.** The players publicly identify a pair $\{P_i, P_j\}$ of players, where at least one of them is corrupted and has deviated from the protocol, and who are not yet registered in Δ. Then we set $\Delta \leftarrow \Delta \cup \{P_i, P_j\}$ and restart the current segment.

2.3   Three-Level Secret-Sharing

We use three different levels of secret-shadings, all based on Shamir's sharing [Sha79], ameliorated with dispute control. The weakest level, called *1D-sharing*, is a polynomial sharing scheme, where the shares of players who are in dispute with the dealer (implicitly) receive a fixed-0 share, called *Kudzu-share*. In order to *1D-share* a value s, the dealer P_D selects a random degree-t polynomial $f(x)$ with $f(0) = s$ and $f(i) = 0$ for every $\{P_D, P_i\} \in \Delta$, and sends the shares $s_i = f(i)$ to every $P_i \in \mathcal{P}$ (the Kudzu-shares are not really sent; instead, the receiver sets his share to 0). A protocol VSS1D for verifiably 1D-share a bunch of values will be given in Section 3.2. Note that 1D-sharings are not robust; reconstruction requires that all players (except those with Kudzu-shares) cooperate.

However, they are detectable in the sense that it can be decided whether or not the reconstruction was successful.

The middle level of secret sharing, called *2D-sharing*, is a two-level polynomial sharings scheme: The share s_i of each player $P_i \in \mathcal{P}$ is 1D-shared among the players (for dealer P_i). More precisely, a value s is 2D-shared when there exists degree-t polynomials $f, f_1, \ldots, f_n$ with $f(0) = s$ and, for $i = 1, \ldots, n$, $f_i(0) = f(i)$ and $\forall P_j \in \mathcal{P} : \{P_i, P_j\} \in \Delta \to f_i(j) = 0$. Every player $P_i \in \mathcal{P}$ holds a share $s_i = f(i)$ of s, the polynomial $f_i(x)$ for sharing s_i, and a share-share $s_{ji} = f_j(i)$ of the share s_j of every player $P_j \in \mathcal{P}$. We say that P_i *owns* the 1D-sharing of s_i, which means in particular that players who are in dispute with P_i hold 0 as share-share of s_i. We will never have a dealer 2D-share a value; instead, we will upgrade 1D-sharings (or rather sums of 1D-sharings) to 2D-sharings, using protocol Upgrade1Dto2D. Note that also 2D-sharings are not robust.

The strongest level of secret sharing, called *2D*-sharing*, is a 2D-sharing, where in addition, the share-shares are secured with information checking (see Section 3.5). More precisely, for each share-share s_{ij} (which is not a Kudzu-share, i.e., $\{P_i, P_j\} \notin \Delta$), the owner P_i of the sharing has provided authentication tags for every verifier $P_V \in \mathcal{P}$ who is neither in dispute with the owner P_i nor the recipient P_j, i.e., $\{P_V, P_i\} \notin \Delta$ and $\{P_V, P_j\} \notin \Delta$. These authentication tags allow P_V in the reconstruction to verify the correctness of the received share-shares; hence, 2D*-sharings are robust. Actually, P_i does not distribute authentication tags for every single share-share s_{ij}, but rather for huge collections of many share-shares $s_{ij}^{(1)}, \ldots, s_{ij}^{(\ell)}$, and P_V can only verify the correctness of all share-shares at once. Also 2D*-sharings are never distributed by a dealer; instead, we will upgrade collections of 2D-sharings to 2D*-sharings, using protocol Upgrade2Dto2D*.

2.4 Main Protocol

The main protocol proceeds in three phases (each making use of segmentation and dispute control):

Preparation phase: The preparation phase uses the circuit-randomization technique of Beaver [Bea91a]: A number of so-called multiplication triples (a, b, c) with $c = ab$ are generated and shared among the players. These triples will then be used in the computation phase for efficiently multiplying shared values. Furthermore, a number of random values are generated and shared, which will be used as outputs of random gates.

Input phase: In the input phase, every player with input shares his input among the players.

Computation phase: In the computation phase, the circuit is evaluated gate by gate (level by level), with help of the prepared multiplication triples and the random values. Given the sharings of the multiplication triples, the random values, and the inputs, the computation phase is fully deterministic. Indeed, the computation phase can be seen as a sequence of reconstructions of known linear combinations of shared values.

Each phase uses dispute control. We initialize the dispute set $\Delta = \{\}$ and enter the first segment of the preparation phase. Then we evaluate segment by segment, and with each segment that fails and is to be repeated, the dispute set Δ grows. Once all segments of the preparation phase have succeeded, the players move on to the first segment of the input phase. Also in this phase, segments can fail and have to be repeated. This allows corrupted players to change their inputs. However, as the adversary obtains no information about whatsoever in the input phase, this does not affect the independence of the inputs. Once all input segments have succeeded, the players move on to the first segment of the computation phase. In this phase, the players (and hence also the adversary) do obtain information about their outputs; however, the computation stage is fully deterministic. Even when a segment fails (and is repeated) *after* the adversary has learned some output, he cannot influence the outputs of the honest players anymore.

In the preparation phase and in the input phase, the private computation is highly parallelized. All proposed sub-protocols process many inputs at once, producing many outputs. This helps reducing the costs for the fault detection and localization, as for all parallel instances, only one single fault-handling procedure is executed. Often, instead of verifying single instances of some test data, we will verify a random linear combination of many instances. Note that the protocols themselves do not use broadcast, but fault handling does.

3   Sub-protocols

All sub-protocols have a private (dispute aware) computation, a fault detection and a fault localization. They can succeed or fail and the players always agree (using broadcast) on what is the case. In case of a failure the public output of the sub-protocol is a (new) pair of players $E = \{P_i, P_j\} \notin \Delta$ such as either P_i or P_j (or both) are corrupted. If some invoked sub-protocol fails with $E = \{P_i, P_j\}$ then the invoking sub-protocol fails with $E = \{P_i, P_j\}$ and is aborted (this abort will be handled in the main protocol).

3.1   Dispute-Control Broadcast

The protocol DC-Broadcast allows every sender $P_S \in \mathcal{P} \setminus \mathcal{X}$ to distribute a vector of ℓ values $s^{(1,S)}, \ldots, s^{(\ell,S)}$ among the players in $\mathcal{P} \setminus \mathcal{X}$, such that it is guaranteed that all honest recipients receive the same vectors (or the protocol fails).

This protocol is rather simple: Every sender directly transmits his vector to the players he is not in dispute with, and via another player to those players he is in dispute with. Then the players pairwisely compare their vectors by using universal hash functions [CW79]. As universal hash with key $k \in \mathcal{F}$, we use the function $U_k : \mathcal{F}^\ell \to \mathcal{F}, (s^{(1)}, \ldots, s^{(\ell)}) \mapsto s^{(1)} + s^{(2)}k + \ldots + s^{(\ell)}k^{\ell-1}$. The probability that two different vectors map to the same hash value for a uniformly chosen key is at most $\ell/|\mathcal{F}|$, which is negligible in our setting with $\mathcal{F} = GF(2^\kappa)$.

Protocol DC-Broadcast
1. PRIVATE COMPUTATION: The following steps are executed in parallel for every sender $P_S \in \mathcal{P} \setminus \mathcal{X}$:
   1.1 P_S sends $s^{(1,S)}, \ldots, s^{(\ell,S)}$ to every P_i with $\{P_S, P_i\} \notin \Delta$.
   1.2 For every P_i with $\{P_S, P_i\} \in \Delta$ (but $P_i \notin \mathcal{X}$), the smallest player $P_{i'}$ with $\{P_S, P_{i'}\} \notin \Delta$ and $\{P_{i'}, P_i\} \notin \Delta$ forwards $s^{(1,S)}, \ldots, s^{(\ell,S)}$ to P_i.[2] We call $P_{i'}$ the *proxy* of P_i.
2. FAULT DETECTION: The following steps are executed in parallel for every verifier $P_V \in \mathcal{P} \setminus \mathcal{X}$:
   2.1 P_V selects a key $k_V \in_R \mathcal{F}$ for a universal hash function U_k and sends it to every P_i with $\{P_V, P_i\} \notin \Delta$.
   2.2 Every P_i with $\{P_V, P_i\} \notin \Delta$ sends the values $h_{S,i} = U_{k_V}(s^{(1,S)}, \ldots, s^{(\ell,S)})$ for every P_S to P_V.
   2.3 P_V broadcasts a bit "accept" or "reject", indicating whether for every $P_S \in \mathcal{P} \setminus \mathcal{X}$, the hash values $h_{S,i}$ of each P_i with $\{P_V, P_i\} \notin \Delta$ are equal.

   If every verifier $P_V \in \mathcal{P} \setminus \mathcal{X}$ broadcasts "accept" in Step 2.3, then the protocol succeeds and terminates.
3. FAULT LOCALIZATION: The following steps are executed for the smallest $P_V \in \mathcal{P} \setminus \mathcal{X}$ reporting a fault.
   3.1 P_V selects S, i, j such that $P_S \notin \mathcal{X}$, $\{P_V, P_i\} \notin \Delta$, and $\{P_V, P_j\} \notin \Delta$, and $h_{S,i} \neq h_{S,j}$, and broadcasts $S, i, j, h_{S,i}, h_{S,j}$, and $k = k_V$.
   3.2 We denote the proxies of P_i and P_j by $P_{i'}$ and $P_{j'}$, respectively (if no proxy exists, we set $i' = i$, respectively $j' = j$). The players $P_S, P_i, P_j, P_{i'}, P_{j'}$ all compute and broadcast a hash value with key k of their vector $s^{(1,S)}, \ldots, s^{(\ell,S)}$, denoted as $h_S, h_i, h_j, h_{i'}, h_{j'}$, respectively. The protocol fails with E being the first pair (P_V, P_i), $(P_i, P_{i'})$, $(P_{i'}, P_S)$, $(P_S, P_{j'})$, $(P_{j'}, P_j)$, or (P_j, P_V), where $h_{S,i} \neq h_i$, $h_i \neq h_{i'}$, $h_{i'} \neq h_S$, $h_S \neq h_{j'}$, $h_{j'} \neq h_j$, or $h_j \neq h_{S,j}$, respectively.

Lemma 1. *If* DC-Broadcast *succeeds, then with overwhelming probability, for each sender* $P_S \in \mathcal{P} \setminus \mathcal{X}$, *all honest players in* $\mathcal{P}$ *hold the same vector* $s^{(1,S)}, \ldots, s^{(\ell,S)}$, *which is the vector of* P_S *if honest. If the protocol fails, a new dispute pair* E *is localized. The protocol communicates* $\mathcal{O}(\ell n^2 + n^3)$ *and broadcasts* $\mathcal{O}(n)$ *field elements.*

Proof. In order to prove that all honest players output the same vector $s^{(1,S)}, \ldots, s^{(\ell,S)}$ when the protocol succeeds, consider two honest players P_i and P_j. As both P_i and P_j are honest, $\{P_i, P_j\} \notin \Delta$ holds, and P_i and P_j have mutually exchanged universal hash values in Step 2. Hence, with overwhelming probability, a difference in the vectors would have been detected and the protocol would have failed. It follows immediately from the protocol that when P_S is

---

[2] The existence of such a player $P_{i'}$ for $P_{S'} \notin \mathcal{X}$ and $P_i \notin \mathcal{X}$ follows by a counting argument.

honest and the protocol succeeds, then all honest players receive the vector directly from P_S. When the protocol fails with dispute pair E, then one can verify by inspection that the two players in E disagree on a value they have privately exchanged, hence either of the players must be faulty. And as players in dispute do not communicate with each other, the localized dispute pair is new.   □

3.2   Verifiable Secret-Sharing

The protocol VSS1D allows every dealer $P_D \in \mathcal{P} \setminus \mathcal{X}$ to verifiably 1D-share ℓ values $s^{(1,D)}, \ldots, s^{(\ell,D)}$ resulting in each player $P_i \in \mathcal{P} \setminus \mathcal{X}$ holding the shares $s_i^{(1,D)}, \ldots, s_i^{(\ell,D)}$ for each dealer P_D. The correctness of these sharings is verified by letting every player take on the role of a verifier P_V and inspect a random linear combination of the sharings of each dealer P_D. For privacy reasons, each such random linear combination is blinded with a random 1D-sharing, i.e., every dealer P_D 1D-shares additional n blinding values $s^{(\ell+1,D)}, \ldots, s^{(\ell+n,D)}$.

Protocol VSS1D

1. PRIVATE COMPUTATION: Every dealer $P_D \in \mathcal{P} \setminus \mathcal{X}$ selects n random blindings $s^{(\ell+1,D)}, \ldots, s^{(\ell+n,D)}$. Then, P_D 1D-shares $s^{(1,D)}, \ldots, s^{(\ell+n,D)}$, i.e., for every $m = 1, \ldots, \ell + n$, P_D picks a random polynomial $f^{(m,D)}(x)$ with $f^{(m,D)}(0) = s^{(m,D)}$ and $f^{(m,D)}(i) = 0$ for every i with $\{P_D, P_i\} \in \Delta$ (the Kudzu-shares), and sends the share $f^{(m,D)}(i)$ to every player P_i with $\{P_D, P_i\} \notin \Delta$; every player P_i with $\{P_D, P_i\} \in \Delta$ sets his share $s_i^{(m,D)} = 0$.

2. FAULT DETECTION: Every verifier $P_V \in \mathcal{P} \setminus \mathcal{X}$ selects a random challenge vector $(r^{(1,V)}, \ldots, r^{(\ell,V)})$. Then, DC-Broadcast is invoked to let every verifier $P_V \in \mathcal{P} \setminus \mathcal{X}$ distribute his vector among the players $P_i \in \mathcal{P} \setminus \mathcal{X}$. Then the following steps are executed for every verifier $P_V \in \mathcal{P} \setminus \mathcal{X}$ (we suppress the index V and denote the challenge vector $(r^{(1)}, \ldots, r^{(\ell)})$):

   2.1 For every dealer P_D, the random linear combination $f^{(*,D)}(x)$ of his 1D-sharings is defined as $f^{(*,D)}(x) = \sum_{m=1}^{\ell} r^{(m)} f^{(m,D)}(x) + f^{(\ell+V,D)}(x)$. Accordingly, for every dealer P_D, every player P_i with $\{P_i, P_D\} \notin \Delta$ and $\{P_i, P_V\} \notin \Delta$ sends to P_V his share $s_i^{(*,D)}$ on $f^{(*,D)}(x)$, i.e., $s_i^{(*,D)} = \sum_{m=1}^{\ell} r^{(m)} s_i^{(m,D)} + s_i^{(\ell+V,D)}$.

   2.2 For each dealer $P_D \in \mathcal{P} \setminus \mathcal{X}$, the verifier P_V checks whether the received shares $s_i^{(*,D)}$ define a correct 1D-sharing for P_D, i.e., whether there exists a degree-t polynomial $\widetilde{f}^{(*,D)}(x)$ with $\widetilde{f}^{(*,D)}(i) = s_i^{(*,D)}$ for every i with $\{P_V, P_i\} \notin \Delta$ and $\{P_D, P_i\} \notin \Delta$, and $\widetilde{f}^{(*,D)}(i) = 0$ for every i with $\{P_D, P_i\} \in \Delta$ (Kudzu).[3] P_V broadcasts a bit "accept" or "reject", indicating whether or not the the above checks succeed for all dealers.

   If all verifiers $P_V \in \mathcal{P} \setminus \mathcal{X}$ broadcast "accept" the protocol succeeded and terminates.

3. FAULT LOCALIZATION: The following steps are executed for the smallest P_V reporting a fault in Step 2.2.

---

[3] Note that any linear combination of Kudzu-shares is Kudzu.

3.1 P_V broadcasts the index D of P_D whose polynomial $\widetilde{f}^{(*,D)}(x)$ does not define a correct 1D-sharing.

3.2 Every player P_i with $\{P_i, P_D\} \notin \Delta$ and $\{P_i, P_V\} \notin \Delta$ broadcasts his share $s_i^{(*,D)}$.

3.3 If the broadcasted shares define a 1D-sharing for dealer P_D, then P_V broadcasts the index i of a player P_i with $\{P_i, P_V\} \notin \Delta$ and $\{P_i, P_D\} \notin \Delta$ who has broadcasted a different share $s_i^{(*,D)}$ in Step 3.2 than he has privately sent to P_V in Step 2.1, and the protocol fails with $E = \{P_V, P_i\}$. Otherwise, when the broadcasted shares do not define a correct 1D-sharing for dealer P_D, then the dealer broadcasts the index i of a player P_i with $\{P_i, P_D\} \notin \Delta$ who has broadcasted a wrong share $s_i^{(*,D)}$ and the protocol fails with $E = \{P_D, P_i\}$.

Lemma 2. *If VSS1D succeeds, then with overwhelming probability, the values $s^{(1,D)}, \ldots, s^{(\ell,D)}$ of each dealer $P_D \in \mathcal{P} \backslash \mathcal{X}$ are correctly 1D-shared. If the protocol fails, then the localized pair $E = \{P_i, P_j\}$ is new (i.e., $E \notin \Delta$) and either P_i or P_j (or both) are corrupted. The privacy of the inputs of the honest players is guaranteed through the whole protocol (even if the protocol fails). The protocol communicates $\mathcal{O}(\ell n^2 + n^3)$ and broadcasts $\mathcal{O}(n)$ field elements.*

Proof. In order to prove the correctness, first consider a dealer P_D, an honest verifier P_V, the (by P_D supposedly correct 1D-shared) values $s^{(1,D)}, \ldots, s^{(\ell,D)}$ and the blinding value $s^{(\ell+V,D)}$. Assume that the sharing of one of the values is not a correct 1D-sharing, i.e., the shares of the honest players (including the Kudzu shares) lie on a polynomial of degree higher than t. Then there are at most $2^{\kappa(\ell-1)}$ (out of $2^{\kappa\ell}$) challenge vectors $(r^{(1)}, \ldots, r^{(\ell)}) \in \mathcal{F}^\ell$ such that the sharing of $s^{(*,D)} = \sum_{m=1}^\ell r^{(m)} s^{(m,D)} + s^{(\ell+V,D)}$ is a correct 1D-sharing, i.e. the polynomial defined by the shares of the honest players is of degree t. As the verifier P_V chooses his challenge vector uniformly at random and gets the correctly linearly combined shares from all honest players (an honest verifier is in dispute with no honest player), the probability of him not detecting the fault is at most $2^{\kappa(\ell-1)}/2^{\kappa\ell} = 1/2^\kappa$. Thus the probability that the protocol succeeds in case of at least one faulty sharing (from any dealer) is negligible.

The privacy of the inputs of the honest players follows from the fact that up to t shares give no information about the secret and from the fact that the reconstructed linear combinations are blinded with a random value chosen by the dealer himself (for every verifier a different one) and are so (for every honest dealer) statistically independent from the dealers secret.

If the protocol fails, then the localized dispute pair consists of two players who have publicly disagreed on a value they have privately exchanged in some previous step (or a value computed from such values), therefore it is obvious, that at least one of them is corrupted. As only players who are not in dispute with each other communicate privately, the localized dispute is a new one.   □

We present a protocol for reconstructing sums of correct 1D-sharings. Consider a set $\mathcal{P}_D \subseteq \mathcal{P} \backslash \mathcal{X}$ of dealers and a set $\mathcal{P}_R \subseteq \mathcal{P} \backslash \mathcal{X}$ of recipients and the actual dispute set Δ. Every dealer $P_D \in \mathcal{P}_D$ has verifiably 1D-shared (with the actual Δ) ℓ summands $s^{(1,D)}, \ldots, s^{(\ell,D)}$ with the polynomials $f^{(1,D)}(x), \ldots, f^{(\ell,D)}(x)$. We denote the share of $f^{(m,D)}(x)$ for player $P_i \in \mathcal{P}$ by $s_i^{(m,D)}$. Note that $s_i^{(m,D)} = 0$ when $\{P_D, P_i\} \in \Delta$ (Kudzu). The values $s^{(1)}, \ldots, s^{(\ell)}$ to be reconstructed are defined as the sums of the above summands, i.e., $s^{(m)} = \sum_{P_D \in \mathcal{P}_D} s^{(m,D)}$. Each of these values is implicitly shared (as Shamir-sharing, not as 1D-sharing) with the polynomial $f^{(m)}(x) = \sum_{P_D \in \mathcal{P}_D} f^{(m,D)}(x)$; we denote the (implicitly defined) share of each player $P_i \in \mathcal{P}$ by $s_i^{(m)} = f^{(m)}(i)$.

Protocol Reconstruct1D

1. PRIVATE COMPUTATION: For every $m = 1, \ldots, \ell$, every player $P_i \in \mathcal{P}$ computes his sum share $s_i^{(m)} = \sum_{P_D \in \mathcal{P}_D} s_i^{(m,D)}$, and sends it to every $P_R \in \mathcal{P}_R$ with $\{P_i, P_R\} \notin \Delta$. Every $P_R \in \mathcal{P}_R$ checks for each $m = 1, \ldots, \ell$ whether the received shares lie on a polynomial $\widetilde{f}^{(m)}(x)$ of degree t. If so, it follows that $\widetilde{f}^{(m)}(x) = f^{(m)}(x)$, and P_R reconstructs $s^{(m)} = \widetilde{f}^{(m)}(0)$.

2. FAULT DETECTION: Every $P_R \in \mathcal{P}_R$ broadcasts "accept" or "reject", indicating whether he could reconstruct all values $s^{(m)}$ for $m = 1, \ldots, \ell$ in Step 1. If all recipients broadcast "accept", then the protocol succeeds and terminates.

3. FAULT LOCALIZATION: The following steps are executed for the smallest complaining recipient $P_R \in \mathcal{P}_R$.

   3.1 P_R broadcasts the index m of the polynomial $\widetilde{f}^{(m)}(x)$ he could not reconstruct.

   3.2 Every player P_i with $\{P_i, P_R\} \notin \Delta$ sends to P_R his summand shares $s_i^{(m,D)}$ for every dealer $P_D \in \mathcal{P}_D$ with $\{P_i, P_D\} \notin \Delta$.

   3.3 P_R verifies for every P_i with $\{P_i, P_R\} \notin \Delta$ that the provided summand shares add up to the previously provided sum share, i.e., $\sum_{P_D : \{P_i, P_D\} \notin \Delta} s_i^{(m,D)} = s_i^{(m)}$.[4] In case of a fault, P_R broadcasts the index i of the bad player P_i, and the protocol fails with $E = \{P_i, P_R\}$.

   3.4 P_R broadcasts the index D of a dealer $P_D \in \mathcal{P}_D$ such that the received shares $s_i^{(m,D)}$ do not define a correct 1D-sharing for dealer P_D, i.e., there is no degree-t polynomial $f(x)$ with $f(i) = s_i^{(m,D)}$ for every i with $\{P_i, P_R\} \notin \Delta$ and $\{P_i, P_D\} \notin \Delta$, and $f(i) = 0$ (Kudzu) for every i with $\{P_i, P_R\} \notin \Delta$ and $\{P_i, P_D\} \in \Delta$.

   3.5 Every player P_i with $\{P_i, P_R\} \notin \Delta$ and $\{P_i, P_D\} \notin \Delta$ broadcasts his summand share $s_i^{(m,D)}$.

   3.6 If the broadcasted summand shares define a correct 1D-sharing for dealer P_D, then P_R broadcasts the index i of a player P_i who has broadcasted a different value $s_i^{(m,D)}$ in Step 3.5 than he has privately sent to P_R in Step 3.2, and the protocol fails with $E = \{P_i, P_R\}$. Otherwise, when the broadcasted summand shares do not define a

---

[4] Note that the Kudzu-shares $s_i^{(*,D)}$ with $\{P_i, P_D\} \in \Delta$ are 0 and do not contribute to the sum.

correct 1D-sharing for P_D, then P_D broadcasts the index i of a player P_i who has broadcasted a wrong share $s_i^{(m,D)}$, and the protocol fails with $E = \{P_i, P_D\}$.

Lemma 3. *If the values $s^{(1,D)}, \ldots, s^{(\ell,D)}$ of each $P_D \in \mathcal{P}_D$ are correctly 1D-shared (for the actual Δ), then the following holds: If Reconstruct1D succeeds, then the privacy is guaranteed and every value reconstructed towards an honest recipient lies on the degree t polynomial defined by the (at least $t + 1$) shares of the honest players. If the protocol fails then the localized pair $E = \{P_i, P_j\}$ is new and contains at least one corrupted player. The protocol communicates $\mathcal{O}(\ell n^2)$ and broadcasts $\mathcal{O}(n)$ field elements.*

Proof. As an honest verifier is not in dispute with any other honest player, he will receive at least $t + 1$ shares of the honest players, which uniquely define a degree t polynomial. If the shares received from the corrupted players lie on this polynomial, he will reconstruct the right secret, otherwise the interpolated polynomial will be of degree higher then t and the protocol will fail. The rest follows (along the lines of proof of Lemma 2) from inspection of the protocol. □

3.3 Generating Random Challenges

The following protocol allows the players to generate a publicly known (i.e., to the players in $\mathcal{P} \setminus \mathcal{X}$) challenge vector $s^{(1)}, \ldots, s^{(\ell)}$, or the protocol fails, if one of the sub-protocols fails, and outputs a new dispute pair $E = \{P_i, P_j\}$:

Protocol GenerateChallenges
1. Every player $P_k \in \mathcal{P} \setminus \mathcal{X}$ selects a random summand vector $s^{(1,k)}, \ldots, s^{(\ell,k)}$.
2. Invoke VSS1D to let every P_k verifiably 1D-share his summand vector.
3. Invoke the protocol Reconstruct1D (with $\mathcal{P}_D = \mathcal{P}_R = \mathcal{P} \setminus \mathcal{X}$) to reconstruct the sum sharings $\sum_{P_k \in \mathcal{P}_D} s^{(1,k)}, \ldots, \sum_{P_k \in \mathcal{P}_D} s^{(\ell,k)}$ towards every $P_j \in \mathcal{P}_R$.

Lemma 4. *If GenerateChallenges succeeds, then with overwhelming probability, the generated values are uniformly distributed. If the protocol fails, then the localized dispute pair $E = \{P_i, P_j\}$ is new and contains at least one corrupted player. The protocol communicates $\mathcal{O}(\ell n^2 + n^3)$ and broadcasts $\mathcal{O}(n)$ field elements.*

3.4 Upgrading 1D-Sharings to 2D-Sharings

We present a protocol for upgrading sums of 1D-sharings to 2D-sharings. The given 1D-sharings must be for the actual Δ; the correctness of these sharings is implicitly verified in the upgrade protocol and must not be a priori guaranteed. The protocol outputs correct 2D-sharings or it fails with a new dispute pair E.

Formally, we consider a set $\mathcal{P}_D \subseteq \mathcal{P} \setminus \mathcal{X}$ of dealers, where each dealer $P_D \in \mathcal{P}_D$ has (for the actual Δ) 1D-shared ℓ summands $s^{(1,D)}, \ldots, s^{(\ell,D)}$ with the polynomials $f^{(1,D)}(x), \ldots, f^{(\ell,D)}(x)$. We denote the share of $f^{(m,D)}(x)$ for player

$P_i \in \mathcal{P}$ by $s_i^{(m,D)}$. Note that $s_i^{(m,D)} = 0$ when $\{P_D, P_i\} \in \Delta$ (Kudzu). The values $s^{(1)}, \ldots, s^{(\ell)}$ to be 2D-shared are defined as the sums of the above summands, i.e., $s^{(m)} = \sum_{P_D \in \mathcal{P}_D} s^{(m,D)}$. Each of these values is implicitly shared (as Shamir-sharing, not as 1D-sharing) with the polynomial $f^{(m)}(x) = \sum_{P_D \in \mathcal{P}_D} f^{(m,D)}(x)$; we denote the (implicitly defined) share of each player $P_i \in \mathcal{P}$ by $s_i^{(m)} = f^{(m)}(i)$.

Protocol Upgrade1Dto2D

1. PRIVATE COMPUTATION: The players first jointly generate a sharing of an additional randomly chosen value $s^{(\ell+1)}$. Then, all $\ell+1$ sharings are upgraded to 2D-sharings, and the correctness is verified with destroying the privacy of this blinding value.

   1.1 Every dealer $P_D \in \mathcal{P}_D$ picks a random summand $s^{(\ell+1,D)}$ and 1D-shares it among the players with polynomial $f^{(\ell+1,D)}(x)$, resulting in every player P_i holding a share $s_i^{(\ell+1,D)}$.

   1.2 For every $m = 1, \ldots, \ell+1$, every player $P_i \in \mathcal{P} \setminus \mathcal{X}$ computes his sum share $s_i^{(m)} = \sum_{P_D \in \mathcal{P}_D} s_i^{(m,D)}$, and 1D-shares it with the polynomial $f_i^{(m)}(x)$, such that $f_i^{(m)}(j) = 0$ for $\{P_i, P_j\} \in \Delta$ (Kudzu). We denote the share-shares as $s_{ij}^{(m)}$. The 1D-sharing of detected players $P_i \in \mathcal{X}$ is the constant-0 sharing (all share-shares are Kudzu).

2. FAULT DETECTION: In order to verify the correctness of the resulting sharings, the players jointly generate a random challenge vector $(r^{(1)}, \ldots, r^{(\ell)}) \in \mathcal{F}^\ell$ using the protocol GenerateChallenges. Then, the correctness of the 2D-sharing of the random linear combination $\sum_{m=1}^{\ell} r^{(m)} s^{(m)} + s^{(\ell+1)}$ will be verified (in parallel) by every player $P_V \in \mathcal{P} \setminus \mathcal{X}$. We denote the linearly combined polynomials by $f(x) = \sum_{m=1}^{\ell} r^{(m)} f^{(m)}(x) + f^{(\ell+1)}(x)$, respectively $f_i(x) = \sum_{m=1}^{\ell} r^{(m)} f_i^{(m)}(x) + f_i^{(\ell+1)}(x)$.
   The following steps are performed in parallel for every verifier $P_V \in \mathcal{P} \setminus \mathcal{X}$:

   2.1 Every P_j with $\{P_V, P_j\} \notin \Delta$ computes and sends to P_V the following linear combinations of his share-shares for every $i = 1, \ldots, n$ with $\{P_i, P_j\} \notin \Delta$: $s_{ij} = \sum_{m=1}^{\ell} r^{(m)} s_{ij}^{(m)} + s_{ij}^{(\ell+1)}$.

   2.2 P_V checks for each $i = 1, \ldots, n$, whether the received share-shares s_{ij} define a valid 1D-sharing for dealer P_i, i.e., there exists a polynomial $\widetilde{f}_i(x)$ with $\widetilde{f}_i(j) = s_{ij}$ for every j with $\{P_V, P_j\} \notin \Delta$ and $\{P_i, P_j\} \notin \Delta$, and $\widetilde{f}_i(j) = 0$ (i.e., Kudzu) for every j with $\{P_i, P_j\} \in \Delta$,[5] and broadcasts a bit "accept" or "reject".

   2.3. P_V checks that the first-level sharing $\widetilde{f}_1(0), \ldots, \widetilde{f}_n(0)$ is a valid Shamir-sharing of degree t and broadcasts "accept" or "reject".

   If all verifiers P_V broadcast "accept" in Steps 2.2 and 2.3, the protocol succeeded and terminates.

3. FAULT LOCALIZATION: The following steps are executed for the smallest complaining verifier P_V.

---

[5] Observe that in this case $\widetilde{f}_i(x) = f_i(x)$.

3.1 If the reported fault was in Step 2.2, i.e., P_V observed that one of the second-level sharings is not a correct 1D-sharing, the following steps are executed:

   3.1.1 P_V broadcasts the index i of the invalid second-level sharing.

   3.1.2 Every P_j with $\{P_j, P_V\} \notin \Delta$ and $\{P_j, P_i\} \notin \Delta$ broadcasts s_{ij}.

   3.1.3 If the broadcasted shares define a correct 1D-sharing, then there exists a player P_j with $\{P_j, P_V\} \notin \Delta$ who has broadcasted a different value than he has privately sent to P_V in Step 2.1. P_V broadcasts his index j, and the protocol fails with $E = \{P_V, P_j\}$. If the broadcasted shares do not define a correct 1D-sharing, the owner P_i of this second-level sharing broadcasts the index j of a player P_j (with $\{P_i, P_j\} \notin \Delta$) who has broadcasted a wrong share $s_{ij} \neq f_i(j)$, and the protocol fails with $E = \{P_i, P_j\}$.

3.2 If the observed fault was in Step 2.3, i.e., P_V could correctly interpolate each second-level sharing $\tilde{f}_1(x), \ldots, \tilde{f}_n(x)$, but the interpolated values $\tilde{f}_1(0), \ldots, \tilde{f}_n(0)$ do not define a valid (first-level) Shamir-sharing of degree t,[6] then the following steps are executed.

   3.2.1 For every dealer P_D, the random linear combination $f^{(*,D)}(x)$ of his 1D-sharings is defined as $f^{(*,D)}(x) = \sum_{m=1}^{\ell} r^{(m)} f^{(m,D)}(x) + f^{(\ell+1,D)}(x)$. Accordingly, for every dealer P_D, every player P_i with $\{P_i, P_D\} \notin \Delta$ and $\{P_i, P_V\} \notin \Delta$ sends to P_V his share $s_i^{(*,D)}$ on $f^{(*,D)}(x)$, i.e., $s_i^{(*,D)} = \sum_{m=1}^{\ell} r^{(m)} s_i^{(m,D)} + s_i^{(\ell+1,D)}$.

   3.2.2 P_V checks for every player P_i with $\{P_V, P_i\} \notin \Delta$ that $\sum_{P_D : \{P_i, P_D\} \notin \Delta} s_i^{(*,D)} = f_i(0)$.[7] If the check fails for some P_i, then P_V broadcasts i, and the protocol fails with $E = \{P_V, P_i\}$.

   3.2.3 P_V broadcasts the index D of P_D such that the received shares $s_i^{(*,D)}$ (for every i with $\{P_i, P_D\} \notin \Delta$ and $\{P_i, P_V\} \notin \Delta$) do not define a correct 1D-sharing.

   3.2.4 Every $P_i \in \mathcal{P}$ with $\{P_i, P_V\} \notin \Delta$ and $\{P_i, P_D\} \notin \Delta$ broadcasts his share $s_i^{(*,D)}$.

   3.2.5 If the broadcasted shares define a correct 1D-sharing for dealer P_D, then P_V broadcasts the index i of the player P_i with $\{P_V, P_i\} \notin \Delta$ who has broadcast a different share $s_i^{(*,D)}$ than he has privately sent to P_V in Step 3.2.1, and the protocol fails with $E = \{P_V, P_i\}$. If the broadcasted shares do not define a correct 1D-sharing for dealer P_D, then P_D broadcasts the index i of a player P_i with $\{P_D, P_i\} \notin \Delta$ who broadcasted a wrong share $s_i^{(*,D)}$, and the protocol fails with $E = \{P_D, P_i\}$.

---

[6] Note that $\tilde{f}_i(0) = f_i(0)$ for every i, i.e., $\tilde{f}_i(0)$ is the linear combination of the values that P_i did indeed 1D-share as his shares $s_i^{(m)}$ in Step 1.

[7] Note that the Kudzu-shares $s_i^{(*,D)}$ with $\{P_i, P_D\} \in \Delta$ are 0 and do not contribute to the sum.

Lemma 5. *If* Upgrade1Dto2D *succeeds, then with overwhelming probability, the upgraded sharings are correct 2D-sharings. If the protocol fails, then the localized pair* $E = \{P_i, P_j\}$ *is new and contains at least one corrupted player. The privacy of the shared values is guaranteed through the whole protocol (even if it fails). The protocol communicates* $\mathcal{O}(\ell n^2 + n^3)$ *and broadcasts* $\mathcal{O}(n)$ *field elements.*

Proof. Along the lines of the proof of Lemma 2.   □

3.5   Information Checking with Dispute Control

An *information-checking (IC)* scheme allows a sender to deliver a message to a recipient in such a way that the recipient can later forward the message and prove its authenticity to a designated verifier. More precisely, an IC-scheme for a sender P_S, recipient P_R, and verifier P_V, consists of two protocols:[8]

IC-Distr: The sender P_S delivers the message m and some authentication tag y to P_R and some checking tag z to P_V.

IC-Reveal: The recipient P_R forwards m and y to P_V, who uses z to verify the authenticity of m, and either accepts or rejects m.

Our information-checking protocol is a variant of the information-checking protocol of [CDD+99] with two modifications. First, our IC-Distr protocol may fail in case of a fault; then, a dispute among two of the three players is identified.[9] Second, our protocol supports authenticating long messages $m = (m_1, \ldots, m_\ell) \in \mathcal{F}^\ell$ without additional costs.[10]

For authenticating $m = (m_1, \ldots, m_\ell)$, a random degree-$\ell$ polynomial $f(x)$ with $f(i) = m_i$ for $i = 1, \ldots, \ell$ is chosen, then the authentication tag is $y = f(0)$ and the verification tag is a random point $z = (u, v)$ with $f(u) = v$ and $u \geq \ell$. One can easily verify that this approach satisfies completeness, secrecy, and correctness (with error probability $\ell/(|\mathcal{F}| - \ell - 1)$) as long as the tags are computed as indicated. In order to ensure that the sender computes the tags correctly, we use a cut-and-choose proof: The sender generates and distributes κ independent tags, and the verifier hands half of them to the recipient, who checks them. The concrete protocols are given in the sequel:

Protocol IC-Distr

1. PRIVATE COMPUTATION: The sender P_S, holding message $m = (m_1, \ldots, m_\ell)$, selects uniformly at random κ authentication tags $y_1, \ldots, y_\kappa \in_R \mathcal{F}^\kappa$, κ elements $u_1, \ldots, u_\kappa \in_R (\mathcal{F} \setminus \{0, \ldots, \ell\})^\kappa$, and computes $v_1, \ldots, v_\kappa$ such that for each $i \in \{1, \ldots, \kappa\}$, the $\ell + 2$ points $(0, y_i), (1, m_1), \ldots, (\ell, m_\ell), (u_i, v_i)$ lie on a polynomial of degree ℓ. P_S sends the message m and the authentication tags $y_1, \ldots, y_\kappa$ to P_R and the verification tags $z_1 = (u_1, v_1), \ldots, z_\kappa = (u_\kappa, v_\kappa)$ to P_V.

---

[8] In [RB89, CDD+99], a different notation is used. They denote the sender as "dealer", the recipient as "intermediary", and the verifier as "receiver".

[9] In our context, the IC-scheme will be used only by triples of players with no a priori dispute among them, so the identified dispute will be a new one.

[10] The costs in the scheme of [CDD+99] grow linearly with the size of the message.

2. FAULT DETECTION:

2.1 P_V partitions the index set $\{1, \ldots, \kappa\}$ into two partitions I and $\bar{I}$ of (almost) equal size, and sends I, $\bar{I}$, and z_i for every $i \in I$ to P_R.

2.2 P_R checks whether for *every* $i \in I$, the points $(0, y_i), (1, m_1), \ldots, (\ell, m_\ell), z_i$ lie on a polynomial of degree ℓ, and broadcasts either "accept" (and the protocol succeeded) or "reject".

3. FAULT LOCALIZATION: If P_R broadcasted "reject", the protocol fails and:

3.1 P_R selects $i \in I$ such that the verification tag z_i received from P_V does not match with the message m and the authentication tag y_i received from P_S, and broadcasts i and z_i.

3.2 P_S and P_V broadcast z_i.

3.3 If the z_i-s broadcasted by P_S and P_V differ, then $E = \{P_S, P_V\}$. Otherwise, if the z_i-s broadcasted by P_R and P_V differ, then $E = \{P_R, P_V\}$. Otherwise, $E = \{P_S, P_R\}$.

Protocol IC-Reveal

1. The recipient P_R sends the message m and the authentication tags y_i for $i \in \bar{I}$ to the verifier P_V.

2. The verifier with verification tags $z_1, \ldots, z_\ell$ accepts $m = (m_1, \ldots, m_\ell)$ if for *any* $i \in \bar{I}$, the points $(0, y_i), (1, m_1), \ldots, (\ell, m_\ell), z_i$ form a polynomial of degree ℓ; otherwise, he rejects m.

Lemma 6. *If IC-Distr succeeds and P_V, P_R are honest, then with overwhelming probability P_V accepts the message m in IC-Reveal (completeness). If IC-Distr fails, then the localized pair E contains at least one corrupted player. If P_S and P_V are honest, then with overwhelming probability, P_V rejects any fake message $m' \neq m$ in IC-Reveal (correctness). If P_S and P_R are honest, then P_V obtains no information about m in IC-Distr (even if it fails) (privacy).*

Proof. Completeness: If the cut-and-choose proof is successful, then the probability that at least one of the remaining authentication tags is valid is at least $1 - \kappa/2^\kappa$. Correctness: The probability that an corrupted receiver can produce at least one correct tag for a message $m' \neq m$ is equal to the probability, that he can guess at least one verification point z_i, which is less than $\kappa/(2^\kappa - \ell - 1)$. Privacy follows from the fact that the verification tag is statistically independent from the message. □

3.6 Upgrading 2D-Sharings to 2D*-Sharings

The following protocol upgrades ℓ 2D-sharings to 2D*-sharings. We denote the 2D-shared values by $s^{(m)}$ (for $m = 1, \ldots, \ell$), the shares of each player $P_i \in \mathcal{P}$ by $s_i^{(m)}$, and P_j's share-share of $s_i^{(m)}$ by $s_{ij}^{(m)}$.

Protocol Upgrade2Dto2D*

1. For every triple of players $P_i, P_j, P_k \in \mathcal{P}$ with no dispute among them (i.e., $\{P_i, P_j\} \notin \Delta$, $\{P_i, P_k\} \notin \Delta$, $\{P_j, P_k\} \notin \Delta$), the protocol IC-Distr is invoked for the message $m = (s_{ij}^{(1)}, \ldots, s_{ij}^{(\ell)})$ with sender P_i, receiver P_j and verifier P_k. The message is not really sent, as P_j already holds it. Furthermore, these up to n^3 parallel invocations are merged when it comes to fault-detection and fault-localization: Every player P_j broadcasts one single bit in the fault-detection, indicating whether he observed a fault in one of the instances he acted as recipient. Then, the smallest player P_j that reported a fault, broadcasts i and k, indicating the instance i, j, k in which he observed the fault, and fault-localization is invoked only for this instance.

Lemma 7. *If the 2D-sharings to be upgraded are correct (for the actual Δ) and the protocol* Upgrade2Dto2D* *succeeds, then the upgraded 2D*-sharings are with overwhelming probability correct. If the protocol fails, then the output pair E is new and contains at least one corrupted player. The privacy of the shared values is guaranteed through the whole protocol (even if it fails). The protocol communicates $\mathcal{O}(n^3 \kappa)$ and broadcasts $\mathcal{O}(n)$ field elements.*

3.7　ABC-Protocol

The following protocol allows every player $P_k \in \mathcal{P} \setminus \mathcal{X}$ to prove that for every $m = 1, \ldots, \ell$, the (for the actual Δ correctly) 1D-shared value $c^{(m,k)}$ is the product of the (for the actual Δ correctly) 1D-shared values $a_k^{(m)}$ and $b_k^{(m)}$. This ABC-protocol is inspired by the corresponding protocol of [CDD+99].

The intuition of the ABC protocol is the following (where we denote the factors as a and b and the product as c): The prover shares a random $\bar{a}$ and $\bar{c} = \bar{a}b$, i.e., $(\bar{a}, b, \bar{c})$ is a multiplication triple, and proves for a random challenge r, that the shared triple $(ra + \bar{a}, b, rc + \bar{c})$ is a correct multiplication triple. This is achieved by first reconstructing $\tilde{a} = ra + \bar{a}$, and then verifying that $z = \tilde{a}b - rc - \bar{c}$ is a sharing of 0. For the sake of efficiency, we parallelize this ABC-proof for many triples and amortize the verification. Instead of reconstructing the sharing of each $\tilde{a}$, we ask the prover to send the (alleged) values $\tilde{a}$ to every player; who then verify that a random linear combination of these sharings reconstructs to the linear combination of the alleged values. Analogously, instead of verifying each z to be zero, the players reconstruct a random linear combination of these values, which must be zero.

Protocol ABC

1. Every player $P_k \in \mathcal{P} \setminus \mathcal{X}$ selects for each $m = 1, \ldots, \ell$ a random $\bar{a}_k^{(m)}$ and computes $\bar{c}^{(m,k)} = \bar{a}_k^{(m)} b_k^{(m)}$.

2. Invoke VSS1D to let every $P_k \in \mathcal{P} \setminus \mathcal{X}$ verifiably 1D-share $\bar{a}_k^{(m)}$ and $\bar{c}^{(m,k)}$ for $m = 1, \ldots, \ell$.

3. Invoke GenerateChallenges to generate one random challenge r.

4. Every $P_k \in \mathcal{P} \setminus \mathcal{X}$ sends $\widetilde{a}_k^{(m)} = r a_k^{(m)} + \overline{a}_k^{(m)}$ for $m = 1, \ldots, \ell$ to every $P_i \in \mathcal{P}$ with $\{P_k, P_i\} \notin \Delta$.

5. Invoke GenerateChallenges to generate ℓ challenges $r^{(1)}, \ldots, r^{(\ell)}$.

6. Invoke Reconstruct1D with $\mathcal{P}_R = \mathcal{P} \setminus \mathcal{X}$ to publicly reconstruct $\widehat{a}_k = \sum_{m=1}^{\ell} r^{(m)} \left(r a_k^{(m)} + \overline{a}_k^{(m)} \right)$ for $k = 1, \ldots, n$.[11]

7. Every $P_i \in \mathcal{P} \setminus \mathcal{X}$ checks for every P_k with $\{P_i, P_k\} \notin \Delta$ whether $\widehat{a}_k = \sum_{m=1}^{\ell} r^{(m)} \widetilde{a}_k^{(m)}$, and broadcasts the index k of a player P_k for whom the check failed, respectively $\bot$ if all checks succeed. If at least one player P_i broadcasts k with $\{P_i, P_k\} \notin \Delta$, then the protocol fails with $E = \{P_i, P_k\}$ for the smallest such P_i (and the accused P_k).

8. Invoke Reconstruct1D with $\mathcal{P}_R = \mathcal{P} \setminus \mathcal{X}$ to reconstruct $z^{(k)} = \sum_{m=1}^{\ell} r^{(m)} \left(\widetilde{a}_k^{(m)} b_k^{(m)} - r c^{(m,k)} - \overline{c}^{(m,k)} \right)$ for $k = 1, \ldots, n$. Note that $\widetilde{a}_k^{(m)}$ is a constant known to all players P_i with $\{P_i, P_k\} \notin \Delta$,[12] hence $z^{(k)}$ is a linear combination of 1D-shared values, as required by Reconstruct1D. Note that when this reconstruction succeeds, then every player $P_V \in \mathcal{P} \setminus \mathcal{X}$ reconstructs the same vector $(z^{(1)}, \ldots, z^{(n)})$.

9. Every player $P_V \in \mathcal{P} \setminus \mathcal{X}$ checks whether the reconstructed values $z^{(k)} = 0$ for every $P_k \in \mathcal{P} \setminus \mathcal{X}$. If this check fails, then P_k is corrupted, and the protocol fails with $E = \{P_i, P_k\}$ for all $P_i \in \mathcal{P}$ (i.e., P_k is in dispute with every player).

Lemma 8. *If all triples $(a_k^{(m)}, b_k^{(m)}, c^{(m,k)})$ are correctly 1D-shared for the actual Δ, then the following holds with overwhelming probability: If ABC succeeds, then the checked triples $(a_k^{(m)}, b_k^{(m)}, c^{(m,k)})$ are correct multiplication triples, i.e. $c^{(m,k)} = a_k^{(m)} b_k^{(m)}$ for every $m = 1, \ldots \ell$, and their privacy is preserved. If the protocol fails, then it localizes a new dispute pair E containing at least one corrupted player (respectively localizes single player who is corrupted). The protocol communicates $\mathcal{O}(\ell n^2 + n^3)$ and broadcasts $\mathcal{O}(n)$ field elements.*

Proof. In order to prove correctness, assume that there is at least one (incorrect) triple $(a_k^{(m)}, b_k^{(m)}, c^{(m,k)})$ (of player P_k) such that $c^{(m,k)} \neq a_k^{(m)} b_k^{(m)}$. Then there is at most one (out of 2^κ) challenge $r \in F$ such that $(r a_k^{(m)} + \overline{a}_k^{(m)}) b_k^{(m)} - r c^{(m,k)} - \overline{c}^{(m,k)} = 0$. If $(r a_k^{(m)} + \overline{a}_k^{(m)}) b_k^{(m)} - r c^{(m,k)} - \overline{c}^{(m,k)} \neq 0$ then there are at most $2^{\kappa(\ell-1)}$ (out of $2^{\kappa\ell}$) challenge vectors $(r^{(1)}, \ldots, r^{(\ell)}) \in \mathcal{F}^\ell$ such that the sum $z^{(k)} = \sum_{m=1}^{\ell} r^{(m)} \left(\left(r a_k^{(m)} + \overline{a}_k^{(m)} \right) b_k^{(m)} - r c^{(m,k)} - \overline{c}^{(m,k)} \right) = 0$. So provided that the values $a_k^{(m)}, b_k^{(m)}, c^{(m,k)}, \overline{a}_k^{(m)}, \overline{c}^{(m,k)}$ for $m = 1, \ldots, \ell$ are correctly 1D-shared, the challenges are random, and in Step 4., player P_k sent the correct

---

[11] Note that the 1D-sharing $\widehat{a}_k$ belongs to dealer P_k. Formally, Reconstruct1D requires every value to be reconstructed to be the *sum* of one 1D-sharing of each dealer in $\mathcal{P}_D$; hence, we implicitly assume constant-0 1D-sharings for the other dealers, and set $\mathcal{P}_D = \mathcal{P} \setminus \mathcal{X}$.

[12] Note that P_k is the owner of the 1D-sharing of $z^{(k)}$; hence, the share of every player P_i with $\{P_i, P_k\} \in \Delta$ is Kudzu, and he does not need to know the constant $\widetilde{a}_k^{(m)}$.

$\widetilde{a}_k^{(m)} = ra_k^{(m)} + \overline{a}_k^{(m)}$ for $m = 1, \ldots, \ell$ to every $P_i \in \mathcal{P}$ with $\{P_k, P_i\} \notin \Delta$, the probability of the false triple not being detected is at most $2/2^\kappa$, which is negligible. As with overwhelming probability the values $a_k^{(m)}, b_k^{(m)}, c^{(m,k)}, \overline{a}_k^{(m)}, \overline{c}^{(m,k)}$ for $m = 1, \ldots, \ell$ are correctly 1D-shared and the challenges are random, it is now sufficient to show that the probability of P_k sending at least one false $\widetilde{a}_k^{(m)} \neq ra_k^{(m)} + \overline{a}_k^{(m)}$ to at least one honest verifier P_i in Step 4 and not being detected (by P_i) in Step 7 is negligible. This holds because for a false $\widetilde{a}_k^{(m)}$ there are at most $2^{\kappa(\ell-1)}$ (out of $2^{\kappa\ell}$) challenge vectors for which the check in Step 7 does not fail.   □

4   Preparation Phase

The goal of this phase is to generate c_M random 2D*-shared multiplication triples (a, b, c) (one for each multiplication gate) and c_R random 2D*-shared values (one for each random gate). We wastefully generate $c_M + c_R$ random multiplication triples and use only the first factor for the random gates.

The generation of the $c_M + c_R$ multiplication triples is divided into n^2 segments, each of length $L = \lceil (c_M + c_R)/n^2 \rceil$. The computation is non-robust, and its correctness is verified at the end of the segment. In fact, the segment will consist of several stages, each with a private computation and fault-detection. As soon as a fault is reported in a fault-detection procedure, the corresponding fault-localization is used to localize a new dispute to be registered in Δ, and the whole segment has failed and is repeated.

Protocol PreparationPhase

Set $\Delta := \{\}$ and $\mathcal{X} = \{\}$, and for each segment (of length L) do the following steps. If any of the invoked sub-protocols fails, then include the localized pair $E = \{P_i, P_j\}$ in Δ, i.e., $\Delta \leftarrow \Delta \cup \{P_i, P_j\}$, and repeat the failed segment.

1. Generate $2L$ correct random 2D-sharings $\left(a^{(1)}, b^{(1)}\right), \ldots, \left(a^{(L)}, b^{(L)}\right)$:

    1.1. Every player $P_k \in \mathcal{P} \setminus \mathcal{X}$ 1D-shares L randomly selected pairs $\left(a^{(1,k)}, b^{(1,k)}\right), \ldots, \left(a^{(L,k)}, b^{(L,k)}\right) \in \mathcal{F}^2$ among the players. We denote the distributed shares of $a^{(m,k)}$ by $a_1^{(m,k)}, \ldots, a_n^{(m,k)}$.

    1.2. Invoke Upgrade1Dto2D with $\mathcal{P}_D = \mathcal{P} \setminus \mathcal{X}$ and $\ell = L$ to upgrade the implicitly defined sum sharings of $\sum_{P_k \in \mathcal{P}_D} a^{(1,k)}, \ldots, \sum_{P_k \in \mathcal{P}_D} a^{(L,k)}$ to 2D-sharings, resulting in L correctly 2D-shared random values $a^{(1)}, \ldots, a^{(\ell)}$. The same for b.

2. Multiply the L pairs $\left(a^{(1)}, b^{(1)}\right), \ldots, \left(a^{(L)}, b^{(L)}\right)$, resulting in L correctly 2D-shared products $c^{(1)}, \ldots, c^{(L)}$:

    2.1. Every player $P_k \in \mathcal{P} \setminus \mathcal{X}$ computes for every $m = 1, \ldots, L$ the product $c^{(m,k)}$ of his shares $a_k^{(m)}$ and $b_k^{(m)}$. Note that the product $c^{(m)} = a^{(m)}b^{(m)}$ can be computed as a weighted sum of these values $c^{(m,k)}$ (namely Lagrange interpolation); accordingly, we will compute a sharing of $c^{(m)}$ as weighted sum of sharings of $c^{(m,1)}, \ldots, c^{(m,n)}$.

2.2. Invoke VSS1D to let every player $P_k \in \mathcal{P} \setminus \mathcal{X}$ verifiably 1D-share his values $c^{(1,k)}, \ldots, c^{(L,k)}$.

2.3. Invoke the protocol ABC to have every player $P_k \in \mathcal{P} \setminus \mathcal{X}$ prove that for every $m = 1, \ldots, L$, the value $c^{(m,k)}$ he shared in Step 2 is indeed the product of his shares $a_k^{(m)}$ and $b_k^{(m)}$, which are implicitly 1D-shared as part of the 2D-sharings of $a^{(m)}$ and $b^{(m)}$, respectively.

2.4. Invoke the protocol Upgrade1Dto2D with $\mathcal{P}_D = \mathcal{P} \setminus \mathcal{X}$ to upgrade the sharings of the weighted sums $\sum_{P_k \in \mathcal{P}_D} \lambda_k c^{(1,k)}, \ldots, \sum_{k=1}^{n} \lambda_k c^{(L,k)}$ to 2D-sharings, where λ_k denotes the Lagrange coefficients.[13]

3. Invoke Upgrade2Dto2D* to upgrade all $3L$ 2D-sharings to 2D*-sharings.

Lemma 9. *With overwhelming probability, the protocol* PreparationPhase *generates* $c_M + c_R$ *correctly 2D*-shared random multiplication triples* (a, b, c) *with* $c = ab$; *the secrecy of the triples is preserved. The protocol communicates* $\mathcal{O}((c_M + c_R)n^2 + n^5 \kappa)$ *and broadcasts* $\mathcal{O}(n^3)$ *field elements.*

Proof. In order to show the correctness first consider one execution of the Steps 1.–3. for one segment of length L. (Note that the dispute set Δ remains unchanged through Steps 1.–3.) If the execution succeeds, then with overwhelming probability, the triples $(a^{(1)}, b^{(1)}, c^{(1)}), \ldots, (a^{(L)}, b^{(L)}, c^{(L)})$ are correctly 2D*-shared (because of Lemma 2, 5, and 7), and $c = ab$ holds because of Lemma 8 for each triple (a, b, c). As there are n^2 segments and the adversary can provoke less than n^2 executions to fail (in total), he has less then $2n^2$ attempts to introduce a segment with a false triple. Because n is at most polynomial in κ, the probability that a false triple is not detected is negligible.

Privacy follows from the privacy of the invoked sub-protocols. Some of them do not guarantee privacy in case of a failure, but in such case all generated values are discarded and completely new shared values will be generated.   □

5 Input Phase

The goal of the input phase is to provide 2D*-sharings of c_I inputs.

We set the upper bound on the number of input gates of a segment to $L = \lceil \frac{c_I}{n^2} \rceil$ and limit each segment to contain only input gates of the same player.

Protocol InputPhase
For each segment, the following steps are executed to let the dealer $P_D \in \mathcal{P} \setminus \mathcal{X}$ verifiably 2D*-share his L inputs $s^{(1)}, \ldots, s^{(L)}$.[14] If any of the invoked sub-protocols fails, include the localized pair $E = \{P_i, P_j\}$ in Δ, i.e., $\Delta \leftarrow \Delta \cup \{P_i, P_j\}$, and repeat the segment.

---

[13] Note that the sharings of detected players $P_D \in \mathcal{X}$ are not considered in the Lagrange interpolation; however, as their shares are 0 (Kudzu), this omission does not falsify the outcome.

[14] If the dealer P_D is detected, i.e., $P_D \in \mathcal{X}$, then the players take the all-zero sharing of 0, i.e., every share is 0 and every share-share is 0 (Kudzu). Note that no authentication tags are needed because all share-shares are Kudzu.

1. P_D (unverifiably) 1D-shares the input values $s^{(1)}, \ldots, s^{(L)}$.
2. Invoke Upgrade1Dto2D with $\mathcal{P} = \{P_D\}$ to upgrade the 1D-sharings of $s^{(1)}, \ldots, s^{(L)}$ to 2D-sharings.
3. Invoke Upgrade2Dto2D* to upgrade the 2D-sharings of $s^{(1)}, \ldots, s^{(L)}$ to 2D*-sharings.

Lemma 10. *With overwhelming probability, the protocol* InputPhase *computes correct 2D*-sharings of c_I inputs, where the privacy of the inputs of the honest players is preserved. The protocol communicates $\mathcal{O}(c_I n^2 + n^5 \kappa)$ and broadcasts $\mathcal{O}(n^3)$ field elements.*

Proof. In one execution of Steps 1.–3., the probability of success in spite of a false sharing is negligible. As there are at most $n^2 + n$ segments and less than n^2 repetitions, the adversary has at most $2n^2 + n$ independent attempts to introduce a segment with a false sharing, hence his success probability is negligible. The privacy is guaranteed even in case of failure (and repetition) of some segment. □

6   Computation Phase

The computation of the circuit proceeds gate-by-gate. First, to every random and every multiplication gate, a prepared 2D*-shared random triple is assigned.

Given the 2D*-sharings of the multiplication triples and of the inputs, all values to be computed (and to be opened) in the computation stage are completely determined. We therefore call the values shared in the preparation phase and in the input phase the *base values* of the computation. All base values are robustly shared with 2D*-sharings.

It turns out that the value of each gate can be computed as linear combination of such base values. This is trivial as long as the circuit only consists of addition and random gates. For a multiplication gate, the players publicly reconstruct two sharings (both linear combinations of base values), such that the value of the multiplication gate is a linear combination of base values, where the coefficients of the linear combination depend on the two reconstructed values [Bea91a]. Hence, the whole computation phase consists only of a sequence of reconstructions of publicly known linear combinations of base sharings. More precisely, the gates are evaluated as follows:

Input Gate: Assign the corresponding 2D*-sharing of the input to the gate.
Random Gate: Assign the 2D*-sharing of a of the assigned multiplication triple (a, b, c) to the gate.
Addition Gate: To both summands, a linear combination of base sharings was assigned. Assign to the gate the sum of these two linear combinations (which is again a linear combination of base sharings).
Multiplication Gate: To both factors, a linear combination of base sharings was assigned. We denote the corresponding values by x and y, and denote the assigned multiplication triple by (a, b, c). The players reconstruct $d_x = x - a$

and $d_y = y - b$ towards every player in $\mathcal{P}$ (both d_x and d_y are represented as known linear combination of base sharings), and assign to the gate the linear combination $d_x d_y + d_x b + d_y a + c$ (i.e., a linear combination of the 2D*-sharings of a, b, and c, all three of them base sharings).

Output Gate: The players reconstruct the assigned linear combination of base sharings towards the designated output player.

Now, we are left with the problem of opening known linear combinations of base values towards designated players. For every multiplication gate, we need $2n$ reconstructions (one towards every player), and for every output gate, we need 1 reconstruction. Hence, in total we need to reconstruct $2nc_M + c_O$ linear combinations of 2D*-sharings. This job is, as usual, divided into n^2 segments, each with at most $L = \lceil (2nc_M + c_O)/n^2 \rceil$ reconstructions. Each reconstruction is processed non-robustly, and at the end of the segment, the players verify that no fault has occurred. In the non-robust reconstruction the receiver either obtains the right value, or he observes a fault, stops the further processing of this segment and only joins again in the fault handling procedure.

Protocol ComputationPhase

For each segment with L reconstructions, the following steps are executed. If in a segment a fault is detected in Step 2., then Step 3 is executed to localize a new dispute pair E, which is included in Δ, i.e., $\Delta \leftarrow \Delta \cup \{E\}$, and the failed segment is repeated.

1. PRIVATE COMPUTATION: Execute the following for each output operation.[15]
   Denote the designated output player with P_k, the publicly known linear combination for the output operation with $\mathcal{L}$, and the 2D*-shared base values used in the linear combination with $s^{(1)}, s^{(2)}, \ldots$. Furthermore, we denote the share and shares-shares of P_i by $s_i^{(m)}, s_{1i}^{(m)}, \ldots, s_{ni}^{(m)}$, respectively, and the polynomial used for the second-level sharing of $s_i^{(m)}$ by $f_i^{(m)}(x)$.

   1.1 Every P_i with $\{P_i, P_k\} \notin \Delta$ sends his linearly combined share $s_i = \mathcal{L}(s_i^{(1)}, s_i^{(2)}, \ldots)$ to P_k, who receives a message in $\mathcal{F} \cup \{\epsilon\}$.[16]

   1.2 If P_k received *all* shares s_i he was supposed to get (i.e., there was no empty message ϵ), and the received shares lie on a polynomial $f(x)$ of degree t, he computes the output value as $s = f(0)$; otherwise P_k observes a fault and aborts the segment, i.e., for the rest of the segment, P_k only sends empty messages.

2. FAULT DETECTION: Every player $P_i \in \mathcal{P} \setminus \mathcal{X}$ broadcasts the index q_i of the first failed reconstruction operation, respectively $\perp$ if he successfully completed the segment. If all players broadcast $\perp$, then the evaluation of the current segment succeeded

---

[15] All output operations at the same level in the circuit can be executed in parallel.

[16] It is legal for an honest player P_i to send the empty message ϵ to P_k, namely when P_i has observed a fault in an earlier gate. Hence, P_k must accept the empty message as valid.

3. FAULT LOCALIZATION: Execute the following steps for the player P_k with the smallest q_k, for the failed reconstruction operation with index q_k:

   3.1 Every player P_i with $\{P_k, P_i\} \notin \Delta$ sends the polynomial $f_i(x) = \mathcal{L}(f_i^{(1)}, f_i^{(2)}, \ldots)$ and all share-shares $s_{ji}(x) = \mathcal{L}(s_{ji}^{(1)}, s_{ji}^{(2)}, \ldots)$ to P_k.

   3.2 If for some P_i with $\{P_k, P_i\} \notin \Delta$, P_k did not receive s_i in Step 1.1, or the provided polynomial $f_i(x)$ is inconsistent with s_i (i.e., $f_i(0) \neq s_i$), then P_k broadcasts i, and the fault localization terminates with $E = \{P_k, P_i\}$.

   3.3 P_k identifies two players P_i, P_j with $\{P_k, P_i\} \notin \Delta$ and $\{P_k, P_j\} \notin \Delta$, such that $f_i(j) \neq s_{ij},$[17] and broadcasts $(i, j, s_{ij}, f_i(j))$.

   3.4 Both P_i and P_j broadcast a bit indicating whether or not they agree with the values broadcasted by P_k. If P_i (respectively P_j) disagrees, the fault localization terminates with $E = \{P_k, P_i\}$ (respectively $E = \{P_k, P_j\}$).

   3.5 As both P_i and P_j agree with s_{ij} respectively $f_i(j)$ as broadcasted by P_k, and as $f_i(j) \neq s_{ij}$, either P_i or P_j delivered a wrong value to P_k. P_j can use the information checking scheme to prove to P_k the correctness of s_{ij}. However, there are no authentication tags for s_{ij} itself, but s_{ij} is computed as a publicly known linear combination $\mathcal{L}$ of base sharings, for which authentication tags exist (one authentication tag for all share-shares x_{ij} of each segment), respectively which are Kudzu and hence publicly known. Hence, P_j executes the protocol IC-Reveal for revealing the provably correct share-shares x_{ij} of every base sharing x, and if P_k accepts all invocations and the linear combination on the share-shares yields s_{ij}, then P_k broadcasts i and $E = \{P_k, P_i\}$, otherwise, P_k broadcasts j and $E = \{P_k, P_j\}$.

Lemma 11. *If all base values are correctly 2D*-shared and all multiplication triples are correct and random, then with overwhelming probability, the circuit evaluation as described above is correct, robust and private. The protocol communicates* $\mathcal{O}((c_I n^2 + c_M n^2 + c_R n^2 + c_O n + n^4)\kappa)$ *and broadcasts* $\mathcal{O}(n^3)$ *field elements.*

Proof. Once the base values are correctly 2D*-shared, the computation phase is purely deterministic. An honest player will never reconstruct a wrong secret: He receives shares from all players he is not in dispute with (otherwise he does not reconstruct at all), hence there are at least $t + 1$ correct shares from the honest players which prevent him from reconstructing a wrong value. Hence, the adversary cannot falsify the outputs of honest players, he can only prevent them from reconstructing. In this case, a fault is detected, a new dispute is localized and included in Δ, and the segment is repeat till eventually all honest players reconstruct all their outputs.

In order to argue about the privacy of the protocol, we observe that share-shares x_{ij} are revealed only when P_i and P_j disagree on some value s_{ij}, hence either P_i or P_j is corrupted. By revealing these values, the adversary obtains no additional information.   □

---

[17] The existence of such a pair (P_i, P_j) is guaranteed due to the correctness of the base 2D*-sharings.

7   The New MPC Protocol and Conclusions

The new MPC protocol consists of the three described phases:

Protocol MPC

1. Invoke PreparationPhase to prepare $c_M + c_R$ random 2D*-shared multiplication triples.
2. Invoke InputPhase to provide 2D*-sharings of the c_I inputs.
3. Invoke ComputationPhase to compute and reconstruct the outputs towards the specified players.

Theorem 1. *A set of n players communicating over a secure synchronous network, can evaluate an agreed function of their inputs securely against an unbounded active adaptive adversary corrupting up to $t < n/2$ of the players with communicating $\mathcal{O}(c_I n^2 + c_M n^2 + c_R n^2 + c_O n + n^5 \kappa)$ field elements and broadcasting $\mathcal{O}(n^3)$ field elements, where c_I, c_M, c_R, c_O denote the number of input gates, multiplication gates, random gates, and output gates, respectively.*

Note that for large enough circuits, the costs for simulating the $\mathcal{O}(n^3)$ broadcast invocations are dominated by the normal communication costs, such that the overall communication complexity is (up to a constant factor) the same as the one of passively secure MPC protocols [BGW88].

However, for very small circuits, the $\mathcal{O}(n^3)$ broadcasts are dominating the overall costs. Note that even in this case, our protocol is substantially more efficient than the most efficient previously known protocol for the same model [CDD+99], which broadcasts $\Omega(n^5)$ field elements *per multiplication*.

Acknowledgments

We would like to thank Micha Riser for the fruitful discussions, and the anonymous referees for their helpful comments.

References

[Bea91a]   D. Beaver. Efficient multiparty protocols using circuit randomization. In *CRYPTO '91*, LNCS 576, pp. 420–432, 1991.

[Bea91b]   D. Beaver. Secure multiparty protocols and zero-knowledge proof systems tolerating a faulty minority. *Journal of Cryptology*, pp. 75–122, 1991.

[BGP92]   P. Berman, J. A. Garay, and K. J. Perry. Bit optimal distributed consensus. *Computer Science Research*, pp. 313–322, 1992. Preliminary version in Proc. 21st STOC, 1989.

[BGW88]   M. Ben-Or, S. Goldwasser, and A. Wigderson. Completeness theorems for non-cryptographic fault-tolerant distributed computation. In *Proc. 20th STOC*, pp. 1–10, 1988.

[CCD88]   D. Chaum, C. Crépeau, and I. Damgård. Multiparty unconditionally secure protocols (extended abstract). In *Proc. 20th STOC*, pp. 11–19, 1988.

[CDD+99]   R. Cramer, I. Damgård, S. Dziembowski, M. Hirt, and T. Rabin. Efficient multiparty computations secure against an adaptive adversary. In *EUROCRYPT '99*, LNCS 1592, pp. 311–326, 1999.

[CDG87]   D. Chaum, I. Damgård, and J. van de Graaf. Multiparty computations ensuring privacy of each party's input and correctness of the result. In *CRYPTO '87*, LNCS 293, pp. 87–119, 1987.

[CW79]   L. Carter and M. N. Wegman. Universal classes of hash functions. *Journal of Computer and System Sciences*, 18(4):143–154, 1979. Preliminary version in Proc. 9st STOC, 1977.

[CW92]   B. A. Coan and J. L. Welch. Modular construction of a Byzantine agreement protocol with optimal message bit complexity. *Information and Computation*, 97(1):61–85, 1992. Preliminary version in Proc. 8th PODC, 1989.

[DS82]   D. Dolev and H. R. Strong. Polynomial algorithms for multiple processor agreement. In *Proc. 14th STOC*, pp. 401–407, 1982.

[GHY87]   Z. Galil, S. Haber, and M. Yung. Cryptographic computation: Secure fault-tolerant protocols and the public-key model. In *CRYPTO '87*, LNCS 293, pp. 135–155, 1987.

[GMW87]   O. Goldreich, S. Micali, and A. Wigderson. How to play any mental game — a completeness theorem for protocols with honest majority. In *Proc. 19th STOC*, pp. 218–229, 1987.

[HM01]   M. Hirt and U. Maurer. Robustness for free in unconditional multi-party computation. In *CRYPTO '01*, LNCS 2139, pp. 101–118, 2001.

[HMP00]   M. Hirt, U. Maurer, and B. Przydatek. Efficient secure multi-party computation. In *ASIACRYPT '00*, LNCS 1976, pp. 143–161, 2000.

[PSL80]   M. Pease, R. Shostak, and L. Lamport. Reaching agreement in the presence of faults. *Journal of the ACM*, 27(2):228–234, Apr. 1980.

[PW92]   B. Pfitzmann and M. Waidner. Unconditional Byzantine agreement for any number of faulty processors. In *Proc. 9th STACS*, LNCS 577, 1992.

[RB89]   T. Rabin and M. Ben-Or. Verifiable secret sharing and multiparty protocols with honest majority. In *Proc. 21st STOC*, pp. 73–85, 1989.

[Sha79]   A. Shamir. How to share a secret. *Communications of the ACM*, 22:612–613, 1979.

[Yao82]   A. C. Yao. Protocols for secure computations. In *Proc. 23rd FOCS*, pp. 160–164, 1982.

Round-Optimal and Efficient
Verifiable Secret Sharing

Matthias Fitzi[1,*], Juan Garay[2,**], Shyamnath Gollakota[3,* * *],
C. Pandu Rangan[3,†], and Kannan Srinathan[4]

[1] Department of Computer Science, Aarhus University, Denmark
fitzi@daimi.au.dk
[2] Bell Labs – Lucent Technologies, 600 Mountain Ave., Murray Hill, NJ 07974
garay@research.bell-labs.com
[3] Department of Computer Science and Engineering, IIT Madras, India
shyam@cse.iitm.ernet.in, rangan@iitm.ernet.in
[4] International Institute of Information Technology, Hyderabad, India
srinathan@iiit.ac.in

Abstract. We consider perfect verifiable secret sharing (VSS) in a synchronous network of n processors (players) where a designated player called the *dealer* wishes to distribute a secret s among the players in a way that no t of them obtain any information, but any $t + 1$ players obtain full information about the secret. The round complexity of a VSS protocol is defined as the number of rounds performed in the sharing phase. Gennaro, Ishai, Kushilevitz and Rabin showed that three rounds are necessary and sufficient when $n > 3t$. Sufficiency, however, was only demonstrated by means of an inefficient (i.e., exponential-time) protocol, and the construction of an efficient three-round protocol was left as an open problem.

In this paper, we present an efficient three-round protocol for VSS. The solution is based on a three-round solution of so-called *weak verifiable secret sharing* (WSS), for which we also prove that three rounds is a lower bound. Furthermore, we also demonstrate that one round is sufficient for WSS when $n > 4t$, and that VSS can be achieved in $1 + \varepsilon$ amortized rounds (for any $\varepsilon > 0$) when $n > 3t$.

1 Introduction

Secret sharing [2, 9] is one of the most important primitives used for the construction of secure multi-party protocols. In secret sharing, a "dealer" wants to share a secret s among a set of n players such that no set of t players will be

---

* Supported by SECOQC, Secure Communication based on Quantum Cryptography, under the Information Societies Technology Programme of the European Commission, IST-2003-506813.
** Work partly done while visiting the Centre de Recerca Matemàtica, Barcelona.
*** Work partly done at Bell Labs India, Bangalore
† Work partly done while visiting Bell Labs, Murray Hill, supported by DIMACS.

S. Halevi and T. Rabin (Eds.): TCC 2006, LNCS 3876, pp. 329–342, 2006.

able to reconstruct the secret while any set of $t + 1$ or more players will be able to reconstruct the secret by combining their shares.

Verifiable secret sharing [4] (VSS) extends ordinary secret sharing for the use in presence of active corruption where an adversary may corrupt up to t players in an arbitrary way. In VSS, it is required that no t players get any information about the secret whereas the n players together can reliably reconstruct the secret even if t of them deliver wrong information.

Prior Work. Secret sharing was introduced in [2, 9] together with a perfectly secure solution for any number $n > t$ of players in the presence of passive corruption, i.e., where no t players get any Shannon information about secret s and any $t + 1$ players get full information about s.

On the other hand, perfectly secure VSS is (efficiently) achievable if and only if $n > 3t$ [1]. When additionally given a broadcast channel among the players, unconditionally secure VSS (with negligible error) can be achieved if $n > 2t$ [8]. As a building block for the VSS protocol in [8], a "degraded" variant of VSS is introduced called *weak* verifiable secret sharing (WSS), where the reconstructed value may also be some default value, in case the dealer is corrupted.

VSS has been extensively studied. Of relevance to our work is the study of the problem's *round complexity* by Gennaro, Ishai, Kushilevitz and Rabin [5], who give tight bounds for perfectly secure VSS. Specifically, it is shown that for $n > 4t$ one round is sufficient when $t = 1$ and that two rounds is a tight bound for general t. For the optimal $n > 3t$, it is shown that three rounds is sufficient as well as necessary; the protocol achieving it, however, requires exponential time. The existence of efficient three-round protocols was left as an open problem.

Our Contributions. In this paper, we solve this open problem by presenting an efficient three-round protocol for VSS perfectly secure for $n > 3t$. The solution is based on a three-round protocol for WSS which we demonstrate to be round optimal itself, by first showing three-round optimality of a problem that we call *weak secure multicast* (WSM), and then showing a reduction to WSS. Furthermore, we show that perfectly secure WSS is efficiently achievable in one round when $n > 4t$ (and $t > 1$). Finally, we present a simple protocol for perfectly secure VSS with amortized $1 + \varepsilon$ rounds for any $\varepsilon > 0$ when $n > 3t$ — which is of special interest for secure multi-party computation [1,3], where a large number of VSS protocols are run sequentially.

Organization of the Paper. We start in Section 2 by presenting the model and definitions of the secret sharing problems we are considering. Section 3 is dedicated to WSS, where we present round-optimal protocols for the cases $n > 3t$ and $n > 4t$. We derive the efficient round-optimal and player-optimal protocol for VSS in Section 4. The amortized $(1 + \varepsilon)$-round protocol is described in Section 5. We conclude with some final remarks in Section 6. For ease of readability, the round optimality proof for player-optimal WSS based on WSM is presented in the appendix.

2   Model and Definitions

We assume a set $\mathcal{P} = \{P_1, P_2, \cdots, P_n\}$ of n players including dealer D, say, $D = P_1$, and assume the standard model of a fully connected network of pairwise secure channels, plus a common broadcast channel, which can be used to force a player to send the same message to all the other players. Furthermore, we assume the presence of an active adversary who may corrupt up to t of the players in an arbitrarily malicious way. Such a corrupted player is called *dishonest* whereas an uncorrupted player is called *honest*. The adversary is modeled to be *rushing* (i.e., it can base the dishonest players' messages for round r on the honest players' messages of the same round), *adaptive* (the adversary can adaptively corrupt players as the protocol proceeds), but *non-mobile* (over the whole period, the adversary corrupts at most t different players). We call such an adversary a "t-adversary." We demand perfect security, i.e., that the resulting protocol has zero error and that no Shannon information is leaked to the adversary.

We consider several forms of secret sharings with different security properties. As in [5], the protocols for all of them have the same following two-phase structure: In a primary phase, the dealer D distributes a secret s, while in a second, later phase, the players cooperate in order to retrieve it. More specifically, the structure is as follows:

SHARING PHASE: The dealer initially holds secret $s \in \mathcal{K}$ where $\mathcal{K}$ is a finite field of sufficient size; and each player P_i finally holds some private information v_i (possibly consisting of several field elements).

RECONSTRUCTION PHASE: In this phase, each player P_i reveals (some of) his private information v_i. Then, on the revealed information v_i' (a dishonest player may reveal $v_i' \neq v_i$), a reconstruction function is applied in order to compute the secret, $s = \mathrm{Rec}(v_1', \cdots, v_n')$.

The sharing phase as well as the reconstruction phase may consist of several communication rounds. We model communication along the lines of [5] where, in each round, a player can privately send messages to other players and/or broadcast a message to all players. With respect to this model, the round complexity of a secret-sharing protocol is defined as the number of such communication rounds that the protocol requires *in the sharing phase*.

Common Requirements. The following requirements have to be satisfied by all secret-sharing protocols we discuss in this paper.

PRIVACY: If D is honest, then the adversary's view during the sharing phase reveals no information about s. More formally, the adversary's view is identically distributed under all different values of s.

CORRECTNESS: If D is honest, then the reconstructed value is equal to the secret s.

Depending on the particular "strength" of the secret-sharing protocol, different commitment properties are required.

Verifiable Secret Sharing (VSS). An n-player protocol is called a (perfect) (n,t)-VSS protocol if, for any t-adversary, the following condition holds in addition to the privacy and correctness conditions:

COMMITMENT: After the sharing phase, a unique value s^* is determined which will be reconstructed in the reconstruction phase; i.e., $s^* = \text{Rec}(v'_1, \cdots, v'_n)$ regardless of the views provided by the dishonest players.

Weak Verifiable Secret Sharing (WSS). An n-player protocol is called a (perfect) (n,t)-WSS protocol if, for any t-adversary, the following condition holds in addition to the privacy and correctness conditions:

WEAK COMMITMENT: After the sharing phase there is a unique value $s^* \in \mathcal{K}$ such that either s^* or default value $\perp \notin \mathcal{K}$ will be reconstructed in the reconstruction phase; i.e., $\text{Rec}(v'_1, \cdots, v'_n) \in \{\perp, s^*\}$ regardless of the views provided by the dishonest players.

Round Complexity and Efficiency. As in [5], we define the round complexity of a secret-sharing protocol as the number of communication rounds in its *sharing phase* — reconstruction can always be done in a single round by having each player reveal all the information he has. A VSS protocol is *efficient* if the total computation and communication performed by all honest players is polynomial in n and the size of the secret.

3 Round-Optimal WSS

We begin by giving a three-round (n,t)-WSS protocol for $n > 3t$, which is optimal, followed by a one-round (n,t)-WSS protocol for $n > 4t$.

3.1 Round-Optimal WSS for n > 3t

The protocol is based on the four-round (n,t)-VSS protocol for $n > 3t$ given in [5]; essentially, it consists of that protocol's first three rounds, with a modified reconstruction phase. Unlike the protocol in [5], where inconsistencies between the shares of honest players are eliminated by using error correcting codes, we use a different technique to detect the dishonest players who deliver false information in the reconstruction phase.

We now present the protocol. The secret s is assumed to be taken from a finite field $\mathcal{K}$, $|\mathcal{K}| > n$; additionally, $1, 2, ..., n$ are interpreted as (arbitrary) distinct non-zero field elements. We call this protocol $(\frac{n}{3})$-WSS.[1]

Sharing Phase. The sharing phase consists of the following three rounds:

1.   — D chooses a random bivariate polynomial $F \in \mathcal{K}[x,y]$ of degree at most t in each variable, satisfying $F(0,0) = s$. D sends to each player P_i the (univariate) polynomials $f_i(x) = F(x,i)$ and $g_i(y) = F(i,y)$.

---

[1] For simplicity; strictly speaking, it should be "$\lfloor \frac{n-1}{3} \rfloor$-WSS."

- Player P_i sends to each player P_j an independent random "pad" r_{ij} picked uniformly from $\mathcal{K}$.
2. Player P_i broadcasts:
   - $a_{ij} = f_i(j) + r_{ij}$ (r_{ij} is the pad P_i sent to P_j)
   - $b_{ij} = g_i(j) + r_{ji}$ (r_{ji} is the pad P_i received from P_j)
3. For each pair $a_{ij} \neq b_{ji}$, the following happens:
   - P_i broadcasts $\alpha_{ij} = f_i(j)$
   - P_j broadcasts $\beta_{ji} = g_j(i)$
   - D broadcasts $\gamma_{ij} = F(j, i)$

   A player is said to be *unhappy* if the value which he broadcast does not match the dealer's value. If there are more than t unhappy players, disqualify the dealer and stop.[2]   ◇

Reconstruction Phase. Every happy player P_i broadcasts his polynomials $f_i(x) = F(x, i)$ and $g_i(y) = F(i, y)$.

Each player P_i now constructs a *consistency graph* G over the set of happy players such that there exists an edge between P_j and P_k in G if and only if $f_j(k) = g_k(j)$ and $g_j(k) = f_k(j)$. Since these polynomials are broadcast, every player P_i constructs the same graph G.

Now each player P_i constructs a set $CORE$ of players as follows. Initially, all the players in G whose node degree is at least $n - t$ are inserted into the set. Next, players in $CORE$ consistent with less than $n - t$ other players in $CORE$ are removed. This process continues until no more players can be removed from the set. If the resulting $CORE$ set contains less than $n - t$ elements then P_i outputs $\bot$ — otherwise, P_i reconstructs the polynomial $F^*(x, y)$ defined by any $t + 1$ players in $CORE$, and the secret $s^* = F^*(0, 0)$ is reconstructed.   ◇

That finishes the description of protocol $(\frac{n}{3})$-WSS. We now show that it is a (n, t)-WSS protocol for $n > 3t$.

As suggested by the construction of graph G above, we say that (the polynomials of) two players P_i and P_j are consistent if the corresponding values of their polynomials (as opened in the reconstruction phase) match, i.e., if $f_i(j) = g_j(i)$ and $g_i(j) = f_j(i)$. Similarly, we say a player P_i is consistent with bivariate polynomial $F(x, y)$ if $f_i(x)$ and $g_i(y)$ lie on $F(x, y)$, i.e., $f_i(x) = F(x, i)$ and $g_i(y) = F(i, y)$. We first prove the following about players in $CORE$.

Lemma 1. *If $|CORE| \geq n - t$, then all the players in CORE are consistent with a polynomial fixed at the end of the sharing phase.*

Proof. At the end of the sharing phase, all the honest happy players are consistent with each other and their shares define a unique bivariate polynomial $F^H(x, y)$ with degree at most t in both variables. To be in $CORE$, every player P_i must be consistent with (at least) $n - t$ players in $CORE$. Moreover, every player in $CORE$ is happy. So there are at least $n - 2t \geq t + 1$ honest players in

---

[2] If necessary, the secret can be assigned a public default value when the dealer gets disqualified.

$CORE$ with whom P_i is consistent. These $t + 1$ players define a unique polynomial $f_i(x)$ of degree at most t for P_i, which is in turn consistent with $F^H(x, y)$. Thus, the polynomial provided by P_i must be $f_i(x)$. Therefore, every player in $CORE$ is consistent with $F^H(x, y)$.   □

Theorem 1. *Protocol $(\frac{n}{3})$-WSS is an efficient, three-round (n, t)-WSS protocol for $n > 3t$.*

Proof. Number of rounds and efficiency are evident. We prove the WSS properties in turn.

PRIVACY: We only need to consider the case when D is honest. Since D distributes consistent information, any pair P_i and P_j of honest players publishes the same mutual padded values. Thus, due to the randomness of the pads, the adversary's view is indistinguishable under different secrets.

CORRECTNESS: If D is honest then all (at least $n - t$) honest players will be happy, and D will not be disqualified in the sharing phase. Since all honest players are mutually consistent, they all end up in set $CORE$ whereas a dishonest player can only be in $CORE$ by revealing his correct polynomials. Thus the information revealed by the players in $CORE$ is consistent with polynomial F and $s^* = F^*(0, 0) = F(0, 0) = s$ is reconstructed.

WEAK COMMITMENT: We need only consider the case when D is dishonest. If $|CORE| < n - t$ then all the players compute $s^* = \bot$ and weak commitment is satisfied. On the other hand, consider $|CORE| \geq n - t$. In this case, it directly follows from Lemma 1 that the secret constructed is the free term of $F^H(x, y)$.   □

We now state a property of the above protocol which will be used in the correctness proof for our VSS protocol in the next section.

Lemma 2. *If the dealer is not disqualified in the reconstruction phase of $(\frac{n}{3})$-WSS, then the polynomial $F^*(x, y)$ reconstructed in that phase is consistent with all the honest happy players.*

Proof. As proved in Lemma 1, the polynomial reconstructed at the end of the reconstruction phase is $F^H(x, y)$. This $F^H(x, y)$ is defined as the polynomial constructed by any $t + 1$ honest happy players. Thus the polynomial constructed is consistent with all the honest happy players.   □

Round Optimality. The proof of the following theorem is given in Appendix A.

Theorem 2. *For $n \leq 4t$ ($t > 1$), there is no perfect (n, t)-WSS protocol requiring less than three rounds.*

3.2   Round-Optimal WSS for n > 4t

When $n > 4t$, perfectly secure WSS can be efficiently achieved in one round as follows.

Sharing Phase. D chooses a random bivariate polynomial $F \in \mathcal{K}[x, y]$ of degree at most t in each variable satisfying $F(0, 0) = s$ and sends to each player P_i the polynomials $f_i(x) = F(x, i)$ and $g_i(y) = F(i, y)$.   □

Reconstruction Phase. Player P_i broadcasts the polynomials $F(x, i)$ and $F(i, y)$ he received in the sharing phase. Player P_i constructs a consistency graph G and a set $CORE$ as in protocol $(\frac{n}{3})$-WSS. Finally, if $|CORE| < n - t$, P_i computes $\perp$; otherwise, $s^* = F^*(0, 0)$, where $F^*(0, 0)$ is the unique bivariate polynomial of degree at most t in both variables defined by any $t + 1$ players in $CORE$.   □

Theorem 3. *Perfectly secure WSS is efficiently achievable in one round when $n > 4t$.*

Proof. We prove that the above protocol achieves the three conditions of WSS.

PRIVACY: Privacy is obvious since the adversary only gets information about at most t players' shares.

CORRECTNESS: If the dealer D is honest then he sends correct shares to all the players. Thus, at the end of the reconstruction phase, set $CORE$ contains (at least) $n - t$ honest players, D is not disqualified, and the secret s is reconstructed since any other secret s^* can be consistent with at most $2t < n - t$ players.

WEAK COMMITMENT: We need only consider the case when D is dishonest. If $|CORE| < n - t$, then all the players compute $\perp$ and weak commitment is satisfied. On the other hand, assume that $|CORE| \geq n - t$. This implies that there is a set $\mathcal{C}$ of at least $n - 2t$ consistent honest players defining a unique secret s^*. Out of set $\mathcal{C}$ at most t players can be consistent with a polynomial defining a different secret $s' \neq s^*$. Thus at most $|\mathcal{P} \setminus \mathcal{C}| + t \leq n - (n - 2t) + t = 3t < n - t$ players overall can be consistent with secret s' — implying weak commitment on s^*.   □

4   Round-Optimal VSS for n > 3t

We now present an efficient three-round (n, t)-VSS protocol for $n > 3t$. Its round optimality follows from the lower bound in [5].

We first give some of the intuition behind our protocol. Overall, we follow the approach in [5] (and in the previous section), where the dealer first hides the secret in a bivariate polynomial $F(x, y)$, and each player P_i gets the respective univariate polynomials $F(x, i)$ and $F(i, y)$ as his secret information. Then, every pair of players compare their common shares by "blinding" them with a random pad and then broadcasting them. In the reconstruction phase the random pads are revealed, allowing the players to compute the shares and finally reconstruct the secret. However, our twist is as follows. In order to guarantee that each player P_i's random pads get revealed consistently, P_i shares a random field element using a round-optimal, player-optimal (n, t)-WSS protocol — namely, protocol $(\frac{n}{3})$-WSS from the previous section, and chooses his pads as *points on the respective polynomial*, as opposed to independently at random as in [5] and in the previous section for WSS. Players whose $(\frac{n}{3})$-WSS protocol instance fails, also get disqualified from the main protocol; on the other hand, players whose protocol instance succeeds enable the reconstruction of all the pads, and in turn the computation of the main shares. Using these multiple instances of an (n, t)-WSS protocol also replaces the need for explicit error correcting codes, as required by some of the VSS protocols (the efficient ones) in [5].

We now present our VSS protocol in detail. We will use superscript "W" to denote the quantities corresponding to the $(\frac{n}{3})$-WSS protocols that are run in order to WSS the players' random pads. We call the resulting VSS protocol $(\frac{n}{3})$-VSS.

Sharing Phase. The sharing phase consists of the following three rounds:

1. – Dealer D chooses a random bivariate polynomial $F \in \mathcal{K}[x, y]$ of degree at most t in each variable satisfying $F(0, 0) = s$. D sends to P_i the polynomials $f_i(x) = F(x, i)$ and $g_i(y) = F(i, y)$.

 – Player P_i, $i = 1, \ldots, n$, selects a random value r_i and starts an instance of $(\frac{n}{3})$-WSS acting as a dealer in order to share r_i by means of bivariate polynomial $F_i^W(x, y)$ $(F_i^W(0, 0) = r_i)$. We call this instance $(\frac{n}{3})$-WSS$_i$. Round 1 of $(\frac{n}{3})$-WSS$_i$ is run.

2. Player P_i broadcasts the following:

 – $a_{ij} = f_i(j) + F_i^W(0, j)$

 – $b_{ij} = g_i(j) + F_j^W(0, i)$

Concurrently, round 2 of $(\frac{n}{3})$-WSS$_i$, $i = 1, \ldots, n$, also takes place.

3. For each pair $a_{ij} \neq b_{ji}$ the following happens:

 – P_i broadcasts $\alpha_{ij} = f_i(j)$

 – P_j broadcasts $\beta_{ji} = g_j(i)$

 – D broadcasts $\gamma_{ij} = F(j, i)$

Concurrently, round 3 of $(\frac{n}{3})$-WSS$_i$, $i = 1, \ldots, n$, also takes place.

A player is said to be *unhappy* if the value that he broadcast does not match the dealer's value. If there are more than t unhappy players, disqualify D and stop.

Local computation:

– Let $\mathcal{H}$ denote the set of happy players. Remove from $\mathcal{H}$ each player P_i who gets disqualified as the dealer in protocol instance $(\frac{n}{3})$-WSS$_i$. Now, if $|\mathcal{H}| < n - t$ then disqualify D and stop.

– For the remaining players, let $\mathcal{H}_i^W$ denote the set of happy players in instance $(\frac{n}{3})$-WSS$_i$. For each player $P_i \in \mathcal{H}$, check that there exist at least $n - t$ players in $\mathcal{H}$ who are also in $\mathcal{H}_i^W$; if not, remove P_i from $\mathcal{H}$. Let us call this final set $CORE_{Sh} := \mathcal{H}$. If $|CORE_{Sh}| < n - t$ then disqualify D and stop.   ♦

Reconstruction Phase. For each $P_i \in CORE_{Sh}$, run the reconstruction phase of $(\frac{n}{3})$-WSS_i, concurrently.

Local computation: Now each player P_i constructs a set $CORE_{Rec}$ as follows. Initially, $CORE_{Rec} := CORE_{Sh}$.

- Remove from $CORE_{Rec}$ every player P_i such that the outcome of $(\frac{n}{3})$-WSS_i equals $\perp$.
- For every $P_i \in CORE_{Rec}$, use the values a_{ij} he broadcast in round two of the sharing phase to compute

$$f_i(j) = a_{ij} - F_i^W(0,j), \quad 1 \leq j \leq n. \tag{1}$$

- Interpolate these points. Check that the resulting polynomial $f_i(x)$ is a polynomial of degree at most t. If not, remove P_i from $CORE_{Rec}$.
- Reconstruct the secret by taking any $t+1$ polynomials $f_i(x)$, $P_i \in CORE_{Rec}$, to obtain $F^*(x,y)$, and compute $s^* = F^*(0,0)$. $\diamond$

Lemma 3. *If D is honest, then $CORE_{Sh}$ contains all the honest players.*

Proof. First, since D is honest, all honest players are happy with respect to $F(x,y)$. Thus, initially, $\mathcal{H}$ contains all the honest players. Similarly, the set of happy players corresponding to $(\frac{n}{3})$-WSS_i started by a honest player P_i will contain all the honest players. Thus $|\mathcal{H}_i^W| \geq n-t$ and all the honest players will be in $\mathcal{H}$. Also, since all honest players are mutually consistent, an honest player P_i is consistent with $n-t$ players in $\mathcal{H}$ and thus $P_i \in CORE_{Sh}$. $\square$

Lemma 4. *If D does not get disqualified in the sharing phase then all the honest players in $CORE_{Sh}$ are consistent with each other and, when $|CORE_{Sh}| \geq n-t$, consistently define a unique polynomial $F^H(x,y)$ of degree at most t in each variable. Furthermore, when D is honest, $F^H(x,y) = F(x,y)$.*

Proof. Since the honest players use their pads faithfully there are no inconsistencies between honest players in $CORE_{Sh}$. Furthermore, if $|CORE_{Sh}| \geq n-t$, then there are at least $t+1$ honest players in $CORE_{Sh}$ defining a unique polynomial $F^H(x,y)$. Finally, in case the dealer is honest, it holds that $F^H(x,y) = F(x,y)$. $\square$

Lemma 5. *If D does not get disqualified in the sharing phase then, at the end of the reconstruction phase, there are at least $t+1$ honest players in $CORE_{Rec}$.*

Proof. In the reconstruction phase a player P_i gets removed from $CORE_{Rec}$ in only two cases: 1) the reconstruction phase of $(\frac{n}{3})$-WSS_i results in $\perp$, or 2) the reconstruction phase of $(\frac{n}{3})$-WSS_i succeeds but the resulting polynomial $f_i(x)$ is of degree larger than t. By the properties of WSS, both cannot happen with respect to a honest player, and thus at least $n-2t > t$ honest players in $CORE_{Sh}$ remain in $CORE_{Rec}$. $\square$

Lemma 6. *If D does not get disqualified in the sharing phase, then any $t+1$ players in $CORE_{Rec}$ define the same bivariate polynomial.*

Proof. If a dishonest player P_i remains in $CORE_{Rec}$, then the reconstruction phase of $(\frac{n}{3})$-WSS_i has succeeded. By Lemma 2 this implies that the reconstructed polynomial $F_i^W(x, y)$ is consistent with all the happy honest players with respect to $(\frac{n}{3})$-WSS_i. By Lemma 5 there are at least $t + 1$ honest players in $CORE_{Rec}$ who, by Lemma 4, define a unique polynomial $F^H(x, y)$ of degree at most t in both variables. Thus, every player remaining in $CORE_{Rec}$ is consistent with $F^H(x, y)$, and the lemma follows. $\qquad\square$

Theorem 4. *Protocol $(\frac{n}{3})$-VSS is an efficient, perfectly secure three-round (n, t)-VSS protocol for $n > 3t$.*

Proof (sketch). We only have to consider the case when D is honest. The number of rounds and polynomial-time computation are immediate. We prove the three VSS properties in turn.

PRIVACY: Assume that, at the end of the reconstruction phase, the players in $\mathcal{A} \subset \mathcal{P}$, $|\mathcal{A}| \leq t$, are corrupted. Let $\text{View}_{\mathcal{A}}^k$, $1 \leq k \leq 3$, denote the adversary's view after step k of the sharing phase. Note that, for all $P_a \in \mathcal{A}$, the polynomials $F_a^W(x, y)$ are exclusively used in order to blind values already known to the adversary, and therefore we can ignore these polynomials.

After step 2 of the sharing phase, the adversary holds (at most) the following polynomials: $F(x, a)$, $F(a, y)$, $F_i^W(x, a)$, and $F_i^W(a, y)$, and it holds that, for all $P_i, P_j \notin \mathcal{A}$, $H\left(F(x, i) | \text{View}_{\mathcal{A}}^2\right) = \log |\mathcal{K}|$, $H\left(F_i^W(j, x) | \text{View}_{\mathcal{A}}^2\right) = H\left(F_i^W(0, x) | \text{View}_{\mathcal{A}}^2\right) = \log |\mathcal{K}|$. Furthermore, for $P_i, P_j \notin \mathcal{A}$, $i \neq j$, the polynomials $F_i^W(x, y)$ and $F_j^W(x, y)$ are independent.

In step 3, in addition, the polynomials $S_i(x) = F(x, i) + F_i^W(0, x)$ get revealed. That is, each $F(x, i)$ is blinded with an independent polynomial $F_i^W(0, x)$ where $H\left(F_i^W(0, x) | \text{View}_{\mathcal{A}}^2\right) = \log |\mathcal{K}|$. Thus, it is still the case that for any $P_i \notin \mathcal{A}$, $H\left(F(x, i) | \text{View}_{\mathcal{A}}^3\right) = \log |\mathcal{K}|$, and therefore, $H\left(F(0, 0) | \text{View}_{\mathcal{A}}^3\right) = \log |\mathcal{K}|$; hence, privacy follows.

CORRECTNESS: We only consider the case when D is honest. By Lemma 3, all the honest players will be in $CORE_{Sh}$, thus $|CORE_{Sh}| \geq n - t$, and the dealer is not disqualified in the sharing phase. By Lemma 4, the shares of the honest players in $CORE_{Sh}$ define the dealer's original polynomial $F^H(x, y) = F(x, y)$. Obviously, all honest players remain in $CORE_{Rec}$, and by Lemma 6, $s = F(0, 0)$ gets reconstructed from the shares of any $t + 1$ players in $CORE_{Rec}$.

COMMITMENT: If D is dishonest and does not get disqualified in the sharing phase, then $|CORE_{Sh}| \geq n - t$ and, by Lemma 5, at least $t + 1$ honest players from $CORE_{Sh}$ remain in $CORE_{Rec}$. By Lemma 4, all honest players in $CORE_{Sh}$ consistently define the same polynomial $F^H(x, y)$ after the sharing phase. Thus, the $t + 1$ honest players in $CORE_{Sh} \cap CORE_{Rec}$ still uniquely define $F^H(x, y)$ and, by Lemma 6, $s^* = F^H(0, 0)$ gets reconstructed from the shares of any $t + 1$ players in $CORE_{Rec}$. $\qquad\square$

5   VSS in $(1 + \varepsilon)$ Rounds

Depending on the particular application, minimizing the round complexity of a stand-alone protocol might not always be the best way to optimize. In multi-party computation, for example, where a large number of VSS protocols are executed sequentially, it is useful to minimize the overall *amortized* round complexity of the VSS instances.

A number m of sequential (n,t)-VSS executions can be easily achieved in $1 + O(\frac{1}{m})$ amortized rounds by "deferring" the commitment as follows. Suppose we have a k-round (n,t)-VSS protocol, and we need to execute m instances of it. In an initial phase, dealer D (or all future dealers in the application, respectively) shares (in parallel) a set of random elements $r_1, \ldots, r_m$ using the given (n,t)-VSS protocol. The sharing phase of the j-th execution of the (n,t)-VSS protocol, $j = 1, \ldots, m$ then simply consists of the dealer broadcasting a correction term $c_j = s_j - r_j$, where s_j is the secret to be shared in this instance. The correction term c_j can be handled in two different ways:

1. c_j is incorporated in the reconstruction phase. That is, after the reconstruction of random element r_j, each player locally computes $s_j = r_j + c_j$; or

2. the sharing is immediately "corrected" at the end of the sharing phase, by having every player P_i compute $F'_k(x, i) = F_k(x, i) + c_k$ and $F'_k(i, y) = F_k(i, y) + c_k$.

Theorem 5. *Any number m of sequential VSS protocols for $n > 3t$ is efficiently achievable in $m + 2$ rounds, thus implying $1 + \varepsilon$ amortized rounds per instance for any $\varepsilon > 0$ when m is sufficiently large.*

Proof. Using any k-round (n,t)-VSS protocol the above approach results in $m + k - 1$ rounds overall, or $1 + \frac{k-1}{m}$ rounds per VSS. In particular, using the round-optimal protocol from Section 4 results in $1 + \frac{2}{m}$ rounds per VSS instance. Thus, in order to achieve $1 + \varepsilon$ amortized rounds, it is sufficient to choose $m \geq \frac{2}{\varepsilon}$. $\square$

6   Summary

In this paper we gave efficient three-round protocols for perfectly secure WSS and VSS when $n > 3t$, and showed that there is no (n,t)-WSS protocol involving less than three rounds when $n \leq 4t$. Furthermore, we gave an efficient one-round protocol for perfectly secure WSS when $n > 4t$, and demonstrated that perfectly secure VSS can be achieved in $(1 + \varepsilon)$ rounds when $n > 3t$.

The following table summarizes the tight bounds on the round complexity of perfectly secure WSS and VSS as given in [5] and in this paper — where round optimality is always achieved efficiently. ("—" stands for impossibility.)

Protocol	Threshold	Number of rounds
WSS	$n \leq 3t$	—
	$3t < n \leq 4t$	3
	$4t < n$	1
VSS	$n \leq 3t$	—
	$3t < n \leq 4t$	3
	$4t < n \ (t > 1)$	2
	$4t < n \ (t = 1)$	1

Note that, same as some (but not all) of the protocols in [5], although our solution for VSS fulfills the standard VSS definition, it is not powerful enough to allow for general multi-party computation. In particular, multiplication of shared secrets is not directly possible since the sharing phase of two different VSS invocations may end up in different $CORE_{Sh}$ sets.

Furthermore, note that, also as the protocols in [5], our VSS protocol satisfies the stronger VSS definition in [7] (Definition 3.3.13) requiring that any set of $t + 1$ honest players be able to reconstruct the shared secret. This condition is satisfied because any set of $t + 1$ honest players can reconstruct the WSS-shared secrets of all players in $CORE_{Sh}$.

Finally, it can be easily seen that our protocols also work with respect to a (possibly corrupted) external dealer while still tolerating t corrupted players among the "share holders."

Acknowledgements

We thank the anonymous reviewers for TCC '06 for their many helpful comments.

References

1. M. Ben-Or, S. Goldwasser, and A. Wigderson. Completeness theorems for non-cryptographic fault-tolerant distributed computation. In *Proceedings of the 20th Annual ACM Symposium on Theory of Computing (STOC '88)*, pages 1–10, 1988.
2. G. R. Blakley. Safeguarding cryptographic keys. In *1979 National Computer Conference*, volume 48 of *AFIPS Conference proceedings*, pages 313–317. AFIPS Press, 1979.
3. D. Chaum, C. Crépeau, and I. Damgård. Multiparty unconditionally secure protocols (extended abstract). In *Proceedings of the 20th Annual ACM Symposium on Theory of Computing (STOC '88)*, pages 11–19. ACM Press, 1988.
4. B. Chor, S. Goldwasser, S. Micali, and B. Awerbuch. Verifiable secret sharing and achieving simultaneity in the presence of faults. In *Proceedings of the 26th Annual IEEE Symposium on Foundations of Computer Science (FOCS '85)*, pages 383–395, 1985.
5. R. Gennaro, Y. Ishai, E. Kushilevitz, and T. Rabin. The round complexity of verifiable secret sharing and secure multicast. In *Proceedings of the 33rd Annual ACM Symposium on Theory of Computing (STOC '01)*, pages 580–589, 2001.

6. R. Gennaro, M. O. Rabin, and T. Rabin. Simplified VSS and fast-track multiparty computations with applications to threshold cryptography. In *Proceedings of the 17th ACM Symposium on Principles of Distributed Computing (PODC '98)*, pages 101–111, 1998.
7. O. Goldreich. Secure multi-party computation, final (incomplete) draft, version 1.4, Oct. 2002.
8. T. Rabin and M. Ben-Or. Verifiable secret sharing and multiparty protocols with honest majority. In *Proceedings of the 21st Annual ACM Symposium on Theory of Computing (STOC '89)*, pages 73–85, 1989.
9. A. Shamir. How to share a secret. *Commun. ACM*, 22:612–613, 1979.

A   Proof of Theorem 2

We show that for $n \leq 4t$, perfect WSS is not possible in less than three rounds. We do this along the lines of the impossibility proof for two-round VSS in [5]. We first introduce the problem of *weak secure multicast* (WSM) and show that perfectly secure WSM is impossible in less than three rounds when $n \leq 4t$. Finally, we show that r-round WSS implies r-round WSM, thus proving the theorem.

Weak Secure Multicast (WSM). Consider an n-player protocol among player set $\mathcal{P} = \{P_1, \ldots, P_n\}$ wherein *sender* $D \in \mathcal{P}$ holds an input m and each player in *multicast set* $M \subseteq \mathcal{P}$ $(D \in M)$ finally computes an output. Such a protocol is called a (perfect) *WSM protocol* if, for any t-adversary, the following conditions hold:

PRIVACY: If all players in M are honest then the adversary learns no information about D's input m.

CORRECTNESS: If D is honest then all honest players in M output m.

WEAK AGREEMENT: Even if D is dishonest, all dishonest players in M output a value in $\{m^*, \perp\}$, where m^* is an unique element in $\mathcal{K}$ and a distinguished value $\perp \notin \mathcal{K}$.

Similarly to VSS, WSM is the "weak" variant of the *secure multicast* (SM) problem formalized in [5], where the Agreement condition, demanding that all the honest players output the same value even if the sender is dishonest, is replaced by Weak Agreement above.

The proof of Theorem 2 follows by proving the impossibility of the following problem and subsequently reducing it to related problems, the last one being the existence of a two-round WSS protocol.

Lemma 7. *There is no deterministic 3-player protocol satisfying the following requirements:*

1. *The protocol is a $(3,1)$-WSM protocol with M being the set of all players.*
2. *The protocol has three communication rounds, where only D speaks in the first round.*

3. If all players are honest then the broadcast messages are independent of D's message m.

The proof of this lemma is identical to the proof of Lemma 7 in [5] for the non-existence of a $(3,1)$-SM protocol satisfying similar requirements.

Lemma 8. *There is no two-round perfect $(4,1)$-WSM protocol with $M=\{P_1, P_2, P_3\}$ (and $D = P_1$).*

Proof (sketch). The existence of such a protocol would imply the existence of the protocol specified in Lemma 7. The proof is almost identical to that of Lemma 6 for the impossibility of a two-round $(4,1)$-SM protocol in [5]. The minor modification is that it is based on our Lemma 7 (instead of their Lemma 7), which accounts for the alternative output $\bot$ of WSM; even though this outcome avoids violation of weak agreement, it still violates correctness.   □

Lemma 9. *There is no two-round perfect $(4,1)$-WSS protocol.*

Proof (sketch). Again, the proof of similar Lemma 3 (and thus of Lemma 5) of [5], which reduces the impossibility of a two-round (n,t)-VSS protocol to the impossibility of a two-round (n,t)-SM protocol, can be based on our Lemma 8, directly implying this stronger lemma.   □

Finally, the proof of Theorem 2 follows from Lemma 9 by a standard player partitioning and simulation argument.

Generalized Environmental Security from Number Theoretic Assumptions

Tal Malkin[1,*], Ryan Moriarty[2,**], and Nikolai Yakovenko[3,***]

[1] Department of Computer Science, Columbia University
tal@cs.columbia.edu
[2] Department of Computer Science, UCLA
ryan@cs.ucla.edu
[3] Google, Inc
yakovenko@google.com

Abstract. We address the problem of realizing concurrently composable secure computation without setup assumptions. While provably impossible in the UC framework of [Can01], Prabhakaran and Sahai had recently suggested a relaxed framework called generalized Environmental Security (gES) [PS04], as well as a restriction of it to a "client-server" setting based on monitored functionalities [PS05]. In these settings, the impossibility results do not apply, and they provide secure protocols relying on new non-standard assumptions regarding the existence of hash functions with certain properties.

In this paper, we first provide gES protocols for general secure computation, based on a new, concrete number theoretic assumption called the relativized discrete log assumption (rDLA). Second, we provide secure protocols for functionalities in the (limited) client-server framework of [PS05], replacing their hash function assumption with the *standard* discrete log assumption. Both our results (like previous work) also use (standard) super-polynomially strong trapdoor permutations.

We believe this is an important step towards obtaining positive results for efficient secure computation in a concurrent environment based on well studied assumptions. Furthermore, the new assumption we put forward is of independent interest, and may prove useful for other cryptographic applications.

1 Introduction

1.1 Background and Motivation

Since its beginnings a few decades ago, theoretical cryptography has developed by formalizing the intuitive notions of security, and basing the strength of protocols realizing these definitions on widely accepted complexity assumptions.

---

* Supported by NSF Early Career Development (CAREER) Grant CCF-0347839.
** This work was done while the author was at the department of Computer Science, Columbia University.
*** This work was done while the author was at the department of Computer Science, Columbia University.

S. Halevi and T. Rabin (Eds.): TCC 2006, LNCS 3876, pp. 343–359, 2006.
© Springer-Verlag Berlin Heidelberg 2006

Much success was achieved in defining and realizing secure multi-party computation, arguably the most general and important task in cryptography, in various 'stand-alone' settings. As our understanding develops side by side with new emerging needs and applications for cryptography in uncontrolled, distributed environments such as the Internet, new goals and challenges arise. An important current direction in cryptography is to model and realize secure protocols operating in such open settings, requiring concurrent composition. Intuitively, one would like to have protocols that will remain secure even if they are composed arbitrarily with other protocols. Such general composability is often what is required in practical settings. It is important to continue to base this new developing theory on well studied and scrutinized complexity assumptions.

The UC/ES Framework. Perhaps the most well known definition of security in a composable setting is the Universally Composable (UC) Security paradigm of Canetti [Can01] (an alternative paradigm was proposed by Pfitzmann et. al. [PW00, BPW04]).

The UC security notion is based on the (by now standard) ideal world / real world simulation paradigm. Very roughly, an ideal world is defined where functions are computed by a trusted party. For a protocol to be secure, we require that for every adversary A operating in the real world under a certain environment, there exists an ideal world adversary S (a simulator), working in time polynomial in that of A, that can simulate everything that happened in the real world under A. This should hold under any environment (which models anything else going on in the world, provides inputs to all parties, watches their interactions, etc). Hence, the UC security notion is also referred to as the Environmental Security (ES) notion.

A major advantage of this framework is that, according to the UC Theorem [Can01], protocols that are secure in this model remain secure even when composed concurrently and arbitrarily (hence the name universally composable security). In particular, consider an arbitrary protocol π which uses some ideal calls (using a trusted party) to compute certain functions. Replacing the ideal calls by UC-secure protocols computing the functions is safe in the sense that whatever an adversary A can achieve, can be simulated in polynomial time within the ideal calls model.

Unfortunately, while this UC/ES framework is very appealing and strong in terms of the provided security guarantees, it turned out to be *too* strong. Indeed, many of the most basic cryptographic tasks (such as commitment or secure computation) were proven impossible to realize in this framework, unless additional "trust" assumptions are being made [Can01, CF01, DG03, CKL03, Lin03, Lin04] (e.g., an honest majority, or a common random string available to all parties and selected by a trusted party).

The gES Framework. Recently, Prabhakaran and Sahai [PS04] introduced a new model of security, generalized Environmental Security (gES). Roughly, this model relaxes the security requirements of the UC/ES setting so as to avoid the impossibility results, while still rendering the model meaningful enough that a protocol secure in this model intuitively implies meaningful guarantees on its actual security if employed "in real life". We discuss the meaningfulness of these guarantees below. This framework is exciting and promising, as it allowed, for the first time, to realize multi-party computation of general functionalities without any setup assumptions, while maintaining security under a pretty general form of composability (see discussion below).

Their idea, roughly, was to perform a thought experiment where the adversary in the ideal world (the simulator) is given super-polynomial computational power (following the approach suggested by Pass [Pas03]). To allow for secure composability, the super-polynomial power in the ideal world is given through a super polynomial angel (oracle), which can answer queries based on its knowledge of who the corrupted parties are. This is the gES notion of security. We refer the reader to [PS04] for more details, but remind that the angel is only required as a tool for the security proof, and is not needed for protocol execution. For a particular angel Γ, the resulting security model is called Γ-ES. Using this model, [PS04] showed how to achieve secure multi-party computation of any functionality, against a static adversary (one who cannot corrupt parties adaptively), and without any setup assumptions. More specifically, it is shown in [PS04] that

1. For every Imaginary Angel Γ, Γ-ES protocols are universally composable.
2. There exists an Imaginary Angel Ψ (under new complexity assumptions) such that there are Ψ-ES protocols for commitment, ZK proofs and any PPT functionality.

This result is very important towards the ultimate goal of reasonable, concurrently secure protocols, without setup assumptions. However, the result of [PS04] is based on a new, non-standard assumption, requiring the existence of a hash function with certain properties regarding distributions of collisions on inputs with the same prefixes (see further discussion in Section 4.1).

Perhaps the most important open problem left in [PS04] is to realize such secure computation without setup assumptions relying on simpler, more standard, easier to analyze, complexity assumptions. Our work provides a big step in this direction.

Very recently (and independently of our work), Barak and Sahai [BS05] have also addressed this problem, and showed how to realize such secure computation (again under a relaxed model where the simulator is super polynomial in the real-world adversary), using reasonably standard assumptions. Namely, assuming the existence of a hash function collection that is collision resistant with respect to super polynomial adversaries, and trapdoor permutations secure against super polynomial adversaries. The main advantage of our solutions, as we shall see, is their simplicity, which will hopefully be useful towards practical implementations, and more importantly, towards distilling a better understanding of this security model, it's meaning, advantages, and limitations.

Finally, we touch upon one more recent framework. Recently, Prabhakaran and Sahai [PS05] suggested a relaxation of gES, introducing Monitored functionalities and Client-Server Computation. This relaxation aims at achieving secure computation (alas, of a limited class of functionalities) with weaker assumptions, simpler, and more efficient protocols. We do not describe or motivate this framework here, except to note that the assumptions used are still non-standard and hard to work with. On the other hand, the functionalities that can be realized, while limited, avoid the impossibility results in the standard UC model, and thus provide an interesting advance.

Why is Security in the gES Framework Meaningful? Before presenting our results, we discuss the meaning of security and composability in the gES framework which we use (the same discussion applies to the framework of [BS05] which also uses super polynomial simulation).

In terms of secure computation of a given function, it can be argued (see [PS04, BS05]) that for most applications of secure computation, the ideal model is still "ideal enough" (or "secure enough") even when the adversary is allowed to run in time that is bounded by a specific super polynomial function (depending on the hardness assumption used).[1] Thus, proving that any adversary in the real world can be simulated by such a (strong) ideal model adversary, still provides a meaningful notion of security.

However, it is important to understand the implications of what exactly is guaranteed (and not guaranteed) by the theorem proving universal composability in this framework (be it Γ-security for some angel Γ as used in [PS04] and in this paper, or the security notion suggested by [BS05]). What *is* guaranteed is that, given any protocol secure according to the notion at hand, the protocol remains secure even when composed in an arbitrary manner with any other arbitrary protocols. But, no guarantees are made for protocols that are *not* secure according to the notion at hand.

In particular, consider a protocol π which is not secure according to this notion, but enjoys some other weaker security features (e.g., the protocol has some security guarantee in the ideal calls model when the adversary is polynomial time bounded). Now, composing this protocol with other protocols within the new framework (e.g., replacing ideal calls to a function f with a Γ-ES secure protocol for f), may break the original (weaker) security guarantee that π had. Indeed, all we know is that anything that happens with adversary A can be simulated in the ideal calls model with an adversary S that has access to the (super-polynomial) Γ. While the protocol for f was Γ-ES secure, Γ (which is used only as a tool in the analysis) may help "break" some other sub-protocol in π.[2] In this sense, the notion of composability guaranteed by the general theorems, is not completely "general composition" as defined by [Lin03].

An argument can be made that one cannot maintain "all possible weak security properties" of insecure protocols under composition (it's not even clear how

---

[1] In fact, in many applications, the ideal model is such that even a computationally unbounded adversary cannot cause damage.

[2] In fact, π may have been designed specifically with this goal in mind, following a "chosen protocol attack" [KSW97]. For example, if Γ is the one used in this paper, namely providing discrete logs (breaking DLA) relative to some primes, π could contain a part that relies on the DLA for an appropriate prime.

to define this), and that the (or a) right notion of security is one that guarantees security to those who use it, while insecure protocols naturally will remain insecure under composition. Moreover, one can argue that this is also the case for the standard UC-security framework: security under composition is only guaranteed when composing protocols that were UC-secure to begin with.[3] On the other hand, it seems clear (historically, intuitively, and practically), that security in an ideal calls model against a polynomial time bounded adversary (the notion of security used for UC) is an extremely natural and important security notion. It may be reasonable to assume that anything that is less secure than that is completely insecure (and thus nothing needs to be preserved under composition for protocols that are not secure according to this notion). For sure, it would be desirable to maintain this security property for protocols even when composed with other protocols in a stronger model such as Γ-ES security. It is important to note that this is, unfortunately, not the case.

At this point we leave the philosophical discussion about the general direction that [PS04] initiated, and [BS05] and the current work follow, and continue to describe our results.

1.2 Our Results

We provide an important step towards realizing gES with standard number theoretic assumptions that are concrete, natural, easier to study and analyze, or indeed refute.

First, we provide an instantiation of the assumptions used by [PS04], based on a new assumption, which we call the relativized discrete log assumption (rDLA), as well as a standard (strong) assumption of trapdoor permutations (TDP) secure against super polynomial adversaries. The details of these assumptions are discussed later in the paper. In particular, we obtain Ω-ES protocols for arbitrary functionalities against static adversaries, with no setup assumptions (here Ω is our imaginary angel, and the security is based on the rDLA). While non-standard, the rDLA is simple to state (intuitively, it says that the DLA over a certain group holds even in the presence of oracles breaking the DLA for other groups), and strictly algebraic/number-theoretic in nature (we work over subgroups of prime order of safe primes, although our assumption and protocols could be considered over other groups). We believe the assumption is easy to understand and think about, and it seems quite reasonable.[4] Since this assumption can be framed as an instantiation of the original [PS04] assumption (through a realization of their hash function), our work simplifies and 'cleans' the previous

---

[3] For example, if a protocol π in the ideal calls model maintained some (weak) notion of security against adversaries bounded by quadratic running time, replacing an ideal call to some f by a UC-secure protocol for f may break that property, as the ideal model adversary in the UC framework is allowed polynomial running time.

[4] We do not claim to be experts in number theory, although several other people that we asked also found the assumption reasonable. Further, the same somewhat philosophical arguments on the plausibility of the assumptions made in [PS04] apply here as well, except that our assumptions are easier to try to attack.

construction, and helps bring the gES model and the [PS04] protocols under more scrutiny, hopefully towards helping to strengthen our belief in its security. We also note that while the [BS05] construction relies on more standard assumptions, and in this sense subsumes our results, our resulting protocols are cleaner, simpler, and hopefully a step towards practically efficient concurrently composable protocols.

Second, we provide an instantiations of the assumptions used by [PS05], based on the standard discrete log assumption (DLA), as well as a standard (strong) assumption of TDP secure against super polynomial adversaries. This allows us to obtain, like [PS05], simpler protocols for a limited class of "server-client" functionalities. Perhaps more importantly, this yields the first results for concurrently secure computation without setup assumptions, under standard computational assumptions.

In sum, our work addresses the arguably most important problem left open by the recent [PS04] pioneering work, and we hope that it provides a useful step towards achieving, or at least understanding, the "holy grail" in this field. Moreover, it provides a significant improvement of the [PS05] results, replacing a new and unstudied assumption by the completely standard DLA. Finally, we believe rDLA, our new assumption, is worth studying independently of the current context, and is likely to find other cryptographic applications.

2 Preliminaries

We do not provide here formal definitions of the gES and related models, and refer the reader to the original papers [PS04, PS05] for definitions (as well as further justifications regarding the meaningfulness of the security model).

In all our protocols we use k as the security parameter, and consider functionalities of up to a polynomial n number of parties.

We assume all the parties have unique IDs, which may be adversarially chosen as long as they adhere to the legal format. In this paper, the IDs are of the form q where q is a safe prime, namely $q = 2p + 1$ for a prime p (p is called a Sophie Germain prime).

All adversaries considered in this work are PPT, non-uniform, and static (namely choose the set of parties to corrupt at the onset of computation).

For two distributions $\mathcal{X}$ and $\mathcal{Y}$ we write $X \approx Y$ to denote that they are indistinguishable by PPT circuits (with respect to the security parameter k).

3 Our Assumptions

In this section we summarize the assumptions that we will use for our results. Section 3.2 describes the rDLA assumption that is new to our work. Section 3.1 describes other assumptions that we will use, which are standard assumptions. The assumptions previously used by [PS04] and [PS05] appear in Section 4.1 and Section 5.1.

The assumptions we use are related to the discrete log assumption (DLA), which is commonly assumed for different groups. We state our assumptions for subgroups of prime order of Z_q^*, specifically for the subgroup G of the quadratic residues, when $q = 2p + 1$ is a safe prime (this can be somewhat extended).

3.1 Standard Assumptions

We sketch these assumptions without full formal details.

The Discrete Log Assumption (DLA). For any PPT adversary A, consider the following probabilistic experiment: choose a random safe prime $q = 2p + 1$ of length k, let G be the subgroup of order p (all quadratic residues) of Z_q^*, let g be a generator of G, and choose a random $y \in G$. Then, the probability that $A(q, g, y) = x \in Z_p^*$ such that $y = g^x$ is negligible.

The Discrete-Log-Safe Trapdoor Permutation Assumption (DLS-TDP). For n polynomial in the security parameter k, there exists a family of trapdoor permutations over $\{0, 1\}^n$, that remain secure against adversaries with access to an oracles solving discrete logarithms in the subgroups of Z_q^*, for safe primes q of size k.

Note that the above assumption is implied by the (more standard assumption of) existence of trapdoor permutations secure against adversaries with super polynomial power 2^{n^ϵ}. Indeed, if such strong TDP exist, we can choose $n = k^{1/\epsilon}$. Then in time 2^k the discrete log problem can be solved, but the TDP remains secure, implying the DLS-TDP assumption.

3.2 The Relativizing Discrete Log Assumption (rDLA)

Let $q = 2p + 1$ be a safe prime of size k, and let G be the subgroup of size p. Then the discrete log problem over G is hard (i.e., no PPT adversary can compute discrete logs with non-negligible probability), even when the adversary has access to oracles that solve the discrete log problem for any input in any other group defined by a safe prime $q' \neq q$ of size k.

Intuitively, the rDLA assumes some sort of "non-malleability" among different groups, asserting that being able to take discrete logs in all other groups of the same size, will not help an adversary take discrete logs in the given group.

4   Achieving gES

In this section we will show how to use rDLA and DLS-TDP to realize secure multi-party computation (for static adversaries) in the gES framework, without any setup assumptions. We do that by showing an instantiation of the assumed hash function in [PS04], proving it satisfies the required properties, and presenting the resulting Angel and protocols for general secure multi-party computation.

4.1   The Assumptions and Angel of [PS04]

The constructions of [PS04] rely on the following assumptions[5]:

Assume there exists a hash function $\mathcal{H} : \{0,1\}^k \to \{0,1\}^l$. The input to $\mathcal{H}$ is of the form $(\mu, r, x, b) \in \mathcal{J} \times \{0,1\}^{k_1} \times \{0,1\}^{k_2} \times \{0,1\}$, where $\mathcal{J}$ is the set of IDs of the parties (each party is assumed to have a unique ID, possibly chosen adversarially), and k_1, k_2, and l are all polynomial in k. It is assumed that $\mathcal{H}$ satisfies the following.

A1 (Collisions and Indistinguishably): For every $\mu \in \mathcal{J}$ and $r \in \{0,1\}^{k_1}$, there is a distribution $\mathcal{D}_r^\mu$ over $\{(x,y,z) | \mathcal{H}(\mu, r, x, 0) = \mathcal{H}(\mu, r, y, 1) = z\} \neq \phi$, such that

$$\{(x,z)|(x,y,z) \leftarrow \mathcal{D}_r^\mu\} \approx \{(x,z)|x \leftarrow \{0,1\}^{k_2}, z = \mathcal{H}(\mu, r, x, 0)\}$$
$$\{(y,z)|(x,y,z) \leftarrow \mathcal{D}_r^\mu\} \approx \{(y,z)|y \leftarrow \{0,1\}^{k_2}, z = \mathcal{H}(\mu, r, y, 1)\}$$

Further, even if the distinguisher is given sampling access to the set of distributions $\{\mathcal{D}_{r'}^{\mu'} | \mu' \in \mathcal{J}, r' \in \{0,1\}^{k_1}\}$, these distributions still remain indistinguishable. (Intuitively, this assumption states that there are collisions in the hash function, which are indistinguishable from a random hash of a 0 or a 1).

A2 (Difficult to find collisions with same prefix): For all PPT circuits M and every id $\mu \in \mathcal{J}$, for a random $r \leftarrow \{0,1\}^{k_1}$, probability that $M(r)$ outputs (x,y) such that $\mathcal{H}(\mu, r, x, 0) = \mathcal{H}(\mu, r, y, 1)$ is negligible. This remains true even when M is given sampling access to the set of distributions $\{\mathcal{D}_{r'}^{\mu'} | \mu' \neq \mu, r' \in \{0,1\}^{k_1}\}$. (Note that without this last requirement, insisting that finding collisions remains difficult even when given sampling access to collisions for other μ', a hash function satisfying these properties could be constructed under standard assumptions).

Additionally, [PS04] also rely on the following assumption:

A3 There exists a family of trapdoor permutations $\mathcal{T}$ over $\{0,1\}^n$, which remains secure even if the adversary has sampling access to $\mathcal{D}_r^\mu$ for all μ and r.

As discussed above (and in [PS04]), A3 can be replaced by the (stronger, but more standard and natural looking) assumption of TDP secure against superpolynomial adversaries.

The Angel Ψ. [PS04] use the following imaginary angel Ψ. On query (μ, r), Ψ checks whether μ is one of the corrupted parties. If so, Ψ outputs a sample from D_μ^r. If not, Ψ returns $\bot$.

4.2   Our Hash Function

We propose the following hash function $\mathcal{H}_0$, point its correspondence to the [PS04] hash function, and prove that (under our assumptions) it realizes their required assumptions A1,A2,A3.

---

[5] This text is extracted almost verbatim from [PS04].

Defining $\mathcal{H}_0$. $\mathcal{H}_0 : \{0,1\}^k \rightarrow \{0,1\}^l$ is defined as follows. The input to $\mathcal{H}_0$ is of the form (q, g_0, g_1, x, b), where:

- $q = 2p + 1$ is a safe prime (namely, p is a prime as well) of length k_1, where k_1 is polynomially related to k. (This corresponds to the party ID μ).[6]
- g_0, g_1 are generators of $G = QR(Z_q^*)$. Equivalently, each of g_0, g_1 is a quadratic residue not equal to 1 in Z_q^*. (This corresponds to r).
- $x \in Z_p^*$
- $b \in \{0,1\}$.

The output of $\mathcal{H}_0$ is then defined as:

$$\mathcal{H}_0(q, g_0, g_1, x, b) = g_b^x \bmod q.$$

We note that it is easy to efficiently check whether the input is of the correct form, by using primarily testing, and testing whether g_i is a quadratic residue by computing its Legendre symbol. It is also easy to generate inputs to $\mathcal{H}_0$; Choosing random generators g_0, g_1 of the QR subgroup can be done by simply choosing a random element of Z_q^* and squaring it. Choosing a safe prime q is easy assuming that safe primes (or Sophie Germain primes) are dense.

Satisfying A1, A2 and A3. We next show that, instantiating $\mathcal{H}$ with $\mathcal{H}_0$, A1 is satisfied unconditionally, A2 is satisfied if rDLA holds, and A3 is satisfied if the DLS-TDP assumption holds.

Lemma 1. $\mathcal{H}_0$ satisfies A1.

Proof. We need to show that for all safe primes q of size k_1 and all $g_0, g_1 \neq 1$ quadratic residues in Z_q^*, there exists a distribution $\mathcal{D}_{g_0,g_1}^q$ over $\{(x, y, z) | g_0^x = g_1^y = z \bmod q\}$, such that

$$\{(x, z) | (x, y, z) \leftarrow \mathcal{D}_{g_0,g_1}^q\} \approx \{(x, z) | x \leftarrow Z_p^*, z = g_0^x \bmod q\}$$
$$\{(y, z) | (x, y, z) \leftarrow \mathcal{D}_{g_0,g_1}^q\} \approx \{(y, z) | y \leftarrow Z_p^*, z = g_1^y \bmod q\}$$

We take $\mathcal{D}_{g_0,g_1}^q$ to be a distribution that outputs (x, y, z) where $x \in Z_p^*$ is chosen at random, $z = g_0^x \bmod q$, and $y \in Z_p^*$ is the unique element satisfying $z = g_1^y \bmod q$. Then, it is clear that the above distributions are identical (and in particular indistinguishable for any distinguisher, even if given sampling access to other $\mathcal{D}_{g_0',g_1'}^{q'}$). $\square$

Lemma 2. If rDLA holds, then $\mathcal{H}_0$ satisfies A2.

Proof. We need to prove that, informally, for every safe prime q of size k_1, and for randomly generated g_0, g_1, it is hard to output a collision (x, y) such that $g_0^x = g_1^y$. This should hold even for an adversary with access to such collision

---

[6] As usual, we consider the party IDs to be unique, and possibly adversarially chosen, subject to the required format.

distributions for other safe primes q' of length k_1. Indeed, any PPT circuit M that outputs collisions for q (given random g_0, g_1) with non-negligible probability, can be converted to a PPT M' which finds the discrete log in the subgroup G of quadratic residues with non-negligible probability. Specifically, given a generator g and a random element $z \in G$, M' can proceed by choosing $g' = g^r$ for a random r, and running $M(g', z)$. If M outputs a collision (x, y) such that $g'^x = z^y$, M' can output $rxy^{-1} \bmod p$. If such an M exists, this contradicts the DLA for the corresponding subgroup of Z_q^*. If there is such an M that uses access to distributions $\mathcal{D}_{g_0', g_1'}^{q'}$ for $q' \neq q$, since these distributions can easily be simulated using an oracle that provides discrete logarithms in the QR subgroup of $Z_{q'}^*$, this contradicts the rDLA.   □

Lemma 3. *If the DLS-TDP assumption holds, then A3 holds.*

Proof. This follows immediately from the fact that for all safe primes q and subgroup generators g_0, g_1, providing collisions (x, y, z) such that $g_0^x = g_1^y = z$ is equivalent to providing the discrete logarithm of g_0 with respect to g_1.   □

4.3   Our Angel Ω

Following the $\Gamma - ES$ Angel model, we use a super-polynomial imaginary angel we call Ω, that breaks the security of already corrupted parties. Specifically, Ω is the following. On a query (q, g_0, g_1) (of the usual format), Ω checks whether q is the ID of one of the corrupted parties. If so, Ω returns a such that $g_0 = g_1^a$ mod q (that is, return the discrete log of g_0 with respect to g_1). If q is the ID of a party that is not corrupted, Ω returns $\perp$.

We remark that we could have used the imaginary angel Ψ from [PS04], instantiated with our hash function $\mathcal{H}_0$. Then, the resulting imaginary angel would have outputted a distribution of collisions for the given q, g_0 and g_1, instead of the discrete logarithm a. These outputs are clearly equivalent, and we chose to present our angel Ω as above since it is a somewhat simpler, cleaner choice.

4.4   Putting the Pieces Together: Secure Multi-party Computation in the Ω-ES Model

We have shown how to realize the hash function, angel, and assumptions required for the constructions of [PS04] using the number theoretic assumption rDLA, and the DLS-TDP (or TDP secure against super polynomial adversaries). This allows the main result from [PS04], namely general secure multi-party computation secure against static adversaries without any setup assumptions, to go through in our setting.

Theorem 1. *If rDLA and the DLS-TDP assumption hold, there is a protocol that Ω-ES realizes any multi-party functionality against static adversaries.*

Functionality $\mathcal{F}_{\widetilde{\text{COM}}}$

The parties are sender C and receiver R, with adversary $\mathcal{S}^{\Omega}$. The security parameter is k, and q_C, q_R are safe primes of size k_1 polynomial in k.

COMMIT PHASE:

1. $\mathcal{F}_{\widetilde{\text{COM}}}$ picks random quadratic residues $g_0, g_1 \leftarrow QR(q_R)$ and sends them to C.
2. $\mathcal{F}_{\widetilde{\text{COM}}}$ receives c from C.
3. $\mathcal{F}_{\widetilde{\text{COM}}}$ sends the message COMMIT to R.

REVEAL PHASE:

1. $\mathcal{F}_{\widetilde{\text{COM}}}$ receives (b, x) from C.
2. $\mathcal{F}_{\widetilde{\text{COM}}}$ checks if $c = g_b^x$. If so, then send message (REVEAL,b) to R and adversary $\mathcal{S}^{\Omega}$.

Fig. 1. The basic commitment functionality $\mathcal{F}_{\widetilde{\text{COM}}}$

Proof. This follows immediately from the previous sections, together with Theorem 3 in [PS04]. All of the protocols necessary for the proof are instantiated directly and the proofs follow from the realization of the assumptions. □

For illustration, and some self-containment, we sketch the general outline of the secure multi-party computation protocol, and present the resulting construction of the basic building blocks, the functionality $\mathcal{F}_{\widetilde{\text{COM}}}$ (in Figure 1) and the protocol BCOM (in Figure **??**) that realizes it, within our framework.

Basic Commitment Semi-Functionality $\mathcal{F}_{\widetilde{\text{COM}}}$. Following the construction from [PS04], we first implement the basic IDEAL commitment semi-functionality $\mathcal{F}_{\widetilde{\text{COM}}}$[7]. We instantiate the REAL protocol BCOM from [PS04] with our hash function and then prove that BCOM Ω-ES realizes the semi-functionality $\mathcal{F}_{\widetilde{\text{COM}}}$. Since $\mathcal{F}_{\widetilde{\text{COM}}}$ is not fully ideal, we need to prove that the commitment is binding separately.

With the help of the Ω angel, for every PPT adversary A^{Ω}, we can demonstrate a PPT simulator S^{Ω} such that no PPT environment can distinguish between the REAL interaction with A^{Ω}, and the IDEAL interaction with S^{Ω}. This proves that BCOM is an Ω-ES realization of $\mathcal{F}_{\widetilde{\text{COM}}}$.

To show that the semi-functionality $\mathcal{F}_{\widetilde{\text{COM}}}$ is binding for a corrupt sender C for any environment, we rely upon the rDLA. We show a polynomial reduction from the problem in which a machine M breaks the binding of the commitment scheme $\mathcal{F}_{\widetilde{\text{COM}}}$, to a problem where an adversary uses M to break the security of the rDLA. By assumption, the later is impossible with better than negligible probability, so the commitment is binding.

---

[7] A semi-functionality is one that is not fully ideal, therefore its ideal properties must be proved separately. See [PS04]

Protocol BCOM

The parties are sender (committer) C and receiver R. The security parameter is k, and $q_C = 2p_C + 1, q_R = 2p_R + 1$ are safe primes of size k_1 polynomial in k.

COMMIT PHASE:

1. R picks random quadratic residues g_0, $g_1 \leftarrow QR(q_R)$ and sends them to C.
2. C chooses $x \leftarrow Z^*_{p_R}$ and computes $c = g_b^x$. C requests $\mathcal{F}_{ENC}$ to send c to R.
3. R receives c from $\mathcal{F}_{ENC}$ and accepts the commitment.

REVEAL PHASE:

1. C requests $\mathcal{F}_{ENC}$ to send (b,x) to R, and R receives.
2. R checks if $c = g_b^x$. If so, he accepts b as the reveal.

Fig. 2. The basic commitment protocol BCOM that Ω-ES realizes $\mathcal{F}_{\widetilde{COM}}$

Building the Rest Of the Tools. We follow [PS04] directly to build the rest of the tools need for secure multi-party computation. In the $\mathcal{F}_{\widetilde{COM}}$-hybrid model we build a multi-bit commitment semi-functionality $\mathcal{F}^*_{\widetilde{COM}}$ [8] and a zero knowledge semi-functionality $\mathcal{F}_{\widetilde{ZK}}$, with corresponding realizations BCOM* and BZK. Proving that that these protocols realize their functionalities does not require the use of angels. The angel Ω is only only to prove the realization of the basic commitment semi-functionality.

The protocols BCOM* and BZK are then used to build the protocol COM, realizing the standard (fully ideal) commitment functionality $\mathcal{F}_{COM}$. In order to realize $\mathcal{F}_{COM}$, however, we need to use the DLS-TDP assumption. In the $\mathcal{F}_{COM}$-hybrid model, there are known protocols for zero knowledge and for a broadcast channel. The zero knowledge functionality $\mathcal{F}_{ZK}$ has a realization due to Canetti and Fischlin [CF01]. The broadcast channel functionality $\mathcal{F}_{BC}$ is due to Goldwasser and Lindell in [GL02]. The proof for both of these realizations are information theoretic, and so they hold in the Ω-ES model, and do not rely on angels.

With the help of $\mathcal{F}_{COM}$, $\mathcal{F}_{ZK}$ and $\mathcal{F}_{BC}$, we can now realize the protocol OM-CP for the ideal functionality for one-to-many commit-and-prove, $\mathcal{F}^{1:M}_{CP}$. Again, the proof of this construction does not rely upon the existence of angels.

Now we have all of the tools needed to perform general multiparty computation against static adversaries.

---

[8] In our model we could also use a Pedersen commitment for the multi-bit commitment, but we felt that our single bit commitment was easier to understand and that our proofs are much simpler if we just instantiate the hash function from [PS04], rather than create completely new protocols. Using the Pedersen commitment could be of independent interest.

Secure Multi-Party Computation. We can now build general MPC following [CLOS02, PS04]. The proofs in [CLOS02] are information theoretic and will therefore carry over to the Ω-ES model.

Following the result of Lemma 2 in [PS04], using the DLS-TDP assumption, we can create a protocol for any functionality $\mathcal{F}$ that is secure against all semi-honest static adversaries. A full explanation appears in [CLOS02].

Now following Lemma 3 in [PS04], we know there exists a compiler that can turn any protocol secure against semi-honest static adversaries into a protocol secure against all static adversaries. The proof relies on having access to a functionality for one-to-make commit-and-prove. Since we have the protocol OM-CP that Ω-ES realizes one-to-many commit-and-prove functionality $\mathcal{F}_{CP}^{1:M}$, we now have a way of performing any multi-party computation.

5   Monitored Functionalities and Client-Server Computation Based on DLA

In [PS05] Prabhakaran and Sahai were able to show how to do any multi-party computation in a "client server model" called "client-server computation". This work is done in the UC/ES framework, but there are many limitations on this model. In the model one party is dedicated as the "server" and all the other parties are "clients". The client receives as output a function of its input and the server's input, while the server receiver as output the client's input. There is, however, the extra security limitation that the server's input to the function is not necessarily independent of the client's input, unless the client had never used that input previously (for more precise details, the reader is referred to [PS05]).

While "client-server computation" is implied by our earlier results in this paper of any multi-party computation we are able to achieve this on much simpler assumptions. With just the standard Discrete Log Assumption and DLS-TDP Assumption against non-uniform adversaries we can show how to do any type of client-server computation (as before, the DLS-TDP assumption is implied by the more standard TDP against super-polynomial adversaries).

The significance of this work is not in the power of the model, but in the fact that we were able to work past inherent restrictions in the UC model using such widely accepted assumptions.

We achieve these results in a similar fashion as in Section 4. We instantiation the hash function used in [PS05], but this time we need only rely on the DLA to do so.

5.1   The Assumptions and Angel of [PS05]

The assumptions made by [PS05] and their imaginary angel, are similar but weaker versions of those used by [PS04]. In particular, the ID of the party is not necessary as an input to the hash function, nor needed by the angel. Intuitively, the reason is that the role played by a certain party (e.g., a server vs a client) does not change across different executions. The imaginary angel can thus decide whether to answer the query in a useful manner based on the identity of the corrupted parties, without requiring the ID. Details follow.

The constructions of [PS05] rely on the following assumptions.

Assume there exists a hash function $\mathcal{H} : \{0,1\}^k \rightarrow \{0,1\}^l$. The input to $\mathcal{H}$ is of the form $(r, x, b) \in \{0,1\}^{k_1} \times \{0,1\}^{k_2} \times \{0,1\}$, where k_1, k_2, and l are all polynomial in k. It is assumed that $\mathcal{H}$ satisfies the following properties:

A′1 (Collisions and Indistinguishably): For every $r \in \{0,1\}^{k_1}$, there is a distribution $\mathcal{D}_r$ over $\{(x, y, z)|\mathcal{H}(r, x, 0) = \mathcal{H}(r, y, 1) = z\} \neq \phi$, such that

$$\{(x, z)|(x, y, z) \leftarrow \mathcal{D}_r\} \approx \{(x, z)|x \leftarrow \{0,1\}^{k_2}, z = \mathcal{H}(r, x, 0)\}$$
$$\{(y, z)|(x, y, z) \leftarrow \mathcal{D}_r\} \approx \{(y, z)|y \leftarrow \{0,1\}^{k_2}, z = \mathcal{H}(r, y, 1)\}$$

Further, given sampling access to $\mathcal{D}_r$ to a distinguisher, these distributions still remain indistinguishable.

A′2 (Difficult to find collisions with same prefix): For all PPT circuits M, for a random $r \leftarrow \{0,1\}^{k_1}$, the probability that $M(r)$ outputs (x, y) such that $\mathcal{H}(r, x, 0) = \mathcal{H}(r, y, 1)$ is negligible.[9]

Additionally, for most of their results, [PS05] also use the following assumption:

A′3 There exists a family of trapdoor permutations which remains secure even when the adversary is given sampling access to $\mathcal{D}_r$ for all r.

Note that each of these assumptions is weaker than (i.e., implied by) the corresponding assumption from [PS04] described in Section 4.1.

The Angel Γ: [PS04] use the following imaginary angel Γ. On query r, Γ checks whether the server is corrupted or not. If so, Γ returns $\bot$. If not, Γ outputs a sample from $\mathcal{D}_r$.

5.2   Our Hash Function

Here, we instantiate the hash function using exactly the same hash function $\mathcal{H}_0 : \{0,1\}^k \rightarrow \{0,1\}^l$ as we defined in Section 4.2, that is,

$$\mathcal{H}_0(q, g_0, g_1, x, b) = g_b^x \bmod q.$$

However, this time (q, g_0, g_1) correspond to r from the [PS05] constructions (rather than q corresponding to an ID). This means, for example, that assumption A′2 (difficulty of collision finding) is required to hold when q (as well as the generators) is chosen randomly (not necessarily for every q).

---

[9] Notice that here, unlike the corresponding assumption A2 from Section 4.1, M does not get access to oracles for the collision finding distributions for with other parameters. This is what will allow us to realize this requirement relying only on DLA, and not rDLA.

Satisfying A′1, A′2, A′3

We show that using our $\mathcal{H}_0$ to instantiate the hash function, A′1 is satisfied unconditionally, A′2 is satisfied if (a standard) DLA holds, and A′3 is satisfied if DLS-TDP holds. This will follow easily using the same arguments as we used in Section 4.2.

Lemma 4. $\mathcal{H}_0$ *satisfies A′1.*

Proof. We showed in Section 4.2 that $\mathcal{H}_0$ satisfies A1. Since A1 implies A′1 (in fact, they are equivalent here), it immediately follows that $\mathcal{H}_0$ satisfies A′1.   □

Lemma 5. *If DLA holds, then $\mathcal{H}_0$ satisfies A′2.*

Proof. Similarly to the proof of Lemma 2, if a PPT circuit M outputs collisions for a random q, g_0, g_1, it can be converted to a PPT circuit M' that computes discrete logs random elements of the quadratic residue subgroup of randomly chosen primes. This contradicts DLA.[10]   □

Lemma 6. *If the DLS-TDP assumption holds, then A′3 holds.*

Proof. This is identical to Lemma 3.   □

5.3   Our Angel Δ

Our imaginary angel Δ will first check if the server $\mathcal{S}$ is corrupted. If $\mathcal{S}$ is corrupted Δ will return $\perp$ on any query. If $\mathcal{S}$ is not corrupted then on input (q, g_0, g_1) Δ will compute the discrete logarithm of g_0 with respect to g_1, namely return a such that $g_0 = g_1^a \mod q$.

5.4   Putting the Pieces Together: Main Theorems for This Model

By proving that our hash function realizes the properties required by [PS05], all their results automatically translate to hold under our angel Δ, using our assumptions. This allows to achieve the first protocols in a (partially) composable model under very standard and widely acceptable assumptions such as the DLA, without any trusted setup, and avoiding the impossibility results of the UC model.

Below we briefly discuss the resulting theorems, and refer the reader to [PS05] for definitions and in-depth discussion of the functionalities and semi-functionalities achieved, as well as the security model. In the full version of the paper we will explicitly present the resulting protocols.

---

[10] Note that the last step in the proof of Lemma 2, dealing with the access M has to collisions for other $Z_{q'}^*$ is not necessary here, due to the weakened assumption A′2.

Monitored Commitment, Zero Knowledge Proof, and Commit and Prove under DLA. In [PS05] Prabhakaran and Sahai achieve protocols that Γ-ES-realize the monitored functionalities $\langle \mathcal{F}_{\widetilde{COM}} \rangle$, $\langle \mathcal{F}_{\widetilde{ZK}} \rangle$ and $\langle \mathcal{F}_{\widetilde{CAP}} \rangle$ given in [PS05] under assumptions A'1 and A'2. Thus we can achieve the protocols for these monitored functionalities with just the Discrete Log Assumption. While these are a means to achieve "client-server computation" they are also of independent interest. The protocols will remain the same as in [PS05] except with $\mathcal{H}$ instantiated with our DL-based $\mathcal{H}_0$.

Theorem 2. *Under the Discrete Log Assumption, protocols COM, ZK and CAP Δ-ES-realize monitored functionalities $\langle \mathcal{F}_{\widetilde{COM}} \rangle$, $\langle \mathcal{F}_{\widetilde{ZK}} \rangle$ and $\langle \mathcal{F}_{\widetilde{CAP}} \rangle$.*

Client-Server Computation Under DLA and DLS-TDP Assumption. In [PS05] Prabhakaran and Sahai achieve "client-server computation" under A'1, A'2 and A'3 and angel Γ. Thus we can achieve "client-server computation" with the Discrete Log Assumption and the DLS-TDP Assumption (or the stronger assumption of TDP secure against super polynomial adversaries). The Client-Server Computation Protocol (CSC) is the same as [PS05].

Theorem 3. *There is a protocol which Δ-ES-realizes monitored functionality $\langle \mathcal{F}_{\widetilde{CSC}} \rangle$ against static adversaries, under the Discrete Log Assumption and the DLS-TDP Assumption.*

Acknowledgments

We are grateful to Boaz Barak, Ran Canetti, Yehuda Lindell, Manoj Prabhakaran, and Amit Sahai for useful discussions. In particular, it was during a conversation with Boaz, Amit, and Manoj that Manoj suggested we try to present our results through an instantiation of the [PS04] hash function, rather than reproving our protocols from our assumptions. We also thank Zeph Grunschlag, Stephen Miller, Rafi Ostrovsky, and Carl Pomerance for helpful discussions and pointers regarding the feasibility of rDLA. Finally, we thank the anonymous referees for suggestions about the presentation.

References

[BPW04] Michael Backes, Birgit Pfitzmann, and Michael Waidner. A general composition theorem for secure reactive systems. In *Theory of Cryptography – TCC 2004*, pages 336–354, 2004.

[BS05] Boaz Barak and Amit Sahai. How to play almost any mental game over the net – concurrent composition via super-polynomial simulation. In *Proc. of the 46th Annu. IEEE Symp. on Foundations of Computer Science*, 2005. To appear.

[Can01] Ran Canetti. Universally composable security: A new paradigm for cryptographic protocols. In *Proc. of the 42nd Annu. IEEE Symp. on Foundations of Computer Science*, pages 136–145, 2001.

[CF01]   Ran Canetti and Marc Fischlin. Universally composable commitments. In *Advances in Cryptology – CRYPTO 2001*, pages 19–40, 2001.

[CKL03]   Ran Canetti, Eyal Kushilevitz, and Yehuda Lindell. On the limitations of universally composable two-party computation without set-up assumptions. In *Advances in Cryptology – EUROCRYPT 2003*, pages 68–86, 2003.

[CLOS02]   Ran Canetti, Yehuda Lindell, Rafail Ostrovsky, and Amit Sahai. Universally composable two-party and multi-party secure computation. In *Proc. of the 34th Annu. ACM Symp. on the Theory of Computing*, pages 494–503, 2002.

[DG03]   Ivan Damgård and Jens Groth. Non-interactive and reusable non-malleable commitment schemes. In *Proc. of the 35th Annu. ACM Symp. on the Theory of Computing*, pages 426–437, 2003.

[GL02]   Shafi Goldwasser and Yehuda Lindell. Secure computation without agreement. In *Proc. of the 16th International Conference on Distributed Computing (DISC)*, pages 17–32, 2002.

[KSW97]   John Kelsey, Bruce Schneier, and David Wagner. Protocol interactions and the chosen protocol attack. In *Proc. of 5th International Security Protocols Workshop*, volume 1361 of *Lecture Notes in Computer Science*, pages 91–104. Springer, 1997.

[Lin03]   Yehuda Lindell. General composition and universal composability in secure multi-party computation. In *Proc. of the 44rd Annu. IEEE Symp. on Foundations of Computer Science*, pages 394–403, 2003.

[Lin04]   Yehuda Lindell. Lower bounds for concurrent self composition. In *Theory of Cryptography – TCC 2004*, pages 203–222, 2004.

[Pas03]   Rafael Pass. Simulation in quasi-polynomial time, and its application to protocol composition. In *Advances in Cryptology – EUROCRYPT 2003*, pages 160–176, 2003.

[PS04]   Manoj Prabhakaran and Amit Sahai. New notions of security: achieving universal composability without trusted setup. In *Proc. of the 36th Annu. ACM Symp. on the Theory of Computing*, pages 242–251, 2004.

[PS05]   Manoj Prabhakaran and Amit Sahai. Relaxing environmental security: Monitored functionalities and client-server computation. In *Theory of Cryptography – TCC 2005*, pages 104–127, 2005.

[PW00]   Birgit Pfitzmann and Michael Waidner. Composition and integrity preservation of secure reactive systems. In *Proc. of the 7th Annu. ACM Conference on Computer and Communications Security (CCS '03)*, pages 245–254, 2000.

Games and the Impossibility of Realizable Ideal Functionality

Anupam Datta[1], Ante Derek[1], John C. Mitchell[1],
Ajith Ramanathan[1], and Andre Scedrov[2]

[1] Stanford University
{danupam, aderek, jcm, ajith}@cs.stanford.edu
[2] University of Pennsylvania
scedrov@math.upenn.edu

Abstract. A cryptographic primitive or a security mechanism can be specified in a variety of ways, such as a condition involving a game against an attacker, construction of an ideal functionality, or a list of properties that must hold in the face of attack. While game conditions are widely used, an ideal functionality is appealing because a mechanism that is indistinguishable from an ideal functionality is therefore guaranteed secure in any larger system that uses it. We relate ideal functionalities to games by defining the *set* of ideal functionalities associated with a game condition and show that under this definition, which reflects accepted use and known examples, bit commitment, a form of group signatures, and some other cryptographic concepts do not have any realizable ideal functionality.

1 Introduction

Many security conditions about cryptographic primitives are expressed using a form of game. For example, the condition that an encryption scheme is semantically secure against chosen ciphertext attack (IND-CCA2) [1] may be expressed naturally by saying that no adversary has better than negligible probability to win a certain game against a challenger. In this definition, the game itself clearly identifies the information and actions available to the adversary, and the condition required to win the game identifies the properties that must be preserved in the face of attack. Another way of specifying security properties uses ideal functionalities [2,3,4,5]. In this approach, usually referred to as Universal Composability [3] (UC) or Reactive Simulatability [6] an idealized way of achieving some goal is presented, possibly using mechanisms such as authenticated channels and trusted third parties that are not basic primitives in practice. An implementation is then considered secure if no feasible attacker can distinguish the implementation from the ideal functionality, in any environment. An advantage of this approach is that indistinguishability from an ideal functionality leads to composable notions of security [3,5,7]. In contrast, if a mechanism satisfies a game condition, there is no guarantee regarding how the mechanism will respond to interactions that do not arise in the specified game.

S. Halevi and T. Rabin (Eds.): TCC 2006, LNCS 3876, pp. 360–379, 2006.
© Springer-Verlag Berlin Heidelberg 2006

In this paper, we develop a framework for comparing game specifications and ideal functionalities, and prove some negative results about the existence of ideal functionalities in certain settings. While most known primitives have game-based definitions (see, e.g., [8]), it has proven difficult to develop useful ideal functionalities for some natural primitives. Some interesting issues are explored in [9, 10], which describe a series of efforts to develop a suitable ideal functionality for digital signatures. In brief, there is a widely accepted game condition for digital signatures, existential unforgeability against chosen message attacks, formulated in [11]. However, there are many possible ideal functionalities that are consistent with this game condition. For example, a functionality could either explicitly disclose information about messages that were signed in the past, or not disclose this information. More generally, given a game condition, it is often feasible to formulate various functionalities that satisfy the game condition yet reveal varying kinds of "harmless" information that does not seem relevant to the goals of the mechanism.

If we have a game or set of games that define a concept like secure encryption, digital signature, or bit-commitment, then we would like to identify precisely the set of possible ideal functionalities associated with each game condition. Since an ideal functionality is intended to be evidently secure by construction, we propose that an *ideal* functionality must satisfy the given game condition on information-theoretic grounds, rather than as a result of computational complexity arguments. Applied to encryption, for example, this means that an ideal functionality for encryption must not provide *any* information about bits of the plaintext to the adversary. Our definition of *ideal functionality* for a set of game conditions is consistent with all examples we have found in the literature, and reflects the useful idea that it should be easier to reason about systems that use an ideal functionality than about systems that use a real protocol. Using our definition, we show that while bit-commitment may be specified using games, there is no realizable ideal functionality for bit-commitment. This may be seen as a negative result about specification using ideal functionality, since there are constructions of bit-commitment protocols that are provably correct under modest cryptographic assumptions (see, e.g., [12]). We also show that there is no realizable ideal functionality for other reasonable and implementable cryptographic primitives, including a form of group signatures and a form of symmetric encryption with integrity guarantees, under certain conditions that allow the encryption key to be revealed after it is used.

The intuition behind our impossibility result is relatively simple. Illustrated using bit-commitment, a good commitment scheme must have two properties: the commitment token must not reveal any information about the chosen bit, while subsequent decommitment must reveal a verifiable relationship between the chosen bit and the commitment token. These are contradictory requirements because the first condition suggests that tokens must be chosen randomly, while the second implies that they are not. Similar "decommitment" issues arise in symmetric encryption or keyed hash, if the encryption key is revealed after some messages using the key have been sent on the visible network. At a more technical level, our proof by contradiction works by showing that if there was a realization of the ideal functionality for bit-commitment, it could be transformed into a protocol for bit-commitment that achieves perfect hiding and binding without using a trusted third party. However, it is well known that such a protocol does not exist [12]. While impossibility results for group signatures and symmetric encryption could be proved by instantiating the general proof method, we present simpler proofs by reducing bit-commitment to these primitives.

In a previous study of ideal functionality for bit commitment, Canetti and Fischlin show that a particular ideal functionality for bit-commitment is not realized by any real protocol [13]. In related work, Canetti [3] shows that particular functionalities for ideal coin tossing, zero-knowledge, and oblivious transfer are not realizable. Canetti et al [14] show that a class of specific functionalities for secure multi-party computation are not realizable, while Canetti and Krawczyk [15] compare indistinguishability-based and simulatability-based definitions of security in the context of key-exchange protocols. Our results are more general since we prove that, given a *game* definition of a primitive, there is *no* realizable ideal functionality associated with that game condition. In addition, our proof is different in that it provides a reduction to a previous negative result independent of universal composability [12], and appears to apply immediately to many primitives. A related issue is the choice of so-called "setup assumptions," such as public-key infrastructure, and common reference string. Our negative results hold under some setup assumptions, such as the absence of shared private information, or the presence of a trusted certificate authority (or PKI), and fail for other setup assumptions, such as the assumption of a common reference string. This is expected, since [13] construct a realizable ideal functionality in the common reference string model. We have yet to characterize precisely the set of possible setup assumptions under which our negative results hold.

While our general proof could be carried out using a number of computational models, we adopt a setting based on a form of process calculus. One advantage of this setting over interacting Turing machines [11,12,3] is a straightforward way of modularizing games that use a functionality. This is useful for defining primitives that are protocols, as opposed to local functions, by games. In principle, some version of our proof could be carried out using some version of Turing machines, augmented with separate function-call-and-return tapes for interacting with some form of oracle that performs public communication visible to the adversary.

Preliminary definitions are presented in Section 2, followed by definitions of bit-commitment functionalities and the main impossibility proof in Section 3. Reductions from other primitives are given in Section 4, with concluding remarks in Section 5.

2   Preliminaries

2.1   Probabilistic Process Calculus

Process calculus is a standard framework for studying concurrency [16,17] that has proved useful for reasoning about security protocols [2,18]. This is more of a "software" model than a "machine" model, since process calculus expressions are a form of program defining a concurrent system. Two main organizing ideas in process calculus are actions and channels. *Actions* occur on channels and are used to model communication flows. *Channels* provide an abstraction of the communication medium. In practice, channels might represent an IP address and port number in distributed computing, or a region of shared memory in a parallel processor.

A probabilistic polynomial-time process calculus (PPC) for security protocols is developed in [19,20,21] and updated in more recent papers [22,18]. The syntax consists of a set of *terms* that represent local sequential probabilistic polynomial-time computation and do not perform any communication with other processes, process *expressions* that can communicate with other processes, and *channels* that are used for communication. Terms contain variables that receive values over channels. There is also a special variable η called the *security parameter*. Each expression defines a set of *processes*, one for each choice of value for the security parameter. Each channel name has a bandwidth polynomial in the security parameter associated with it. The bandwidth ensures that no message gets too large and, thus, ensures that any expression can be evaluated in time polynomial in the security parameter.

Syntax of PPC: Expressions of PPC are constructed from the following grammar.

$$\mathcal{P} ::= \oslash \mid \nu(c)\mathcal{P} \mid \text{in}(c,x).(\mathcal{P}) \mid \text{out}(c,\text{T}).(\mathcal{P}) \mid [\text{T}].(\mathcal{P}) \mid (\mathcal{P} \mid \mathcal{P}) \mid !_{q(\eta)}(\mathcal{P})$$

Intuitively, $\oslash$ is the *empty process* taking no action. A process $\text{in}(c,x).\mathcal{P}$ with an *input* operator waits until it receives a value for input variable x on the channel c and then proceeds with process $\mathcal{P}$. Similarly, an *output* $\text{out}(c,\text{T}).\mathcal{P}$ transmits that value of the term T on the channel c and then proceeds with $\mathcal{P}$. Channel names that appear in an input or an output operation can be either public or private, with a channel being private if it is bound by the *private-binding* operator, ν and public otherwise. Actions on a private channel bound by a ν are not observable outside the scope of the ν operator. Hence private channels can be used to provide a form of secure communication. The *match* operator [T], a form of "if", executes the expression that follows it iff T evaluates to 1. The *parallel composition* operator, $\mid$, applied to two expressions allows them to evaluate concurrently, possibly communicating over any shared channels. The *bounded replication* operator has bound determined by the polynomial q affixed as a subscript. The expression $!_{q(\eta)}(\mathcal{P})$ is expanded to the $q(\eta)$-fold parallel composition $\mathcal{P} \mid \cdots \mid \mathcal{P}$ before evaluation. There is also a syntactic notion of context in PPC. A *context* $\mathcal{C}[\,\cdot\,]$ is an expression with a hole $[\,\cdot\,]$ such that we can substitute any expression into the hole and obtain a well-formed expression. Contexts may be used to represent the environment or adversary that interacts with a protocol or process.

Evaluating PPC expressions: To evaluate an expression in PPC we choose a probabilistic scheduler that selects communication steps. We then evaluate every term and match that is not in the scope of an input expression. When we can no longer evaluate terms and matches, we select a pair of input and output expressions on the same channel according to the scheduler, erase the output expression and substitute the value transmitted by the output (truncated suitably by the bandwidth of the channel) for the variable bound by the input. This procedure is repeated until no communication steps are possible. Further discussion, and explanation of a number of issues related to probabilistic scheduling, are explained in [18, 22, 23, 24].

Equivalence relations over PPC: Two equivalence relations over PPC will prove useful for studying security issues. The first relation, *computational observational equivalence*, written $\cong$, relates two expressions just when, in any context, the difference between the distributions they induce on observable behavior (messages over public channels) is negligible in the security parameter η. Formally $\mathcal{P} \cong \mathcal{Q}$ just when $\forall$ contexts $\mathcal{C}[\ \cdot\ ].\forall$ observables o:

$$\text{Prob}\,[\mathcal{C}[\mathcal{P}] \text{ produces } o] - \text{Prob}\,[\mathcal{C}[\mathcal{Q}] \text{ produces } o] \text{ is negligible in } \eta$$

Since the evaluation of all expressions and contexts in PPC are guaranteed to terminate in polynomial-time, $\cong$ is a natural way to state that two expressions are computationally indistinguishable to a poly-time attacker. The second relation, *information-theoretic observational equivalence*, written $=$, relates two expressions just when they induce exactly the same distribution on observable behavior in all contexts. Formally $\mathcal{P} = \mathcal{Q}$ just when $\forall$ contexts $\mathcal{C}[\ \cdot\ ].\forall$ observables o:

$$\text{Prob}\,[\mathcal{C}[\mathcal{P}] \text{ produces } o] - \text{Prob}\,[\mathcal{C}[\mathcal{Q}] \text{ produces } o] = 0$$

As a consequence, we can use $=$ to state that two expressions are indistinguishable even to unbounded attackers.

2.2   Function Calls and Returns

Process calculus allows processes to be programmed in a modular way, with one process relying on another for certain computations or actions. For example, one process P might wish to send a number bit-by-bit on a channel d. This can be done by writing another process Q that handles all the communication on channel d for P. This process Q receives a number n on some channel c used only for communication between P and Q, and then sends the bits of n on a channel d as required. If P wants a return value, such as notification that Q has finished sending the message, then P can execute an input action on channel c immediately after sending the number n to Q. This pattern of sends and receives essentially works like an ordinary remote procedure call and return. If the channel c is private, we can think of this as a remote procedure call between one process and another on the same processor, through a loopback interface, or a remote procedure call between two processors behind a firewall that makes LAN traffic invisible to an external attacker.

We will refer to the pattern of sends and receive just described for processes P and Q as a *function call and return*. Function calls and returns turn out to be a very useful concept in structuring games that specify properties of cryptographic primitives. To give a relatively concise notation, we will write $\mathsf{Call}^\eta(\langle \mathsf{params} \rangle, \mathbb{C})$ returns $\langle \mathsf{vars} \rangle. \mathcal{P}$ for a call that sends (outputs) parameters params on calling channels $\mathbb{C}$, and then waits to receive (input) return values $\langle \mathsf{vars} \rangle$ before executing process $\mathcal{P}$. To emphasize that a function call and return hides the structure of Q from the calling process P, we sometimes refer to this as a *black-box call*. Since PPC provides private channels, a function call and return will always be done on a private channel to avoid exposing the parameters or return values to an adversary.

For every function call and return to proceed, there must be a process that waiting to receive the call and then send a return value. Rather than write out all the input and output actions associated with responding to a remote procedure call, we will simply write $\mathsf{Impl}[\mathbb{C}, \mathbb{D}]$ for a process that responds to blackbox calls on channels $\mathbb{C}$, possibly using channels in $\mathbb{D}$ for some other purpose. For example, the process Q described above has the form $\mathsf{Impl}[c, d]$. since it receives function calls on channel c and performs public communication on channel d.

2.3 Interfaces and Cryptographic Primitives

In this paper, a *cryptographic primitive* is defined by an interface and a set of required security or correctness conditions that are expressible using the interface. The *interface* is the set of actions defined and applicable to the primitive, expressed as a set of function calls and returns. For example, the interface to an encryption primitive consists of calls to three probabilistic functions: key-generation, encryption, and decryption. A correctness condition for encryption is that the decryption of an encryption under the correct key returns the message encrypted. A semantically-secure encryption primitive must also satisfy a security condition stating that no probabilistic polynomial-time adversary can win a game that involves guessing which of two messages has been encrypted.

A *protocol* for a primitive is a process that responds to a set of function calls and supplies the associated returns, without using any additional private communication. For example, RSA can be formulated as an encryption protocol that implements key-generation, encryption, and decryption. A *functionality* for a primitive similarly supports the given interface, but may use additional private communication (such as used for a trusted third party; see Section 3.3). These restrictions on private communication are meant to prevent abusing the security associated with private channels, which are not a realistic primitive on the open public network. However, there are no restrictions on the way a functionality can communicate or reveal information to the adversary. For example, a functionality for signatures [10] could let the attacker choose the bitstrings for signatures.

An *ideal functionality* for an interface and a set of game conditions is a functionality that satisfies the correctness conditions with high probability and satisfies the security conditions in an information-theoretic way (i.e., against an unbounded adversary).

2.4 Universal Composability

Universal composability [3, 13, 25, 26, 14] involves a protocol to be evaluated, an ideal functionality, two adversaries, and an environment. The protocol realizes the ideal functionality if, for every attack on the protocol, there exists an attack on the ideal functionality, such that the observable behavior of the protocol under attack is the same as the observable behavior of the idealized functionality under attack. Each set of observations is performed by the same environment. The intuition here is that the ideal functionality 'obviously' possess a desired security property, possibly because the ideal functionality is constructed using a central authority, trusted third party, or private channels. Therefore, if a protocol is indistinguishable from an ideal functionality, the protocol must have the desired security property. In previous work, that which makes an ideal functionality "ideal" appears not to have been characterized precisely.

Universal composability can be expressed as a relation in process calculus [23, 24]. To give a form appropriate for the present paper, let $\mathcal{P}_1, \ldots, \mathcal{P}_n$ be n principals. We will assume that for some k, every principal $\mathcal{P}_i$ $(i > k)$ is in collusion with the adversary. Given an expression $\mathcal{P}$, we will write $\mathcal{P}[\mathbb{C}]$ to denote an instance of P running over the channels in $\mathbb{C}$. We say that an implementation Impl *securely realizes* a functionality $\mathcal{F}$ just when for any real world adversary $\mathcal{A}$, there exists a simulator $\mathcal{S}$ such that for any environment $\mathcal{E}$:

$$\nu(\mathbb{C}_1, \ldots, \mathbb{C}_k)(\mathcal{P}_1[\mathbb{C}_1, \mathbb{D}] \mid \cdots \mid \mathcal{P}_n[\mathbb{C}_n, \mathbb{D}] \mid \mathsf{Impl}[\mathbb{C}_1, \mathbb{D}] \mid \cdots \mid \mathsf{Impl}[\mathbb{C}_k, \mathbb{D}]) \mid$$
$$\mathcal{A}[\mathbb{C}_{k+1}, \ldots, \mathbb{C}_n, \mathbb{D}] \mid \mathcal{E}$$
$$\cong \nu(\mathbb{C}_1, \ldots, \mathbb{C}_k)(\mathcal{P}_1[\mathbb{C}_1, \mathbb{D}] \mid \cdots \mid \mathcal{P}_n[\mathbb{C}_n, \mathbb{D}] \mid \mathcal{F}[\mathbb{C}_1, \ldots, \mathbb{C}_k, \mathbb{D}]) \mid$$
$$\mathcal{S}[\mathbb{C}_{k+1}, \ldots, \mathbb{C}_n, \mathbb{D}] \mid \mathcal{E}$$

Here the first[1] k principals are assumed to be honest, and the remainder are assumed to be dishonest and acting in collusion with the adversary. To prevent the adversary/simulator from unfairly interfering with communications between the honest principals and the implementations (real or ideal), we make the links between the honest principals and the implementations private. Specifically, participant $\mathcal{P}_i$ uses private channels $\mathbb{C}_i$ to communicate with the implementation (real or ideal). The set of network channels $\mathbb{D}$ is used for communication between different participants. Both the adversary and the simulator have access to these channels.

Secure realisability requires that if we replace the real implementations Impl with an ideal implementation $\mathcal{F}$ (the functionality), there exists a simulator (that can interact with $\mathcal{F}$) which makes the ideal and real configurations indistinguishable. Another way to state this is that every real attack can be translated, using the simulator, into an attack on the functionality. We note that the principals that act in collusion with the attacker execute arbitrary programs and, in the ideal world, interact directly with the simulator (which mounts the ideal attack). Example configurations with two honest participants $\mathcal{I}$ and $\mathcal{R}$ are given in Figure 1.

---

[1] Since parallel composition is associative, the order in which we write the processes does not matter, and we may assume without loss of generality that the k honest principals occur first in the list.

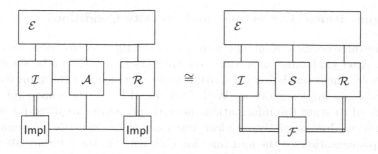

Fig. 1. Real and ideal configurations with two honest participants

3   Functionalities for Bipartite Bit-Commitment

A bipartite bit-commitment protocol allows a principal A to commit on a bit b to the principal B. However, B gains no information about the bit b until A later opens the commitment. We therefore formulate bit-commitment using four function calls, one call for each principal in each phase of the commitment. After defining the interface for bipartite bit-commitment, we define the game conditions for bit commitment and prove that no ideal functionality for these game conditions is realizable. We stress that the game conditions for bit-commitment as formulated in this paper are equivalent to standard security notions [27, 12], and that they can be realized using standard cryptographic assumptions such as the existence of pseudorandom functions [27].

3.1   Commitment Interface

A bipartite bit-commitment scheme provides four function calls:

$$\mathsf{SendCommit}^\eta(b, \mathbb{C}) \text{ returns } \langle \sigma \rangle \qquad \mathsf{GetCommit}^\eta(\mathbb{C}) \text{ returns } \langle \sigma \rangle$$
$$\mathsf{Open}^\eta(\sigma, \mathbb{C}) \text{ returns } \emptyset \qquad \mathsf{Verify}^\eta(\sigma, \mathbb{C}) \text{ returns } \langle \mathbf{r} \rangle$$

The initiator A commits to a bit using the call $\mathsf{SendCommit}^\eta(b, \mathbb{C})$ returns $\langle \sigma \rangle$, which communicates the commitment value over the channels in $\mathbb{C}$. Some state information σ is generated that can, amongst other things, be used to differentiate between different commitments and is needed to open the corresponding commitment. A responder B may receive a commitment from A by executing a call $\mathsf{GetCommit}^\eta(\mathbb{C})$ returns $\langle \sigma \rangle$ over the channels in $\mathbb{C}$, which may also returns some state information σ.

In the decommitment phase, the initiator A may open the commitment using the function call $\mathsf{Open}^\eta(\sigma, \mathbb{C})$ returns $\emptyset$, which uses the state information from the initial call to indicate which commitment is to be opened. The responder B can then verify the committed value by making the call $\mathsf{Verify}^\eta(\sigma, \mathbb{C})$ returns $\langle \mathbf{r} \rangle$. If verification succeeds, $\mathbf{r}$ contains the value of the committed bit. Otherwise, $\mathbf{r}$ is a symbol $\perp$ indicating failure.

3.2   Commitment Correctness and Security Conditions

There are three conditions on bit-commitment [27,12] — correctness, hiding, and binding. After explaining each condition, we show that each can be stated as an equivalence. The equivalences are written using $\cong$, which give the game condition required of any implementation. With $\cong$ replaced by $=$, the same equivalences can be used to state the information-theoretic properties required for an ideal functionality. More precisely, an *ideal functionality for bipartite bit-commitment* is an implementation for the four function calls listed in the interface above such that the correctness property below is satisfied with high probability, and the hiding and binding properties of bipartite bit-commitment below are satisfied with an information-theoretic equivalence. It is easy to verify that the concrete functionality considered in [13] is an instance of the ideal functionality for bipartite bit-commitment.

Given a game condition, there is a canonical way of writing it as an indistinguishability between expressions. The basic idea is that, since $\cong$ quantifies over all contexts, any successful attack on the game condition can be translated into a similarly successful context that distinguishes between the two sides of the equivalence. Conversely, since all expressions and contexts in PPC are guaranteed to evaluate in polynomial time and since the class of terms is precisely the class of probabilistic poly-time functions, every successfully distinguishing context can be translated into a successful attack on the corresponding game conditions.

Hiding: An implementation Impl is *hiding* if for an honest initiator, no adversary can gain, with non-negligible advantage, information about the committed bit. In other words, probability P_{Adv} that the attacker Adv, after interacting with an honest initiator committing to a randomly chosen bit b, successfully guesses the bit b should be close to a half. Writing this property as an equivalence yields:

$$\nu(\mathbb{C}, c).(\mathsf{Impl}[\mathbb{C}, \mathbb{D}] \mid \mathsf{out}(c, \mathsf{rand}) \mid \mathsf{in}(c, b).$$
$$\mathsf{SendCommit}^\eta(b, \mathbb{C}) \text{ returns } \langle \sigma \rangle.\mathsf{in}(d, b').\mathsf{out}(dec, b' \stackrel{?}{=} b))$$
$$\cong \nu(\mathbb{C}, c).(\mathsf{Impl}[\mathbb{C}, \mathbb{D}] \mid \mathsf{out}(c, \mathsf{rand}) \mid \mathsf{in}(c, b).$$
$$\mathsf{SendCommit}^\eta(b, \mathbb{C}) \text{ returns } \langle \sigma \rangle.\mathsf{in}(d, b').\mathsf{out}(dec, \mathsf{rand}))$$

Both expressions select a random bit and commit to it. The adversary (expressed as a context) interacts with the commitment protocol and tries to guess the committed-to value. The difference between the two expressions is that the LHS tests, over the channel dec, whether the adversary's guess matches the chosen bit, while the RHS assumes, again over the channel dec, that the adversary fails with probability 1/2. Clearly, any successfully distinguishing context must guess the bit with non-negligible advantage, thereby proving the existence of an adversary that violates the hiding property. Hence, we can naturally express the hiding condition that for all Adv, the probability $P_{Adv} - \frac{1}{2}$ is negligible in η as a process calculus equivalence. To say that an implementation is perfectly or information-theoretically secure we require that $\forall Adv\colon P_{Adv} - \frac{1}{2} = 0$, which is the same as replacing $\cong$ by $=$ in the equivalence above.

Binding: The *binding* property is that no adversary can open a commitment to an arbitrary value. This condition can be restated using a game in which the adversary commits to a challenger (an honest responder), who then picks a random b and challenges the adversary to open the commitment to b. As an equivalence, it is stated as:

$\nu(\mathbb{C}, d)(\mathsf{GetCommit}^\eta(\mathbb{C})$ returns $\langle\sigma\rangle.\mathsf{out}(d, \mathrm{rand}) \mid \mathsf{in}(d, b).\mathsf{out}(c, b)$.
$\quad\mathsf{Verify}^\eta(\sigma, \mathbb{C})$ returns $\langle\mathbf{r}\rangle.\mathsf{out}(dec, \mathbf{r} \stackrel{?}{=} b) \mid \mathsf{Impl}[\mathbb{C}, \mathbb{D}])$
$\cong \nu(\mathbb{C}, d)(\mathsf{GetCommit}^\eta(\mathbb{C})$ returns $\langle\sigma\rangle.\mathsf{out}(d, \mathrm{rand}) \mid \mathsf{in}(d, b).\mathsf{out}(c, b)$.
$\quad\mathsf{Verify}^\eta(\sigma, \mathbb{C})$ returns $\langle\mathbf{r}\rangle.\mathsf{out}(dec, \text{if } \mathbf{r} \stackrel{?}{=} \perp \text{ then false else rand}) \mid \mathsf{Impl}[\mathbb{C}, \mathbb{D}])$

Here both expressions wait for a commitment, and then challenge the adversary to open the commitment to a randomly chosen bit. The LHS tests whether the adversary successfully does so, whilst the RHS assumes that if the attempt to open does not fail (i.e., the result of Verify is not $\perp$) the adversary fails with probability $1/2$. Perfect binding is expressed by replacing $\cong$ with $=$ in the equivalence above.

Correctness: An implementation Impl is *correct* if an honest responder is able to verify an opened commitment by an honest initiator with overwhelming probability. This correctness property may be expressed as the process calculus equivalence.

$\nu(\mathbb{C}, c)(\mathsf{out}(c, \mathrm{rand}) \mid \mathsf{in}(c, b).\mathsf{SendCommit}^\eta(b, \mathbb{C})$ returns $\langle\sigma_I\rangle$
$\quad.\mathsf{Open}^\eta(\sigma_I, \mathbb{C})$ returns $\emptyset.\mathsf{in}(d, b').\mathsf{out}(dec, b \stackrel{?}{=} b') \mid \mathsf{Impl}[\mathbb{C}, \mathbb{D}]) \mid$
$\quad\nu(\mathbb{C}')(\mathsf{GetCommit}^\eta(\mathbb{C}')$ returns $\langle\sigma_R\rangle.\mathsf{Verify}^\eta(\sigma_R, \mathbb{C}')$ returns $\langle\mathbf{r}\rangle.\mathsf{out}(d, r) \mid \mathsf{Impl}[\mathbb{C}', \mathbb{D}])$
$\cong \nu(\mathbb{C}, c)(\mathsf{out}(c, \mathrm{rand}) \mid \mathsf{in}(c, b).\mathsf{SendCommit}^\eta(b, \mathbb{C})$ returns $\langle\sigma_I\rangle$
$\quad.\mathsf{Open}^\eta(\sigma_I, \mathbb{C})$ returns $\emptyset.\mathsf{in}(d, b').\mathsf{out}(dec, \text{true}) \mid \mathsf{Impl}[\mathbb{C}, \mathbb{D}]) \mid$
$\quad\nu(\mathbb{C}')(\mathsf{GetCommit}^\eta(\mathbb{C}')$ returns $\langle\sigma_R\rangle.\mathsf{Verify}^\eta(\sigma_R, \mathbb{C}')$ returns $\langle\mathbf{r}\rangle.\mathsf{out}(d, r) \mid \mathsf{Impl}[\mathbb{C}', \mathbb{D}])$

Here, both expressions pick a random bit, commit to it, and then try to open it. The LHS checks whether the verifier obtained the correct value for the bit, whilst the RHS assumes that the verifier gets the right value all the time.

3.3   Impossibility of Bit-Commitment

In this section, we show that no ideal functionality for bit-commitment can be realized. This generalizes the impossibility result for one particular functionality given in [13]. Other plausible bit-commitment functionalities can be constructed by adjusting the level of information and possible actions provided to the attacker by the functionality. For example, the functionality may let the attacker change the identity of the committer, hence making the commitment unauthenticated. Alternatively, the functionality may let the attacker change the committed bit if the attacker manages to correctly guess an internal secret of the functionality (since this is a low probability event, correctness still holds). Our proof shows that all of these variants (as well as further variants discussed in [13]) and their combinations are not realizable. Although we have not yet obtained a general characterization, our theorem applies under some setup assumptions, and fails in the common reference string model in accordance with the construction given in [13].

Our proof by contradiction roughly works as follows: given a real protocol P that realizes an ideal functionality F for bit-commitment, we construct another real protocol Q which provides the same correctness guarantee. However, in protocol Q all calls to the bit-commitment interface by principals are handled by copies of F. As a consequence, Q provides perfect hiding and binding, which is a contradiction.

In order to state the theorem formally, we require some definitions. We say that P is a *real protocol* if each instance of P only communicates with one principal over a set of private channels. Intuitively, since it cannot communicate with two separate parties over private channels hidden from the adversary, a real protocol cannot act as a secure trusted third party. We say that a protocol P for bit-commitment is *terminating* when the following expression will, with high probability, produce the messages "go" and "done".

$$\nu(\mathbb{C})(\mathsf{SendCommit}^\eta(b, \mathbb{C}) \text{ returns } \langle\sigma_I\rangle.\mathsf{in}(c, z).\mathsf{Open}^\eta(\sigma_I, \mathbb{C}) \text{ returns } \emptyset.\mathsf{in}(d, z) \mid P[\mathbb{C}, \mathbb{D}] \mid$$
$$\nu(\mathbb{C}')(\mathsf{GetCommit}^\eta(\mathbb{C}') \text{ returns } \langle\sigma_R\rangle.\mathsf{out}(c, \text{"go"}).$$
$$\mathsf{Verify}^\eta(\sigma_R, \mathbb{C}') \text{ returns } \langle\mathbf{r}\rangle.\mathsf{out}(d, \text{"done"}) \mid P[\mathbb{C}', \mathbb{D}])$$

Intuitively, if the function calls are implemented with P, in the absence of the attacker two honest parties should be able to first finish the commitment stage, synchronize, and then finish the decommitment stage.

Theorem 1. *If F is an ideal functionality for bilateral bit-commitment, then there does not exist a terminating real protocol P that securely realizes F.*

Before giving the proof, the following two lemmas will be useful. The first lemma states the well known fact [12] that perfect hiding and binding protocols for bit-commitment do not exist without a trusted third party. We omit the proof here. The second lemma states that any realization of F will also be correct for bit-commitment. The proof sketch is in Appendix A. Similarly, any realization of F will enjoy complexity-theoretic hiding and binding guarantees; however, we do not require this fact for the impossibility result.

Lemma 1. *There does not exist a terminating real protocol P which is correct with high probability, and both perfectly hiding and perfectly binding.*

Lemma 2. *If P is a terminating real protocol that securely realizes F, then P is correct with high probability.*

Proof (Proof of Theorem 1). We assume that P securely realizes F. It follows that for any configuration involving principals making use of P, there exists a simulator S such that replacing the calls to P with calls to the simulator in conjunction with the functionality yields an indistinguishable configuration.

Consider the following real configuration when the environment plays the role of the responder honestly. It selects a bit and sends that bit to the initiator. The initiator then commits to that bit using a copy of the implementation P_I. The responder is corrupted by the adversary to simply forward messages to the environment. After corrupting the responder, the adversary simply forwards messages. The environment then honestly plays the responder's role using a copy

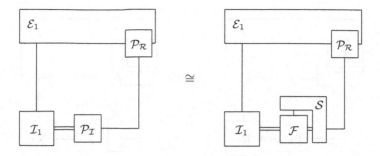

Fig. 2. Configurations for the first step

of the real implementation $\mathcal{P_R}$. At the conclusion of the commitment phase, the environment initiates decommitment by instructing the initiator to open. The environment then verifies the initiator's attempt to open, and then decides if the bit the initiator opened to was the bit the environment selected at the start of the run. The programs of the four principals are given below, where $\mathsf{Forward}(\mathbb{C} \leftrightarrow \mathbb{D})$ is an expression that forwards in an order-preserving way messages received on the channels $\mathbb{C}$ to channels $\mathbb{D}$ and vice versa:

$\mathcal{E}_1 \equiv \nu(\mathbb{C}, c)(\mathsf{out}(c, \mathsf{rand}) \mid \mathsf{in}(c, b).\mathsf{out}(\mathrm{IO}_I, b).\mathsf{GetCommit}^\eta(\mathbb{C})$ returns $\langle \sigma \rangle.\mathsf{out}(\mathrm{IO}_I, \mathsf{open}).$

$\qquad \mathsf{Verify}^\eta(\mathbf{r}, \mathbb{C})$ returns $\langle \sigma \rangle.\mathsf{out}(dec, b \overset{?}{=} \mathbf{r}))$

$\mathcal{I}_1 \equiv \nu(\mathbb{C}')(\mathsf{in}(\mathrm{IO}_I, b).\mathsf{SendCommit}^\eta(b, \mathbb{C}')$ returns $\langle \sigma' \rangle.\mathsf{in}(\mathrm{IO}_I, x).\mathsf{Open}^\eta(\sigma', \mathbb{C}')$ returns $\emptyset)$

$\mathcal{A}_1 \equiv \mathsf{Forward}(\mathrm{Net}_I \leftrightarrow \mathrm{Net}_R)$

$\mathcal{R}_1 \equiv \mathsf{Forward}(\mathrm{IO}_R \leftrightarrow \mathrm{Net}_R)$

This real configuration and its corresponding ideal configuration are shown in Figure 2 on the left and right, respectively (omitting the forwarders for clarity). Let us consider the ideal configuration. Here, the initiator uses the ideal functionality $\mathcal{F}$, whilst the environment continues using the real protocol. A simulator $\mathcal{S}$ must exist such that it can "convert" the messages of the functionality into messages that $\mathcal{P_R}$ understands and vice versa. This simulator sits between $\mathcal{P_R}$ and $\mathcal{F}$ and is connected to $\mathcal{F}$ via the bit-commitment interface and the unspecified interface of $\mathcal{F}$. Since $\mathcal{P}$ securely realizes $\mathcal{F}$, it follows that the configurations are indistinguishable. Furthermore, by Lemma 2 the environment in the real configuration must register success with high probability, since the adversary does nothing. Whence the expression $\mathcal{Q}$ consisting of $\mathcal{F}$ and $\mathcal{S}$ wired in the way that they are must be able to commit to $\mathcal{P_R}$ and, then, successfully open the commitment.

Let us now consider another real configuration (Figure 3) where the initiator is corrupted to be a forwarder but the responder is honest. As before, the adversary, after corrupting the initiator, does nothing. The environment selects a bit and then runs the initiator's role directly. However, instead of using $\mathcal{P}$ to implement the initiator's role, the environment uses the expression $\mathcal{Q}$ from the first part of the argument. To commit, the environment sends the bit to the functionality whose messages are then translated by the simulator into messages suitable for the copy of the implementation $\mathcal{P_R}$ used by the honest responder.

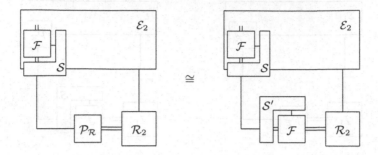

Fig. 3. Configurations for the second step

After committing, the environment waits for a receipt from the responder, before decommitting. It then waits for the responder to send the bit it believes the initiator committed to and the environment checks that the bit it received was the same as the bit to which it committed. The responder, for its part, receives a commitment, sends a receipt to the environment, then verifies a commitment, and forwards the result to the environment. The programs are given below:

$$\mathcal{E}_2 \equiv \nu(\mathbb{C}, c)(\mathsf{out}(c, \mathsf{rand}) \mid \mathsf{in}(c, b).\mathsf{SendCommit}^\eta(b, \mathbb{C}) \text{ returns } \langle\sigma\rangle.\mathsf{in}(\mathrm{IO}_R, x).$$

$$\mathsf{Open}^\eta(\sigma, \mathbb{C}) \text{ returns } \emptyset.\mathsf{in}(\mathrm{IO}_R, b')\mathsf{out}(dec, b \overset{?}{=} b') \mid \mathcal{Q}[\mathbb{C}, \mathrm{IO}_R])$$

$$\mathcal{R}_2 \equiv \nu(\mathbb{C}')(\mathsf{GetCommit}^\eta(\mathbb{C}') \text{ returns } \langle\sigma'\rangle.\mathsf{out}(\mathrm{IO}_R, \mathsf{receipt}).$$

$$\mathsf{Verify}^\eta(\mathbf{r}, \mathbb{C}') \text{ returns } \langle\sigma'\rangle.\mathsf{out}(\mathrm{IO}_R, \mathbf{r}))$$

$$\mathcal{A}_2 \equiv \mathsf{Forward}(\mathrm{Net}_I \leftrightarrow \mathrm{Net}_R)$$

$$\mathcal{I}_2 \equiv \mathsf{Forward}(\mathrm{IO}_I \leftrightarrow \mathrm{Net}_I)$$

In this scenario, the simulator $\mathcal{S}'$ sits between the expression $\mathcal{Q}$ (consisting of simulator $\mathcal{S}$ and functionality) and the functionality $\mathcal{F}$. Again, from secure realizability, Lemma 2, and the fact that $\mathcal{Q}$ looks like an initiator running the implementation $\mathcal{P}$, we know that in the real configuration, the environment will, with high probability, register a success. Therefore, so must the ideal configuration, whence the expression $\mathcal{Q}'$ consisting of $\mathcal{F}$ and $\mathcal{S}'$ must correctly play the role of the responder running the implementation $\mathcal{P}$.

If we look at the ideal configuration, we notice that the functionality is no longer working as a trusted third party. Every message is run through the simulators $\mathcal{S}$ and $\mathcal{S}'$. Thus, we have an implementation of bit-commitment that is a real protocol. The initiator executes the code given by the expression $\mathcal{Q}$ while the responder executes the code given by the expression $\mathcal{Q}'$. From the above argument it follows that the implementation $\mathcal{Q} \mid \mathcal{Q}'$ is a correct implementation. Furthermore, the $\mathcal{Q} \mid \mathcal{Q}'$ has to be information-theoretically hiding and binding because of the way they make use of the functionality. For example, to commit to a bit, the caller passes the bit to the functionality which, by definition, reveals no information about the bit regardless of the other parties in the configuration until the open step. Thus, we have a correct with high probability, and information-theoretically hiding and binding implementation of the bit-commitment interface that does not make use of trusted third parties. This contradicts Lemma 1.

4   Generalization of the Impossibility Result and Other Examples

In this section state a more general impossibility result: if $\mathcal{G}$ is a functionality and P is a protocol which uses $\mathcal{G}$ to achieve bit-commitment with perfect hiding and binding, then the functionality $\mathcal{G}$ cannot be realized. Intuitively, the functionality $\mathcal{G}$ together with protocol P constitutes an ideal functionality for bit-commitment $\mathcal{F}$, and any realization of $\mathcal{G}$ will lead to the realization of $\mathcal{F}$. Therefore, we would expect that all primitives that can be used to build bit-commitment are not realizable as functionalities. We illustrate this by showing that certain (rather strong) variants of symmetric encryption and group signatures cannot have realizable ideal functionalities. Due to space constrains, the security definitions are mostly informal and proof sketches have been moved to Appendix A.

Hybrid Protocols: We will consider implementations of primitives which, in addition to public channels, may use a particular functionality. Let $\mathcal{G}$ be any functionality, a $\mathcal{G}$-*hybrid protocol* P for a primitive is an implementation of the primitive's interface which does not make use of the trusted third party except maybe by making calls to $\mathcal{G}$'s interface. We will write $P[\mathcal{Q}]$ to denote an instance of P where calls to $\mathcal{G}$'s interface are handled by the implementation $\mathcal{Q}$ (real or ideal).

Theorem 2. *If $\mathcal{G}$ is a functionality and P is a terminating $\mathcal{G}$-hybrid protocol for bit-commitment which is correct with high probability and provides perfect hiding and perfect binding, then no protocol realizes functionality $\mathcal{G}$.*

Symmetric Encryption: Symmetric encryption primitive is defined by the standard interface for key generation, encryption and decryption.

KeyGen$^{\eta}(\mathbb{C})$ returns $\langle K \rangle$ Encrypt$^{\eta}(K, p, \mathbb{C})$ returns $\langle c \rangle$ Decrypt$^{\eta}(K, c, \mathbb{C})$ returns $\langle p \rangle$

In addition to the obvious correctness property, we assume, as in [28], that the encryption scheme is CCA-secure and that it provides ciphertext integrity. Provably secure schemes with respect to these two properties exist under reasonable assumptions [29]. Informally, we can describe the properties as follows:

- *CCA-security* means that it is hard for an adaptive attacker with access to the decryption oracle to distinguish the plaintext from a random value of the same length given the ciphertext. *Perfect CCA-security* means that the probability of success is exactly half.
- *Integrity of ciphertexts* means that it is hard for an attacker to find a ciphertext c which will successfully decrypt unless that ciphertext has been produced by the encryption algorithm for some key and plaintext. *Perfect integrity of ciphertexts* means that the probability of an attacker finding such a ciphertext is zero.

Corollary 1. *If $\mathcal{F}$ is a functionality for symmetric encryption providing perfect CCA-security and perfect integrity of ciphertext then $\mathcal{F}$ cannot be realized.*

Group Signatures: Group signature primitive is defined by the interface for key generation, group signing, group signature verification and opening. For simplicity we will assume that the group is always of size two.

GKeyGen$^\eta$($\mathbb{C}$) returns $\langle gpk, gmsk, gsk_0, gsk_1 \rangle$ GSign$^\eta$($m, gsk, \mathbb{C}$) returns $\langle sig \rangle$
GVerify$^\eta$($gpk, m, sig, \mathbb{C}$) returns $\langle result \rangle$ GOpen$^\eta$($gmsk, m, sig, \mathbb{C}$) returns $\langle identity \rangle$

In addition to the obvious correctness properties, we assume that the group signature scheme provides anonymity and traceability even against dishonest group managers. This is a stronger security requirement than the version principally considered in [30] (though [30] does briefly discuss this variant); [30] also shows that schemes with these properties exist if trapdoor permutations exist. Informally, we can describe the properties as follows:

- *Anonymity* means that it is hard for an adaptive attacker with access to an opening oracle to recover the identity of the signer given a signature and a message, even if the attacker has all the signing keys. *Perfect anonymity* means that the probability of success is exactly half (assuming, as we do, only two possible signers).
- *Traceability* means that it is hard for an attacker that adaptively corrupts a coalition of signers and has access to a signing oracle to produce a valid message-signature pair that opens to a signer not in the coalition, even when the group manager is dishonest. *Perfect traceability* means that the probability of an attacker forging such a signature is zero.

Corollary 2. *If $\mathcal{F}$ is a functionality for group signatures providing perfect anonymity and perfect traceability then $\mathcal{F}$ cannot be realized.*

5   Conclusion and Future Directions

We articulate accepted practice in the literature by giving a precise definition of an *ideal* functionality satisfying any given game specification: An ideal functionality must be a process or a set of processes that realize the game conditions in an information-theoretic, rather than computational complexity, sense. Using this definition we show that bit commitment, group signatures, and other cryptographic concepts that are definable using games do not have any realizable ideal functionality.

This proof appears applicable to other functionalities, and to a range of so-called "setup assumptions." However, we have not yet characterized the applicable setup assumptions precisely. Some examples of setup assumptions are pre-shared keys, certificate authorities, random oracles or common reference strings. These additional assumptions can be captured by a hybrid model where parties have access to an additional ideal functionality "implementing" the assumption, such as a trusted certificate authority. As demonstrated in [13], there are realizable ideal functionalities for bit-commitment when parties have access to one form of common reference string functionality. One possibility is to restrict attention to functionalities that are used only in the initialization phase. Our intuition suggests that a similar impossibility proof can be constructed for this case as long as these setup functionalities are global, i.e. all honest parties can access them. This does not contradict the common reference string construction in [13] since a fresh instance of the common reference string functionality is required for each pair of participants engaged in a session and cannot be accessed by other honest parties. We hope to develop this idea more fully in future work.

An appealing property of indistinguishability-based specifications is the connection with composability: if a real security mechanism is indistinguishable from an ideal one, then any larger system using the real mechanism will behave in the same way as the same system using the ideal mechanism instead. In light of the limitations on indistinguishability-based specifications explored in this paper, there are several modifications to the basic theory that might provide useful forms of composability. One direction is to relax or modify the requirements for ideal functionality. For example, information-theoretic equivalence could be replaced with the indistinguishability of random systems in the sense of [31]. This would allow adaptive, computationally unbounded distinguishers to query the system at most polynomially many times in the security parameter. Another possible direction involves the modification of the Universal Composability framework recently considered in [32, 33], which allows a commitment functionality. In the modified framework, parties are typed in a certain way, and the typing must be respected by the simulator. On the other hand, since the intuition for some of these directions is not clear, it may be more productive to develop methods for stating and proving *conditional* forms of composability. More precisely, primitives and protocols could be guaranteed to operate securely only in environments that satisfies certain conditions. Games currently provide a very limited form of conditional composability, since a game condition provides guarantees for any system whose actions can be regarded as (or reduced to) moves in a relevant game. We also consider the work on protocol composition logic [34, 35] a potentially relevant form of conditional composability, since protocols or primitives proved correct in that framework carry guarantees that apply to any environment respecting certain invariants expressed explicitly in the logic.

References

1. Bellare, M., Boldyreva, A., Micali, S.: Public-key encryption in a multi-user setting: Security proofs and improvements. In: Advances in Cryptology - EUROCRYPT 2000, International Conference on the Theory and Application of Cryptographic Techniques, Proceeding. Volume 1807 of Lecture Notes in Computer Science., Springer-Verlag (2000) 259–274

2. Abadi, M., Gordon, A.D.: A calculus for cryptographic protocols: the spi calculus. Information and Computation **143** (1999) 1–70 Expanded version available as SRC Research Report 149 (January 1998).
3. Canetti, R.:   Universally composable security: A new paradigm for crypto-graphic protocols.   In: FOCS '01: Proceedings of the 42nd IEEE sympo-sium on Foundations of Computer Science. (2001) 136 Full version available at http://eprint.iacr.org/.
4. Lincoln, P., Mitchell, J.C., Mitchell, M., Scedrov, A.: A probabilistic poly-time framework for protocol analysis. In: ACM Conference on Computer and Commu-nications Security. (1998) 112–121
5. Pfitzmann, B., Waidner, M.: Composition and integrity preservation of secure reactive systems. In: ACM Conference on Computer and Communications Security. (2000) 245–254
6. Backes, M., Pfitzmann, B., Waidner, M.:   A composable cryptographic library with nested operations. In: CCS '03: Proceedings of the 10th ACM conference on Computer and communications security, ACM Press (2003) 220–230
7. Backes, M., Pfitzmann, B., Waidner, M.: A general composition theorem for secure reactive systems.   In: TCC '04: Proceedings of the 1st Theory of Cryptography Conference. Volume 2951 of Lecture Notes in Computer Science., Springer-Verlag (2004) 336–354
8. Shoup, V.: Sequences of games: a tool for taming complexity in security proofs. Cryptology ePrint Archive, Report 2004/332 (2004) http://eprint.iacr.org/2004/332.
9. Backes, M., Hofheinz, D.:   How to break and repair a universally composable signature functionality. In: Information Security, 7th International Conference, ISC 2004, Proceedings. Volume 3225 of Lecture Notes in Computer Science., Springer-Verlag (2004) 61–72
10. Canetti, R.: Universally composable signature, certification, and authentication. In: CSFW '04: Proceedings of the 17th IEEE Computer Security Foundations Workshop, IEEE Computer Society (2004) 219–233
11. Goldwasser, S., Micali, S., Rackoff, C.: The knowledge complexity of interactive proof systems. SIAM Journal on Computing **18**(1) (1989) 186–208
12. Goldreich, O.: Foundations of Cryptography: Basic Tools. Cambridge University Press (2000)
13. Canetti, R., Fischlin, M.: Universally composable commitments. In: Advances in Cryptology - CRYPTO 2001, 21st Annual International Cryptology Conference, Proceedings. Volume 2139 of Lecture Notes in Computer Science., Springer-Verlag (2001) 19–40
14. Canetti, R., Kushilevitz, E., Lindell, Y.: On the limitations of universally com-posable two-party computation without set-up assumptions.   In: Advances in Cryptology - EUROCRYPT 2003, International Conference on the Theory and Applications of Cryptographic Techniques, Proceedings. Volume 2656 of Lecture Notes in Computer Science., Springer-Verlag (2003) 68–86
15. Canetti, R., Krawczyk, H.: Analysis of key-exchange protocols and their use for building secure channels. In: Advances in Cryptology - EUROCRYPT 2001, Inter-national Conference on the Theory and Application of Cryptographic Techniques, Proceeding. Volume 2045 of Lecture Notes in Computer Science., Springer-Verlag (2001) 453–474
16. Milner, R.: Communication and Concurrency. International Series in Computer Science. Prentice Hall (1989)

17. van Glabbeek, R.J., Smolka, S.A., Steffen, B.: Reactive, generative, and strati-
    fied models of probabilistic processes. International Journal on Information and
    Computation **121**(1) (1995)
18. Ramanathan, A., Mitchell, J.C., Scedrov, A., Teague, V.: Probabilistic bisimula-
    tion and equivalence for security analysis of network protocols. In: Foundations
    of Software Science and Computation Structures, 7th International Conference,
    FOSSACS 2004, Proceedings. Volume 2987 of Lecture Notes in Computer Science.,
    Springer-Verlag (2004) 468–483
19. Mitchell, J.C., Mitchell, M., Scedrov, A.: A linguistic characterization of bounded
    oracle computation and probabilistic polynomial time. In: FOCS '98: Proceedings
    of the 39th Annual IEEE Symposium on the Foundations of Computer Science,
    IEEE Computer Society (1998) 725–733
20. Lincoln, P.D., Mitchell, J.C., Mitchell, M., Scedrov, A.: Probabilistic polynomial-
    time equivalence and security protocols. In: Formal Methods World Congress, vol.
    I. Number 1708 in Lecture Notes in Computer Science, Springer-Verlag (1999)
    776–793
21. Mitchell, J.C., Ramanathan, A., Scedrov, A., Teague, V.: A probabilistic
    polynomial-time calculus for the analysis of cryptographic protocols (preliminary
    report). In: 17th Annual Conference on the Mathematical Foundations of Program-
    ming Semantics. Volume 45., Electronic notes in Theoretical Computer Science
    (2001)
22. Ramanathan, A., Mitchell, J.C., Scedrov, A., Teague, V.: Probabilistic bisimula-
    tion and equivalence for security analysis of network protocols. Unpublished, see
    http://www-cs-students.stanford.edu/~ajith/ (2003)
23. Datta, A., Küsters, R., Mitchell, J.C., Ramanathan, A., Shmatikov, V.: Unifying
    equivalence-based definitions of protocol security. In: 2004 IFIP WG 1.7, ACM
    SIGPLAN and GI FoMSESS Workshop on Issues in the Theory of Security (WITS
    2004). (2004)
24. Datta, A., Küsters, R., Mitchell, J.C., Ramanathan, A.: On the relationships
    between notions of simulation-based security. In: TCC '05: Proceedings of the 2nd
    Theory of Cryptography Conference. Volume 3378 of Lecture Notes in Computer
    Science., Springer-Verlag (2005) 476–494
25. Canetti, R., Krawczyk, H.: Universally composable notions of key exchange and
    secure channels. In: Advances in Cryptology - EUROCRYPT 2002, International
    Conference on the Theory and Application of Cryptographic Techniques, Proceed-
    ing. Volume 2332 of Lecture Notes in Computer Science., Springer-Verlag (2002)
    337–351
26. Canetti, R., Lindell, Y., Ostrovsky, R., Sahai, A.: Universally composable two-
    party and multi-party secure computation. In: STOC '02: Proceedings of the 34th
    annual ACM symposium on Theory of computing, ACM Press (2002) 494–503
27. Naor, M.: Bit commitment using pseudorandomness. Journal of Cryptology **4**(2)
    (1991) 151–158
28. Backes, M., Pfitzmann, B.: Symmetric encryption in a simulatable Dolev-Yao style
    cryptographic library. In: CSFW '04: Proceedings of the 17th IEEE Computer
    Security Foundations Workshop, IEEE Computer Society (2004) 204–218
29. Rogaway, P., Bellare, M., Black, J., Krovetz, T.: OCB: A block-cipher mode of
    operation for efficient authenticated encryption. In: CCS '01: Proceedings of the
    8th ACM Conference on Computer and Communications Security, ACM Press
    (2001) 196–205

30. Bellare, M., Micciancio, D., Warinschi, B.: Foundations of group signatures: Formal definitions, simplified requirements, and a construction based on general assumptions. In: Advances in Cryptology - EUROCRYPT 2003, International Conference on the Theory and Applications of Cryptographic Techniques, Proceedings. Volume 2656 of Lecture Notes in Computer Science., Springer-Verlag (2003) 614–629
31. Maurer, U.M.: Indistinguishability of random systems. In: Advances in Cryptology - EUROCRYPT 2002, International Conference on the Theory and Application of Cryptographic Techniques, Proceeding. Volume 2332 of Lecture Notes in Computer Science., Springer-Verlag (2002) 110–132
32. Prabhakaran, M., Sahai, A.: New notions of security: Achieving universal composability without trusted setup. In: STOC '04: Proceedings of the 36th annual ACM symposium on Theory of computing, ACM Press (2004) 242–251
33. Prabhakaran, M., Sahai, A.: Relaxing environmental security: Monitored functionalities. In: TCC '05: Proceedings of the 2nd Theory of Cryptography Conference. Volume 3378 of Lecture Notes in Computer Science., Springer-Verlag (2005) 104–127
34. Datta, A., Derek, A., Mitchell, J.C., Pavlovic, D.: A derivation system and compositional logic for security protocols. Journal of Computer Security **13** (2005) 423–482
35. He, C., Sundararajan, M., Datta, A., Derek, A., Mitchell, J.C.: A modular correctness proof of TLS and IEEE 802.11i. In: ACM Conference on Computer and Communications Security. (2005)

A   Proof Sketches

Proof (Proof sketch of Lemma 2). Consider a configuration consisting of an honest initiator running the implementation P, an honest responder running the implementation P, and an adversary that does nothing. The initiator waits for a bit from the environment and then commits to that bit. It then waits for a message from the environment and then opens its commitment. The responder, after receiving a commitment, sends a receipt to the environment. After a successful verification of an attempt to open the commitment, it sends the opened-to value to the environment. The environment selects a bit, sends it to the initiator and waits for a receipt from the responder. Once it gets this message, it instructs the initiator to open its commitment, and then waits for the responder to reveal the bit to which the initiator committed. If that bit matches the bit the environment selects at the start of the run, the environment registers success. Otherwise it registers failure.

By the terminating property of P, we know that this run will complete. The ideal configuration has the initiator talking to the functionality which talks directly to the responder. Though a simulator exists in the ideal configuration, it can do nothing since both the initiator and responder are connected directly to the functionality. By virtue of the functionality's correctness, we know that in the ideal configuration the environment will register success with high probability. Since P securely realizes $\mathcal{F}$, the environment must register success in the real configuration with high probability. Whence the correctness of P is established.

Proof (Proof sketch of Theorem 2). Assume that P is a terminating $\mathcal{G}$-hybrid protocol for bit-commitment, which is correct with high probability and provides perfect hiding and binding. A functionality $\mathcal{F} = P[\mathcal{G}]$ is clearly an ideal functionality for bit-commitment. Let Q be a real protocol which is a realization of $\mathcal{G}$, consider a real protocol $R = P[Q]$ in which all the calls of P to the functionality $\mathcal{G}$ are implemented with Q. We claim that R is a secure realization of $\mathcal{F}$. Choose any real configuration for R, consisting of an attacker A, and parties $P_1, \ldots, P_n$. We need to show that there is a simulator S such that for any environment E this configuration is indistinguishable from one where parties call functionality $\mathcal{F}$ instead of R. This configuration is also a real configuration for the protocol Q. Therefore, there is a simulator such that when all calls to Q are replaced with calls to $\mathcal{G}$, the two configurations are indistinguishable for any environment. Since this ideal configuration is exactly the ideal configuration for $\mathcal{F}$ we are done with the proof, because by Theorem 1 there can be no protocol realizing any ideal functionality for bit-commitment.

Proof (Proof sketch of Corollary 1). Assume that $\mathcal{F}$ is an ideal functionality for symmetric encryption and construct a $\mathcal{F}$-hybrid protocol for bit-commitment providing perfect hiding and binding. The initiator can commit to b by generating a new key, encrypting b and sending the ciphertext via public channel. To open the commitment, initiator sends the key. This protocol provides perfect hiding because of the perfect CCA-security provided $\mathcal{F}$, and provides perfect binding because of the perfect integrity of ciphertexts provided by $\mathcal{F}$. By Theorem 2, functionality $\mathcal{F}$ cannot be realized.

Proof (Proof sketch of Corollary 2). Construct a $\mathcal{F}$-hybrid protocol for bit-commitment providing perfect hiding and binding. The initiator can commit to b by generating all the group keys, signing a random message with b's signing key, and then sending, as the signature, the tuple consisting of the b's signature, the message, the group public key, and all the signing keys. To open the commitment, the initiator sends the group manager's secret key. This protocol provides perfect hiding because of the perfect anonymity provided $\mathcal{F}$, and provides perfect binding because of the perfect traceability provided by $\mathcal{F}$. By Theorem 2, functionality $\mathcal{F}$ cannot be realized.

Universally Composable Symbolic Analysis of Mutual Authentication and Key-Exchange Protocols*

(Extended Abstract)

Ran Canetti[1],** and Jonathan Herzog[2],***

[1] IBM Research
[2] The MITRE Corporation

Abstract. Symbolic analysis of cryptographic protocols is dramatically simpler than full-fledged cryptographic analysis. In particular, it is simple enough to be automated. However, symbolic analysis does not, by itself, provide any cryptographic soundness guarantees. Following recent work on cryptographically sound symbolic analysis, we demonstrate how Dolev-Yao style symbolic analysis can be used to assert the security of cryptographic protocols within the universally composable (UC) security framework. Consequently, our methods enable security analysis that is completely symbolic, and at the same time cryptographically sound with strong composability properties.

More specifically, we concentrate on mutual authentication and key-exchange protocols. We restrict attention to protocols that use public-key encryption as their only cryptographic primitive and have a specific restricted format. We define a mapping from such protocols to Dolev-Yao style symbolic protocols, and show that the symbolic protocol satisfies a certain symbolic criterion if and only if the corresponding cryptographic protocol is UC-secure. For mutual authentication, our symbolic criterion is similar to the traditional Dolev-Yao criterion. For key exchange, we demonstrate that the traditional Dolev-Yao style symbolic criterion is insufficient, and formulate an adequate symbolic criterion.

Finally, to demonstrate the viability of our treatment, we use an existing tool to automatically verify whether some prominent key-exchange protocols are UC-secure.

1 Introduction

The analysis of cryptographic protocols is a complex and subtle business. One main reason is the need to capture an adversary that is very powerful in terms of commu-

---

* This work was first presented at the DIMACS workshop on protocol security analysis, June 2004. Most of the research was done while both authors were at CSAIL, MIT.

** Supported by NSF CyberTrust Grant #0430450.

*** The author's affiliation with The MITRE Corporation is provided for identification purposes only, and is not intended to convey or imply MITRE's concurrence with, or support for, the positions, opinions or viewpoints of the author.

S. Halevi and T. Rabin (Eds.): TCC 2006, LNCS 3876, pp. 380–403, 2006.

nication, while being computationally bounded. Furthermore, security typically holds only in a probabilistic sense, and only under computational intractability assumptions. Indeed, developing adequate mathematical models and formulations of security properties has been a main endeavor in modern cryptography from its early stages, beginning with the notions of pseudo-randomness and semantic security of encryption [13, 14, 41, 25], through zero-knowledge, non-malleability, and general cryptographic protocols, e.g. [27, 28, 23, 26, 42, 9, 16, 49, 17]. Consequently, we now have a variety of mathematical models where one can represent cryptographic protocols, specify the security requirements of cryptographic tasks, and then potentially prove that (the mathematical representation of) a protocol meets the specification in a way that is believed to faithfully represent the security of actual protocols in actual systems.

However, the models above are complex and delicate, even for relatively simple protocols for simple tasks. In particular, they directly represent adversaries as resource-bounded and randomized entities, and directly bound their success probabilities with a function of the consumed resources. This entails either asymptotic formalisms or alternatively parameterized notions of concrete security. Furthermore, since these notions are typically satisfied only under some underlying hardness assumptions, proofs of security typically require a reduction to the underlying hard problem. Coming up with such reductions typically requires "human creativity" which is hard to mechanize. Consequently, full-fledged cryptographic analysis of even moderately complex cryptographic systems is a daunting prospect.

Several alternatives to this "computational" approach to protocol security analysis have been proposed, such as the Dolev-Yao model [24] and its many derivatives (e.g. [54, 22]), the BAN logic [15], and a number of process calculi and other models, e.g. [2, 33, 34, 37]. In these approaches, cryptographic primitives are represented as symbolic operations which guarantee a set of idealized security properties by fiat. (For instance, transmission of encrypted data is modeled as communication that is inaccessible to the adversary, e.g. [15], or as a symbolic operation that completely hides the message, e.g. [24].) Consequently, the model becomes dramatically simpler. There is no need for computational assumptions; randomization can be replaced by non-determinism; and protocols can be modeled by simple finite constructs without asymptotics. Indeed, protocol analysis in these models is much simpler, more mechanical, and amenable to automation (see e.g. [36, 39, 53, 46, 11]). These are desirable properties when attempting to analyze large-scale systems. Until recently, however, there has been no concrete justification for this high level of abstraction. Thus, these models could not be used to prove that protocols remain secure when the abstract security primitives are realized by actual algorithms.

Within the past few years, however, there have been several efforts towards devising symbolic models that enjoy the relative simplicity of "abstract cryptography" while maintaining cryptographic soundness. One attractive approach towards this goal was introduced in the ground-breaking work of Abadi and Rogaway [4] in the context of passive security of symmetric encryption schemes.

Essentially, they showed that proving indistinguishability of distribution ensembles of a certain class can be done by translating these ensembles to symbolic forms and verifying a symbolic criterion on these forms. This work has been extended several times ([44, 3, 31, 5]). Of particular importance is the work of Micciancio and Warinschi [45] who extend this approach to include active adversaries. Specifically, they provide a formal criteria for two-party protocols, and show that symbolic protocols which satisfy this criteria achieve *mutual authentication* (as defined in [10]) if they are implemented with public-key encryption secure against chosen-ciphertext attacks (as in [51, 23]).

An alternative approach was taken in the works of Backes, Pfitzmann and Waidner, and Canetti (e.g. [49, 17, 6, 7, 18]). Here, the idea is to define idealized abstractions of cryptographic primitives directly in a full-fledged cryptographic model. These abstractions are realizable by actual concrete protocols in a cryptographic setting, but can at the same time be used as abstract primitives by higher-level protocols. Soundness for this style of abstraction provided via a strong composition theorem. This approach is attractive due to its generality and the strong compositional security properties it guarantees when the protocol runs together with arbitrary other protocols. Furthermore, the analysis of the higher-level protocols becomes more straightforward and mechanical when the lower-level primitives are replaced by their abstractions. However, this model still requires the analyst to directly reason about protocols within a full-fledged cryptographic model, with its asymptotics, error probabilities etc., and so this approach retains much of the original complexity of the problem.

Our Approach. This work demonstrates how formal and symbolic reasoning in a simple finite model can be used to simplify analyzing the security of protocols within a full-fledged cryptographic model with strong composability properties. Specifically, we use the universally composable (UC) security framework [17]. The overall approach follows that of [4, 45]: We want to assert whether a given concrete, fully specified protocol satisfies a concrete security property. (In our case, the concrete property is realizing a given ideal functionality within the UC framework.) Instead of proving this assertion directly, we proceed as follows. The first step is to abstract out from the cryptographic primitives, and use instead ideal functionalities that represent these primitives in an idealized manner. This step is done still within the UC framework, and its soundness comes from secure composability. The next step is to translate this semi-abstract protocol to a simpler, symbolic ("Dolev-Yao style") protocol. Now, standard tools for symbolic protocol analysis are used to prove that the symbolic protocol satisfies a certain symbolic criterion. Finally, we show that this implies that the concrete protocol satisfies the concrete security property (i.e., realizes the given ideal functionality). The main gain here is that all steps, except for one, can be done once and for all. Only the symbolic analysis of the protocol at hand needs to be done per protocol. This analysis typically considerably simpler than full-fledged cryptographic analysis within the UC framework. The approach is summarized in Figure 1.

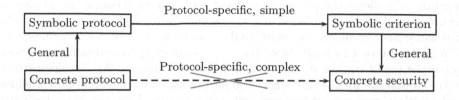

Fig. 1. Using symbolic (formal) analysis to simplify cryptographic analysis. Instead of directly proving that a given concrete protocol π realizes a concrete ideal functionality F, translate π to a symbolic protocol, verify that the symbolic protocol satisfies a simple symbolic criterion, and use this to show that π realizes F. The first and third steps are general and proven once and for all. Only the second step needs to be repeated per protocol.

In a way, this approach takes the best of the two approaches described above: On the one hand, it guarantees the strong security and composition properties of the ideal-functionality approach. On the other hand, we end up with a relatively simple, finite symbolic model, and symbolic criteria to verify within that model. In fact, our analysis is even simpler than current ones: The strong compositional security properties of the UC framework allow us to specify and analyze protocols in terms of a *single instance* of the protocol in question. Security in a setting, where an unbounded number of instances of a protocol may run concurrently with each other and with arbitrary other protocols, is guaranteed via the UC and UC with joint state theorems [17, 21]. In contrast, existing symbolic models (e.g. [24, 54, 22]) directly address the more complex multi-session case, even in the symbolic model. Consequently, our symbolic modeling involves fewer runtime states and thus lends to more effective mechanical analysis.

This Work. In this work, we apply the above approach to the problems of mutual-authentication and key-exchange protocols. In particular, we progress as follows.

First, we translate concrete protocols to their symbolic counterparts. We follow the approach of [45], and concentrate on a restricted class of concrete cryptographic protocols. We call such protocols **simple protocols**. Simple protocols use public-key encryption as their only cryptographic operations and conform to a restrictive format, or "programming language." (The reason to use a restricted class of protocols is that existing symbolic models can only handle such protocols. Indeed, any enrichment in the symbolic model would translate to an analogous enrichment in the definition of simple protocols, while preserving the validity of the treatment.) We note that, while restricted, this format is still very meaningful; in particular, it allows expressing known 'benchmark' protocols such as several variants of the Needham-Schroeder-Lowe (NSL) protocol [47, 35, 36], and the Dwork-Dolev-Naor [23] protocol.

In order to further simplify the treatment, we require simple protocols to use an "ideal encryption functionality" rather than directly using some concrete encryption scheme. This ideal functionality, denoted $\mathsf{F}_{\mathrm{CPKE}}$ (for "certified public-key encryption") allows the parties to encrypt and decrypt messages in an ideally secure way. Using F_{CPKE} instead of concrete encryption simplifies the analysis in two ways: first, it allows the entire analysis to be done in terms of a single session of the protocol at hand. Next, the entire analysis is unconditional, and does not make use of computational bounds on the adversaries. As we demonstrate below, soundness for the case where the parties use concrete encryption schemes and multiple instances of the protocol run concurrently is guaranteed via the UC and UC with joint state theorems.

Next, we consider the symbolic and computational criteria for mutual authentication and key exchange. The computational criteria are expressed in terms of UC functionalities. The symbolic criterion for mutual authentication is traditional and drawn from the existing literature, but the symbolic criterion for key-exchange is not. In fact, we demonstrate that the traditional symbolic criterion for key-exchange is strictly weaker than the computational one. This is done via by example: We show that a natural way to extend the NSL authentication protocol to key exchange results in a protocol that satisfies the traditional symbolic secrecy criterion for the key, but whose computational counterpart can be easily broken. In particular, the computational counterpart is not UC-secure. We thus define a new symbolic criterion for key exchange which is closer in spirit to traditional computational criteria.

Finally we show that:

1. The original, concrete protocol realizes the mutual authentication functionality in the UC framework *if and only if* its translation into the symbolic model fulfills the symbolic mutual authentication criterion, and
2. The original, concrete protocol realizes the key exchange functionality in the UC framework *if and only if* its translation into the symbolic mode fulfills our new symbolic criterion.

We stress that, as in [4], the symbolic criterion is formulated in terms of a finite and relatively simple model, whereas the concrete criterion (realizing an ideal functionality) is formulated in the standard asymptotic terms of cryptographic security. Still, equivalence holds.

As a result, both vertical arrows of Figure 1 are firmly established for mutual authentication and key exchange. The only work that remains is to show that particular symbolic protocols fulfill the symbolic criteria. However, this last step is protocol specific and rather mechanical. In particular, as we demonstrate, it can be readily automated.

Automated Analysis: Proof of Concept. Since our symbolic key secrecy criterion is not standard, one might wonder whether this criterion retains the main advantage of the symbolic model, namely amenability to automation. We demonstrate that it does, by applying the ProVerif [12] automated verification tool to verify whether our symbolic key exchange criterion is satisfied by known protocols.

Specifically, we consider two very natural extensions of the NSL protocol to key exchange: In one extension the output key is the nonce generated by the initiator, and in the other extension the output key is the nonce generated by the responder. As we demonstrate within, one extension is demonstrably insecure, while the other extension seems secure. The automated analysis supports these impressions: The insecure extension fails the automates test, while the seemingly secure extension passes the test. In fact, the tool now provides a proof of security for this extension.

1.1   Related Work

Pfitzmann and Waidner [49] provide a general definition of integrity properties and prove that such properties are preserved under protocol composition in their framework. Our symbolic mutual authentication criterion can be cast as such an integrity property. In addition, Backes, Pfitzmann and Waidner [7], building on the idealized cryptographic library in [6], demonstrate that several known protocols satisfy a property that is similar to our symbolic mutual authentication criterion. However, these results do not answer the question which is the focus of this work, namely whether a given concrete cryptographic protocol realizes an ideal functionality (say, the mutual authentication functionality) in a cryptographic model (say, the UC framework.) Furthermore, since the [6] abstraction is inherently multi-session, the [7] analysis has to directly address the more complex multi-session case.

Our results for mutual authentication protocols follow the lines of Micciancio and Warinschi [45]. However, since we use the UC abstraction of idealized encryption, our characterization results are unconditional (rather than based on computational assumptions), can be meaningfully stated in the simpler terms of a single session, and provide the stronger security guarantees of the UC framework.

Laud [32] investigates the concrete cryptographic properties guaranteed by certain symbolic secrecy criteria for protocols using *symmetric* encryption. He also shows how these symbolic criteria can be automatically verified. However, these criteria are different from the ones discussed here. Specifically, following the traditional symbolic formulation, it is only required that the adversary obtains no information about the key during the course of the protocol, and "real-or-random secrecy" against active adversaries is not considered. Consequently, these criteria do not guarantee secure key exchange, nor are they preserved under composition.

Concurrently to this work, Cortier and Warinschi [52] formulate another symbolic secrecy criterion for key exchange protocols, demonstrate how to automatically verify this criterion, and show that this criterion implies a cryptographic secrecy criterion in the style of "real-or-random" secrecy against active adversaries. However, also in that work the symbolic criterion follows the tradition of only requiring that the adversary obtains no information on the secret key. Consequently, their cryptographic criterion falls short of guaranteeing secrecy in a general protocol setting, as exhibited in [50, 20]. In particular, their criterion admits the above-mentioned buggy extension of the NSL protocol to key exchange.

Blanchet [11] provides a symbolic criterion (cast in a variant of the spi-calculus [1]) that captures a secrecy property, called "strong secrecy", that is similar to our symbolic secrecy criterion for the exchanged key. Essentially, the criterion says that the view of any adversarial environment remains unchanged (modulo renaming of variables) when the symbol representing the secret key is replaced by a fresh symbol that's unrelated to the protocol execution. In addition, an automated tool for verifying this criterion is provided. Indeed, Blanchet's tool is the one we use for the automated analysis reported above.

Concurrently to this work, Backes and Pfitzmann [8] propose an abstract secrecy criterion for key-exchange protocols that use their cryptographic library, and demonstrate that this criterion suffices for guaranteeing cryptographically sound secrecy. However, their criterion is still formulated within their full-fledged cryptographic framework, rather than in a simplified symbolic model as done here. Furthermore, it does not carry any secure composability guarantees.

Herzog, Liskov and Micali [30] provide an alternative cryptographic realization of the Dolev-Yao abstraction of public-key encryption. Their realization makes stronger cryptographic requirements from encryption scheme in use (namely, they require "plaintext aware encryption"), and assumes a model where both the sender and the receiver have public keys. Herzog later relaxes this requirement to standard CCA-2 security [29], but that work (lacking any composition theorems) still considers the multi-session case. Furthermore, it only connects executions of protocols in the concrete setting to executions of protocols in the symbolic setting. It does not investigate whether security in the symbolic setting implies or is implied by security in the concrete setting, which the main focus of this work.

Micciancio and Panjwani [43] study computationally sound symbolic analysis of group key agreement protocols with adaptively changing membership. However, both their symbolic and concrete secrecy criteria are very different than the ones here. In particular, their symbolic criterion is a trace property, rather than a "real-or-random" style criterion as the one here.

Patil [48] extends the present work to handle also mutual authentication and key exchange protocols that use digital signatures *in addition to* public-key encryption. That work demonstrates the flexibility and modularity of the approach initiated here.

Organization. This paper is an extended abstract of the work presented in [19]. Section 2 contains a very high-level overview of the UC framework and the Dolev-Yao style symbolic model. Section 3 defines simple protocols and presents two variants of the NSL protocol, written as simple protocols. It also sketches how simple protocols can be instantiated using concrete, fully-specified encryption schemes. Section 4 presents the *Mapping lemma,* which translates between traces of simple protocols and symbolic protocols. This lemma plays a central role in our analysis.

We then turn to the specific tasks of mutual authentication and key-exchange. Due to lack of space, we omit the treatment of mutual authentication from this abstract. (Full treatment appear in [19].) The treatment of key exchange is sketched in Section 5. This includes a discussion of the inadequacy of the traditional symbolic secrecy criterion, the new symbolic criterion, and the equivalence with the UC notion of realizing the ideal key exchange functionality. We conclude by discussing future research directions.

2   Background

2.1   The UC Framework

The UC framework provides a general way for specifying the security requirements of cryptographic tasks, and asserting whether a given protocol realizes the specification. A salient property of this framework is that it provides strong composability guarantees: A protocol that realizes the specification continues to realize the specification regardless of the activity in the rest of the network, without "unexpected side-effects".

The security requirements of tasks are specified by envisioning an "ideal process" where the participants can hand their inputs for the task to an imaginary "trusted party", who locally computes the desired outputs and hands them back to the parties. The code run by the trusted party is called an ideal functionality. This code is intended to capture the security and correctness requirements of the cryptographic task at hand.

Deciding whether a protocol π UC-realizes an ideal functionality F (namely, whether π is a secure protocol for the corresponding task) is done in three steps, as follows. We first formulate a model for executing the protocol. This model consists of the parties running the protocol, plus two adversarial entities: the environment Z, which generates the inputs for the parties and reads their outputs, and the adversary A, which reads the outgoing messages generated by the parties, and delivers incoming messages to the parties. The adversary and the environment can interact freely during the protocol execution. (In fact, in this model one can treat them as a single entity without losing generality.)

Next, we consider an "ideal process" for realizing the given ideal functionality (i.e., the task). This process is similar to the process of executing the protocol π, with two important exceptions. First, the inputs that the environment generates for the parties running the protocol are given to a trusted party that executes the code of the ideal functionality F. Similarly, the outputs generated by F are given to the environment as the outputs coming from the parties. Second, the adversary A for interacting the protocol is replaced by an adversary S that does not interact directly with the parties; instead, S interacts directly with F (in a way specified by F). The communication between S and Z remains arbitrary.

Finally, we say that π UC-realizes functionality F if for *any* adversary A there exists an adversary S such that no (polytime) environment Z can tell with non-negligible probability whether it is interacting with and execution of π with adversary A, or alternatively with the ideal process for F and adversary S. This in particular means that the I/O behavior of the good parties in the protocol execution is essentially the same as that of the ideal functionality; in addition, the information that the environment learns from A on the execution of π can be generated (or, "simulated") by S, given only the information that it learns legally from interacting with F.

The following universal composition theorem holds in this framework. Let π is a protocol that UC-realizes functionality F, and let ρ be a protocol that makes calls to (multiple instances of) the trusted party running F. Let ρ^{π} be the "composed protocol" which is identical to ρ except that calls to F are replaced by calls to π. Then, protocol ρ^{π} behaves in an indistinguishable way from the original ρ. In particular, if ρ UC-realizes some ideal functionality G then so does ρ^{π}.

An additional theorem that will be useful for substantiating our treatment is universal composition with joint state (JUC) [21]. Notice that the UC theorem only applies to protocols ρ^{π} where the honest parties maintain completely disjoint local states for the different instances of π. In contrast, the JUC theorem applies to cases where the different instances of π have some joint state. Specifically, let $\hat{\pi}$ be a protocol that, in one instance, UC-realizes multiple instances of ideal functionality F. (Formally, let $\hat{F}$ be the ideal functionality that exhibits, in a single instance, the behavior of multiple instances of F. Then $\hat{\pi}$ is a protocol that UC-realizes $\hat{F}$.) Let ρ be an arbitrary protocol that uses multiple instances of F, and let $\pi^{[\hat{\pi}]}$ be the composed protocol where each party runs a single instance of ρ plus a *single* instance of $\hat{\pi}$, and where all the inputs provided by π to all the instances of F are forwarded to the instance of $\hat{\pi}$. Similarly, the outputs of the single instance of $\hat{\pi}$ are given to ρ as coming from the various instances of F. Then, the JUC theorem states that protocol $\rho^{[\pi]}$ UC-emulates the original protocol ρ. Then, the JUC theorem states that protocol $\rho^{[\pi]}$ behaves in an indistinguishable way from the original ρ. In particular, if ρ UC-realizes some ideal functionality G then so does ρ^{π}.

2.2   The Symbolic Model

The symbolic model (also called the "Dolev-Yao" model) is a simplified model for analyzing protocols that use cryptographic primitives. In this model, messages are represented as strings of symbols, explicitly describing their parse trees, and encryption is represented as an abstract operation. Thus, $Enc(M; K)$ is not the application of an algorithm to a pair of bit-strings, but the sequence of characters "$Enc(M; K)$" (or the parse-tree created when the encryption constructor is applied to the sub-trees M and K). Because of its simplicity, this model allows the analyst to focus on the *structure* of protocols independently of the specific algorithms used to implement them. While the full-fledged Dolev-Yao model includes a variety of primitives such as symmetric encryption and signatures, we focus on a sub-model which includes only asymmetric encryption.

The symbolic mode has several components. Firstly, the model uses a symbolic algebra $\mathcal{A}$ to represent messages of a protocol. The *atomic* elements of the algebra are used to represent primitive structures such as party identifiers, public encryption keys, random challenges ("nonces"), and secret keys. (The party identifiers and public keys can be either honest, or corrupted.) The two operations of the algebra represent abstracted pairing (or concatenation) and encryption. Thus, the *compound* elements of the algebra (*i.e.*, those messages produced by the operations) represent those messages that pair or encrypt primitive messages (or other, simpler, compound messages). Lastly, the algebra is defined to be *free*: each message has exactly one representation. Put another way, the algebra admits no equalities other than identity: two distinct parse-trees will always represent two distinct messages.

Secondly, *symbolic protocols* are defined simply as sets of roles, which are themselves defined by a state transition table. Participants that engage in a role must maintain their current state. (For convenience, this state is defined to be the sequence of messages they have seen and sent so far.) Then, when a participant receives a message, it cross-indexes that message and its current state in the state transition table to discover (1) the message it must then transmit, and (2) possibly a message to output locally. It then updates its internal state accordingly. Here all inputs, outputs, and messages are compound messages from the algebra.

Definition 1. *A role R in a symbolic protocol $\mathcal{P}$ is a mapping from the set of states $\mathcal{S} = (\mathcal{A})^*$, an element in the algebra A representing the incoming message, and a name from the set of names $\mathcal{M}$ representing the name of the participant, to a pair of values from $\mathcal{A}$ representing values to transmit and (locally) output respectively, and a new state (which is the old state with the addition of the new incoming message). That is:*

$$R : \mathcal{S} \times \mathcal{A} \times \mathcal{M} \to \mathcal{A} \times \mathcal{A} \times \mathcal{S}.$$

Thirdly, the symbolic model considers a very restricted **symbolic adversary**. In particular, the adversary is defined in two parts: its initial knowledge (a set of symbolic messages), and the **adversary operations** it can use to deduce new messages from known ones. (These known messages can include not only its initial knowledge, but also the messages sent during the protocol execution.) These adversary operations are extremely limited. Specifically, the adversary can concatenate messages, de-concatenate elements of a message, encrypt a message with a given public key, or decrypt a given symbolic ciphertext if the corresponding public key is corrupted. Note that this list of adversary operations implicitly defines the strength of "ideal" encryption: it is strong enough to prevent the adversary from performing any other operations to ciphertexts. The adversary may, however, combine these basic operations in any way that it pleases. This gives rise to the definition of closure: the closure of a message (or a set of messages) is the set of all messages that the adversary can potentially derive from the given message (or set). That is, the closure operation defines the messages which the adversary can create and transmit at any point.

With this, the symbolic model defines (in the straightforward way) how symbolic protocols **execute** in the presence of a symbolic adversary. That is, an execution consists of a sequence of events where each event is either:

- The delivery of a message to a participant and the participant replying in accordance with its role, or
- The adversary intercepting a message and replacing it with a message drawn from the closure.

The **trace** of an execution is the sequence of these events. The **security properties** of the symbolic model are typically (but not always) predicates on sets of traces: a protocol satisfies such a security property if the predicate is satisfied by the set of that protocol's possible (or valid) traces.

The true power of the symbolic model comes from the fact that so many aspects of execution (such as complexity-classes and probabilities) are simply abstracted away, allowing the analyst to focus on "structural flaws". Also, because the symbolic adversary is easily described as a simple, non-deterministic machine, it becomes possible to create specialized algorithms to analyze protocols in this setting. Examples of this abound (*e.g.* [36, 40, 46, 38]) and later in this work we use one such automated tool to perform a symbolic analysis which we have (by then) shown to be computationally sound in the UC model.

3   Simple Protocols

Although the UC framework and the symbolic model were both designed for the purpose of security analysis, they differ in some very important ways. For example, the symbolic model does not explicitly represent the internal workings of the honest participants, and therefore makes no guarantee that the transition tables of the honest participants can be efficiently computed. The UC framework, on the other hand, imposes very little structure on the format of messages and allows participants to create messages using computations that cannot be modeled in the symbolic model.

Thus, to reconcile these two frameworks, we limit our attention to a particular class of protocols called simple protocols. These protocols are still sets of roles (for our purposes, the two roles of initiator and responder) but these roles are programs written in the programming language of Figure 2.

The language of simple protocols is defined in terms of the UC framework. Still, the commands of this language reflect the structure of the symbolic model. Furthermore, the encryption operation of this language is defined in terms of the abstract UC 'certified public-key encryption' functionality F_{CPKE} in Figure 4. This functionality captures, in an idealized way, the properties of public-key encryption in the case where parties know the public keys of each other in advance. In [19] we show how F_{CPKE} can be realized given a certification authority plus any encryption scheme that is secure against chosen ciphertext attacks.

To demonstrate the expressive power of the programming language that defines simple protocols, we express in this language two protocols. One protocol is the Dolev-Dwork-Naor authentication protocol which was originally presented in concrete cryptographic terms [23]. The other protocol is the Needham-Schroeder-Lowe (NSL) protocol, which is traditionally presented in symbolic form [47, 35, 36]. In fact, we extend the traditional description of NSL (which treat the protocol as a mutual authentication protocol) to a key exchange protocol. That is, we prescribe a way for the parties to locally output a key. Furthermore, we present two alternative methods for computing the output key. While these two methods look very similar, they turn out to have very different security properties. See more details in Section 5. These two variants of NSL are shown in Figure 3.

Π ::= BEGIN; STATEMENTLIST

BEGIN ::= input(SID, PID_0, PID_1, RID);

(Store $\langle\text{"sid"}, \text{SID}\rangle$, $\langle\text{"pid"}, \text{PID}_0\rangle$, $\langle\text{"pid"}, \text{PID}_1\rangle$, $\langle\text{"role"}, \text{RID}\rangle$) in local variables MySID, MyName, PeerName and MyRole)

STATEMENTLIST ::= STATEMENT STATEMENTLIST

| FINISH

STATEMENT ::= newrandom(v)

(generate a k-bit random string r and store $\langle\text{"random"}, r\rangle$ in v)

| encrypt(v1, v2, v3)

(Send (Encrypt, $\langle\text{PID}, \text{SID}\rangle$, v2) to F_{CPKE} where $v1 = \langle\text{"pid"}, \text{PID}\rangle$, receive c, and store $\langle\text{"ciphertext"}, c, \langle\text{PID}_1, \text{SID}\rangle\rangle$ in v3)

| decrypt(v1, v2)

(If the value of v1 is $\langle\text{"ciphertext"}, c'\rangle$ then send (Decrypt, $\langle\text{PID}_0, \text{SID}\rangle$, c') to F_{CPKE} instance $\langle\text{PID}_0, \text{SID}\rangle$, receive some value m, and store m in v2. Otherwise, end.)

| send(v)

(Send value of variable v)

| receive(v)

(Receive message, store in v)

| output(v)

(Send value of v to local output)

| pair(v1, v2, v3)

(Store $\langle\text{"pair"}, \sigma_1, \sigma_2\rangle$ in v3, where σ_1 and σ_2 are the values of v1 and v2, respectively.)

| separate(v1, v2, v3)

(If the value of v1 is $\langle\text{"join"}, \sigma_1, \sigma_2\rangle$, store σ_1 in v2 and σ_2 in v3 (else end))

| if (v1 == v2 then STATEMENTLIST else STATEMENTLIST

(where v1 and v2 are compared by value, not reference)

FINISH ::= output($\langle\text{"finished"}, v\rangle$); end.

The symbols v, v1, v2 and v3 represent program variables. It is assumed that $\langle\text{"pair"}, \sigma_1, \sigma_2\rangle$ encodes the bit-strings σ_1 and σ_2 in such a way that they can be uniquely and efficiently recovered. A party's input includes its own PID and the PID of its peer. Recall that the SID of an instance of F_{CPKE} is an encoding $\langle\text{SID}, \text{PID}\rangle$ of the PID and SID of the legitimate recipient.

Fig. 2. The grammar of simple protocols

3.1 From Simple Protocols to Fully-Specified Protocols

Simple protocols are by themselves somewhat abstract, in that they use F_{CPKE} rather than some fully-specified public-key encryption. This abstraction is justified as follows. In [19] we show how F_{CPKE} can be realized using functionality F_{PKE} (which represents the basic properties of public-key encryption schemes) and functionality F_{REG} (which represents some basic properties of a certification service). Furthermore, it is shown in [17] how to realize F_{PKE} given any public-key encryption scheme that is secure against chosen ciphertext attacks.

In the standard notation of the symbolic model, the protocol is usually written as:

$$1.\ A \to B : Enc(A\,N_a; K_B)$$
$$2.\ B \to A : Enc(b\,N_a\,N_b; K_A)$$
$$3.\ A \to B : Enc(N_b; K_B)$$

where $A \to B : M$ indicates that A sends the message M to B, N_a and N_b are random values (generated by A and B respectively, and K_A and K_B are the public encryption keys of A and B respectively. In Version 1 of the protocol, the parties output N_a as their secret key. In Version 2, As a simple protocol, the parties output N_b as the secret key. Written as a simple protocol, the protocol involves two roles, as follows:

On input $(p1 : PID; r1 : RID; s : SID), (p2 : PID; r2 : RID)$, do:

Initiator (M_{init}):

```
send((p1; r1; s), (p2; r2));
newrandom(na);
pair(p1, na, a_na);
encrypt(p2, s, r2, a_na, a_na_enc);
send(a_na_enc);
receive(b_na_nb_enc);
decrypt(b_na_nb_enc, b_na_nb);
separate(b_na_nb, b, na_nb);
if (b == p2) then
  separate(na_nb, na2, nb);
  if (na == na2) then
    encrypt(p2, s, r2, nb, nb_enc);
    send(nb_enc);
    pair(p1, p2, a_b);
    pair(a_b, [x], output);
    output(⟨"finished", output⟩);
  end.
  else send(⟨"finished", ⊥⟩); end.
else send(⟨"finished", ⊥⟩); end.
```

Responder (M_{resp}):

```
receive(a_na_enc);
decrypt(a_na_Enc(, ; a) _na);
separate(a_na, a, na);
if (b == p2) then
  newrandom(nb);
  pair(p1, na, b_na);
  pair(b_na, nb, b_na_nb);
  encrypt(p2, s, r2, b_na_nb, b_na_nb_enc);
  send(b_na_nb_enc);
  receive(nb_enc);
  decrypt(nb_enc, nb2);
  if (nb == nb2) then
    pair(p1, p2, b_a);
    pair(b_a, [x], output);
    output(⟨"finished", output⟩);
  end.
  else send(⟨"finished", ⊥⟩); end.
else send(⟨"finished", ⊥⟩); end.
```

Version 1: x=na (Initiator's nonce output as secret key)
Version 2: x=nb (Responder's nonce output as secret key)

Fig. 3. The Needham-Schroeder-Lowe (NSL) protocol

These facts, combined with the UC theorem, provide a straightforward way of instantiating simple protocols, while preserving security: replace each instance of F_{CPKE} by an instance of a CCA-secure encryption scheme, and use the certification authority to publicize the public keys. However, this method results in highly inefficient protocols, where each instance of the instantiated simple protocol uses its own instance of the public-key encryption scheme. Instead, we would like to obtain a protocol where each party uses a single instance of the public-key encryption scheme for multiple instances of the instantiated simple protocol.

Functionality F$_{\text{CPKE}}$

F$_{\text{CPKE}}$ proceeds as follows, when parameterized by message domain M, a probabilistic function E with domain M and range $\{0,1\}^*$, and a probabilistic function D of domain $\{0,1\}^*$ and range $M \cup \text{error}$. The SID is assumed to consist of a pair SID $= (\text{PID}_{owner}, \text{SID}')$, where PID$_{owner}$ is the identity of a special party, called the **owner** of this instance.

Encryption: Upon receiving a value (Encrypt, SID, m) from a party P proceed as follows:
   1. If $m \notin M$ then return an error message to P.
   2. If $m \in M$ then:
      – If party PID$_{owner}$ is corrupted, then let $ciphertext \leftarrow \mathsf{E}_k(m)$.
      – Otherwise, let $ciphertext \leftarrow \mathsf{E}_k(1^{|m|})$.
      Record the pair (m,c), and return c.
Decryption: Upon receiving a value (Decrypt, SID, c) from the owner of this instance, proceed as follows. (If the input is received from another party then ignore.)
   1. If there is a recorded pair (c,m) for some m, then hand m to P. (If there is more than a single recorded pair for c entry then return an error message.)
   2. Otherwise, compute $m = \mathsf{D}(c)$, and hand m to P.

Fig. 4. The certified public-key encryption functionality, F$_{\text{CPKE}}$

One way to do that would be to consider the entire multi-session interaction as a single instance of a more complex protocol. That protocol can now use a single instance of F$_{\text{CPKE}}$ per party. But this approach would force us to directly analyze the more complex multi-session protocols as a single unit. Instead we would like to be able to specify and analyze simple protocols n terms of a single instance (e.g., a single exchange of a key in the case of key-exchange), while making sure that the instantiated protocol uses only a single instance of F$_{\text{CPKE}}$ per party. This can be obtained using the UC with joint state theorem, along with an additional simple technique from [21].

We first observe that the following protocol realizes, in a single instance, multiple instances of F$_{\text{CPKE}}$ (which has the same decryptor), using only a single instance of F$_{\text{CPKE}}$: Whenever some party asks to encrypt a message m for an instance of F$_{\text{CPKE}}$ with session identifier sid, the protocol encrypts the pair (m, sid). Whenever some party asks to decrypt a ciphertext c for an instance sid, the protocol decrypts c, verifies that the decrypted value is of the form (m, sid') for some m, verifies that $sid' = sid$, and returns m. If $sid' \neq sid$ then an error value is returned. Denote this protocol by ES, for "encrypt the session ID". (This protocol and its analysis are analogous to the [21] protocol for realizing multiple instances of an ideal signature functionality using a single instance.)

Now, consider some protocol Π that involves multiple instances of a simple protocol π. (Protocol Π may simply describe an adversarially-controlled invocation of multiple instances of π, or alternatively Π may be geared towards realizing some other ideal functionality, potentially calling other protocols as subroutines.) In Π, each party uses a different instance of $\mathsf{F}_{\mathrm{CPKE}}$ per instance of π. We can now use the JUC theorem to assert that the protocol $\Pi^{[\mathrm{ES}]}$ behaves in the same way as Π. Furthermore, in $\Pi^{[\mathrm{ES}]}$ each party uses a single instance of $\mathsf{F}_{\mathrm{CPKE}}$ throughout the interaction.

4 The Mapping Lemma

While simple protocols are concrete protocols within the UC framework and are expressed in terms of interactive Turing machines, etc., they can be can be thought of as lying in the *intersection* of the UC framework and the symbolic model. This intuition is formalized via a protocol mapping which translates a concrete simple protocol p into a symbolic protocol $symb(\mathsf{p})$. The variables of the 'program' are interpreted as elements of the symbolic message algebra $\mathcal{A}$. Symbols are used instead of values for names and fresh randomness. Instead of using the functionality $\mathsf{F}_{\mathrm{CPKE}}$ for encryption and decryption, the symbolic constructor is applied or removed. Lastly, the symbolic pairing operator is applied or removed in the place of bit-string concatenation or separation.

We proceed as follows. First, we define the trace of an execution of a simple protocol in the presence of an adversarial environment within the UC framework. The trace provides a global view of the execution, including the views of the environment and the participants. It consists of a sequence of input, outputs, messages, and local variables (represented in strings). It also contains the participants' calls to $\mathsf{F}_{\mathrm{CPKE}}$, thus capturing their internal cryptographic operations. Similarly, we define the trace of an execution of a symbolic protocol within the symbolic model. Again, the trace represents a global view of the (now symbolic) execution. Here, the trace consists of a sequence of expressions from the underlying symbolic algebra, but as opposed to concrete traces the internal cryptographic operations of participants are not represented.

Next, we define a trace mapping, also denoted $symb()$, which translates a trace of a concrete simple protocol into a symbolic trace. This mapping is straightforward except that the calls to $\mathsf{F}_{\mathrm{CPKE}}$ in the concrete trace do not map to events in the symbolic trace, but are instead used as intermediate values in the mapping.

Finally, we show that this mapping provides soundness to trace properties in the symbolic protocol. That is, $symb()$ almost always translates traces of a concrete simple protocol to a trace of the corresponding symbolic protocol that is valid (meaning: one that could have been produced by the symbolic adversary and symbolic protocol). That is, we prove the following mapping lemma:

Lemma 1. *For all simple protocols* p, *adversaries* A, *environments* Z, *and inputs z of length polynomial in the security parameter* k, *the probability*

$$\Pr\left[\mathsf{t} \leftarrow \mathrm{TRACE}_{\mathsf{p},\mathsf{A},\mathsf{Z}}(k, z) : symb(\mathsf{t}) \text{ is not a valid DY trace for } symb(\mathsf{p})\right]$$

is negligible.

Thus, the adversary in the UC setting can do nothing with its general computational power that the symbolic adversary cannot also do (except with negligible probability).

We note that the statement of the mapping lemma is unconditional. Furthermore, it applies even to computationally unbounded environments and adversaries. In fact, the only source or error in the mapping is in cases where the environment in the concrete model "guesses" the value of some nonce. Since nonces are chosen at random from a large enough domain, the probability of error is negligible (in fact, it is exponentially small in the security parameter).

The mapping lemma is a central technical tool in our proofs of equivalence of the symbolic and concrete security criteria for mutual authentication and key exchange. Indeed, mutual authentication follows almost immediately from this lemma. (One can interpret this lemma as saying that trace properties of the symbolic protocol must also be trace properties of the original simple protocol, and mutual authentication is a trace property.) The lemma also seems to be of general interest beyond the rest of this work.

Finally, we note that the approach of mapping computational traces to symbolic ones comes from [45]. However, there the mapping holds only for computationally bounded adversaries and only under computational hardness assumptions.

5   Key Exchange

Key-exchange protocols require two security guarantees: an *agreement* property, establishing that the two parties share a common key, and a *secrecy* property for the agreed key. That is, the agreement property requires that if two parties P and P' obtain keys and associate these keys with each other, then the two keys are equal. The secrecy requirement requires that in this case the joint key should be "unknown" to the adversary.

In the UC model, these requirements are both captured in the ideal functionality F_{2KE} (Figure 5). This functionality waits to receive requests from two parties to exchange a key with each other, and then hands a secretly chosen random key to the parties. (Each party gets the output key only when the adversary instructs. Furthermore, the key is guaranteed to be random and secret only if both parties are uncorrupted.[1])

Providing a sound symbolic security criterion for key-exchange turns out be a more delicate task. The traditional symbolic criterion for key exchange requires these two properties in a straightforward way. Agreement is represented as a trace property: in any valid trace where both parties output a key symbol, it must be the same key symbol which is output. Secrecy, on the other hand, is represented by the separate trace property that there be no valid trace in which the adversary transmits the shared key 'in the clear.' Because the symbolic adversary is able to transmit any message it can derive, such a requirement implies that the session key will never be something which the symbolic adversary can learn.

However, notice that this symbolic secrecy property differs in flavor from the standard definitional approach of the computational model. The traditional sym-

---

[1] The present formulation of F_{2KE} is slightly different than the formulation in [17]. But the difference only affects the expected order of receiving the initial inputs from the parties, and does not affect the secrecy and authenticity properties of the exchange.

Functionality F_{2KE}

F_{2KE} proceeds as follows, running with security parameter k. At the first activation, choose and record a value $\kappa \xleftarrow{R} \{0,1\}^k$. Next:

1. Upon receiving an input $(\texttt{EstablishSession}, \textsf{SID}, \textsf{P}, \textsf{P}', \textsf{RID})$ from some party $\textsf{P}$ send this input to the adversary. In addition, if no pair is recorded, or the pair $(\textsf{P}', \textsf{P})$ is recorded, then record $(\textsf{P}, \textsf{P}')$. (Note that at most two pairs are ever recorded, and if there are two pairs, then they consist of the same party identities in reverse order.)
2. Upon receiving a request $(\texttt{SessionKey}, \textsf{SID}, \textsf{P}'', \tilde{k})$ from the adversary, do:
   (a) If a tuple $(\textsf{P}'', \textsf{P}''')$ is recorded, and $\textsf{P}'''$ is corrupted, then output $(\texttt{Finished}, sid, \tilde{k})$ to $\textsf{P}''$. (Here the adversary determines the key.)
   (b) If $\textsf{P}'''$ is uncorrupted, then Output $(\texttt{Finished}, \textsf{SID}, \kappa)$ to $\textsf{P}''$.
   (c) If no tuple $(\textsf{P}'', \textsf{P}''')$ is recorded, then ignore the request.

Fig. 5. The Key Exchange functionality

bolic definition requires only that the adversary be unable to derive the value of the key. However, the UC definition (following other computational definitions, e.g. [20]) require that the adversary be unable to distinguish between the real key and a random key even when given the candidate value *during the protocol execution*. It is tempting at first to believe that, since in the symbolic model the security guarantees are "all or nothing" in flavor, the ability to symbolically generate a secret and the ability to distinguish it from random should be equivalent. However, it turns out that this is not the case. That is, there exists a protocol which provably secure in the sense of the traditional symbolic definition, , but is insecure when instantiated by real cryptographic primitives. In particular, it does not realize the functionality F_{2KE}.

Consider the NSL protocol from Figure 3. This protocol was originally proposed for mutual authentication only, but it has long been recognized that either of the two random values used in the protocol (N_a and N_b in the symbolic notation, na and nb in the simple protocols) could be regarded as a secret session key. Furthermore, it has been proven many times (*e.g.*, [36, 54]) that both of these values are secret in the sense of the symbolic definition. However, as seen by the attack below, the NSL variant which outputs nb as the shared key (version 1 of Figure 3) is insecure in any reasonable protocol setting. In particular, it does *not* implement F_{2KE}.

Consider an execution of the NSL protocol that proceeds normally until the initiator has sent the third message, but before the responder receives that message. At this point, the responder is expecting to receive the random value nb, encrypted in his public encryption key. However, initiator has already completed the protocol and terminated, and so the attacker has already received the value nb and must distinguish it from a random value. Rephrased in terms of the UC framework, the environment has received the local output from one of the participants but it doesn't know if this is the real key nb (as in the protocol setting) or a random key (as in the functionality and simulator setting). The third message of the protocol provides an straightforward way of distinguishing these two cases.

We provide a detailed specification of the attack in terms of the UC framework. We stress however that the attack is quite generic, and does not depend on the specific formulation of one model or another.

The adversary flips a coin to choose a value. If the coin is 'heads,' the adversary chooses the candidate key. If the coin is 'tails,' on the other hand, the adversary chooses a new random key of the same length. In either case, the adversary encrypts the chosen value in the public key of the responder and sends it to that party.

- If the adversary is in the protocol setting, then the responder will be able to distinguish between the candidate key (which is the actual key) and a new random value, and progress accordingly.
- If the adversary is in the functionality/simulator setting, the simulator (who must simulate the responder's behavior) does not see the session key produced by F_{2KE}. This it will be unable to determine the coin-flip of the adversary. Thus, the simulator will be able to accurately simulate the responder with only 50% probability.

The salient point here is that, while the protocol never explicitly leaks the key, it give the adversary an opportunity to verify candidate values for the key. Thus, this protocol cannot fulfill the UC definition of key-secrecy, even though it has been shown to fulfill the traditional symbolic definition. Thus, computational soundness against the UC framework requires a new symbolic definition of secrecy.

The New Symbolic Criterion. Unlike the traditional symbolic criterion, our new definition is not expressed as a predicate on valid traces. Instead, it translates into the symbolic model the intuition behind the real-or-random secrecy criterion from cryptographic definitions of secrecy. To do that, we formalize the notion of a symbolic **adversary strategy**.

Definition 2 (Adversary Strategy). *Let an* adversary strategy *be a sequence of adversary events that respect the Dolev-Yao assumptions. That is, a strategy Ψ is a sequence of instructions $I_1, I_2 \ldots I_n$, where each I_i has one of the following forms, where i, j, k are integers:*

$$[\text{"receive"}, i] \quad [\text{"enc"}, j, k, i] \quad [\text{"dec"}, j, k, i]$$
$$[\text{"pair"}, j, k, i] \; [\text{"extract-l"}, j, i] \; [\text{"extract-r"}, j, i]$$
$$[\text{"random"}, i] \quad [\text{"name"}, i] \quad [\text{"pubkey"}, i]$$
$$[\text{"deliver"}, j, P_i]$$

When executed against protocol $\mathcal{P}$, a strategy Ψ produces the following Dolev-Yao trace $\Psi(\mathcal{P})$. Go over the instructions in Ψ one by one, and:

- *For each ["receive", i] instruction, if this is the first activation of party P_i, or P_i was just activated with a delivered message m, then add to the trace* a participant event (P'_j, L, m) which is consistent with the protocol $\mathcal{P}$. Else output the trace $\perp$.

– For any other instruction, add the corresponding event to the trace, where the index i is replaced by m_i, the message expression in the ith event in the trace so far. (If adding the event results in an invalid trace then output the trace $\perp$.)

We also need to define the 'observable portion' of a trace, which we do using public-key patterns (due originally to Abadi and Rogaway [4].)

Definition 3 (Public-key pattern[4, 29]). Let $T \subseteq \mathcal{K}_{Pub}$ (public keys) and $\mathsf{m} \in \mathcal{A}$. We recursively define the function $p(\mathsf{m}, T)$ to be:

– $p(K, T) = K$ if $K \in \mathcal{K}$ (public keys)
– $p(A, T) = A$ if $A \in \mathcal{M}$ (names/party identifiers)
– $p(N, T) = N$ if $N \in \mathcal{R}$ (random challenges/nonces)
– $p(N_1|N_2, T) = p(N_1, T)|p(N_2, T)$ (pairing)
– $p(Enc(\mathsf{m}; K), T) = \begin{cases} Enc(p(\mathsf{m}, T); K) & \text{if } K \in T \\ \langle\!|T|\!\rangle_K & \text{(where } T \text{ is the type tree of } \mathsf{m}) \text{ o.w.} \end{cases}$

Then $pattern_{pk}(\mathsf{m}, T)$, the public-key pattern of an Dolev-Yao message m relative to the set T, is

$$p(\mathsf{m}, \mathcal{K}_{Pub} \cap C[\{\mathsf{m}\} \cup T]).$$

If $t = H_1, H_2, \ldots H_n$ is a Dolev-Yao trace where event H_i contains message m_i then $pattern_{pk}(t, T)$ is exactly the same as t except that each m_i is replaced by $p(\mathsf{m}_i, \mathcal{K}_{Pub} \cap C[S \cup T])$ where $S = \{\mathsf{m}_1, \mathsf{m}_2, \ldots \mathsf{m}_n\}$. The base pattern of a message m, denoted $pattern(\mathsf{m})$, is defined to be $pattern_{pk}(\mathsf{m}, \emptyset)$, and $pattern(t)$ is defined to be $pattern_{pk}(t, \emptyset)$.

Our new symbolic definition of secure key-exchange requires that, for all adversary strategies, when a given strategy is applied to the protocol, the observable portion of the resulting trace looks the same when the shared key is the output of the protocol and when it is a fresh key symbol (representing a fresh random key).

Definition 4 (Variable Renaming). Let R_1, R_2 be random-strings symbols, and let t be an expression in the algebra $\mathcal{A}$. Then $t_{[R_1 \mapsto R_2]}$ is the expression where every instance of R_1 is replaced by R_2.

Definition 5 (Symbolic Criterion for Key Exchange). A Dolev-Yao protocol $\mathcal{P}$ provides Dolev-Yao two-party secure key exchange (DY-2SKE) if

1. (Agreement) For all P_0 and $P_1 \notin \mathcal{M}_{Adv}$ and Dolev-Yao traces valid for $\mathcal{P}$ in which P_0 outputs $\langle Starting|P_0|P_1|\mathsf{m}\rangle$ and P_1 outputs $\langle Starting|P_1|P_0|\mathsf{m}'\rangle$, if participant P_0 produces output message $\langle Finished|\mathsf{m}_0\rangle$ and participant P_1 produces output message $\langle \texttt{finished}|\mathsf{m}_1\rangle$, then $\mathsf{m}_0 = P_0|P_1|R$ and $\mathsf{m}_1 = P_1|P_0|R$ for some $R \in \mathcal{R}$.

2. *(Real-or-random secrecy) Let $\mathcal{P}_f$ be the protocol $\mathcal{P}$ except that a fresh fake key R_f is output by terminating participants in place of the real key R_r. Then for every adversary strategy Ψ,*

$$pattern\,(\Psi(\mathcal{P})) = pattern\,\big(\Psi(\mathcal{P}_f)_{[R_f \mapsto R_r]}\big)$$

Finally, we demonstrate that the new symbolic security criterion for key exchange is equivalent to the UC criterion. (Again, equivalence holds unconditionally.)

Theorem 1. *Let* p *be a simple protocol. Then* p *UC-realizes* $\mathsf{F}_{2\text{KE}}$ *if and only if* $symb(\mathsf{p})$ *achieves Dolev-Yao secure key-exchange.*

To demonstrate the "only if" part (namely, the completeness of the symbolic condition) we show how to turn any symbolic trace of $symb(\mathsf{p})$ that violates the symbolic key exchange criterion into a strategy of a concrete environment for distinguishing between an execution of p and the ideal process for $\mathsf{F}_{2\text{KE}}$.

The "if" part (namely, the soundness of the symbolic condition) is proven as follows. Given a simple protocol p, we construct a general strategy for a simulator (i.e, an ideal-process adversary) within the UC framework. We then show that, except with negligible probability, any environment that distinguishes between real and ideal executions can be turned into a (single) symbolic trace of $symb(\mathsf{p})$ that violates the symbolic key exchange criterion.

This sketch omits many details however; the proof is rather delicate. In particular, demonstrating the second property with respect to the symbolic secrecy criterion requires some work.

6   Future Research

This work demonstrates that completely symbolic analysis of security properties within a simulation-based, compositional cryptographic framework is possible. Furthermore, the chosen symbolic framework is one that is very close to the language of known automated verification tools. As such, it opens the door to a number of questions and challenges. For example one might wish to generalize our results to a richer and less restrictive "programming language" for protocols. One direction is to enlarge the set of allowed operations and to incorporate other cryptographic primitives, while retaining the ability to analyze only a single session of the protocol in question. Natural candidates include the Diffie-Hellman exchange, signatures schemes, pseudo-random functions, and message authentication codes. Other generalizations include adaptive security (i.e. security against adversaries that corrupt parties throughout the computation), and protocols where even their symbolic counterparts are randomized.

A second direction is to apply a similar analytical methodology to other cryptographic tasks, and even tasks that were never before addressed using formal tools. For instance, it may be possible to come up with a symbolic representation of, say, two-party protocols that use commitment schemes, and provide a symbolic criterion for when such protocols are zero-knowledge protocols (e.g., satisfy the ideal zero-knowledge functionality). Similarly, one can potentially come up with symbolic criteria as to when a protocol UC-realizes an arbitrary given ideal functionality.

Acknowledgments

We thank Shai Halevi and Akshay Patil for very useful comments and discussions. In particular, Shai discovered a bug in a previous version of the proof of Theorem 1, and Akshay discovered a bug in a previous formulation of F_{CPKE}.

References

1. Martín Abadi and Bruno Blanchet. Analyzing security protocols with secrecy types and logic programs. In *Conference Record of POPL 2002: The 2pth SIGPLAN-SIGACT Symposium on Principles of Programming Languages*, pages 33–44, January 2002.
2. Martín Abadi and Andrew Gordon. A calculus for cryptographic protocols: the spi calculus. *Information and Computation*, 148(1):1–70, 1999.
3. Martín Abadi and Jan Jürjens. Formal eavesdropping and its computational interpretation. In Naoki Kobayashi and Benjamin C. Pierce, editors, *Proceedings, 4th International Symposium on Theoretical Aspects of Computer Software TACS 2001*, volume 2215 of *Lecture Notes in Computer Science*, pages 82–94. Springer, 2001.
4. Martín Abadi and Phillip Rogaway. Reconciling two views of cryptography (the computational soundness of formal encryption). *Journal of Cryptology*, 15(2):103–127, 2002.
5. Pedro Adao, Gergei Bana, Jonathan Herzog, and Andre Scedrov. Soundness of abadi-rogaway logics in the presence of key-cycles. In *Proceedings of the 10th European Symposium On Research In Computer Security (ESORICS 2005)*. Springer, September 2005.
6. M. Backes, B. Pfitzmann, and M. Waidner. A composable cryptographic library with nested operations (extended abstract). In *Proceedings, 10th ACM conference on computer and communications security (CCS)*, October 2003. Full version available at http://eprint.iacr.org/2003/015/.
7. Michael Backes and Birgit Pfitzmann. A cryptographically sound security proof of the Needham-Schroeder-Lowe public-key protocol. In *Proceedings of the 23rd Conference on Foundations of Software Technology and Theoretical Computer Science – FSTTCS*, volume 2914 of *Lecture Notes in Computer Science*, pages 140–152. Springer-Verlag, December 2003.
8. Michael Backes and Birgit Pfitzmann. Relating symbolic and cryptographic secrecy. Cryptology ePrint Archive, Report 2004/300, November 2004. http://eprint.iacr.org/.
9. Donald Beaver. Secure multiparty protocols and zero-knowledge proof systems tolerating a faulty minority. *Journal of Cryptology*, 4(2):75–122, 1991.
10. Mihir Bellare and Phillip Rogaway. Entity authentication and key distribution. In D. Stinson, editor, *Advances in Cryptology - CRYPTO 1993*, volume 773 of *Lecture Notes in Computer Science*, pages 232–249. Springer-Verlag, August 1993. Full version of paper available at http://www-cse.ucsd.edu/users/mihir/.

11. Bruno Blanchet. Automatic proof of strong secrecy for security protocols. In proceedings of the 2004 IEEE Symposium on Security and Privacy (S&P), Oakland, CA, USA, May 2004. IEEE.
12. Bruno Blanchet. ProVerif automatic cryptographic protocol verifier user manual. Available at *http://www.di.ens.fr/ blanchet/crypto-eng.html*, November 2004.
13. Manual Blum and Silvio Micali. How to generate cryptographically strong sequences of pseudo random bits. In *Proceedings, 22th Annual Syposium on Foundations of Computer Science (FOCS 1982)*, pages 112–117, 1982.
14. Manual Blum and Silvio Micali. How to generate cryptographically strong sequences of pseudo-random bits. *SIAM Journal on Computing*, 13(4):850–864, 1984.
15. Michael Burrows, Martín Abadi, and Roger Needham. A logic of authentication. *ACM Transactions in Computer Systems*, 8(1):18–36, February 1990.
16. Ran Canetti. Security and composition of multiparty cryptographic protocols. *Journal of Cryptology*, 13(1):143–202, 2000.
17. Ran Canetti. Universal composable security: A new paradigm for cryptographic protocols. In *42nd Annual Syposium on Foundations of Computer Science (FOCS 2001)*, pages 136–145. IEEE Computer Society, October 2001.
18. Ran Canetti. Universally composable signature, certification, and authentication. In *Proceedings of the 17th IEEE Computer Security Foundations Workshop (CSFW 16)*, pages 219–233. IEEE Computer Society, June 2004.
19. Ran Canetti and Jonathan Herzog. Universally composable symbolic analysis of cryptographic protocols (the case of encryption-based mutual authentication and key exchange). Cryptology ePrint Archive, Report 2004/334, 2004.
20. Ran Canetti and Hugo Krawczyk. Analysis of key-exchange protocols and their use for building secure channels. In Birgit Pfitzmann, editor, *Advances in Cryptology - Eurocrypt 2001*, volume 2045 of *Lecture Notes in Computer Science*, pages 453–474. Springer-Verlag, May 2001.
21. Ran Canetti and Tal Rabin. Universal composition with joint state. In Advances in Cryptology - CRYPTO 2003, LNCS 2729, 2003, pages 265–281.
22. I. Cervesato, N. A. Durgin, P. D. Lincoln, J. C. Mitchell, and A. Scedrov. A meta-notion for protocol analysis. In Proceedings of the 12th IEEE Computer Security Foundations Workshop (CSFW 12). IEEE Computer Society, June 1999.
23. D. Dolev, C. Dwork, and M. Naor. Non-malleable cryptography. *SIAM Journal of Computing*, 30(2):391–437, 2000.
24. D. Dolev and A. Yao. On the security of public-key protocols. *IEEE Transactions on Information Theory*, 29:198–208, 1983.
25. S. Goldwasser and S. Micali. Probabilistic encryption. *Journal of Computer and System Sciences*, 28(2):270–299, 1984.
26. Shafi Goldwasser and Leonid Levin. Fair computation of general functions in presence of immoral majority. In Alfred Menezes and Scott A. Vanstone, editors, *CRYPTO*, volume 537 of *Lecture Notes in Computer Science*, pages 77–93. Springer, August 1990.
27. Shafi Goldwasser, Silvio Micali, and Charles Rackoff. The knowledge complexity of interactive proof systems. *SIAM Journal on Computing*, 18(1):186–208, 1989.
28. Shafi Goldwasser, Silvio Micali, and Ronald L. Rivest. A digital-signature scheme secure against adaptive chosen-message attacks. *SIAM J. Computing*, 17(2):281–308, April 1988.
29. Jonathan Herzog. A computational interpretation of dolev-yao adversaries. *Theoretical Computer Science*, 340:57–81, June 2005.

30. Jonathan Herzog, Moses Liskov, and Silvio Micali. Plaintext awareness via key registration. In Advances in Cryptology - CRYPTO 2003, LNCS 2729, 2003, pages 548–564.
31. O. Horvitz and V. Gligor. Weak key authenticity and the computational completeness of formal encryption. In Advances in Cryptology - CRYPTO 2003, LNCS 2729, 2003, pages 530–547.
32. P. Laud. Symmetric encryption in automatic analyses for confidentiality against active adversaries. In proceedings of the 2004 IEEE Symposium on Security and Privacy (S&P), Oakland, CA, USA, May 2004. IEEE.
33. P. D. Lincoln, J. C. Mitchell, M. Mitchell, and A. Scedrov. A probabilistic poly-time framework for protocol analysis. In *Proceedings of the 5th ACM Conference on Computer and Communication Security (CCS '98)*, pages 112–121, November 1998.
34. P. D. Lincoln, J. C. Mitchell, M. Mitchell, and A. Scedrov. Probabilistic polynomial-time equivalence and security protocols. In Jeannette M. Wing, Jim Woodcock, and Jim Davies, editors, *World Congress on Formal Methods*, volume 1708 of *Lecture Notes in Computer Science*, pages 776–793. Springer, September 1999.
35. Gavin Lowe. An attack on the Needham–Schroeder public-key authentication protocol. *Information Processing Letters*, 56:131–133, 1995.
36. Gavin Lowe. Breaking and fixing the Needham–Schroeder public-key protocol using FDR. In Margaria and Steffen, editors, *Tools and Algorithms for the Construction and Analysis of Systems*, volume 1055 of *Lecture Notes in Computer Science*, pages 147–166. Springer–Verlag, 1996.
37. Nancy Lynch. I/O automaton models and proofs for shared-key communication systems. In proceedings of the 12th IEEE Computer Security Foundations Workshop (CSFW 12). IEEE Computer Society, June 1999.
38. P. Maggi and R. Sisto. Using SPIN to verify security protocols. In *Proceedings of the 9th International SPIN Workshop on Model Checking of Software*, number 2318 in Lecture Notes in Computer Science, pages 187–204, 2002.
39. Catherine Meadows. Applying formal methods to the analysis of a key management protocol. *The Journal of Computer Security*, 1(1), January 1992.
40. Catherine Meadows. The nrl protocol analyzer: An overview. *J. Log. Program.*, 26(2):113–131, 1996.
41. Silvio Micali, Charles Rackoff, and Bob Sloan. The notion of security for probabilistic cryptosystems. *SIAM Journal on Computing*, 17(2):412–426, April 1988.
42. Silvio Micali and Phillip Rogaway. Secure computation (abstract). In Joan Feigenbaum, editor, *CRYPTO*, volume 576 of *Lecture Notes in Computer Science*, pages 392–404. Springer, August 1991.
43. Daniele Micciancio and Saurabh Panjwani. Adaptive security of symbolic encryption. In *Theory of cryptography conference - Proceedings of TCC 2005*, volume 3378 of *LNCS*, pages 169–187. Springer-Verlag, 2005.
44. Daniele Micciancio and Bogdan Warinschi. Completeness theorems for the Abadi-Rogaway logic of encrypted expressions. Workshop on Issues in the Theory of Security (WITS '02), January 2002.
45. Daniele Micciancio and Bogdan Warinschi. Completeness theorems for the Abadi-Rogaway logic of encrypted expressions. *Journal of Computer Security*, 12(1):99–129, 2004.
46. John C. Mitchell, Mark Mitchell, and Ulrich Stern. Automated analysis of cryptographic protocols using Murφ. In *Proceedings, 1997 IEEE Symposium on Security and Privacy*, pages 141–153. IEEE, Computer Society Press of the IEEE, 1997.

47. Roger Needham and Michael Schroeder. Using encryption for authentication in large networks of computers. *Communications of the ACM*, 21(12):993–999, 1978.
48. Akshay Patil. On symbolic analysis of cryptographic protocols. Master's thesis, Massachusetts Institute of Technology, May 2005.
49. Birgit Pfitzmann and Michael Waidner. Composition and integrity preservation of secure reactive systems. In *Proceedings of the 7th ACM Conference on Computer and Communication Security (CCS 2000)*, pages 245–254. ACM Press, November 2000.
50. C. Rackoff. Personal communication. 1995.
51. C. Rackoff and D. Simon. Noninteractive zero-knowledge proof of knowledge and the chosen-ciphertext attack. In *Advances in Cryptology– CRYPTO 91*, number 576 in Lecture Notes in Computer Science, pages 433–444, 1991.
52. Shmuel Sagiv, editor. *Computationally Sound, Automated Proofs for Security Protocols.*, volume 3444 of *Lecture Notes in Computer Science*. Springer, April 2005.
53. D. Song. Athena, an automatic checker for security protocol analysis. In proceedings of the 12th IEEE Computer Security Foundations Workshop (CSFW 12). IEEE Computer Society, June 1999.
54. F. Javier THAYER Fábrega, Jonathan C. Herzog, and Joshua D. Guttman. Strand spaces: Proving security protocols correct. *Journal of Computer Security*, 7(2/3):191–230, 1999.

Resource Fairness and Composability
of Cryptographic Protocols

Juan Garay[1], Philip MacKenzie[2], Manoj Prabhakaran[3], and Ke Yang[2]

[1] Bell Labs – Lucent Technologies
garay@research.bell-labs.com
[2] Google
philmac@gmail.com, yangke@google.com
[3] Computer Science Department, University of Illinois at Urbana-Champaign
mmp@uiuc.edu

Abstract. We introduce the notion of *resource-fair* protocols. Informally, this property states that if one party learns the output of the protocol, then so can all other parties, as long as they expend roughly the same amount of resources. As opposed to similar previously proposed definitions, our definition follows the standard simulation paradigm and enjoys strong composability properties. In particular, our definition is similar to the security definition in the universal composability (UC) framework, but works in a model that allows any party to request additional resources from the environment to deal with dishonest parties that may prematurely abort.

In this model we specify the ideally fair functionality as allowing parties to "invest resources" in return for outputs, but in such an event offering all other parties a fair deal. (The formulation of fair dealings is kept independent of any particular functionality, by defining it using a "wrapper.") Thus, by relaxing the notion of fairness, we avoid a well-known impossibility result for fair multi-party computation with corrupted majority; in particular, our definition admits constructions that tolerate arbitrary number of corruptions. We also show that, as in the UC framework, protocols in our framework may be arbitrarily and concurrently composed.

Turning to constructions, we define a "commit-prove-fair-open" functionality and design an efficient resource-fair protocol that securely realizes it, using a new variant of a cryptographic primitive known as "time-lines." With (the fairly wrapped version of) this functionality we show that some of the existing secure multi-party computation protocols can be easily transformed into resource-fair protocols while preserving their security.

1 Introduction

Secure multi-party computation (MPC) is one of the most fundamental problems in cryptography, and has been investigated thoroughly over many years [54,55,38,7,18,37]. Defining security is one of the first challenges in achieving this [37,13,48,14,40,43,3,50]. The *universal composability* (UC) framework

S. Halevi and T. Rabin (Eds.): TCC 2006, LNCS 3876, pp. 404–428, 2006.

of Canetti [14] is among the models that provide perhaps the strongest security guarantees. A protocol π that is secure in this framework is guaranteed to remain secure when arbitrarily composed with other protocols, by means of a "composition theorem."

In this paper we investigate a less studied aspect of multiparty computation, namely *fairness*. Informally, a protocol is fair if either all the parties learn the output of the function, or no party learns anything (about the output).[1] Clearly, fairness is a very desirable property for secure MPC protocols, and in fact, many of the security definitions cited above imply fairness. (See [40] for an overview of different types of fairness, along with their corresponding histories.) Here we briefly describe some known results about (complete) fairness. Let n be the total number of participating parties and t be the number of corrupted parties. It is known that if $t < n/3$, then fairness can be achieved without any set-up assumptions, both in the information-theoretic setting [7,18] and in the computational setting [38,37] (assuming the existence of trapdoor permutations). If $t < n/2$, one can still achieve fairness if all parties have access to a broadcast channel; this also holds both information theoretically [51] and computationally [38,37].

Unfortunately, the above fairness results no longer hold when $t \geq n/2$, i.e., when a majority of the parties are corrupted. In fact, it was proved that there do not exist fair MPC protocols in this case, even when parties have access to a broadcast channel [19,37]. Intuitively, this is because the adversary, controlling a majority of the corrupted parties, can abort the protocol prematurely and always gain some unfair advantage. This impossibility result easily extends to the *common reference string* (CRS) model (where there is a common string drawn from a prescribed distribution available to all the parties).

Nevertheless, fairness is still important (and necessary) in many applications in which at least half the parties may be corrupted. One such application is contract signing (or more generally, the fair exchange of signatures) by two parties [8]. To achieve some form of fairness, various approaches have been explored. One such approach adds to the model a trusted third party, who is essentially a judge that can be called in to resolve disputes between the parties. (There is a large body of work following this approach; see, e.g., [2,12] and references therein.) This approach requires a trusted external party that is constantly available. Another recent approach adds an interesting physical communication assumption called an "envelope channel," which might be described as a "trusted postman" [42].

A different approach that avoids the available trusted party requirement uses a mechanism known as "gradual release," where parties take turns to release their secrets in a "bit by bit" fashion. Therefore, if a corrupted party aborts prematurely, it is only a little "ahead" of the honest party, and the honest party can "catch up" by investing an amount of time that is comparable to (and maybe greater than) the time spent by the adversary. (Note that this is

---

[1] This property is also known as "complete fairness," and can be contrasted with "partial fairness," where fairness is achieved only when there are certain restrictions on corruption of parties [40].

basically an *ad hoc* notion of fairness.) Early works in this category include [8,27,30,39,4,23]. More recent work has focused on making sure — under the assumption that there exist problems, such as modular exponentiation, that are not well suited for parallelization[2] — that this "unfairness" factor is bounded by a small constant [11,36,49]. As we discuss below, our constructions also use a gradual release mechanism secure against parallel attacks.

Resource fairness. In this paper we propose a new notion of fairness with a rigorous simulation-based security definition (without a trusted third party), that allows circumvention of the impossibility result discussed above in the case of corrupted majorities. We call this new notion *resource fairness*. In a nutshell, resource fairness means that if any party learns the output of a function, then all parties *will be able to learn the output of the function by expending roughly the same amount of resources.* (In our case, the resource will be time.) In order to model this, we allow honest parties in our framework (both in the real world and in the ideal process) to request resources from the environment, and our definition of resource fairness relates the amount of requested resources to the amount of resources available to corrupted parties.

Slightly more formally, a resource-fair functionality can be described in two steps. We start with the most natural notion for a fair functionality $\mathcal{F}$. A critical feature of a fair functionality is the following:

– There are certain messages that $\mathcal{F}$ sends to multiple parties such that all of them must receive the message in the same round of communication. (For this it is necessary that the adversary in the ideal process cannot block messages from $\mathcal{F}$ to the honest parties.[3])

Then we modify it using a "wrapper" to obtain a functionality $\mathcal{W}(\mathcal{F})$. The wrapper allows the adversary to make "deals" of roughly the following kind:

– Even if $\mathcal{F}$ requires a message to be simultaneously delivered to all parties, the adversary can "invest" computational resources and obtain the message from $\mathcal{W}(\mathcal{F})$ in an earlier communication round.
– However, in this case, $\mathcal{W}(\mathcal{F})$ will offer a "fair deal" to the honest parties: each of them will be given the option of obtaining its message by investing (at most) the same amount of computational resources as invested by the adversary.

Once we define $\mathcal{W}(\mathcal{F})$ as our ideal notion of a fair functionality, we need to define when a *protocol* is considered to be as fair as $\mathcal{W}(\mathcal{F})$. We follow the same paradigm as used in the UC framework for defining security: A protocol π is said

---

[2] Indeed, there have been considerable efforts in finding efficient exponentiation algorithms (e.g., [1,53]) and still the best methods are sequential.

[3] In the original formulation of the UC framework [14], the adversary in the ideal process could block the outputs from the ideal functionality to all the parties. Thus, the ideal process itself is already completely unfair, and therefore discussing fair protocols is not possible. The new version [15] also has "immediate functionalities" as the default—see Section 2.1.

to be as fair as $\mathcal{W}(\mathcal{F})$ if for every real adversary $\mathcal{A}$ there exists an ideal adversary (simulator) $\mathcal{S}$ such that no environment can distinguish between interacting with $\mathcal{A}$ and parties running a protocol π (the real world), and interacting with $\mathcal{S}$ and parties talking to $\mathcal{W}(\mathcal{F})$ (the ideal world). But in addition we require that $\mathcal{S}$ cannot invest much more resources than $\mathcal{A}$ has.

This last condition is crucial for the notion of resource fairness. To see this, note the following:

- In the ideal world, in the event of the adversary $\mathcal{S}$ obtaining a message by investing some amount of resources, an honest party can be required to invest the same amount of resources to get its message.
- By the indistinguishability condition, this is the same as the amount of resources required by the honest parties in the real world. Thus, the resources required by the honest parties in the real world can be as much as that invested by the adversary $\mathcal{S}$ in the ideal world.

Recall that the (intuitive) notion of resource fairness requires that the resources required by an honest party in the real world should be comparable to what the adversary $\mathcal{A}$ (in the real world) expends, to obtain its output. Thus, to achieve the notion, we must insist that the amount of resources invested by the ideal world adversary $\mathcal{S}$ is comparable to what the real world adversary $\mathcal{A}$ expends.

Note that for these comparisons, *the resources in the ideal world must be measured using the same units as in the real world*. However, these invested resources do not have a physical meaning in the ideal world: it is just a "currency" used to ensure that the fairness notion is correctly reflected in the ideal world process.

The only resource we shall consider in this work is computation time.

Fairness through gradual release. Our definition is designed to capture the fairness guarantees offered by the method of *gradual release*. The gradual release method by itself is not new, but our simulation-based definition of fairness is.

Typical protocols using gradual release consist of a "computation" phase, where some computation is carried out, followed by a "revealing" phase, where the parties gradually release their private information towards learning a result y. Our simulation-based definition requires one to be able to simulate both the computation phase and the release phase. In contrast, previous *ad hoc* security definitions did not require this, and consisted, explicitly or implicitly, of the following three conditions:

1. The protocol must be completely simulatable up to the revealing phase.
2. The revealing phase must be completely simulatable *if the simulator knows y*.
3. If the adversary aborts in the revealing phase and computes y by brute force in time t, then all the honest parties can compute y in time comparable to t.[4]

While carrying some intuition about security and fairness, we note that these definitions are not fully simulation-based. To see this, consider a situation where

---

[4] As we discussed before, an honest party typically will spend *more* time than the adversary in this case.

an adversary $\mathcal{A}$ aborts early on in the revealing phase, such that it is still infeasible for $\mathcal{A}$ to find y by brute force. At this time, it is also infeasible for the honest parties to find y by brute force. Now, how does one simulate $\mathcal{A}$'s view in the revealing phase? Notice that the revealing phase is simulatable *only if the ideal adversary $\mathcal{S}$ knows y*. However, since nobody learns y in the real world, they should not learn y in the ideal world, and, in particular, $\mathcal{S}$ should not learn y. Thus, the above approach gives no guarantee that $\mathcal{S}$ can successfully simulate $\mathcal{A}$'s view. In other words, by aborting early in the revealing phase, $\mathcal{A}$ might gain some unfair advantage. This can become an even more serious security problem when protocols are composed.

Environment's role. In our formulation of fairness, if a protocol is aborted, the honest parties get the *option* of investing resources and recovering a message from the functionality. However, the decision of whether to exercise this option is not specified by the protocol itself, but left to the environment. Just being provided with this option is considered fair.[5] The fairness guarantee is that the amount of resources that need to be invested by the adversary to recover the message will be comparable to what the honest party requires. Whether the adversary actually makes that investment or not is not known to the honest parties.

Leaving the recovery decision to the environment has the consequence that our notion of fairness becomes a robust "relative" notion. In some environments the execution might be (intuitively) unfair if, for instance, the environment refuses to grant any requests for resources. However, this is analogous to the situation in the case of security: Some environments can choose to reveal all the honest parties' inputs to the adversary. The protocol's guarantee is limited to mimicking the ideal functionality (which by definition is secure and fair). We do not seek to incorporate absolute guarantees of fairness (or security) into the protocol, as they are dependent on the environment.

Our results. We now summarize the main results presented in this paper.

1. **A fair multi-party computation framework.** We start with a framework for fair multi-party computation (FMPC), which is a variation of the UC framework, but with modifications so that it is possible to design functionalities such that the ideal process is (intuitively) fair. We then present a generic *wrapper* functionality, denoted $\mathcal{W}(\cdot)$, that converts a fair functionality into one that allows for a resource-fair realization in the real world. We then present definitions for resource-fair protocols that securely realize functionalities in this framework. We emphasize that these definitions

---

[5] In a previous version of this work [35], we insisted that the protocol itself must decide whether or not to invest computational resources and recover a message from an aborted protocol. Further, for being fair, we required that if the adversary could have obtained its part of the message, then the protocol *must* carry out the recovery. This leads to the unnatural requirement that the protocol must be aware of the computational power of the adversary (up to a constant).

are in the (standard) simulation paradigm[6] and admit protocols that tolerate an arbitrary number of corruptions. Finally, we prove a *composition theorem* similar to the one in the UC framework.

2. **The "commit, prove and fair-open" functionality.** We define a *commit-prove-fair-open* functionality $\mathcal{F}_{CPFO}$ in the FMPC framework. This functionality allows all parties to each commit to a value, prove relations about the committed value, and more importantly, open all committed values simultaneously to all parties. This functionality (more specifically, a wrapped version of it) lies at the heart of our constructions of resource-fair MPC protocols. We then construct an efficient resource-fair protocol GradRel that securely realizes $\mathcal{F}_{CPFO}$, assuming static corruptions. Our protocol uses a new variant of a cryptographic primitive known as *time-lines* [31], which enjoys a property that we call *strong pseudorandomness*. In turn, the construction of time-lines hinges on a refinement of the *generalized BBS* assumption [11], which has broader applicability.

3. **Efficient and resource-fair MPC protocols.** By using the $\mathcal{W}(\mathcal{F}_{CPFO})$ functionality, many existing secure MPC protocols can be easily transformed into resource-fair protocols while preserving their security. In particular, we present two such constructions. The first construction converts the universally composable MPC protocol by Canetti *et al.* [17] into a resource-fair MPC protocol that is secure against static corruptions in the CRS model in the FMPC framework. Essentially, the only thing we need to do here is to replace an invocation of a functionality in the protocol called "commit-and-prove" by our $\mathcal{W}(\mathcal{F}_{CPFO})$ functionality.

    The second construction turns the efficient MPC protocol by Cramer *et al.* [21] into a resource-fair one in the "public key infrastructure" (PKI) model in a similar fashion. The resulting protocol becomes secure and resource fair (assuming static corruptions) in the FMPC framework, while preserving the efficiency of the original protocol — an additive overhead of only $O(\kappa^2 n)$ bits of communication and an additional $O(\kappa)$ rounds, for κ the security parameter.

Organization of the paper. The paper has two main components: the formalization of the notion of resource-fairness, and protocol constructions satisfying this notion. In Section 2 we present the new notion, and Section 3 is dedicated to explaining the protocol constructions. Within Section 2, we describe the FMPC framework, describe "wrapped" functionalities, give security and fairness definitions and finally state a composition theorem. In Section 3 we present the $\mathcal{F}_{CPFO}$ functionality and show a protocol that realizes a wrapped version of it, which we then use to achieve resource-fair MPC. Due to space limitations, proofs, detailed remarks and extensions are omitted from this extended abstract and can be found in the full version of the paper [32].

---

[6] Indeed, as explained in Section 2.4, our definition of resource fairness subsumes the UC definition of security.

2   FMPC Framework and Resource Fairness

2.1   The FMPC Framework

We now define the new framework used in our paper, which we call the *fair multi-party computation* (FMPC) framework. It is similar to the universal composability (UC) framework [14,15]. In particular, there are n parties, $P_1, P_2, ..., P_n$, a real-world adversary $\mathcal{A}$, an ideal adversary $\mathcal{S}$, an ideal functionality $\mathcal{F}$, and an environment $\mathcal{Z}$. However, FMPC contains some modifications so that fairness becomes possible. We stress that the FMPC framework still inherits the strong security of UC, and we shall prove a composition theorem in the FMPC framework similar to UC.

Instead of describing the FMPC framework from scratch, we only discuss its most relevant features and differences from the UC framework. Refer to [15] for a detailed presentation of the UC framework. The critical features of the FMPC framework are:

Interactive circuits/PRAMs. Instead of interactive Turing machines, we assume the computation models in the FMPC framework are non-uniform interactive PRAMs (IPRAMs).[7] This is a non-trivial distinction, since we will work with exact time bounds in our security definition, and the "equivalence" between various computation models does not carry over there. The reason to make this modification is that, we will need to model machines *that allow for simulation and subroutine access with no significant overhead.* Thus, if we have two protocols, and one calls the other as a black-box, then the total running time of the two protocols together will be simply the sum of their running times. Obviously, Turing machines are not suitable here.

We say an IPRAM is *t-bounded* if it runs for a total of at most t steps.[8] We always assume that t is a polynomial of the security parameter κ, though for simplicity we do not explicitly write $t(\kappa)$. We can view a t-bounded IPRAM as a "normal" IPRAM with an explicit "clock" attached to it that terminates the execution after a total number of t cumulative steps (notice that an IPRAM is reactive: i.e., it maintains state across activations).

Synchronous communication with rounds. In the UC framework, the communication is asynchronous, and controlled by the adversary, and further there is no notion of time. This makes fair MPC impossible, since the adversary may, for example, choose not to deliver the final protocol message to an uncorrupted party P_i. In this case, P_i will never obtain the final result *because it is never activated again.* What is needed is to let parties be able to *time out* if they do not receive an expected message within some time bound. However, instead of incorporating a full-fledged notion of time into the model, for simplicity we shall work in a "synchronous model." Specifically, in the FMPC framework there will

---

[7] IPRAMs are simply extensions to the PRAM machines with special read-only and write-only memories for interacting with each other.

[8] For simplicity, we assume that an IPRAM can compute a modular squaring operation (i.e., compute $x^2 \bmod M$ on input (x, M)) in constant time.

be synchronous rounds of communication in both the real world and the ideal process. (See [41,45] for other synchronous versions of the UC framework.)

In each round we allow the adversary to see the messages sent by other parties in that round, before generating its messages (i.e., we use a *rushing adversary* model.

Note that this model of communication is used in both the real *and* ideal worlds used for defining security. (As we shall see later, a resource-fair ideal functionality is designed to be aware of this round structure. This is necessary because the amount of resources required by an honest party to retrieve messages that the adversary blocks, is directly related to the number of communication rounds in the protocol that pass prior to that.) This allows also the environment to be aware of the round structure.

We stress that in our protocols, we use the synchronous communication model only as a substitute for having time-outs on messages (which are sequentially numbered). Our use of the synchronous model is only that if a message does not arrive in a communication round in which it is expected, then the protocol can specify an action to take.

For simplifying our protocols, we also incorporate an authenticated broadcast capability into our communication model. (This is not essential for the definitions and composition theorem.) The broadcast can be used to ensure that all parties receive the same message; however no fairness guarantee is assumed: some parties may not receive a message broadcast to them. Indeed, such a broadcast mechanism can be replaced by resorting to, for instance, the broadcast protocol from [40] (with a slight modification to the ideal abstraction of broadcasting, to allow for the round structure in our synchronous model).

Guaranteed-round message delivery from functionalities. Following the revised formulation of the UC framework [15], in our model the messages from an ideal functionality $\mathcal{F}$ are forwarded directly to the uncorrupted parties and cannot be blocked by $\mathcal{S}$.[9] (Note that this is not guaranteed by the previous specification regarding synchronous communication.) Specifically, $\mathcal{F}$ may output (fairdeliver, sid, $msg\text{-}id$, $\{(\mathsf{msg}_1, P_{i_1}), \ldots, (\mathsf{msg}_m, P_{i_m})\}, j)$, meaning that each message msg_i will be delivered to the appropriate party P_i at round j. We will call this feature *guaranteed-round message delivery*.

Resource requests. Typically, an honest party's execution time (per activation) is bounded *a priori* by a polynomial in the security parameter. But in our model, an honest party can "request" the environment to allow it extra computation time. If the request is granted, then the party can run for longer in its activations, for as many computation steps as granted by the environment. More formally, an honest party in the real-world execution can send a message of the form (dealoffer, sid, $msg\text{-}id$, β) to the environment; if the environment responds

---

[9] In the original UC formulation, messages from the ideal functionality $\mathcal{F}$ were forwarded to the uncorrupted parties by the ideal adversary $\mathcal{S}$, who may block these messages and never actually deliver them. The ability of $\mathcal{S}$ to block messages from $\mathcal{F}$ makes the ideal process inherently unfair.

Functionality $\mathcal{F}_f$

$\mathcal{F}_f$ proceeds as follows, running with security parameter κ, parties $P_1, \ldots, P_n$, and an adversary $\mathcal{S}$.

— Upon receiving a value (input, sid, v) from P_i, set $x_i \leftarrow v$.

— As soon as inputs have been received from all parties, compute $y \leftarrow f(x_1, \ldots, x_n)$.

— Wait to receive message (deliverat, sid, s) from $\mathcal{S}$. As soon as the message is received, output (fairdeliver, $sid, 0, \{((\text{output}, y), P_i)\}_{1 \leq i \leq n}, s)$, that is, set up a fair delivery of message (output, sid, y) to all parties for delivery in the sth round.

Fig. 1. The SFE functionality for evaluating an n party function f

to this with (dealaccept, $sid, msg\text{-}id$), then the party gets a "credit" of β extra computational steps (which added to the credits it accumulated before). In a hybrid model, these credits may also be used to accept deals offered by sub-functionality instances. Note that the environment can decide to grant a request or not, depending on the situation.

2.2   A Fair SFE Functionality

Before we introduce the notion of "wrapped functionalities," it is useful to note that in the model described above, we can construct a functionality that can be considered a fair *secure function evaluation* functionality $\mathcal{F}_f$. This functionality is similar to the homonymous functionality in the UC framework [14], except for (1) the fact that there is no reference to the number of corrupted parties, as in our case it may be arbitrary, (2) the output is a single public value, instead of private outputs to each party[10], (3) the added round structure—in particular, the adversary specifies the round at which the outputs are to be produced (deliverat message)[11], and (4) the use of the fair delivery mechanism of the FMPC framework.

We emphasize that in the FMPC framework, and because $\mathcal{F}_f$ uses the fair delivery mechanism, it is easy to see that in the ideal model, the functionality $\mathcal{F}_f$ satisfies the intuitive definition of fairness for secure function evaluation. (This is called "complete fairness" in [40].) Specifically, if one party receives the output, all parties receive the output.

---

[10] This can be easily extended to the case where each party receives a different private output, since y may contain information for each individual party, encrypted using a one-time pad. In fact, the framework developed here accommodates interactive func-tionalities with even more general fairness requirements, where different messages from the functionality can be fairly delivered to different sets of parties at multiple points in the execution.

[11] Alternatively, the functionality could take the number of rounds as a parameter.

2.3   Wrapped Functionalities

As we have stated previously, according to the result of Cleve [19], it is impossible to construct fair protocols, and thus there is no protocol that could realize the functionality $\mathcal{F}_f$ describe above. Therefore we will create a relaxation of $\mathcal{F}_f$ that can be realized, and that will be amenable to analysis in terms of resource fairness. To do this, we will actually construct a more general *wrapper functionality* which provides an interface to any functionality and will be crucial to defining resource fairness. We denote the wrapper functionality as $\mathcal{W}()$, and a wrapped functionality as $\mathcal{W}(\mathcal{F})$.[12]

The wrapper operates as follows. For ease of explanation, assume the functionality $\mathcal{F}$ schedules a single fair delivery to all parties with the same message. Basically, the wrapper handles this fair delivery by storing the message internally until the specified round for delivery, and then outputing the message to be delivered immediately to each party. It also allows the adversary $\mathcal{S}$ to *invest* resources and obtain the message in advance. (Of course, in the ideal process, this investment is simply notational - the adversary does not actually expend any resources.) It will still deliver the message to each party at the specified round unless $\mathcal{S}$ offers a deal to a party to "expend" a certain amount of resources. If that party does not take the deal, then the wrapper will not deliver the message at any round. The wrapper enforces the condition that it only allows $\mathcal{S}$ to offer a deal for at most the amount of resources that $\mathcal{S}$ itself invested. Except for the messages discussed above, all communication to and from $\mathcal{F}$ are simply forwarded directly to and from $\mathcal{F}$.

The formal definition of $\mathcal{W}(\mathcal{F})$ is given in Figure 2. Here we provide some intuition behind some of the labels and variables. Let $F(msg\text{-}id)$ denote a fairdeliver message record (containing message-destination pairs (msg_i, P_i) and $(msg_{\mathcal{S}}, \mathcal{S})$), with identifier $msg\text{-}id$. Associated with any such record is a round number, which specifies the communication round in which the messages in that record will be delivered to all the parties and $\mathcal{S}$. Initially each such record is marked unopened to signify that no party has received any of the messages yet. At any round the adversary $\mathcal{S}$ has the option of obtaining its messages (i.e., messages for the corrupt players and $\mathcal{S}$) by investing $\alpha_{msg\text{-}id}$ amount of resources.[13] If it does so, then the record is marked opened. Once a message is marked opened, $\mathcal{W}(\mathcal{F})$ will ensure that each honest party is offered a fair deal. For each honest party P_i this can happen in one of two ways: either the adversary offers a deal to the honest party to obtain its message msg_i by investing at most $\alpha_{msg\text{-}id}$ amount of resources (in which case the pair (msg_i, P_i) is marked dealt), or if the adversary

---

[12] Assuming $\mathcal{F}$ is a fair functionality, one could say that $\mathcal{W}(\mathcal{F})$ is a "resource-fair" functionality. However, there is an important distinction: a protocol that securely realizes F would be called a "fair" protocol, while a protocol that securely realizes F would not be called a "resource-fair" protocol unless it satisfies an additional requirement, as is discussed below.

[13] This simply means that the adversary sends a message (invest, sid, $msg\text{-}id$, $\alpha_{msg\text{-}id}$) to $\mathcal{W}(\mathcal{F})$, and the amount $\alpha_{msg\text{-}id}$ is counted towards the total amount of resources invested by $\mathcal{S}$.

Wrapper functionality $\mathcal{W}(\mathcal{F})$

$\mathcal{W}(\mathcal{F})$ proceeds as follows, running with parties $P_1, \ldots, P_n$, and an adversary $\mathcal{S}$: It internally runs a copy of $\mathcal{F}$.

- Whenever it receives an incoming communication, which is not one of the special messages (invest, noinvest, dealoffer and dealaccept), it immediately passes this message on to $\mathcal{F}$.
- Whenever $\mathcal{F}$ outputs any message *not* marked for fair delivery, output this message (i.e., pass it on to its destination, allowing the adversary to block this message[a]).
- Whenever $\mathcal{F}$ outputs a record (fairdeliver, sid, msg-id, $\{(\mathsf{msg}_1, P_{i_1}), \ldots, (\mathsf{msg}_m, P_{i_m}), (\mathsf{msg}_S, \mathcal{S})\}, j)$,[b] $\mathcal{W}(\mathcal{F})$ stores this for future delivery (in communication round j). The message record is marked unopened to indicate that the adversary has not yet obtained this message. Also all the pairs (msg_i, P_i) in the record are marked undealt to indicate that no deal has been offered to the party P_i for obtaining this message.
- If a record with ID msg-id is marked as unopened and the adversary sends a message (noinvest, sid, msg-id), then that record is erased (and the messages in it will not be delivered to any party).
- If msg-id is marked as unopened and the adversary $\mathcal{S}$ sends a message (invest, sid, msg-id, α), then
    - the record with ID msg-id is marked as opened, and α is stored as $\alpha_{msg\text{-}id}$. For each corrupt party P_i, if the record contains the message (msg, P_i), that message is delivered to $\mathcal{S}$ immediately (even if the round j has not yet been reached). If the record contains $(\mathsf{msg}_S, \mathcal{S})$ then that message is also delivered to $\mathcal{S}$ at this point.
- At any round in which a fairdeliver record (marked unopened or opened) is stored for delivery at that round, for every pair (msg, P) in that record marked undealt, msg is output for immediate delivery to P (i.e., using the fair delivery mechanism). Then that record is erased.
- If a record msg-id is marked as opened and the adversary sends (dealoffer, sid, msg-id, P_i, β) for some honest party P_i, then
    - $\mathcal{W}(\mathcal{F})$ marks the pair (msg_i, P_i) in the record msg-id as dealt, and sends (dealoffer, sid, msg-id, β') to P_i, where $\beta' = \min(\beta, \alpha_{msg\text{-}id})$.
- If an honest party P_i responds to (dealoffer, sid, msg-id, β) with (dealaccept, sid, msg-id, β), then the stored message msg_i is immediately delivered to P_i, and erased from the stored record.

---

[a] In a typical fair functionality, all messages from $\mathcal{F}$ could be marked for fair delivery. However we allow for non-fair message delivery also in the model.

[b] A message record is identified using the ID msg-id, which $\mathcal{F}$ will ensure is unique for each record.

Fig. 2. The wrapper functionality $\mathcal{W}(\mathcal{F})$

makes no such offer, then P_i receives the message at the specified round without having to make any investment at all.

The following fact is easy to verify.

Fact 1. If the adversary obtains a message that was set for fair delivery with message ID *msg-id*, every honest party that is set to receive a message in the fair delivery with message ID *msg-id* will either receive it at the specified round, or will be offered a deal for at most the amount invested by the adversary.

Conventions. Below we clarify some of the conventions in the new framework.

- **Using resource-requesting subroutines.** A protocol interfaces with a resource-requesting subroutine in a natural way. When a protocol ρ uses a subroutine π which makes resource requests (for instance, if π accesses a wrapped functionality $\mathcal{W}(\mathcal{F})$, or if π securely realizes a wrapped functionality $\mathcal{W}(\mathcal{F})$), it is for ρ to decide when to grant resource requests made by π. ρ can grant resource requests only using resources it already has (which is either part of its running time, or part of resources granted to it by *its* environment). In the cases we consider, the outer protocol ρ will simply transfer resource requests it receives to *its* environment, and will transfer the resources granted to it back to the subroutine.

- **Resource requests granted by the environment.** We do not impose any restriction on the amount of resources that the environment can grant to the honest parties. In particular, the environment could grant a super-polynomial amount of resources to an honest party. This allows a wider class of environments for which the security guarantee holds. Jumping ahead, we point out that this does not render the system insecure, because of an extra condition that the entire system be simulatable in polynomial time, independent of the amount of resources granted by the environment. This requirement is captured in the definition of security using a device called the *full simulator* (see Definition 1).

- **Dummy honest parties in the ideal world.** An honest party in the ideal world is typically a "dummy" party. In the original UC framework this means that it acts as a transparent mediator in the communication between the environment and the ideal functionality. In our framework too this is true, but now the interaction also involves dealoffer and dealaccept messages.

- **$\mathcal{A}$'s resources in a hybrid model.** When working in $\mathcal{W}(\mathcal{F})$-hybrid model, the convention regarding bounding the resources of the adversary $\mathcal{A}$ needs special attention: any amount of resources that $\mathcal{A}$ sends as investment to $\mathcal{W}(\mathcal{F})$ gets counted towards its running time. That is, if $\mathcal{A}$ is a t-bounded IPRAM, then the total amount invested by it plus the total number of steps it runs is at most t.

2.4   Security and Fairness Definitions

So far, we have described the ideal world notion of fairness. As mentioned in Section 1, for a protocol to be resource-fair, for each real world adversary $\mathcal{A}$, the ideal world adversary $\mathcal{S}$ built to simulate the protocol should be such that the amount of resources $\mathcal{S}$ invests is not much more than that available to $\mathcal{A}$.

Below we shall quantify the resource fairness of a protocol by the ratio of the amount of resources that S invests to the actual resources available to A (which technically also includes those available to the environment).

The typical order of quantifiers in the simulation-based security definitions allows the ideal-world adversary to depend on the real-world adversary that it simulates, but it should be independent of the environment (i.e., $\forall A \exists S \forall Z$). A stronger definition of security (which all current constructions in the UC framework satisfy) could require the ideal-world adversary to be a "black-box" simulator which depends on A only by making black-box invocations of A. We employ a slight weakening of this definition: we pass S a bound t on the running times of A and Z, as an input parameter. More formally we model A and Z as bounded IPRAMs. Our security definition will use the order of quantifiers $\exists S$ $\forall t$-bounded A and Z, and it will refer to $S^A(t)$. Now recall that we allow the ideal-world adversary to invest resources with an ideal functionality. An ideal-world adversary S with input parameter t (see above) is said to be λ-restricted if there is a polynomial $\zeta(\kappa)$ such that the sum of all investments sent by S to the ideal functionality is bounded by $\lambda t + \zeta(\kappa)$.

The definition of security and fairness using the simulator captures the intuitive requirements of these notions. However, this by itself does not give us universal composability. We shall strengthen the definition as described below to guarantee universal composition as well.

The full simulator. The strengthening is by requiring that (in addition to the security requirement above) there should exist a "full simulator" which can replace A *and the honest parties* running the protocol in the real world, without an environment being able to detect the change. We call it a full simulator because it simulates all of the execution of a session to the environment, in contrast to a simulator which does not control the honest parties. In this new scenario, since there are no more honest parties involved in the execution, there is no ideal functionality involved. Such a full simulation would be trivial, because the full simulator has access to all the inputs of A as well as of the honest parties, and it can simply execute the code of these parties in its simulation. The non-triviality comes from another requirement: the running time of full simulator should be bounded by a fixed polynomial, *independent of the resource-requests granted by Z.*

We shall denote the random variable corresponding to the output produced by Z on interaction with a full simulator X by $\text{FSIM}_{X^A, Z}$.

Definition 1 (Securely Realizing Functionalities). *Let W_1 and W_2 be two functionalities. We say a protocol π securely realizes the functionality W_1 in the W_2-hybrid model if there exist an ideal world adversary S and a full simulator X, such that for all t-bounded A and Z*

1. $\text{HYB}_{\rho, A, Z}^{W_2} \approx \text{IDEAL}_{W_1, S^A(t), Z}$, *and*
2. $\text{HYB}_{\rho, A, Z}^{W_2} \approx \text{FSIM}_{X^A, Z}$.

Furthermore, if S is λ-restricted, then π securely realizes W_1 with λ-investment (in the W_2-hybrid model).

Although the definition above is stated with respect to general functionalities (and this will be useful in proving our composition theorem), this notion of realizing a functionality with λ-investment will be particularly relevant in the case when $\mathcal{W}_1$ is a wrapped functionality, and specifically a wrapped "fair" functionality. To elaborate, let us consider the case where $\mathcal{W}_1$ is $\mathcal{W}(\mathcal{F})$ for some $\mathcal{F}$. (The functionality $\mathcal{W}_2$ can be a wrapped or non-wrapped functionality, i.e., $\mathcal{W}_2$ above can be a non-wrapped functionality like $\mathcal{F}_{\text{CRS}}$, or it can be a wrapped functionality which we use as a module in a larger protocol.) Then we make the following definition.

Definition 2. *Let π be a protocol that securely realizes $\mathcal{W}(\mathcal{F})$ with λ-investment. Then π λ-fairly realizes $\mathcal{F}$.*

Let us give some intuition behind this definition. First, by Fact 1, $\mathcal{W}$ guarantees that any time a corrupted party (or in particular, the ideal adversary that has corrupted that party) receives its fairdeliver message, then every honest party is at least offered a deal to receive its fairdeliver message, and this deal is bounded by the amount that the ideal adversary invests. Second, by the definition above, the ideal adversary invests an amount within a factor of λ to the resources available to the real adversary. Thus, by expending resources at most a factor λ more than the amount available to the real adversary, an honest party in the ideal world may obtain its message. Since the ideal world is indistinguishable from the real world, the honest party in the real world may also obtain the message expending that amount of resources.

To summarize, we use the term λ-fairly to denote "resource fairness" where an honest party may need to spend at most a factor of λ more resources (i.e., time) than an adversary in order to keep the fair deliveries "fair." Now we consider the case where $\mathcal{F}$ is in fact the fair SFE functionality $\mathcal{F}_f$, and formally define resource fairness and (standard) fairness.

Definition 3. *Let π be a protocol that securely realizes $\mathcal{W}(\mathcal{F}_f)$ with λ-investment. Then we say π is λ-fair. If $\lambda = O(n)$, then we say π is resource fair, and if $\lambda = 0$, then we say π is fair.*

Note that in a "fair" protocol, only a fixed polynomial investment is made by the ideal adversary, and thus all deals are bounded by a fixed polynomial. This could simply be incorporated into the protocol, and thus no deals would need to be made. Thus the protocol would actually securely realize $\mathcal{F}_f$. (Of course, as discussed above, if the adversary may corrupt more than a strict minority of parties, then no such protocol exists.)

On choosing $\lambda = O(n)$. The intuition behind the choice of $\lambda = O(n)$ for resource-fair protocols is as follows. As discussed before, since corrupted parties can abort and gain unfair advantage, an honest party needs more time to catch up. In the worst case, there can be $(n-1)$ corrupted parties against one honest party. Since the honest party may need to invest a certain amount of work against every corrupted party, we expect that the honest party would run about $(n-1)$ times as long as the adversary. Thus, we believe that $O(nt)$ is the "necessary" amount of time an honest party needs for a t-bounded adversary. On the other hand, as we show in the sequel, there exist $O(n)$-fair protocols in the FMPC framework, and thus $\lambda = O(n)$ is also sufficient.

Security of resource-fair protocols. Our definition of resource fairness subsumes the UC definition of security. First of all, if a protocol π λ-fairly realizes $\mathcal{F}$, then, by definition it is also a secure realization of $\mathcal{W}(\mathcal{F})$. However it is not a secure realization of $\mathcal{F}$ itself, because $\mathcal{W}(\mathcal{F})$ offers extra features. But note that for adversaries which never use the feature of sending an invest message, $\mathcal{F}$ and $\mathcal{W}(\mathcal{F})$ behave identically. In fact, $\mathcal{F}$ in the original (unfair) UC model of [14] can be modeled using a rigged wrapper: consider $\mathcal{W}'(\mathcal{F})$ which behaves like $\mathcal{W}(\mathcal{F})$ except that it does not offer any deals to the honest parties (but interacts with the adversary in the same way: in particular, it allows the adversary to obtain its outputs by "investing" any amount of resources). Except for the round structure we use, $\mathcal{W}'(\mathcal{F})$ is an exact modeling of $\mathcal{F}$ in the original UC framework. Clearly $\mathcal{W}(\mathcal{F})$, is intuitively as secure as $\mathcal{W}'(\mathcal{F})$ (but is also fair).

2.5   A Composition Theorem

We now examine the composition of protocols. It turns out that the composition theorem of the UC framework does not automatically imply an analog in the FMPC framework. The main reason for this is that the running time of a resource-requesting protocol is not bounded *a priori*, as there is no bound on the amount of time the environment may decide to grant it in response to a request. This is the reason we introduced the full simulator, whose running time *is* bounded by a polynomial, independent of the environment, and added the extra requirement concerning the full simulator in our definition of security. Using this extra requirement, we are able to prove the composition theorem below.

For simplicity, we shall modify Definition 1, so that the simulator $\mathcal{S}$ is passed t which is a bound on the *sum* of the running times of the environment $\mathcal{Z}$ and the adversary $\mathcal{A}$ (rather than on the maximum of these two). We state the composition theorem accordingly. This makes a difference of at most a constant factor in the parameters below.

Theorem 2 (Universal Composition of Resource-Fair Protocols). *Let* $\mathcal{W}_2$ *be an ideal functionality. Let* π *be a protocol in the* $\mathcal{W}_2$-*hybrid model, which uses atmost ℓ sessions of* $\mathcal{W}_2$. *Let* ρ *be a protocol that securely and λ-fairly realizes* $\mathcal{W}_2$. *Then there exists a λ'-restricted black-box hybrid-mode adversary* $\mathcal{H}$, *such that for all t, for any t_1-bounded real-world adversary* $\mathcal{A}$ *and t_2-bounded environment* $\mathcal{Z}$ *such that $t_1 + t_2 \leq t$, we have*

$$\mathrm{REAL}_{\pi^\rho, \mathcal{A}, \mathcal{Z}} \approx \mathrm{HYB}^{\mathcal{W}_2}_{\pi, \mathcal{H}^{\mathcal{A}}(t), \mathcal{Z}}, \tag{1}$$

where $\lambda' = \lambda\ell$.

Corollary 1. *Let* $\mathcal{W}_1$ *and* $\mathcal{W}_2$ *be ideal functionalities. Let* π *be a protocol that securely realizes* $\mathcal{W}_1$ *with λ-investment in the* $\mathcal{W}_2$-*hybrid model. Let* ρ *be a protocol that securely realizes* $\mathcal{W}_2$ *with λ'-investment. Then the protocol* π^ρ *securely* *realizes* W_1 *with* λ''-*investment. Here, if* ℓ *is an upperbound on the number of sessions of* W_2 *used by* π, *then* $\lambda'' = \lambda(\ell(\lambda' + 1))$.

3   Resource-Fair Protocols

3.1   The Commit-Prove-Fair-Open Functionality

We first present the "commit-prove-fair-open" functionality $\mathcal{F}_{\text{CPFO}}$, and then show how to construct a protocol, GradRel, that securely realizes $W(\mathcal{F}_{\text{CPFO}})$ with $O(n)$-investment using "time-lines." Functionality $\mathcal{F}_{\text{CPFO}}$ is described below.

Functionality $\mathcal{F}_{\text{CPFO}}^R$

$\mathcal{F}_{\text{CPFO}}^R$ is parameterized by a polynomial-time computable binary relation R. It proceeds as follows, running with parties P_1, P_2, ..., P_n and an adversary $\mathcal{S}$.

Round 1 – commit phase: Receive message (commit, sid, x_i) from every party P_i and broadcast (RECEIPT, sid, P_i) to all parties and $\mathcal{S}$.

Round 2 – prove phase: Receive message (prove, sid, y_i) from every party P_i, and if $R(y_i, x_i) = 1$, broadcast (PROOF, sid, P_i, y_i) to all parties and $\mathcal{S}$.

Oopen phase: Wait to receive message (open, sid) from party P_i, $1 \leq i \leq n$, and a message (deliverat, sid, s) from $\mathcal{S}$. As soon as all n open messages and the deliverat message are received, output (fairdeliver, $sid, 0, \{((\text{DATA}, (x_1, x_2, ..., x_n)), P_i)\}_{1 \leq i \leq n} \cup \{((\text{DATA}, (x_1, x_2, ..., x_n)), \mathcal{S})\}, s)$.

Fig. 3. The commit-prove-fair-open functionality $\mathcal{F}_{\text{CPFO}}$ with relation R

Functionality $\mathcal{F}_{\text{CPFO}}$ is similar to the "commit-and-prove" functionality $\mathcal{F}_{\text{CP}}$ in [17] in that both functionalities allow a party to commit to a value v and prove relations about v. Note that although $\mathcal{F}_{\text{CP}}$ does not provide an explicit "opening" phase, the opening of v can be achieved by proving an "equality" relation. However, while $\mathcal{F}_{\text{CP}}$ is not concerned with fairness, $\mathcal{F}_{\text{CPFO}}$ is specifically designed to enforce fairness in the opening. In the open phase, $\mathcal{F}_{\text{CPFO}}$ does not require the outputs to be handed over to the parties as soon as the parties request an opening. Instead, it specifies (to $W(\mathcal{F}_{\text{CPFO}})$) a round s in the future when the outputs are to be handed over. We allow the adversary to determine this round by sending a deliverat message to $\mathcal{F}_{\text{CPFO}}$. (Implicitly we assume that if the round number in the deliverat message is less than the current round number, then the functionality will ignore it.)

Later in the paper, we shall see that by replacing some invocations to the $\mathcal{F}_{\text{CP}}$ functionality by invocations to $W(\mathcal{F}_{\text{CPFO}})$, we can convert the MPC protocol by Canetti *et al.* (which is completely unfair) into a resource-fair protocol.

Before showing a protocol that securely realizes $W(\mathcal{F}_{\text{CPFO}})$, we present a variant of a cryptographic primitive known as "time-lines" [31] that will play an essential role in the construction of resource-fair protocols. Before doing that, we present the assumptions used by these protocols.

Preliminaries for protocol constructions. Let κ be the cryptographic security parameter. A function $f : \mathbb{Z} \to [0,1]$ is negligible if for all $\alpha > 0$ there exists an $\kappa_\alpha > 0$ such that for all $\kappa > \kappa_\alpha$, $f(\kappa) < |\kappa|^{-\alpha}$. All functions we use in this paper will include a security parameter as input, either implicitly or explicitly, and we say that these functions are negligible if they are negligible in the security parameter. (They will be polynomial in all other parameters.) Furthermore, we assume that n, the number of parties, is polynomially bounded by κ as well.

A prime p is *safe* if $p' = (p-1)/2$ is also a prime. A Blum integer is a product of two primes, each equivalent to 3 modulo 4. We will be working with a special class of Blum integers $N = p_1 p_2$ where p_1 and p_2 are both safe primes. We call such numbers *safe Blum integers*.

The assumptions used in this paper are the *composite decisional Diffie-Hellman* assumption (CDDH) [10], the *decision composite residuosity* assumption (DCRA) [46], and a further refinement of the *generalized Blum-Blum-Shub* assumption (GBBS) [11], which we now state.[14]

Given security parameter κ, let $N = p_1 p_2$ be a safe Blum integer with $|p_1| = |p_2| = \kappa$, and let k be an integer bounded from below by κ^c for some positive c. Let $\boldsymbol{a}$ be an arbitrary ℓ-dimensional vector where $0 = a[1] < a[2] < \cdots < a[\ell] < 2^k$, and x be an integer between 0 and 2^k such that $\mathsf{Dist}(x, \boldsymbol{a}) = S$, where $\mathsf{Dist}(x, \boldsymbol{a})$ denotes the minimal absolute difference between x and elements in $\boldsymbol{a}$. (Note that, in particular, we have $x \geq S$, since $a[1] = 0$.) Let g be a random element in $\mathbb{Z}_N^*$; define the "repeated squaring" function as $\mathsf{RepSq}_{N,g}(x) = g^{2^x} \bmod N$. Let $\boldsymbol{u}$ be an ℓ-dimensional vector such that $u[i] = \mathsf{RepSq}_{N,g}(a[i])$, for $i = 1, ..., \ell$.

Now let $\mathcal{A}$ be a PRAM algorithm whose running time is bounded by $\delta \cdot S$ for some constant δ, and let R be a random element in $\mathbb{Z}_N^*$. The *GBBS assumption* states that there exists a negligible function $\epsilon(\kappa)$ such that for any $\mathcal{A}$,

$$\left| \Pr[\mathcal{A}(N, g, \boldsymbol{a}, \boldsymbol{u}, x, \mathsf{RepSq}_{N,g}(x)) = 1] - \Pr[\mathcal{A}(N, g, \boldsymbol{a}, \boldsymbol{u}, x, R^2) = 1] \right| \leq \epsilon(\kappa).$$

(2)

In this paper we present protocols that work in the CRS model and in the PKI model. In the CRS model, there is a common reference string (CRS) generated from a prescribed distribution accessible to all parties at the beginning of the protocol. The $\mathcal{F}_{\mathrm{CRS}}$ functionality simply returns the CRS. The public key infrastructure (PKI) model is stronger. Upon initial activation, a PKI functionality, $\mathcal{F}_{\mathrm{PKI}}$, generates a public string as well as a private string for each party. We note that both models can be defined in the UC and the FMPC frameworks.

Time-lines. We present a definition of a time-line suitable for our purposes, followed by an efficient way to generate them (according to this definition), the security of which relies on GBBS and CDDH-QR.

---

[14] Refer to [32] for remarks on the differences between the version presented here and the original one.

Definition 4. *Let κ be a security parameter. A decreasing time-line is a tuple $L = \langle N, g, u \rangle$, where $N = p_1 p_2$ is a safe Blum integer where both p_1 and p_2 are κ-bit safe primes, g is an element in $\mathbb{Z}_N^*$, and u is a κ-dimensional vector defined as $u[i] = \mathsf{RepSq}_{N,g}(2^\kappa - 2^{\kappa-i})$ for $i = 1, 2, ..., \kappa$. We call N the time-line modulus, g the seed, the elements of u the points in L, and $u[\kappa]$ the end point in L.*

To randomly generate a time-line, one picks a random safe Blum integer N along with $g \xleftarrow{R} \mathbb{Z}_N^*$ as the seed, and then produces the points. Naturally, one can compute the points by repeated squaring: By squaring the seed g $2^{\kappa-1}$ times, we get $u[1]$, and from then on, we can compute $u[i]$ by squaring $u[i-1]$; it is not hard to verify that $u[i] = \mathsf{RepSq}_{N,u[i-1]}(2^{\kappa-i})$, for $i = 2, ..., \kappa$. Obviously, using this method to compute all the points would take exponential time. However, if one knows the factorization of N, then the time-line can be efficiently computed [11].

Alternatively, and assuming one time-line is already known, Garay and Jakobsson [31] suggested the following way to efficiently generate additional time-lines. Given a time-line L, one can easily *derive* a new time-line from L, by raising the seed and every point in L to a fixed power α. Clearly, the result is a time-line with the same modulus.

Definition 5. *Let $L = \langle N, g, u \rangle$ and $L' = \langle N, h, v \rangle$ be two lines of identical modulus. We say that time-line L' is derived from L with shifting factor α if there exists an $\alpha \in \mathbb{Z}_{[1, \frac{N-1}{2}]}$ such that $h = g^\alpha \bmod N$. We call L the master time-line.*

Note that the cost of derivation is just one exponentiation per point, and there is no need to know the factorization of N. In fact, without knowing the master time-line L, if an adversary $\mathcal{A}$ of running time $\delta \cdot 2^\ell$ sees only the seed and the *last* $(\ell + 1)$ points of a derived time-line L', the previous point (which is at distance 2^ℓ away) appears pseudorandom to $\mathcal{A}$, assuming that the GBBS assumption holds. Obviously, this pseudorandomness is no longer true if $\mathcal{A}$ also knows the entire master time-line L and the shifting factor α, since it can then use the deriving method to find the previous point (in fact, any point) on L' efficiently. Nevertheless, as we state in the following lemma, assuming CDDH and GBBS, this pseudorandomness remains true if $\mathcal{A}$ knows L, but not the shifting factor α.

Lemma 1 (Strong Pseudorandomness). *Let $L = \langle N, g, u \rangle$ be a randomly generated decreasing time-line and $L' = \langle N, h, v \rangle$ be a time-line derived from L with random shifting factor α. Let κ and δ be as in the GBBS assumption. Let w be the vector containing the last $(\ell+1)$ elements in v, i.e., $w = (v[\kappa - \ell], v[\kappa - \ell + 1], ..., v[\kappa])$. Let $\mathcal{A}$ be a PRAM algorithm whose running time is bounded by $\delta \cdot 2^\ell$ for some constant δ. Let R be a random element in $\mathbb{Z}_N^*$. Then, assuming CDDH and GBBS hold, there exists a negligible function $\epsilon(\cdot)$ such that, for any $\mathcal{A}$,*

$$\left| \Pr[\mathcal{A}(N, g, u, h, w, v[\kappa - \ell - 1]) = 1] - \Pr[\mathcal{A}(N, g, u, h, w, R^2) = 1] \right| \leq \epsilon(\kappa).$$
$$(3)$$

Realizing $\mathcal{W}(\mathcal{F}_{\text{CPFO}})$: *Protocol* GradRel. Now we construct a protocol, GradRel, that securely realizes wrapped functionality $\mathcal{W}(\mathcal{F}_{\text{CPFO}})$ in the $(\mathcal{F}_{\text{CRS}}, \hat{\mathcal{F}}_{\text{ZK}})$-hybrid model using the time-lines introduced above. We use the multi-session version of the "one-to-many" $\hat{\mathcal{F}}_{\text{ZK}}$ functionality from [17], which is shown in Figure 4.[15] In particular, we need the $\hat{\mathcal{F}}_{\text{ZK}}$ functionality for the following relations.

Functionality $\hat{\mathcal{F}}_{\text{ZK}}^{R}$

$\hat{\mathcal{F}}_{\text{ZK}}^{R}$ proceeds as follows, running parties $P_1, \ldots, P_n$, and an adversary $\mathcal{S}$:

- Upon receiving (zk-prove, $sid, ssid, x, w$) from P_i: If $R(x, w)$ does not hold, ignore. Otherwise, request $\mathcal{S}$ for permission to send (ZK-PROOF, sid, $ssid, P_i, x$) to each of P_j ($j \neq i$). Send the messages as permissions are granted.

Fig. 4. The (multi-session) zero-knowledge functionality for relation R

Discrete log. $\text{DL} = \{((M, g, h), \alpha) \mid h = g^\alpha \bmod M\}$.

Diffie-Hellman quadruple. $\text{DH} = \{((M, g, h, x, y), \alpha) \mid h = g^\alpha \bmod M \land y = x^\alpha \bmod M\}$.

Blinded relation. Given a binary relation $R(y, x)$, we define a "blinded" relation $\hat{R}$ as: $\hat{R}((M, g, h, w, z, y), \alpha) = (h = g^\alpha \bmod M) \land R(y, z/w^\alpha \bmod M)$. Intuitively, $\hat{R}$ "blinds" the witness x using the Diffie-Hellman tuple $(g, h, w, z/x)$. Obviously $\hat{R}$ is an NP relation if R is.

We now describe protocol GradRel informally. The CRS in GradRel consists of a master time-line $L = \langle N, g, \boldsymbol{u} \rangle$. To commit to a value x_i, party P_i derives a new time-line $L_i = \langle N, g_i, \boldsymbol{v_i} \rangle$, and uses the tail of L_i to "blind" x_i. More precisely, P_i sends $z_i = v_i[\kappa] \cdot x_i$ as a "timeline-commitment" to x_i together with a zero-knowledge proof of knowledge (through $\hat{\mathcal{F}}_{\text{ZK}}^{\text{DL}}$) that it knows L_i's shifting factor, and thus, x_i. Note that any party can *force-open* the commitment by performing repeated squaring from points in the time-line. However, forced opening can take a long time, and in particular, since $v_i[\kappa]$ is $(2^\kappa - 1)$ steps away from the seed g_i, it appears pseudorandom to the adversary.

The prove phase is directly handled by the $\hat{\mathcal{F}}_{\text{ZK}}^{\hat{R}}$ functionality. The opening phase consists of κ rounds. In the i-th round, all parties reveal the ith point in their derived time-lines, followed by a zero-knowledge proof that this point is valid (through $\hat{\mathcal{F}}_{\text{ZK}}^{\text{DH}}$), for $i = 1, 2, \ldots \kappa$. If at any time in the gradual opening stage, an uncorrupted party does not receive a ZK-PROOF message in a round when it is expected (possibly because the adversary blocked it, or a corrupted party did not send a proper zk-prove message to an $\hat{\mathcal{F}}_{\text{ZK}}$ functionality) then it enters the *panic mode*. In this mode, an uncorrupted party requests time from the environment to force-open the commitments of all other parties. If the environment accepts, the party forces-open the commitment; otherwise it aborts.

---

[15] In [17] the framework used is that originally presented in [14]. However, since we are using the modified version from [15], we modify the functionality $\hat{\mathcal{F}}_{\text{ZK}}$ by explicitly allowing the adversary to block messages from the functionality to the parties.

Protocol GradRel^R

Set-up: The CRS consists of a master time-line $L = \langle N, g, u \rangle$.

Round 1 (commit phase) For each party P_i, $1 \le i \le n$, upon receiving input (commit, sid, x_i), do:

    1.   Pick $\alpha_i \xleftarrow{R} [1, \frac{N-1}{2}]$, set $g_i \leftarrow g^{\alpha_i} \bmod N$, and compute from L a derived time-line $L_i = \langle N, g_i, v_i \rangle$.

    2.   Set $z_i \leftarrow v_i[\kappa] \cdot x_i = (u[\kappa])^{\alpha_i} \cdot x_i \bmod N$ and broadcast message (COMMIT, sid, P_i, g_i, z_i).

    3.   Send message (zk-prove, sid, 0, (N, g, g_i), α_i) to the $\hat{\mathcal{F}}_{\mathsf{ZK}}^{\mathsf{DL}}$ functionality.

All parties output (RECEIPT, sid, P_i) after receiving (ZK-PROOF, sid, 0, P_i, (N, g, g_i)) from $\hat{\mathcal{F}}_{\mathsf{ZK}}^{\mathsf{DL}}$.

Round 2 (prove phase) For each party P_i, $1 \le i \le n$, upon receiving input (prove, sid, y_i), do:

    1.   Send message (zk-prove, sid, 0, $(N, g, g_i, u[\kappa], z_i, y_i)$, α) to the $\hat{\mathcal{F}}_{\mathsf{ZK}}^{\hat{R}}$ functionality.

    2.   After receiving messages (ZK-PROOF, sid, 0, P_i, $(N, g, g_i, u[\kappa], z_i, y_i)$) from $\hat{\mathcal{F}}_{\mathsf{ZK}}^{\hat{R}}$, all parties output (PROOF, sid, P_i, y_i).

Round $r = 3, \ldots, (\kappa + 2)$ (open phase) Let $\ell = r - 2$. For each party P_i, $1 \le i \le n$, do:

    1.   Broadcast (RELEASE, sid, $v_i[\ell]$) and send message (zk-prove, sid, r, $(N, g, g_i, u[\ell], v_i[\ell])$, α_i) to ideal functionality $\hat{\mathcal{F}}_{\mathsf{ZK}}^{\mathsf{DH}}$.

    2.   After receiving all n RELEASE and ZK-PROOF messages, proceed to the next round. Otherwise, if any of the broadcast messages is missing, go to panic mode.

At the end of round $(\kappa + 2)$, compute $x_j = z_j \cdot (v_j[\kappa])^{-1} \bmod N$, for $1 \le j \le n$, output (DATA, sid, $x_1, x_2, \ldots, x_n$) and terminate.

Panic mode: For each party P_i, $1 \le i \le n$, do:

– Send (dealoffer, sid, $\emptyset$, $n\delta \cdot 2^{\kappa - \ell + 1}$) to the environment.

– If the environment responds with (dealaccept, sid, $\emptyset$), for $j = 1, 2, \ldots, n$, and use $v_j[\ell - 1]$ from the previous round to directly compute x_j committed by P_j as $x_j = z_j \cdot \left(\mathsf{RepSq}_{N, v_j[\ell - 1]}(2^{\kappa - \ell + 1} - 1) \right)^{-1} \bmod N$. Then output (DATA, sid, $x_1, x_2, \ldots, x_n$) in round $(\kappa + 2)$ and terminate.

– Otherwise, output $\perp$ in round $(\kappa + 2)$ and terminate.

Fig. 5. Protocol GradRel, running in the CRS model in $(\kappa + 2)$ rounds

The detailed description of the protocol is given in Figure 5. The security of this protocol is based on CDDH, DCRA, and GBBS. The δ in the protocol is the constant δ from the GBBS assumption. As a technical note, GradRel assumes that all the committed values are quadratic residues in $\mathbb{Z}_N^*$. In [32] we discuss how this assumption can be removed. Clearly, protocol GradRel uses $O(\kappa^2 n)$ bits of communication. As mentioned in Section 2.1, the protocol employs a broadcast channel for convenience.

We can show an ideal adversary for $\mathcal{W}(\mathcal{F}_{\text{CPFO}}^R)$ that invests nt/δ and produces a simulation indistinguishable from GradRel. Therefore, GradRel securely realizes $\mathcal{W}(\mathcal{F}_{\text{CPFO}}^R)$ with n/δ-investment.

Theorem 3. *Assume that GBBS and CDDH hold. Then protocol* GradRel *securely realizes the ideal functionality* $\mathcal{W}(\mathcal{F}_{\text{CPFO}}^R)$ *with* $O(n)$-*investment in the* $(\mathcal{F}_{\text{CRS}}, \hat{\mathcal{F}}_{\text{ZK}}^{\text{DL}}, \hat{\mathcal{F}}_{\text{ZK}}^{\text{DH}}, \hat{\mathcal{F}}_{\text{ZK}}^{\hat{R}})$-*hybrid model, assuming static corruptions.*

Refer to [32] for the proof of this theorem. Here we sketch the essential new elements involving the wrapper. In constructing a simulator $\mathcal{S}$, the most interesting aspect is the simulation of the fair-open phase. Note that the opening takes place in rounds, with the value released in each round being "closer" to the value to be revealed.

- $\mathcal{S}$ internally runs the adversary $\mathcal{A}$, and simulates to it the protocol messages from the honest parties. Initially $\mathcal{S}$ uses random values to simulate the values released by the honest parties in each round.
- However, once the released value gets sufficiently close to the final value, $\mathcal{S}$ can no longer use random values, because even a t-bounded adversary and environment can distinguish between that and the values released by the honest party in an actual execution. So, before that point, $\mathcal{S}$ will *invest* sufficient amount of time with $\mathcal{W}(\mathcal{F}_{\text{CPFO}})$ and obtain the value to be opened. (The "sufficient" amount is the same as what an honest party entering the panic mode at this point would have requested the environment.) Further rounds in the simulation are carried out using the value obtained from $\mathcal{W}(\mathcal{F}_{\text{CPFO}})$ (and hence in those rounds the simulation is perfect).
- At this point a deal is still not offered by $\mathcal{W}(\mathcal{F}_{\text{CPFO}})$ to any honest party. But if in a future round, the adversary $\mathcal{A}$ causes a RELEASE or a ZK-PROOF message not to reach an honest party P (which in the real execution would prompt P to enter the panic mode), at that point $\mathcal{S}$ would request $\mathcal{W}(\mathcal{F}_{\text{CPFO}})$ to send a deal to P, with investment required from P being the actual time that the protocol would request the environment then. This amount will be no more than what $\mathcal{S}$ invested.
- In the ideal world protocol, if P receives a deal offer from $\mathcal{W}(\mathcal{F}_{\text{CPFO}})$, then it would pass it on to the environment, and if the deal is accepted by the environment, then P will invest the amount of time specified in the deal, and obtain the committed value from $\mathcal{W}(\mathcal{F}_{\text{CPFO}})$. In the real world protocol, if P enters the panic mode it will send the deal offer to the environment, and if the deal is accepted by the environment, then P will use the amount of time specified in the deal offer to force-open the computed value. In either, case the environment sees the same behavior from P.

To show that this simulation is good, we depend on the fact that the values released in the initial rounds of the actual execution are pseudorandom, and that in the simulation $\mathcal{S}$ switches to the actual values before this pseudorandomness ceases to hold. The $O(n)$ factor in the amount invested by $\mathcal{S}$ is because of the fact that $\mathcal{S}$ has to make the advance investment for commitments by all honest parties (at most n), whereas the adversary $\mathcal{A}$ might choose to attack any one of them. The $O(n)$ factor also includes (in the constant) the factor δ from the GBBS assumption.

To prove the theorem we must also show a full simulator. A full simulator is essentially a faithful execution of the adversary and the honest parties. The only non-triviality resides in that its running time should not depend on the amount of resources granted by the environment. This is not a problem, since the full simulator will *know* the committed values and need not extract it as the honest parties do in the protocol.

By "plugging in" the UCZK protocol from [17] into protocol GradRel, we have the following corollary.

Corollary 2. *Assume GBBS and CDDH hold, and that enhanced trapdoor permutations exist. Then there exists a protocol that securely realizes $\mathcal{W}(\mathcal{F}_{\mathrm{CPFO}}^R)$ with $O(n)$-investment in the $\mathcal{F}_{\mathrm{CRS}}$-hybrid model, assuming static corruptions.*

3.2 Resource-Fair Multi-party Computation

We show how to construct resource-fair protocols that securely realize the (wrapped) SFE functionality in the FMPC framework. At a high level, our strategy is very simple. Typical secure multi-party protocols (e.g., [21,17,25]) contain an "output" phase, in which every party reveals a secret value, and once all secret values are revealed, every party computes the output of the function. We modify the output phase to have the parties invoke the $\mathcal{W}(\mathcal{F}_{\mathrm{CPFO}})$ functionality. A bit more concretely, assuming each party P_i holds a secret value v_i to reveal, each P_i first commits to v_i and then proves its correctness. Finally $\mathcal{W}(\mathcal{F}_{\mathrm{CPFO}})$ opens all the commitments simultaneously.

In the full paper, we present two constructions that convert the MPC protocols of Canetti *et al.* [17] and Cramer *et al.* [21] into resource-fair MPC protocols. Here we state the results.

Theorem 4. *Assuming the existence of enhanced trapdoor permutations, for any polynomial-time computable function f, there exists a polynomial-time protocol that securely realizes $\mathcal{W}(\mathcal{F}_f)$ with $O(n)$-investment in the $(\mathcal{F}_{\mathrm{CRS}}, \mathcal{W}(\mathcal{F}_{\mathrm{CPFO}}))$-hybrid model in the FMPC framework, assuming static corruptions.*

Corollary 3. *Assuming GBBS, CDDH, and the existence of enhanced trapdoor permutations, for any polynomial-time computable function f, there exists a resource-fair protocol that securely realizes $\mathcal{W}(\mathcal{F}_f)$ in the $\mathcal{F}_{\mathrm{CRS}}$-hybrid model in the FMPC framework, assuming static corruptions.*

Theorem 5. *Assuming GBBS, CDDH, DCRA, and strong RSA, for any polynomial-time computable function f, there exists a resource-fair protocol that securely realizes $\mathcal{W}(\mathcal{F}_f)$ in the $(\mathcal{F}_{\mathrm{PKI}}, \mathcal{W}(\mathcal{F}_{\mathrm{CPFO}}))$-hybrid model in the FMPC framework, assuming static corruptions. Furthermore, this protocol has communication complexity $O(\kappa n|C| + \kappa^2 n)$ bits and consists of $O(d + \kappa)$ rounds.*

Acknowledgements. We thank Amit Sahai for helpful discussions on the formulation of the notion of resource fairness, and Yehuda Lindell and Jesper Nielsen, as well as the anonymous reviewers for *TCC '06* for their many helpful comments.

References

1. L. Adleman and K. Kompella. Using smoothness to achieve parallelism. In *20th STOC*, pp. 528–538, 1988.
2. N. Asokan, V. Shoup, and M. Waidner. Optimistic Fair Exchange of Digital Signatures (Extended Abstract). In *EUROCRYPT 1998*, pp. 591–606, 1998.
3. M. Backes, B. Pfitzmann, and M. Waidner. A general composition theorem for secure reactive systems. In *1st Theory of Cryptography Conference (TCC)*, LNCS 2951, pp. 336-354, 2004.
4. D. Beaver and S. Goldwasser. Multiparty Computation with Faulty Majority. In *30th FOCS*, pages 503–513, 1990.
5. J. Benaloh and M. de Mare. One-Way Accumulators: A Decentralized Alternative to Digital Signatures. In *Eurocrypt 1993*, LNCS 765, pp. 274–285, 1994.
6. M. Ben-Or, O. Goldreich, S. Micali and R. Rivest. A Fair Protocol for Signing Contracts. *IEEE Transactions on Information Theory* 36(1):40–46, 1990.
7. M. Ben-Or, S. Goldwasser, and A. Wigderson. Completeness theorems for non-cryptographic fault-tolerant distributed computation. In *20th STOC*, pp. 1–10, 1988.
8. M. Blum. How to exchange (secret) keys. In *ACM Transactions on Computer Systems*, 1(2):175–193, May 1983.
9. L. Blum, M. Blum, and M. Shub. A simple unpredictable pseudo-random number generator. *SIAM Journal on Computing*, 15(2):364–383, May 1986.
10. D. Boneh. The decision Diffie-Hellman problem. In *Proceedings of the Third Algorithmic Number Theory Symposium*, LNCS 1423, pp. 48–63, 1998.
11. D. Boneh and M. Naor. Timed commitments (extended abstract). In *Advances in Cryptology—CRYPTO '00*, LNCS 1880, pp. 236–254, Springer-Verlag, 2000.
12. C. Cachin and J. Camenisch. Optimistic Fair Secure Computation. In *Advances in Cryptology—CRYPTO '00*, LNCS 1880, pp. 93–111, Springer-Verlag, 2000.
13. R. Canetti. Security and Composition of Multiparty Cryptographic Protocols. *Journal of Cryptology*, 13(1):143-202, Winter 2000.
14. Ran Canetti. Universally composable security: A new paradigm for cryptographic protocols. Electronic Colloquium on Computational Complexity (ECCC) TR01-016, 2001. Previous version "A unified framework for analyzing security of protocols" availabe at the ECCC archive TR01-016. Extended abstract in FOCS 2001.
15. Ran Canetti. Universally composable security: A new paradigm for cryptographic protocols. Cryptology ePrint Archive, Report 2000/067, 2005. Revised version of [14].
16. R. Canetti and M. Fischlin. Universally composable commitments. In *CRYPTO 2001*, LNCS 2139, pp. 19–40, 2001.
17. R. Canetti, Y. Lindell, R. Ostrovsky, and A. Sahai. Universally Composable Two-party and Multi-party Secure Computation. In *34th STOC*, 2002.
18. D. Chaum, C. Crépeau, and I. Damgård. Multiparty unconditionally secure protocols. In *20th STOC*, pp. 11–19, 1988.

19. R. Cleve. Limits on the security of coin flips when half the processors are faulty. In *Proceedings of the 18th Annual ACM Symposium on Theory of Computing (STOC 1986)*, pp. 364-369, 1986.
20. R. Cramer. Modular Design of Secure yet Practical Cryptographic Protocols. Ph.D. Thesis. CWI and University of Amsterdam, 1997.
21. R. Cramer, I. Damgård, and J. Nielsen. Multiparty Computation from Threshold Homomorphic Encryption In *Advances in Cryptology - EuroCrypt 2001 Proceedings*, LNCS 2045, pp. 280–300, Springer-Verlag, 2001.
22. R. Cramer, I. Damgård, and B. Schoenmakers. Proofs of partial knowledge and simplified design of witness hiding protocols. In *Advances in Cryptology - CRYPTO '94*, LNCS 839, pp. 174–187, 1994.
23. I. Damgård. Practical and Provably Secure Release of a Secret and Exchange of Signatures. In *Journal of Cryptology* 8(4), pp. 201–222, 1995.
24. I. Damgård and M .Jurik. Efficient protocols based probabilistic encryptions using composite degree residue classes. In *Research Series RS-00-5*, BRICS, Department of Computer Science, University of Aarhus, 2000.
25. I. Damgård, and J. Nielsen. Universally Composable Efficient Multiparty Computation from Threshold Homomorphic Encryption. In *Advances in Cryptology - CRYPTO '03*, 2003.
26. D. Dolev, C. Dwork and M. Naor. Non-malleable cryptography. *SIAM J. on Comput.*, 30(2):391–437, 2000. An earlier version appeared in *23rd ACM Symp. on Theory of Computing*, pp. 542–552, 1991.
27. S. Even, O. Goldreich, and A. Lempel. A randomized protocol for signing contracts. *Commun. ACM*, 28(6):637–647, June 1985.
28. M. Fitzi, D. Gottesman, M. Hirt, T. Holenstein and A. Smith. Detectable Byzantine Agreement Tolerating Faulty Majorities (from scratch). In *21st PODC*, pp. 118–126, 2002.
29. P. Fouque, G .Poupard, and J. Stern. Sharing decryption in the context of voting or lotteries. In *Proceedings of Financial Crypto 2000*, 2000.
30. Z. Galil, S. Haber, and M. Yung. Cryptographic Computation: Secure Fault-tolerant Protocols and the Public-Key Model. In *CRYPTO'87*, pp. 135–155, 1988.
31. J. Garay and M. Jakobsson. Timed Release of Standard Digital Signatures. In *Financial Cryptography '02*, LNCS 2357, pp. 168–182, Springer-Verlag, 2002.
32. J. Garay, P. MacKenzie, M. Prabhakaran and K. Yang. Resource Fairness and Composability of Cryptographic Protocols. In Cryptology ePrint Archive, http://eprint.iacr.org/2005/370.
33. J. Garay, P. MacKenzie and K. Yang. Strengthening Zero-Knowledge Protocols using Signatures. In *Advances in Cryptology - Eurocrypt 2003*, LNCS 2656, pp.177-194, 2003. Full version in Cryptology ePrint Archive, http://eprint.iacr.org/2003/037, 2003. To appear in *Journal of Cryptology*.
34. J. Garay, P. MacKenzie and K. Yang. Efficient and Universally Composable Committed Oblivious Transfer and Applications. In *1st Theory of Cryptography Conference (TCC)*, LNCS 2951, pp. 297-316, 2004.
35. J. Garay, P. MacKenzie and K. Yang. Efficient and Secure Multi-Party Computation with Faulty Majority and Complete Fairness. In Cryptology ePrint Archive, http://eprint.iacr.org/2004/019.
36. J. Garay and C. Pomerance. Timed Fair Exchange of Standard Signatures. In *Financial Cryptography 2003*, LNCS 2742, pp. 190–207, Springer-Verlag, 2003.
37. O. Goldreich. Secure Multi-Party Computation (Working Draft, Version 1.2), March 2000. Available from http://www.wisdom.weizmann.ac.il/~oded/pp.html.

38. O. Goldreich, S. Micali, and A. Wigderson. How to Play any Mental Game – A Completeness Theorem for Protocols with Honest Majority. In *19th ACM Symposium on the Theory of Computing*, pp. 218–229, 1987.
39. S. Goldwasser and L. Levin. Fair computation of general functions in presence of immoral majority, In *CRYPTO '90*, pp. 77-93, Springer-Verlag, 1991.
40. S. Goldwasser and Y. Lindell. Secure Computation Without Agreement. In *Journal of Cryptology*, 18(3), pp. 247-287, 2005.
41. D. Hofheinz and J. Müller-Quade. A Synchronous Model for Multi-Party Computation and Incompleteness of Oblivious Transfer. In Cryptology ePrint Archive, http://eprint.iacr.org/2004/016, 2004.
42. M. Lepinski, S. Micali, C. Peikert, and A. Shelat. Completely fair SFE and coalition-safe cheap talk. In *23rd PODC*, pp. 1–10, 2004.
43. Y. Lindell. General Composition and Universal Composability in Secure Multi-Party Computation.In *FOCS 2003*.
44. P. MacKenzie and K. Yang. On Simulation Sound Trapdoor Commitments. In *Advances in Cryptology–Eurocrypt '04*, pp.382–400, 2004.
45. J. B. Nielsen. On Protocol Security in the Cryptographi Model. Ph.D. Thesis. Aarhus University, 2003.
46. P. Paillier. Public-key cryptosystems based on composite degree residue classes. In *Advances in Cryptology–Eurocrypt '99*, pp.223–238, 1999.
47. T. P. Pedersen. Non-Interactive and Information-Theoretic Secure Verifiable Secret Sharing. In *Advances in Cryptology – CRYPTO '91*, LNCS 576, 129–140, Springer-Verlag, 1991.
48. B. Pfitzmann and M. Waidner. Composition and Integrity Preservation of Secure Reactive Systems. In *ACM Conference on Computer and Communications Security (CSS)*, pp. 245–254, 2000.
49. B. Pinkas. Fair Secure Two-Party Computation. In *Eurocrypt 2003*, pp. 87–105, 2003.
50. M. Prabhakaran and A. Sahai. New notions of security: Achieving universal composability without trusted setup. Cryptology ePrint Archive, Report 2004/139. Extended abstract in Proc. 36th STOC, pp. 242–251, 2004.
51. T. Rabin and M. Ben-Or. Verifiable Secret Sharing and Multiparty Protocols with Honest Majority. In *21st STOC*, pp. 73–85, 1989.
52. V. Shoup. A Computational Introduction to Number Theory and Algebra. *Preliminary book, available at* http://shoup.net/ntb/.
53. J. Sorenson. A Sublinear-Time Parallel Algorithm for Integer Modular Exponentiation. Available from http://citeseer.nj.nec.com/sorenson99 sublineartime.html.
54. A. Yao. Protocols for Secure Computation. In *FOCS 1982*, pp. 160–164, 1982.
55. A. Yao. How to generate and exchange secrets. In *FOCS 1986*, pp. 162–167, 1986.

Finding Pessiland*

Hoeteck Wee

Computer Science Division,
University of California, Berkeley
hoeteck@cs.berkeley.edu

Abstract. We explore the minimal assumptions that are necessary for non-trivial argument systems, such as Kilian's argument system for NP with poly-logarithmic communication complexity [K92]. We exhibit an oracle relative to which there is a 2-round argument system with poly-logarithmic communication complexity for some language in NP, but no one-way functions. The language lies outside $\mathsf{BPTime}(2^{o(n)})$, so the relaxation to computational soundness is essential for achieving sublinear communication complexity. We obtain as a corollary that under black-box reductions, non-trivial argument systems do not imply one-way functions.

1 Introduction

Pessiland, coined by Impagliazzo [I95], is a world in which there are hard-on-average languages in NP but no one-way functions. In Pessiland, generating hard instances of NP-languages is easy, but we do not know of a way of exploiting these hard-on-average problems in cryptography. In fact, Impagliazzo and Luby [IL89] proved that most cryptographic applications, including bit commitment, private-key encryption and digital signatures, require one-way functions (which allow us to generate hard instances of NP-languages along with a witness) and are therefore impossible to realize in Pessiland.

Recently, Barak's construction of (non-black-box) zero-knowledge arguments [B01] renewed interest in the round complexity and the minimal assumptions necessary for the existence of non-trivial argument systems for NP and NEXP [K92, M00, BG02, W05]. We consider an argument system for NP or NEXP to be non-trivial if the communication complexity is subpolynomial in the length of the witness. Currently, the best construction for NEXP is a 4-round protocol based on the existence of (standard) collision-resistant hash functions [BG02]. If we could relax the assumption to one-way functions, then Barak's construction would yield a constant-round zero-knowledge argument for NP under the same assumption. On the other hand, we do not even know if one-way functions are necessary for non-trivial argument systems. For 2-round argument systems, it is known that a relaxation of hard-on-average languages in NP is necessary [W05] (also, Appendix A.2).

---

* Work supported by US-Israel BSF Grant 2002246. Presently visiting Tsinghua University, Beijing, China.

S. Halevi and T. Rabin (Eds.): TCC 2006, LNCS 3876, pp. 429–442, 2006.
© Springer-Verlag Berlin Heidelberg 2006

1.1 Main Results

In this work, we establish a connection between the two problems: we provide a relativized construction of Pessiland which contains a non-trivial 2-round argument system for a language in NP.

Theorem 1. *There exists an oracle relative to which there exists a strongly hard-on-average language in* NP $\cap$ coNP, *but no one-way functions. Furthermore, there is a 2-round public-coin argument system with poly-logarithmic communication complexity for a language that lies within* NP *but outside* $\mathsf{BPTime}(2^{o(n)})$.

It is important that our argument system is for a language outside $\mathsf{BPTime}(2^{o(n)})$, as it means that the relaxation to computational soundness is essential for achieving sublinear communication complexity. This rules out trivial 2-round argument systems with poly-logarithmic communication complexity for languages in BPP or $\mathsf{NTime}(\log^2 n)$. In particular, a relativizing argument in [GH98] implies that languages outside $\mathsf{BPTime}(2^{o(n)})$ do not have interactive proof systems with sublinear (total) communication complexity, regardless of the number of rounds, and even if the verifier is allowed a polynomial amount of private randomness.

As a corollary, we deduce that there does not exist a black-box construction (such as those used in [V04, W05]) of one-way functions or collision-resistant hash functions from non-trivial 2-round argument systems. This partially explains why we have not been able to prove a statement of the form "if there exists a non-trivial 2-round argument system, then there exists one-way functions". In particular, a proof of this statement must use a non-relativizing argument or make some stronger assumptions on the underlying language. On the other hand, we do not expect to disprove this statement. Suppose non-trivial 2-round argument systems do not exist (which is quite plausible); then, the statement is vacuously true.

The black-box construction of primitives from interactive protocols in [V04, W05] only yields auxiliary-input primitives, as the input instance for the protocol is hard-wired into the algorithm computing the primitive. As such, one would ideally like to rule out auxiliary-input one-way functions (that is, we only require that the function be computable by a nonuniform polynomial-time algorithm) while exhibiting a non-trivial argument system. At this point, we are only able to achieve a much weaker result:

Theorem 2. *There exists an oracle relative to which there exists a strongly hard-on-average language in* NP, *but no auxiliary-input one-way functions.*

The analysis of our first construction is fairly straight-forward apart from some subtle details, and uses several techniques from previous work (such as [IR89, GT00]); the insight lies in the construction and in establishing a connection between Pessiland and non-trivial argument systems. Our second construction, on the other hand, requires a more intricate and novel analysis.

1.2 Perspective and Related Works

Round-efficient argument systems. All previous constructions of non-trivial argument systems (in the standard model) [K92, BG02] require 4 rounds and the existence of collision-resistant hash functions. Micali [M00] gave the first relativized construction of a non-trivial 2-round argument system, by using a random oracle to instantiate collision-resistant hash functions and the Fiat-Shamir paradigm in Kilian's 4-round protocol [K92]. While these previous constructions were for either NP-complete or NEXP-complete languages, our relativized construction (which does not require one-way functions or collision-resistant hash functions) is for a language in NP but possibly not NP-complete. We stress that previous work [W05] deducing hard-on-average problems in NP from non-trivial argument systems for NP (and NEXP) does not exploit the structure of NP in any way; it merely uses the fact NP does not have a proof system with the same communication complexity as the underlying argument system under standard complexity assumptions.

Relationships between cryptographic primitives. Starting with the work of Impagliazzo and Rudich [IR89], the study of relationships between cryptographic primitives has focused on the impossibility of basing complex primitives on simpler ones, particularly one-way functions and one-way permutations. Our main result goes in the reverse direction: it shows the impossibility of constructing simpler primitives from a specific cryptographic application (in a black-box manner). It also provides an example of a cryptographic application (for a contrived language, unfortunately) which may be based on weaker assumptions than the existence of one-way functions. In an unpublished work, Impagliazzo and Rudich gave the first[1] relativized construction of Pessiland, which yields a black-box separation between hard-on-average languages in NP and one-way functions.

2   Preliminaries

We use Π_ℓ to denote the set of all permutations on $\{0,1\}^\ell$, $\mathcal{F}_{n,\ell}$ to denote the set of all functions from $\{0,1\}^n$ to $\{0,1\}^\ell$, and U_n to denote the uniform distribution over $\{0,1\}^n$. A negligible function is a function of the form $n^{-\omega(1)}$. In the context of describing probability distributions, we write $x \sim U_n$ to denote choosing x according to the distribution U_n; we also use $x \in S$ to denote choosing an element x from the set S uniformly at random. We use $\cdot$ to denote the standard dot product of binary strings, and $H(\cdot)$ to denote the Shannon entropy function, namely, $H(p) = -p \log p - (1-p) \log(1-p)$, for $p \in [0,1]$.

2.1   Models of Computation

A circuit has AND and OR gates where each gate has in-degree 2 and out-degree 1, and is labeled with a bit that indicates whether its value should be negated. The size of a circuit is the number of gates. A nonuniform polynomial-time algorithm refers to a

---

[1] We only learnt about the work of Impagliazzo and Rudich after independently arriving at the same construction. We also clarify that *finding* in the title alludes to the search for constructions of Pessiland with stronger cryptographic implications (and a positive result for exploiting average-case hardness) than a mere separation between hard-on-average languages and one-way functions.

family of polynomial-size circuits; specifically, we may consider the polynomial-time algorithm as being circuit evaluation and the nonuniformity being the corresponding circuit. An oracle circuit has 3 types of gates: AND, OR and oracle gates. The in/out-degree of the oracle gate matches the input/output length of the oracle. It is easy to see that an oracle circuit of size s having input/output length n and oracle access to a function $f : \{0,1\}^n \to \{0,1\}$ can be encoded using $O(sn\log(sn))$ bits. A nonuniform oracle polynomial-time algorithm refers to a family of polynomial-size oracle circuits.

2.2   Average-Case Hardness and One-Way Functions

Definition 1. *For any $\alpha \in [0,1/2]$, a function $f : \{0,1\}^n \to \{0,1\}$ is α-hard for size s if every circuit of size s fails to compute f on an α fraction of inputs.*

Definition 2. *For any function $\alpha : \mathbb{N} \to [0,1/2]$, a function $f : \{0,1\}^* \to \{0,1\}$ is α-hard if for every nonuniform polynomial-time algorithm A, for all sufficiently large n's,*

$$\Pr_{x \sim U_n}[A(x) \neq f(x)] > \alpha(n)$$

A function f is weakly hard-on-average *(resp.* strongly hard-on-average*) if f is α-hard for some $\alpha(n) = n^{-c}$ where $c > 0$ is a constant (resp. some $\alpha(n) = 1/2 - n^{-\omega(1)}$). A language L is α-hard if the characteristic function for L is α-hard. We also extend the notions of weakly and strongly hard-on-average to languages.*

Definition 3. *For any function $\alpha : \mathbb{N} \to [0,1]$, a function $f : \{0,1\}^* \to \{0,1\}^*$ is α-one-way (resp.* auxiliary-input α-one-way*) if f is computable in polynomial time (resp. by a nonuniform polynomial-time algorithm) and if for every nonuniform polynomial-time algorithm A, and all sufficiently large n's,*

$$\Pr_{x \sim U_n}[A(f(x)) \notin f^{-1}(f(x))] > \alpha(n)$$

A function f is weakly one-way *(resp.* strongly one-way*) if f is α-one-way for some $\alpha(n) = n^{-c}$ where $c > 0$ is a constant (resp. some $\alpha(n) = 1 - n^{-\omega(1)}$).*

All of these notions extend naturally to the setting of oracle nonuniform polynomial-time algorithms (and oracle circuits). We will often appeal to the following technical lemma from [GT00] stating that random permutations are strongly one-way. We will also use the fact that the proof relativizes.

Lemma 1 ([GT00]). *For all sufficiently large ℓ, with probability $1 - 2^{-2^{\ell/2}}$ over $\pi \in \Pi_\ell$, for all oracle circuits A of size $2^{\ell/5}$,*

$$\Pr_{x \sim U_\ell}[A^\pi(\pi(x)) \neq x] > 1 - 2^{-\ell/5}$$

2.3   Interactive Proofs and Argument Systems

For a relation $R \subseteq \{0,1\}^* \times \{0,1\}^*$, the *language associated with R* is $L_R = \{x : \exists y \, (x,y) \in R\}$.

Definition 4. *An interactive protocol* (P,V) *is an* interactive proof system *for a language L if there is a relation R such that* $L = L_R$, *and functions* $c,s : \mathbb{N} \to [0,1]$ *such that* $1 - c(n) > s(n) + 1/poly(n)$ *and the following holds:*

- *(efficiency): the length of all the messages are bounded by a polynomial in the length of the common input x, and V is computable in probabilistic polynomial time.*
- *(completeness): for all* $(x,w) \in R$, *then V accepts in* $(P(w),V)(x)$ *with probability at least* $1 - c(|x|)$,
- *(soundness): for all* $x \notin L$, *then for every* P^*, *V accepts in* $(P^*,V)(x)$ *with probability at most* $s(|x|)$.

We call $c(\cdot)$ the *completeness error* and $s(\cdot)$ the *soundness error*. We say that (P,V) has *negligible error* if both c and s are negligible. We say that it has *perfect completeness* if $c = 0$. P is an *efficient prover* if $P(w)$ is computable by a probabilistic polynomial-time algorithm when $(x,w) \in R$. The *communication complexity* of the proof system is the total length of all the messages exchanged by both parties, and the *round complexity* is the total number of messages exchanged by both parties (in both directions).

Definition 5. *An* argument system (P,V) *is defined in the same way as an interactive proof system, with the following modification:*

- *The soundness condition is replaced with* computational soundness: *For every nonuniform polynomial-time machine* P^* *and for all sufficiently long* $x \notin L$, *the verifier V accepts in* $(P^*,V)(x)$ *with probability at most* $s(|x|)$.

In this paper, we focus on public-coin argument systems with perfect completeness, negligible soundness error, and an efficient prover.

2.4 Relativization and Black-Box Reductions

In each of our relativized constructions, we consider a family of oracles $\mathcal{O} = \{\mathcal{O}_n\}_{n \geq 1}$, with an oracle for each input length. For simplicity, we will only present our results for the model where an oracle Turing machine (respectively an oracle circuit) on an input of length m only queries $\mathcal{O}_n$ for a single value of n, where $n = n(m)$ is polynomially related to m. This is already sufficient to capture most black-box reductions and transformations used in cryptography.

For black-box constructions of cryptographic primitives from interactive protocols, we require that the construction uses oracle access to the efficiently computable entities in the protocol, such as the verifier, the efficient prover (if one exists), and the simulator (in the case of zero-knowledge). An example is the construction of one-way functions from zero-knowledge proof systems in [V04], where the function is computed using black-box access to the simulator and the verifier for the underlying proof system. Such constructions usually only yield auxiliary-input cryptographic primitives because we need to hardwire the instance used in the protocol into the algorithm for computing the primitive. We omit a formal definition of black-box constructions used in this work (as a sufficiently general framework will be fairly involved without yielding any additional insight); instead, we refer the reader to [RTV04] for a formal treatment of black-box constructions and reductions.

3   The Impagliazzo-Rudich Construction

We begin by reviewing the relativized construction of Pessiland due to Impagliazzo and Rudich (unpublished). We use some of the ideas and proofs in our main constructions.

Theorem 3 (Impagliazzo-Rudich). *There exists an oracle relative to which there exists a strongly hard-on-average language in* $\mathsf{NP} \cap \mathsf{coNP}$, *but no one-way functions.*

For any $f \in \mathcal{F}_{n,n}$ (namely, a function from $\{0,1\}^n$ to $\{0,1\}^n$), we define a verification oracle for f:

$$V_f(x,y) = \begin{cases} 1 & \text{if } f(x) = y \\ 0 & \text{otherwise} \end{cases}$$

The construction used in the proof of Theorem 3 is as follows:

Construction 1. *For each* $n \in \mathbb{N}$, *we have an oracle* V_π, *for some permutation* $\pi \in \Pi_n$ *(specifically, one that satisfies the condition in Lemma 1 and that in Lemma 2 below). In addition, we provide access to a* PSPACE *oracle.*

We choose π by sampling a random permutation on $\{0,1\}^n$. If π is strongly one-way, then the NP-relation $\{(x,w) \mid \pi(w) = x\}$ yields a hard-on-average search problem (with a unique witness), and upon applying the Goldreich-Levin transformation [GL89], we obtain a strongly hard-on-average language in $\mathsf{NP} \cap \mathsf{coNP}$. Furthermore, a polynomial-time oracle Turing machine M makes a query to V_π of the form $(x, \pi(x))$ with negligible probability, so M^Z agrees with M^{V_π} on almost all inputs. Here, $Z : \{0,1\}^* \to \{0,1\}$ denotes the function that evaluates to 0 everywhere. Using the PSPACE oracle, we may then invert M^Z everywhere and thus M^{V_π} almost everywhere.

Lemma 2. *Fix* $T(n) = n^{\log n}$ *and an encoding of oracle Turing machines. For all sufficiently large* n, *with probability at least* $1/2n^2$ *over* $\pi \in \Pi_n$, *for all oracle Turing machines* M *that can be described using at most* $\log n$ *bits and makes at most* $T(n)$ *oracle queries,*

$$\Pr_{x \sim U_n} \left[M^{V_\pi}(x) = M^Z(x) \right] \geq 1 - \frac{1}{2T(n)}$$

Proof. Fix an oracle Turing machine M. By linearity of expectations, we have

$$\mathbb{E}_{\pi \in \Pi_n} \left[|\{x \in \{0,1\}^n : M^{V_\pi}(x) \neq M^Z(x)\}| \right] \leq 2^n \cdot \frac{T(n)}{2^n - T(n)}$$

By Markov's inequality,

$$\Pr_{\pi \in \Pi_n} \left[|\{x \in \{0,1\}^n : M^{V_\pi}(x) \neq M^Z(x)\}| \geq \frac{2^n}{2T(n)} \right] \leq \frac{2T(n)^2}{2^n - T(n)} < \frac{1}{4n^3}$$

This allows us to take a union bound over all oracle Turing machines M with description at most $\log n$ bits (there are at most $2n$ of them). $\square$

Remark 1. As stated, the above lemma only allows us to rule out one-way functions computed by oracle Turing machines M that on an input of length n, only queries V_π corresponding to a permutation on $\{0,1\}^n$. To handle the case where M queries oracles corresponding to permutations on different input lengths, we choose $\pi \in \Pi_n$ to allow for a union bound over all oracle Turing machines M that can be described using at most $\log n$ bits and makes at most $T(n)$ queries to V_π on some input of length $m(n)$ where $m(n)$ is polynomially related to n (instead of only considering $m(n) = n$).

Lemma 3 ([LTW05]). *Let $f, g : \{0,1\}^n \to \{0,1\}^n$ be functions that agree on an ε fraction of inputs. Let $A()$ be the probabilistic procedure that, for every $y \in \{0,1\}^n$, $A(y)$ outputs $\perp$ if $f^{(-1)}(y) = \emptyset$, and a uniformly random element of $f^{(-1)}(y)$ otherwise. Then, the probability that $A(g(x)) \in g^{(-1)}(g(x))$ is at least ε^2, when taken over the uniform choice of $x \in \{0,1\}^n$ and over the internal coin tosses of A.*

Remark 2. Since we also provide access to a PSPACE oracle, we should say that with overwhelming probability over π, $M^{Z,\text{PSPACE}}$ agrees with $M^{V_\pi,\text{PSPACE}}$ almost everywhere. This is true since the proof of Lemma 2 relativizes. With a PSPACE oracle, we may uniformly sample pre-images for $M^{Z,\text{PSPACE}}$ in probabilistic polynomial time, which together with Lemma 3, is sufficient to rule out one-way functions.

Lemma 4 ([GT00, GL89]). *For all sufficiently large n, with probability $1 - o(1/n^2)$ over $\pi \in \Pi_n$, the function $f : \{0,1\}^{2n} \to \{0,1\}$ given by $f(y,r) = \pi^{-1}(y) \cdot r$ is $(1/2 - n^{-\log n})$-hard against oracle circuits of size $n^{\log n}$ with oracle access to π.*

4 Our First Pessiland

We present our construction that establishes Theorem 1. Fix n and $\ell = 100 \log^2 n$. For each $f \in \mathcal{F}_{n,3n}$ and a collection of permutations $\{\pi_y \in \Pi_\ell \mid y \in \{0,1\}^{3n}\}$, we define a 3-tuple (V_π, V_f, T) where V_π and V_f are verification oracles for checking the relations induced by $\{\pi_y\}$ and f, and T is a trapdoor permutation oracle for computing π_y and π_y^{-1} if given (w, y) such that $f(w) = y$.

Our 2-round protocol for the language $L_f = \{y \mid \exists w : f(w) = y\}$ is shown in Fig 1. On input $y \in \{0,1\}^{3n}$, the prover is asked to invert π_y on a random input, and the verifier checks the answer using the verification oracle V_π. The trapdoor permutation oracle yields an efficient prover for the YES instances. For the NO instances, generating an accepting response is as hard as inverting a random permutation.

$$V_\pi(y, \alpha, \beta) = \begin{cases} 1 & \text{if } \pi_y(\alpha) = \beta \\ 0 & \text{otherwise} \end{cases}$$

$$V_f(w, y) = \begin{cases} 1 & \text{if } f(w) = y \\ 0 & \text{otherwise} \end{cases}$$

$$T(w, y, b, z) = \begin{cases} \pi_y(z) & \text{if } f(w) = y \text{ and } b = 0 \\ \pi_y^{-1}(z) & \text{if } f(w) = y \text{ and } b = 1 \\ \perp & \text{otherwise} \end{cases}$$

Common input: An instance $y \in \{0,1\}^{3n}$.
Prover's private input: A witness $w \in \{0,1\}^n$.

$V \to P$:   Send $\beta \xleftarrow{R} \{0,1\}^{O(\log^2 n)}$.
$P \to V$:   Send $\alpha = T(w,y,\beta)$.

Verification: V accepts if $V_\pi(y,\alpha,\beta) = 1$ (that is, $\pi_y(\alpha) = \beta$).

Fig. 1. 2-round public-coin protocol prot for the language $L_f = \{y \mid \exists w : f(w) = y\}$

Construction 2. *For each $n \in \mathbb{N}$, we have an oracle (V_π, V_f, T), for some appropriate choices of $f \in \mathcal{F}_{n,3n}$ and $\{\pi_y \in \Pi_{O(\log^2 n)} \mid y \in \{0,1\}^{3n}\}$. In addition, we provide access to a PSPACE oracle.*

We begin with an overview of the analysis for our construction.

Computational soundness. A successful cheating prover is one that inverts π_y on a noticeable fraction of inputs, for some $y \notin L_f$. However, for each $y \notin L_f$, the random permutation π_y is one-way against oracle circuits of size $n^{\log n}$ with probability $1 - 2^{-n^{\log n}}$ (Lemma 1). This holds even if the circuit is given oracle access to V_f, π_y and $(\pi_{y'}, \pi_{y'}^{-1})$ for all $y' \neq y$ (which are sufficient to simulate the oracles (V_π, V_f, T)), because $\pi_{y'}$ and f are chosen independently of π_y. We can then take a union bound to ensure that every permutation in the collection $\{\pi_y\}$ is strongly one-way, as shown in Lemma 5.

Ruling out low-communication proof systems. A 2-round argument system for L_f with communication complexity $\ell(n)$ is only interesting if we could rule out 2-round interactive proof systems for the language L_f with the same communication complexity. We prove in Lemma 6 that there is no subexponential-size oracle circuits for deciding L_f, given oracle access to V_f and to $\{(\pi_y, \pi_y^{-1})\}_{y \in \{0,1\}^{3n}}$, which is sufficient to simulate oracle access to (V_π, V_f, T). This implies $L_f \notin \mathsf{BPTime}(2^{o(n)})$. Note that an algorithm running in time $\mathsf{BPTime}(2^{O(\ell(n))})$ can compute and invert the permutations π_y everywhere given oracle access to V_π. It is therefore essential to our proof that the collection of permutations $\{\pi_y\}$ is defined independently of f.

Ruling out one-way functions. The analysis is virtually identical to that for the Impagliazzo-Rudich Pessiland, since a polynomial-time oracle Turing machine is unlikely to query (V_π, V_f, T) at any input where the answer is neither 0 nor $\bot$. Note that in order to satisfy the efficient prover condition (for YES instances), it suffices to provide oracle access to $\pi_{f(w)}^{-1}$ in T. By incorporating oracle access to $\pi_{f(w)}$ into T, we also rule out the trivial auxiliary-input one-way permutation given by $\pi_{f(w)}^{-1}$. However, we do not know how to rule out every auxiliary-input one-way function for this construction.

A strongly hard-on-average language. We can construct the language from the strongly hard-on-average function given by $g : \{0,1\}^{3n+2\ell} \to \{0,1\}$ where $g(y, \beta, r) = \pi_y^{-1}(\beta) \cdot r$.

Lemma 5. *For all sufficiently large n, for every $f \in \mathcal{F}_{n,3n}$, with probability $1 - 2^{-\Omega(n^{\log n})}$ over $\{\pi_y\}_{y \in \{0,1\}^{3n}} \in \Pi_\ell^{2^{3n}}$, for all $y \in \{0,1\}^{3n}$ and for all oracle circuits A of size $n^{\log n}$,*

$$\Pr_{x \sim U_\ell}[A^{V_f, \pi_y, \{(\pi_{y'}, \pi_{y'}^{-1}) | y' \neq y\}}(\pi_y(x)) = x] < 2^{-n^{\log n}}$$

Proof. By Lemma 1 (and the fact that it relativizes), if we fix a sufficiently large n, along with any $f \in \mathcal{F}_{n,3n}$, any $y \in \{0,1\}^{3n}$, and any $\pi_{y'} \in \Pi_\ell$ for all $y' \neq y$, we know that with probability $2^{-\Omega(n^{\log n})}$ over $\pi_y \in \Pi_\ell$, for all oracle circuits A of size $n^{\log n}$,

$$\Pr_{x \sim U_\ell}[A^{V_f, \pi_y, \{(\pi_{y'}, \pi_{y'}^{-1}) | y' \neq y\}}(\pi_y(x)) = x] < 2^{-n^{\log n}}$$

The lemma follows from taking a union bound over all $y \in \{0,1\}^{3n}$. □

Lemma 6. *For all sufficiently large n, for every collection of permutations $\{\pi_y\}_{y \in \{0,1\}^{3n}}$, with probability $1 - 2^{-\Omega(2^n)}$ over $f \in \mathcal{F}_{n,3n}$, there is no oracle circuit of size $2^{n/5}$ that given oracle access to V_f and to $\{(\pi_y, \pi_y^{-1})\}_{y \in \{0,1\}^{3n}}$ decides L_f.*

Proof. We establish this result following the counting argument in [GT00]. We may neglect oracle access to $\{(\pi_y, \pi_y^{-1})\}_{y \in \{0,1\}^{3n}}$ since the argument relativizes. The idea is to show that any function f for which there is an oracle circuit A that given oracle access to V_f decides L_f has a "short" description (given A). There are very few such functions, so a random f satisfies the hardness property with overwhelming probability.

Formally, fix an oracle circuit $A : \{0,1\}^{3n} \to \{0,1\}$ of size $2^{n/5}$ and suppose A on oracle access to V_f decides L_f for some $f \in \mathcal{F}_{n,3n}$. We simulate A on every input in $\{0,1\}^{3n}$ in lexicographic order and observe the queries that A makes to V_f. WLOG, assume A never makes the same query twice on a given input. Define $X \subseteq \{0,1\}^n$ to be all x such that A queries V_f on $(x, f(x))$.

CASE 1: $|X| \leq \frac{3}{4} \cdot 2^n$. Given the set X and $f|_X$, we may simulate A on all inputs without oracle access to V_f, thereby recovering the set $f(\{0,1\}^n)$. We may then specify f on each input outside X using just n bits (instead of $3n$ bits) since we only need n bits to specify an element in the set $f(\{0,1\}^n)$.

CASE 2: $|X| > \frac{3}{4} \cdot 2^n$. Over all possible inputs, A makes at most $2^{3n} \cdot 2^{n/5}$ queries to V_f. Therefore, there are at most $\frac{1}{4} \cdot 2^n$ values of x for which A makes more than $4 \cdot 2^{2n} \cdot 2^{n/5}$ queries to V_f of the form $(x, \cdot)$. In particular, there is a subset X' of X with $\frac{1}{2} \cdot 2^n$ elements, and for each $x \in X'$, A makes at most $4 \cdot 2^{2n} \cdot 2^{n/5}$ queries to V_f of the form $(x, \cdot)$. Given the circuit A, the set X' and $f|_{\{0,1\}^n \setminus X'}$, we may specify f on each input in X' using $11n/5 + 2$ bits (instead of $3n$ bits) since we only need to specify i such that the i'th query A makes of the form $(x, \cdot)$ returns 1.

In both cases, given A, we may specify f with $2^n(2n/5 - 2)$ less bits (relative to the $2^n \cdot 3n$ bits required to specify a function in $\mathcal{F}_{n,3n}$). It takes an additional $O(2^{n/5}n^2)$ bits to specify A. □

5   A Second Pessiland

We present our next construction that establishes Theorem 2. It is similar to the Impagliazzo-Rudich Pessiland except we provide a verification oracle for a random function instead of a random permutation.

Construction 3. *For each $n \in \mathbb{N}$, we have an oracle V_f, for some appropriate choice of $f \in \mathcal{F}_{n,n}$. In addition, we provide access to a PSPACE oracle.*

First, we show that for most $f \in \mathcal{F}_{n,n}$, the language $L_f = \{y \mid \exists x : f(x) = y\}$ is weakly hard-on-average (Lemma 7); the proof is an extension of that for Lemma 6, except more involved because we are establishing average-case hardness instead of worst-case hardness. Since the main technical result from [HVV04] on hardness amplification within NP relativizes, we may deduce that there is a strongly hard-on-average language L'_f in NP/poly, obtained by applying some monotone transformation to some padded variant of L_f. We provide an additional oracle that on input 1^n, outputs the nonuniformity needed to compute L'_f in NP. To rule out auxiliary-input one-way functions, it suffices to show that the function computed by any small oracle circuit may be approximated by the function computed by a standard circuit with a polynomial blow-up in size (Lemma 8).

Lemma 7. *For all sufficiently large n, with probability $1 - 2^{-\Omega(n^2)}$ over $f \in \mathcal{F}_{n,n}$, the language $L_f = \{y \mid \exists x : f(x) = y\}$ is 0.01-hard against oracle circuits of size $2^{o(n)}$ with oracle access to V_f.*

Proof (sketch). A standard "balls in bins" analysis (e.g. [MR95, Theorem 4.18]) tells us that with probability $1 - 2^{-\Omega(2^n)}$ over $f \in \mathcal{F}_{n,n}$, $|f(\{0,1\}^n)|$ is bounded from above by $\frac{2}{3} \cdot 2^n$ (we may replace $\frac{2}{3}$ by any constant larger than $1 - \frac{1}{e}$). We may then simply focus on f such that $|f(\{0,1\}^n)| < \frac{2}{3} \cdot 2^n$, and proceed as in the proof of Lemma 6. Again, we consider an oracle circuit $A : \{0,1\}^n \to \{0,1\}$ that solves L_f on at least a 0.99 fraction of inputs and we define X to be all x such that A queries V_f on $(x, f(x))$.

CASE 1: $|X| \leq 0.02 \cdot 2^n$. Let $Y = \{y \mid A(y) \neq L_f(y)\}$, that is, the subset of inputs on which A is wrong. Given $f|_X$ and the sets X, Y (which may be specified using $(0.02n + H(0.02) + H(0.01) + o(1))2^n$ bits), we may simulate A on all inputs without oracle access to V_f, thereby recovering the set $f(\{0,1\}^n)$. We may then specific f on inputs outside X using $\log(\frac{2}{3} \cdot 2^n)$ bits. Therefore, given the circuit A, we may specify f using $2^n n - (0.98 \log \frac{3}{2} - H(0.01) - H(0.02) - o(1))2^n < 2^n(n - 0.35)$ bits.

CASE 2: $|X| > 0.02 \cdot 2^n$. We argue that there is a subset X' of X with $0.01 \cdot 2^n$ elements, and for each $x \in X'$, A makes at most $100 \cdot 2^{o(n)}$ queries to V_f of the form $(x, \cdot)$. Given the circuit A, we may then specify f using $(0.99 + o(1))2^n n$ bits. $\square$

To facilitate the proof of the next lemma, we introduction an additional notation: for any $f \in \mathcal{F}_{n,n}$ and any subset Q of $\{0,1\}^n$, we define:

$$V_{f,Q}(x,y) = \begin{cases} 1 & \text{if } f(x) = y \text{ and } x \in Q \\ 0 & \text{otherwise} \end{cases}$$

Lemma 8. *For all sufficiently large n, with probability $1 - 2^{-\Omega(n^2)}$ over $f \in \mathcal{F}_{n,n}$, for all oracle circuits C of size s where $n \leq s \leq 2^{n/10}$ and for all $\varepsilon \geq 2^{-n/10}$, there exists a circuit C' of size $O(s^4 n^3 / \varepsilon^2)$ such that C^{V_f} and C' agree on a $1 - \varepsilon/2$ fraction of inputs.*

To see why the naive approach of setting $C' = C^Z$ (as in Lemma 2) fails, consider an oracle circuit C that independent of its input, outputs $V_f(0^n, 1^n)$. Then, with probability $1 - 2^{-n}$, C' and C agree on all inputs, and with probability 2^{-n}, disagree on all inputs. This is not sufficient for a union bound over all polynomial-size circuits. To work around this, we hardwire into C' information about f. Specifically, we show that with overwhelming probability over $f \in \mathcal{F}_{n,n}$, for all C of size s, there exists a set $Q \subseteq \{0,1\}^n$ of size $O(s^4 n^2 / \varepsilon^2)$ such that the circuit $C^{V_{f,Q}}$ agrees with C^{V_f} on a $1 - \varepsilon/2$ fraction of inputs. Note that we allow Q to depend on f. We may specify $f|_Q$ using $|Q| n$ bits of nonuniformity, so $C^{V_{f,Q}}$ may be computed by a circuit C' of size $O(s^4 n^3 / \varepsilon^2)$ (without oracle access to V_f).

Here is an outline of the analysis. Let us examine the first oracle query made by the circuit C on different inputs, and we define Q_1 to be all x such that the first query C makes to V_f matches $(x, \cdot)$ on more than a $\varepsilon^3 / s^3 n^2$ fraction of inputs. Therefore, $|Q_1| = \text{poly}(s, n, 1/\varepsilon)$. Now, consider the oracle circuit C_1 that behaves like C, except the first oracle query is made to V_{f,Q_1} instead of V_f. Suppose C and C_1 differs on a $\varepsilon/2s$ fraction of inputs. This must be because for a $\varepsilon/2s$ fraction of inputs, the first query C makes to V_f matches $(x, f(x))$, for some $x \notin Q_1$. For a random f and a fixed x, this happens with probability 2^{-n}. Moreover, this must happen for at least $s^2 n^2 / \varepsilon^2$ different values of x not in Q_1 (since each $x \notin Q_1$ accounts for at most a $\varepsilon^3 / s^3 n^2$ fraction of inputs). For a random f, the evaluation of f on each of these x values are independent. Thus, the probability (over f) that C and C_1 differs on a $\varepsilon/2s$ fraction of inputs is roughly $2^{-\Omega(ns^2)}$.

Proof. Formally, fix $f \in \mathcal{F}_{n,n}$. We define oracle circuits $C_0, C_1, \ldots, C_s$ and subsets $Q_0, Q_1, \ldots, Q_s$ of $\{0,1\}^n$ inductively as follows:

- $Q_0 = \emptyset$ and $C_0 = C$.
- Q_i is union of Q_{i-1} and the set

$$\left\{ x \in \{0,1\}^n \, \middle| \, \Pr_z \left[i\text{'th oracle query for computing } C_{i-1}^{V_f}(z) \text{ matches } (x, \cdot) \right] \geq \varepsilon^2 / s^3 n^2 \right\}$$

- C_i on input z and oracle access to V_f simulates the computation of $C^{V_f}(z)$ except for $j = 1, 2, \ldots, i$, the j'th oracle query is answered using V_{f,Q_j} instead of V_f. We will hardwire the description of the sets $Q_1, \ldots, Q_i$ into C_i, so upon oracle access to V_f, C_i may simulate the oracles $V_{f,Q_j}, j = 1, \ldots, i$.

Claim. For all $i = 1, 2, \ldots, s$, $\quad \Pr_{f \in \mathcal{F}_{n,n}} \left[\Pr_z \left[C_{i-1}^{V_f}(z) \neq C_i^{V_f}(z) \right] < \varepsilon/2s \right] \geq 1 - 2^{-\Omega(sn^2)}$

It follows readily from the claim that

$$\Pr_{f \in \mathcal{F}_{n,n}} \left[\Pr_z \left[C^{V_f}(z) \neq C_s^{V_f}(z) \right] < \varepsilon/2 \right] \geq 1 - s \cdot 2^{-\Omega(sn^2)}$$

This implies that with overwhelming probability over f, C^{V_f} and $C^{V_{f,Q_s}}$ agree on a $1 - \varepsilon/2$ fraction of inputs. We may bound $|Q_s|$ by $s^4 n^2/\varepsilon^2$ since $|Q_i| \le |Q_{i-1}| + s^3 n^2/\varepsilon^2$. Hence, $C^{V_{f,Q_s}}$ may be computed by a circuit C' of size $O(s^4 n^2/\varepsilon^2)$. The lemma then follows from taking a union bound over all circuits of size s, all s between n and $2^{n/10}$, and all $1/\varepsilon$ between 2 and $2^{n/10}$. $\square$

Now, we provide the proof of the above claim.

Proof (of claim). We start with the case $i = 1$. Note that the definition of Q_1 does not depend on f. Consider any input z to C^{V_f}. If the first oracle query made by C^{V_f} corresponds to an element in Q_1, then $\Pr_f[C_1^{V_f}(z) = C^{V_f}(z)] = 1$. Otherwise, $\Pr_f[C_1^{V_f}(z) = C^{V_f}(z)] = 1 - 2^{-n}$. For each $x \in \{0,1\}^n$, we define

$$\alpha_x = \begin{cases} \Pr_z[\text{first oracle query for } C^{V_f}(z) \text{ matches } (x, \cdot)] & \text{if } x \notin Q_1 \\ 0 & \text{otherwise} \end{cases}$$

(note that α_x is independent of f) and Y_x to be the random variable (where the randomness is over $f \in \mathcal{F}_{n,n}$) for the probability

$$\Pr_z[\text{first oracle query for } C^{V_f}(z) \text{ matches } (x, \cdot) \text{ and } C^{V_f}(z) \neq C_1^{V_f}(z)]$$

Hence, we have $\sum_x \alpha_x \le 1$ and for all $x \in \{0,1\}^n$:

$$0 \le Y_x \le \alpha_x \le \varepsilon^2/s^3 n^2 \text{ and } E_f[Y_x] = \alpha_x 2^{-n}$$

In addition,

$$\Pr_{f,z}[C_1^{V_f}(z) \neq C^{V_f}(z)] = E_f\left[\sum_x Y_x\right]$$

By convexity, we have $\sum_x \alpha_x^2 \le \varepsilon^2/s^3 n^2$. Applying the Hoeffding bound [H63] yields:

$$\Pr_f\left[\sum_x Y_x - 2^{-n} \ge \varepsilon/4s\right] \le e^{-2(\varepsilon/4s)^2/\sum_x \alpha_x^2} \le e^{-sn^2/8}$$

In the general case, we fix an assignment to $f|_{Q_{i-1}}$, so the set Q_i is also fixed. As before, we define

$$\alpha_x = \begin{cases} \Pr_z[i\text{'th oracle query for } C_{i-1}^{V_f}(z) \text{ matches } (x, \cdot)] & \text{if } x \notin Q_i \\ 0 & \text{otherwise} \end{cases}$$

(here, α_x is independent of $f|_{\{0,1\}^n \setminus Q_{i-1}}$) and Y_x to be the random variable (where the randomness is over $f|_{\{0,1\}^n \setminus Q_{i-1}}$) for the probability

$$\Pr_z[i\text{'th oracle query for } C_{i-1}^{V_f}(z) \text{ matches } (x, \cdot) \text{ and } C_{i-1}^{V_f}(z) \neq C_i^{V_f}(z)]$$

Again, the Hoeffding bound yields:

$$\Pr_{f|_{\{0,1\}^n \setminus Q_{i-1}}}\left[\sum_x Y_x - 2^{-n} \ge \varepsilon/4s\right] \le e^{-sn^2/8}$$

This holds for all $f|_{Q_{i-1}}$. Averaging over all possible assignments of $f|_{Q_{i-1}}$, we have:

$$\Pr_f\left[\Pr_z\left[C_{i-1}^{V_f}(z) \neq C_i^{V_f}(z)\right] \geq \varepsilon/4s + 2^{-n}\right] \leq e^{-sn^2/8}$$

This completes the proof of the technical claim.   □

Acknowledgements

I am grateful towards Salil Vadhan for sharing his insightful observations which led me towards the problems addressed in this work, and Luca Trevisan for his help with the proofs in Section 5. In addition, I thank Lance Fortnow and Russell Impagliazzo for pointing out previous constructions of Pessiland, and the anonymous referees for their helpful and constructive feedback.

References

[B01]   B. Barak. How to go beyond the black-box simulation barrier. In *Proc. 42nd FOCS*, 2001.

[BG02]   B. Barak and O. Goldreich. Universal arguments and their applications. In *Proc. 17th CCC*, 2002.

[GH98]   O. Goldreich and J. Håstad. On the complexity of interactive proofs with bounded communication. *IPL*, 67(4):205–214, 1998.

[GL89]   O. Goldreich and L. Levin. Hard-core predicates for any one-way function. In *Proc. 21st STOC*, 1989.

[GT00]   R. Gennaro and L. Trevisan. Lower bounds on efficiency of generic cryptographic constructions. In *Proc. 41st FOCS*, 2000.

[H63]   W. Hoeffding. Probability inequalities for sums of bounded random variables. *Journal of the American Statistical Association*, 58:13–30, 1963.

[HVV04]   A. Healy, S. Vadhan, and E. Viola. Using nondeterminism to amplify hardness. In *Proc. 36th STOC*, 2004.

[I95]   R. Impagliazzo. A personal view of average-case complexity. In *Proc. 10th Structure in Complexity Theory Conference*, 1995.

[IL89]   R. Impagliazzo and M. Luby. One-way functions are essential for complexity based cryptography. In *Proc. 30th FOCS*, 1989.

[IR89]   R. Impagliazzo and S. Rudich. Limits on the provable consequences of one-way permutations. In *Proc. 21st STOC*, 1989.

[K92]   J. Kilian. A note on efficient zero-knowledge proofs and arguments. In *Proc. 24th STOC*, 1992.

[LTW05]   H. Lin, L. Trevisan, and H. Wee. On hardness amplification of one-way functions. In *Proc. 2nd TCC*, 2005.

[M00]   S. Micali. Computationally sound proofs. *SICOMP*, 30(4):1253–1298, 2000.

[MR95]   R. Motwani and P. Raghavan. *Randomized Algorithms*. Cambridge University Press, 1995.

[RTV04]   O. Reingold, L. Trevisan, and S. Vadhan. Notions of reducibility between cryptographic primitives. In *Proc. 1st TCC*, 2004.

[V04]   S. Vadhan. An unconditional study of computational zero knowledge. In *Proc. 45th FOCS*, 2004.

[W05]   H. Wee. On round-efficient argument systems. In *Proc. 32nd ICALP (Track C)*, 2005.

A Appendix

A.1 The Hoeffding Bound

We state the concentration result for sum of independent bounded random variables (with possibly arbitrary distributions) used in the proof of Lemma 8.

Lemma 9 ([H63]). *If $X_1, \ldots, X_n$ are independent random variables such that $a_i \leq X_i \leq b_i$, $i = 1, 2, \ldots, n$, then for all $t > 0$,*

$$\Pr[X - \mathrm{E}[X] \geq t] \leq e^{-2t^2/\Sigma_i (b_i - a_i)^2}$$

where $X = X_1 + \ldots X_n$.

A.2 Necessity of Hardness Assumptions

For ease of reference, we reproduce the proof from [W05] (with a minor improvement in the result) that a 2-round argument system for NP with subpolynomial communication complexity implies hard-on-average search problems in NP. Under complexity assumptions, such a protocol cannot be a proof system [GH98]. Hence, there exists infinitely many NO instances that are merely "computationally sound", from which we may construct hard-on-average search problems in NP. We stress that the construction of hard-on-average search problems uses the underlying verifier in a black-box manner.

Lemma 10 ([W05]). *Suppose a promise problem $\Pi = (\Pi_Y, \Pi_N)$ has a 2-round public-coin argument system (P, V) with communication complexity $m(n)$, perfect completeness and negligible soundness error. Then, there exists a subset $I \subset \Pi_N$ such that:*

- *Ignoring inputs in I, Π has a 2-round public-coin proof system with communication complexity $m(n)$, perfect completeness and soundness error less than 1. This implies $(\Pi_Y, \Pi_N \setminus I) \in \mathsf{DTime}(2^{O(m(n))})$.*
- *When $x \in I$, the predicate $V(x, \cdot, \cdot)$ induces a distribution over hard-on-average search instances in NP. That is, for every $x \in I$:*

$$\Pr_r[\exists y : V(x, r, y) = 1] = 1,$$

but for every n, every $x \in I \cap \{0, 1\}^n$ and every nonuniform polynomial-time algorithm A, there exists a negligible function $\varepsilon(n)$ such that

$$\Pr_r[V(x, r, A(r)) = 1] < \varepsilon(n)$$

Theorem 4 ([W05]). *Suppose NP has a 2-round public-coin argument system (P, V) with communication complexity $n^{o(1)}$, perfect completeness and negligible soundness error. Then, (at least) one of the following is true:*

- $\mathsf{NP} \subseteq \mathsf{DTime}(2^{n^{o(1)}})$
- *There exists an infinite set I such that for all $x \in I$, the predicate $V(x, \cdot, \cdot)$ induces a distribution over hard-on-average search instances in NP (as formalized in Lemma 10). This yields an auxiliary-input samplable distribution over satisfiable instances in NP where the search problem is infinitely-often strongly hard-on-average.*

Pseudorandom Generators from One-Way Functions: A Simple Construction for Any Hardness

Thomas Holenstein

ETH Zurich, Department of Computer Science, CH-8092, Zurich
thomahol@inf.ethz.ch

Abstract. In a seminal paper, Håstad, Impagliazzo, Levin, and Luby showed that pseudorandom generators exist if and only if one-way functions exist. The construction they propose to obtain a pseudorandom generator from an n-bit one-way function uses $\mathcal{O}(n^8)$ random bits in the input (which is the most important complexity measure of such a construction). In this work we study how much this can be reduced if the one-way function satisfies a stronger security requirement. For example, we show how to obtain a pseudorandom generator which satisfies a standard notion of security using only $\mathcal{O}(n^4 \log^2(n))$ bits of randomness if a one-way function with exponential security is given, i.e., a one-way function for which no polynomial time algorithm has probability higher than 2^{-cn} in inverting for some constant c.

Using the uniform variant of Impagliazzo's hard-core lemma given in [7] our constructions and proofs are self-contained within this paper, and as a special case of our main theorem, we give the first explicit description of the most efficient construction from [6].

1 Introduction

A pseudorandom generator is a deterministic function which takes a uniform random bit string as input and outputs a longer bit string which cannot be distinguished from a uniform random string by any polynomial time algorithm. This concept, introduced in the fundamental papers of Yao [16] and Blum and Micali [1] has many uses. For example, it immediately gives a semantically secure cryptosystem: the input of the pseudorandom generator is the key of the cryptosystem, and the output is used as a one-time pad. Other uses of pseudorandom generators include the construction of pseudorandom functions [2], pseudorandom permutations [11], statistically binding bit commitment [13], and many more.

Such a pseudorandom generator can be obtained from an arbitrary one-way function, as shown in [6]. The given construction is not efficient enough to be used in practice, as it requires $\mathcal{O}(n^8)$ bits of input randomness (for example, if one would like to have approximately the security of a one-way function with $n = 100$ input bits, the resulting pseudorandom generator would need several petabits of input, which is clearly impractical). On the other hand, it is possible

S. Halevi and T. Rabin (Eds.): TCC 2006, LNCS 3876, pp. 443–461, 2006.

to obtain a pseudorandom generator very efficiently from an arbitrary one-way *permutation* [4] or from an arbitrary regular one-way function [3] (see also [5]), i.e., a one-way function where every image has the same number of preimages. In other words, if we have certain guarantees on the *combinatorial structure* of the one-way function, we can get very efficient reductions.

In this paper we study the question whether a pseudorandom generator can be obtained more efficiently under a stronger assumption on the *computational difficulty* of the one-way function. In particular, assume that the one-way function is harder to invert than usually assumed. In this case, one single invocation of the one-way function could be more useful, and fewer invocations might be needed. We will see that is indeed the case, even if the pseudorandom generator is supposed to inherit a stronger security requirement from the one-way function, and not only if it is supposed to satisfy the standard security notion.

2   Overview of the Construction

The construction given in [6] uses several stages: first the one-way function is used to construct a false entropy generator, i.e., a function whose output is computationally indistinguishable from a distribution with more entropy. (This is the technically most difficult part of the construction and the security proof can be significantly simplified by using the uniform hard-core lemma from [7].) Next, the false entropy generator is used to construct a pseudoentropy generator (a function whose output is computationally indistinguishable from a distribution which has more entropy than the input), and finally a pseudorandom generator is built on top of that. If done in this way, their construction is very inefficient (requiring inputs of length $\mathcal{O}(n^{34})$), but it is also sketched in [6] how to "unroll" the construction in order to obtain an $\mathcal{O}(n^{10})$ construction. Similarly it is mentioned that an $\mathcal{O}(n^8)$ construction is possible (by being more careful).

In this work we explicitly describe an $\mathcal{O}(n^8)$ construction (in an unrolled version the construction we describe is the one sketched in [6]). Compared to [6] we choose a different way of presenting this construction; namely we use a two-step approach (see Figure 1). First, (in Section 4) we use the one-way function to construct a pair (g, P) where g is an efficiently evaluable function and P is a predicate. The pair will satisfy that predicting $P(x)$ from $g(x)$ is computationally

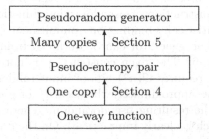

Fig. 1. Overview of our construction

difficult (in particular, more difficult than it would be information theoretically). In [5] the term *pseudo-entropy pair* is coined for such a pair and we will use this term as well. In a second step we use many instances of such a pseudo-entropy pair to construct a pseudorandom generator.

Further, we generalize the construction to the case where stronger security guarantees on the one-way function are given. This enables us to give more efficient reductions under stronger assumptions.

Indepenently of this work, Haitner, Harnik, and Reingold [5] give a better method to construct a pseudo-entropy pair from a one-way function. Their construction has the advantage that the entropy of $P(x)$ given $g(x)$ can be estimated, which makes the construction of the pseudorandom generator from the pseudo-entropy pair more efficient.

3 Definitions and Result

3.1 Definitions and Notation

Definition 1. *A one-way function with security $s(n)$ against $t(n)$-bounded inverters is an efficiently evaluable family of functions $f : \{0,1\}^n \to \{0,1\}^m$ such that for any algorithm running in time at most $t(n)$*

$$\Pr_{x \leftarrow_R \{0,1\}^n}[f(A(f(x))) = f(x)] < \frac{1}{s(n)}$$

for all but finitely many n.

For example the standard notion of a one-way function is a function which is one-way with security $p(n)$ against $p(n)$-bounded inverters for all polynomials $p(n)$.

In [15] it is shown that a random permutation is $2^{n/10}$-secure against $2^{n/5}$-bounded inverters, and also other reasons are given why it is not completely unreasonable to assume the existence of one-way permutations with exponential security. In our main theorem we can use one-way functions with exponential security, a weaker primitive than such permutations.

Definition 2. *A pseudorandom-generator with security $s(\ell)$ against $t(\ell)$-bounded distinguishers is an efficiently evaluable family of (expanding) functions $h : \{0,1\}^\ell \to \{0,1\}^{\ell+1}$ such that for any algorithm running in time at most $t(\ell)$*

$$\left| \Pr_{x \leftarrow_R \{0,1\}^\ell}[A(h(x)) = 1] - \Pr_{u \leftarrow_R \{0,1\}^{\ell+1}}[A(u) = 1] \right| \le \frac{1}{s(\ell)},$$

for all but finitely many ℓ.

The standard notion of a pseudorandom generator is a pseudorandom generator with security $p(\ell)$ against $p(\ell)$-bounded distinguishers, for all polynomials $p(\ell)$.

As mentioned above, we use pseudo-entropy pairs as a step in our construction. For such a pair of functions we first define the advantage an algorithm A has in predicting $P(w)$ from $g(w)$ (by convention, we use the letter w to denote the input here).

Definition 3. *For any algorithm A, any function $g : \{0,1\}^n \to \{0,1\}^m$ and any predicate $P : \{0,1\}^n \to \{0,1\}$, the advantage of A in predicting P given g is*

$$\text{Adv}^A(g, P) := 2\left(\Pr_{w \leftarrow_R \{0,1\}^n}[A(g(w)) = P(w)] - \frac{1}{2}\right).$$

The following definition of a pseudo-entropy pair contains (somewhat surprisingly) the conditioned entropy $H(P(W)|g(W))$; we give an explanation below.

Definition 4. *A pseudo-entropy pair with gap $\phi(n)$ against $t(n)$-bounded predictors is a pair (g, P) of efficiently evaluable functions $g : \{0,1\}^n \to \{0,1\}^m$ and $P : \{0,1\}^n \to \{0,1\}$ such that for any algorithm A running in time $t(n)$*

$$\text{Adv}^A(g, P) \leq 1 - H(P(W)|g(W)) - \phi,$$

for all but finitely many n (where W is uniformly distributed over $\{0,1\}^n$).

The reader might think that it would be more natural if we used the best advantage for computationally unbounded algorithms (i.e., the information theoretic advantage), instead of $1 - H(P(W)|g(W))$. Then ϕ would be the gap which comes from the use of $t(n)$-bounded predictors. We quickly explain why we chose the above definition. First, to get an intuition for the expression $1 - H(P(W)|g(W))$, assume that the pair (g, P) has the additional property that for every input w, $g(w)$ either fixes $P(w)$ completely or does not give any information about it, i.e., for a fixed value v either $H(P(W)|g(W) = v) = 1$ or $H(P(W)|g(W) = v) = 0$ holds. Then, a simple computation shows that $1 - H(P(W)|g(W))$ is a tight upper bound on the advantage of computationally unbounded algorithms, i.e., in this case our definition coincides with the above "more natural definition". We mention here that the pairs (g, P) we construct will be close to pairs which have this property. If there are values v such that $0 < H(P(W)|g(W) = v) < 1$, the expression $1 - H(P(W)|g(W))$ is *not* an upper bound anymore and in fact one might achieve significantly greater advantage than $1 - H(P(W)|g(W))$. Therefore in this case, Definition 4 requires something stronger than the "more natural definition", and, consequently, constructing a pseudorandom generator from a pseudo-entropy pair becomes easier.[1]

We use $\|$ to denote concatenation of strings, ax denotes the multiplication of bitstrings a and x over $\text{GF}(2^n)$ (with an arbitrary representation), and $x|_\lambda$ denotes the first $\lfloor \lambda \rfloor$ bits of the bit string x. For fixed x and $\bar{x}$, $x \neq \bar{x}$, the probability that $(ax)|_i$ equals $(a\bar{x})|_i$ for uniformly chosen a can be computed as

$$\Pr_{a \leftarrow \{0,1\}^n}\left[(ax)|_i = (a\bar{x})|_i\right] = \Pr_{a \leftarrow \{0,1\}^n}\left[(a(x - \bar{x}))|_i = 0^i\right] = 2^{-i}, \quad (1)$$

an expression we will use later.

---

[1] In fact, we do not know a direct way to construct a pseudorandom generator from a pseudo-entropy pair with the "more natural definition".

For bitstrings x and r of the same length n we use $x \odot r := \bigoplus_{i=1}^{n} x_i r_i$ for the inner product. We use the convention that $f^{-1}(y) := \{x \in \{0,1\}^n | f(x) = y\}$, i.e., f^{-1} returns a set.

For two distributions P_{X_0} and P_{X_1} over $\mathcal{X}$ the statistical distance is

$$\Delta(X_0, X_1) := \frac{1}{2} \sum_{x \in \mathcal{X}} |P_{X_0}(x) - P_{X_1}(x)|.$$

We also say that a distribution is ε-close to another distribution if the statistical distance of the distributions is at most ε. For a distribution P_X over $\mathcal{X}$ the min-entropy is $H_\infty(X) := -\log(\max_{x \in \mathcal{X}} P_X(x))$. For joint distributions P_{XY} over $\mathcal{X} \times \mathcal{Y}$ the conditional min-entropy is defined with $H_\infty(X|Y) := \min_{y \in \mathcal{Y}} H_\infty(X|Y = y)$.

Finally, we define $[n] := \{1, \ldots, n\}$.

3.2   Result

We give a general construction of a pseudorandom generator from a one-way function. The construction is parametrized by two parameters ε and ϕ. The parameter ε should be chosen such that it is smaller than the target indistinguishability of the pseudorandom generator: an algorithm which distinguishes the output of the pseudorandom generator with advantage less than ε will not help us in inverting f. The second parameter ϕ should be chosen such that the given one-way function cannot be inverted with probability more than about $2^{-n\phi}$ (as an example, for standard security notions choosing $\phi = \frac{1}{n}$ and $\varepsilon = 2^{-n}$ would be reasonable – these should be considered the canonical choices).

Theorem 1. *Let functions* $f : \{0,1\}^n \to \{0,1\}^m$, $\phi : \mathbb{N} \to [0,1]$, $\varepsilon : \mathbb{N} \to [0,1]$ *be given, computable in time* $\mathrm{poly}(n)$, *and satisfying* $2^{-n} \le \varepsilon \le \frac{1}{n} \le \phi$.

There exists an efficient to evaluate oracle function $h_{\varepsilon,\phi}^f$ *with the following properties:*

- $h_{\varepsilon,\phi}^f$ *is expanding,*
- $h_{\varepsilon,\phi}^f$ *has input of length* $\mathcal{O}(\frac{n^4}{\phi^4} \log(\frac{1}{\varepsilon}))$, *and*
- *an algorithm* A *which distinguishes the output of* $h_{\varepsilon,\phi}^f$ *from a uniform bit string with advantage* γ *can be used to get an oracle algorithm which inverts* f *with probability* $\mathcal{O}(\frac{1}{n^3}) 2^{-n\phi}$, *using* $\mathrm{poly}(n, \frac{1}{\gamma - \varepsilon})$ *calls to* A.

For example, if we set $\phi := \log(n)/n$ and $\varepsilon := n^{-\log(n)} = 2^{-\log^2(n)}$ and use a standard one-way function in the place of f, then $h_{\varepsilon,\phi}^f$ will be a standard pseudorandom generator, using $\mathcal{O}(n^8)$ bits[2] of randomness.

Corollary 1. *Assume that* $f : \{0,1\}^n \to \{0,1\}^m$ *is a one-way function with security* $p(n)$ *against* $p(n)$-*bounded inverters, for all polynomials* $p(n)$. *Then there exists a pseudorandom generator* $h : \{0,1\}^\ell \to \{0,1\}^{\ell+1}$ *with security* $p(\ell)$ *against* $p(\ell)$-*bounded distinguishers, for all polynomials* $p(\ell)$. *The construction calls the one-way function for one fixed* n *dependent of* ℓ *and satisfies* $\ell \in \mathcal{O}(n^8)$.

---

[2] This could be insignificantly reduced by choosing ε slightly bigger.

Alternatively if we have a much stronger one-way function which no polynomial time algorithm can invert with better probability than 2^{-cn} for some constant c, we can set ϕ to some appropriate small constant and $\varepsilon := n^{-\log(n)}$, which gives us a pseudorandom generator using $\mathcal{O}(n^4 \log^2(n))$ bits of input:

Corollary 2. *Assume that $f : \{0,1\}^n \to \{0,1\}^m$ is a one-way function with security 2^{-cn} against $p(n)$-bounded inverters, for some constant c and all polynomials $p(n)$. Then there exists a pseudorandom generator $h : \{0,1\}^\ell \to \{0,1\}^{\ell+1}$ with security $p(\ell)$ against $p(\ell)$-bounded distinguishers, for all polynomials $p(\ell)$. The construction calls the one-way function for one fixed n dependent of ℓ, and satisfies $\ell \in \mathcal{O}(n^4 \log^2(n))$.*

If we want a pseudorandom generator with stronger security we set ε smaller in our construction. For example, if a one-way function f has security 2^{cn} against 2^{cn} bounded distinguishers, we set ϕ (again) to an appropriate constant and $\varepsilon := 2^{-n}$. With these parameters our construction needs $\mathcal{O}(n^5)$ input bits, and, for an appropriate constant d, an algorithm with distinguishing advantage 2^{-dn}, and running in time 2^{dn}, can be used to get an inverting algorithm which contradicts the assumption about f. (A corollary similar to the ones before could be formulated here).

The proof of Theorem 1 is in two steps (see Figure 1). In Section 4 we use the Goldreich-Levin Theorem and two-universal hash-functions to obtain a pseudo-entropy pair. In Section 5 we show how such a pair can be used to obtain a pseudorandom generator.

3.3   Extractors

Informally, an extractor is a function which can extract a uniform bit string from a random string with sufficient min-entropy. The following well known left-over hash lemma from [10] shows that multiplication over $\mathrm{GF}(2^n)$ with a randomly chosen string a and then cutting off an appropriate number of bits can be used to extract randomness. For completeness we give a proof (adapted from [12]).

Lemma 1 (Left-over hash lemma). *Let $x \in \{0,1\}^n$ be chosen according to any source with min-entropy λ. Then, for any $\varepsilon > 0$, and uniform random a, the distribution of $\big((ax)|_{\lambda-2\log(\frac{1}{\varepsilon})} \,\|\, a\big)$ is $\frac{\varepsilon}{2}$-close to a uniform bit string of length $\lfloor \lambda - 2\log(\frac{1}{\varepsilon}) \rfloor + n$.*

Proof. Let $m := \lfloor \lambda - 2\log(\frac{1}{\varepsilon}) \rfloor$, and P_{VA} be the distribution of $(ax)|_m \| a$. Further, let P_U be the uniform distribution over $\{0,1\}^{m+n}$. Using the Cauchy-Schwartz inequality $(\sum_{i=1}^k a_i)^2 \leq k \sum_{i=1}^k a_i^2$ we obtain for the statistical distance in question

$$\Delta(VA, U) = \frac{1}{2} \sum_{v \in \{0,1\}^m, a \in \{0,1\}^n} \left| P_{VA}(v,a) - \frac{1}{2^n 2^m} \right|$$

$$\leq \frac{1}{2}\sqrt{2^n\,2^m}\sqrt{\sum_{v,a} P_{VA}^2(v,a) - 2\sum_{v,a}\frac{P_{VA}(v,a)}{2^n\,2^m} + \sum_{v,a}\left(\frac{1}{2^n\,2^m}\right)^2}$$

$$= \frac{1}{2}\sqrt{2^n\,2^m}\sqrt{\sum_{v,a} P_{VA}^2(v,a) - \frac{1}{2^n\,2^m}}\,.\qquad(2)$$

Let now X_0 and X_1 be independently distributed according to P_X (i.e., the source with min-entropy λ). Further, let A_0 and A_1 be independent over $\{0,1\}^n$. The collision probability of the output distribution is

$$\Pr\Big[((X_0A_0)|_m \,\|\, A_0) = ((X_1A_1)|_m \,\|\, A_1)\Big] = \sum_{v,a} P_{VA}^2(v,a).$$

Thus we see that equation (2) gives an un upper bound on $\Delta(VA, U)$ in terms of the collision probability of two independent invocations of the hash-function on two independent samples from the distribution P_X. We can estimate this collision probability as follows:

$$\Pr\Big[((X_0A_0)|_m \,\|\, A_0) = ((X_1A_1)|_m \,\|\, A_1)\Big]$$
$$= \Pr[A_0 = A_1]\,\Pr[(X_0A_0)|_m = (X_1A_0)|_m]$$
$$\leq \Pr[A_0 = A_1]\Big(\Pr[X_0 = X_1] + \Pr\big[(X_0A_0)|_m = (X_1A_0)|_m \,\big|\, X_0 \neq X_1\big]\Big)$$
$$\leq \frac{1}{2^n}\Big(\frac{1}{2^{m+2\log(1/\varepsilon)}} + \frac{1}{2^m}\Big) = \frac{1+\varepsilon^2}{2^n\,2^m},\qquad(3)$$

where we used (1) in the last inequality. We now insert (3) into (2) and get $\Delta(VA, U) \leq \frac{\varepsilon}{2}$. □

Using the usual definition of an extractor, the above lemma states that multiplying with a random element of $GF(2^n)$ and then cutting off the last bits is a strong extractor. Consequently, we will sometimes use the notation $\text{Ext}_m(x, a)$ to denote the function $\text{Ext}_m(x, a) := (ax)|_m \,\|\, a$, extracting $\lfloor m \rfloor$ bits from x.

Further we use the following proposition on independent repetitions from [8], which is a quantitative version of the statement that for k independent repetitions of random variables, the min-entropy of the resulting concatenation is roughly k times the (Shannon-)entropy of a single instance (assuming k large enough and tolerating a small probability that something improbable occured). A similar lemma with slightly weaker parameters is given in [10] (the latter would be sufficient for our application, but the expression from [8] is easier to use).

Proposition 1. Let $(X_1, Y_1), \ldots, (X_k, Y_k)$ i.i.d. according to P_{XY}. For any ε there exists a distribution $P_{\overline{X}\,\overline{Y}}$ which has statistical distance at most $\frac{\varepsilon}{2}$ from $(X_1, \ldots, X_k, Y_1, \ldots, Y_k)$ and satisfies

$$H_\infty(\overline{X}|\overline{Y}) \geq kH(X|Y) - 6\sqrt{k\log(1/\varepsilon)}\log(|\mathcal{X}|).$$

We can combine the above propositions as follows:

Lemma 2. *Let k, ε with $k > \log(1/\varepsilon)$ be given. Let $(X_1, Y_1), \ldots, (X_k, Y_k)$ i.i.d. according to P_{XY} over $\mathcal{X} \times \mathcal{Y}$ with $\mathcal{X} \subseteq \{0,1\}^n$. Let A be uniform over $\{0,1\}^{kn}$. Then,*

$$\text{Ext}_{kH(X|Y) - 8\log(|\mathcal{X}|)\sqrt{k\log(1/\varepsilon)}}(X_1\|\cdots\|X_k, A)\|Y_1\|\cdots\|Y_k$$

is ε-close to $U \times Y^k$, where U is an independent uniform chosen bitstring of length $\lfloor kH(X|Y) - 8\log(|\mathcal{X}|)\sqrt{k\log(1/\varepsilon)}\rfloor + kn$.

Proof. Combine Lemma 1 and Proposition 1. □

4   A Pseudo-Entropy Pair from Any One-Way Function

The basic building block we use to get a pseudo-entropy pair is the following theorem by Goldreich and Levin [4] (recall that $x \odot r = x_1 r_1 \oplus \cdots \oplus x_n r_n$ is the inner product of x and r):

Proposition 2 (Goldreich-Levin). *There is an oracle algorithm $B^{(\cdot)}$ such that for any $x \in \{0,1\}^n$ and any oracle A satisfying*

$$\Pr_{r \leftarrow_R \{0,1\}^n} [A(r) = x \odot r] \geq \frac{1}{2} + \gamma$$

B^A does $\mathcal{O}(\frac{n}{\gamma^2})$ queries to A and then efficiently outputs a list of $\mathcal{O}(\frac{1}{\gamma^2})$ elements such that x is in the list with probability $\frac{1}{2}$.

This proposition implies that for any one-way function f, no efficient algorithm will be able to predict $x \odot r$ from $f(x)$ and r much better than random guessing, as otherwise the one-way function can be broken.

This suggests the following method to get a pseudo-entropy pair: if we define $g(x, r) := f(x)\|r$ and $P(x, r) := x \odot r$, then predicting $P(x, r)$ from $g(x, r)$ is computationally hard. The problem with this approach is that since $f(x)$ may have many different preimages, $H(P(X, R)|g(X, R)) \approx 1$ is possible. In this case, $P(x, r)$ would not only be *computationally* unpredictable, but also *information theoretically* unpredictable, and thus (g, P) will not be a pseudo-entropy pair.

The solution of this problem (as given in [6]), is that one additionally extracts some information of the input x to f; the amount of information extracted is also random. The idea is that in case one is lucky and extracts roughly $\log(|f^{-1}(f(x))|)$ bits, then these extracted bits and $f(x)$ fix x in an information theoretic way, but computationally $x \odot r$ is still hard to predict because of Proposition 2.

Thus, we define functions $g : \{0,1\}^{4n} \to \{0,1\}^{m+4n}$ and $P : \{0,1\}^{4n} \to \{0,1\}$ as follows (where $i \in [n]$ is a number[3], x, a, and r are bitstrings, and we ignore padding which should be used to get $(ax)|_i$ to length n)

---

[3] Technically, we should choose i as a uniform number from $[n]$. We can use an n bit string to choose a uniform number from $[2^n]$ and from this we can get an "almost" uniform number from $[n]$ (for example by computing the remainder when dividing by n). This only gives an exponentially small error which we ignore from now on.

$$g(x, i, a, r) := f(x) \| i \| a \| (ax)|_i \| r \tag{4}$$

$$P(x, i, a, r) := x \odot r. \tag{5}$$

We will alternatively write $g(w)$ and $P(w)$, i.e., we use w as an abbreviation for (x, i, a, r). We will prove that (g, P) is a pseudo-entropy pair in case f is a one-way function. Thus we show that no algorithm exceeds advantage $1 - H(P(W)|g(W)) - \phi$ in predicting $P(w)$ from $g(w)$ (the gap ϕ does not appear in the construction, but the pair will have a bigger gap if the one-way function satisfies as stronger security requirement, as we will see).

We first give an estimate on $H(P(W)|g(W))$. The idea is that we can distinguish two cases: either $i \geq \log(|f^{-1}(f(x))|)$, in which case $H(P(W)|g(W)) \approx 0$, since $(ax)|_i$, a, and $f(x)$ roughly fix x, or $i < \log(|f^{-1}(f(x))|)$, in which case $H(P(W)|g(W)) \approx 1$.

Lemma 3. *For the functions g and P as defined above*

$$H(P(W)|g(W)) \leq \frac{\mathrm{E}_{x \leftarrow_R \{0,1\}^n} [\log(|f^{-1}(f(x))|)] + 2}{n}$$

Proof. From (1) and the union bound we see that if $i > \log(|f^{-1}(y)|)$ the probability that x is not determined by the output of g is at most $2^{-(i-\log(|f^{-1}(y)|))}$. This implies $H(P(W)|g(W), f(X) = y, I = i) \leq 2^{-(i-\log(|f^{-1}(y)|))}$, and thus

$$H(P(W)|g(W)) = \frac{1}{2^n} \sum_{x \in \{0,1\}^n} H\big(P(W)|g(W), f(X) = f(x)\big)$$

$$= \frac{1}{2^n} \sum_{x \in \{0,1\}^n} \frac{1}{n} \sum_{i=1}^{n} H\big(P(W)|g(W), f(X) = f(x), I = i\big)$$

$$\leq \frac{1}{2^n} \sum_{x \in \{0,1\}^n} \left(\frac{\log(|f^{-1}(f(x))|)}{n} \right.$$

$$\left. + \frac{1}{n} \sum_{i=\lceil \log(|f^{-1}(f(x))|)\rceil}^{n} 2^{-(i-\log(|f^{-1}(f(x))|))} \right)$$

$$\leq \frac{1}{2^n} \sum_{x \in \{0,1\}^n} \frac{\log(|f^{-1}(f(x))|) + 2}{n}$$

$$= \frac{\mathrm{E}_{x \leftarrow_R \{0,1\}^n} [\log(|f^{-1}(f(x))|)] + 2}{n}.$$

$\square$

We can now show that (g, P) is a pseudo-entropy pair. For this, we show that any algorithm which predicts P from g with sufficient probability can be used to invert f. Recall that ϕ is usually $\frac{1}{n}$.

Lemma 4. *Let $f : \{0,1\}^n \to \{0,1\}^m$ and $\phi : \mathbb{N} \to [0,1]$ be computable in time $\mathrm{poly}(n)$. Let functions g and P be as defined above. There exists an oracle* *algorithm* $B^{(\cdot)}$ *such that, for any* A *which has advantage* $\mathrm{Adv}^A(g^f, P^f) \geq 1 - H(P^f(W)|g^f(W)) - \phi$ *in predicting* P^f *from* g^f, B^A *inverts* f *with probability* $\Omega(\frac{1}{n^3})2^{-n\phi}$ *and* $\mathcal{O}(n^3)$ *calls to* A.

We find it convenient to present our proof using random experiments called "games", similar to the method presented in [14].

Proof. Assume that a given algorithm $A(y, i, a, z, r)$ has an advantage exceeding the bound in the lemma in predicting P from g. To invert a given input $y = f(x)$, we will choose i, a, and z uniformly at random. Then we run the Goldreich-Levin algorithm using $A(y, i, a, z, \cdot)$, i.e., the Goldreich-Levin calls $A(y, i, a, z, r)$ for many different r, but always using the same y, i, a, and z. This gives us a list L containing elements from $\{0,1\}^n$. For every $\overline{x} \in L$ we check whether $f(\overline{x}) = y$. If at least one $\overline{x} \in L$ satisfies this we succeeded in inverting f.

In order to see whether this approach is successful, we first define α to be the advantage of A for a fixed y, i, a and z in predicting $\overline{x} \odot r$ for a preimage $\overline{x}$ of y:

$$\alpha(y, i, a, z) := \max_{\overline{x} \in f^{-1}(y)} \left(2 \Pr_{r \leftarrow \{0,1\}^n}[A(y, i, a, z, r) = \overline{x} \odot r] - 1\right).$$

We maximize over all possible $\overline{x} \in f^{-1}(y)$, since it is sufficient if the above method finds *any* preimage of y. We will set the parameters of the algorithm such that it succeeds with probability $\frac{1}{2}$ if $\alpha(y, i, a, z) > \frac{1}{4n}$ (i.e., with probability $\frac{1}{2}$ the list returned by the algorithm contains $\overline{x}$). It is thus sufficient to show for uniformly chosen x, i, a, and z the inequality $\alpha(f(x), i, a, z) > \frac{1}{4n}$ is satisfied with probability $\Omega(\frac{1}{n^3})2^{-n\phi}$.

Together with Lemma 3, the requirement of this lemma implies that in the following Game 0 the expectation of the output is at least $1 - H(P^f(W)|g^f(W)) - \phi \geq 1 - \frac{1}{n} \mathrm{E}_x[\log(|f^{-1}(f(x))|)] - \frac{2}{n} - \phi$ (this holds even *without* the maximization in the definition of α and using $\overline{x} = x$ instead – clearly, the maximization cannot reduce the expected output of Game 0).

> **Game 0:**
> $x \leftarrow_R \{0,1\}^n$, $y := f(x)$, $i \leftarrow_R [n]$
> $a \leftarrow_R \{0,1\}^n$, $z := (ax)|_i$
> **output** $\alpha(y, i, a, z)$

Note that even though we can approximate $\alpha(y, i, a, z)$ we do not know how to compute the exact value in reasonable time. However, we do not worry about finding an efficient implementation of our games.

If i is much larger than $\log(|f^{-1}(y)|)$ then predicting $P(w)$ from $g(w)$ is not very useful in order to invert f, since $(ax)|_i$ gives much information about x which we do not have if we try to invert y. Thus, we ignore the cases where i is much larger than $\log(|f^{-1}(y)|)$ in Game 1.

> **Game 1:**
> $x \leftarrow_R \{0,1\}^n$, $y := f(x)$, $i \leftarrow_R [n]$
> **if** $i \leq \log(|f^{-1}(y)|) + n\phi + 3$ **then**
> $a \leftarrow_R \{0,1\}^n$, $z := (ax)|_i$

    output $\alpha(y, i, a, z)$
  fi
  output 0

It is not so hard to see that the probability that the if clause fails is at most $1 - \frac{1}{n} E_x[\log(|f^{-1}(f(x))|)] - \frac{3}{n} - \phi$. Thus, in Game 1 the expectation of the output is at least $\frac{1}{n}$ (because the output only decreases in case the if clause fails, and in this case by at most one).

 In Game 2, we only choose the first j bits of z as above, where j is chosen such that these bits will be $\frac{1}{4n}$-close to uniform (this will be used later). We fill up the rest of z with the best possible choice; clearly, this cannot decrease the expectation of the output.

 Game 2:
   $x \leftarrow_R \{0,1\}^n, \ y := f(x), \ i \leftarrow_R [n]$
   if $i \le \log(|f^{-1}(y)|) + n\phi + 3$ **then**
    $j := \min(\lfloor \log(|f^{-1}(y)|) - 2\log(4n) \rfloor, i)$
    $a \leftarrow_R \{0,1\}^n, \ z_1 := (ax)|_j$
    set $z_2 \in \{0,1\}^{j-i}$ such that $\alpha(y, i, a, z_1 \| z_2)$ is maximal
    output $\alpha(y, i, a, z_1 \| z_2)$
  fi
  output 0

We now chose z_1 uniformly at random. Lemma 1 implies that the statistical distance of the previous distribution of z_1 to the uniform distribution (given a, i, and y but not x) is at most $\frac{1}{4n}$. Thus, the expecation of the output is at least $\frac{1}{2n}$.

 Game 3:
   $x \leftarrow_R \{0,1\}^n, \ y := f(x), \ i \leftarrow_R [n]$
   if $i \le \log(|f^{-1}(y)|) + n\phi + 3$ **then**
    $j := \min(\lfloor \log(|f^{-1}(y)|) - 2\log(4n) \rfloor, i)$
    $a \leftarrow_R \{0,1\}^n, \ z_1 \leftarrow_R \{0,1\}^j$
    set $z_2 \in \{0,1\}^{j-i}$ such that $\alpha(y, i, a, z_1 \| z_2)$ is maximal
    output $\alpha(y, i, a, z_1 \| z_2)$
  fi
  output 0

As mentioned above, we will be satisfied if we have values $y, i, a, (z_1 \| z_2)$ such that $\alpha(y, i, a, z_1 \| z_2) \ge \frac{1}{4n}$. In Game 4, we thus do not compute the expectation of α anymore, but only output success if this is satisfied, and fail otherwise.

 Game 4:
   $x \leftarrow_R \{0,1\}^n, \ y := f(x), \ i \leftarrow_R [n]$
   if $i \le \log(|f^{-1}(y)|) + n\phi + 3$ **then**
    $j := \min(\lfloor \log(|f^{-1}(y)|) - 2\log(4n) \rfloor, i)$
    $a \leftarrow_R \{0,1\}^n, \ z_1 \leftarrow_R \{0,1\}^j$
    set $z_2 \in \{0,1\}^{j-i}$ such that $\alpha(y, i, a, z_1 \| z_2)$ is maximal
    if $\alpha(y, i, a, z_1 \| z_2) > \frac{1}{4n}$
    output success

$$\text{fi}$$
$$\text{fi}$$
output fail

The usual Markov style argument shows that the probability that the output is success is at least $\frac{1}{4n}$ (this is easiest seen by assuming otherwise and computing an upper bound on the expectation of the output in Game 3: it would be less than $\frac{1}{2n}$).

In Game 5, we choose all of z uniformly at random.

Game 5:
$$x \leftarrow_R \{0,1\}^n, \ y := f(x), \ i \leftarrow_R [n]$$
$$\text{if } i \leq \log(|f^{-1}(y)|) + n\phi + 3 \text{ then}$$
$$a \leftarrow_R \{0,1\}^n, \ z \leftarrow_R \{0,1\}^i$$
$$\text{if } \alpha(y,i,a,z) > \tfrac{1}{4n}$$
$$\text{output success}$$
$$\text{fi}$$
$$\text{fi}$$
output fail

In Game 5, we can assume that z is still chosen as $z_1 \| z_2$. For z_1, the distribution is the same as in Game 4, for z_2, we hope that we are lucky and choose it exactly as in Game 4. The length of z_2 is at most $2\log(4n) + n\phi + 3$, and thus this happens with probability at least $\frac{1}{128n^2} 2^{-n\phi}$. Thus, in Game 4, with probability at least $\frac{1}{512n^3} 2^{-n\phi}$ the output is success. As mentioned at the start of the proof, in this case running the Goldreich-Levin algorithm with parameter $\frac{1}{4n}$ will invert f with probability $\frac{1}{2}$, which means that in total we have probability $\Omega(\frac{1}{n^3}) 2^{-n\phi}$ in inverting f. $\qquad\Box$

5 A Pseudorandom Generator from a Pseudo-Entropy Pair

We now show how we can obtain a pseudorandom generator from a pseudo-entropy pair (g, P) as constructed in the last section. The idea here is that we use many (say k) parallel copies of the function g. We can then extract about $kH(g(W))$ bits from the concatenated outputs of g, about $kH(W|g(W)P(W))$ bits from the concatenated inputs, and about $k(H(P(W)|g(W)) + \phi)$ bits from the concatenated outputs of P. Using the identity $H(g(W)) + H(P(W)|g(W)) + H(W|g(W)P(W)) = H(W)$, we can see that this will be expanding, and we can say that the $k\phi$ bits of pseudorandomness from P are used to get the expanding property of h.

The key lemma in order to prove the security of the construction is the following variant of Impagliazzo's hard-core lemma [9] proven in [7][4]. For a set T

---

[4] The proposition here is slightly stronger then the corresponding lemma in [7], as we do not require γ to be noticeable. It is easy to see that the proof in [7] works in this case as well.

let χ_T be the characteristic function of T:

$$\chi_T(x) := \begin{cases} 1 & x \in T \\ 0 & x \notin T. \end{cases}$$

Proposition 3 (Uniform Hard-Core Lemma). *Assume that the given functions $g : \{0,1\}^n \to \{0,1\}^m$, $P : \{0,1\}^n \to \{0,1\}$, $\delta : \mathbb{N} \to [0,1]$ and $\gamma : \mathbb{N} \to [0,1]$ are computable in time poly(n), where δ is noticeable and $\gamma > 2^{-n/3}$.*

Further, assume that there exists an oracle algorithm $A^{(\cdot)}$ such that, for infinitely many n, the following holds: for any set $T \subseteq \{0,1\}^n$ with $|T| \geq \delta 2^n$, A^{χ_T} outputs a circuit C satisfying

$$\mathrm{E}\Big[\Pr_{x \leftarrow_R T}[C(g(x)) = P(x)]\Big] \geq \frac{1+\gamma}{2}$$

(where the expectation is over the randomness of A).

Then, there is an algorithm B which calls A as a black box poly$(\frac{1}{\gamma}, n)$ times, such that

$$\mathrm{Adv}^B(g, P) \geq 1 - \delta$$

for infinitely many n. The runtime of B is bounded by poly$(\frac{1}{\gamma}, n)$ times the runtime of A.

The advantage of using Proposition 3 is as follows: in order to get a contradiction, we will use a given algorithm A as oracle to contradict the hardness of a pseudo-entropy pair, i.e., we will give B such that $\mathrm{Adv}^B(g, P) \geq 1 - H(P(W)|g(W)) - \phi$. Proposition 3 states that for this it is sufficient to show how to get circuits which perform slightly better than random guessing on a fixed set of size $2^n(H(P(W)|g(W)) + \phi)$, given access to a description of this set. Often, this is a much simpler task.

In the following construction of a pseudorandom generator from a pseudo-entropy pair we assume that parameters ε and ϕ are provided (thus they reappear in Theorem 1). The parameter ε describes how much we lose in the indistinguishability (by making our extractors imperfect), while ϕ is the gap of the pseudo-entropy pair.

Further we assume that parameters α and β are known which give certain information about the combinatorial structure of the given predicate. We will get rid of this assumption later by trying multiple values for α and β such that one of them must be correct.[5]

Lemma 5. *Let g and P be efficiently evaluable functions, $g : \{0,1\}^n \to \{0,1\}^m$, $P : \{0,1\}^n \to \{0,1\}$, $\varepsilon : [0,1] \to \mathbb{N}$, and $\phi : [0,1] \to \mathbb{N}$ be computable in*

---

[5] Haitner, Harnik, and Reingold [5] construct a pseudo-entropy pair for which $H(P(W)|g(W)) = \frac{1}{2}$ is fixed. Because of this, they are able to save a factor of n in the seed length under standard assumptions (they do not need to try different values for α).

polynomial time, $\phi > \frac{1}{n}$. Assume that parameters α and β are such that

$$\alpha \le H(P(W)|g(W)) \le \alpha + \phi/4$$
$$\beta \le H(g(W)) \le \beta + \phi/4.$$

There is an efficient to evaluate oracle function $h^g_{\alpha,\beta,\varepsilon,\phi}$ with the following properties:

- *$h^g_{\alpha,\beta,\varepsilon,\phi}$ is expanding,*
- *$h^g_{\alpha,\beta,\varepsilon,\phi}$ has inputs of length $\mathcal{O}(n^3 \frac{1}{\phi^2} \log(\frac{1}{\varepsilon}))$, and*
- *any algorithm A which distinguishes the output of $h^g_{\alpha,\beta,\varepsilon,\phi}$ from a uniform bit string with advantage γ can be used to get an oracle algorithm B^A satisfying $\mathrm{Adv}^B(g, P) \ge 1 - H(P(W)|g(W)) - \phi$ which does $\mathrm{poly}(\frac{1}{\gamma-\varepsilon}, n)$ calls to A.*

Proof. Let $k := 4096 \cdot (\frac{n}{\phi})^2 \cdot \log(\frac{3}{\varepsilon})$ be the number of repetitions (this is chosen such that

$$\frac{k\phi}{8} = 512\frac{n^2}{\phi}\log\left(\frac{3}{\varepsilon}\right) = 8n\sqrt{k\log\left(\frac{3}{\varepsilon}\right)}, \tag{6}$$

which we use later). To simplify notation we set $\lambda := n - \alpha - \beta - \phi/2$. Using the notation $w^k := w_1\|\ldots\|w_k$, $g^{(k)}(w^k) := g(w_1)\|\ldots\|g(w_k)$ and $P^{(k)}(w^k) := P(w_1)\|\ldots\|P(w_k)$, the function $h_{\alpha,\beta,\varepsilon,\phi}$ is defined as

$$h_{\alpha,\beta,\varepsilon,\phi}(w^k, s_1, s_2, s_3) :=$$

$$\mathrm{Ext}_{k(\beta-\phi/8)}\big(g^{(k)}(w^k), s_1\big) \, \big\| \, \mathrm{Ext}_{k(\alpha+7\phi/8)}\big(P^{(k)}(w^k), s_2\big) \, \big\| \, \mathrm{Ext}_{k(\lambda-\phi/8)}\big(w^k, s_3\big).$$

Clearly, the input length is $\mathcal{O}(n^3 \frac{1}{\phi^2} \log(\frac{1}{\varepsilon}))$. We further see by inspection that, excluding the additional randomness s_1, s_2, and s_3, the function h maps kn bits to at least $k(\alpha + \beta + \lambda) + 5k\frac{\phi}{8} - 3 = k(n - \frac{\phi}{2}) + k\frac{5\phi}{8} - 3 = k(n + \frac{\phi}{8}) - 3 > kn$ bits. Since the additional randomness is also completely contained in the output, $h_{\alpha,\beta,\varepsilon,\phi}$ is expanding for almost all n.

We now show that an algorithm A which has advantage γ in distinguishing $h_{\alpha,\beta,\varepsilon,\phi}(w^k, s_1, s_2, s_3)$ from a uniform bit string of the same length can be used to predict $P(w)$ given $g(w)$ as claimed above. Per definition the probability that the output is **true** in the following game is at least $\frac{1+\gamma}{2}$.

Game 0:

$$(w_1, \ldots, w_k) \leftarrow_R \{0,1\}^{nk}$$
$$b \leftarrow_R \{0,1\}$$
if $b = 0$ **then**   *(Run A with the output of h)*
$$\quad s_1 \leftarrow_R \{0,1\}^{mk}, \; v_1 := \mathrm{Ext}_{k(\beta-\phi/8)}\big(g^{(k)}(w^k), s_1\big)$$
$$\quad s_2 \leftarrow_R \{0,1\}^k, \; v_2 := \mathrm{Ext}_{k(\alpha+7\phi/8)}\big(P^{(k)}(w^k), s_2\big)$$
$$\quad s_3 \leftarrow_R \{0,1\}^{nk}, \; v_3 := \mathrm{Ext}_{k(\lambda-\phi/8)}\big(w^k, s_3\big)$$
else   *(Run A with uniform randomness)*
$$\quad v_1 \leftarrow_R \{0,1\}^{mk+k(\beta-\phi/8)}$$

$$v_2 \leftarrow_R \{0,1\}^{k+k(\alpha+7\phi/8)}$$
$$v_3 \leftarrow_R \{0,1\}^{nk+k(\lambda-\phi/8)}$$
fi
output $b = A(v_1 \| v_2 \| v_3)$

We now make two transition based on statistical indistinguishability. First, we replace the last part v_3 in the *if*-clause of Game 0 with uniform random bits. Because $H(W|g(W)P(W)) = H(W) - H(g(W)) - H(P(W)|g(W)) \geq n - \alpha - \beta - \frac{\phi}{2} = \lambda$, Lemma 2 implies that conditioned on the output of $g^{(k)}$ and $P^{(k)}$ (and thus also conditioned on the extracted bits of those outputs) $\text{Ext}_{k\lambda-k\phi/8}(w^k, s_3) = \text{Ext}_{k\lambda-8n\cdot\sqrt{k\log(\frac{3}{\varepsilon})}}(w^k, s_3)$ is $\frac{\varepsilon}{3}$-close to the uniform distribution (here we used (6)). Thus this only loses $\varepsilon/3$ of the advantage γ in distinguishing.

Second, we replace v_1 in the *else*-clause with $\text{Ext}_{k(\beta-\phi/8)}(g^{(k)}(w^k), s_1)$. Since $H(g(W)) \geq \beta$, Lemma 2 implies that we only lose $\varepsilon/3$ in the advantage again. In total, in the following Game 1 we have advantage at least $\gamma - 2\varepsilon/3$ over random guessing.

Game 1:

$(w_1, \ldots, w_k) \leftarrow_R \{0,1\}^{nk}$
$b \leftarrow_R \{0,1\}$
if $b = 0$ **then**

$s_1 \leftarrow_R \{0,1\}^{mk}, v_1 := \text{Ext}_{k(\beta-\phi/8)}(g^{(k)}(w^k), s_1)$
$s_2 \leftarrow_R \{0,1\}^k, v_2 := \text{Ext}_{k(\alpha+7\phi/8)}(P^{(k)}(w^k), s_2)$
$v_3 \leftarrow_R \{0,1\}^{nk+k(\lambda-\phi/8)}$

else

$s_1 \leftarrow_R \{0,1\}^{mk}, v_1 :- \text{Ext}_{k(\beta-\psi/8)}(g^{(k)}(w^k), s_1)$
$v_2 \leftarrow_R \{0,1\}^{k+k(\alpha+7\phi/8)}$
$v_3 \leftarrow_R \{0,1\}^{nk+k(\lambda-\phi/8)}$

fi
output $b = A(v_1 \| v_2 \| v_3)$

We would like to ignore the parts which are the same in case $b = 0$ and $b = 1$. It is easy to see that A' in Game 2 can be designed such that it calls A with the same distribution as in Game 1.

Game 2:

$(w_1, \ldots, w_k) \leftarrow_R \{0,1\}^{nk}$
$b \leftarrow_R \{0,1\}$
if $b = 0$ **then**

$s \leftarrow_R \{0,1\}^k, v := \text{Ext}_{k(\alpha+7\phi/8)}(P^{(k)}(w^k), s)$

else

$v \leftarrow_R \{0,1\}^{k+k(\alpha+7\phi/8)}$

fi
output $b = A'(g^{(k)}(w^k) \| v)$

Later we want to use Proposition 3. Thus we will have an oracle χ_T which implements the characteristic function of a set T of size at least $(\alpha + \phi)2^n$. From now on we will use the oracle implicitly in the games by testing whether $w \in T$.

In Game 3 it is easy to check that in case $b = 0$ the distribution with which A' is called does not change from Game 2. On the other hand, if $b = 1$, then (since $|T| \geq 2^n(\alpha + \phi)$) the p_i contain independent random variables with entropy at least $\alpha + \phi$ (where the entropy is conditioned on $g(w_i)$). Using Lemma 2 we see that in this case v is $\frac{\varepsilon}{3}$-close to uniform, implying that in Game 3 the advantage of A' in predicting b is still $\gamma - \varepsilon$.

Game 3:

$(w_1, \ldots, w_k) \leftarrow_R \{0,1\}^{nk}$
$b \leftarrow_R \{0,1\}$
for $i \in [n]$ **do**
    if $w_i \in T \wedge b = 1$ **then**
        $p_i \leftarrow_R \{0,1\}$
    else
        $p_i := P(w_i)$
    fi
od
$s \leftarrow_R \{0,1\}^k$, $v := \mathrm{Ext}_{k(\alpha + 7\phi/8)}(p^k, s)$
output $b = A'(g^{(k)}(w^{(k)}) \| v)$

From Game 3, we will now apply a standard hybrid argument to get a predictor for a single position. For this, consider Game 4.

Game 4:

$(w_1, \ldots, w_k) \leftarrow_R \{0,1\}^{nk}$
$j \leftarrow_R [n]$
for $i \in \{1, \ldots, j-1\}$ **do**
    if $w_i \in T$ **then** $p_i \leftarrow_R \{0,1\}$ **else** $p_i := P(w_i)$ **fi**
od
for $i \in \{j+1, \ldots, n\}$ **do** $p_i := P(w_i)$ **od**
$b \leftarrow_R \{0,1\}$
if $w_j \in T \wedge b = 1$ **then** $p_j \leftarrow_R \{0,1\}$ **else** $p_j := P(w_j)$ **fi**
$s \leftarrow_R \{0,1\}^k$, $v := \mathrm{Ext}_{k(\alpha + 7\phi/8)}(p^k, s)$
output $b = A'(g^{(k)}(w^{(k)}) \| v)$

The distribution A' is called in Game 4 in case $b = 0$ and $j = 1$ is the same as in Game 3 in case $b = 0$; the distribution used in Game 4 in case $b = 1$ and $j = n$ is the same as in Game 3, in case $b = 1$. Further, the distribution in Game 4 does not change if b is set from 1 to 0 and j is increased by one. This implies that the advantage of A' in predicting b is $(\gamma - \varepsilon)/k$.

In Game 5, we replace A' with A'' which does all the operations common in case $b = 0$ and $b = 1$ (the w chosen in Game 5 corresponds to w_j in Game 4, and A'' chooses the value of j, and all other w_i before calling A').

Game 5:
$w \leftarrow_R \{0,1\}^n$
$b \leftarrow_R \{0,1\}$
if $w \in \mathcal{T} \wedge b = 1$ **then**
$\quad p \leftarrow_R \{0,1\}$
$\quad$**output** $A''(g(w)\|p) = b$
else
$\quad$**output** $A''(g(w)\|P(w)) = b$
fi

An easy calculation now yields that for $w \leftarrow_R \mathcal{T}$ and $p \leftarrow_R \{0,1\}$ the probabillity that

$$1 \oplus p \oplus A''(g(w)\|p) = P(w)$$

is at least $\frac{1}{2} + \frac{\gamma - \varepsilon}{k}$. Since this works for any $\mathcal{T}$ with $|\mathcal{T}| \geq (\alpha + \phi)2^n$, and thus for every $\mathcal{T}$ with $|\mathcal{T}| \geq (H(P(W)|g(W)) + \phi)2^n$, we can apply Proposition 3 and get the lemma. $\qquad\square$

With this lemma, we can now prove Theorem 1.

Proof (of Theorem 1). Given ε and ϕ, we use the construction of Lemma 4 to get a predicate which we use in the construction of Lemma 5 for $\frac{16n}{\phi^2}$ different values of α and β (note that $0 \leq H(g(W)) \leq n$), such that for at least one of those choices the requirements of Lemma 5 hold. Further, in those applications we use $\varepsilon' := \Omega(\varepsilon \frac{\psi^4}{n^5})$ in place of ε. Since $\varepsilon' = \Omega(\varepsilon^{10})$, this satisfies $\mathcal{O}(\log(\frac{1}{\varepsilon})) = \mathcal{O}(\log(\frac{1}{\varepsilon'}))$.

For every choice of α and β we concatenate $h_{\alpha,\beta,\varepsilon',\phi} : \{0,1\}^\ell \to \{0,1\}^{\ell+1}$ with itself, in order to obtain a function $h'_{\alpha,\beta,\varepsilon',\phi} : \{0,1\}^\ell \to \{0,1\}^{16n\phi^{-2}\ell+1}$, i.e., the first part of the output of $h_{\alpha,\beta,\varepsilon',\phi}$ is used to call $h_{\alpha,\beta,\varepsilon',\phi}$ again, and this process is repeated $16n\phi^{-2}\ell \in \mathcal{O}(n^5\frac{1}{\phi^4})$ times, and every time we get one more bit of the final output.

The function $h_{\varepsilon,\phi} : \{0,1\}^{16n\phi^{-2}\ell} \to \{0,1\}^{16n\phi^{-2}\ell+1}$ divides its input into $\frac{16n}{\phi^2}$ blocks of length ℓ, calls the functions $h'_{\alpha,\beta,\varepsilon',\phi}$ with seperate blocks, and XORs the outputs.

Assume now that an algorithm A can distinguish the output of $h_{\varepsilon,\phi}$ from a unifrom random string with advantage γ. For every choice of α and β (and in particular the choice which satisfies the requirements of Lemma 5) we try the following to invert f. First, since we can simulate the other instances, we see that we have advantage γ in distinguishing the output of $h'_{\alpha,\beta,\varepsilon',\phi}$ from a random string. We can use the hybrid argument to get an algorithm which has advantage $\gamma' := \Omega(\gamma\phi^4 n^{-5})$ in distinguishing the output of $h_{\alpha,\beta,\varepsilon',\phi}$ from a random string. From Lemma 5 we get an algorithm which predicts P from g with advantage at least $1 - H(P(W)|g(W)) - \phi$, and the number of calls is bounded by $\mathrm{poly}(\frac{1}{\gamma'-\varepsilon'}, n) = \mathrm{poly}(\frac{1}{\gamma-\varepsilon}, n)$. Finally, Lemma 4 implies that we can get an inverter with the claimed complexity and success probability. $\qquad\square$

Acknowledgments

I would like to thank Ueli Maurer, Krzysztof Pietrzak, Dominik Raub, Renato Renner, Johan Sjödin, and Stefano Tessaro for helpful comments and discussions. Also, I would like to thank the anonymous referees who provided helpful criticism about the presentation of this paper. I was supported by the Swiss National Science Foundation, project no. 200020-103847/1.

References

1. Manuel Blum and Silvio Micali. How to generate cryptographically strong sequences of pseudo-random bits. *Siam Journal on Computation*, 13(4):850–864, 1984.
2. Oded Goldreich, Shafi Goldwasser, and Silvio Micali. How to construct random functions. *Journal of the ACM*, 33(4):792–807, 1986.
3. Oded Goldreich, Hugo Krawczyk, and Michael Luby. On the existence of pseudorandom generators. *Siam Journal on Computation*, 22(6):1163–1175, 1993.
4. Oded Goldreich and Leonid A. Levin. A hard-core predicate for all one-way functions. In *Proceedings of the Twenty First Annual ACM Symposium on Theory of Computing*, pages 25–32, 1989.
5. Iftach Haitner, Danny Harnik, and Omer Reingold. On the power of the randomized iterate. Technical Report TR05-135, Electronic Colloquium on Computational Complexity (ECCC), 2005.
6. Johan Håstad, Russell Impagliazzo, Leonid A. Levin, and Michael Luby. A pseudorandom generator from any one-way function. *Siam Journal on Computation*, 28(4):1364–1396, 1999.
7. Thomas Holenstein. Key agreement from weak bit agreement. In *Proceedings of the Thirty Seventh Annual ACM Symposium on Theory of Computing*, pages 664–673, 2005.
8. Thomas Holenstein and Renato Renner. On the smooth Rényi entropy of independently repeated random experiments. In preparation, 2005.
9. Russell Impagliazzo. Hard-core distributions for somewhat hard problems. In *The 36th Annual Symposium on Foundations of Computer Science*, pages 538–545, 1995.
10. Russell Impagliazzo, Leonid A. Levin, and Michael Luby. Pseudo-random generation from one-way functions (extended abstract). In *Proceedings of the Twenty First Annual ACM Symposium on Theory of Computing*, pages 12–24, 1989.
11. Michael Luby and Charles Rackoff. How to construct pseudorandom permutations from pseudorandom functions. *Siam Journal on Computation*, 17(2):373–386, 1988.
12. Michael Luby and Avi Wigderson. Pairwise independence and derandomization. Technical Report ICSI TR-95-035, International Computer Science Institute, Berkeley, CA, 1995.
13. Moni Naor. Bit commitment using pseudorandomness. *Journal of Cryptology*, 4(2):151–158, 1991.
14. Victor Shoup. Sequences of games: a tool for taming complexity in security proofs. Technical Report 332, http://eprint.iacr.org/2004/332, 2004.

15. Hoeteck Wee. On obfuscating point functions. In *Proceedings of the Thirty Seventh Annual ACM Symposium on Theory of Computing*, pages 523–532, 2005.
16. Andrew C. Yao. Theory and applications of trapdoor functions (extended abstract). In *The 23rd Annual Symposium on Foundations of Computer Science*, pages 80–91, 1982.

On the Complexity of Parallel Hardness Amplification for One-Way Functions

Chi-Jen Lu[*]

Institute of Information Science, Academia Sinica, Taipei, Taiwan
cjlu@iis.sinica.edu.tw

Abstract. We prove complexity lower bounds for the tasks of hardness amplification of one-way functions and construction of pseudo-random generators from one-way functions, which are realized non-adaptively in black-box ways.

First, we consider the task of converting a one-way function $f : \{0,1\}^n \to \{0,1\}^m$ into a harder one-way function $\bar{f} : \{0,1\}^{\bar{n}} \to \{0,1\}^{\bar{m}}$, with $\bar{n}, \bar{m} \leq \text{poly}(n)$, in a black-box way. The hardness is measured as the fraction of inputs any polynomial-size circuit must fail to invert. We show that to use a constant-depth circuit to amplify hardness beyond a polynomial factor, its size must exceed $2^{\text{poly}(n)}$, and to amplify hardness beyond a $2^{o(n)}$ factor, its size must exceed $2^{2^{o(n)}}$. Moreover, for a constant-depth circuit to amplify hardness beyond an $n^{1+o(1)}$ factor in a security preserving way (with $\bar{n} = O(n)$), it size must exceed $2^{n^{o(1)}}$.

Next, we show that if a constant-depth polynomial-size circuit can amplify hardness beyond a polynomial factor in a weakly black-box way, then it must basically embed a hard function in itself. In fact, one can derive from such an amplification procedure a highly parallel one-way function, which is computable by an NC^0 circuit (constant-depth polynomial-size circuit with bounded fan-in gates).

Finally, we consider the task of constructing a pseudo-random generator $G : \{0,1\}^{\bar{n}} \to \{0,1\}^{\bar{m}}$ from a strongly one-way function $f : \{0,1\}^n \to \{0,1\}^m$ in a black-box way. We show that any such a construction realized by a constant-depth $2^{n^{o(1)}}$-size circuit can only have a sublinear stretch (with $\bar{m} - \bar{n} = o(\bar{n})$).

1 Introduction

One of the most fundamental notions in cryptography is that of one-way functions. Informally speaking, a one-way function is a function which is easy to compute but hard to invert. The adversaries we consider here are polynomial-size circuits, which are non-uniform versions of polynomial-time algorithms. We measure the hardness of a one-way function as the fraction of n-bit inputs on

[*] This work was supported in part by the National Science Council under the Grant NSC 94-2213-E-001-015, and by the Taiwan Information Security Center(TWISC), National Science Council under the Grants NSC 94-3114-P-001-001-Y and NSC 94-3114-P-011-001.

S. Halevi and T. Rabin (Eds.): TCC 2006, LNCS 3876, pp. 462–481, 2006.

which such adversaries must fail to invert. A one-way function with hardness larger than $1 - 1/\text{poly}(n)$ is called a strongly one-way function, which is known to be sufficient for building a large number of cryptographical primitives. Can we further weaken the hardness assumption? Can we start from a one-way function which is only hard to invert in a worst-case sense (with hardness 2^{-n})? This has been a long-standing open problem in cryptography.

It is known that one can start from a weakly one-way function, a one-way function with hardness at least $1/\text{poly}(n)$. The transformation from a weakly one-way function to a strongly one-way function was first discovered by Yao [21], using the so-called direct product approach. The direct product approach has the advantage of being extremely simple and highly parallel. However, the drawback is that it blows up the input length and thus degrades the security (the hardness of the new function is now measured against much smaller circuits). Ideally, one would like to have a security preserving hardness amplification, in which the new function's input length is only increased by a constant factor. Goldreich et al. [7] gave the first security preserving hardness amplification which transforms any weakly one-way *permutation* to a strongly one-way *permutation* of the same input length. Their approach is based on taking random walks on expander graphs and is much more involved than the direct product approach. Moreover, the transformation requires a higher complexity and seems sequential in nature. Therefore, even if the initial function can be evaluated efficiently in parallel, it is not clear if the resulting function will be so. This raises the following question: can a security preserving hardness amplification be carried out in parallel or in a low complexity class?

Another fundamental primitive in cryptography is pseudo-random generator, which stretches a short random seed into a longer random-looking string. A celebrated result due to Håstad et al. shows that a pseudo-random generator can be constructed from any strongly one-way function [9]. A crucial parameter of a pseudo-random generator $G : \{0,1\}^r \to \{0,1\}^{r+s}$ is its stretch s. In several cryptographical applications, we need the stretch to be at least linear. The pseudo-random generator construction in [9] only has a sublinear stretch. In particular, the hard-core function approach can only extract $O(\log n)$ pseudo-random bits from a one-way function. Given a pseudo-random generator of sublinear stretch, one can increase the stretch to linear, but the known construction appears inherently sequential. In [20], Viola asked the question: can the construction of pseudo-random generators with linear stretch from one-way functions be realized efficiently in parallel?

In fact, a more general question is: can cryptographic constructions (or reductions) be realized in a low complexity class? Very little is known for the questions we raised above. For the task of hardness amplifications and pseudo-random generator constructions, there has been no success in realizing them in a low complexity class. Could they be impossible tasks? We would like to say so by showing that they basically all require a high complexity. However, it is not clear what this means. For example, suppose there indeed exists a strongly one-way function computed by a low-complexity procedure, then it gives a trivial hardness amplification procedure of low complexity: just ignore the initial weakly one-way function and compute the strongly one-way function from scratch.

Black-Box Constructions. One important paradigm of cryptographic constructions is the so-called black-box constructions [12], in which one cryptographic primitive is used as a black box to construct another cryptographic primitive. Call a hardness amplification for one-way functions a *black-box* one if the following two conditions hold. First, the initial function f is given as a black-box to construct the new function $\bar{f}$. That is, there is an oracle algorithm AMP such that $\bar{f} = \text{AMP}^f$, so $\bar{f}$ only uses f as an oracle and does not depend on the internal structure of f. Second, the hardness of the new function $\bar{f}$ is proved in a black-box way. That is, there is an oracle Turing machine DEC, such that given any A breaking the hardness of $\bar{f}$, DEC using A as an oracle can break the hardness of f. Again, DEC only uses A as an oracle and does not depend on the internal structure of A. We assume that the procedure DEC makes only a polynomial number of queries to the oracle, and we will study the complexity needed to realize the procedure AMP. In fact, all previous hardness amplification results (and almost all cryptographic reductions) were done in such a black-box way, so it is important to understand its limitation.

A hardness amplification is called a *weakly black-box* one if only the first condition above is required while the second is dropped, namely, without requiring the hardness of the new one-way function to be guaranteed in a black-box way. Note that it seems difficult to obtain negative results for weakly black-box constructions, because one could always build the function $\bar{f}$ from scratch if it exists (without relying on the function f). Therefore, showing that this is indeed the case is usually the best one could expect.

Similarly, one can also define the notion of black-box construction of pseudorandom generators from one-way functions.

Previous Lower Bound Results. Lin, Trevisan, and Wee [14] provided complexity lower bounds for black-box hardness amplification of one-way functions. They showed that to amplify a δ-hard function to an $(1 - \varepsilon)$-hard function in a blackbox way, the procedure AMP must make $q = \Omega((1/\delta)\log(1/\varepsilon))$ queries to the oracle, and the resulting new function must have an input length longer than that of the initial function by $\Omega(\log(1/\varepsilon)) - O(\log q)$ bits. They also showed that if there exists a weakly black-box transformation from a δ-hard *permutation* to an $(1 - \varepsilon)$-hard *permutation* beating this lower bound, then one-way permutations exist unconditionally.

Viola [20] provided a complexity lower bound for black-box construction of pseudo-random generators from strongly one-way functions. He introduced the notion of *parallel* black-box construction, in which the procedure AMP works in the following way. Given an input $\bar{x} \in \{0,1\}^n$, AMP first generates *nonadaptive* queries $x_1, \ldots, x_t \in \{0,1\}^n$ and an AC^0 (constant-depth polynomialsize) circuit A, then accesses the oracle f at these t places to obtain the values $y_1 = f(x_1), \ldots, y_t = f(x_t)$, and finally computes the value $A(y_1, \ldots, y_t)$ as its output. He then showed that if the procedure AMP is realized in this way, then the resulting pseudo-random generator can only have a sublinear stretch.

In a different setting, Lu, Tsai, and Wu [15] considered the hardness of computing Boolean functions instead of inverting one-way functions. They provided complexity lower bounds for procedures which amplify this kind of hardness.

Our Results. We adopt Viola's model [20] and consider hardness amplifications and pseudo-random generator constructions realized in a parallel (non-adaptive) way. Our first result shows that any black-box hardness amplification realized by a low-complexity procedure can not increase the hardness substantially. More precisely, consider any black-box hardness amplification which maps any ε-hard function $f : \{0,1\}^n \to \{0,1\}^m$ to an $\bar{\varepsilon}$-hard function $\bar{f} : \{0,1\}^{\bar{n}} \to \{0,1\}^{\bar{m}}$ with $\bar{n}, \bar{m} \leq \text{poly}(n)$. We show that a constant-depth circuit of $2^{\text{poly}(n)}$ size cannot amplify the hardness to any $\bar{\varepsilon} > \varepsilon \cdot \text{poly}(n)$, and a constant-depth circuit of $2^{2^{o(n)}}$ size cannot amplify the hardness to any $\bar{\varepsilon} > \varepsilon \cdot 2^{o(n)}$. This implies that a procedure in polynomial hierarchy (PH) cannot amplify hardness beyond a polynomial factor, and an alternating Turing machine with constant alternations and $2^{o(n)}$ time ($\text{ATIME}(O(1), 2^{o(n)})$) cannot amplify hardness beyond a $2^{o(n)}$ factor. As a result, a procedure in PH cannot transform a one-way function with hardness lower than $1/\text{poly}(n)$ into a one-way function with constant hardness (let alone a strongly one-way function), and a procedure in $\text{ATIME}(O(1), 2^{o(n)})$ cannot transform a one-way function with worst-case hardness into a weakly one-way function (let alone a strongly one-way function). Note that not only do we rule out the possibility of using a polynomial-time procedure for doing such hardness amplifications (as is usually hoped for in cryptography), we show that even a procedure in a high complexity class, such as PH (or $\text{ATIME}(O(1), 2^{o(n)})$), can not do the job. This just demonstrates how difficult the task is. Moreover, we show that to have $\bar{n} = O(n)$, a constant-depth circuit of $2^{n^{o(1)}}$ size cannot amplify the hardness to any $\bar{\varepsilon} > \varepsilon \cdot n^{1+o(1)}$. This explains why the security preserving hardness amplification procedures of [7, 4] are sequential while the parallel hardness amplification procedure by direct product [21] blows up the input length: they are all done in a black-box way.

Our second result shows that if a parallel weakly black-box hardness amplification can increase the hardness substantially, then it must basically embed a one-way function in itself. More precisely, consider any weakly black-box hardness amplification which maps any ε-hard function $f : \{0,1\}^n \to \{0,1\}^m$ to an $\bar{\varepsilon}$-hard function $\bar{f} : \{0,1\}^{\bar{n}} \to \{0,1\}^{\bar{m}}$. We show that if an AC^0 circuit can amplify the hardness to $\bar{\varepsilon} > \sqrt{\varepsilon} \cdot \text{poly}(n)$, then one can derive from it a one-way function computable in NC^1 with hardness roughly $\bar{\varepsilon}$. From [2], this implies the existence of a one-way function computable in NC^0. This is interesting in the following sense. Consider one-way functions which are computed in polynomial time or even in a higher complexity class. It is possible for a low-complexity procedure, say in AC^0, to amplify hardness for such functions, for example using the direct product approach [21]. However, if it amplifies hardness beyond a polynomial factor, we can derive from such an amplification procedure a one-way function which is computable in NC^0, an extremely low complexity class.

Our third result extends Viola's lower bound for black-box constructions of pseudo-random generators [20]. We show that any black-box construction of pseudo-random generators from strongly one-way functions realized by a constant-depth circuit can only have a sublinear stretch unless the circuit size is exponential. This improves the super-polynomial lower bound of Viola [20].

Our Techniques. We follow the approach of Viola [20], which relies on the fact that applying random restrictions on the input of AC^0 circuits are likely to make their output bits biased since such circuits are insensitive to noise on their input [13, 3]. A similar idea was also used in [15]. However, since our setting is different, we have different problems to solve.

Assume that an AC^0 circuit can amplify hardness beyond a certain bound (the idea can be generalized to a larger class of circuits). It is known that a random function f is likely to be one-way. As shown in [20], it is still likely to be so even with a random restriction ρ applied to its output bits, as long as ρ gives each output bit the symbol $\star$ (leave the bit free) at a rate above some threshold. On the other hand, AC^0 circuits are likely to become biased after applying a random restriction on its input. As the rate of $\star$ decreases, the effect a random f on $\mathrm{AMP}^{f\restriction\rho}(\bar{x})$ becomes smaller, for any input $\bar{x}$. If the rate of $\star$ is small enough, the functions $\mathrm{AMP}^{f\restriction\rho}$'s for most f become close to each other (agreeing with each other on most inputs). As a result, they are close to some fixed function (depending on ρ) which can then be used as an oracle to invert $f\restriction\rho$. This would lead to a contradiction, and we could conclude that such hardness amplification cannot be realized by AC^0 circuits.

However, there is an obstacle in front us. In order to guarantee that the functions $\mathrm{AMP}^{f\restriction\rho}$'s for most f are close to each other, we need the random restriction to give $\star$ in a very low rate. Had we applied a conventional random restriction, say from [5, 8] (as was done in [20]), we would end up having too few free bits left in $f(x)$ for almost every x, and consequently $f\restriction\rho$ would not be one-way for most f. To overcome this problem, we would like the $\star$'s to appear in a somewhat clustered fashion: for any x, either $f(x)$ has no $\star$ at all, or it has a sufficient number of $\star$'s. This motivates us to consider a new kind of random restriction (described in Section 3), and we show that it also makes the output bits of AC^0 circuits highly biased.

This new kind of random restriction also helps us improve the result of Viola. In [20], a super-polynomial size lower bound was shown for black-box constructions of pseudo-random generators from one-way functions. What prevents the argument there from getting a better bound is exactly the same obstacle we just discussed above. Namely, to guarantee $f\restriction\rho$ being one-way using a conventional random restriction, the rate of $\star$ cannot be too low, which fails to make the output bits of larger circuits biased enough. With the help of our new random restriction, we are able to overcome this problem and obtain an exponential lower bound.

Another technical contribution of ours is in the derivation of one-way functions from weakly black-box hardness amplification procedures. In the different setting of Boolean functions, if a function $f : \{0,1\}^n \to \{0,1\}$ agrees on most inputs with a hard-to-compute function $f' : \{0,1\}^n \to \{0,1\}$ (any adversary fails to compute f' correctly on a large portion of inputs), then f itself must also be hard enough, which can be proved in a black-box way. However, this does not seem to be the case for one-way functions. That is, even though a function $f : \{0,1\}^n \to \{0,1\}^m$ is close to a hard-to-invert function $f' : \{0,1\}^n \to \{0,1\}^m$, it is not clear if f itself must also be hard to invert. In fact, this cannot be proved in a black-box way (more in Section 5). The technique in [14] faces the same problem, and the result there is only on weakly hardness amplification which produces one-way *permutations*, since the injective condition makes the problem disappear. As we consider a more restricted type of hardness amplification, that realizable in parallel, we are able overcome this difficulty and obtain results for weakly hardness amplification which produces general one-way *functions*.

2 Preliminaries

For any $n \in \mathbb{N}$, let $[n]$ denote the set $\{1, 2, \ldots, n\}$ and let $\mathcal{U}_n$ denote the uniform distribution over the set $\{0,1\}^n$. When sampling from a finite set, the default distribution we use is the uniform one. For a string $x \in \Sigma^n$, let x_i, for $i \in [n]$, denote the entry in the i'th dimension of x, and let x_I, for $I \subseteq [n]$, denote the substring of x which is the projection of x onto those dimensions in I.

We will consider functions computed by Boolean circuits of AND/OR/NOT gates. Let NC^i denote the class of functions computed by circuits of depth $O(\log^i n)$ and size $\mathrm{poly}(n)$ with *bounded* fan-in gates. Let $\mathsf{AC}(d, s)$ denote the class of functions computed by circuits of depth d and size s with *unbounded* fan-in gates. Let $\mathsf{AC}^0(s)$ denote the class $\mathsf{AC}(O(1), s)$, and note that $\mathsf{AC}^0(\mathrm{poly}(n))$ corresponds to the standard complexity class AC^0. Let $\mathsf{ATIME}(d, t)$ denote the class of functions computed by alternating Turing machines in time t with d alternations. The class $\mathsf{ATIME}(O(1), \mathrm{poly}(n))$ corresponds to the polynomial-time hierarchy PH. More information about complexity classes can be found in standard textbooks, such as [18].

Next, we will introduce the notion of one-way functions and pseudo-random generators. Informally speaking, a function is called a one-way function if it is easy to compute but hard to invert. For a many-to-one function f, we say that an algorithm M inverts $f(x)$ if $M(f(x))$ is in the preimage of $f(x)$, namely, $f(M(f(x))) = f(x)$. When we mention a function $f : \{0,1\}^n \to \{0,1\}^m$, we usually mean a sequence of functions $(f : \{0,1\}^n \to \{0,1\}^{m(n)})_{n \in \mathbb{N}}$, and when we make a statement about f, we usually mean that it holds for any sufficiently large $n \in \mathbb{N}$.

Definition 1. *A function $f : \{0,1\}^n \to \{0,1\}^m$ is (n, m, ε)-hard, or ε-hard for short, if for any polynomial-size circuit M, $\Pr_{x \in \mathcal{U}_n}[M^f$ fails to invert $f(x)] \geq \varepsilon$. A function $f : \{0,1\}^n \to \{0,1\}^m$ is an (n, m, ε)-OWF, or ε-OWF for short, if it can be computed in polynomial time but is ε-hard to invert.*

A pseudo-random generator is a function which stretches a short random seed into a longer random-looking string.

Definition 2. *A function $M : \{0,1\}^m \to \{0,1\}$ ε-distinguishes a function $g : \{0,1\}^n \to \{0,1\}^m$ if $|\Pr_{x \in \mathcal{U}_n}[M(g(x)) = 1] - \Pr_{y \in \mathcal{U}_m}[M(y) = 1]| > \varepsilon$. A function $g : \{0,1\}^n \to \{0,1\}^m$, with $n < m$, is an (n, m, ε)-PRG, or ε-PRG for short, if it can be computed in polynomial time, but no polynomial-size circuit can ε-distinguish g.*

2.1 Black-Box Constructions

Next, we introduce the notion of black-box hardness amplification.

Definition 3. *A black-box hardness amplification from* (n, m, ε)-*hard functions to* $(\bar{n}, \bar{m}, \bar{\varepsilon})$-*hard functions consists of two oracle algorithms* AMP *and* DEC *satisfying the following two conditions. First, for any* $f : \{0,1\}^n \to \{0,1\}^m$, AMP^f *is a function from* $\{0,1\}^{\bar{n}}$ *to* $\{0,1\}^{\bar{m}}$. *Second,* DEC *makes at most* poly(n) *oracle queries, and for any* $f : \{0,1\}^n \to \{0,1\}^m$ *and* $\bar{M} : \{0,1\}^{\bar{m}} \to \{0,1\}^{\bar{n}}$, *if* $\Pr_{\bar{x} \in \mathcal{U}_{\bar{n}}}[\bar{M} \ inverts \ \text{AMP}^f(\bar{x})] > 1 - \bar{\varepsilon}$,[1] *then* $\Pr_{x \in \mathcal{U}_n}[\text{DEC}^{\bar{M}, f} \ inverts \ f(x)] > 1 - \varepsilon$.

Here the transformation of the initial function f into a harder function is done in a black-box way, as the harder function AMP^f only uses f as an oracle. Furthermore, the hardness of AMP^f is also guaranteed in a black-box way, in the sense that any algorithm $\bar{M}$ breaking the hardness condition of AMP^f can be used as an oracle for DEC to break the hardness condition of f. We call AMP the *encoding procedure* and DEC the *decoding procedure*.

A weaker notion is the following weakly black-box hardness amplification, in which only the encoding is required to be done in a black-box way.

Definition 4. *A weakly black-box hardness amplification from* (n, m, ε)-*hard functions to* $(\bar{n}, \bar{m}, \bar{\varepsilon})$-*hard functions consists of an oracle algorithm* AMP *such that* AMP^f *is* $(\bar{n}, \bar{m}, \bar{\varepsilon})$-*hard given any* (n, m, ε)-*hard function* f.

Following [20], we consider the notion of *parallel* black-box hardness amplification. In [20], only the case with $d = O(1)$ and $s \leq$ poly(n) was considered, but here we allow arbitrary d and s. This makes our impossibility results stronger, since we rule out a larger class of hardness amplification procedures.

Definition 5. *We say that a black-box hardness amplification is realized by* AC(d, s) *if the following additional condition holds. Given any* $\bar{x} \in \{0,1\}^{\bar{n}}$, AMP *first produces an* AC(d, s) *circuit* A *and makes* $t \leq$ poly(s) *non-adaptive queries* $x_1, \ldots, x_t \in \{0,1\}^n$ *to the oracle to obtain answers* $y_1, \ldots, y_t \in \{0,1\}^m$, *and then computes its output as* $A(y_1, \ldots, y_t)$.

Note that $x_1, \ldots, x_t$ and A only depend on $\bar{x}$ and are independent of the oracle f. For the black-box case, no complexity constraint is placed on the part of generating the queries and the circuit, which again makes our impossibility results stronger. For the weakly black-box case, we need this part to be computed by an AC(d, s) circuit too, since we want to derive from the procedure AMP an efficiently computable one-way function. Similarly, one can define the notion of black-box construction of pseudo-random generators from hard functions, which is omitted here and can be found in [20].

2.2   Limited Independence

A sequence of random variables is called k-wise independent if any k of them are independent. It is well known that such a space can be sampled in a randomness-efficient way.

---

[1] Here we consider the case that $\bar{M}$ does not query AMP^f. This makes such hardness amplification easier to find and our impossibility results stronger.

Fact 1. *Any k-wise independent random variables $X_1, \ldots, X_N \in V$ can be generated in polynomial time using a seed of length $O(k(\log N + \log |V|))$.*

A sequence of variables is called (k, δ)-wise independent if any k of them together has a statistical distance at most δ to the uniform distribution. We need efficient constructions of such a space from [16, 1]. From this, we can obtain the following, whose proof is omitted due to the space constraint.

Lemma 1. *Suppose $b \geq t^2/\varepsilon^3$. Then there exists a family $\bar{\mathcal{H}}$ of hash functions from $\{0, 1\}^n$ to $[b]$ which can be sampled using a seed of length $r_0 = O(\log n + \log b + \log(1/\varepsilon))$ and satisfies the following two properties.*

1. *For any distinct $x_1, \ldots, x_t \in \{0, 1\}^n$, the probability over $h \in \bar{\mathcal{H}}$ that $h(x_i) = h(x_j)$ for some $i \neq j$ is at most $o(\varepsilon)$.*
2. *For any $S \subseteq [b]$ of size $3\varepsilon b$, the probability over $h \in \bar{\mathcal{H}}$ that $h(x) \in S$ for less than 2ε fraction of x is at most $o(\varepsilon)$.*

2.3 Fourier Analysis

As in [20], we will apply Fourier analysis on Boolean functions. For $N \in \mathbb{N}$ and $I \subseteq [N]$, define the function $\chi^I : \{-1, 1\}^n \to \{-1, 1\}$ as $\chi^I(x) = \prod_{i \in I} x_i$ for any $x \in \{-1, 1\}^N$. For any $C : \{-1, 1\}^N \to \{-1, 1\}$ and any $I \subseteq [N]$, let $\hat{C}(I) = \mathbb{E}_{x \in \{-1,1\}^N}[C(x) \cdot \chi^I(x)]$. Here are some useful facts.

Fact 2. *For any $C : \{-1, 1\}^N \to \{-1, 1\}$ and for any $x \in \{-1, 1\}^N$, $C(x) = \sum_I \hat{C}(I) \cdot \chi^I(x)$.*

Lemma 2. *[19] For any $C : \{-1, 1\}^N \to \{-1, 1\} \in \mathsf{AC}(d, s)$, $\sum_I \hat{C}(I)^2 (1 - 2\delta)^{|I|} \geq 1 - O(\delta \log^{d-1} s)$.*

3 Random Restriction

We will need the notion of random restriction [5, 8]. A restriction ρ on m variables is an element of $\{0, 1, \star\}^m$, or seen as a function $\rho : [m] \to \{0, 1, \star\}$. A variable is fixed by ρ if it receives a value in $\{0, 1\}$ while a variable remains free if it receives the symbol $\star$. For a string $y \in \{0, 1\}^m$ and a restriction $\rho \in \{0, 1, \star\}^m$, let $y\restriction_\rho \in \{0, 1\}^m$ be the restriction of y with respect to ρ: for $i \in [m]$, the i'th bit of $y\restriction_\rho$ is y_i if $\rho_i = \star$ and is ρ_i if $\rho_i \in \{0, 1\}$. For a string $z \in \{0, 1, \star\}^m$, let $\#_\star(z)$ denote the number of i's such that $z_i = \star$.

As in [20], we will consider applying a random restriction to a function $f : \{0, 1\}^n \to \{0, 1\}^m$ in the following sense. Take a restriction $\rho \in \{0, 1, \star\}^{2^n m}$, seen as a function $\rho : \{0, 1\}^n \to \{0, 1, \star\}^m$, let $f\restriction_\rho$ be the function from $\{0, 1\}^n$ to $\{0, 1\}^m$ such that for $x \in \{0, 1\}^n$, $f\restriction_\rho(x) = f(x)\restriction_{\rho(x)}$, the result of applying the restriction $\rho(x) \in \{0, 1, \star\}^m$ on $f(x) \in \{0, 1\}^m$.

Let $\mathcal{R}^m_\delta$ denote the random restriction (distribution over restrictions) on m variables such that each variable independently receives the symbol $\star$ with probability δ, the value 1 with probability $(1 - \delta)/2$, and the value 0 with probability $(1 - \delta)/2$. For our purpose later, we will need a new kind of random restriction.

Definition 6. *Let $\mathcal{R}_{\alpha,\beta}^{1,m}$ be the random restriction on m variables defined as* $\mathcal{R}_{\alpha,\beta}^{1,m} = \alpha \cdot \mathcal{R}_\beta^m + (1-\alpha) \cdot \mathcal{R}_0^m$. *That is, $\mathcal{R}_{\alpha,\beta}^{1,m}$ distributes as $\mathcal{R}_\beta^m$ with probability α and as $\mathcal{R}_0^m = \mathcal{U}_m$ with probability $1 - \alpha$. Let $\mathcal{R}_{\alpha,\beta}^{t,m}$ be the random restriction on tm variables, defined as $\mathcal{R}_{\alpha,\beta}^{t,m} = (\mathcal{R}_{\alpha,\beta}^{1,m})^t$, namely, t independent copies of $\mathcal{R}_{\alpha,\beta}^{1,m}$.*

It is known that AC^0 circuits are insensitive to noise and (standard kind of) random restrictions are likely to make their output values highly biased [13, 3, 20]. We show that this is still true with respect to our new kind of random restrictions.

Lemma 3. *For any $C : \{0,1\}^{tm} \to \{0,1\} \in \mathsf{AC}(d,s)$, the probability over $\rho \in \mathcal{R}_{\alpha,\beta}^{t,m}$ and $y, y' \in \mathcal{U}_{tm}$ that $C(y\restriction_\rho) \neq C(y'\restriction_\rho)$ is at most $O(\alpha\beta \log^{d-1} s)$.*

Proof. We would like to apply Fourier analysis on C, so for now let us use $\{-1,1\}$ for the binary values $\{0,1\}$. Partition the tm input positions evenly into t parts $B_1, \ldots, B_t$ of size m, with $B_i = \{(i-1)m+1, \ldots, im\}$.

We know that $\Pr_{\rho;y,y'}[C(y\restriction_\rho) \neq C(y'\restriction_\rho)] = \frac{1}{2}(1 - \mathrm{E}_{\rho;y,y'}[C(y\restriction_\rho) \cdot C(y'\restriction_\rho)])$. From Fact 2, $\mathrm{E}_{\rho;y,y'}[C(y\restriction_\rho) \cdot C(y'\restriction_\rho)]$ is equal to

$$\mathop{\mathrm{E}}_{\rho;y,y'}\left[\left(\sum_{I \subseteq [tm]} \hat{C}(I)\chi^I(y\restriction_\rho)\right) \cdot \left(\sum_{J \subseteq [tm]} \hat{C}(J)\chi^J(y'\restriction_\rho)\right)\right]$$

$$= \sum_{I,J \subseteq [tm]} \hat{C}(I) \cdot \hat{C}(J) \cdot \mathop{\mathrm{E}}_{\rho;y,y'}\left[\chi^I(y\restriction_\rho) \cdot \chi^J(y'\restriction_\rho)\right].$$

To bound the expectation $\mathrm{E}_{\rho;y,y'}\left[\chi^I(y\restriction_\rho) \cdot \chi^J(y'\restriction_\rho)\right]$, consider two cases.

Case 1: $I \neq J$. There must exist some block B_i such that $B_i \cap I \neq B_i \cap J$. Observe that $\mathrm{E}_{\rho;y,y'}[\chi^I(y\restriction_\rho) \cdot \chi^J(y'\restriction_\rho)]$ is qual to

$$\mathop{\mathrm{E}}_{\rho;y,y'}\left[(\chi^{I \cap B_i}(y\restriction_\rho) \cdot \chi^{J \cap B_i}(y'\restriction_\rho))(\chi^{I \setminus B_i}(y\restriction_\rho) \cdot \chi^{J \setminus B_i}(y'\restriction_\rho))\right]$$

$$= \mathop{\mathrm{E}}_{\rho;y,y'}\left[\chi^{I \cap B_i}(y\restriction_\rho) \cdot \chi^{J \cap B_i}(y'\restriction_\rho)\right] \mathop{\mathrm{E}}_{\rho;y,y'}\left[\chi^{I \setminus B_i}(y\restriction_\rho) \cdot \chi^{J \setminus B_i}(y'\restriction_\rho)\right],$$

where the second equality is because $\chi^{I \cap B_i}(y\restriction_\rho) \cdot \chi^{J \cap B_i}(y'\restriction_\rho)$ and $\chi^{I \setminus B_i}(y\restriction_\rho) \cdot \chi^{J \setminus B_i}(y'\restriction_\rho)$ are distributed independently. Note that

$$\mathop{\mathrm{E}}_{\rho;y,y'}\left[\chi^{I \cap B_i}(y\restriction_\rho) \cdot \chi^{J \cap B_i}(y'\restriction_\rho)\right] = \mathop{\mathrm{E}}_{\rho_i;y_i,y_i'}\left[\chi^{I \cap B_i}(y_i\restriction_{\rho_i}) \cdot \chi^{J \cap B_i}(y_i'\restriction_{\rho_i})\right],$$

with $\rho_i \in \mathcal{R}_{\alpha,\beta}^{1,m} = (1-\alpha) \cdot \mathcal{R}_0^m + \alpha \cdot \mathcal{R}_\beta^m$ and $y_i, y_i' \in \mathcal{U}_m$, so the expectation is

$$(1-\alpha) \cdot \mathop{\mathrm{E}}_{\rho_i \in \mathcal{R}_0^m;y_i,y_i'}\left[\chi^{I \cap B_i}(y_i\restriction_{\rho_i}) \cdot \chi^{J \cap B_i}(y_i'\restriction_{\rho_i})\right]$$

$$+ \alpha \cdot \mathop{\mathrm{E}}_{\rho_i \in \mathcal{R}_\beta^m;y_i,y_i'}\left[\chi^{I \cap B_i}(y_i\restriction_\rho) \cdot \chi^{J \cap B_i}(y_i'\restriction_{\rho_i})\right],$$

which is $0 + 0 = 0$. This implies that $\mathrm{E}_{\rho;y,y'}[\chi^I(y\restriction_\rho) \cdot \chi^J(y'\restriction_\rho)] = 0$ when $I \neq J$.

Case 2: $I = J$. Partition I into t parts $I_1, \ldots, I_t$ where $I_i = I \cap B_i$. Then,

$$
\mathop{\mathrm{E}}_{\rho;y,y'} \left[\chi^I(y\!\restriction_\rho) \cdot \chi^I(y'\!\restriction_\rho) \right] = \mathop{\mathrm{E}}_{\rho;y,y'} \left[\prod_{i\in[t]} \chi^{I_i}(y_i\!\restriction_{\rho_i}) \cdot \chi^{I_i}(y_i'\!\restriction_{\rho_i}) \right]
$$

$$
= \prod_{i\in[t]} \mathop{\mathrm{E}}_{\rho_i;y_i,y_i'} \left[\chi^{I_i}(y_i\!\restriction_{\rho_i}) \cdot \chi^{I_i}(y_i'\!\restriction_{\rho_i}) \right]
$$

$$
= \prod_{i\in[t]} \left((1-\alpha)\cdot 1 + \alpha \cdot (1-\beta)^{|I_i|} \right)
$$

$$
\geq \prod_{i\in[t]} (1-\alpha\beta)^{|I_i|}
$$

$$
= (1-\alpha\beta)^{|I|},
$$

where the inequality follows from Jensen's inequality.[2]

Combining the two cases, we have $\mathrm{E}_{\rho;y,y'} \left[C(y\!\restriction_\rho) \cdot C(y'\!\restriction_\rho) \right]$ equal to

$$
\sum_I \hat{C}(I)^2 \cdot \mathop{\mathrm{E}}_{\rho;y,y'} \left[\chi^I(y\!\restriction_\rho) \cdot \chi^I(y'\!\restriction_\rho) \right] \geq \sum_I \hat{C}(I)^2 \cdot (1-\alpha\beta)^{|I|},
$$

which equals to $1 - O(\alpha\beta \log^{d-1} s)$ by Lemma 2. Then,

$$
\mathop{\mathrm{Pr}}_{\rho;y,y'} [C(y\!\restriction_\rho) \neq C(y'\!\restriction_\rho)] = \frac{1}{2}\left(1 - \mathop{\mathrm{E}}_{\rho;y,y'} [C(y\!\restriction_\rho) \cdot C(y'\!\restriction_\rho)] \right) = O(\alpha\beta \log^{d-1} s). \quad \square
$$

Note that a random restriction from $\mathcal{R}_{\alpha,\beta}^{1,m}$ can be sampled using a seed of length $\ell_1 + m\ell_2$ consisting of $m+1$ parts. The first part of the seed has length $\ell_1 = O(\log(1/\alpha))$ and is used to determine whether the restriction $\mathcal{R}_\beta^m$ or $\mathcal{R}_0^m$ is applied. The remaining m parts of the seed, each of length $\ell_2 = O(\log(1/\beta))$, are used to generate the m symbols in $\{0,1,\star\}$. For simplicity, we use a longer seed of length $\ell = (m+1)\ell_0$ and let each part have the same length $\ell_0 = \max(\ell_1, \ell_2)$.

Furthermore, there is an $\mathrm{AC}^0(\mathrm{poly}(\ell))$ circuit W which given such a random seed of length ℓ produces the random restriction $\mathcal{R}_{\alpha,\beta}^{1,m}$. Thus, a random restriction from $\mathcal{R}_{\alpha,\beta}^{b,m}$ can be sampled using a seed of length $b\ell$ and produced by an $\mathrm{AC}^0(\mathrm{poly}(b\ell))$ circuit W^b, the concatenation of b independent copies of W.

4   Black-Box Hardness Amplification

In this section, we study black-box hardness amplification from (n, m, ε)-hard functions to $(\bar{n}, \bar{m}, \bar{\varepsilon})$-hard functions. We will show that no such hardness amplification realized by $\mathrm{AC}^0(2^{\mathrm{poly}(n)})$ can amplify the hardness to *any* $\bar{\varepsilon} > \varepsilon \cdot \mathrm{poly}(n)$ while keeping the function's output or input length to $\mathrm{poly}(n)$. Our main technical result is the following.

---

[2] Consider the function $f(x) = (1 - \beta x)^k$, which is convex for x in the interval $[0, 1]$.
Then $(1-\alpha)\cdot 1 + \alpha \cdot (1-\beta)^k = (1-\alpha)\cdot f(0) + \alpha \cdot f(1) \geq f((1-\alpha)\cdot 0 + \alpha \cdot 1) = (1-\alpha\beta)^k$.

Theorem 1. *No black-box hardness amplification from (n, m, ε)-hard functions to $(\bar{n}, \bar{m}, \bar{\varepsilon})$-hard functions can be realized by $\mathsf{AC}(d, s)$ with $\varepsilon \leq \bar{\varepsilon} \cdot \gamma$, for any $\gamma \leq o(m/(\bar{m} \log^{d+1} s))$ and any $s \geq \mathrm{poly}(n)$.*

Since any $\mathsf{ATIME}(d, t)$ computation with an oracle can be simulated by an $\mathsf{AC}(O(d), 2^{O(dt)})$ circuit with oracle answers given as part of its input, we have the following. In particular, with $\bar{m} \leq \mathrm{poly}(m)$, no such hardness amplification can be realized in PH for any $\bar{\varepsilon} \geq \varepsilon \cdot n^{\omega(1)}$, and nor can it be realized in $\mathsf{ATIME}(O(1), 2^{o(n)})$ for any $\bar{\varepsilon} \geq \varepsilon \cdot 2^{\Omega(n)}$.

Corollary 1. *No black-box hardness amplification from (n, m, ε)-hard functions to $(\bar{n}, \bar{m}, \bar{\varepsilon})$-hard functions can be realized in $\mathsf{ATIME}(d, t)$ with $\varepsilon \leq \bar{\varepsilon} \cdot m/(\bar{m} \cdot t^{cd})$ for some constant c.*

Theorem 1 states that a low-complexity procedure cannot amplify the hardness substantially without blowing up the output length. Next, we show that one cannot avoid blowing up the input length either. In particular, no $\mathsf{AC}^0(2^{n^{o(1)}})$ circuit can amplify hardness beyond an $n^{1+o(1)}$ factor in a security preserving way (with $\bar{n} = O(n)$).

Theorem 2. *No black-box hardness amplification from (n, m, ε)-hard functions to $(\bar{n}, \bar{m}, \bar{\varepsilon})$-hard functions can be realized by $\mathsf{AC}(d, s)$ with $\varepsilon \leq \bar{\varepsilon} \cdot \gamma$, for any $\gamma \leq o(n/(\bar{n} \log^{2d+1} s))$ when $s \geq 2^{\Omega(n^{1/(d-1)})}$, or for any $\gamma \leq o(n/(\bar{n} n^{(2d+1)/(d-1)}))$ when $s \leq 2^{O(n^{1/(d-1)})}$.*

4.1 Proof of Theorem 1

Assume that such a hardness amplification exists, with AMP realized by $\mathsf{AC}(d, s)$ and $\varepsilon = o(\bar{\varepsilon} \cdot m/(\bar{m} \log^{d+1} s))$. We will show that this leads to a contradiction. The idea is the following. First, we show that for a random function f and a suitable random restriction ρ, the resulting function $f\!\restriction_\rho$ is likely to be one-way. The key is to show that for a sufficient number of x, ρ leaves enough bits in $f(x)$ free. Next, we show that such a random restriction is likely to kill off the effect of a random function f on $\mathrm{AMP}^{f\!\restriction_\rho}$ so that the functions $\mathrm{AMP}^{f\!\restriction_\rho}$'s for most f's are close to each other. The key is to show that an $\mathsf{AC}(d, s)$ circuit is likely to become highly biased after such a random restriction. This yields a way to invert $\mathrm{AMP}^{f\!\restriction_\rho}$ well for most f's, which can then be used as an oracle to invert $f\!\restriction_\rho$, and we have a contradiction. To make sure that both conditions above hold, we need the random restriction to give $\star$'s at a very small rate but in a clustered way: $f(x)$ receives no $\star$ at all for most x, but gets an enough number of $\star$'s for the rest. This motivates us to consider the new random restriction $\mathcal{R}^{b,m}_{\alpha,\beta}$ introduced in Section 3.

As in [20], we would like to make sure that a restriction does not give away too much information about the input, so that the function $f\!\restriction_\rho$ is one-way even given ρ. Therefore we will hash the input from the space $\{0, 1\}^n$ down to a smaller space $[b]$ before applying the restriction from $\mathcal{R}^{b,m}_{\alpha,\beta}$. Here we choose the following parameters:

$$\alpha = 2\varepsilon, \quad \beta = (\log^2 s)/m, \quad \text{and } b = t^2/\varepsilon^3.$$

Let $\mathcal{H}$ denote the set of functions from $\{0,1\}^n$ to $[b]$. Then define our random restriction $\mathcal{R}$ as the uniform distribution over the set of restrictions $\sigma \circ h :$ $\{0,1\}^n \to \{0,1,\star\}^m$, with $h \in \mathcal{H}$ and $\sigma \in \mathcal{R}_{\alpha,\beta}^{b,m}$. Let $\mathcal{F}$ denote the set of functions from $\{0,1\}^n$ to $\{0,1\}^m$.

Definition 7. *We call a restriction* $\rho : \{0,1\}^n \to \{0,1,\star\}^m$ *good if both of the following two conditions hold:*

1. $\Pr_{x \in \mathcal{U}_n}[\#_\star(\rho(x)) \geq \beta m/2] \geq (2/3)\alpha.$
2. $\Pr_{\bar{x} \in \mathcal{U}_{\bar{n}}; f, f' \in \mathcal{F}}[\mathrm{AMP}^{f\restriction\rho}(\bar{x}) \neq \mathrm{AMP}^{f'\restriction\rho}(\bar{x})] = o(\bar{\varepsilon}).$

Note that if we use a traditional random restriction (of [5, 8]) as in [20], it is unlikely to have both conditions hold at the same time, because the second condition requires a low rate of $\star$ (lower than $\bar{\varepsilon}/(\bar{m}\log^{d-1} s)$) which makes the first condition unlikely to hold. On the other hand, using our new random restriction, we can have both conditions hold with high probability.

Lemma 4. $\Pr_{\rho \in \mathcal{R}}[\rho \text{ is not good}] = o(1).$

Due to the space limitation, we defer the proof to the journal version and only sketch the idea here. To show that the first condition fails with a small probability, note that about α fraction of x's are turned "on" in the sense that it receives the restriction from $\mathcal{R}_\beta^m$ and should have $\#_\star(\rho(x))$ about βm, so large deviation from this has a small probability. To show that the second condition fails with a small probability, note that for any $\bar{x} \in \{0,1\}^{\bar{n}}$, most $\rho \in \mathcal{R}$ can kill off the effect of a random function f so that the value $\mathrm{AMP}^{f\restriction\rho}(\bar{x})$ is the same for most $f \in \mathcal{F}$, which is guaranteed by Lemma 3, with $\alpha\beta = O((\varepsilon \log^2 s)/m) = o(\bar{\varepsilon}/(\bar{m}\log^{d-1} s)).$

Next, we show that for a good ρ, the function $f\restriction_\rho$ is ε-hard for most $f \in \mathcal{F}$. In fact, as will be needed later, we prove hardness against slightly stronger algorithms: algorithms which can depend on ρ and have arbitrarily high complexity but make only a polynomial number of queries to $f\restriction_\rho$.

Lemma 5. *For any good* ρ, *for any* M_ρ *making at most* $\mathrm{poly}(n)$ *oracle queries,* $\Pr_{x \in \mathcal{U}_n, f \in \mathcal{F}}[M_\rho^{f\restriction_\rho} \text{ inverts } f\restriction_\rho(x)] \leq 1 - \varepsilon.$

Due to space limitation, we defer the proof to the journal version. The argument is somewhat standard, which can be modified, say, from [20, 6].

This implies that for any good ρ, the function $\bar{A}_\rho$, defined by $\bar{A}_\rho(\bar{x}) = \max \arg_z \Pr_{f \in \mathcal{F}}[\mathrm{AMP}^{f\restriction_\rho}(\bar{x}) = z]$, is close to $\mathrm{AMP}^{f\restriction_\rho}$ for most f, because

$$\Pr_{\bar{x},f}\left[\bar{A}_\rho(\bar{x}) \neq \mathrm{AMP}^{f\restriction_\rho}(\bar{x})\right] \leq \Pr_{\bar{x},f,f'}\left[\mathrm{AMP}^{f\restriction_\rho}(\bar{x}) \neq \mathrm{AMP}^{f'\restriction_\rho}(\bar{x})\right] = o(\bar{\varepsilon}).$$

This then provides us a way to invert the function $\mathrm{AMP}^{f\restriction_\rho}$.

Lemma 6. *For any good* ρ, *there exists a function* $\bar{M}_\rho : \{0,1\}^{\bar{m}} \to \{0,1\}^{\bar{n}}$ *such that* $\Pr_{\bar{x} \in \mathcal{U}_{\bar{n}}, f \in \mathcal{F}}[\bar{M}_\rho \text{ inverts } \mathrm{AMP}^{f\restriction_\rho}(\bar{x})] \geq 1 - o(\bar{\varepsilon}).$

Proof. Fix any good ρ, and let $\bar{M}_\rho$ be the function which on input $\bar{y}$ outputs a random element in the set $\bar{A}_\rho^{-1}(\bar{y})$. Then $\Pr_{\bar{x},f}[\bar{M}_\rho$ fails to invert $\text{Amp}^{f\restriction_\rho}(\bar{x})]$ is

$$\Pr_{\bar{x},f}\left[\text{Amp}^{f\restriction_\rho}(\bar{M}_\rho(\text{Amp}^{f\restriction_\rho}(\bar{x}))) \neq \text{Amp}^{f\restriction_\rho}(\bar{x})\right]$$

$$\leq \Pr_{\bar{x},f}\left[\text{Amp}^{f\restriction_\rho}(\bar{M}_\rho(\bar{A}_\rho(\bar{x}))) \neq \bar{A}_\rho(\bar{x})\right] + \Pr_{\bar{x},f}\left[\bar{A}_\rho(\bar{x}) \neq \text{Amp}^{f\restriction_\rho}(\bar{x})\right]$$

$$< \sum_{\bar{y}} \Pr_{\bar{x}}\left[\bar{A}_\rho(\bar{x}) = \bar{y}\right] \cdot \Pr_{\bar{x},f}\left[\text{Amp}^{f\restriction_\rho}(\bar{M}_\rho(\bar{y}))) \neq \bar{y} \mid \bar{A}_\rho(\bar{x}) = \bar{y}\right] + o(\bar{\varepsilon})$$

$$= \sum_{\bar{y}} \Pr_{\bar{x}}\left[\bar{A}_\rho(\bar{x}) = \bar{y}\right] \cdot \Pr_{\bar{x},\bar{x}',f}\left[\text{Amp}^{f\restriction_\rho}(\bar{x}') \neq \bar{y} \mid \bar{A}_\rho(\bar{x}) = \bar{A}_\rho(\bar{x}') = \bar{y}\right] + o(\bar{\varepsilon})$$

$$= \sum_{\bar{y}} \Pr_{\bar{x}}\left[\bar{A}_\rho(\bar{x}) = \bar{y}\right] \cdot \Pr_{\bar{x}',f}\left[\text{Amp}^{f\restriction_\rho}(\bar{x}') \neq \bar{A}_\rho(\bar{x}') \mid \bar{A}_\rho(\bar{x}') = \bar{y}\right] + o(\bar{\varepsilon})$$

$$= \Pr_{\bar{x},f}\left[\bar{A}_\rho(\bar{x}) \neq \text{Amp}^{f\restriction_\rho}(\bar{x})\right] + o(\bar{\varepsilon})$$

$$= o(\bar{\varepsilon}).$$ □

From Lemma 6 and Definition 3, for any good ρ, a Markov's inequality implies that for most $f \in \mathcal{F}$, the function $M_\rho^{f\restriction_\rho} = \text{Dec}^{\bar{M}_\rho,f\restriction_\rho}$ can achieve $\Pr_x[M_\rho^{f\restriction_\rho}$ inverts $f\restriction_\rho(x)] > 1 - \varepsilon$. This contradicts Lemma 5 since Dec makes at most a polynomial number of queries to the oracle. Therefore, no such hardness amplification is possible, which proves Theorem 1.

4.2   Proof of Theorem 2

Let $\bar{\mathcal{H}}$ denote the family of hash functions from $\{0,1\}^m$ to $\{0,1\}^{3n}$ derived from a $(2, 2^{-3n})$-wise independent space. We will use the construction of [1], based on finite fields of characteristic two, with each function in the family specified by $O(n)$ bits. Then using ideas from [11,10], given the specification of a function $h \in \bar{\mathcal{H}}$ and an input $x \in \{0,1\}^n$, one can compute $h(x)$ by an $\text{AC}(d, 2^{O(n^{1/(d-1)})})$ circuit.

The key to the theorem is the following, which says that one can transform a hard function $f : \{0,1\}^n \to \{0,1\}^m$ with any $m \leq \text{poly}(n)$ into a hard function $f' : \{0,1\}^{n'} \to \{0,1\}^{m'}$ with $n', m' = O(n)$.

Lemma 7. *A black-box hardness amplification from (n, m, ε)-hard functions to $(\bar{n}, \bar{m}, \bar{\varepsilon})$-hard functions can be realized in $\text{AC}(d, 2^{O(n^{1/(d-1)})})$ with $\bar{\varepsilon} = \varepsilon - 2^{-n+1}$, $\bar{n} = O(n)$, and $\bar{m} = O(n)$.*

Proof. Given any ε-hard function $f : \{0,1\}^n \to \{0,1\}^m$, define the function $f' = \text{Amp}^f : \{0,1\}^{n'} \to \{0,1\}^{m'}$ as

$$f'(x, h) = (h(f(x)), h),$$

with $x \in \{0,1\}^n$ and $h \in \bar{\mathcal{H}}$. Thus, $n' = n + O(n) = O(n)$ and $m' = 3n + O(n) = O(n)$. From the discussion at the beginning, Amp can be realized in $\text{AC}(d, 2^{O(n^{1/(d-1)})})$.

Next, we prove the hardness of f' in a black-box way. Suppose M' is a function which inverts f' with probability more than $1-(\varepsilon-2^{-n+1})$. Consider the function $M = \mathrm{DEC}^{M'}$, which on input $y \in \{0,1\}^m$ generates a random $h \in \bar{\mathcal{H}}$, calls $M'(h(y), h)$, and outputs the first component from the answer. We will show that M inverts f with probability more than $1 - \varepsilon$. Let M_h denote the function M with the random choice h. Call $h \in \bar{\mathcal{H}}$ *colliding* if there exist x, x' with $f(x) \neq f(x')$ and $h(f(x)) = h(f(x'))$. Then, $\Pr_{x \in \mathcal{U}_n}[M$ inverts $f(x)]$ is

$$\Pr_{x \in \mathcal{U}_n, h \in \bar{\mathcal{H}}} [f(M_h(f(x))) = f(x)]$$

$$\geq \Pr_{x \in \mathcal{U}_n, h \in \bar{\mathcal{H}}} [h(f(M_h(f(x)))) = h(f(x)) \wedge h \text{ is not colliding}]$$

$$\geq \Pr_{x \in \mathcal{U}_n, h \in \bar{\mathcal{H}}} [f'(M(f'(x, h))) = f'(x, h)] - \Pr_{h \in \bar{\mathcal{H}}} [h \text{ is colliding}]$$

$$> 1 - (\varepsilon - 2^{-n+1}) - 2^{2n}(2^{-3n} + 2^{-3n})$$

$$= 1 - \varepsilon.$$

This proves the lemma.   □

Consider any black-box hardness amplification from (n, m, ε)-hard functions to $(\bar{n}, \bar{m}, \bar{\varepsilon})$-hard functions realized by $\mathsf{AC}(d, s)$, with $\varepsilon \leq \bar{\varepsilon} \cdot \gamma$. Assume we have $s \geq 2^{\Omega(n^{1/(d-1)})}$ and $\gamma \leq o(n/(\bar{n} \log^{2d+1} s))$. Then by combining this with Lemma 7, we get a black-box hardness amplification from (n, m, ε)-hard functions to $(\bar{n}', \bar{m}', \bar{\varepsilon}')$-hard functions realized by $\mathsf{AC}(2d, s')$, with $\bar{m}' = O(\bar{n})$, $s' = O(s)$, and $\varepsilon \leq \bar{\varepsilon}' \cdot \gamma'$, for $\gamma' \leq o(m/(\bar{m}' \log^{2d+1} s'))$, which contradicts Theorem 1. Therefore, no such hardness amplification can exist. Next, assume we have $s \leq 2^{O(n^{1/(d-1)})}$ and $\gamma \leq o(n/(\bar{n} \cdot n^{(2d+1)/(d-1)}))$. Combining this with Lemma 7, we get a black-box hardness amplification from (n, m, ε)-hard functions to $(\bar{n}', \bar{m}', \bar{\varepsilon}')$-hard functions realized by $\mathsf{AC}(2d, s')$, with $\bar{m}' = O(\bar{n})$, $s' \leq 2^{O(n^{1/(d-1)})}$, and $\varepsilon \leq \bar{\varepsilon}' \cdot \gamma'$, for $\gamma' \leq o(m/(\bar{m}'n^{(2d+1)/(d-1)})) = o(m/(\bar{m}' \log^{2d+1} s'))$, which contradicts Theorem 1. Thus, no such hardness amplification can exist either. This completes the proof of Theorem 2.

5   Weakly Black-Box Hardness Amplification

In this section, we consider weakly black-box hardness amplifications from (n, m, ε)-hard functions to $(\bar{n}, \bar{m}, \bar{\varepsilon})$-hard functions. Suppose such an amplification procedure, consisting of both the query-generation part and the answer-combination part, can be computed in AC^0. We will show that if it can amplify the hardness beyond a polynomial factor, then one can derive from it a highly-parallel one-way function. To simplify the presentation, we do not attempt to derive the strongest possible result here.

Theorem 3. *Suppose a weakly black-box hardness amplification from (n, m, ε)-hard functions to $(\bar{n}, \bar{m}, \sqrt{\bar{\varepsilon}})$-hard functions can be computed in AC^0 with $\varepsilon \leq \bar{\varepsilon} \cdot \gamma$, for $\gamma < m/(\bar{m} \cdot \mathrm{poly}(\log n))$ and $\bar{\varepsilon} \geq 1/\mathrm{poly}(n)$. Then one can obtain from it a $(1 - o(1))\sqrt{\bar{\varepsilon}}$-OWF computable in NC^0.*

We will give the proof of Theorem 3 in Section 5.2. It will rely on a derandomized version of the random restriction $\mathcal{R}$ used in the previous section, which is discussed next.

5.1 Pseudo-Random Restriction

Set the parameters $\alpha = 2\varepsilon, \beta = (\log^2 s)/m, b = t^2/\varepsilon^3$ as in the previous section, and suppose $\varepsilon < \bar{\varepsilon} \cdot m/(\bar{m} \cdot \text{poly}(\log n))$. Now we describe our choice of pseudo-random restriction $\bar{\rho} : \{0,1\}^n \to \{0,1,\star\}^m$. Again, we will first hash $\{0,1\}^n$ down to a smaller space $[b]$. Following [20], we would like to replace the random hash function by a pseudo-random one, but a more careful choice is needed. Here we use the family $\bar{\mathcal{H}}$ of hash functions in Lemma 1. Then we would like to replace the random restriction $\mathcal{R}_{\alpha,\beta}^{b,m}$ by a pseudo-random one, such that it is still good with high probability. For this, we need the following two constructions. (Recall from Section 3 that a random restriction from $\mathcal{R}_{\alpha,\beta}^{b,m}$ can be generated by a circuit $W^b \in$ AC$^0 : \{0,1\}^{b\ell} \to (\{0,1,\star\}^m)^b$ using a random seed of length $b\ell = b(m+1)\ell_0$.)

- Let IND $: \{0,1\}^{r_1} \to \{0,1\}^{b\ell}$ be the generator defined as follows, with $r_1 = \text{poly}(\log n)$. First, use the input as the seed for the generator in Fact 1 to produce b random variables over $\{0,1\}^{O(\ell_0+\log m)}$ that are pairwise independent. Next, take each variable as the seed for the generator in Fact 1 to generate $m+1$ new random variables over $\{0,1\}^{\ell_0}$ that are 3-wise independent. The output of IND is the concatenation of these $b(m+1)$ new random variables over $\{0,1\}^{\ell_0}$.
- Let NIS $: \{0,1\}^{r_2} \to \{0,1\}^{b\ell}$ be Nisan's $o(\bar{\varepsilon})$-PRG for AC^0 circuits [17], with $r_2 = \text{poly}(\log n)$.

Our pseudo-random restriction $\bar{\mathcal{R}}$ is the uniform distribution over the set of restrictions $\bar{\rho}_{h,z_1,z_2}$, with $(h, z_1, z_2) \in \{0,1\}^{r_0} \times \{0,1\}^{r_1} \times \{0,1\}^{r_2}$, defined as

$$\bar{\rho}_{h,z_1,z_2}(x) = W^b \left(\text{IND}(z_1) \oplus \text{NIS}(z_2)\right)_{h(x)}.$$

Recall the definition of a good restriction from the previous section. The following says that such a pseudo-random restriction is still likely to be good.

Lemma 8. $\Pr_{\bar{\rho} \in \bar{\mathcal{R}}}[\bar{\rho} \text{ is not good}] = o(1)$.

Due to the space limitation, we defer the proof to the journal version. The idea is similar to that of Lemma 4. Now we use the generators IND and NIS, respectively, to guarantee that the two conditions of being good also fail with a small probability.

5.2 Proof of Theorem 3

Suppose there exists such a weakly black-box hardness amplification with $\varepsilon < \bar{\varepsilon} \cdot m/(\bar{m} \cdot \text{poly}(\log n))$ and $\bar{\varepsilon} \geq 1/\text{poly}(n)$. We will show how to obtain from it a hard function. The idea is the following. From Section 4, we know that for most ρ and f the function $\text{AMP}^{f\restriction\rho}$ is hard (to invert), but we do not know which ρ and f give a hard function. Our first step is to replace the random restriction ρ by a pseudo-random one $\bar{\rho}$ so that the function $\text{AMP}^{f\restriction\bar{\rho}}$ is still likely to be hard. Then we show that by replacing the random function f by a pseudo-random one $\bar{f}$, the resulting function $\text{AMP}^{\bar{f}\restriction\bar{\rho}}$ is likely to be close to $\text{AMP}^{f\restriction\bar{\rho}}$. However, having $\text{AMP}^{\bar{f}\restriction\bar{\rho}}$ close to a hard function $\text{AMP}^{f\restriction\bar{\rho}}$ does not seem sufficient to guarantee that $\text{AMP}^{\bar{f}\restriction\bar{\rho}}$ is hard. The problem is that on input $\text{AMP}^{\bar{f}\restriction\bar{\rho}}(\bar{x}) = \text{AMP}^{f\restriction\bar{\rho}}(\bar{x})$, an inverter might output $\bar{x}'$ such that $\text{AMP}^{\bar{f}\restriction\bar{\rho}}(\bar{x}) = \text{AMP}^{\bar{f}\restriction\bar{\rho}}(\bar{x}') \neq \text{AMP}^{f\restriction\bar{\rho}}(\bar{x}')$. Thus, one might succeed in inverting $\text{AMP}^{\bar{f}\restriction\bar{\rho}}$ but not $\text{AMP}^{f\restriction\bar{\rho}}$ for many such $\bar{x}$'s. We will come up with a carefully designed function that avoids this problem.

First, similar to Lemma 5, we have the following. We omit the proof here due to space limitation.

Lemma 9. *For any good $\bar{\rho} \in \mathcal{R}$, $\Pr_f[\text{AMP}^{f\restriction\bar{\rho}}$ is not $\sqrt{\bar{\varepsilon}}$-hard$] = o(\bar{\varepsilon})$.*

Next, we want to replace the random function by the following pseudo-random one. Let $\bar{\mathcal{F}}$ be the class of functions $\bar{f}_{h,z_3}$, with $h \in \bar{\mathcal{H}}$ and $z_3 \in \{0,1\}^{r_3}$, defined as

$$\bar{f}_{h,z_3}(x) = \text{NIS}'(z_3)_{h(x)},$$

where $\text{NIS}' : \{0,1\}^{r_3} \to (\{0,1\}^m)^b$ is Nisan's $o(\bar{\varepsilon})$-PRG for AC^0, with $r_3 = \text{poly}(\log n)$. One can show that it has a similar effect as the random one in the sense that for any $\bar{x} \in \{0,1\}^{\bar{n}}$, $\bar{\rho} \in \mathcal{R}$, and $\bar{y} \in \{0,1\}^{\bar{m}}$,

$$\left| \Pr_{f\in\mathcal{F}}\left[\text{AMP}^{f\restriction\bar{\rho}}(\bar{x}) = \bar{y}\right] - \Pr_{\bar{f}\in\bar{\mathcal{F}}}\left[\text{AMP}^{\bar{f}\restriction\bar{\rho}}(\bar{x}) = \bar{y}\right]\right| = o(\bar{\varepsilon}). \tag{1}$$

This is because NIS' can fool such a test.

For any good $\bar{\rho} \in \mathcal{R}$, we know by definition that there is a large subset $B \subseteq \{0,1\}^{\bar{n}}$ of inputs such that for each input in B, the output of AMP is the same for most $f \in \mathcal{F}$, and by (1), for most $\bar{f} \in \bar{\mathcal{F}}$. We would like our function to output this corresponding value for each input in B, and to output a value different from all these values for inputs not in B. We use $\bar{f}^p = (\bar{f}_1, \ldots, \bar{f}_p) \in \bar{\mathcal{F}}^p$, with $p = n^c$ for some large enough constant c, to locate one such set of inputs. Let $\text{MAJ}_{\bar{\rho},\bar{f}^p}(\bar{x})$ be the majority value in $\{\text{AMP}^{\bar{f}_1\restriction\bar{\rho}}(\bar{x}), \ldots, \text{AMP}^{\bar{f}_p\restriction\bar{\rho}}(\bar{x})\}$. Let

$$B_{\bar{\rho},\bar{f}^p} = \left\{\bar{x} \in \{0,1\}^{\bar{n}} : \Pr_{i\in[p]}\left[\text{AMP}^{\bar{f}_i\restriction\bar{\rho}}(\bar{x}) \neq \text{MAJ}_{\bar{\rho},\bar{f}^p}(\bar{x})\right] < \sqrt{\bar{\varepsilon}}\right\}.$$

Now for $\bar{\rho} \in \mathcal{R}$, $\bar{f}^p \in \bar{\mathcal{F}}^p$, and $\bar{y} \in \{0,1\}^{\bar{m}}$, define the function $\bar{A}_{\bar{\rho},\bar{f}^p,\bar{y}} : \{0,1\}^{\bar{n}} \to \{0,1\}^{\bar{m}}$ as

$$\bar{A}_{\bar{\rho},\bar{f}^p,\bar{y}}(\bar{x}) = \begin{cases} \text{MAJ}_{\bar{\rho},\bar{f}^p}(\bar{x}) & \text{if } \bar{x} \in B_{\bar{\rho},\bar{f}^p}, \\ \bar{y} & \text{otherwise.} \end{cases}$$

Call $(\bar{\rho}, \bar{f}^p, \bar{y}) \in \mathcal{R} \times \bar{\mathcal{F}}^p \times \{0,1\}^{\bar{m}}$ nice if $\bar{\rho}$ is good and the following three conditions all hold:

(a) $|B_{\bar{\rho},\bar{f}^p}| \geq (1 - o(\sqrt{\bar{\varepsilon}}))2^{\bar{n}}$.

(b) For any $\bar{x} \in B_{\bar{\rho},\bar{f}^p}$, $\Pr_{f\in\mathcal{F}}\left[\bar{A}_{\bar{\rho},\bar{f}^p,\bar{y}}(\bar{x}) \neq \text{AMP}^{f\restriction\bar{\rho}}(\bar{x})\right] = o(\bar{\varepsilon})$.

(c) For any $\bar{x} \notin B_{\bar{\rho},\bar{f}^p}$ and $\bar{x}' \in B_{\bar{\rho},\bar{f}^p}$, $\bar{A}_{\bar{\rho},\bar{f}^p,\bar{y}}(\bar{x}) \neq \bar{A}_{\bar{\rho},\bar{f}^p,\bar{y}}(\bar{x}')$.

The following lemma says that a randomly chosen $(\bar{\rho}, \bar{f}^p, \bar{y})$ is likely to be nice. Due to the space limitation, we omit the proof here.

Lemma 10. $\Pr_{\bar{\rho} \in \mathcal{R}, \bar{f}^p \in \bar{\mathcal{F}}^p, \bar{y} \in \mathcal{U}_{\bar{m}}}[(\bar{\rho}, \bar{f}^p, \bar{y}) \text{ is not nice}] = o(1).$

The following shows that a nice $(\bar{\rho}, \bar{f}^p, \bar{y})$ gives a hard function.

Lemma 11. For any nice $(\bar{\rho}, \bar{f}^p, \bar{y})$, the function $\bar{A}_{\bar{\rho}, \bar{f}^p, y}$ is $(1 - o(1))\sqrt{\bar{\varepsilon}}$-hard.

Proof. Fix any nice $(\bar{\rho}, \bar{f}^p, \bar{y})$. Consider any polynomial-size circuit $\bar{M}$ which tries to invert $\bar{A}_{\bar{\rho}, \bar{f}^p, y}$. For notational convenience, let us write $\hat{A}$ for $\bar{A}_{\bar{\rho}, \bar{f}^p, \bar{y}}$, A^f for $\text{AMP}^{f \restriction \bar{\rho}}$, and B for $B_{\bar{\rho}, \bar{f}^p}$. Suppose we sample $\bar{x}$ uniformly from $\{0,1\}^{\bar{n}}$ and f uniformly from $\mathcal{F}$. Let E be the event that $\bar{M}$ inverts $\hat{A}(\bar{x})$. Clearly, E is the union of the two events $E_1 : (\bar{M} \text{ inverts } \hat{A}(\bar{x})) \wedge (\hat{A}(\bar{x}') = A^f(\bar{x}'))$ and $E_2 : (\bar{M} \text{ inverts } \hat{A}(\bar{x})) \wedge (\hat{A}(\bar{x}') \neq A^f(\bar{x}'))$, where $\bar{x}' = \bar{M}(\hat{A}(\bar{x}))$.

First, note that the event E_1 is contained in the union of the two events $E_{1,1} : \hat{A}(\bar{x}) \neq A^f(\bar{x})$ and $E_{1,2} : \bar{M}$ inverts $A^f(\bar{x})$. From items (a) and (b), we have $\Pr_{\bar{x}, f}[E_{1,1}] \leq \Pr_{\bar{x}}[\bar{x} \notin B] + \Pr_{\bar{x}, f}[\hat{A}(\bar{x}) \neq A^f(\bar{x}) \mid \bar{x} \in B] = o(\sqrt{\bar{\varepsilon}})$. Then by Lemma 9, $\Pr_{\bar{x}, f}[E_{1,2}]$ is at most

$$\Pr_f \left[A^f \text{ is not } \sqrt{\bar{\varepsilon}}\text{-hard} \right] + \Pr_{\bar{x}, f} \left[\bar{M} \text{ inverts } A^f(\bar{x}) \mid A^f \text{ is } \sqrt{\bar{\varepsilon}}\text{-hard} \right] \leq o(\varepsilon) + 1 - \sqrt{\bar{\varepsilon}}.$$

Next, note that the event E_2 is contained in the union of the two events $E_{2,1} : \bar{x} \notin B$ and $E_{2,2} : (\bar{x} \in B) \wedge (\bar{M} \text{ inverts } \hat{A}(\bar{x})) \wedge (\hat{A}(\bar{x}') \neq A^f(\bar{x}'))$. From item (a), $\Pr_{\bar{x}}[E_{2,1}] = o(\sqrt{\bar{\varepsilon}})$. Observe that the event $E_{2,2}$ implies that $(\bar{x}' \in B) \wedge (\hat{A}(\bar{x}') \neq A^f(\bar{x}'))$, so by item (b), $\Pr_{\bar{x}, f}[E_{2,2}] = o(\varepsilon)$.

Combining these bounds together, we get $\Pr_{\bar{x}, f}[E] \leq 1 - \sqrt{\bar{\varepsilon}} + o(\sqrt{\bar{\varepsilon}})$, which proves the lemma. $\qquad \square$

Finally, define the function $\bar{A} : \{0,1\}^{\bar{n}} \times \mathcal{R} \times \bar{\mathcal{F}}^p \times \{0,1\}^{\bar{m}} \to \{0,1\}^{\bar{m}} \times \mathcal{R} \times \bar{\mathcal{F}}^p \times \{0,1\}^{\bar{m}}$ as

$$\bar{A}(\bar{x}, \bar{\rho}, \bar{f}^p, \bar{y}) = (\bar{A}_{\bar{\rho}, \bar{f}^p, \bar{y}}(\bar{x}), \bar{\rho}, \bar{f}^p, \bar{y}).$$

Note that the input length of $\bar{A}$ is at most $\text{poly}(n)$, since each $\bar{\rho} \in \mathcal{R}$ can be specified by $\text{poly}(\log n)$ bits and each $\bar{f}^p \in \bar{\mathcal{F}}^p$ can be specified by $\text{poly}(n)$ bits.

Lemma 12. The function $\bar{A}$ is $(1 - o(1))\bar{\varepsilon}$-hard.

Proof. Consider any polynomial-size circuit $\bar{M}$ which attempts to invert $\bar{A}$. Then $\Pr_{\bar{x}, \bar{\rho}, \bar{f}^p, \bar{y}}[M \text{ fails to invert } \bar{A}(\bar{x}, \bar{\rho}, \bar{f}^p, \bar{y})]$ is at least

$$\Pr_{\bar{\rho}, \bar{f}^p, \bar{y}} \left[(\bar{\rho}, \bar{f}^p, \bar{y}) \text{ nice} \right] \cdot \Pr_{\bar{x}, \bar{\rho}, \bar{f}^p, \bar{y}} \left[M \text{ fails to invert } \bar{A}(\bar{x}, \bar{\rho}, \bar{f}^p, \bar{y}) \mid (\bar{\rho}, \bar{f}^p, \bar{y}) \text{ nice} \right],$$

which by Lemma 10 & 11 is at least $(1 - o(1)) \cdot (1 - o(1))\bar{\varepsilon} = (1 - o(1))\bar{\varepsilon}$. $\qquad \square$

Since Nisan's PRG, the generator IND, and functions in $\mathcal{H}$ all can be computed in NC^1, the function $\bar{A}$ can be computed in NC^1 too. From [2], this yields a OWF in NC^0, which proves the theorem.

6   Black-Box Construction of PRG from OWF

In this section, we study the complexity-quality tradeoff for black-box constructions of pseudo-random generators from strongly one-way functions. Our result is the following.

Theorem 4. *No black-box construction of $(\bar{n}, \bar{m}, 1/5)$-PRGs from $(n, m, 1 - n^{-\log n})$-hard functions can be realized by $\mathsf{AC}(d, s)$ with $\bar{m} > \bar{n}(1 + (\log^{d+5} s)/m)$ and $s \leq 2^{m^{o(1/d)}}$. In particular, with $d = O(1)$, such construction of PRG can only have a sublinear stretch unless $s \geq 2^{m^{\Omega(1)}}$.*

Proof. Assume for the sake of contradiction that such a black-box construction realized by $\mathsf{AC}(d, s)$ exists with $\bar{m} \geq \bar{n}(1 + (\log^{d+5} s)/m)$ and $s \leq 2^{m^{o(1/d)}}$. We will show that this leads to a contradiction. The idea is similar to that in Section 4. First, we will show that for a random restriction ρ and a random function f, the function $f\restriction_\rho$ is weakly hard, and the function derived from it using direct product is strongly hard. On the other hand, suppose we have such a PRG construction. Then we will show that a random restriction can reduce the effect of a random function, and consequently there exists a distinguisher which breaks the PRG. This can then be used to invert the strongly-hard function well, and we reach a contradiction.

Let PRG be the encoding procedure and DEC the decoding procedure. Let $k = c_0 \log^{d+3} s$ for a large enough constant c_0, let $n' = n/k$ and $m' = m/k$. Note that $n', m' \geq \text{poly}(n)$ since $s \leq 2^{n^{o(1/d)}}$ and $k \leq n^{o(1)}$. Now we replace the parameters n and m in the previous sections by n' and m', and consider sampling function $f : \{0,1\}^{n'} \to \{0,1\}^{m'}$ and restriction $\rho : \{0,1\}^{n'} \to \{0,1,\star\}^{m'}$. Set the parameters:

$$\alpha = 1/\log^{d+1} s, \quad \beta = (\log^2 s)/m', \quad \text{and } b = t^2 m'.$$

Similar to Lemma 5, one can show that the function $f\restriction_\rho$ is $\Omega(\alpha)$-hard with high probability (using an almost identical proof). If $f\restriction_\rho$ is $\Omega(\alpha)$-hard, the function $f_\rho^k : \{0,1\}^{kn'} \to \{0,1\}^{km'}$ defined as $f_\rho^k(x_1, \ldots, x_k) = (f\restriction_\rho(x_1), \ldots, f\restriction_\rho(x_k))$ is $(1 - n^{-\log n})$-hard, according to [21]. Thus we have the following.

Lemma 13. *For most $\rho \in \mathcal{R}$, for any oracle algorithm M_ρ making at most $\text{poly}(n)$ oracle queries, for most $f \in \mathcal{F}$, $\Pr_{x \in \mathcal{U}_n}[M_\rho^{f_\rho^k} \text{ inverts } f_\rho^k(x)] \leq n^{-\log n}$.*

For $x, x' \in \{0,1\}^n$, let $\triangle(x, x') = |\{i \in [n] : x_i \neq x'_i\}|/n$, their relative Hamming distance. Then as in Section 4, one can show that the random restriction can reduce the effect of the random function on PRG.

Lemma 14. *For most $\rho \in \mathcal{R}$, there exists a function $\bar{G}_\rho : \{0,1\}^{\bar{n}} \to \{0,1\}^{\bar{m}}$ such that for most $f \in \mathcal{F}$, $\mathrm{E}_{\bar{x}}[\triangle(\bar{G}_\rho(\bar{x}), \mathrm{PRG}^{f_\rho^k}(\bar{x})] = \mu$ for some $\mu = O(1/m')$.*

Form such a function $\bar{G}_\rho$, one can construct a distinguisher $\bar{D}_\rho : \{0,1\}^{\bar{m}} \to \{0,1\}$ for $\mathrm{PRG}^{f_\rho^k}$, defined by $\bar{D}_\rho(\bar{y}) = 1$ if and only if there exists some $\bar{y}'$ in the image of $\bar{G}_\rho$ such that $\triangle(\bar{y}, \bar{y}') \leq 5\mu$. The we have the following, whose proof is omitted due to space limitation.

Lemma 15. *For most $\rho \in \mathcal{R}$, there exists a distinguisher $\bar{D}_\rho : \{0,1\}^{\bar{m}} \to \{0,1\}$ such that for most $f \in \mathcal{F}$, $\bar{D}_\rho$ can $1/5$-distinguish $\mathrm{PRG}^{f_\rho^k}$.*

According to the lemma, for most $\rho \in \mathcal{R}$ and $f \in \mathcal{F}$, the function $M_\rho = \mathrm{DEC}^{\bar{D}_\rho}$ achieves $\mathrm{Pr}_x[M_\rho^{f_\rho^k} \text{ inverts } f_\rho^k(x)] > n^{-\log n}$. This contradicts Lemma 13, since DEC makes at most a polynomial number of queries to the oracle. Thus we have the theorem.   □

References

1. Noga Alon, László Babai, Johan Håstad, and Rene Peralta. Some constructions of alomost k-wise independent random variables. *Random Structures and Algorithms*, 3(3), pages 289–304, 1992.
2. Benny Applebaum, Yuval Ishai, and Eyal Kushilevitz. Cryptography in NC^0. In *Proceedings of the 45th Annual IEEE Symposium on Foundations of Computer Science*, pages 166–175, 2004.
3. Ravi B. Boppana. The average sensitivity of bounded-depth circuits. *Information Processing Letters*, 63(5), pages 257–261, 1997.
4. Giovanni Di Crescenzo and Russell Impagliazzo. Security-preserving hardness-amplification for any regular one-way function. In *Proceedings of the 31st Annual ACM Symposium on Theory of Computing*, pages 169–178, 1999.
5. Merrick L. Furst, James B. Saxe, and Michael Sipser. Parity, circuits, and the polynomial-time hierarchy. *Mathematical Systems Theory*, 17(1), pages 13–27, 1984.
6. Rosario Gennaro and Luca Trevisan. Lower bounds on the efficiency of generic cryptographic constructions. In *Proceedings of the 41st Annual IEEE Symposium on Foundations of Computer Science*, pages 305–313, 2000.
7. Oded Goldreich, Russell Impagliazzo, Leonid A. Levin, Ramarathnam Venkatesan, and David Zuckerman. Security preserving amplification of hardness. In *Proceedings of the 31st Annual IEEE Symposium on Foundations of Computer Science*, pages 318–326, 1990.
8. Johan Håstad. *Computational limitations for small depth circuits*. PhD thesis, MIT Press, 1986.
9. Johan Håstad, Russel Impagliazzo, Leonid A. Levin, and Michael Luby. A pseudorandom generator from any one-way function. *SIAM Journal on Computing*, 28(4), pages 1364–1396, 1999.
10. Alexander Healy and Emanuele Viola. Constant-depth circuits for arithmetic in finite fields of characteristic two. *Electronic Colloquium on Computational Complexity*, TR05-087, 2005.
11. William Hesse, Eric Allender, and David A. M. Barrington. Uniform constant-depth threshold circuits for division and iterated multiplication. *Journal of Computer and System Sciences*, 65(4), pages 695–716, 2002.
12. Russell Impagliazzo and Steven Rudich. Limits on the provable consequences of one-way permutations. In *Proceedings of the 21st Annual ACM Symposium on Theory of Computing*, pages 44–61, 1989.
13. Nathan Linial, Yishay Mansour, and Noam Nisan. Constant depth circuits, Fourier transform, and learnability. *Journal of the ACM*, 40(3), pages 607–620, 1993.

14. Henry Lin, Luca Trevisan, and Hoeteck Wee. On hardness amplification of one-way functions. In *Proceedings of the 2nd Theory of Cryptography Conference*, pages 34–49, 2005.
15. Chi-Jen Lu, Shi-Chun Tsai, and Hsin-Lung Wu. On the complexity of hardness amplification. In *Proceedings of the 20th Annual IEEE Conference on Computational Complexity*, pages 170–182, 2005.
16. Joseph Naor and Moni Naor. Small-bias probability spaces: efficient constructions and applications. *SIAM Journal on Computing*, 22(4), pages 838–856, 1993.
17. Noam Nisan. Pseudorandom bits for constant depth circuits. *Combinatorica*, 11(1), pages 63–70, 1991.
18. Christos Papadimitriou. *Computational Complexity*. Addison-Wesley, 1994.
19. Emanuele Viola. The complexity of constructing pseudorandom generators from hard functions. *Computational Complexity*, 13(3-4), pages 147–188, 2005.
20. Emanuele Viola. On constructing parallel pseudorandom generators from one-way functions. In *Proceedings of the 20th Annual IEEE Conference on Computational Complexity*, pages 183–197, 2005.
21. Andrew Chi-Chih Yao. Theory and applications of trapdoor functions. In *Proceedings of the 23rd Annual IEEE Symposium on Foundations of Computer Science*, pages 80–91, 1982.

On Matroids and Non-ideal Secret Sharing

Amos Beimel* and Noam Livne

Dept. of Computer Science, Ben-Gurion University, Beer-Sheva 84105, Israel

Abstract. Secret-sharing schemes are a tool used in many cryptographic protocols. In these schemes, a dealer holding a secret string distributes shares to the parties such that only authorized subsets of participants can reconstruct the secret from their shares. The collection of authorized sets is called an access structure. An access structure is *ideal* if there is a secret-sharing scheme realizing it such that the shares are taken from the same domain as the secrets. Brickell and Davenport (J. of Cryptology, 1991) have shown that ideal access structures are closely related to matroids. They give a necessary condition for an access structure to be ideal – the access structure must be induced by a matroid. Seymour (J. of Combinatorial Theory B, 1992) showed that the necessary condition is not sufficient: There exists an access structure induced by a matroid that does not have an ideal scheme.

In this work we continue the research on access structures induced by matroids. Our main result in this paper is strengthening the result of Seymour. We show that in any secret sharing scheme realizing the access structure induced by the Vamos matroid with domain of the secrets of size k, the size of the domain of the shares is at least $k + \Omega(\sqrt{k})$. Our second result considers non-ideal secret sharing schemes realizing access structures induced by matroids. We prove that the fact that an access structure is induced by a matroid implies lower and upper bounds on the size of the domain of shares of subsets of participants even in non-ideal schemes (this generalized results of Brickell and Davenport for ideal schemes).

1 Introduction

Secret sharing schemes are a tool used in many cryptographic protocols. A secret sharing scheme involves a dealer who has a secret, a finite set of n participants, and a collection $\mathcal{A}$ of subsets of the set of participants called the access structure. A secret-sharing scheme for $\mathcal{A}$ is a method by which the dealer distributes shares to the parties such that: (1) any subset in $\mathcal{A}$ can reconstruct the secret from its shares, and (2) any subset not in $\mathcal{A}$ cannot reveal any partial information about the secret in the information theoretic sense. A secret sharing scheme can only exist for monotone access structures, i.e. if a subset A can reconstruct the secret, then every superset of A can also reconstruct the secret. Given any monotone

---

* Partially supported by the David and Lucile Packard Foundation grant of Matthew Franklin, and by the Frankel Center for Computer Science.

S. Halevi and T. Rabin (Eds.): TCC 2006, LNCS 3876, pp. 482–501, 2006.

access structure, Ito, Saito, and Nishizeki [22] show how to build a secret sharing scheme that realizes the access structure. Even with more efficient schemes presented since, e.g. in [5, 41, 10, 25, 44, 21], most access structures require shares of exponential size: if the domain of the secrets is binary, the shares are strings of length $2^{\Theta(n)}$, where n is the number of participants

Certain access structures give rise to very economical secret sharing schemes. A secret sharing scheme is called *ideal* if the shares are taken from the same domain as the secrets. For example, Shamir's threshold secret sharing scheme [40] is ideal. An access structure is called ideal if there is an ideal secret sharing scheme which realizes the access structure over some finite domain of secrets. Ideal access structures are interesting for a few reasons: (1) they are the most efficient secret sharing schemes as proved by [26], (2) they are most suitable for composition of secret sharing schemes, and (3) they have interesting combinatorial structure, namely, they have a matroidial structure, as proved by [11] and discussed in the next paragraph.

Brickell and Davenport [11] have shown that ideal access structures are closely related to matroids over a set containing the participants and the dealer. They give a necessary condition for an access structure to be ideal – the access structure must be induced by a matroid – and a somewhat stronger sufficient condition – the matroid should be representable over some finite field. The question of an exact characterization of ideal access structures is still open. Seymour [39] has shown that the necessary condition is not sufficient: there exists an access structure induced by a matroid that does not have an ideal scheme. The following natural open question arises: How far from ideal can access structures induced by matroids be? Is there an upper bound on the shares' size implied by being an access structure induced by a matroid? There is no better known upper bound on the share size than the $2^{O(n)}$ bound for general access structures. Most known secret sharing schemes are linear (see discussion in [2]). On one hand, the number of linear schemes with n participants, binary domain of secrets, and shares of size poly(n) is $2^{\text{poly}(n)}$. On the other hand, the number of matroids with n points is $\exp(2^{\Theta(n)})$ (see [47]) and every matroid induces at least one access structure. Thus, for most access structures induced by matroids, the size of the shares in linear secret-sharing schemes is super-polynomial. This gives some evidence that access structures induced by matroids do not have efficient secret sharing schemes for a reasonable size of domain of secrets.

Our Results. In this work we continue the research on access structures induced by matroids. Seymour [39] showed that any access structure induced by the Vamos matroid [46] is not ideal. Our main result is strengthening this result. We consider an access structure induced by the Vamos matroid and show that in any secret sharing scheme realizing this access structure with domain of the secret of size k, the size of the domain of the shares is at least $k + \Omega(\sqrt{k})$ (compared to the lower bound of $k + 1$ implied by [39]). Towards proving this lower bound, we needed to strengthen some results of [11] to non-ideal secret sharing schemes realizing access structures induced by matroids. We then needed to generalize Seymour's ideas to obtain our lower bound. We note that the upper-bound on the size of the domain of shares in a secret sharing scheme realizing the access structure induced by the Vamos matroid is poly(k), thus our work still leaves open the question of the minimal-size share domain for this access structure.

Brickell and Davenport [11] proved that the size of the domain of shares of a subset of participants in an ideal scheme is exactly determined by the size of the domain of secrets and the rank of the subset in the matroid inducing the access structure. We consider non-ideal secret sharing schemes realizing access structures induced by matroids. We prove that the fact that an access structure is induced by a matroid implies lower and upper bounds on the size of the domain of shares of subsets of participants even in non-ideal schemes. These lower and upper bounds, beside being interesting for their own, are used to prove our main result. We need both the lower bounds and the upper bounds to prove our main result – the lower bound on the size of the domain of shares in the Vamos matroid.

We prove two incomparable versions of such bounds. The first version, in Section 3, contains somewhat weaker bounds; however, this is the version we can use in the proof of our main result. The second version, in Section 5, contains bounds on the entropy of shares of subsets of participants. Entropy arguments have been used to give bounds on the size of shares in secret sharing schemes starting with [26, 12]. Specifically, entropy arguments have been used for ideal secret sharing schemes in [27]. We were not able to use the bounds we proved via entropy in the proof of our main result for technical reasons. We include them in this paper since we believe that they are interesting for their own sake. Furthermore, they might be useful in proving stronger bounds than the lower bound proved here, either for the matroid induced by the Vamos matroid, or for access structures induced by other matroids. See discussion in Example 4 at the end of this paper.

Historical Background. Secret sharing schemes were introduced by Blakley [6] and Shamir [40] for the threshold case, that is, for the case where the subsets that can reconstruct the secret are all the sets whose cardinality is at least a certain threshold. Secret sharing schemes for general access structures were introduced by Ito, Saito, and Nishizeki in [22]. More efficient schemes were presented in, e.g., [5, 41, 10, 25, 44, 21]. Originally motivated by the problem of secure information storage, secret-sharing schemes have found numerous other applications in cryptography and distributed computing, e.g., Byzantine agreement [38], secure multiparty computations [4, 13, 15], threshold cryptography [19], and access control [33].

Several lower bounds on the share size of secret-sharing schemes were obtained [5, 12, 7, 20, 18, 17]. The strongest current bound is $\Omega(n^2/\log n)$ [17] for the total size of the shares of all the participants, where n is the number of participants in the system. However, there is a huge gap between these lower bounds and the best known upper bounds of $2^{O(n)}$ for general access structures. The question of super-polynomial lower bounds on the size of shares for some (explicit or random) access structures is still open.

Ideal secret sharing schemes and ideal access structures have been first considered in [10] and have been studied extensively thereafter, e.g. in

[1, 3, 8, 11, 23, 24, 27, 29, 30, 31, 32, 34, 35, 37, 42, 43, 45, 16]. There are two common definitions for ideal access structures in the secret sharing literature. The first, that will also be used here, can be found implicitly in [11] and explicitly in [31, 1, 34, 35, 3]. The second can be found in [29, 30, 32]. Livne [28] pointed that these definitions are not necessarily equivalent. Furthermore, he proposed a candidate access structure that is ideal according to one definition but possibly is not ideal according to the stronger definition.

Organization. In Section 2 we present basic definitions of secret sharing schemes and matroids, and discuss the relation between them. In Section 3 we prove some technical lemmas concerning weak secret sharing schemes; these lemmas are used to prove our main result. In Section 4 we prove a lower bound on the size of shares in any secret sharing realizing the access structure induced by the Vamos matroid. Finally, in Section 5 we prove upper and lower bounds on the entropy of shares of subsets of participants in secret sharing schemes realizing matroid induced access structures. In Appendix A we supply some background results on the entropy function.

2   Preliminaries

In this section we define weak secret sharing schemes, review some background on matroids, and discuss the connection between secret sharing schemes and matroids.

Definition 1 (Access Structure). *Let P be a finite set of participants. A collection $\mathcal{A} \subseteq 2^P$ is* monotone *if $B \in \mathcal{A}$ and $B \subseteq C \subseteq P$ imply that $C \in \mathcal{A}$. An* access structure *is a monotone collection $\mathcal{A} \subseteq 2^P$ of non-empty subsets of P. Sets in $\mathcal{A}$ are called* authorized, *and sets not in $\mathcal{A}$ are called* unauthorized. *A set B is called a* minterm *of $\mathcal{A}$ if $B \in \mathcal{A}$ and for every $C \subsetneq B$, the set C is unauthorized. A participant is called* redundant *if there is no minterm that contains it. An access structure is called* connected *if it has no redundant participants.*

In this section we only give a relaxed definition of secret sharing scheme, which we call a *weak secret sharing scheme.* The formal definition of (strong) secret sharing scheme appears in Section 5. While in the definition of secret sharing schemes it is required that the uncertainty of the secret given the shares of an unauthorized subset of participants is the same as the a-priory uncertainty of the secret (in the information theoretic sense), here we require merely that no value of the secret could be ruled out, i.e. that each value of the secret has probability greater than zero. In particular, every secret sharing scheme is a weak secret sharing scheme. Thus, in the proof of our main result we prove lower bounds on the size of shares in weak secret sharing schemes.

Definition 2 (Weak Secret-Sharing Scheme and Weakly Ideal Access Structure). *Let P be a set of participants, and let K be a finite set of secrets. A* weak secret sharing scheme *with domain of secrets K is a matrix M whose columns are indexed by $P \cup \{p_0\}$, where $p_0 \notin P$, and with all entries in column*

p_0 from K. When the dealer wants to distribute a secret $s \in K$, it chooses a row $r \in M$ such that $M_{r,p_0} = s$, and privately communicates to each participant $p \in P$ the value $M_{r,p}$. We refer to $M_{r,p}$ as the share of participant p. Given a vector of shares $\mathbf{K_A}$, denote by $K(p_0|\mathbf{K_A})$ the possible values of the secret given that the participants in A receive the vector of shares $\mathbf{K_A}$.

We say that M realizes a weak secret sharing scheme for the access structure $\mathcal{A} \subseteq 2^P$ if the following two requirements hold:

CORRECTNESS. The secret can be reconstructed by any authorized set of participants: $|K(p_0|\mathbf{K_A})| = 1$ for any $A \in \mathcal{A}$ and every possible vector of shares $\mathbf{K_A}$ for the set A.

WEAK PRIVACY. Given a vector of shares of an unauthorized set of participants, none of the values of the secret can be ruled out: $K(p_0|\mathbf{K_A}) = K$ for any $A \notin \mathcal{A}$ and every possible vector of shares $\mathbf{K_A}$ for the set A.

If an access structure has a weak secret sharing scheme with shares' domain of every participant equal to the domain of the secret for some finite domain of secrets, we say that the access structure is weakly ideal.

Example 1. As an example, consider Shamir's threshold scheme [40]. Denote $P = \{1, \ldots, n\}$, let $t \leq n$, and define the threshold access structure $\mathcal{A}_t = \{A \subseteq P : |A| \geq t\}$. We choose some prime number $q \geq n$, and define a secret sharing scheme with domain of secrets of size q as follows. In order to distribute a secret $s \in \{0, \ldots, q - 1\}$, the dealer randomly chooses, with uniform distribution, a polynomial p of degree $t - 1$ over $\mathrm{GF}(q)$ such that $p(0) = s$. The dealer then distributes to each participant $p_i \in P$ the share $p(i)$. When an authorized subset of participants (of size at least t) wants to reconstruct the secret, it has at least t distinct points of the polynomial p, therefore it can determine p, and it can calculate $p(0)$. An unauthorized subset cannot eliminate any value of the secret. In this scheme, the matrix M contains q^t rows; a row $\langle p(0), p(1), \ldots, p(n) \rangle$ for every polynomial p of degree $t - 1$ over $\mathrm{GF}(q)$.

We next give some notations concerning weak secret sharing schemes. Given $A, B \subseteq P \cup \{p_0\}$ and $\mathbf{K_B} \in K(B)$, denote by $K(A|\mathbf{K_B})$ the set of combinations of shares the participants in A can receive given that the participants in B received the vector of shares $\mathbf{K_B}$. That is, if M' is the restriction of M to the rows such that the values in the columns in B are $\mathbf{K_B}$, then $K(A|\mathbf{K_B})$ is the set of the distinct rows in the restriction of M' to the columns in A. Given $\mathbf{K_A} \in K(A|\mathbf{K_B})$, we say that $\mathbf{K_A}$ coincides with $\mathbf{K_B}$ (that is, there is a row in M that gives to the participants in A the shares in $\mathbf{K_A}$ and to the participants in B the shares in $\mathbf{K_B}$). Of course, this relation is symmetric. We denote $K(\{v_{i_1}, v_{i_2}, \ldots, v_{i_\ell}\})$ by $K(v_{i_1}, v_{i_2}, \ldots, v_{i_\ell})$. Given sets of participants $A, B_1, \ldots, B_\ell \subseteq V$, and vectors of shares $\mathbf{K_{B_i}} \in K(B_i)$ for $1 \leq i \leq \ell$, we also denote $K(A|\mathbf{K_{B_1}}, \ldots, \mathbf{K_{B_\ell}})$ as the set of vectors of shares the (ordered) set of participants A can receive given that the participants of B_i received the shares $\mathbf{K_{B_i}}$ for $1 \leq i \leq \ell$. Given two sets of participants $A, B \subseteq V$, and a set $X_B \subseteq K(B)$ we denote $K(A|X_B) \stackrel{\text{def}}{=} \bigcup_{\mathbf{K_B} \in X_B} K(A|\mathbf{K_B})$.

2.1   Matroids

A matroid is an axiomatic abstraction of linear independence. There are several equivalent axiomatic systems to describe matroids: by independent sets, by bases, by the rank function, or, as done here, by circuits. For more background on matroid theory the reader is referred to [47, 36].

Definition 3 (Matroid). *A matroid $M = \langle V, C \rangle$ is a finite set V and a collection C of subsets of V that satisfy the following three axioms:* (C0) $\emptyset \notin C$. (C1) *If $X \neq Y$ and $X, Y \in C$, then $X \nsubseteq Y$.* (C2) *If C_1, C_2 are distinct members of C and $x \in C_1 \cap C_2$, then there exists $C_3 \in C$ such that $C_3 \subseteq (C_1 \cup C_2) \setminus \{x\}$. The elements of V are called* points, *or simply* elements, *and the subsets in C are called* circuits.

For example, let $G = (V, E)$ be an undirected graph and C be the collection of simple cycles in G. Then, (E, C) is a matroid.

Definition 4. *A subset of V is* dependent *in a matroid M if it contains a circuit. If a subset is not dependent, it is* independent. *The* rank *of a subset $A \subseteq V$, denoted* rank(A), *is the size of a maximal independent subset of A. A matroid is* connected *if for every pair of elements x, y there is a circuit containing x and y.*

The following lemma shows that a stronger statement than (C2) can be made about the circuits of a matroid. Its proof can be found, e.g., in [47, 36].

Lemma 1. *If C_1, C_2 are distinct members of C and $x \in C_1 \cap C_2$, then for any element $y \in C_1 \triangle C_2$ there exists $C_3 \in C$ such that $y \in C_3$ and $C_3 \subseteq (C_1 \cup C_2) \setminus \{x\}$.*

The following lemma, whose proof can be found in [47, 36], states that if a matroid is connected then the set of circuits through a fixed point uniquely determines the matroid.

Lemma 2. *Let e be an element of a connected matroid M and let C_e be the set of circuits of M that contain e. For $C_1, C_2 \in C_e$ define:*

$$I_e(C_1, C_2) \stackrel{def}{=} \bigcap \{C_3 : C_3 \in C_e, C_3 \subseteq C_1 \cup C_2\}$$

and

$$D_e(C_1, C_2) \stackrel{def}{=} (C_1 \cup C_2) \setminus I_e(C_1, C_2).$$

Then, all of the circuits of M that do not contain e are the minimal *sets of the form $D_e(C_1, C_2)$ where C_1 and C_2 are distinct circuits in C_e.*

2.2   Matroids and Secret Sharing

We next define the access structures induced by matroids. This definition is used to give a necessary condition for ideal access structures.

Definition 5. *Let* $\mathcal{M} = \langle V, \mathcal{C} \rangle$ *be a matroid and* $p_0 \in V$. *The* induced access structure *of* $\mathcal{M}$ *with respect to* p_0 *is the access structure* $\mathcal{A}$ *on* $P = V \setminus \{p_0\}$, *where*

$$\mathcal{A} \overset{def}{=} \{A : \text{ there exists } C_0 \in \mathcal{C} \text{ such that } p_0 \in C_0 \text{ and } C_0 \setminus \{p_0\} \subseteq A\}.$$

That is, a set is a minterm of $\mathcal{A}$ *if by adding* p_0 *to it, it becomes a circuit of* $\mathcal{M}$. *We think of* p_0 *as the dealer. We say that an access structure is* induced *from* $\mathcal{M}$, *if it is obtained by setting some arbitrary element of* $\mathcal{M}$ *as the dealer. In this case, we say that* $\mathcal{M}$ *is the* appropriate matroid *of* $\mathcal{A}$.

If a connected access structure has an appropriate matroid, then this matroid is also connected. Thus, by Lemma 2, if a connected access structure has an appropriate matroid, then this matroid is unique. Of course, not every access structure has an appropriate matroid.

We now quote some results concerning weak secret sharing schemes. Since every secret sharing scheme is, in particular, a weak secret sharing scheme, these results hold for the regular case as well. The following fundamental result, which is proved in [11], connects matroids and secret sharing schemes.

Theorem 1 ([11]). *If an access structure is weakly ideal, then it has an appropriate matroid.*

The following result, which is implicit in [11], shows the connection between the rank function of the appropriate matroid and the size of the domain of shares of sets of participants.

Lemma 3 ([11]). *Assume that the access structure* $\mathcal{A} \subseteq 2^P$ *is weakly ideal, and let* $\langle P \cup \{p_0\}, \mathcal{C} \rangle$ *be its appropriate matroid where* $p_0 \notin P$. *Let* M *be an ideal weak secret sharing scheme realizing* $\mathcal{A}$ *with domain of secrets (and shares)* K. *Then* $|K(X)| = |K|^{\mathrm{rank}(X)}$ *for any* $X \subseteq P \cup \{p_0\}$, *where* $\mathrm{rank}(X)$ *is the rank of* X *in the matroid.*

Remark 1. A corollary of Lemma 3 is that M can realize a secret sharing scheme for *any* access structure induced from $\mathcal{M}$ (i.e., with every element set as the dealer).

Example 2. Consider the threshold access structure $\mathcal{A}_t$ and Shamir's scheme [40] realizing it (see Example 1). The appropriate matroid of $\mathcal{A}_t$ is the matroid with $n + 1$ points, whose circuits are the sets of size $t + 1$ and $\mathrm{rank}(X) = \min\{|X|, t\}$. Since every t points determine a unique polynomial of degree $t - 1$, in Shamir's scheme $|K(X)| = |K|^{\min\{|X|, t\}}$, as implied by Lemma 3.

3 Secret Sharing Schemes Realizing Matroid-Induced Access Structures

We now prove some lemmas concerning weak secret sharing schemes and matroid-induced access structures with arbitrary size of shares domain. The next lemma gives a lower bound on the size of the shares of certain subsets of participants. This lemma holds for every access structure.

Lemma 4. *Let* $\mathcal{A} \subseteq 2^P$ *be an access structure,* $A, B \subseteq P$, *and* $b \in B \setminus A$ *such that* $A \cup B \in \mathcal{A}$ *and* $A \cup B \setminus \{b\} \notin \mathcal{A}$. *Denote the dealer* p_0 *and define* $K \stackrel{def}{=} K(p_0)$ *(that is,* K *is the domain of secrets). Then,* $|K(b|\mathbf{K_A})| \geq |K|$ *for any* $\mathbf{K_A} \in K(A)$.

Proof. Since $A \cup B \setminus \{b\} \notin \mathcal{A}$, by the privacy requirement, for any $\mathbf{K_{A \cup B \setminus \{b\}}} \in K(A \cup B \setminus \{b\})$,

$$K(p_0|\mathbf{K_{A \cup B \setminus \{b\}}}) = K. \tag{1}$$

Since $A \cup B$ is authorized, by the correctness requirement, for any $\mathbf{K_{A \cup B}} \in K(A \cup B)$,

$$|K(p_0|\mathbf{K_{A \cup B}})| = 1. \tag{2}$$

Furthermore, $K(p_0|\mathbf{K_{A \cup B \setminus \{b\}}}) = \bigcup_{\mathbf{K_b} \in K(b|\mathbf{K_{A \cup B \setminus \{b\}}})} K(p_0|\mathbf{K_{A \cup B \setminus \{b\}}}, \mathbf{K_b})$ for any $\mathbf{K_{A \cup B \setminus \{b\}}} \in K(A \cup B \setminus \{b\})$. Since, by (2), every set in this union is of size one, and since, by (1), the size of the union is $|K|$, there are at least $|K|$ sets in the union. Hence $|K(b|\mathbf{K_{A \cup B \setminus \{b\}}})| \geq |K|$. Define $\mathbf{K_A}$ as the restrictions of the vector $\mathbf{K_{A \cup B \setminus \{b\}}}$ to the set A. Since $K(b|\mathbf{K_{A \cup B \setminus \{b\}}}) \subseteq K(b|\mathbf{K_A})$, the lemma follows.   □

Lemma 5. *Let* $\mathcal{M} = \langle P \cup \{p_0\}, \mathcal{C} \rangle$ *be the appropriate matroid of an access structure* $\mathcal{A} \subseteq 2^P$, *and let* $C \in \mathcal{C}$ *such that* $p_0 \in C$. *Let* $A \subseteq P \cup \{p_0\}$ *and* $D \subseteq P$ *such that* $A \cap D = \emptyset$. *If* $A \cup D \subsetneq C$, *then* $|K(A|\mathbf{K_D})| \geq |K|^{|A|}$ *for every* $\mathbf{K_D} \in K(D)$.

Proof. We will prove the lemma by induction on $|A|$. If $|A| = 0$, the claim is trivial. For the induction step, let $a \in A$. Since $A \cup D \subsetneq C$, we have $A \cup D \setminus \{a\} \subsetneq C$. By the induction hypothesis, $|K(A \setminus \{a\}|\mathbf{K_D})| \geq |K|^{|A|-1}$. Therefore, it is sufficient to prove that $|K(A|\mathbf{K_D})| \geq |K||K(A \setminus \{a\}|\mathbf{K_D})|$ for some $a \in A$. If $p_0 \in A$, then we choose $a = p_0$. Note that $A \cup D \setminus \{p_0\}$ is unauthorized. If this is not the case, then $A \cup D$ contains a circuit C_0 which contains p_0. But since $A \cup D$ is properly contained in C, it follows that C_0 is properly contained in C, a contradiction to Axiom (C1) of the matroids. Now since $A \cup D \setminus \{p_0\}$ is unauthorized, by the privacy requirement, $|K(p_0|\mathbf{K_{A \setminus \{p_0\}}}, \mathbf{K_D})| = |K|$ for any $\mathbf{K_{A \setminus \{p_0\}}} \in K(A \setminus \{p_0\})$. Therefore, $|K(A|\mathbf{K_D})| = |K||K(A \setminus \{a\}|\mathbf{K_D})|$, which concludes this case.

If $p_0 \notin A$, then we choose an arbitrary $a \in A$. Now $A \cup D \setminus \{a\}$ is unauthorized. Otherwise $(A \cup D \setminus \{a\}) \cup \{p_0\}$ contains a circuit C_0 which contains p_0. But since $A \cup D$ is properly contained in C, it follows that C_0 is properly contained in C, a contradiction. Moreover, $A \cup D \subseteq C \setminus \{p_0\}$, and $C \setminus \{p_0\}$ is authorized. Therefore, by Lemma 4, $|K(a|\mathbf{K_{A \setminus \{a\}}}, \mathbf{K_D})| \geq |K|$ for any $\mathbf{K_{A \setminus \{a\}}} \in K(A \setminus \{a\})$. It follows that $|K(A)| \geq |K||K(A \setminus \{a\}|\mathbf{K_D})|$, which concludes the proof.   □

In the ideal case, by Lemma 3 we have an upper bound on the share domain of every subset of participants that form a circuit in the appropriate matroid. In the non-ideal case we cannot apply Lemma 3. Lemma 6 will be used to overcome this difficulty. To prove Lemma 6, we need the following claim.

Claim. Let N and K be 2 finite sets, where $|N| = m, |K| = k$, and $m \geq k$. Let f_1, f_2 be functions from a subset of N onto K. Then

$$| \{\langle x_1, x_2 \rangle : x_1, x_2 \in N, f_1(x_1) = f_2(x_2)\} | \leq k - 1 + (m - k + 1)^2.$$

Proof. Without loss of generality, assume $K = \{1, 2, \ldots, k\}$. For $1 \leq i \leq k$ define $a_i \overset{\text{def}}{=} |f_1^{-1}(i)|$ and $b_i \overset{\text{def}}{=} |f_2^{-1}(i)|$. Then $\sum_{1 \leq i \leq k} a_i \leq m$ and $\sum_{1 \leq i \leq k} b_i \leq m$, since both these sums are the size of the domains of the functions. Moreover, since both these functions are onto K, we have $a_i \geq 1$ and $b_i \geq 1$ for all $1 \leq i \leq k$. Thus, $1 \leq a_i \leq m - k + 1$ and $1 \leq b_i \leq m - k + 1$ for every $1 \leq i \leq k$. From the definitions $| \{\langle x_1, x_2 \rangle : x_1, x_2 \in N, f_1(x_1) = f_2(x_2)\} | = \sum_{1 \leq i \leq k} a_i b_i$. Assume without loss of generality that a_1 is maximal in $a_1, a_2, \ldots, a_k$. Then

$$\sum_{i=1}^{k} a_i b_i \leq \sum_{i=1}^{k} (a_i + a_1(b_i - 1)) \leq a_1(m - k) + m \leq k - 1 + (m - k + 1)^2.$$

We note that this claim is tight as shown in the following simple example: $f_1(i) = f_2(i) = i$ for $1 \leq i \leq k$ and $f_1(i) = f_2(i) = 1$ for $k + 1 \leq i \leq m$. □

Lemma 6. *Let $\mathcal{A}$ be an access structure, and denote the dealer by p_0. Let $A \subseteq P$ and $b_1, b_2 \in P$ such that $A \notin \mathcal{A}$, $A \cup \{b_1\} \in \mathcal{A}$, and $A \cup \{b_2\} \in \mathcal{A}$. Consider a weak secret sharing scheme realizing $\mathcal{A}$ in which the size of the domain of the secret is k, and the size of the domain of the shares of each participant is bounded by m. Then $|K(b_1, b_2|\mathbf{K_A})| \leq k - 1 + (m - k + 1)^2$ for any $\mathbf{K_A} \in K(A)$.*

Proof. Fix some $\mathbf{K_A} \in K(A)$. Since $A \cup \{b_1\} \in \mathcal{A}$, given $\mathbf{K_A}$, any $\mathbf{K_{b_1}} \in K(b_1|\mathbf{K_A})$ determines the secret. Moreover, since $A \notin \mathcal{A}$, given $\mathbf{K_A}$ any value of the secret is possible. Therefore, $\mathbf{K_A}$ induces a function from $K(b_1|\mathbf{K_A})$ onto $K(p_0)$. Formally, the set of 2-vectors $K(b_1, p_0|\mathbf{K_A})$ viewed as a set of ordered pairs form a function with $K(b_1|\mathbf{K_A})$ as its domain and $K(p_0)$ as its image. Denote this function by f_1. Similarly $\mathbf{K_A}$ also induces a function from $K(b_2|\mathbf{K_A})$ onto $K(p_0)$. Denote this function by f_2.

Given $\mathbf{K_A}$, consider any $\langle x_1, x_2 \rangle \in K(b_1, b_2|\mathbf{K_A})$. There is a row r in M that gives to the participants in A the values in $\mathbf{K_A}$, and to b_1, b_2 the values x_1, x_2 respectively. However, $M_{r,p_0} = f_1(x_1) = f_2(x_2)$. Informally, given $\mathbf{K_A}$, the shares x_1 and x_2 must "agree" on the secret. Thus, $f_1(x_1) = f_2(x_2)$ for every $\langle x_1, x_2 \rangle \in K(b_1, b_2|\mathbf{K_A})$. Since both f_1 and f_2 are onto $K(p_0)$, and since the domain of both functions is bounded by m, Claim 3 implies that $|K(b_1, b_2|\mathbf{K_A})| \leq k - 1 + (m - k + 1)^2$. □

4   Secret Sharing and the Vamos Matroid

In this section we prove lower bounds on the size of shares in secret sharing schemes realizing an access structure induced by the Vamos matroid. The Vamos matroid [46] is the smallest known matroid that is non-representable over any field, and is also non-algebraic (for more details on these notions see [47, 36]).

Definition 6 (The Vamos Matroid). *The Vamos matroid $\mathcal{V}$ is defined on the set $V = \{v_1, v_2, \ldots, v_8\}$, and its independent sets are all the sets of cardinality ≤ 4 except for five, namely $\{v_1, v_2, v_3, v_4\}$, $\{v_1, v_2, v_5, v_6\}$, $\{v_3, v_4, v_5, v_6\}$, $\{v_3, v_4, v_7, v_8\}$, and $\{v_5, v_6, v_7, v_8\}$.*

Note that these 5 sets are all the unions of two pairs from $\{v_1, v_2\}$, $\{v_3, v_4\}$, $\{v_5, v_6\}$, and $\{v_7, v_8\}$, excluding $\{v_1, v_2, v_7, v_8\}$. The five sets listed in Definition 6 are circuits, a fact that will be used later. Seymour [39] proved that any access structure induced by the Vamos matroid is non-ideal. In this section we strengthen this result.

Definition 7 (The Access Structure V_8). *The access structure V_8 is the access structure induced by the Vamos matroid with respect to v_8.[1] That is, in this access structure, a set of participants is a minterm, if this set together with v_8 is a circuit in $\mathcal{V}$.*

Example 3. We next give examples of authorized and non-authorized sets in V_8. The set $\{v_3, v_4, v_7\}$ is authorized, since $\{v_3, v_4, v_7, v_8\}$ is a circuit. The circuit $\{v_1, v_2, v_3, v_4\}$ is unauthorized, since the set $\{v_1, v_2, v_3, v_4, v_8\}$ does not contain a circuit that contains v_8. To check this, we first note that this 5-set itself cannot be a circuit, since it contains the circuit $\{v_1, v_2, v_3, v_4\}$. Second, the only circuit it contains is $\{v_1, v_2, v_3, v_4\}$, which does not contain v_8. The set $\{v_1, v_2, v_3, v_4, v_5\}$ is authorized, since $\{v_1, v_2, v_3, v_5, v_8\}$ is a circuit (as well as $\{v_1, v_2, v_4, v_5, v_8\}$, $\{v_1, v_3, v_4, v_5, v_8\}$, and $\{v_2, v_3, v_4, v_5, v_8\}$).

For a given secret sharing scheme realizing V_8, assume $|K(v_8)| = k$, and $|K(v_i)| < m$ for $1 \leq i \leq 7$, i.e., the size of the domain of the secrets is k and the size of the domain of the shares of each participant is upper bounded by m. By [26], for every secret sharing scheme, the size of the domain of shares of each non redundant participant is at least the size of the domain of secrets, that is, $m \geq k$. Seymour [39] proved that the Vamos access structure is not ideal, that is, $m \geq k + 1$. We next strengthen this result. To achieve the lower bound on m here, we fix an arbitrary $\langle x_1, x_2 \rangle \in K(v_1, v_2)$ and calculate an upper bound on the size of $K(v_7, v_8 | x_1, x_2)$ as a function of m and k. By Lemma 5 the size of this set is at least k^2, and thus, we achieve a lower bound on m.

Fix some arbitrary $\langle x_1, x_2 \rangle \in K(v_1, v_2)$, and define $A \overset{\text{def}}{=} K(v_5, v_6 | x_1, x_2)$ (see Fig. 1). Our goal is to count the possible shares $\{v_7, v_8\}$ can receive given $\langle x_1, x_2 \rangle$. We upper bound this value by considering all the possible shares $\{v_5, v_6\}$ can receive given $\langle x_1, x_2 \rangle$ (namely, the set A), and considering the union of all the sets $K(v_7, v_8 | y_5, y_6)$ for all the vectors $\langle y_5, y_6 \rangle$ in A. We first bound the size of A.

---

[1] There are two non-isomorphic access structures induced by the Vamos matroid. The access structure V_8 is isomorphic to the access structure obtained by setting v_1, v_2, or v_7 as the dealer. The other access structure is obtained by setting v_3, v_4, v_5, or v_6 as the dealer.

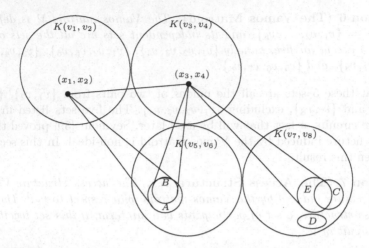

Fig. 1. Sets in the proof of Theorem 2. Circles denote sets, and points denote elements in the sets. Two elements are connected if they coincide. A line connects an element and a subset, if the subset is the set of all elements that coincide with the element. For example, $\langle x_1, x_2 \rangle$ and A are connected with lines because A is the set of elements in $K(v_5, v_6)$ that can coincide with $\langle x_1, x_2 \rangle$.

Lemma 7. $|A| \leq m^{\frac{k-1+(m-k+1)^2}{k}}$.

Proof. Fix an arbitrary $x_3 \in K(v_3|x_1, x_2)$. The set $\{v_1, v_2, v_3\}$ is unauthorized (since $\{v_1, v_2, v_3, v_8\}$ is independent). Since $\{v_1, v_2, v_3, v_5, v_8\}$ is a circuit, $\{v_1, v_2, v_3, v_5\}$ is authorized. Similarly, the set $\{v_1, v_2, v_3, v_6\}$ is authorized too. Since $|K(v_5)| \leq m$, and $|K(v_6)| \leq m$, by Lemma 6,

$$|K(v_5, v_6|x_1, x_2, x_3)| \leq k - 1 + (m - k + 1)^2. \tag{3}$$

We now bound the size of $K(v_3, v_5, v_6|x_1, x_2)$. Notice that

$$K(v_3, v_5, v_6|x_1, x_2) = \bigcup_{y_3 \in K(v_3|x_1, x_2)} \{\langle y_3, y_5, y_6 \rangle : \langle y_5, y_6 \rangle \in K(v_5, v_6|x_1, x_2, y_3)\}.$$

That is, we count all the y_3's that coincide with $\langle x_1, x_2 \rangle$, and for each such y_3 we count all the $\langle y_5, y_6 \rangle$'s that coincide with $\langle x_1, x_2, y_3 \rangle$. Since (3) is true for any $y_3 \in K(v_3|x_1, x_2)$, the size of each set in the union is at most $k - 1 + (m - k + 1)^2$, and since $|K(v_3|x_1, x_2)| \leq |K(v_3)| \leq m$, there are at most m sets in the union. Therefore,

$$|K(v_3, v_5, v_6|x_1, x_2)| \leq m \left(k - 1 + (m - k + 1)^2\right). \tag{4}$$

On the other hand,

$$K(v_3, v_5, v_6|x_1, x_2) = \bigcup_{\langle y_5, y_6 \rangle \in A} \{\langle y_3, y_5, y_6 \rangle : \langle y_3 \rangle \in K(v_3|x_1, x_2, y_5, y_6)\}.$$

Since $\{v_1, v_2, v_5, v_6\}$ is unauthorized, but $\{v_1, v_2, v_3, v_5, v_6\}$ is authorized, by Lemma 4 each set in this union is of size at least k. Since all these sets are disjoint, and by (4), there are at most $\frac{m}{k}(k-1+(m-k+1)^2)$ sets in this union. We conclude that $|A| \le m(k-1+(m-k+1)^2)/k$. $\square$

In addition to x_1, x_2, fix an arbitrary vector $\langle x_3, x_4 \rangle \in K(v_3, v_4 | x_1, x_2)$. We define, in addition to A, a set of vectors $B \stackrel{\text{def}}{=} K(v_5, v_6 | x_1, x_2, x_3, x_4)$. That is, the set A is the shares $\{v_5, v_6\}$ can receive given $\langle x_1, x_2 \rangle$, and B is the shares $\{v_5, v_6\}$ can receive given $\langle x_1, x_2, x_3, x_4 \rangle$. Clearly $B \subseteq A$.

To count the vectors in $K(v_7, v_8 | A)$, we define two sets $C \stackrel{\text{def}}{=} K(v_7, v_8 | B) = \bigcup_{\langle y_5, y_6 \rangle \in B} K(v_7, v_8 | y_5, y_6)$, and $D \stackrel{\text{def}}{=} K(v_7, v_8 | A \setminus B)$.

Lemma 8. $|C| + |D| \le m - k^2 + \left(\frac{k-1+(m-k+1)^2}{k} \right) m^2$.

Proof. First we show that $|C| \le |B|(m-k) + m$. Define $E \stackrel{\text{def}}{=} K(v_7, v_8 | x_3, x_4)$. Informally, we will show that E is small and for any $\langle y_5, y_6 \rangle \in B$ the set E contains a large portion of $K(v_7, v_8 | y_5, y_6)$.

Since $\{v_3, v_4, v_7\}$ is authorized and by the correctness requirement, given $\langle x_3, x_4 \rangle$ any $y_7 \in K(v_7 | x_3, x_4)$ determines the secret, therefore

$$|E| = |K(v_7, v_8 | x_3, x_4)| = |K(v_7 | x_3, x_4)| \le |K(v_7)| \le m. \tag{5}$$

Since $\{v_3, v_4, v_5, v_6\}$ is unauthorized, for any $\langle y_5, y_6 \rangle \in K(v_5, v_6 | x_3, x_4)$, and in particular for any $\langle y_5, y_6 \rangle \in B$, we have $|K(v_8 | x_3, x_4, y_5, y_6)| = k$. Therefore,

$$|K(v_7, v_8 | x_3, x_4, y_5, y_6)| \ge k$$

for any $\langle y_5, y_6 \rangle \in B$. Clearly, $K(v_7, v_8 | x_3, x_4, y_5, y_6) \subseteq E$ for any $\langle y_5, y_6 \rangle \in B$. Since $K(v_7, v_8 | x_3, x_4, y_5, y_6) \subseteq K(v_7, v_8 | y_5, y_6)$ we conclude that for any $\langle y_5, y_6 \rangle \in B$,

$$|K(v_7, v_8 | y_5, y_6) \cap E| \ge k. \tag{6}$$

That is, given any $\langle y_5, y_6 \rangle \in B$, at least k elements from $K(v_7, v_8 | y_5, y_6)$ are in E. We now upper bound the number of elements of $K(v_7, v_8 | y_5, y_6)$ not in E. To do this, we bound the total number of elements in $K(v_7, v_8 | y_5, y_6)$ for any $\langle y_5, y_6 \rangle$. Since $\{v_5, v_6, v_7\}$ is authorized, by the correctness requirement, given $\langle y_5, y_6 \rangle$ any $y_7 \in K(v_7 | y_5, y_6)$ determines the secret, therefore for any $\langle y_5, y_6 \rangle \in K(v_5, v_6 | x_1, x_2)$,

$$|K(v_7, v_8 | y_5, y_6)| = |K(v_7 | y_5, y_6)| \le |K(v_7)| \le m. \tag{7}$$

With (6), we conclude that for any $\langle y_5, y_6 \rangle \in B$,

$$|K(v_7, v_8 | y_5, y_6) \setminus E| \le m - k. \tag{8}$$

That is, given any $\langle y_5, y_6 \rangle \in B$, at most $m - k$ elements from $K(v_7, v_8 | y_5, y_6)$ are not in E. Thus, by (5),

$$|C| \le |E| + |B|(m-k) \le m + |B|(m-k). \tag{9}$$

Furthermore, by (7), given any element in $A \setminus B$, the number of possible shares for $\{v_7, v_8\}$ is at most m. Therefore,

$$|D| \leq |A \setminus B|m. \tag{10}$$

Finally, since $\{v_1, v_2, v_3, v_4\}$ is unauthorized, but $\{v_1, v_2, v_3, v_4, v_5\}$ is authorized, by Lemma 4 we have $|K(v_5|x_1, x_2, x_3, x_4)| \geq k$, and therefore

$$|B| = |K(v_5, v_6|x_1, x_2, x_3, x_4)| \geq |K(v_5|x_1, x_2, x_3, x_4)| \geq k. \tag{11}$$

We now complete the proof of the lemma:

$$|C| + |D| \leq m + |B|(m - k) + |A \setminus B|m = m - k|B| + |A|m$$
$$\leq m - k^2 + \left(\frac{k - 1 + (m - k + 1)^2}{k}\right)m^2.$$

The first inequality follows (9) and (10). The equality is implied by the fact that $B \subseteq A$. The last inequality follows (11) and Lemma 7.   □

Lemma 9. *For every* $\langle x_1, x_2 \rangle \in K(v_1, v_2)$

$$K(v_7, v_8|x_1, x_2) \leq m - k^2 + m^2\frac{k + (m - k + 1)^2}{k}.$$

Proof. We first show that $K(v_7, v_8|x_1, x_2) \subseteq K(v_7, v_8|A)$. Take any $\langle y_7, y_8 \rangle \in K(v_7, v_8|x_1, x_2)$. The vector $\langle x_1, x_2, y_7, y_8 \rangle$ can be extended to a vector

$$\langle x_1, x_2, y_5, y_6, y_7, y_8 \rangle \in K(v_1, v_2, v_5, v_6, v_7, v_8).$$

Thus, $\langle y_5, y_6, y_7, y_8 \rangle \in K(v_5, v_6, v_7, v_8)$ and $\langle y_5, y_6 \rangle \in K(v_5, v_6|x_1, x_2) = A$, and so $\langle y_7, y_8 \rangle \in K(v_7, v_8|A)$. Consequently,

$$|K(v_7, v_8|x_1, x_2)| \leq |K(v_7, v_8|A)| \leq |C| + |D|$$
$$= m - k^2 + m^2\frac{k - 1 + (m - k + 1)^2}{k}$$
$$< m - k^2 + m^2\frac{k + (m - k + 1)^2}{k}.   □$$

Theorem 2. *For any* $0 < \lambda < 1$ *there exists* $k_0 \in \mathbb{N}$, *such that for any secret sharing scheme realizing* V_8, *with the domain of the secret of size* $k > k_0$, *the size of at least one share domain is larger then* $k + \lambda\sqrt{k}$.

Proof. Let $0 < \lambda < 1$, and assume $m \leq k + \lambda\sqrt{k}$. Since $\{v_1, v_2, v_3, v_7, v_8\}$ is a circuit in the Vamos matroid and $\{v_1, v_2, v_7, v_8\} \subseteq \{v_1, v_2, v_3, v_7, v_8\}$, by Lemma 5, $|K(v_7, v_8|x_1, x_2)| \geq k^2$ for every $\langle x_1, x_2 \rangle \in K(v_1, v_2)$ in any secret sharing scheme realizing V_8. Combining this with Lemma 9, we have that if m is an upper bound on the size of the domain of the shares, then the following inequality must hold:

$$\left(m - k^2 + m^2\frac{k + (m - k + 1)^2}{k}\right) \geq k^2. \tag{12}$$

Since the left side of Inequality (12) increases as m increases, and since $m \leq k + \lambda\sqrt{k}$, we can substitute m with $k + \lambda\sqrt{k}$. After rearranging we have:

$$k^2 \leq k + \lambda\sqrt{k} - k^2 + (k^2 + \lambda^2 k + 2\lambda k\sqrt{k})\frac{k + (\lambda\sqrt{k} + 1)^2}{k} = \lambda^2 k^2 + p_\lambda(k),$$

where $p_\lambda(k)$ is a polynomial of degree 1.5 in k. Thus, $1 - \lambda^2 \leq \frac{p_\lambda(k)}{k^2}$. Since $1 - \lambda^2 > 0$ and since $\lim_{k \to \infty} \frac{p_\lambda(k)}{k^2} = 0$, we conclude that there exists some $k_0 \in \mathbb{N}$, such that for any $k \geq k_0$, Inequality (12) does not hold. We conclude that for any $k \geq k_0$, at least one participant must have domain of shares larger than $k + \lambda\sqrt{k}$. □

5   Upper and Lower Bounds for Matroid Induced Access Structures

In this section we define secret sharing schemes using the entropy function, as done in [26, 12], and then use some tools from information theory to prove lower and upper bounds on sizes of shares' domains of subsets of participants in matroid induced access structures. The purpose of these lemmas is to generalize Lemma 3 of [11] to non-ideal secret sharing schemes for matroid induced access structures. These lemmas were not used in the proof of Theorem 2, but they might be used to prove a stronger bound than the lower bound proved here. For a review on the notions from information theory, see Appendix A. We start by defining (strong) secret sharing schemes using the entropy function.

Definition 8 (Distribution Scheme). *Let P be a set of participants, and $p_0 \notin P$ be a special participant called the dealer. Furthermore, let K be a finite set of secrets. A distribution scheme Σ with domain of secrets K is a pair $\langle \{M^s\}_{s \in K}, \{\Pi_s\}_{s \in K} \rangle$, where $\{M^s\}_{s \in K}$ is a family of matrices whose columns are indexed by P, and Π_s is a probability distribution on the rows of M^s for each $s \in K$. When the dealer wants to distribute a secret $s \in K$, it chooses according to the probability distribution Π_s on M^s, a row $r \in M^s$, and privately communicates to each participant $p \in P$ the value $M^s_{r,p}$. We refer to $M^s_{r,p}$ as the share of participant p.*

Let A be an access structure whose set of participants is P, and denote the dealer by p_0. Assume that Σ is a distribution scheme for A. Any probability distribution on the domain of secrets, together with the scheme Σ, induces a probability distribution on $K(A)$, for any subset $A \subseteq P$. We denote the random variable taking values in $K(A)$ according to this probability distribution by S_A, and denote the random variable taking values in K according to the probability distribution on the secrets by S. Note that the random variable taking values in $K(A \cup B)$ can be written either as $S_{A \cup B}$ or as $S_A S_B$.

Definition 9 (Secret Sharing Scheme). *A distribution scheme is a secret sharing scheme realizing an access structure A if the following two requirements hold:*

CORRECTNESS. *The secret can be reconstructed by any authorized set.*

$$A \in \mathcal{A} \implies H(S|S_A) = 0. \tag{13}$$

PRIVACY. *Every unauthorized set can learn nothing about the secret (in the information theoretic sense) from its shares. Formally,*

$$A \notin \mathcal{A} \implies H(S|S_A) = H(S). \tag{14}$$

5.1   Lower Bounds on the Entropy of Shares of Subsets

Let $p_0 \in V$ and let $\langle V, \mathcal{C} \rangle$ be the appropriate matroid of an access structure $\mathcal{A} \subseteq 2^{V \setminus \{p_0\}}$. In Theorem 3 we prove a lower bound on the entropy of the shares of any subset of V. To prove Theorem 3 we prove two lemmas. The first lemma, which generalizes Lemma 4, makes no use of the fact that $\mathcal{A}$ has an appropriate matroid; it is proven for any access structure.

Lemma 10. *Let $A, B \subseteq V \setminus \{p_0\}$ and $b \in B \setminus A$ such that $A \cup B \in \mathcal{A}$ and $A \cup B \setminus \{b\} \notin \mathcal{A}$. Then, $H(S_b|S_A) \geq H(S)$.*

Proof.

$$
\begin{aligned}
H(S_b|S_A) &\geq H(S_b|S_A S_{B \setminus \{b\}}) && \text{(from (21))} \\
&= H(S|S_{A \cup B}) + H(S_b|S_{A \cup B \setminus \{b\}}) && \text{(since } H(S|S_{A \cup B}) = 0 \text{ by (13))} \\
&= H(S_b S|S_{A \cup B \setminus \{b\}}) && \text{(from (22))} \\
&= H(S_b|S_{A \cup B \setminus \{b\}} S) + H(S|S_{A \cup B \setminus \{b\}}) && \text{(from (22))} \\
&\geq H(S) && \text{(from (19) and (14), and because } A \cup B \setminus \{b\} \notin \mathcal{A}) \quad \square
\end{aligned}
$$

A consequence of Lemma 10 is that if $I \subseteq A$ for a minterm A and $i \in I$, then $H(S_i|S_{I \setminus \{i\}}) \geq H(S)$. Combining this with (20), we get by induction that $H(S_I) \geq |I|H(S)$. We now generalize this claim for every independent set. We next prove a lemma on matroids that will be used to prove this generalization. The next lemma, intuitively, states that in every independent set of participants there is a participant that is needed in order to reveal the secret. That is, there is a minterm (minimal authorized set) such that omitting this participant from the union of the independent set and the minterm results in an unauthorized set. Define $\mathcal{C}_0 \stackrel{\text{def}}{=} \{C \in \mathcal{C} : p_0 \in C\}$.

Lemma 11. *For every independent set $I \subseteq V \setminus \{p_0\}$, there exists $i \in I$ and $C \in \mathcal{C}_0$ such that $i \in C$ and there is no $C_1 \in \mathcal{C}_0$ such that $C_1 \subseteq C \cup I \setminus \{i\}$.*

Proof. For every $i \in I$ there exists a circuit $C \in \mathcal{C}_0$ such that $i \in C$ (since $\mathcal{M}$ is connected). Choose an $i \in I$ and $C \in \mathcal{C}_0$ such that $i \in C$ and for every $C' \in \mathcal{C}_0$

$$I \cap C' \neq \emptyset \implies C' \setminus I \text{ is not properly contained in } C \setminus I. \tag{15}$$

(Note that not necessarily every i can be chosen.) We claim that such i and C satisfy the conditions of the lemma, namely, there is no $C_1 \in \mathcal{C}_0$ such that

$C_1 \subseteq C \cup I \setminus \{i\}$. Assume towards contradiction that this is not the case, and choose $C_1 \in C_0$ such that

$$C_1 \subseteq C \cup I \setminus \{i\}. \tag{16}$$

We have $C_1 \cap I \neq \emptyset$, otherwise $C_1 \subsetneq C$ in a contradiction to Axiom (C1) of the matroids. Therefore, by (15) and (16), $C \setminus I = C_1 \setminus I$. Let $c \in C \setminus I = C_1 \setminus I$. Such c exists, otherwise we have $C_1 \subseteq I$ and so I is not independent. Since $c \in C \cap C_1$, by Axiom (C2) there exists a circuit $C_2 \subseteq C \cup C_1 \setminus \{c\}$. We have $C_2 \cap I \neq \emptyset$ (otherwise $C_2 \subsetneq C$), and so $p_0 \notin C_2$ (otherwise we have a contradiction to (15)), and so $p_0 \in C \setminus C_2$. Moreover, $C_2 \setminus I \neq \emptyset$, otherwise $C_2 \subseteq I$ contradicting the independence of I. So there exists $c' \in C_2 \setminus I$, where $c' \neq c$. Since $C_2 \setminus I \subseteq C \setminus I$ we have that $c' \in C \setminus I$, so $c' \in C_2 \cap C$, and therefore there is a circuit $C_3 \in C_0$ such that $C_3 \subseteq C_2 \cup C \setminus \{c'\}$ (from Lemma 1). Since $c' \in C \setminus C_3$, we have $C_3 \setminus I \subsetneq C \setminus I$. Moreover $C_3 \cap I \neq \emptyset$ (otherwise $C_3 \subsetneq C$), and therefore C_3 is a contradiction to the minimality of $C \setminus I$ (defined in (15)), so C and i satisfy the conditions of the lemma.   □

Theorem 3. *For every* $A \subseteq V, H(S_A) \geq \mathrm{rank}(A)H(S)$.

Proof. From the definition of the rank function and (20), it is sufficient to show that the statement holds for any independent set $I \subseteq V$. Since every subset of an independent set in a matroid is independent, by induction, it is sufficient to show that for every independent set I there exists $i \in I$ such that $H(S_I) \geq H(S) + H(S_{I \setminus \{i\}})$. If $p_0 \in I$ then since I is independent it contains no circuit, and particularly no circuit which contains p_0. Therefore, $I \setminus \{p_0\}$ contains no minterm, and we have $I \setminus \{p_0\} \notin \mathcal{A}$. Now by (14) $H(S|S_{I \setminus \{p_0\}}) = H(S)$, and we have $H(S_I) = H(S|S_{I \setminus \{p_0\}}) + H(S_{I \setminus \{p_0\}}) = H(S) + H(S_{I \setminus \{p_0\}})$. Otherwise, by Lemma 11 for every independent set $I \subseteq V \setminus \{p_0\}$, there exists $i \in I$ and $C \in C_0$ such that $i \in C$ and there is no $C_1 \in C_0$ such that $C_1 \subseteq C \cup I \setminus \{i\}$. Therefore, we have $I \cup C \setminus \{i, p_0\} \notin \mathcal{A}$, but $I \cup C \setminus \{p_0\} \in \mathcal{A}$, and so, by Lemma 10, $H(S_i|S_{I \setminus \{i\}}) \geq H(S)$ and we have $H(S_I) = H(S_i|S_{I \setminus \{i\}}) + H(S_{I \setminus \{i\}}) \geq H(S) + H(S_{I \setminus \{i\}})$.   □

5.2   Upper Bounds on the Entropy of Shares of Subsets

In Lemma 15 we prove an upper bound on the entropy of "the last element of a circuit," that is, we prove an upper bound on the entropy of an element in a circuit, given the rest of the elements, and assuming an upper bound on the entropy of the participants. This enables us to prove, in Theorem 4, upper bounds on the entropy of shares of subsets. Let $\mathcal{M}$ and Σ be as above, and assume that, for every $v \in V \setminus \{p_0\}$, $H(S_v) \leq (1 + \lambda)H(S)$ for some $\lambda \geq 0$. Define $C_0 \stackrel{\text{def}}{=} \{C \in \mathcal{C} : p_0 \in C\}$ as above. For lack of space, some proofs in this section are omitted.

Lemma 12. *For every* $C \in C_0$ *and* $c \in C$, $H(S_c|S_{C \setminus \{c\}}) \leq \lambda H(S)$.

Lemma 13. *For every* $C \in \mathcal{C} \setminus C_0$ *and* $c \in C$, *there exists* $C_1, C_2 \in C_0$ *such that* $C = D_{p_0}(C_1, C_2)$, *and* $c \in C_1 \setminus C_2$ *(where* $D_{p_0}(C_1, C_2)$ *is defined in Lemma 2).*

Proof. From Lemma 2 there are $C_1, C_2 \in \mathcal{C}_0$ such that $C = D_{p_0}(C_1, C_2)$. If $c \in C_1 \triangle C_2$ we are done. Otherwise, $c \in C_1 \cap C_2$. By the definition of $D_{p_0}(C_1, C_2)$, there must be some $C_3 \in \mathcal{C}_0$ such that $C_3 \subseteq C_1 \cup C_2 \setminus \{c\}$ (otherwise $c \in I_{p_0}(C_1, C_2)$), and so we have $c \in C_1 \setminus C_3$. We now prove that $C = D_{p_0}(C_1, C_3)$ and this completes the proof. Notice that $C_1 \cup C_3 \subseteq C_1 \cup C_2$, from which we get $I_{p_0}(C_1, C_2) \subseteq I_{p_0}(C_1, C_3)$. Therefore, $D_{p_0}(C_1, C_3) \subseteq D_{p_0}(C_1, C_2)$. By Lemma 2, the circuits which do not contain p_0 are the *minimal* sets of the form $D_{p_0}(C_1, C_2)$ for all $C_1, C_2 \in \mathcal{C}_0$. Thus, since $D_{p_0}(C_1, C_2)$ is a circuit, $D_{p_0}(C_1, C_3) = D_{p_0}(C_1, C_2)$, and therefore $C = D_{p_0}(C_1, C_3)$ as desired. □

Lemma 14. *Let* $C = D_{p_0}(C_1, C_2)$, *and* $I = I_{p_0}(C_1, C_2) \setminus \{p_0\}$. *Then,*

$$H(S_I | S_C) \geq |I| H(S).$$

Lemma 15. *For every* $C \in \mathcal{C} \setminus \mathcal{C}_0$ *such that* $C = D_{p_0}(C_1, C_2)$, *and* $c \in C$ *such that* $c \in C_1 \setminus C_2$, $H(S_c | S_{C \setminus \{c\}}) \leq |I_{p_0}(C_1, C_2)| \lambda H(S)$. *In particular, for every* $C \in \mathcal{C} \setminus \mathcal{C}_0$ *and* $c \in C$, $H(S_c | S_{C \setminus \{c\}}) \leq n \lambda H(S)$.

Theorem 4. *Let* $\mathcal{M} = \langle V, \mathcal{C} \rangle$ *be a connected matroid where* $|V| = n + 1$, $p_0 \in V$ *and let* $\mathcal{A}$ *be the induced access structure of* $\mathcal{M}$ *with respect to* p_0. *Furthermore, let* Σ *be a secret sharing scheme realizing* $\mathcal{A}$, *and let* $\lambda \geq 0$ *be such that* $H(S_v) \leq (1 + \lambda) H(S)$ *for every* $v \in V \setminus \{p_0\}$. *Then, for every* $A \subseteq V$

$$H(S_A) \leq \mathrm{rank}(A)(1 + \lambda)H(S) + (|A| - \mathrm{rank}(A))\lambda n H(S).$$

The previous theorem is useful only when $\lambda \leq 1/(n-1)$ (otherwise the bound $H(S_A) \leq |A|(1 + \lambda)H(S)$ is better). We next show how to apply these results to the Vamos matroid, considered in Section 4. We then compare this bound to the bound we achieve in Section 4.

Example 4. Consider a secret sharing scheme realizing the Vamos access structure V_8. Recall that the set $\{v_1, v_2, v_5, v_6\}$ is a circuit of the Vamos matroid. By Theorem 4, $H(S_{\{v_1, v_2, v_5, v_6\}}) \leq (3 + 10\lambda)H(S)$ (by using Lemma 15 we can get a better dependence of λ). Since $\{v_1, v_2\}$ is independent, by Theorem 3, $H(S_{\{v_1, v_2\}}) \geq 2H(S)$. Thus, by (20), $H(S_{\{v_5, v_6\}} | S_{\{v_1, v_2\}}) = H(S_{\{v_1, v_2, v_5, v_6\}}) - H(S_{\{v_1, v_2\}}) \leq (1 + 10\lambda)H(S)$. Thus, there is a vector of shares $\langle x_1, x_2 \rangle$ such that

$$H\left(S_{\{v_5, v_6\}} | S_{\{v_1, v_2\}} = \langle x_1, x_2 \rangle\right) \leq (1 + 10\lambda)H(S).$$

Now, we consider a specific setting of the parameters. Let us assume that there are k possible secrets distributed uniformly, and the size of the domain of shares of each participant is at most $2k$. Thus, $H(S) = \log k$ and, by (18), $H(S_{v_i}) \leq \log(2k) = H(S) + 1 = (1 + 1/\log k)H(S)$. Thus, there is a vector of shares $\langle x_1, x_2 \rangle$ such that $H(S_{\{v_5, v_6\}} | S_{\{v_1, v_2\}} = \langle x_1, x_2 \rangle) \leq (1 + 10/\log k)H(S)$. This should be compared to the bound of approximately $2H(S)$ we can achieve by Lemma 7 and (18). Notice that in the proof of our main result we prove in Lemma 7 an *upper bound* on the number of possible shares of $\{v_5, v_6\}$ given a vector of shares $\langle x_1, x_2 \rangle$ of $\{v_1, v_2\}$. Here we give a better upper-bound on the entropy of the shares of $\{v_5, v_6\}$ given a vector of shares $\langle x_1, x_2 \rangle$ of $\{v_1, v_2\}$.

We do not know how to use this better bound on the entropy in the proof of the lower bound for the Vamos access structure.

Acknowledgment. We thank Enav Weinreb for very helpful discussions.

References

1. A. Beimel and B. Chor. Universally ideal secret sharing schemes. *IEEE Trans. on Information Theory*, 40(3):786–794, 1994.
2. A. Beimel and Y. Ishai. On the power of nonlinear secret-sharing. *SIAM Journal on Discrete Mathematics*, 19(1):258-280, 2005.
3. A. Beimel, T. Tassa, and E. Weinreb. Characterizing ideal weighted threshold secret sharing. In *TCC 2005*, vol. 3378 of *LNCS*, pages 600–619. 2005.
4. M. Ben-Or, S. Goldwasser, and A. Wigderson. Completeness theorems for non-cryptographic fault-tolerant distributed computations. In *Proc. of the 20th STOC*, pages 1–10, 1988.
5. J. Benaloh and J. Leichter. Generalized secret sharing and monotone functions. In *CRYPTO '88*, vol. 403 of *LNCS*, pages 27–35. 1990.
6. G. R. Blakley. Safeguarding cryptographic keys. In *Proc. of the 1979 AFIPS National Computer Conference*, pages 313–317. 1979.
7. C. Blundo, A. De Santis, L. Gargano, and U. Vaccaro. On the information rate of secret sharing schemes. *Theoretical Computer Science*, 154(2):283–306, 1996.
8. C. Blundo, A. De Santis, D. R. Stinson, and U. Vaccaro. Graph decomposition and secret sharing schemes. *J. of Cryptology*, 8(1):39 64, 1995.
9. C. Blundo, A. De Santis, and A. Giorgio Gaggia. Probability of shares in secret sharing schemes. *Inform. Process. Lett.*, 72:169 175, 1999.
10. E. F. Brickell. Some ideal secret sharing schemes. *Journal of Combin. Math. and Combin. Comput.*, 6:105–113, 1989.
11. E. F. Brickell and D. M. Davenport. On the classification of ideal secret sharing schemes. *J. of Cryptology*, 4(73):123–134, 1991.
12. R. M. Capocelli, A. De Santis, L. Gargano, and U. Vaccaro. On the size of shares for secret sharing schemes. *J. of Cryptology*, 6(3):157–168, 1993.
13. D. Chaum, C. Crépeau, and I. Damgård. Multiparty unconditionally secure protocols. In *Proc. of the 20th STOC*, pages 11–19, 1988.
14. T. M. Cover and J. A. Thomas. *Elements of Information Theory*. John Wiley & Sons, 1991.
15. R. Cramer, I. Damgård, and U. Maurer. General secure multi-party computation from any linear secret-sharing scheme. In *EUROCRYPT 2000*, vol. 1807 of *LNCS*, pages 316–334. 2000.
16. R. Cramer, V. Daza, I. Gracia, J. Jimenez Urroz, G. Leander, J. Marti-Farre, and C. Padro. On codes, matroids and secure multi-party computation from linear secret sharing schemes. In *CRYPTO 2005*, vol. 3621 of *LNCS*, pages 327–343. 2005.
17. L. Csirmaz. The dealer's random bits in perfect secret sharing schemes. *Studia Sci. Math. Hungar.*, 32(3–4):429–437, 1996.
18. L. Csirmaz. The size of a share must be large. *J. of Cryptology*, 10(4):223–231, 1997.
19. Y. Desmedt and Y. Frankel. Shared generation of authenticators and signatures. In *CRYPTO '91*, vol. 576 of *LNCS*, pages 457–469. 1992.

20. M. van Dijk. On the information rate of perfect secret sharing schemes. *Designs, Codes and Cryptography*, 6:143–169, 1995.

21. M. van Dijk. A linear construction of secret sharing schemes. *Designs, Codes and Cryptography*, 12(2):161–201, 1997.

22. M. Ito, A. Saito, and T. Nishizeki. Secret sharing schemes realizing general access structure. In *Proc. of Globecom 87*, pages 99–102, 1987. Journal version: Multiple assignment scheme for sharing secret. *J. of Cryptology*, 6(1):15-20, 1993.

23. W. Jackson and K. M. Martin. Perfect secret sharing schemes on five participants. *Designs, Codes and Cryptography*, 9:267–286, 1996.

24. W. Jackson, K. M. Martin, and C. M. O'Keefe. Ideal secret sharing schemes with multiple secrets. *J. of Cryptology*, 9(4):233–250, 1996.

25. M. Karchmer and A. Wigderson. On span programs. In *Proc. of the 8th Structure in Complexity Theory*, pages 102–111, 1993.

26. E. D. Karnin, J. W. Greene, and M. E. Hellman. On secret sharing systems. *IEEE Trans. on Information Theory*, 29(1):35–41, 1983.

27. K. Kurosawa, K. Okada, K. Sakano, W. Ogata, and S. Tsujii. Nonperfect secret sharing schemes and matroids. In *EUROCRYPT '93*, vol. 765 of *LNCS*, pages 126–141. 1994.

28. N. Livne. On matroids and non-ideal secret sharing. Master's thesis, Ben-Gurion University, Beer-Sheva, 2005.

29. J. Martí-Farré and C. Padró. Secret sharing schemes on access structures with intersection number equal to one. In *SCN '02*, vol. 2576 of *LNCS*, pages 354–363. 2002.

30. J. Martí-Farré and C. Padró. Secret sharing schemes with three or four minimal qualified subsets. *Designs, Codes and Cryptography*, 34(1):17–34, 2005.

31. K. M. Martin. *Discrete Structures in the Theory of Secret Sharing*. PhD thesis, University of London, 1991.

32. P. Morillo, C. Padró, G. Sáez, and J. L. Villar. Weighted threshold secret sharing schemes. *Inform. Process. Lett.*, 70(5):211–216, 1999.

33. M. Naor and A. Wool. Access control and signatures via quorum secret sharing. *IEEE Transactions on Parallel and Distributed Systems*, 9(1):909–922, 1998.

34. S.-L. Ng. A representation of a family of secret sharing matroids. *Designs, Codes and Cryptography*, 30(1):5–19, 2003.

35. S.-L. Ng and M. Walker. On the composition of matroids and ideal secret sharing schemes. *Designs, Codes and Cryptography*, 24(1):49 – 67, 2001.

36. J. G. Oxley. *Matroid Theory*. Oxford University Press, 1992.

37. C. Padró and G. Sáez. Secret sharing schemes with bipartite access structure. *IEEE Trans. on Information Theory*, 46:2596–2605, 2000.

38. M. O. Rabin. Randomized Byzantine generals. In *Proc. of the 24th FOCS*, pages 403–409, 1983.

39. P. D. Seymour. On secret-sharing matroids. *J. of Combinatorial Theory, Series B*, 56:69–73, 1992.

40. A. Shamir. How to share a secret. *Communications of the ACM*, 22:612–613, 1979.

41. G. J. Simmons, W. Jackson, and K. M. Martin. The geometry of shared secret schemes. *Bulletin of the ICA*, 1:71–88, 1991.

42. J. Simonis and A. Ashikhmin. Almost affine codes. *Designs, Codes and Cryptography*, 14(2):179–197, 1998.

43. D. R. Stinson. An explication of secret sharing schemes. *Designs, Codes and Cryptography*, 2:357–390, 1992.

44. D. R. Stinson. Decomposition construction for secret sharing schemes. *IEEE Trans. on Information Theory*, 40(1):118–125, 1994.

45. T. Tassa. Hierarchical threshold secret sharing. In *TCC 2004*, vol. 2951 of *LNCS*, pages 473–490. 2004.
46. P. Vamos. On the representation of independence structures. Unpublished manuscript, 1968.
47. D. J. A. Welsh. *Matroid Theory*. Academic press, London, 1976.

A   Basic Definitions from Information Theory

We review here the basic concepts of Information Theory used in this paper. For a complete treatment of this subject, see [14]. All the logarithms here are of base 2.

Given a probability distribution $\{p(x)\}_{x \in X}$ on a finite set X, we define the *entropy* of X, denoted $H(X)$, as

$$H(X) \overset{\text{def}}{=} - \sum_{x \in X, p(x) > 0} p(x) \log p(x).$$

Given two sets X and Y and a joint probability distribution $\{p(x, y)\}_{x \in X, y \in Y}$ on $X \times Y$, we define the *conditioned entropy of X given Y* as

$$H(X|Y) \overset{\text{def}}{=} - \sum_{y \in Y, p(y) > 0} \sum_{x \in X, p(x|y) > 0} p(y)p(x|y) \log p(x|y).$$

We also define the *conditioned mutual information $I(X; Y|Z)$* between X and Y given Z as

$$I(X; Y|Z) \overset{\text{def}}{=} H(X|Z) - H(X|YZ). \tag{17}$$

For convenience, in the following text, when dealing with the entropy function XY will denote $X \cup Y$. We will use the following properties of the entropy function. Let X, Y, and Z be random variables, and $|X|$ be the size of the support of X (the number of values with probability greater than zero).

$$0 \le H(X) \le \log |X| \tag{18}$$

$$0 \le H(X|Y) \le H(X) \tag{19}$$

$$H(Y) \le H(XY) = H(X|Y) + H(Y) \le H(X) + H(Y) \tag{20}$$

$$H(X|Y) \ge H(X|YZ) \tag{21}$$

$$H(XY|Z) = H(X|YZ) + H(Y|Z) \tag{22}$$

$$I(X; Y|Z) = H(X|Z) - H(X|YZ) = H(Y|Z) - H(Y|XZ) = I(Y; X|Z) \tag{23}$$

Secure Computation with Partial Message Loss*

Chiu-Yuen Koo

Dept. of Computer Science, University of Maryland, College Park, USA
cykoo@cs.umd.edu

Abstract. Existing communication models for multiparty computation (MPC) either assume that *all* messages are delivered eventually or *any* message can be lost. Under the former assumption, MPC protocols guaranteeing output delivery are known. However, this assumption may not hold in some network settings like the Internet where messages can be lost due to denial of service attack or heavy network congestion. On the other hand, the latter assumption may be too conservative. Known MPC protocols developed under this assumption have an undesirable feature: output delivery is not guaranteed even only one party suffers message loss.

In this work, we propose a communication model which makes an intermediate assumption on message delivery. In our model, there is a common global clock and three types of parties: (i) Corrupted parties (ii) Honest parties with connection problems (where message delivery is never guaranteed) (iii) Honest parties that can normally communicate but may lose a small fraction of messages at each round due to transient network problems. We define secure MPC under this model. Output delivery is guaranteed to type (ii) parties that do not abort and type (iii) parties.

Let n be the total number of parties, e_f and e_c be upper bounds on the number of corrupted parties and type (ii) parties respectively. We construct a secure MPC protocol for $n > 4e_f + 3e_c$. Protocols for broadcast and verifiable secret sharing are constructed along the way.

1 Introduction

The study of secure multiparty computation (MPC) was initiated by Yao[26] in the 2-party setting and extended to the multiparty setting by Goldreich, Micali and Wigderson[16]. Roughly speaking, a set of n parties wants to jointly compute a function g of their (private) inputs. However, up to t of them are corrupted by an adversary. The requirements are that (i) non-corrupted parties obtain their outputs and (ii) the adversary learns nothing but the outputs of the corrupted parties.

Several communication models are considered in the current body of work, giving rise to different feasibility results, as follows.

1. *Common global clock and message delivery within bounded time:* This is the synchronous model. Protocol execution consists of rounds. It is assumed that the duration of one round is sufficient for a message to be sent and delivered from one party to another.

---

* Supported by NSF Trusted Computing grant #0310751.

S. Halevi and T. Rabin (Eds.): TCC 2006, LNCS 3876, pp. 502–521, 2006.

If $t < n/3$, then information-theoretically secure MPC protocols exist[3, 11] in a point-to-point network. If we assume the existence of a broadcast channel, then information-theoretically secure MPC protocols exist for $t < n/2$[22, 12, 1].

If the definition of secure MPC is relaxed such that non-corrupted parties are not guaranteed to receive their outputs (i.e., without guarantee on output delivery), then computationally secure MPC protocols exist[16, 17] for any $t < n$.

2. *Eventual message delivery without bound on delivery time:* This is the asynchronous model with eventual message delivery assumption. There is no assumed bound on the network latency. Under this communication model, information-theoretically secure MPC protocols exist for $t < n/3$[2, 4, 7].[1]

3. *No eventual message delivery:* This is known as the "message blocking" model and is the communication model considered in [8, 10]. There is no assumed bound on the network latency and any message can be lost. Assuming a common reference string, an Universally Composable[8] secure multiparty computation protocol (without guarantee on output delivery) exists for any $t < n$ in the computational setting[10].

4. *Local clock with bounded drift and message delivery within bounded time:* This is the timing model considered in [18]. It is assumed that the local clocks of the parties proceed at the same rate and an upper bound is known on the network latency. Under this model, an universally composable secure multiparty computation protocol (without guarantee on output delivery) exists for any $t < n$ in the computational setting[18]. It is worth mentioning that the security of the protocol holds as long as the assumption about local clocks holds. The network latency assumption is used to ensure non-triviality of the protocol.

We note that the results mentioned in models 1-3 hold for an adaptive adversary while the result mentioned in model 4 holds for a static adversary.

1.1   Applicability of Existing Models to General Network Setting

We discuss whether the assumptions made in the existing models are applicable to general network settings like the internet.

– Message delivery within bounded time: In a real network setting, an upper bound on the network latency can be very large. Even worse, as noted in [18], any reasonable bound is unlikely to hold, and hence the security of a protocol can be compromised. Consider the following scenario: n parties execute the protocol by Ben-Or, Goldwasser and Wigderson[3] and the adversary corrupts $n/6+1$ parties. In the first round, parties share their private inputs using a $(n/3+1)$-out-n secret sharing scheme. Suppose an uncorrupted party

---

[1] We note that this result cannot be translated to the synchronous model since the definition of secure asynchronous computation is different from the synchronous counterpart.

p_i suffers network congestion: $n/6$ uncorrupted parties fail to receive their shares of p_i's input in time. These $n/6$ uncorrupted parties will broadcast complaints in the next round and p_i will reveal the corresponding shares in the round after the next round.[2] The adversary will then have enough shares to reconstruct the private input of p_i.

- Eventual message delivery: This is a weaker assumption than the previous one. Under this assumption, we can have secure MPC protocols that guarantee output delivery. However, given the current form of the internet, this assumption may still be too strong. Messages sent to a party can be lost due to denial of service(DoS) attack[19] or heavy network congestion.
- Messages can be blocked: Under the message blocking model, known MPC protocols have an undesirable feature: output delivery cannot be guaranteed when one party suffers message loss (even if all parties are honest). An adversary can then have a simple strategy to prevent parties from receiving the outputs: carrying out a denial of service attack on a chosen party.

  The assumption that *every* message can be lost may be conservative. Depending on the scenarios, it may be reasonable to assume a few, but not many, parties suffer from from DoS attack or network congestion at the same time.

In this work, we propose a communication model that is an intermediate between eventual message delivery model and message blocking model. Under this model, we construct a secure multiparty computation protocol that guarantees output delivery to all parties except those experience severe message loss.

1.2   Our Model

Three assumptions are made in our model:

1. *A common global clock:* Given the current state of art for modern network, we believe it is reasonable to assume the existence of a common global clock[3]. Protocol execution consists of rounds. Every party knows when a round starts and ends.
2. *Three type of parties:* We assume that there are three types of parties in the network:
   - Corrupted parties who are controlled by an adversary.
   - Honest parties with connection problems (where message delivery is never guaranteed). An honest party that fails to contact the common global clock belongs to this category.
   - Honest parties that can normally communicate but at each round they may fail to send/receive a small fraction of messages due to transient network problems.

---

[2] We remark that this is a simplification of what actually happens.

[3] For instance, NIST has provided such a service: http://tf.nist.gov/timefreq/service/its.htm?

From now on, we will address the second type of parties as constrained parties and the third type of parties as fault-free parties. A constrained party does not necessarily realize that it suffers from connection problem.

3. *A time bound related to network latency:* We assume that there is a time bound Δ such that
   - Duration of a round is equal to Δ.
   - Any fault-free party p can successfully communicate with all but δ fraction of fault-free parties in any round (i.e., message transmission from it to another party takes time less than Δ and vice versa). The set of fault-free parties p can communicate with may vary each round.

1.3   Discussions of Our Model and Related Works

The first assumption is also made in the synchronous model.

The second assumption is inspired by the previous work in distributed computing. Thambidurai and Park[24], and independently Garay and Perry[14], introduced the concept of hybrid failure model which allows a mix of different degrees of failures. Our second assumption can be viewed as assuming a mix of omission[21, 20] and Byzantine failures, which is a more general assumption than the previous ones considered in the literature.

In [23, 25, 6], protocols for broadcast and consensus are considered in a communication model where the edges can be faulty (in addition to faulty nodes). In [9], Canetti, Halevi and Herzberg considered the problem of maintaining authenticated communication over untrusted communication channels, where an adversary can corrupt links for transient periods of time. In both lines of work, there is a bound on the number of faulty links connected to an honest party. On the other hand, our model captures the scenario when an honest party suffers from *arbitrary* message loss.

We believe the third assumption is more realizable than assuming a time bound on the maximum latency, and yet it is sufficient to guarantee output delivery of fault-free parties.

If constrained parties are absent, then our model is reduced to the standard synchronous model. On the other hand, if fault-free parties are absent, then our model can be viewed as a message blocking model with time-out.

1.4   Our Results

We define secure multiparty computation under our communication model. We defer the formal definition to the next section, but roughly speaking, we require the followings: (i) the fault-free parties always receive their outputs; (ii) if a constrained party does not realize that it suffers from connection problems, then it will receive its output, otherwise, it aborts; (iii) the adversary learns nothing but the outputs of the corrupted parties.

We consider an adaptive, rushing adversary. The adversary is adaptive in the sense that in any round, it can turn a fault-free party into a constrained party or into a corrupted party. In each round, the adversary has the power to decide the set of messages a constrained party can receive/send and the set of parties a fault-free party can communicate with (subject to the δ constraint). We assume each pair of parties is connected by a secure channel.

Let e_f be an upper bound on the number of corrupted parties; e_c be an upper bound on the number of constrained parties; n be the total number of parties. For $\delta < \frac{1}{6}$, we construct an information-theoretically secure MPC protocol for $n > 4e_f + 3e_c$. We define broadcast and verifiable secret sharing (VSS) under the new communication model along the way. The results are as follows (all results hold for $\delta < \frac{1}{6}$):

- A broadcast protocol for $n > 3e_f + 2e_c$; we also have a different broadcast protocol for $n \geq 3e_f + 2e_c$ if it is known that $e_f, e_c \geq 1$.
- A VSS protocol for $n > 4e_f + 3e_c$.

2   Notations, Definitions and Overview

2.1   Notations

We use two special symbols ϕ and $\perp$ in the paper. ϕ is a special symbol denoting the failure of receiving a valid message. During a protocol execution, if a party p_r fails to receive a valid message from another party p_s, then we say p_r receives ϕ. If p_r is not a corrupted parties and p_r receives ϕ from p_s, assuming $\delta = 0$, then one of the followings must hold:

1. p_s is a corrupted party while p_r is a fault-free party or a constrained party.
2. p_s is a constrained party while p_r is a fault-free party or a constrained party.
3. p_s is a fault-free party while p_r is a constrained party.

If $\delta > 0$, then it is possible that both p_s and p_r are fault-free parties and yet p_r receives ϕ from p_s.

$\perp$ is a special symbol denoting abortion. In any (sub-)protocol execution, if a party outputs $\perp$, then the party aborts the entire execution at that point. We also assume that a constrained party outputs $\perp$ if it fails to contact the common global clock. Our protocols are designed in such a way that *only* a constrained party will output $\perp$. For clarity, when we refer to an uncorrupted party p_i in our proofs, unless otherwise specified, we implicitly assume that p_i is a fault-free party or a constrained party who has not aborted at that point (i.e., we do not consider a constrained party that has already aborted).

2.2   Definition of Secure Computation

We define the secure multiparty computation using the ideal/real world paradigm. We assume the function g is defined in a way such that if the input of an party is ϕ, then the evaluation of g does not depend on the input of that party and the corresponding output for that party is $\perp$. We also assume that if the output of a party is not equal to $\perp$, then its output contains the set of parties which input ϕ. As a warm-up, we will start with the case of a *non-adaptive* adversary.

The Non-adaptive Case

Ideal world: In the ideal world, there is a trusted party (TP) which carries out the evaluation of the function g. The evaluation consists of the following steps:

1. The adversary chooses a set of corrupted parties $\mathcal{P}^f$, modifies their inputs (which can become ϕ) and sends them to the trusted party; the adversary chooses three sets of constrained parties $\mathcal{P}^{c_1}$, $\mathcal{P}^{c_2}$ and $\mathcal{P}^{c_3}$;[4] the trusted party receives the private inputs from parties that are not in $\mathcal{P}^f \cup \mathcal{P}^{c_1}$; the trusted party receives ϕ as the input from the constrained parties in $\mathcal{P}^{c_1}$.
2. The trusted party evaluates g. Let $\mathcal{P}^\phi$ be the set of parties which send ϕ to the trusted party.
3. The adversary receives the outputs of the parties in $\mathcal{P}^f$; parties in $\mathcal{P}^{c_1} \cup \mathcal{P}^{c_2}$ receive $\perp$; other parties receive their outputs (note that the outputs contain the set $\mathcal{P}^\phi$).

Real world: In the real world, the parties execute a protocol Π to evaluate g. Corrupted parties may deviate from the protocol in an arbitrary manner. Messages delivery are controlled by the adversary, subject to the constraints in our communication model.

At the end of the protocol execution, the fault-free parties and the constrained parties output their outputs from Π; the real-world adversary generates an output (which can depend on the information it gathers during the execution of Π).

We say a protocol Π is a secure multiparty computation protocol if the following holds: for every real world adversary $\mathcal{A}$, there exists an ideal world adversary $\mathcal{I}$ with the same set of corrupted parties and same set of constrained parties such that (1) and (2) are indistinguishable:

1. The output of $\mathcal{I}$ and the outputs of fault-free parties and constrained parties in the ideal world.
2. The output of $\mathcal{A}$ and the outputs of fault-free parties and constrained parties in the real world.

The Adaptive Case. The only difference between this case and the non-adaptive case is the definition of the ideal world. In the *ideal world,*

1. The adversary chooses a set of corrupted parties $\mathcal{P}^{f_1}$ (in an adaptive manner), modifies their inputs (which can become ϕ) and sends them to the trusted party; the adversary chooses a set of constrained parties $\mathcal{P}^{c_1}$; the trusted party receives the private inputs from parties that are not in $\mathcal{P}^{f_1} \cup \mathcal{P}^{c_1}$; the trusted party receives ϕ as the input from the constrained parties in $\mathcal{P}^{c_1}$.
2. The trusted party evaluates g. Let $\mathcal{P}^\phi$ be the set of parties who send ϕ to the trusted party.

---

[4] A constrained party in $\mathcal{P}^{c_3}$ is not distinguishable from a fault-free party in the ideal world, the set $\mathcal{P}^{c_3}$ is defined due to a subtle technical point.

3. The adversary receives the outputs of the parties in $\mathcal{P}^{f_1}$; depending on the outputs it received, the adversary can (adaptively) choose to corrupt a new set of parties $\mathcal{P}^{f_2}$ and obtain their outputs; the adversary then chooses two sets of constrained parties $\mathcal{P}^{c_2}$ and $\mathcal{P}^{c_3}$; parties in $\mathcal{P}^{c_1}$ and $\mathcal{P}^{c_2}$ receive $\perp$; other uncorrupted parties receive their outputs.

At the end, the fault-free parties and the constrained parties output what they receive from the trusted party; the ideal-world adversary generates an output.

2.3   Overview

In section 5, we construct a MPC protocol for $n > 4e_f + 3e_c$ which uses broadcast(section 3) and VSS(section 4) as sub-protocols. We note that we assume $\delta = 0$(i.e. a fault-free party can always successfully receive messages from other fault-free parties) in sections 3, 4 and 5. In section 6, we discuss how to extend our results to the case of $\delta < \frac{1}{6}$. In section 7, we conclude and state some open problems.

3   Broadcast

3.1   Definitions

In broadcast, there is a distinguished sender p_s with input v. We can define broadcast using the ideal/real world paradigm by specifying the ideal world as follows:

1. The adversary chooses a set of corrupted parties $\mathcal{P}^f$ and a set of constrained parties $\mathcal{P}^c$.
   - If $p_s \in \mathcal{P}^f$, then the adversary obtains the value v and p_s sends a possibly modified value v' (v' can be equal to ϕ) to the trusted party.
   - If $p_s \in \mathcal{P}^c$, then the adversary sends a flag b to p_s; if b is equal to true, then p_s sends v to the trusted party else p_s sends ϕ.
   - If $p_s \notin \mathcal{P}^f \cap \mathcal{P}^c$, then p_s sends v to the trusted party.
2. The trusted party sends the value it received from p_s to all parties not in $\mathcal{P}^c$; it sends ϕ to all parties in $\mathcal{P}^c$.

However, the above definition is an overkill for our application (as a building block for VSS and MPC). If p_s is a constrained party and it fails to broadcast the message (i.e., p_s receives $b =$ false from the adversary and the honest parties receive ϕ), then the adversary should obtain no knowledge about the message. We need some kind of secret sharing to achieve this in the real world. However, our intention is to construct a VSS protocol using broadcast, not the other way round! To solve this dilemma, we observe that in our applications of broadcast, privacy is not an issue. In more details, what we need are as follows:

- If the sender is corrupted, then all fault-free parties receive the same value v' (v' can be equal to ϕ). A constrained party should receive v' or ϕ.

- If the sender is constrained, then all fault-free parties receive the same value v' (v' has to be equal to v or ϕ). A constrained party should receive v' or ϕ.
- If the sender is fault-free, then all fault-free parties receive the same value $v' = v$. A constrained party should receive v' or ϕ.

It will ease the designing of the VSS protocol if we place a more stringent requirement: if a constrained party does not receive v', then it aborts (i.e. outputting $\perp$).

More formally, we say broadcast is achieved if the followings hold:

- Agreement: If an uncorrupted party outputs $v'(\neq\perp)$[5], then all fault-free parties output v'.
- Correctness: If the sender is uncorrupted and an uncorrupted party outputs $v'(\neq\perp)$, then $v' = v$ or $v' = \phi$. If the sender is fault-free, then all fault-free parties output v.

If the sender is constrained, it is possible that all fault-free parties output ϕ. We assume a constrained sender will abort if it outputs ϕ in the broadcast protocol. We reduce the broadcast problem to the *consensus* problem. In consensus, every party p_i has an input v_i. Consensus is achieved if the followings hold:

- Agreement: All fault-free parties output a common value v. A constrained party either outputs v or $\perp$(abort).
- Persistence: If all fault-free parties have the same input v', then $v = v'$.

For the rest of the section, we focus on the case where the domain of values is restricted to $\{0,1\}$. It is easy to see that if we have a broadcast protocol for a single bit, then we can have broadcast protocol for a ℓ-bit string by running the bit-protocol ℓ-times sequentially. For a bit b, we denote its complement by $\bar{b}$.

3.2   Reducing Broadcast to Consensus

Under our communication model, broadcast cannot be achieved by simply having the sender sending its value to all parties and then running the consensus protocol. The problem is that the sender could be constrained and fault-free parties may not receive the value from the sender. Nevertheless, we show the following:

Lemma 1. *Consensus implies broadcast.*

Proof. We construct a broadcast protocol from any consensus protocol:

1. Sender p_s sends the bit v to all parties. Let b_i be the bit p_i received from p_s. If p_i does not receive anything from the sender, then sets $b_i = 0$.
2. Parties execute the consensus protocol. Each party p_i enters the protocol with input b_i and let b_i^* be the output. If $b_i^* = \perp$, then p_i outputs $\perp$ and aborts.

---

[5] v' can be equal to ϕ.

3. If $v = b_s^*$, then p_s sends 1 to all parties; otherwise, p_s does not send anything.
4. If p_i receives 1 from the sender, then sets $d_i = 1$ else sets $d_i = 0$. Parties execute the consensus protocol again. This time, p_i enters the protocol with input d_i and lets d_i^* be the corresponding output. If $d_i^* = 1$, then p_i outputs b_i^* else if $d_i^* = 0$, then p_i outputs ϕ else p_i outputs $\perp$ ($d_i^* = \phi$ for the last case) .

The consensus protocol is run twice in the construction. Roughly speaking, the first execution establishes a common value among the parties. However, if the sender is constrained, then the established value may be different from v. The second execution is to determine if the sender is "happy" with the established value. If the sender is not happy, then all parties output ϕ. The formal proof proceeds as follows:

Agreement: (i) If an uncorrupted party outputs $b \notin \{\phi, \perp\}$, then by the agreement property of consensus, $d_i^* = 1$ and $b_i^* = b$ for all fault-free parties p_i. Hence all fault-free parties output b. (ii) If an uncorrupted party outputs ϕ, then $d_i^* = 0$ for all fault-free parties p_i. Hence all fault-free parties output ϕ.

Correctness: Consider two cases: (i) a fault-free sender and (ii) a constrained sender. (i) If the sender is fault-free, then all fault-free parties receive v from the sender in step 1. By the persistence property of consensus, all fault-free parties have $b_i^* = v$. Hence all fault-free parties p_i receive 1 from the sender in step 3 and sets $d_i = 1$ in step 4. By the persistence property of consensus again, $d_i^* = 1$. Hence a fault-free party outputs v. (ii) If the sender p_s is constrained, consider two sub-cases: (a) $b_s^* = v$ (b) $b_s^* \neq v$. For case (a), a fault-free party p_i may or may not receive 1 from p_s in step 3 and d_i^* may equal to 0 or 1. However, note that $b_i^* = b_s^* = v$. Therefore, p_i either outputs v or ϕ. For case (b), p_s does not send anything in step 3. Hence all fault-free parties p_i enter the consensus protocol in step 4 with input $d_i = 0$. By the persistence property of consensus, $d_i^* = 0$. Therefore all uncorrupted parties output ϕ.

3.3   Consensus for $n > 3e_f + 2e_c$

Following the principle of Berman et al.[5], the construction of the consensus protocol is done through constructing protocols for weaker consensus variants: weak consensus, graded consensus, king consensus and then consensus. In all these (sub-)protocols, we only need to know the number of fault-free parties $e_{ff} \stackrel{\text{def}}{=} n - e_f - e_c$, but not e_f and e_c; moreover, we only require authenticate (but not secure) point-to-point channels. For all these (sub-)protocols, p_i has an input bit and we denote it as b_i.

Weak Consensus. We say weak consensus is achieved if the following two conditions hold:

- Persistence: If all fault-free parties have the same input bit b, then all fault-free parties output b.
- Agreement: If an uncorrupted party outputs $b \in \{0, 1\}$, then all uncorrupted parties output b or 2.

Protocol WConsensus(p_i,b_i,e_{ff})

1. p_i sends b_i to all parties.
2. Let X_i^0 and X_i^1 be the number of 0 and 1 received by p_i respectively.
   If $X_i^0 \geq e_{ff}$, p_i outputs 0, else if $X_i^1 \geq e_{ff}$, p_i outputs 1 else p_i outputs 2.

Lemma 2. *Protocol WConsensus achieves weak consensus for $n > 3e_f + 2e_c$.*

Proof. Persistence: Note that $e_{ff} > \frac{n}{2}$. If all fault-free parties p_i have the same input bit b, then $X_i^b \geq e_{ff}$ and $X_i^{\bar{b}} < e_{ff}$. Hence p_i will output b. *Agreement:* Suppose there exists two uncorrupted parties p_i and p_j outputting 0 and 1 respectively. Then $|X_i^0 \cap X_j^1| \geq e_{ff} - (e_f + e_c) = n - e_f - e_c - (e_f + e_c) > e_f$. Therefore, there exists more than e_f parties sending different bits to p_i and p_j in round 1. This is a contradiction since there are at most e_f corrupted parties. $\qquad\square$

Graded consensus. In graded consensus, every party p_i outputs a bit along with a grade g_i. Graded consensus is achieved if the following three conditions are satisfied:

- Persistence: If all fault-free parties have the same input bit b, then all fault-free parties output b with $g = 1$.
- Agreement: If an uncorrupted party outputs b with $g = 1$, then all fault-free parties output b, all constrained parties output b or $\bot$.
- Completeness: No fault-free party outputs $\bot$.

Protocol GConsensus(p_i,b_i, e_{ff})

1. p_i sends the output of WConsensus(p_i, b_i, e_{ff}) to all parties.
2. Let X_i^0, X_i^1 and X_i^2 be the number of 0, 1 and 2 received by p_i respectively.
   If $\max\{X_i^0, X_i^1\} + X_i^2 < e_{ff}$, then p_i outputs $\bot$ and abort.
   If $X_i^0 \geq e_{ff}$, p_i outputs 0 with $g_i = 1$,
   else if $X_i^1 \geq e_{ff}$, p_i outputs 1 with $g_i = 1$,
   else if $X_i^0 > X_i^1$, p_i outputs 1 with $g_i = 0$,
   else p_i outputs 0 with $g_i = 0$.

Lemma 3. *Protocol GConsensus achieves graded consensus for $n > 3e_f + 2e_c$.*

Proof. Persistence: If all fault-free parties have the same input bit b, then following the persistence property of weak consensus, they output the same bit b in WConsensus. For a fault-free party p_i, $X_i^b \geq e_{ff}$ and $X_i^{\bar{b}} < e_{ff}$. Therefore p_i outputs b with $g_i = 1$. *Agreement:* If an uncorrupted party p_i outputs b with $g_i = 1$, then $X_i^b \geq e_{ff}$. Hence at least $e_{ff} - e_f$ uncorrupted parties have b as the output of WConsensus. Following the agreement property of weak consensus, the number of uncorrupted parties that output $\bar{b}$ in WConsensus is equal to 0. By counting, the number of uncorrupted parties that output 2 in WConsensus is at most $e_{ff} + e_c - (e_{ff} - e_f) = e_c + e_f$. Assume on contrary that there exists an uncorrupted party p_j outputs $\bar{b}$ in GConsensus. Then $X_j^{\bar{b}} \geq X_j^b$. Note that all $\bar{b}$ p_j received in step 1 are from corrupted parties. Therefore, $X_j^2 \leq e_f + e_c + (e_f - X_j^{\bar{b}})$.

($e_f + e_c$ corresponds to the number of 2 received due to uncorrupted parties; $e_f - X_j^{\bar{b}}$ corresponds to the number of 2 received due to corrupted parties.) But $X_j^{\bar{b}} + X_j^2 \leq X_j^{\bar{b}} + e_f + e_c + (e_f - X_j^{\bar{b}}) = 2e_f + e_c < e_{ff}$ as $n > 3e_f + 2e_c$ and $e_{ff} = n - e_f - e_c$. Therefore, $\max\{X_j^0, X_j^1\} + X_j^2 < e_{ff}$, p_j should output $\perp$ instead. Contradiction. *Completeness:* By the agreement property of weak consensus, for some bit b, each fault-free party has either b or 2 as the output of WConsensus. Therefore, for a fault-free party p_i, $\max\{X_i^0, X_i^1\} + X_i^2 \geq e_{ff}$. Hence no fault-free party outputs $\perp$.

King Consensus. In king consensus, there is a designated party p_k known as the king. King consensus is achieved if the followings hold:

- Persistence: If all fault-free parties have the same input bit b, then all uncorrupted parties that do not abort output b.
- Correctness: If the king p_k is fault-free, then all uncorrupted parties that do not abort output the same bit.
- Completeness: No fault-free party outputs $\perp$.

Protocol KConsensus$_{p_k}(p_i, b_i, e_{ff})$

1. Let (v_i, g_i) be the output of GConsensus(p_i, b_i, e_{ff}). If $(v_i, g_i) = \perp$, then p_i outputs $\perp$.
2. p_k sends v_k to all parties.
3. If $(g_i \neq 1)$ and p_i receives $\check{v}_k$ from p_k and $v_k \neq \phi$, then p_i sets $v_i = v_k$.
4. Let (v_i', g_i') be the output of GConsensus(p_i, v_i, e_{ff}). p_i outputs v_i'.

Lemma 4. *Protocol KConsensus achieves king consensus for $n > 3e_f + 2e_c$.*

Proof. Persistence: If all fault-free parties have the same input bit b, then by the persistence property of graded consensus, all fault-free parties p_i have $(b, 1)$ as the output of the first execution of GConsensus, i.e., $v_i = b$ and $g_i = 1$. Since $g_i = 1$, v_i will not be modified in step 3. All fault-free parties enter the second execution of GConsensus with the same input b. By the persistence and the agreement properties of graded consensus, all uncorrupted parties output the bit b in KConsensus. *Correctness:* Suppose p_k is a fault-free party. Consider two cases: (i) there exists a fault-free party p_i with $g_i = 1$ by the end of step 1. (ii) all fault-free parties p_i have $g_i = 0$ by the end of step 1. For case (i), following the agreement property of graded consensus, all fault-free parties p_j (including p_k) have the same value for v_j (i.e., $v_j = v_i = v_k$) by the end of step 1. It does not matter whether p_j resets its value in step 3. For case (ii), all fault-free parties p_j receive v_k from p_k in step 3 and set $v_j = v_k$. Combining two cases, all fault-free parties enter the second execution of GConsensus with the same input v_k. Following the persistence and agreement properties of graded consensus, all uncorrupted parties output v_k in KConsensus. *Completeness:* Completeness of KConsensus follows the completeness of graded consensus since no fault-free party will output $\perp$ in the executions of GConsensus.

Consensus. We show how to construct a consensus protocol from a king consensus protocol:

Procotol Consensus(p_i, b_i, e_{ff})

1. Set $b'_i = b_i$.
2. for $k = 1$ to $n - e_{ff} + 1$ do:
   (a) Set b'_i to the output of KConsensus$_{p_k}(p_i, b'_i, e_{ff})$.
   (b) If $b'_i = \perp$, then p_i outputs $\perp$ and abort.
3. p_i outputs b'_i.

Theorem 1. *Protocol Consensus achieves consensus for $n > 3e_f + 2e_c$.*

Proof. Persistence: If all fault-free parties enter the protocol Consenus with the same input bit b, then by the persistence property of king consensus, all uncorrupted parties that do not abort output the same bit b. *Agreement:* Note that $n - e_{ff} + 1 = e_f + e_c + 1$. There exists a fault-free party $p_i \in \{p_1, \ldots, p_{e_f + e_c + 1}\}$. By the correctness property of king consensus, all fault-free parties will have the same value for b' after KConsensus$_{p_i}$ is run. Agreement then follows from the persistence property of king consensus.

3.4 Consensus for $n \geq 3e_f + 2e_c$, $e_f, e_c \geq 1$

If the values of e_f and e_c are known a priori, then we can improve the bound in Theorem 1. On a high level, the construction takes two steps. First, based on the consensus protocol we have for $n > 3e_f + 2e_c$, we construct a *weak broadcast* (to be defined) protocol for $n \geq 3e_f + 2e_c$. Second, we convert a weak broadcast protocol into a consensus protocol.

Weak Broadcast. In weak broadcast, there is a sender p_s with an input bit b. Weak broadcast is achieved if the following two conditions hold:

- Agreement: All fault-free parties output a common bit b'.
- Correctness: If the sender is fault-free, then $b = b'$.

Note that in weak broadcast, we do not concern the outputs of constrained parties. Due to lack of the space, we omit the description of the protocol. The details will appear in the full version [6].

From weak broadcast to consensus. Once we have a protocol for weak broadcast, it is easy to construct a consensus protocol:

1. Each party p_i weak-broadcasts the input bit b_i using the protocol WBroadcast.
2. If the majority of the broadcasted bits is 1, then p_i sets $b'_i = 1$ else p_i sets $b'_i = 0$. p_i sends b'_i to all parties.

---

[6] A preliminary full version is available at the author's homepage: http://www.cs.umd.edu/~cykoo

3. Let X_i^0 and X_i^1 be the number of 0 and 1 received by p_i in last round respectively. If $X_i^0 > \frac{1}{2}n$, then p_i outputs 0 else if $X_i^1 > \frac{1}{2}n$, then p_i outputs 1 else p_i outputs $\perp$.

Since all fault-free parties p_i have the same output in weak broadcasts, they will have the same value for b_i'. In particular, if all of them have the same input bit b, then $b_i' = b$. As the majority of parties are fault-free, if an uncorrupted party p_j receives $> \frac{1}{2}$ copies of $b_j' \in \{0,1\}$ in round 2, then $b_j' = b_i'$ for any fault-free party p_i. Therefore both persistence and agreement properties hold. Hence we have the following:

Theorem 2. *There is a consensus protocol for $n \geq 3e_f + 2e_c$, $e_f \geq 1, e_c \geq 1$, assuming the values of e_f and e_c are known a priori.*

4   Verifiable Secret Sharing (VSS)

In verifiable secret sharing (VSS), there is a special party p_d known as the dealer. The dealer holds a secret s. A VSS protocol consists of two phases: a *sharing phase* and a *reconstruction phase*. In the sharing phase, the dealer shares the secret with other parties. Parties may *disqualify* a non fault-free dealer. If the dealer is not disqualified, then in the reconstruction phase, the parties reconstruct a value based on their views in the sharing phase.

In our case, VSS protocol is used as a tool for multiparty computation. Our definition requires a VSS protocol to have the verifiable secret and polynomial sharing property[15]. In this section, we assume the values of e_f and e_c are known a priori. We say a protocol achieves *verifiable secret sharing* if the followings hold:

- Privacy: If the dealer is uncorrupted, then the view of the adversary during the sharing phase reveals no information on s.
- Agreement: If an uncorrupted party disqualifies the dealer, then all uncorrupted parties that do not abort disqualify the dealer.
- Commitment: If the dealer is not disqualified, then there exists a polynomial $h'(x)$ of degree e_f such that at the end of the sharing phase, all fault-free parties p_i (locally) output $h'(i)$; a constrained party p_j which does not abort outputs $h'(j)$. All uncorrupted parties that do not abort output $h'(0)$ in the reconstruction phase.
- Correctness: No fault-free party will abort the protocol. A fault-free dealer will not be disqualified while a constrained dealer may be disqualified. But if an uncorrupted dealer is not disqualified, then $h'(0) = s$.

Theorem 3. *Assuming the values e_f and e_c are known a priori, there exists a VSS protocol for $n \geq 4e_f + 3e_c + 1$.*

Proof. We construct a VSS protocol with the above resilience. The protocol is based on the bivariate solution of Feldman-Micali[13]. We start by giving a high level description of the protocol. In round 1, the dealer shares the secret via a random bivariate polynomial of degree $e_f + 1$. If the dealer is constrained, then a fault-free party may not receive its entitled share. However, unlike [13], the fault-free party cannot take a default value since it will not be on the polynomial and correctness will be violated. Instead, a party broadcasts "receive" in round 2 if it has received its entitled share. Let $\mathcal{G}$ be the group of parties who broadcast "receive". If $|\mathcal{G}|$ is too small, then the dealer is disqualified. Otherwise, the parties within $\mathcal{G}$ proceed to verify if the dealer has shared a valid secret, using a similar approach as in [13](with suitable modifications to tolerate the presence of constrained parties). The parties that are not in $\mathcal{G}$ will not take part in the verification. After the verification, if the dealer is not disqualified, then all parties in $\mathcal{G}$ have shares correspond to a valid secret. The parties outside $\mathcal{G}$ then compute their shares by interpolating the shares from the parties in $\mathcal{G}$ (here we exploit the fact that the secret is shared using a bivariate polynomial).

We assume the secret s is taken from some finite field $\mathcal{F}$. In the following, if the dealer broadcasts a value and the parties receive ϕ as the output, then we implicitly assume that all parties disqualify the dealer.

VSS-share(p_d)

1. The dealer chooses a random bivariate polynomial f of degree at most c_f in each variable such that $f(0,0) = s$. The dealer sends to party p_i the polynomials $g_i(x) \stackrel{\text{def}}{=} f(x,i)$ and $h_i(x) \stackrel{\text{def}}{=} f(i,x)$.
2. p_i broadcasts "0" if it does not receive $g_i(x)$ and $h_i(x)$ from the dealer (or $g_i(x)$ and $h_i(x)$ are not polynomials of degree e_f), otherwise p_i broadcasts "1". Let $\mathcal{G}$ be the group of parties which broadcast "1". If $|\mathcal{G}| < 3e_f + 2e_c + 1$, then the dealer is disqualified. Otherwise, each party $p_i \in \mathcal{G}$ does the following:
   (a) For every party $p_j \in \mathcal{G}$, p_i sends $g_i(j)$ and $h_i(j)$ to p_j. Let $g'_{j,i}$ and $h'_{j,i}$ be the two values received by p_i from $p_j \in \mathcal{G}$. p_i aborts if it receives values from less than $2e_f + e_c + 1$ parties.
   (b) For every party $p_j \in \mathcal{G}$, if $(g'_{j,i} \neq h_i(j)$ and $g'_{j,i} \neq \phi)$ or $(h'_{j,i} \neq g_i(j)$ and $h'_{j,i} \neq \phi)$, then p_i broadcasts "complaint : i,j" else p_i broadcasts "no complaint: i,j". Note that if an uncorrupted p_i broadcasts "complaint: i,j", then p_j or the dealer is corrupted.
   (c) The dealer broadcasts $f_{j,k} \stackrel{\text{def}}{=} f(j,k)$ and $f_{k,j} \stackrel{\text{def}}{=} f(k,j)$ if "complaint: j,k" is broadcasted by a party $p_j \in \mathcal{G}$.
   (d) If there exists a j such that (i) $f_{j,i}$ and $f_{i,j}$ are revealed in last step and (ii) $f_{j,i} \neq g_i(j)$ or $f_{i,j} \neq h_i(j)$, then p_i broadcasts "complaint", otherwise p_i broadcasts "okay". (If an uncorrupted p_i broadcasts "complaint", then the dealer must be corrupted. On the other hand, if the dealer is uncorrupted and p_i broadcasts "complaint", then p_i must be corrupted.)
   (e) If p_i broadcasts "complaint" in last step, then the dealer broadcasts $g_i(x)$ and $h_i(x)$.
   (f) p_i broadcasts "reject" if one of the followings hold:
      – p_i broadcasts "complaint" in step 2(d)
      – There exists a public polynomial $g_k(x)$ and $h_k(x)$ such that $g_k(i) \neq h_i(k)$ or $h_k(i) \neq g_i(k)$

      – The dealer does not respond to the complaints broadcasted in step
        2(b) or step 2(d);

      otherwise, p_i broadcasts "accept".

3. If less than $3e_f + 2e_c + 1$ parties in $\mathcal{G}$ broadcast "accept", then the dealer
   is disqualified. Otherwise, note that two polynomials $g_i(x)$ and $h_i(x)$ are
   associated with each uncorrupted party p_i in $\mathcal{G}$ (If the polynomials are not
   made public in step 2(e), then the polynomials associated with p_i are the
   two polynomials p_i received in step (1)). Each party $p_i \in \mathcal{G}$ sends $h_i(j)$ to
   all parties $p_j \notin \mathcal{G}$.

4. For each party $p_i \notin \mathcal{G}$, p_i constructs a degree e_f polynomial $g_i(x)$ by using
   the Reed-Solomon error-correction interpolation procedure on the values it
   received in last step (If p_i cannot construct such polynomial or p_i receives
   less than $2e_f + e_c + 1$ shares, then p_i aborts).

5. Each party p_i outputs $g_i(0)$.

VSS-reconstruct$(g_i(0))$

1. Party p_i sends $g_i(0)$ to all parties.
2. Let SS^i be the set of secret shares p_i receives in last round. If $|SS^i| < 3e_f + 1$,
   then p_i aborts else p_i reconstructs a polynomial $h_0(x)$ of degree e_f by using
   the Reed-Solomon error-correction interpolation on the set SS^i and outputs
   $h_0(0)$.

We now proceed to prove that the above protocol achieves VSS.

– Privacy: Consider an uncorrupted dealer. If an uncorrupted party p_i broadcasts "complaint: i,j" in step 2(b), then p_j must be a corrupted party. It is
  easy to see that if a party p_i broadcasts a complaint in step 2(d), then p_i
  is a corrupted party. Therefore, all the information broadcasted by an uncorrupted dealer on f, if any, is a subset of the shares the corrupted parties
  entitled to receive in step 1. Since the secret is shared by a random bivariate polynomial of degree e_f, we conclude that the view of the adversary is
  independent of s during the sharing phase.
– Agreement: Note that the decision of disqualifying a dealer is completely
  dependent on the messages broadcasted by the parties. If an uncorrupted
  party does not abort by the end of the sharing phase, then by the agreement
  property of broadcast, the values it received from the broadcasts are same
  as those received by fault-free parties. Hence agreement follows.
– Correctness: We first consider a fault-free dealer. All fault-free parties will be
  in $\mathcal{G}$. Since $n \geq 4e_f + 3e_c + 1$, it follows that $|\mathcal{G}| \geq 3e_f + 2e_c + 1$. In addition,
  all fault-free parties broadcast "accept" in step 2(f). For a constrained party
  p_i that is not in $\mathcal{G}$, it is easy to see that $g_i(x)$ reconstructed in step 3 (if p_i
  does not abort) is equal to $f(x, i)$.

    Next we consider a constrained dealer that is not disqualified. For an
uncorrupted party $p_i \in \mathcal{G}$ that does not abort by step 2(f), it is easy to see
that $g_i(x) = f(x, i)$ and $h_i(x) = f(x, i)$. If the dealer is not disqualified, then
$\geq 3e_f + 2e_c + 1$ parties broadcasts "accept" in step 2(f). Let $\mathcal{G}'$ be the set

of fault-free parties among these $\geq 3e_f + 2e_c + 1$ parties. $|\mathcal{G}'| \geq 2e_f + e_c + 1$.
It then follows that every fault-free party (or constrained party that does
not abort in step 3) p_i that is not in $\mathcal{G}$ can reconstruct $g_i(x) = f(x, i)$. An
uncorrupted party p_i (if it does not abort) will then output $f(0, i)$ in step 4.
- Commitment: We consider the case of a non-disqualified corrupted dealer.
   If a corrupted dealer is not disqualified, then at least $2e_f + e_c + 1$ fault-free
   parties broadcast "accept" in step 2(f). Let $\mathcal{G}'$ be the set of such fault-free
   parties. Following [13, Lemma 2], there exists a bivariate polynomial f' of
   degree e_f in each variable such that for all $p_i \in \mathcal{G}'$, $g_i(x) = f'(x, i)$ and
   $h_i(x) = f'(i, x)$. Now consider an uncorrupted party $p_j \in \mathcal{G}$ but not in $\mathcal{G}'$.
   There are 2 possible scenarios:

   - p_j broadcasts a complaint in step 2(d). $g_j(x)$ and $h_j(x)$ are made public
      in step 2(e). For all $p_i \in \mathcal{G}'$, $g_j(i) = h_i(j)$ and $h_j(i) = g_i(j)$. Hence it
      follows that $g_j(x) = f'(x, j)$ and $h_i(x) = f'(x, i)$.
   - p_j does not broadcast a complaint in step 2(d). If p_j does not abort in
      step 2(b), then $h_j(i) = f_i(j)$ and $f_j(i) = h_i(j)$ for at least $2e_f + e_c + 1 -$
      $(e_f + e_c) = e_f + 1$ parties $p_i \in \mathcal{G}'$. Since f' is a bivariate polynomial of
      degree e_f, it follows that $h_j(x) = f'(j, x)$ and $g_j(x) = f'(x, j)$. Therefore
      if p_j does not broadcast a complaint in step 2(d), it will not broadcast
      "reject" in step 2(f).

We conclude that for all uncorrupted parties $p_i \in \mathcal{G}$ that do not abort by
step 2, $h_i(x) = f'(i, x)$ and $g_i(x) = f'(x, i)$. It is easy to see that if an
uncorrupted party $p_j \notin \mathcal{G}$ does not abort by step 3, p_j can reconstruct
$g_j(x) = f'(x, j)$.
Hence it follows that an uncorrupted party p_i (if it does not abort) outputs
$f'(0, i)$ in step 4.
   It also follows that all uncorrupted parties that do not abort output $h'(0)$
by the end of VSS-reconstruct.

5   MPC

We now construct a MPC protocol following the paradigm in [3]. On a high level,
each party shares its private input, evaluates the circuit gate by gate, and then
reconstructs the outputs.

Input Phase: Every party shares its private input using VSS-share. If a party is
disqualified (when it plays the role of the dealer), then the party is added to the
set $\mathcal{D}$. Note that by the end of the input phase, all uncorrupted parties that do
not abort have the same view on $\mathcal{D}$.

Circuit Evaluation: All parties that do not abort evaluate the circuit $g^{\mathcal{D}}$ gate
by gate. A party who was disqualified in the input phase does not take part in
this phase and all other parties will ignore the messages sent from that party.
It suffices to consider the addition and multiplication gates. The evaluation procedures are very similar to the one in [7, section 4.52] and we omit the details
here.

Output phase: Reconstructing output is easy. For each output wire, each party sends its share to the party who is entitled to receive the output. The corresponding party then reconstructs the output from the shares it received using error correction. A party aborts if it receives less than $3e_f + 1$ entitled shares.

6   Extending to the Case of $\delta < \frac{1}{6}$

We describe how to extend the results from the previous sections (which assume $\delta = 0$) to the case of $\delta < \frac{1}{6}$, at the expense of increasing the round complexity by a factor of 2. More precisely, we show how to compile a protocol Π for $\delta = 0$ into a protocol Π' for $\delta < \frac{1}{6}$.

Our broadcast protocol Π for $\delta = 0$ assumes $n \geq 3e_f + 2e_c$ but does not assume secure channels. If p_i is supposed to send p_j a message m in Π, then the followings are carried out in Π':

- p_i sends m to all parties who then forward the message to p_j.
- If there exists m' such that p_j receives $\geq \frac{4}{3}e_f$ copies of them, then p_j sets $m = m'$ else $m = \phi$.

Consider the following two cases:

1. Both p_i and p_j are fault-free parties: since $n \geq 3e_f + 2e_c$, at least $(2e_f + e_c)(1 - \delta)$ fault-free parties receive m from p_i. The number of copies of m p_j received is at least $(2e_f + e_c)(1 - 2\delta)$ which is greater than $\frac{4}{3}e_f$ if $\delta < \frac{1}{6}$. Hence p_j can receive m from p_i.
2. At least one of the p_i and p_j is a constrained party: suppose p_j receives m' from p_i in Π' and $m' \neq \phi$. Since p_j receives at least $\frac{4}{3}e_f$ copies of m', at least $\frac{1}{3}e_f$ copies are from uncorrupted parties. Hence $m' = m$ (assuming $e_f \geq 3$).

For VSS and MPC protocols, we assume $n \geq 4e_f + 3e_c + 1$ but we also assume secure channels. If p_i is supposed to send p_j a message m in Π, then the following steps are carried out in Π':

- p_i picks a random polynomial $h(x)$ of degree e_f such that $h(0) = m$. p_i sends $h(k)$ to p_k who then forwards the share to p_j.
- Based on the shares p_j received, using the Reed-Solomon error-correction interpolation procedure, p_j constructs a polynomial $h'(x)$ of degree e_f such that at least $2e_f + 1$ shares are on $h'(x)$. If p_j cannot construct such polynomial, then p_j sets $m = \phi$ else $m = h'(0)$.

First we note that if p_i is uncorrupted, then the view of the adversary is independent of m since h is a random polynomial of degree e_f. Second, if both p_i and p_j are uncorrupted and p_j does not set $m = \phi$, then p_j receives $2e_f + 1$ shares that are on $h'(x)$. $e_f + 1$ of these shares are from uncorrupted parties. Hence $h'(x) = h(x)$ since both $h'(x)$ and $h(x)$ are of degree $e_f + 1$. Finally, if both p_i and p_j are fault-free parties, then p_j will receive at least $(3e_f + 2e_c + 1)(1 - 2\delta) \geq 2e_f + 1$ (assuming $\delta < \frac{1}{6}$ and $e_c \geq 1$) correct shares. On the other hand, p_j will receive at most e_f corrupted shares. Hence p_j can always reconstruct $h(x)$ using error-correction.

7   Conclusion and Open Problems

In this paper, we consider a communication model where message delivery is neither always guaranteed nor always in the hands of the adversary. We has developed broadcast and VSS protocols under this model. However, we do not know if the bounds are tight. Another interesting direction is to consider what is achievable if the global clock is removed from the model.

Acknowledgments

The author thanks Omer Horvitz, Jonathan Katz, Ruggero Morselli, Tsuen-Wan "Johnny" Ngan and Ji Sun Shin for helpful comments and encouragement. In particular, discussions on the communication model with Ruggero Morselli and Tsuen-Wan "Johnny" Ngan are very helpful. The author also thanks the anonymous referees for providing useful references and thoughtful comments.

References

1. D. Beaver. Multiparty protocols tolerating half faulty processors. In G. Brassard, editor, *Advances in Cryptology - CRYPTO '89, 9th Annual International Cryptology Conference*, volume 435 of *Lecture Notes in Computer Science*, pages 560–572. Springer, 1989.
2. M. Ben-Or, R. Canetti, and O. Goldreich. Asynchronous secure computation. In *STOC '93: Proceedings of the twenty-fifth annual ACM symposium on Theory of computing*, pages 52–61, New York, NY, USA, 1993. ACM Press.
3. M. Ben-Or, S. Goldwasser, and A.Wigderson. Completeness theorems for non-cryptographic fault-tolerant distributed computations. In *Proceedings of the 20th annual ACM symposium on Theory of computing*, pages 1–10, 1988.
4. M. Ben-Or, B. Kelmer, and T. Rabin. Asynchronous secure computations with optimal resilience (extended abstract). In *PODC '94: Proceedings of the thirteenth annual ACM symposium on Principles of distributed computing*, pages 183–192, New York, NY, USA, 1994. ACM Press.
5. P. Berman, J. A. Garay, and K. J. Perry. Towards optimal distributed consensus (extended abstract). In *Proceedings of the 30th Annual Symposium on Foundations of Computer Science*, pages 410–415. IEEE, 1989.
6. M. Biely. Optimal agreement protocol in malicious faulty processors and faulty links. *IEEE Transactions on Knowledge and Data Engineering*, 4(3):266–280, 1992.
7. R. Canetti. *Studies in Secure Multiparty Computation and Applications*. PhD thesis, Weizmann Institute of Science, Rehovot 76100, Israel, June 1995.
8. R. Canetti. Universally composable security: A new paradigm for cryptographic protocols. In *FOCS '01: Proceedings of the 42nd IEEE symposium on Foundations of Computer Science*, page 136, Washington, DC, USA, 2001. IEEE Computer Society.
9. R. Canetti, S. Halevi, and A. Herzberg. Maintaining authenticated communication in the presence of break-ins. In *PODC '97: Proceedings of the sixteenth annual ACM symposium on Principles of distributed computing*, pages 15–24, New York, NY, USA, 1997. ACM Press.

10. R. Canetti, Y. Lindell, R. Ostrovsky, and A. Sahai. Universally composable two-party and multi-party secure computation. In *STOC '02: Proceedings of the thiry-fourth annual ACM symposium on Theory of computing*, pages 494–503, New York, NY, USA, 2002. ACM Press.

11. D. Chaum, C. Crepeau, and I. Damgard. Multiparty unconditionally secure protocols. In *STOC '88: Proceedings of the twentieth annual ACM symposium on Theory of computing*, pages 11–19, New York, NY, USA, 1988. ACM Press.

12. R. Cramer, I. Damgård, S. Dziembowski, M. Hirt, and T. Rabin. Efficient multiparty computations secure against an adaptive adversary. In J. Stern, editor, *Advances in Cryptology — EUROCRYPT '99*, volume 1592 of *Lecture Notes in Computer Science*, pages 311–326. Springer-Verlag, May 1999.

13. P. Feldman and S. Micali. An optimal probabilistic protocol for synchronous byzantine agreement. *SIAM J. Comput.*, 26(4):873–933, 1997.

14. J. A. Garay and K. J. Perry. A continuum of failure models for distributed computing. In *Proceedings of the 6th International Workshop on Distributed Algorithms*, pages 153–165. Springer-Verlag, 1992. Full version availabe at http://cm.bell-labs.com/who/garay/continuum.ps.

15. R. Gennaro, M. O. Rabin, and T. Rabin. Simplified vss and fast-track multiparty computations with applications to threshold cryptography. In *PODC '98: Proceedings of the seventeenth annual ACM symposium on Principles of distributed computing*, pages 101–111, New York, NY, USA, 1998. ACM Press.

16. O. Goldreich, S. Micali, and A. Wigderson. How to play any mental game. In *Proceedings of the nineteenth annual ACM conference on Theory of computing*, pages 218–229. ACM Press, 1987.

17. S. Goldwasser and Y. Lindell. Secure computation without agreement. In *DISC '02: Proceedings of the 16th International Conference on Distributed Computing*, pages 17–32, London, UK, 2002. Springer-Verlag.

18. Y. T. Kalai, Y. Lindell, and M. Prabhakaran. Concurrent general composition of secure protocols in the timing model. In *STOC '05: Proceedings of the thirty-seventh annual ACM symposium on Theory of computing*, pages 644–653, New York, NY, USA, 2005. ACM Press.

19. A. D. Keromytis, V. Misra, and D. Rubenstein. Sos: secure overlay services. *SIG-COMM Comput. Commun. Rev.*, 32(4):61–72, 2002.

20. P. R. Parvédy and M. Raynal. Optimal early stopping uniform consensus in synchronous systems with process omission failures. In *Proceedings of the sixteenth annual ACM symposium on Parallelism in algorithms and architectures*, pages 302–310. ACM Press, 2004.

21. K. J. Perry and S. Toueg. Distributed agreement in the presence of processor and communication faults. *IEEE Trans. Softw. Eng.*, 12(3):477–482, 1986.

22. T. Rabin and M. Ben-Or. Verifiable secret sharing and multiparty protocols with honest majority. In *STOC '89: Proceedings of the twenty-first annual ACM symposium on Theory of computing*, pages 73–85, New York, NY, USA, 1989. ACM Press.

23. U. Schmid, B. Weiss, and J. Rushby. Formally verified byzantine agreement in presence of link faults. In *ICDCS '02: Proceedings of the 22 nd International Conference on Distributed Computing Systems (ICDCS'02)*, page 608, Washington, DC, USA, 2002. IEEE Computer Society.

24. P. Thambidurai and Y.-K. Park. Interactive consistency with multiple failure modes. In *Proceedings of the 7th Reliable Distributed Systems Symposium*, pages 93–100, 1988.

25. K. Q. Yan, Y. H. Chin, and S. C. Wang. Optimal agreement protocol in malicious faulty processors and faulty links. *IEEE Transactions on Knowledge and Data Engineering*, 4(3):266–280, 1992.
26. A. C.-C. Yao. Protocols for secure computations. In *FOCS '82: Proceedings of the 23rd Symposium on Foundations of Computer Science*, pages 160–164, Los Alamitos, CA, USA, 1982. IEEE Computer Society Press.

Communication Efficient Secure Linear Algebra

Kobbi Nissim* and Enav Weinreb**

Ben Gurion University, Beer-Sheva 84105, Israel
{kobbi, weinrebe}@cs.bgu.ac.il

Abstract. We present communication efficient secure protocols for a variety of linear algebra problems. Our main building block is a protocol for computing Gaussian Elimination on encrypted data. As input for this protocol, Bob holds a $k \times k$ matrix M, encrypted with Alice's key. At the end of the protocol run, Bob holds an encryption of an upper-triangular matrix M' such that the number of nonzero elements on the diagonal equals the rank of M. The communication complexity of our protocol is roughly $O(k^2)$.

Building on Oblivious Gaussian elimination, we present secure protocols for several problems: deciding the intersection of linear and affine subspaces, picking a random vector from the intersection, and obliviously solving a set of linear equations. Our protocols match known (insecure) communication complexity lower bounds, and improve the communication complexity of both Yao's garbled circuits and that of specific previously published protocols.

1 Introduction

Linear algebra plays a central role in computer science in general and in cryptography in particular. Numerous cryptographic applications such as private information retrieval, secret sharing schemes, multi party secure computation, and many more make use of linear algebra. In particular, the ability to efficiently solve a set of linear equations constitutes an important algorithmic and cryptographic tool. In this work we design communication efficient secure protocols for various linear algebraic problems.

The basic linear algebraic problem we focus on is linear subspace intersection. Alice and Bob hold subspaces of F^k for some finite field F, each subspace representing a set of linear equations held by the players. They wish to study different properties of the intersection of their input subspaces, without leaking any information not revealed by the result of the computation. The first variant is a search problem where Alice and Bob wish to *compute* the intersection, while in the second variant they only wish to *decide* whether the intersection is the trivial zero subspace. We also consider the problems of picking a random vector from the intersection, and of affine subspaces intersection.

* Research partially Supported by the Frankel Center for Computer Science.
** Partially supported by a Kreitman Foundation Fellowship and by the Frankel Center for Computer Science.

S. Halevi and T. Rabin (Eds.): TCC 2006, LNCS 3876, pp. 522–541, 2006.

Cramer and Damgård introduced secure protocols for solving various linear algebraic problems [5]. Their work was done in the information theoretical setup, with the main goal of reducing the round complexity to a constant. The communication complexity of their protocols is $\Omega(k^3)$ while the size of the inputs is merely $O(k^2)$. Another approach for designing secure protocols for these linear algebraic problems is to apply the garbled circuit method of Yao [18]. The communication complexity of such protocols is related to the Boolean circuit complexity of the underlying problems. However, as these problems are strongly related to the problem of matrix multiplication, the communication complexity of the resulting protocol is essentially the circuit complexity of the latter. The best known upper bound for this problem is $O(k^\omega)$ [6] for $\omega \cong 2.38$, which is still larger than the input size.

We introduce a protocol for the subspace intersection problem[1] with communication complexity of roughly $O(k^2)$. Even for *insecure* computation, it is shown in [3] that the deterministic communication complexity of the problem is $\Omega(k^2)$. This result agrees with ours up to a polylogarithmic factor. Although determining the *randomized* communication complexity of subspace intersection is an open problem, it serves as an evidence that our upper bound may be tight. Our protocol gives rise to communication efficient secure protocols for problems reducible to linear algebra, e.g., perfect matching, and functions with low span program complexity [11]. Unlike the protocol of [5] and [18], our protocols are not constant round. However, using a combination of our techniques and the general techniques of [18], we achieve a round complexity of $O(k^{1-\frac{1}{\omega-1}}) \approx O(k^{0.275})$.

Techniques. We use public key homomorphic encryption to put the players in the following situation: Alice holds a private key of a public key homomorphic encryption scheme, while Bob holds an encrypted matrix. Bob wants to perform computations on his matrix, such as checking if it has full rank, without leaking any information on it. Specifically, we show how Bob can use Alice's help to securely perform the Gaussian Elimination procedure on the matrix. As the current state of art in homomorphic encryption does not allow an encryption scheme with both homomorphic addition and multiplication, we use standard techniques to make Alice multiply encrypted elements for Bob.

The Gaussian Elimination procedure requires Bob to find a row with a non-zero element in its leftmost coordinate for the elimination of the leftmost column. As the matrix is encrypted, Bob cannot find such a row on his own. We use Alice's help and randomness to overcome this problem. Alice's help in this case may be interpreted as performing an 'if' statement in Bob's code, although the computation is oblivious. To save communication, we use the paradigm of *lazy evaluation*, in which Bob uses Alice as a storage device as well. Instead of instantly sending Bob the results of the computations, Alice keeps an image of Bob's memory, and sends him only the information he needs for the next round of computation. To conclude, Bob uses Alice for calculations, for flow control, and as a storage device, without enclosing any of his data to her.

---

[1] All the bounds mentioned in the introduction are for the case where Alice and Bob hold subspaces of dimension $k/2$. Exact bounds are presented later in the text.

The round complexity of our basic protocol is $O(k)$. We use a combination of the garbled circuit method of Yao, and the techniques described above, to reduce the number of rounds to $O(k^{0.275})$. We use randomness to ensure both correctness and security for our protocols. In the context of finding a random vector in the intersection of the input subspaces, we use a technique of adding random constraints to reduce the solution set into only one solution. This is inspired by the "hashing paradigm" [17] that was employed, e.g., by Bellare et al. [2] for uniformly picking an NP witness. Another use of randomness is achieved via the next basic linear algebraic claim. Take a rank r matrix and multiply it from the left and from the right by random full rank matrices. Then, with constant probability, the top-left $r \times r$ sub-matrix of the resulting matrix is of rank r. This fact enables us to reduce problems on non-square matrices to problems on their square counterparts.

Organization. We start in Section 2 with preliminaries and notation. In Section 3 we present secure protocols for computing subspaces intersection and for deciding if the intersection is trivial. Later, in Section 4, we design our main building block, the Oblivious Gaussian Elimination protocol. In Section 5 we show how to securely pick a random vector from the intersection, and finally, in Section 6, we design secure protocols for analogous problems on affine subspaces.

2   Preliminaries

Notation. Let F be a finite field. We denote by $\boldsymbol{v}$ a row vector in the vector space F^k where $k > 0$ and $\boldsymbol{0}$ denotes the row vector whose entries are all zero. For a matrix M with entries from F, we denote by $\boldsymbol{M_i}$ the ith row of M. For an encryption scheme, we let λ be its security parameter. W.l.o.g, we assume that the result of encrypting a field element is of length $O(\lambda)$. We use $\mathbf{neg}(k)$ to denote a function that is negligible in k, i.e. $\mathbf{neg}(k) = k^{-\omega(1)}$.

Homomorphic encryption schemes. Our constructions use semantically-secure public-key encryption schemes that allow for simple computations on encrypted data. In particular, we use encryption schemes where the following operations can be performed without knowledge of the private key: (i) Given two encryptions $\mathsf{Enc}(m_1)$ and $\mathsf{Enc}(m_2)$, we can efficiently compute $\mathsf{Enc}(m_1 + m_2)$; and (ii) Given an encryption $\mathsf{Enc}(m)$ and $c \in F$, we can efficiently compute $\mathsf{Enc}(cm)$.

Several constructions of homomorphic encryption schemes are known, each with its particular properties (see e.g. [15,10,8,14,16,13,7,1]). These have been in use in a variety of cryptographic protocols. Over $F = GF(2)$, the encryption scheme of Goldwasser and Micali [10], based on quadratic residuosity, is sufficient for our constructions.

For a vector $\boldsymbol{v} \in F^n$, we denote by $\mathsf{Enc}(\boldsymbol{v})$ the coordinate-wise encryption of $\boldsymbol{v}$. That is, if $\boldsymbol{v} = \langle a_1, \dots, a_n \rangle$ where $a_1, \dots, a_n \in F$, then $\mathsf{Enc}(\boldsymbol{v}) = \langle \mathsf{Enc}(a_1), \dots, \mathsf{Enc}(a_n) \rangle$. Similarly, for a matrix $M \in F^{m \times n}$, we denote by $\mathsf{Enc}(M)$ the $m \times n$ matrix such that $\mathsf{Enc}(M)[i,j] = \mathsf{Enc}(M[i,j])$. An immediate consequence of the above properties of homomorphic encryption schemes is the ability to perform the following operations without knowledge of the secret key: (i) Given encryptions of two vectors $\mathsf{Enc}(v_1)$ and $\mathsf{Enc}(v_2)$, we can efficiently compute $\mathsf{Enc}(v_1 + v_2)$, and similarly with matrices. (ii) Given an encryption of a vector $\mathsf{Enc}(v)$ and a constant $c \in F$, we can efficiently compute $\mathsf{Enc}(cv)$. (iii) Given an encryption of a matrix $\mathsf{Enc}(M)$ and a matrix M' of the appropriate dimensions, we can efficiently compute $\mathsf{Enc}(MM')$ and $\mathsf{Enc}(M'M)$.

Adversary model. Our protocols are constructed for the two-party semi-honest adversary model. Roughly speaking, both parties are assumed to act in accordance with their prescribed actions in the protocol. Each party may, however, collect any information he/she encounters during the protocol run, and try to gain some information about the other party's input.

Remark 1. Our protocols achieve information theoretic security for Bob while Alice's security relies on that of the underlying encryption scheme.

Basic Building Blocks. In our protocols Bob holds data encrypted by a public key homomorphic encryption scheme, while Alice holds the private decryption key. Bob uses Alice's help to perform different calculations, without enclosing his data to her. As a simple example of a protocol where Bob uses Alice's help, assume Bob holds $\mathsf{Enc}(a)$ and $\mathsf{Enc}(b)$ and needs to compute $\mathsf{Enc}(ab)$. Let `Multiply` be the following (folklore) solution: (i) Bob chooses random masks $r_a, r_b \in_R F$ and sends $\mathsf{Enc}(a+r_a)$ and $\mathsf{Enc}(b+r_b)$ to Alice; (ii) Alice deciphers these messages and returns $\mathsf{Enc}((a+r_a)(b+r_b))$; (iii) Given $\mathsf{Enc}((a+r_a)(b+r_b))$, Bob computes $\mathsf{Enc}(ab) = \mathsf{Enc}((a+r_a)(b+r_b) - r_b a - r_a b - r_a r_b)$. It is easy to see that neither Alice nor Bob gain any information about a and b (and ab). The communication complexity of this protocol is $O(\lambda)$. This protocol is easily generalized to vectors of length k, the resulting protocol `Vector Multiply` is of communication complexity $O(\lambda k)$.

Linear Algebra. We need the following simple linear algebraic claim.

Claim 1 ([4]). *Let $k_a < k_b$ be positive integers, F be a finite field, and M be a $k_a \times k_b$ matrix over F. Suppose $r \leq \mathrm{rank}(M)$ and let T_A and T_B be $k_a \times k_a$ and $k_b \times k_b$ randomly chosen full rank matrices over F. Let $M' = T_A M T_B$, and denote the top-left $r \times r$ sub-matrix of M' by N'. Then with constant probability $\mathrm{rank}(N') = r$.*

3   Linear Subspace Intersection

Let F be a finite field and k be a positive integer. Alice holds a subspace $V_A \subseteq F^k$ of dimension $k_a \leq k$. The subspace V_A is represented by a $k_a \times k$ matrix A, where the rows of A span V_A. Similarly, Bob's input is a subspace $V_B \subseteq F^k$ of dimension k_b, represented by a $k_b \times k$ matrix B. Letting $V_I = V_A \cap V_B$, Alice and Bob wish to securely study different properties of V_I.

The first variant of the problem is of *computing* the subspace V_I itself. The second is of *deciding* whether V_I is the trivial zero subspace. Ignoring security issues, computing the intersection of the input subspaces is at least as hard as deciding whether they have a non trivial intersection. However, constructing a *secure* protocol for the latter turns to be somewhat easier as the players gain less information from its output.

A common step in solving both variants is the following reduction of computing V_I into solving a homogeneous linear system. Let $V_B^\perp$ be the perpendicular subspace[2] of V_B. Define $k_b' = k - k_b$ and let $B^\perp$ be a $k \times k_b'$ matrix whose columns span exactly the subspace $V_B^\perp$. Finally define the $k_a \times k_b'$ matrix $M = AB^\perp$.

Claim 2. *Let $v \in F^{k_a}$. Then $vA \in V_I$ if and only if $vM = \mathbf{0}$.*

Proof. If $vA \in V_I$ then $vA \in V_B$, and thus $vM = (vA)B^\perp = \mathbf{0}$. For the other direction, if $vM = \mathbf{0}$, then $(vA)B^\perp = \mathbf{0}$, and thus $vA \in V_B$. As vA is a linear combination of the rows of A, we get that $vA \in V_A$, hence $vA \in V_A \cap V_B = V_I$.

3.1   Computing the Intersection

Protocol `Intersection Computation` securely computes V_I in one round of communication. The communication complexity of the protocol is $O(\lambda k_a k)$. The protocol uses homomorphic encryption to enable a multiplication of an encrypted matrix by an open matrix without the knowledge of the private decryption key.

Protocol `Intersection Computation`

INPUT: Alice (resp. Bob) holds a $k_a \times k$ (resp. $k_b \times k$) matrix A (resp. B) over a finite field F representing a subspace $V_A \subseteq F^k$ (resp. $V_B \subseteq F^k$).
OUTPUT: Alice holds a matrix representing $V_I = V_A \cap V_B$.

1. Bob locally computes a $k \times k_b'$ matrix $B^\perp$ that represents the subspace $V_B^\perp$.
2. Alice generates keys for a homomorphic public key encryption system, and sends Bob $\mathsf{Enc}(A)$ and the public key.
3. Bob randomly chooses a $k_b' \times k_b'$ full rank matrix T_B, locally computes $\mathsf{Enc}(M)$, where $M \stackrel{\text{def}}{=} AB^\perp T_B$, and sends $\mathsf{Enc}(M)$ to Alice.
4. Alice decrypts M and computes the subspace $K = ker(M^T)$, that is, $K = \{v : vM = \mathbf{0}\}$.
5. Alice computes the subspace $V_I = \{vA : v \in K\}$.

Correctness and Security. The correctness of the `Intersection Computation` protocol derives[3] from Claim 2. Alice's security immediately follows from the fact she only sends information encrypted in a semantically-secure encryption scheme. To prove Bob's security, we show a simulator for Alice's view. The simulator and its related security proof appear in Appendix A.

---

[2] Recall that $V_B^\perp \stackrel{\text{def}}{=} \{u : \langle u, v \rangle = 0 \text{ for all } v \in V_B\}$, and is of dimension $k_b' \stackrel{\text{def}}{=} k - k_b$.
[3] Note that although in the protocol $M = AB^\perp T_B$, Claim 2 still applies, as the columns of $B^\perp$ and the columns of $B^\perp T_B$ both span the subspace $V_B^\perp$.

3.2 Deciding Whether the Intersection Is Trivial

Let V_A, V_B be as above and V_I their intersection[4]. By Claim 2, there is a non trivial intersection between V_A and V_B if and only if there exist a non-zero vector $v \in F^{k_a}$ such that $vAB^\perp = 0$. This happens only if $rank(AB^\perp) < k_a$, that is, if $AB^\perp$ is not a full rank matrix. Hence, computing $AB^\perp$ seems useful also in deciding whether $V_I = \{0\}$. However, unlike in the Intersection Computation protocol ,we cannot have Bob sending $AB^\perp$ nor any information regarding its dimension to Alice. Such information would compromise the protocol privacy.

As in the Intersection Computation protocol, Alice sends an encryption of her input and a public key to Bob, who computes an encryption of $AB^\perp$. Here, we are only interested in whether $AB^\perp$ is of full rank. Our main building block is a private protocol that transforms the encryption of $AB^\perp$ into an encryption of an upper triangular matrix[5]. In particular, there is a 0 on the main diagonal of the resulting matrix if and only if $AB^\perp$ is of full rank.

Definition 1 (Oblivious Gaussian Elimination Problem). *Input: Alice holds a private key of a public key homomorphic encryption scheme over a finite field F. Bob holds a $k_a \times k_b$ matrix M encrypted by Alice's public key, where $k_a \leq k_b$.*

Output: Suppose $\mathrm{rank}(M) = r$. *In the end of the protocol Bob holds an encryption of a $k_a \times k_b$ matrix M'. With probability $1 - \mathbf{neg}(k)$, the matrix M' is upper triangular and: (i) There are at most r non-zero elements on the main diagonal of M'. (ii) With constant probability there are exactly r non-zero elements on the main diagonal of M'.*

The following theorem summarizes the properties of our protocol for solving the Oblivious Gaussian Elimination Problem. The protocol is described in Section 4.

Theorem 3. *There is a secure protocol with communication complexity $\tilde{O}(\lambda k_a k_b)$ and round complexity $k_a^{0.275}$, that solves the Oblivious Gaussian Elimination Problem.*

Having a secure protocol for solving the problem above, deciding the intersection of the input subspaces is done in two steps. A procedure for deciding the intersection with *one sided constant error probability* is depicted in Protocol Intersection Decision below. To get a protocol with negligible error probability, Alice and Bob run protocol Intersection Decision for $m = \omega(\log k)$ times. Alice and Bob then obliviously compute the logical OR of all the executions. The correctness of the protocol is straight forward assuming the correctness of Oblivious Gaussian Elimination.

Theorem 4. *Protocol Intersection Decision is a secure protocol for the subspace intersection decision problem. The communication complexity of the protocol is $\tilde{O}(\lambda k_a k)$ and the round complexity is $\tilde{O}(k_a^{0.275})$.*

---

[4] W.l.o.g., we assume that $k_a + k_b \leq k$, as otherwise V_A and V_B always have a non-trivial intersection.

[5] For non-square matrices upper triangular means $i > j \Rightarrow M[i, j] = 0$.

Protocol Intersection Decision

INPUT: Alice (resp. Bob) holds a $k_a \times k$ (resp. $k_b \times k$) matrix A (resp. B) over a finite field F representing a subspace $V_A \subseteq F^k$ (resp. $V_B \subseteq F^k$). Let $B^\perp$ be a $k \times k_b'$ matrix that represents the subspace $V_B^\perp$.

OUTPUT: If V_I is not the trivial zero subspace, Bob outputs $\mathsf{Enc}(0)$ with probability 1. Else, with constant probability, Bob outputs $\mathsf{Enc}(r)$ for some non-zero $r \in F$.

1. Alice generates keys for a homomorphic public key encryption system, and sends Bob $\mathsf{Enc}(A)$ and the public key.
2. Bob locally computes $\mathsf{Enc}(M)$, where $M \overset{\text{def}}{=} AB^\perp$. Note that M is a $k_a \times k_b'$ matrix.
3. Alice and Bob run protocol **Oblivious Gaussian Elimination** on $\mathsf{Enc}(M)$. Denote by M' the resulting $k_a \times k_b'$ matrix that Bob holds at the end of the protocol execution.
4. Bob and Alice use the **Multiply** protocol such that Bob eventually locally outputs $\mathsf{Enc}(r)$ where $r \overset{\text{def}}{=} \prod_{i=1}^{k_a} M'[i,i]$.

---

4   Oblivious Gaussian Elimination

In this section we introduce a protocol for the Oblivious Gaussian Elimination problem (See Definition 1), with parameters matching Theorem 3. We first define the Oblivious Gaussian Elimination problem for square matrices. Then we design a protocol for this special case, and finally we reduce the problem on general matrices to the problem on their square counterparts.

Definition 2 (Oblivious Gaussian Elimination Problem for Square Matrices). *Input: Alice holds a private key of a public key homomorphic encryption scheme over a finite field F. Bob holds a $k \times k$ matrix M encrypted by Alice's public key.*

Output: In the end of the protocol Bob holds an encryption of a $k \times k$ matrix M'. With probability $1 - \mathbf{neg}(k)$, the matrix M' is upper triangular and: (i) If M is full rank then with probability $1 - \mathbf{neg}(k)$ all the diagonal entries of M' are non-zero. (ii) If M is not full rank then there is a 0 entry on the diagonal of M'.

There are two differences between this definition and Definition 1. Here, the diagonal of the resulting matrix M' does not reflect the exact rank of M, but rather only whether M is full rank or not. On the other hand, here we require very high success probability, while in Definition 1, the success probability is constant.

4.1   Gaussian Elimination

The Gaussian Elimination algorithm is a well known method for transforming a matrix into a triangular form, while keeping its rank. Consider the following 'textbook' Gaussian Elimination procedure. To simplify the presentation, we assume the underlying field is the unique finite field with two elements, that is, $F = GF(2)$. The generalization of all our protocols to other finite fields of fixed size is straight forward[6].

Input: A $k \times k$ matrix M over $F = GF(2)$:
(1) Find a row M_j, such that the leftmost coordinate in M_j is 1, that is, $M[j, 1] = 1$.
(2) For every $i \neq j$, if $M[i, 1] = 1$, add M_j to M_i, so that the result is 0 in the leftmost coordinate.
(3) Swap the first and the jth rows of M.
(4) If $k > 1$, perform steps (1) – (4) on the lower-right $(k-1) \times (k-1)$ sub-matrix of M.

Consider obliviously running Gaussian Elimination on an encrypted $k \times k$ matrix M over $GF(2)$ held by Bob. In step (1) Bob faces the problem of choosing the row M_j as he cannot distinguish a 0 entry from a 1 entry, and letting Bob (or Alice) learn j may compromise privacy. To go around this problem, we let Alice and Bob eliminate the leftmost column using several rows. For each of the rows they use, if the leftmost entry is 1 then we get the desired elimination. On the other hand, if the leftmost entry is 0, the matrix is not changed at all. We use randomization to guarantee that with high probability, the leftmost entry in at least one of the rows used is 1.

4.2   Column Elimination

Protocol Basic Column Elimination securely eliminates the leftmost column of a matrix using its jth row.

In the second step of the protocol Bob uses Alice's assistance in computing $\mathsf{Enc}(M[i, 1] \cdot M[j, 1] \cdot M_j)$. Note that if $M[j, 1] = 0$ then the result of step 2 is an encryption of $\mathbf{0}$. Therefore, if $M[j, 1] = 0$, Bob adds encryptions of $\mathbf{0}$ to every row, and thus $M' = M$. If $M[j, 1] = 1$, then Bob adds M_j exactly to the rows M_i with $M[i, 1] = 1$, as in the Gaussian Elimination procedure.

The communication complexity of the protocol is $O(\lambda k^2)$, as we run the Vector Multiply protocol for $O(k)$ times. However, in all iterations Bob multiplies an encryption of M_j. Hence, it is enough for Bob to randomly choose r_{M_j} and send Alice $\mathsf{Enc}(M_j + r_{M_j})$ only once. We get that the communication complexity from Bob to Alice is reduced to $O(\lambda k)$ while the communication from Alice to Bob remains $O(\lambda k^2)$. The communication from Alice to Bob will later be reduced as well.

Oblivious Column Elimination. As we noted above, if the leftmost coordinate of the eliminating row M_j is 0, running Basic Column Elimination does not

---

[6] To generalize our protocols to a field F, use the a sub-protocol for the following problem: Bob holds $\mathsf{Enc}(a)$ for $a \in F$, and Alice holds the private decryption key. In the end of the protocol Bob should hold $\mathsf{Enc}(a^{-1})$ if $a \neq 0$ and $\mathsf{Enc}(0)$ if $a = 0$. If $|F|$ is large, this can be done using the garbled circuit method of Yao, without affecting the asymptotic complexity of the protocol.

Protocol **Basic Column Elimination**

INPUT: As in Definition 2
OUTPUT: At the end of the protocol Bob holds an encryption of a matrix M' with the following properties: If $M[j, 1] = 0$ then $M' = M$. Otherwise, $M'_i = M_i$ for every $i \leq j$, and for $i > j$ (i) if $M[i, 1] = 0$ then $M'_i = M_i$, and (ii) if $M[i, 1] = 1$ then $M'_i = M_i + M_j$.

For every $j < i \leq k$ do the following:

1. Alice and Bob run protocol **Multiply**, with Bob's inputs being $\mathsf{Enc}(M[j, 1])$ and $\mathsf{Enc}(M[i, 1])$. As a result, Bob holds $\mathsf{Enc}(M[i, 1] \cdot M[j, 1])$.
2. Alice and Bob run protocol **Vector Multiply**, with Bob's inputs being $\mathsf{Enc}(M[i, 1]M[j, 1])$ and $\mathsf{Enc}(M_j)$. As a result, Bob holds $\mathsf{Enc}(M[i, 1] \cdot M[j, 1] \cdot M_j)$.
3. Bob locally computes $\mathsf{Enc}(M'_i) = \mathsf{Enc}(M_i + M[i, 1] \cdot M[j, 1] \cdot M_j)$.

---

advance the elimination process. Protocol **Oblivious Column Elimination** below uses the upper m rows of M to eliminate the leftmost column. The process is successful if any of these m rows contains 1 in the leftmost coordinate, and the parameter m is chosen such that this happens with high probability. Let $i \in \{1, \ldots, m\}$ be the minimal row index such that $M[i, 1]$ is non-zero. Note that (i) the column elimination process using any of the $i - 1$ upper rows does not change the matrix; (ii) the ith row M_i eliminates the leftmost column of M; (iii) the column elimination process using rows $i + 1$ to m does not effect M anymore. Denote by M' the resulting matrix.

Next, Alice and Bob swap the ith and first rows of M'. However, as the process is run obliviously, Bob does not know what i is. For that, we slightly modify Gaussian Elimination. Note that if the elimination was successful, the ith row in M' is the only row that does not have 0 in the leftmost coordinate. Bob adds the top m rows in M' into the top row of the matrix: $M'_1 = \sum_{j=1}^{m} M'_j$. The result is a leftmost 1 entry in at most two rows of M': the first and ith.

To eliminate the non-zero entry in M' we run **Basic Column Elimination** once more using the top row. If M is a full-rank matrix, and there is a 1 entry in the leftmost column of at least one of the top m rows of M, then in the resulting M' satisfies: (i) $M'[1, 1] = 1$ and (ii) $M'[j, 1] = 0$ for $2 \leq j \leq k$.

We note that Alice and Bob may agree on T by choosing a seed to a pseudorandom generator. Hence, the communication complexity of this protocol is m times that of protocol **Basic Column Elimination**. It is simple to verify that neither Alice nor Bob gain any information about M. Furthermore, $\mathrm{rank}(M') = \mathrm{rank}(M)$ as M is transformed into M' via a sequence of elementary matrix operations. Finally, the following claim shows that the elimination is successful with high probability.

Claim 5. *Let M be a $k \times k$ matrix and T be a random $k \times k$ matrix of full rank, both over $\mathrm{GF}(2)$ and let $m = \omega(\log k)$. If the leftmost column of M is non-zero, then with probability $1 - \mathbf{neg}(k)$, at least one entry in the leftmost column of the top m rows of the matrix TM is non-zero.*

Protocol Oblivious Column Elimination

INPUT: As In Definition 2
OUTPUT: At the end of the execution Bob holds an encryption $\mathsf{Enc}(M')$ of a $k \times k$ matrix such that $\mathrm{rank}(M') = \mathrm{rank}(M)$. Furthermore, if the leftmost column of M is non-zero then with high probability $M'[1,1] = 1$ and $M'[i,1] = 0$ for $2 \leq i \leq k$.

1. Alice and Bob agree on a random non-singular matrix $T \in_R \mathrm{GF}(2)^{k \times k}$. Bob uses the homomorphic properties of the encryption scheme to compute $\mathsf{Enc}(M')$ where $M' = TM$.
2. For every $1 \leq i \leq m(k)$, Alice and Bob run protocol **Basic Column Elimination** with Bob's inputs being $\mathsf{Enc}(M')$ and i.
3. Bob locally assigns $M_1' = \sum_{j=1}^{m} M_j'$ by adding the m upper encrypted rows of M'.
4. Alice and Bob run protocol **Basic Column Elimination** protocol with Bob's inputs being $\mathsf{Enc}(M')$ and 1.

Proof. Denote the leftmost non-zero column of M by c, the m top rows of T by $T_1, \ldots, T_m$, and note that $TM[i,1] = T_i c$. If T was a random matrix, that is $T_1, \ldots, T_m$ were independently randomly chosen vectors, then for every $i \in [m]$ the probability that $TM[i,1] = 0$ would be exactly $1/2$. Hence the probability that $TM[i,1] \neq 0$ for at least one value of i would be $1 - \mathbf{neg}(k)$. As a random matrix has full rank with constant probability [4], it follows that for a random non-singular matrix the probability that such an event occurs is also negligible.

4.3 Oblivious Gaussian Elimination

We now have the ingredients to present our **Oblivious Gaussian Elimination** protocol. On a matrix $M \in \mathrm{GF}(2)^{k \times k}$, the protocol first applies **Oblivious Column Elimination**, to eliminate the leftmost column, and then recurses on the lower-right $(k-1) \times (k-1)$ sub-matrix. For clarity of presentation, we first construct a 'naive' protocol, of communication complexity $\tilde{O}(\lambda k^3)$ and round complexity $\tilde{O}(k)$, and then discuss how to reduce the communication complexity to $\tilde{O}(\lambda k^2)$ and the round complexity to $\tilde{O}(k^{0.275})$.

As before, it is easy to verify that the parties gain no information about the matrix M. The following claim asserts the correctness of the protocol.

Claim 6. *At the end of the execution of the Oblivious Gaussian Elimination Protocol, Bob holds an encryption of an upper triangular matrix M' as required by Definition 2.*

4.4 Reducing Communication Complexity Via Lazy Evaluation

Informally, in the above protocol, Bob uses Alice as a 'calculator' for performing multiplications of encrypted field elements. The communication complexity of protocol **Oblivious Gaussian Elimination** is $O(\lambda m k^3) = \tilde{O}(\lambda k^3)$, by picking $m = \mathrm{polylog}(k)$. We now show that Bob can also use Alice as a storage device,

---

Protocol `Oblivious Gaussian Elimination` (for Square matrices)

INPUT AND OUTPUT: As in Definition 2.

1. Alice and Bob run protocol `Oblivious Column Elimination` on M. Let Bob's output be $\mathsf{Enc}(M')$.
2. Alice and Bob recursively run `Oblivious Gaussian Elimination`, on the lower-right $(k-1) \times (k-1)$ submatrix of M'. Let Bob's output be $\mathsf{Enc}(M'')$.
3. Bob locally outputs
$$\begin{pmatrix} \mathsf{Enc}(M'[1,1]), \; \mathsf{Enc}(M'[1,2]), \ldots, \mathsf{Enc}(M'[1,k]) \\ 0 \\ \vdots \qquad\qquad\qquad \mathsf{Enc}(M'') \\ 0 \end{pmatrix}.$$

---

and by this to reduce the communication complexity by a factor of k. Note that in each round of the protocol, Bob sends to Alice one row and one column of $\mathsf{Enc}(M)$, (masked with random vectors). In return, Alice sends $O(k)$ vectors that Bob adds to the matrix M. Each of these vectors is of size k, resulting in $\tilde{O}(\lambda k^2)$ communication per round.

However, as Bob is not using all the matrix entries in the following round, we can have Alice send him only the single row and column that are needed for completing the next round. We make a simple modification to the protocol, and let Alice maintain a matrix L, where $L[i,j]$ equals the sum of elements Bob needs to add to the entry $M[i,j]$. Alice would then send $\mathsf{Enc}(L[i,j])$ just before the ith row, or the jth column is needed for Bob. Moreover, whenever Bob multiplies his matrix by a full-rank matrix, Alice needs to multiply L by the same matrix, and this is the reason why Alice and Bob choose the random matrices together. This reduces the communication complexity of each round to $\tilde{O}(\lambda k)$, and hence the communication of the entire protocol to $\tilde{O}(\lambda k^2)$.

4.5 Reducing the Round Complexity

The round complexity of our protocol is linear in the matrix dimension, that is $\Omega(k)$. In this section we show how to reduce the round complexity to sub-linear while preserving the low communication complexity. The idea is to combine our communication efficient protocol with the general purpose round efficient protocol of Yao [18]. This idea was used before, in, e.g., [12].

The protocol is still based on Gaussian Elimination, only that here we eliminate a number of columns together in the same round. Let $\ell = k^\epsilon$ where $0 < \epsilon < 1$ is a parameter to be specified later. The first modification we make to `Oblivious Gaussian Elimination` is that Bob multiplies the matrix M by full rank matrices from *both* sides and not only from the left. By Claim 1, if M is a full rank matrix then with constant probability, the top-left $\ell \times \ell$ sub-matrix of M, denoted by N, is of full rank as well.

In this stage Alice and Bob execute a secure sub-protocol base on [18], such that at the end of the protocol Bob holds an encryption of N^{-1} if N is invertible, and an encryption of the 0 matrix if N is not full rank. Following this stage, the protocol is very similar to the original Oblivious Gaussian Elimination protocol. We divide the k rows of M into k/ℓ blocks of ℓ rows each. Denote the block of the top ℓ rows of M by K. The notations are depicted in Figure 1. For every other ℓ-rows block L, Alice and Bob perform the following:

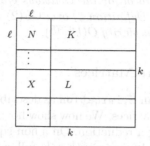

Fig. 1. Notations for the round efficient protocol

Denote the $\ell \times \ell$ left sub-matrix of L by X. Bob uses the help of Alice to compute $L \leftarrow L - XN^{-1}K$. If N is not invertible, then Bob has an encryption of the 0 matrix as N^{-1}, and thus the matrix is left unchanged. Otherwise, this procedure zeros the ℓ leftmost columns of L. As this process succeeds with constant probability, we repeat it a polylogarithmic number of times. Using basic techniques Alice and Bob can make sure that after finding a non-invertible N, no changes are done to the matrix till working on the next block of columns.

We first analyze the communication complexity of the protocol *excluding* the sub-protocol for computing N^{-1}. The communication complexity from Bob to Alice in each round is $\tilde{O}(\lambda \ell k)$ as Bob sends a masking of the top ℓ rows and the leftmost ℓ columns to Alice. Therefore, as there are k/ℓ rounds, the overall communication from Bob to Alice is $\tilde{O}(\lambda k^2)$. The communication complexity from Alice to Bob in each round is large as she needs to send $O(\lambda \ell k)$ bits for every ℓ-rows block. However, as before, we use lazy evaluation. Alice only sends Bob the $O(\lambda \ell k)$ bits he needs for the next block of columns, and keeps a matrix with the changes needed to be made to all the other entries in the matrix of Bob. This makes the overall communication complexity from Alice to Bob $O(\lambda k^2)$, excluding the protocol for computing N^{-1}.

We now analyze the communication complexity of the secure sub-protocol for computing N^{-1}. The communication complexity of securely inverting a matrix using Yao's garbled circuit method is related to the circuit complexity of matrix inversion. As matrix inversion is reducible to matrix multiplication, this can be done using a circuit of size $O(\ell^\omega)$, where the best known upper bound [6] for ω is approximately 2.38.

Therefore, the communication complexity of the sub-protocol is $O(\lambda \ell^\omega)$. As it is executed $\tilde{O}(k/\ell)$ times through the protocol we get that the overall complexity of executing the sub-protocol is:

$$(\lambda k/\ell)\ell^\omega = \lambda k \ell^{\omega-1} = kk^{\epsilon(\omega-1)} = \lambda k^{1+\epsilon\omega-\epsilon}.$$

To get a communication complexity of $\tilde{O}(\lambda k^2)$, we set the value of ϵ such that $1 + \epsilon\omega - \epsilon = 2$, i.e., $\epsilon = 1/(\omega - 1) = 1/1.38 \cong 0.725$. The round complexity of the protocol is $\tilde{O}(k^{1-\epsilon}) = \tilde{O}(k^{0.275})$. Choosing different values for ℓ, one gets a tradeoff between the communication complexity and the round complexity.

Theorem 7. *There is a protocol for the Oblivious Gaussian Elimination Problem for Square Matrices (See Definition 2) over* GF(2) *with communication complexity* $\tilde{O}(\lambda k^2)$ *and round complexity* $\tilde{O}(k^{0.275})$.

4.6   Handling Non-square Matrices

Protocol Oblivious Gaussian Elimination as described above works with very high probability for square matrices. We now show how to generalize the protocol to non-square matrices using a reduction. On a non-square matrix M of dimensions $k_a \times k_b$, Bob first randomly chooses a $k_a \times k_a$ full-rank matrix T_A and a $k_b \times k_b$ full-rank matrix T_B and computes $M^* = T_A M T_B$. Suppose w.l.o.g., that $k_a < k_b$ (otherwise perform the elimination on M^T). Alice and Bob execute the Oblivious Gaussian Elimination protocol on the top left $k_a \times k_a$ of M^*, denoted by N^*. The $k_b - k_a$ right columns of M are updated during the protocol, but are not eliminated. By Claim 2, if rank$(M) \geq r$ then with constant probability, rank$(N^*) = r$, and thus after executing the Oblivious Gaussian Elimination protocol on N^*, Bob holds an encrypted matrix Enc(M') such that M' is upper triangular, and with constant probability M' has exactly r non-zero entries on its diagonal. The communication complexity of the protocol is $\tilde{O}(\lambda k_a k_b)$ and the round complexity remains $\tilde{O}(k_a^{0.275})$. This completes the proof of Theorem 3.

5   Finding a Random Element in the Intersection

As in the previous sections, Alice holds a k_a-dimensional subspace $V_A \subseteq F^k$ represented by a $k_a \times k$ matrix A, while Bob holds holds a k_b-dimensional subspace $V_B \subseteq F^k$ represented by B. Alice and Bob wish to securely compute a uniformly distributed random vector in the subspace $V_A \cap V_B$. The main step in the design of our protocol is the addition of random linear constraints to the linear system created by the input subspaces, to reduce the number of solutions into only *one* random uniformly distributed solution.

We start with a definition of the Oblivious Linear Solve Problem.

Definition 3 (Oblivious Linear Solve Problem). *Input: Alice holds a private key of a public key homomorphic encryption scheme over a finite field F. Bob holds a $k_a \times k_b$ matrix M and a vector $v \in F^{k_b}$, encrypted by Alice's public key.*

Output: (i) If there exists a vector x such that $xM = v$, then with constant probability, Bob holds an encryption of an arbitrary such vector, and with constant probability Bob holds an encryption of $\mathbf{0}$. (ii) Otherwise Bob holds an encryption of $\mathbf{0}$.

In Appendix B we modify protocol Oblivious Gaussian Elimination to get protocol Oblivious Linear Solve whose properties are summarizes in the following claim.

Claim 8. *Protocol* Oblivious Linear Solve *is a secure protocol for the Oblivious Linear Solve Problem. The communication complexity of the protocol is* $\tilde{O}(\lambda k_a k)$ *and the round complexity is* $\tilde{O}(k_a^{0.275})$.

As in our previous protocols, Alice sends Bob $\mathsf{Enc}(A)$, and Bob computes $\mathsf{Enc}(M)$ for $M = AB^{\perp}$. By Claim 2, it is enough for Alice and Bob to find a random solution vector x to the linear system $xM = \mathbf{0}$. However, this linear system may have many solutions and picking an *arbitrary* solution is not satisfactory for our purpose. Therefore, we add random linear constrains to the linear system. That is, we concatenate a matrix R to M from the left, and a vector u to $\mathbf{0}$ and solve the linear system $x(R|M) = (u|\mathbf{0})$. We want to choose R and u so that with high probability, the system has a unique uniformly distributed solution.

The number of constraints needed to be added to the linear system depends on the dimension of the solution space of $xM = \mathbf{0}$. To this end, Alice and Bob first execute the Oblivious Gaussian Elimination protocol on M. By Theorem 3, with constant probability, the number of non-zero elements on the main diagonal of the result matrix M' equals the rank of M. Thus, Alice and Bob add a random linear constraint to R and u for every 0 on the main diagonal of M' and a trivial $x\mathbf{0} = 0$ constraint for every non-zero element on the diagonal of M'. Alice and Bob pick each random constraint by Alice sending the encryption of a random vector to Bob, who adds to it a second random vector. This way neither Alice nor Bob have information regarding the random constraints used. The technical method to add the constraints is depicted in Protocol Random Intersection Vector.

After adding the random constraints, Alice and Bob run the Oblivious Linear Solve protocol to get an encryption of a solution to the system $x(R|M)=(u|\mathbf{0})$. There are three possible cases: (i) The vector $(u|\mathbf{0})$ is not in the row span of the matrix $(R|M)$. In this case we get $x = \mathbf{0}$. (ii) There exists a non-zero vector x such that $x(R|M) = (u|\mathbf{0})$, but x is not unique. In this case it holds that $xM = \mathbf{0}$ but we do not argue that x is a random vector satisfying this requirement. (iii) There exist a unique non-zero vector x such that $x(R|M) = (u|\mathbf{0})$. In this case, by a symmetry argument, the vector x is a random vector satisfying $xM = 0$.

Alice and bob run Linear Equations Solve ℓ times and finally use the sum of the vectors x_j computed in these ℓ executions. The vectors x satisfying $xM = \mathbf{0}$ form a subspace, and hence are closed for addition. Thus, it is enough for one execution of Linear Equations Solve to yield a random solution, as in case (iii) above. To get to case (iii) we need the Oblivious Gaussian Elimination protocol to succeed and we need the linear system $x(R|M) = (u|\mathbf{0})$ to have a unique solution. The first event succeeds with constant probability. The success probability of the second event equals the probability that the sum of two random subspaces $V_1, V_2 \subseteq F^n$ of dimensions s and $n-s$ satisfy $V_1 \oplus V_2 = F^n$. The probability for this event is a constant as well. As both events occur with

---

Protocol Random Intersection Vector

INPUT: Alice (resp. Bob) holds a $k_a \times k$ (resp. $k_b \times k$) matrix A (resp. B) over $GF(2)$ representing a subspace $V_A \subseteq GF(2)^k$ (resp. $V_B \subseteq GF(2)^k$).
OUTPUT: Alice locally outputs a random vector v satisfying $v \in V_A \cap V_B$.

1. Alice generates keys for a homomorphic public key encryption system, and sends Bob $\mathsf{Enc}(A)$ and the public key.
2. Bob locally computes $\mathsf{Enc}(M)$, where $M \overset{\text{def}}{=} AB^{\perp}$.
3. For every $j \in \{1, \ldots, \ell\}$:
   (a) Alice and Bob execute Protocol **Oblivious Gaussian Elimination** on M. Let Bob's output be $\mathsf{Enc}(M')$.
   (b) For every $1 \leq i \leq k_a$, Alice and Bob choose a random vector w_i and set the ith column of the matrix R to be $c_i = (1 - M'[i, i])w_i$. That is, For every 0 on the diagonal of M', the vector c_i is a random vector, and for every 1 on the diagonal it is an encryption of $\mathbf{0}$.
   (c) Bob generates the vector $u \in GF(2)^{k_a}$ in the following way. For $1 \leq i \leq k_a$, if $M'[i, i] = 1$ Bob assigns $u[i] = 0$, while if $M'[i, i] = 0$ Bob randomly assigns $u[i] \in_R \{0, 1\}$. That is, Bob adds a random constraint for every 0 on the diagonal of M'.
   (d) Alice and Bob execute protocol **Linear Equations Solve** on $(R|M)$ and $(u|\mathbf{0})$ to get an encryption of a vector x_j such that $x_j(R|M) = (u|\mathbf{0})$, or $\mathsf{Enc}(\mathbf{0})$ if no such vector exists.
4. Bob's computes $\mathsf{Enc}(x) = \sum_{j=1}^{\ell} \mathsf{Enc}(x_j)$ and sends $\mathsf{Enc}(x)$ to Alice.
5. Alice outputs $v = xA$.

---

constant probability, case (iii) occurs with constant probability, and thus it is enough to run **Linear Equations Solve** $\omega(\log k)$ times, to get a negligible error probability.

Theorem 9. *Protocol* **Random Intersection Vector** *is a secure protocol for computing a random intersection vector. The communication complexity of the protocol is* $\tilde{O}(\lambda k_a k)$ *and the round complexity is* $\tilde{O}(k_a^{0.275})$.

6 Intersection of Affine Subspaces

In the affine subspace intersection problem Alice's input is an affine subspace $v_a + V_A$ where $v_a \in F^k$ and $V_A \subseteq F^k$ is a k_a dimensional linear subspace. Similarly, Bob's input is $v_b + V_B$, where $k_b = \dim(V_B)$. We design secure protocols for several problems concerning $(v_a + V_A) \cap (v_b + V_B)$. Our protocols are based on reductions to problems on linear subspaces. For example, to *compute* the intersection of two affine subspaces, we need both the **Intersection Computation** and the **Random Intersection Vector** protocols on linear subspaces. The following simple claims reduces the problem into computing whether a vector is contained in a subspace.

Claim 10. *There exists a vector $v \in (v_a + V_A) \cap (v_b + V_B)$ if and only if $v_a - v_b \in V_A + V_B$.*

Proof. Assume $v \in (v_a + V_A) \cap (v_b + V_B)$. Then $v = v_a + w_a$ for some $w_a \in V_A$ and $v = v_b + w_b$ for some $w_b \in V_B$. Hence $v_a + w_a = v_b + w_b$, and therefore, $v_a - v_b = w_b - w_a$, which means that $v_a - v_b \in V_A + V_B$. Now assume $v_a - v_b \in V_A + V_B$. Then there exist $w_a \in V_A$ and $w_b \in V_B$ such that $v_a - v_b = w_a + w_b$. Then $z \overset{\text{def}}{=} v_a - w_a = v_b + w_b$ is in the intersection $(v_a + V_A) \cap (v_b + V_B)$.

Claim 11. *Suppose $v_a + w_a = v_b + w_b$ for some $w_a \in V_A$ and $w_b \in V_B$. Then $(v_a + V_A) \cap (v_b + V_B) = (v_a + w_a) + (V_A \cap V_B)$.*

Proof. Let $v \in (v_a + V_A) \cap (v_b + V_B)$. Then there exist $z_a \in V_A$ and $z_b \in V_B$ such that $v = v_a + z_a = v_b + z_b$. As $v_a + w_a = v_b + w_b$, by subtracting equations we get $w_a - z_a = w_b - z_b$. Since $w_a - z_a \in V_A$ and $w_b - z_b \in V_B$, we get that $w_a - z_a \in V_A \cap V_B$. Thus $v = (v_a + w_a) - (w_a - z_a) \in (v_a + w_a) + (V_A \cap V_B)$.

For the other direction, let $v \in (v_a + w_a) + (V_A \cap V_B)$. Thus, $v = v_a + w_a + z = v_b + w_b + z$ where $z \in (V_A \cap V_B)$. As $w_a + z \in V_A$ and $w_b + z \in V_B$ we get $v \in (v_a + V_A) \cap (v_b + V_B)$.

Deciding if $(v_a + V_A) \cap (v_b + V_B)$ is empty. Protocol **Affine Intersection Decision** below is based on Claim 10. I.e., it checks whether $v = v_a - v_b \in \text{span}(V_A + V_B)$. The privacy of the protocol follows from that of protocol **Linear Equations Feasibility**, and the communication complexity is $\tilde{O}(\lambda k (k_a + k_b))$.

Protocol **Affine Intersection Decision**

INPUT: Alice holds a k_a dimensional affine subspace $v_a + V_A$ of $GF(2)^k$. Bob holds a k_b dimensional affine subspace $v_b + V_B$ of $GF(2)^k$.
OUTPUT: The output is 1 if and only if $v_a + V_A \cap v_b + V_B \neq \emptyset$.

1. Alice sends Bob $\mathsf{Enc}(A)$ and $\mathsf{Enc}(v_a)$, where A is a $k_a \times k$ matrix representing V_A.
2. Bob computes $\mathsf{Enc}(B)$ and $\mathsf{Enc}(v_b)$, where B is a $k_b \times k$ matrix representing V_B.
3. Alice and Bob execute Protocol **Linear Equations Feasibility** on the matrix $\begin{pmatrix} \mathsf{Enc}(A) \\ \mathsf{Enc}(B) \end{pmatrix}$ and the vector $\mathsf{Enc}(v_a + v_b)$. Bob sends the outcome of Protocol **Linear Equations Feasibility** to Alice, that decrypts it as the output.

Computing $(v_a + V_A) \cap (v_b + V_B)$. We describe a protocol for computing $(v_a + V_A) \cap (v_b + V_B)$, assuming the intersection is not empty. By Claim 11, it is enough for Alice and Bob to compute $V_A \cap V_B$, and find $w_a \in V_A$ and $w_b \in V_B$ such that $v_a + w_a = v_b + w_b$. We use Protocol **Linear Equation Solve** on the matrix $\begin{pmatrix} \mathsf{Enc}(A) \\ \mathsf{Enc}(B) \end{pmatrix}$ and the vector $\mathsf{Enc}(v_a + v_b)$. In the end of Protocol **Linear Equation Solve** Bob holds an encryption of a vector $c \in GF(2)^{k_a + k_b}$. Bob denotes the k_a

leftmost coordinates of c by w_a. Alice and Bob now execute Protocol Random Intersection Element on V_A and V_B, such that Bob holds an encryption of a vector $r \in_R V_A \cap V_B$. Bob sends $\mathsf{Enc}(v_i) \overset{\text{def}}{=} \mathsf{Enc}(v_a + w_a + r)$ to Alice. Now Alice and Bob execute Protocol Intersection Computation such that Alice learns $V_I \overset{\text{def}}{=} V_A \cap V_B$. Alice outputs $v_i + V_I$.

Random Intersection Vector. Note that if instead of computing V_I in the protocol above, Alice simply outputs v_i, we get a protocol for computing a random intersection element.

Acknowledgments. We thank Amos Beimel, Yinnon Haviv, Benny Pinkas and Lior Zolf for helpful conversations.

References

1. D. Boneh, E. Goh, and K. Nissim. Evaluating 2-DNF Formulas on Ciphertexts. TCC 2005 pages 325–341.
2. M. Bellare, O. Goldreich, and E. Petrank. Uniform Generation of NP-Witnesses Using an NP-Oracle. In Inf. Comput. 163(2): 510-526 (2000).
3. A. Beimel, and E. Weinreb. Separating the Power of Monotone Span Programs over Different Fields. In FOCS 2003: 428-437.
4. A. Borodin, J. von zur Gathen, and J. Hopcroft. Fast parallel matrix and gcd computations. In Information and Control, 52(3):241-256, March 1982.
5. R. Cramer, and I. Damgård. Secure Distributed Linear Algebra in a Constant Number of Rounds. In CRYPTO 2001: 119-136.
6. D. Coppersmith, and S. Winograd. Matrix Multiplication via Arithmetic Progressions. In Proc. 19th ACN Symp. on Theory of Computing, pp. 1–6, 1987.
7. I. Damgård and M. Jurik. A generalization, a simplification and some applications of Paillier's probabilistic public-key system. In K. Kim, editor, *Proceedings of Public Key Cryptography 2001*, volume 1992 of *LNCS*, pages 119–136. Springer, 2001.
8. T. ElGamal. A public key cryptosystem and a signature scheme based on discrete logarithms. *IEEE Transactions on Information Theory*, 31(4):469–472, Jul 1985.
9. O. Goldreich. *The Foundations of Cryptography - Volume 2*. Cambridge University Press, 2004.
10. S. Goldwasser and S. Micali. Probabilistic encryption & how to play mental poker keeping secret all partial information. In *Proceedings of the fourteenth annual ACM symposium on Theory of computing*, pages 365–377. ACM Press, 1982.
11. M. Karchmer and A. Wigderson. On Span Programs In *Proc. of the 8th IEEE Structure in Complexity Theory*, pages 102–111, 1993.
12. Y. Lindell and B. Pinkas. Privacy Preserving Data Mining In *J. Cryptology* 15(3):177–206, 2002.
13. P. Pallier. Public-key cryptosystems based on composite degree residuosity classes. In J. Stern, editor, *Proceedings of Eurocrypt 1999*, volume 1592 of *LNCS*, pages 223–238. Springer-Verlag, May 1999.
14. T. P. Pedersen. A threshold cryptosystem without a trusted party. In D. Davies, editor, *Proceedings of Eurocrypt 1991*, volume 547 of *LNCS*, pages 522–526. Springer, 1991.

15. R. L. Rivest, A. Shamir, and L. M. Adleman. A Method for Obtaining Digital Signatures and Public-Key Cryptosystems. Commun. ACM 21(2): 120–126 (1978).
16. T. Sander, A. Young, and M. Yung. Non-interactive CryptoComputing for NC^1. In *Proceedings of the 40th Symposium on Foundations of Computer Science (FOCS)*, pages 554–567, New York, NY, USA, Oct. 1999. IEEE Computer Society Press.
17. M. Sipser. A Complexity Theoretic Approach to Randomness. In Proc. of the 15th Annual Symp. on the Theory of Computing, 1983.
18. A. C. Yao. Protocols for secure computations. In *Proceedings of the 23rd Symposium on Foundations of Computer Science (FOCS)*, pages 160–164. IEEE Computer Society Press, 1982.

A   Security Proof for the Intersection Computation Protocol

In this section we prove Bob's security in Protocol `Intersection Computation`. Note that the only information Bob sends to Alice is $\mathsf{Enc}(M)$, from which she learns M.

Simulator Alice

INPUT: A $k_i \times k$ matrix C representing V_I, a $k_a \times k$ matrix A representing V_A and an integer $k_b \leq k$.
OUTPUT: A matrix M in the same distribution of M in protocol `Intersection Computation`.

1. Compute a $k_i \times k_a$ matrix D, satisfying $DA = C$. As $V_I \subseteq V_A$ such a matrix exists and is easy to compute.
2. Compute a $k_b \times k_a$ matrix E by adding $k_b - k_i$ zero rows to D.
3. Compute the $k_a \times k_{b'}$ matrix $E^\perp$ whose columns represent the kernel of the matrix E.
4. Randomly choose a $k_b' \times k_b'$ full rank matrix T_E and output $M = E^\perp T_E$.

Claim 12. *M_S is distributed identically to M in* `Intersection Computation`.

Proof. Define the subspace $W = \{v : vA \in A \cap B\}$. The rows of the matrices D and E in the simulator span W. Therefore the columns of the matrix $E^\perp$ span the subspace $W^\perp$. Moreover, according to Claim 2, the columns of the matrix $AB^\perp$ from protocol `Intersection Computation` also span $W^\perp$.

Thus, there exists a full rank matrix T_0 of dimensions $k_b' \times k_b'$ such that $E^\perp T_0 = AB^\perp$. The probability that a matrix M_S is the simulator output is $Pr_{T_E}[M_S = E^\perp T_E]$. For every such choice of T_E take $T_B = T_0^{-1} T_E$ to be the choice of the protocol, to get $M = AB^\perp T_B = AB^\perp T_0^{-1} T_E = E^\perp T_E$. Conversely, for every random choice T_B of the protocol, set $T_E = T_0 T_B$ to get $M_S = E^\perp T_E = E^\perp T_0 T_B = AB^\perp T_B$. Therefore, the distributions are identical.

B   Obliviously Solving Sets of Linear Equations

Let Bob hold an encrypted matrix $\mathsf{Enc}(M)$ and an encrypted vector $\mathsf{Enc}(v)$. We consider the decisional and functional versions of solving the linear system $cM = v$ (i.e., deciding whether exists a vector c satisfying $cM = v$, and finding such c).

Protocol Linear Equations Feasibility

INPUT: Alice holds a private key for a public-key homomorphic encryption scheme over GF(2). Bob holds an encryption $\mathsf{Enc}(M)$ of a $k_a \times k_b$ matrix over $GF(2)$ (we assume $k_a \leq k_b$; the general case is analogous), and an encryption $\mathsf{Enc}(v)$ of a vector $v \in \mathrm{GF}(2)^{k_b}$.

OUTPUT: If a vector $c \in \mathrm{GF}(2)^{k_a}$ exists such that $cM = v$ then Bob locally outputs $\mathsf{Enc}(1)$; Otherwise, he outputs $\mathsf{Enc}(0)$.

1. Bob randomly chooses a non-singular $k_a \times k_a$ matrix T_R, and a non-singular $k_b \times k_b$ matrix T_C, and computes $M' = T_R M T_C$, and $v' = v T_C$.
2. Alice and Bob run protocol Oblivious Gaussian Elimination, on the $(k_a+1) \times k_b$ matrix $\begin{pmatrix} M' \\ v' \end{pmatrix}$, with the following exception: when multiplying the matrix M' by random matrices from the left, Alice and Bob pick a matrix that does not change the lower row of M'. Let Bob's output be $\mathsf{Enc}(M'')$.
3. Alice and Bob use the Multiply protocol to compute an encryption of $\prod_{i=1}^{k_b}(1 - M''[k_a + 1, i])$. This product is 1 if and only if the $k_a + 1$ row of M'' is $\mathbf{0}$.

In the first step of the protocol Bob multiplies M by random operators from the left and from the right to get $M' = T_R M T_C$. The following simple claim shows that it is enough to check if there exists a vector c' such that $c'M' = v'$ to solve the original $cM = v$ system.

Claim 13. *There exists a vector $c \in \mathrm{GF}(2)^{k_a}$ such that $cM = v$ if and only if there exists a vector $c'\mathrm{GF}(2)^{k_a}$ such that $c'M' = v'$.*

Proof. If there exists a vector $c \in \mathrm{GF}(2)^{k_a}$ such that $cM = v$, then the rows of M span v. Multiplying M by T_R from the left does not change the row space of M. Thus, there exists a vector c^* such that $c^* T_R M = v$. Multiplying both sides by T_C from the right results in $c^* M' = c^* T_R M T_C = v T_C = v'$. The other direction follows similarly.

By Claim 1, if M is a rank r matrix, then with constant probability the $r \times r$ top left sub-matrix of M' is of full rank. In the second step, Alice and Bob jointly perform Gaussian Elimination on the matrix $\begin{pmatrix} M' \\ v' \end{pmatrix}$. We run the protocol on the $k_a \times k_a$ top left sub-matrix, letting Basic Column Elimination update the entire matrix. If the rows of the matrix M' span the row v', then by the end of the Gaussian Elimination protocol, the bottom row will be $\mathbf{0}$. Otherwise, the bottom row will not be $\mathbf{0}$. In step 3. we translate a zero vector in the last column to $\mathsf{Enc}(1)$, and a non-zero vector to $\mathsf{Enc}(0)$.

The protocol has a one-sided error. If the answer is NO then Bob will always hold an encryption of 0. If the answer is YES, then if in step (1) the rank of the top-left $k_a \times k_a$ sub-matrix of M' is that of M, Bob will hold an encryption of 1. As this happens with constant probability, Alice and Bob can execute the protocol a polylogarithmic number of times, and OR the results in order to make the error probability negligible.

Solving the Linear System. Note that the computation done in the Gaussian Elimination protocol may be viewed as multiplying M by non-singular matrices from the right and from the left (for column elimination, and for randomizing). That is, at the end of the protocol we get an encryption of M' where $M' = T_1 M T_2$ for some non-singular matrices T_1 and T_2.

To have Bob hold an encryption of a vector c such that $cM = v$, we need Bob to hold an encryption of T_1. We modify the Gaussian Elimination protocol such that any operation done on the rows of the input matrix M is simultaneously performed on the rows of a unit matrix I_{k_a+1}. At the end of this process Bob holds an encryption $T_1 I_{k_a+1} = T_1$.

We now describe protocol **Linear Equations Solve**. We assume that $cM = v$ is feasible, and compute such a solution c. Alice and Bob execute protocol **Linear Equations Feasibility**, using the modified **Oblivious Gaussian elimination** protocol. As a result, Bob holds a matrix T_1 such that $T_1 \begin{pmatrix} M' \\ v' \end{pmatrix} = \begin{pmatrix} M'' \\ \mathbf{0} \end{pmatrix}$.

Denote the lower row of T_1 by t_1. The vector t_1 gives a linear combination of M' and v' that gives the vector $\mathbf{0}$. Moreover, as we modified the Gaussian Elimination protocol not to use the bottom row in the elimination process, the rightmost entry of t_1 must be 1. Thus, denoting the k_a left entries of t_1 by c_1, we get that $c_1 M' + v' = 0$, that is, over $GF(2)$, $c_1 M' = v'$. Recall that $M' = T_R M T_C$ and $v' = v T_C$, and thus $c_1 T_R M T_C = v T_C$. Therefore, $c_1 T_R M = v$, and having an encryption of c_1, Bob can output $\mathsf{Enc}(c_1 T_R)$ as the output of the protocol.

Threshold and Proactive Pseudo-Random Permutations

Yevgeniy Dodis[1,*], Aleksandr Yampolskiy[2,**], and Moti Yung[3]

[1] New York University, Department of Computer Science,
251 Mercer Street, New York, NY 10012, USA
dodis@cs.nyu.edu
[2] Yale University, Department of Computer Science,
51 Prospect Street, New Haven, CT 06520, USA
aleksandr.yampolskiy@yale.edu
[3] RSA Laboratories and Columbia University, Department of Computer Science,
1214 Amsterdam Avenue, New York, NY 10027, USA
moti@cs.columbia.edu

Abstract. We construct a reasonably efficient threshold and proactive pseudo-random permutation (PRP). Our protocol needs only $O(1)$ communication rounds. It tolerates up to $(n-1)/2$ of n dishonest servers in the semi-honest environment. Many protocols that use PRPs (*e.g.*, a CBC block cipher mode) can now be translated into the distributed setting. Our main technique for constructing invertible threshold PRPs is a distributed Luby-Rackoff construction where both the secret keys *and* the input are shared among the servers. We also present protocols for obliviously computing pseudo-random functions by Naor-Reingold [41] and Dodis-Yampolskiy [25] with shared input and keys.

Keywords: Distributed Block Ciphers, Distributed Luby-Rackoff Construction, Oblivious Pseudo-Random Functions, Threshold Cryptography.

1 Introduction

Block ciphers are familiar cryptographic tools, which transform blocks of plaintext into blocks of ciphertext of the same length. The DES (U.S. Data Encryption Standard) is a well-known example of a block cipher, which was, until recently, used by many financial firms to protect online transactions. Traditionally, **pseudo-random permutations** (PRPs) have been used to model secure block ciphers [35]. They map l-bit inputs into unique l-bit outputs that appear random to parties who lack the secret key. A close relative of the PRP is a **pseudo-random function** (PRF), which needs not be invertible, but whose outputs also look like random bit-strings without the secret key [29].

* Supported in part by NSF career award CCR-0133806 and NSF grant CCR-0311095.
** Supported by NSF grants CCR-0098078,ANI-0207399,CNS-0305258, and CNS-0435201.

S. Halevi and T. Rabin (Eds.): TCC 2006, LNCS 3876, pp. 542–560, 2006.
© Springer-Verlag Berlin Heidelberg 2006

MOTIVATION. The security of these functions relies on the owner of the secret key, who has a primary responsibility of keeping the key safe. Alas, it is not always realistic to put all trust into a single party. The area of **threshold cryptography** deals exclusively with sharing the ability to perform cryptographic operations between a set of n servers [20]. A long line of research produced many distributed protocols that are more efficient than generic multi-party solutions when used for public key encryption [44, 17, 45], digital signatures [19, 21, 22, 28], key generation [1, 8, 27], pseudo-random functions [42, 40, 10], and other applications. The extra security and increased availability of constructions justify the added complexity. The pseudo-random permutation is the only primitive that is still missing from this long list.

Several initial attempts [9, 37] gave a very basic sharing structure with many limitations and drawbacks.[1] The question of constructing a threshold PRP was yet left open. In this paper, we resolve this problem. Many protocols are defined for PRPs (block ciphers) and, when needed, can now be readily translated into the distributed setting. This makes sense for sensitive operations like key-encrypting-key in the Key Distribution Center [40]. Applications such as distributed remotely keyed authenticated encryption and CBC encryption mode become possible, since they require a PRP as a building block (regular PRFs do not suffice).

We focus on implementing the Luby-Rackoff construction [35] as a method for building PRPs. It uses the **Feistel permutation** for function F (denoted $\bar{F}$), which sends a $2l$-bit string (x^L, x^R) to a $2l$-bit string $(x^R, x^L \oplus F(x^R))$. Luby and Rackoff showed that a composition of four Feistel permutations (denoted $\Psi(F_1, F_2, F_3, F_4) = \bar{F}_1 \circ \cdots \circ \bar{F}_4$) is a secure $2l$-bit PRP when F_i are independent l-bit PRFs. While a sequential composition of PRFs to build a sequential PRP is generic, there is a major technical difficulty in the distributed Luby-Rackoff construction. Particularly, the difficult part is that if one uses a PRF as an intermediate round function, then not just the secret key, but also the output needs to be kept distributed to assure the security of the entire Luby-Rackoff construction. At the same time, the computation needs to continue and compute on these shares, which means that we need to compute on shared inputs as well.

OUR RESULTS. This paper describes an $O(1)$ round distributed protocol for evaluating $\Psi(F_1, F_2, F_3, F_4)$, which results in a **threshold pseudo-random permutation**. Our protocol invokes the multiplication protocol for the underlying secret sharing scheme $O(mn + m \log m)$ times, where n is the number of servers and m is the maximum input length. It tolerates up to $\tau = \lfloor (n-1)/2 \rfloor$ dishonest servers in the semi-honest model, which is consistent with some prior work on distributed PRFs [40] and multiparty tools [1, 12, 15] used in our constructions.

---

[1] They showed how to build rather inefficient cascade ciphers $E_k(x) = g_{k_m}(\cdots g_{k_2}(g_{k_1}(x)))$, where $g(\cdot)$ is itself a secure cipher, by sharing a sequence of keys in a special way. For τ-out-of-n sharing, the number of keys and composition layers is on the order of $\binom{n}{\tau}$, which is exponential for most $\tau = \omega(1)$.

It can be made robust using standard techniques [46] and, as we show, can be amended to ensure **proactive security** [33].

As we have explained, intermediate Feistel values arising after each round of the Luby-Rackoff construction must be kept secret, yet we must evaluate the PRFs F_i on them. Unfortunately, prior distributed PRF constructions [42,40,10] are inapplicable to our problem, because they require the PRF input to be publicly known. We give two protocols for distributed computation of PRFs by Naor-Reingold [41] and Dodis-Yampolskiy [25] when both the secret keys *and* the input are shared among the servers. Effectively, we implement **oblivious distributed PRFs**, where servers do not learn what the input is, yet blindly compute the PRF value.

We note that, theoretically, we can always use general multi-party techniques [5] to distribute the computation of a PRP. Until recently, this was not a viable option. These techniques either (i) required a linear number of communication rounds (in the circuit depth) [5,48] or (ii) ran in $O(1)$ rounds but used expensive zero-knowledge proofs for each gate of the circuit [4]. A recent improvement by Damgård *et al.* [16] allows to securely evaluate any circuit C in $O(1)$ rounds using $O(|C|n)$ cryptographic operations ($|C|$ is the circuit size). If we distribute the DES circuit (which is believed to be a PRP) using Damgård *et al.*'s techniques, we obtain comparable efficiency to our threshold PRP.[2] Our protocol is thus fairly practical. In addition, it has theoretical value in and of itself and could be of independent interest in other fields.

OVERVIEW OF OUR CONSTRUCTION. In our protocol, servers hold Shamir shares [47] of secret keys SK_i to PRFs F_i used in the Luby-Rackoff (LR) construction of a PRP $\Psi(F_1, F_2, F_3, F_4)$. The untrusted user who wants to compute the PRP's value broadcasts his input x to the servers. Servers somehow verify that the user is entitled to evaluate the PRP and engage in an interactive protocol, which terminates with shares of the PRP's value.

Our round functions F_i are based on a PRF by Dodis-Yampolskiy [25]. We chose this PRF because it possesses useful algebraic properties and can be computed in $O(1)$ rounds. Given an l-bit input $x = x_1 x_2 \ldots x_l$ (which can be viewed as an element of $\mathbb{Z}_Q$) and a secret key $SK \in \mathbb{Z}_Q$, the PRF value is $F_{SK}(x) = g^{1/(x+SK)}$. Here, g is the generator of a group in which the **decisional Diffie-Hellman inversion** (y-DDHI) problem is hard. The y-DDHI problem asks: "given $(g, g^x, \ldots, g^{(x^y)}, R)$ as input, to decide whether $R = g^{1/x}$ or not." It appears hard in a quadratic residues subgroup G_Q of $\mathbb{Z}_P^*$ ($P = 2Q + 1$) for sufficiently large primes P, Q.

Dodis and Yampolskiy showed that $F_{SK}(\cdot)$ is secure only for inputs of small length $l = \omega(\log k)$, which makes it unsuitable for the LR construction, whose

---

[2] In fact, for realistic settings, our algorithm performs better. The full DES circuit contains about $|C| \approx 16000$ Boolean gates [6]. Let the group size be a $m = 1024$ bit prime and the number of servers be $n = 100$. Our protocol performs roughly $(mn + m \log m) \cdot (m^2 n + mn^2 \log n) \approx 1.95 \times 10^{13}$ bit operations, while Damgård *et al.*'s protocol [16] performs $(|C|n) \cdot (mn^2 \log n) \approx 10.9 \times 10^{13}$ operations to compute the DES circuit.

round functions must accept longer $l = \Theta(k)$ bit inputs (k is the security parameter). In this paper, we assume subexponential hardness of the y-DDHI problem. This immediately allows us to support inputs of size $a = \Theta(k^\delta)$ for some small $\delta \approx 1/3$. We can shrink the input to the LR construction from $l = \Theta(k)$ bits down to $a = \Theta(k^\delta)$ bits using an ϵ-universal hash function $h_i(x) = (ix \bmod Q) \bmod 2^a$. We thus get a new PRF $F'_{i,SK}(x) = F_{SK}(h_i(x))$, which can be used in the (centralized) LR construction.

We distribute the LR construction using well-known multiparty tools of addition, multiplication, inversion, etc. [3, 5, 1]. We rely heavily on an $O(1)$ round protocol by Damgård et al. [15], which computes shares of bits of $x \in \mathbb{Z}_P$ from shares of x. This protocol allows us to efficiently perform modular reduction, exponentiation, and truncation of shared values.

We can compute the PRF $F'_{i,SK}(x)$ with shared input x and keys (i, SK) as follows. Computing the ϵ-universal hash $h_i(x) = (ix \bmod Q) \bmod 2^a$ amounts to a single multiparty multiplication, followed by a call to Damgård et al.'s protocol to extract the trailing a bits. We can also distribute the computation of $F_{SK}(x) = g^{1/(x+SK)}$ because it is well-known how to do multiparty addition, inversion, and exponentiation. As a result, we obtain a sharing of $F'_{i,SK}(x)$, a random group element in G_Q, whereas we need a sharing of a random l-bit string. We can use a deterministic extractor $E(x) = (x^{(P+1)/4} \bmod P) \bmod 2^l$ to convert this group element into a random l-bit string. Computing this extractor distributively entails a single distributed exponentiation followed by a call to Damgard et al.'s protocol to extract l bits.

Armed with a protocol for computing the PRF $F'_{i,SK}(\cdot)$, we can distribute a single Feistel permutation, which maps (x^L, x^R) into $(x^R, x^L \oplus F'_{i,SK}(x^R))$. The only missing link is how to XOR shares of PRF's bits with shares of input's bits. Given shares of bits $b_1, b_2 \in \{0, 1\}$, we can get a share of $b_1 \oplus b_2$ by distributively computing $(b_1 + b_2) \cdot (2 - (b_1 + b_2))$. This completes our calculation. We obtain a threshold PRP by iterating the distributed Feistel permutation four times, cross-feeding its outputs to inputs.

PAPER ORGANIZATION. The remainder of the paper is organized as follows. Section 2 reviews some preliminaries and defines our system model. In Section 3, we give distributed protocols for evaluating pseudo-random functions by Naor-Reingold [41] and Dodis-Yampolskiy [25] when keys and input are shared. In Section 4, we present our distributed Luby-Rackoff protocol. Some practical applications of our construction appear in Section 5. We conclude in Section 6. Due to limitations in space, some details have been omitted; they can be found in the full version, available as a Yale CS technical report [26].

2 Preliminaries

In this section, we discuss some basic definitions and assumptions.

2.1 Our Model

Let k be a security parameter. We consider n computationally bounded servers $P_1, \ldots, P_n$, which are connected by secure and authentic channels[3]. Our protocols are secure against a static, honest-but-curious adversary who controls up to $\tau = \lfloor (n-1)/2 \rfloor$ servers. This threshold results from the multiplication protocol by Ben-Or et al. [5], which is used throughout the paper. We prove security in the framework by Canetti [11]. In the honest-but-curious setting, privacy is preserved under non-concurrent modular composition of protocols. This composition theorem will be the main source of our privacy proofs.

2.2 Notation

The notation in this paper is adapted from [1, 15]. We define $\mathbb{Z}_P$ as the set $\{0, \ldots, P-1\}$. We denote additive shares over $\mathbb{Z}_P$ of a value $a \in \mathbb{Z}_P$ by $\langle a \rangle_1^P, \ldots, \langle a \rangle_n^P \in \mathbb{Z}_P$; i.e., $a = \sum_{j=1}^n \langle a \rangle_j^P \bmod P$. Meanwhile, we denote Shamir shares [47] of $a \in \mathbb{Z}_P$ by $[a]_1^P, \ldots, [a]_n^P \in \mathbb{Z}_P$; i.e., $a = \sum_{j=1}^\tau \lambda_j [a]_j^P \bmod P$, where τ is the threshold and λ_j are the Lagrange coefficients.

We denote protocols as follows: the term $[a]_j^P \leftarrow \mathtt{PROTOCOL}([b]_j^P, c)$ means that server P_j executes the protocol $\mathtt{PROTOCOL}$ with local input $[b]_j^P$ and public input c. As a result of the protocol, it gets back local output $[a]_j^P$. In all cases, the local inputs and outputs will be Shamir shares over the appropriate field.

2.3 Building Blocks

We review some standard tools for multiparty computation that are used throughout the paper. All these protocols require $O(1)$ rounds of communication. We measure their running time in terms of bit operations in $m = \lceil \log_2 P \rceil$ (the modulus length) and n (the number of servers). Below, we use $\mathcal{B}$ as a shorthand for $O(nm^2 + mn^2 \log n)$.

Sharing a Secret. To compute a Shamir sharing of $x \in \mathbb{Z}_P$ over $\mathbb{Z}_P$, player P_j chooses random coefficients $\alpha_k \in \mathbb{Z}_P$ for $k = 1, \ldots, \tau$. He then sends $[x]_i^P = x + \sum_{k=1}^\tau \alpha_k \cdot i^k \bmod P$ to player P_i. We denote this protocol by $\mathtt{RANDSS}(x, \mathbb{Z}_P)$; it takes $O(n^2 m \log n)$ bit operations.

Basic Operations. Addition and multiplication of a constant and a Shamir share can be done locally. Hence, $[x]_j^P + c \bmod P$ is a polynomial share of $x + c \bmod P$ and $c \cdot [x]_j^P \bmod P$ is a share of $xc \bmod P$. These operations take $O(m)$ and $O(m^2)$ bit operations, respectively. Similarly, we can compute $[x]_j^P + [y]_j^P \bmod P$, which is a share of $x + y \bmod P$. Addition requires $O(m)$ bit operations.

Multiplication. We note that a product of polynomially many shared secrets $x_1, \ldots, x_s \in \mathbb{Z}_P^*$ can be computed in constant rounds [3, 15]. We denote this protocol by $\mathtt{MUL}([x_1]_j^P, \ldots, [x_s]_j^P)$; it uses $O(s\mathcal{B})$ bit operations.

---

[3] Such channels can be implemented using public-key encryption and digital signatures.

Conversion Between Bit Shares. Given Shamir shares of a single bit $b \in \{0,1\}$ in $\mathbb{Z}_P$, we may need to obtain its shares in $\mathbb{Z}_Q$. We can do this as follows. First, each server P_j locally computes $[b']_j^P \leftarrow -2 \cdot [b]_j^P + 1 \pmod{P}$ to convert the bit from a $0/1$ to a $1/-1$ encoding. Next, P_j chooses a random $b_j \in \{1, -1\}$ and shares it among servers in both $\mathbb{Z}_P$ and $\mathbb{Z}_Q$. He computes $[b'']_j^P \leftarrow \text{MUL}([b']_j^P, [b_1]_j^P, \ldots, [b_n]_j^P)$ and reveals it for all servers to reconstruct b''. Finally, P_j multiplies $b'' \pmod{Q}$ by its share of $\text{MUL}([b_1]_j^Q, \ldots, [b_n]_j^Q)$ and converts the result to a $0/1$ encoding. The protocol requires $O(1)$ rounds and $O(n\mathcal{B})$ bit operations.

Bit Representation. Let $x \in \mathbb{Z}_P$ be a shared secret (written $x_m \ldots x_1$ in binary). In some situations, we will need to obtain Shamir shares of the bits of x. For this, we will use a protocol by Damgård *et al.* [15], denoted $([x_1]_j^P, \ldots, [x_m]_j^P) \leftarrow \text{BITS}([x]_j^P)$, which uses $O((m \log m)\mathcal{B})$ bit operations.

Occasionally, we will need to compute shares of a least significant bits of $x \in \mathbb{Z}_P$ in $\mathbb{Z}_Q$ (rather than in $\mathbb{Z}_P$). We will first run the $\text{BITS}([x]_j^P)$ protocol and then convert each bit share from $\mathbb{Z}_P$ to $\mathbb{Z}_Q$. We denote this protocol by $([x_1]_j^Q, \ldots, [x_a]_j^Q) \leftarrow \text{BITS}([x]_j^P, a, \mathbb{Z}_Q)$. It requires $O(1)$ rounds and $O((an + m \log m)\mathcal{B})$ bit operations.

Given bit-by-bit shares of $x \in \mathbb{Z}_P$, denoted $[x_1]_j^P, \ldots, [x_m]_j^P$, we can easily obtain shares of x by locally computing $[x]_j^P \leftarrow \sum_{i=1}^{m} 2^{i-1} \cdot [x_i]_j^P \bmod P$. This takes $O(m^3)$ bit operations.

Inversion. Let $x \in \mathbb{Z}_P$ be a shared secret. A protocol due to Bar-Ilan and Beaver [3], denoted by $\text{INV}([x]_j^P)$, allows us to compute the shares of $x^{-1} \bmod P$. It takes an expected number of $O(\mathcal{B})$ bit operations.

Generating a Random Number. Occasionally, servers may need to jointly generate shares of a random number. A simple protocol, denoted $\text{JRP}(\mathbb{Z}_P)$, accomplishes this in $O(mn^2 \log n)$ bit operations [1]. There also exists a protocol $\text{JRPZ}(\mathbb{Z}_P)$ to jointly compute a sharing of zero modulo P in $O(mn^2 \log n)$ bit operations.

Exponentiation. Some of our protocols require computing the shares of $x^y \bmod P$ when: (i) the exponent $y \in \mathbb{Z}_Q$ is shared, but the base is fixed; (ii) the base $x \in \mathbb{Z}_P$ is shared and the exponent is fixed; or (iii) both the base and the exponent are shared. We denote protocols for the above scenarios $\text{EXP}_1(x, [y]_j^Q)$, $\text{EXP}_2([x]_j^P, y)$, and $\text{EXP}([x]_j^P, [y]_j^Q)$. They run in $O(1)$ rounds and require, respectively, $O((mn + m \log m)\mathcal{B})$, $O(m^3 + n\mathcal{B})$, and $O(m^4 + (mn + m \log m)\mathcal{B})$ bit operations per player. They appear in the full version of the paper [26].

3 Distributed Pseudo-Random Functions

In this section, we describe two distributed PRF constructions, where both the secret key and the input are shared. This will ensure that unscrupulous servers do not learn the results of intermediate Luby-Rackoff computations. In Section 3.1, we show how to do this for the PRF by Naor and Reingold [41]. Then in Section 3.2, we describe how to do this for the recently introduced PRF by Dodis and Yampolskiy [25].

Let the input size $l : \mathbb{N} \mapsto \mathbb{N}$ be a function computable in poly(k) time. Sometimes, for simplicity, we will write l for $l(k)$. The initial input for all servers is a triple (P, Q, g), where P, Q are large primes such that $P = 2Q + 1$ and $P \equiv 3 \bmod 4$. Here g is a generator of quadratic residues subgroup G_Q of $\mathbb{Z}_P^*$. The group $\mathbb{Z}_P^*$ must be sufficiently large, i.e., $P \gg 2^k$. Such a triple can be publicly chosen without a trusted party by executing Bach's algorithm [2].

Both centralized PRFs take as input an l-bit message x, the secret key SK and output a random group element in G_Q. In our distributed PRFs, each server P_j receives a share of the secret key SK and l shares of bits of x.

3.1 Naor-Reingold PRF

The secret key $SK = (a_0, a_1, \ldots, a_l)$ consists of $l + 1$ random exponents in $\mathbb{Z}_Q$. Given an l-bit input $x = x_1 \ldots x_l$, the PRF $F_{SK}^{NR} : \{0,1\}^l \mapsto G_Q$ is defined as

$$F_{SK}^{NR}(x) = (g^{a_0})^{\prod_{i : x_i = 1} a_i}.$$

This PRF was shown to be secure for polynomially sized inputs, $l(k) = $ poly(k), under the decisional Diffie-Hellman (DDH) assumption: "given (g, g^x, g^y) and $R \in G_Q$, it is hard to determine if $R = g^{xy}$ or not."

We can compute the PRF value recursively. Set $h_0 = g^{a_0}$. Then, for all $i = 1, \ldots, l$,

$$h_i = \begin{cases} h_{i-1}^{a_i} & \text{if } x_i = 1, \\ h_{i-1} & \text{otherwise.} \end{cases} \tag{1}$$

It is easily seen that the PRF value must be equal to h_l. This form is convenient for distributed computation when both the input x and the secret exponents a_i are shared. One problem here is that we need to implement an if-condition on secret input x. We can use a simple trick and rewrite Equation (1) as

$$h_i = h_{i-1}(1 - x_i) + h_{i-1}^{a_i} x_i \text{ for } x_i \in \{0, 1\}. \tag{2}$$

Computing the PRF value distributively amounts to several rounds of distributed multiplication and exponentiation(see Algorithm 1).

Proving security of this protocol is straightforward given the security of its sub-protocols by the composition theorem. The size of the secret key is proportional to the length of the input. What is worse, this protocol requires $O(l)$ rounds of communication. The running time is dominated by l calls to exponentiation protocol in line 6, yielding $O(m^4 l + (mn + m \log m)\mathcal{B}l)$. bit operations per player.

Algorithm 1. A protocol $\text{PRF-NR}(([a_0]_j^Q, \ldots, [a_l]_j^Q), ([x_1]_j^P, \ldots, [x_l]_j^P))$ for distributed computation of $F_{SK}^{NR}(x)$.

---

1 $[0]_j^P \leftarrow \text{JRPZ}(\mathbb{Z}_P)$ ▷Servers jointly generate a sharing of 0 mod P

2 $[h_0]_j^P \leftarrow [0]_j^P + g \bmod P$ ▷And compute a share of generator g.

3 **for** $i \leftarrow 1$ **to** l ▷For all input bits i,

4 **do**

5 $\quad [r]_j^P \leftarrow \text{MUL}([h_{i-1}]_j^P, 1 - [x_i]_j^P)$

6 $\quad [s]_j^P \leftarrow \text{EXP}([h_{i-1}]_j^P, [a_i]_j^Q)$

7 $\quad [t]_j^P \leftarrow \text{MUL}([s]_j^P, [x_i]_j^P)$ ▷we compute shares of Equation (2)

8 $\quad [h_i]_j^P \leftarrow [r]_j^P + [t]_j^P$

9 **end**

10 **return** $[h_l]_j^P$ ▷Return a share of the PRF value.

3.2 Dodis-Yampolskiy PRF

The pseudo-random function $F_{SK}^{DY} : \{0,1\}^{l(k)} \mapsto G_Q$ ($|Q| > l$) is as follows. Given an l-bit input x (which can also be thought of as an element in $\mathbb{Z}_Q$) and the secret key $SK \in \mathbb{Z}_Q$, the function value is $F_{SK}^{DY}(x) = g^{1/(x+SK)}$ [25]. Dodis-Yampolskiy's proof of security relied on an unorthodox q-decisional **Diffie-Hellman inversion** (q-DDHI) assumption: "given the tuple $(g, g^x, \ldots, g^{(x^q)})$ and $R \in G_Q$ as input, it is hard to decide whether $R = g^{1/x}$ or not." Specifically, they showed:

Theorem 1 (Dodis-Yampolskiy). *Suppose an attacker who runs for $s(k)$ steps cannot break the $2^{l(k)}$-DDHI assumption in group G_Q with advantage $\epsilon(k)$. Then no algorithm running in less than $s'(k) = s(k)/(2^{l(k)} \cdot poly(k))$ steps can distinguish $F_{SK}^{DY}(\cdot)$ from a random function with advantage $\epsilon'(k) = \epsilon(k) \cdot 2^{l(k)}$.*

Because the security reduction is rather loose, we can construct PRFs only with small superlogarithmic input $l(k) = \omega(\log k)$. Unfortunately, "as is" this PRF is unsuitable for use in the Feistel transformation. A Feistel transformation uses length-preserving PRFs which map $l(k) = poly(k)$ input bits to $l(k)$ pseudo-random bits. In theory, small inputs are not a problem. We can either (1) shrink the inputs using a collision-resistant hash function [14] or (2) utilize the generic tree construction [29] to extend the input range. However, when we need to distribute the computation of this PRF between different servers, neither of these options becomes acceptable. As of today, we do not know how to efficiently distribute collision-resistant hash functions. And if we decide to utilize the generic tree construction, then we might as well use the Naor-Reingold PRF from the start.

Instead, we assume subexponential hardness of the q-DDHI assumption in G_Q; that is, we suppose that there is no way to break the q-DDHI assumption except by computing the discrete logarithm of g^x in $\mathbb{Z}_P^*$. The fastest algorithm for computing discrete logarithms modulo P runs in time roughly $\exp((1+o(1)) \cdot$

$\sqrt{\log P}\sqrt{\log \log P}$) [13]. It seems reasonable to assume that no algorithm running in time less than $s(k) = 2^{k^{\epsilon_2}}$ (for some small $\epsilon_2 \approx \frac{1}{3}$) can break the q-DDHI assumption. Formally:

Definition 1 (strong DDHI assumption). *We say that the strong DDHI assumption holds in G_Q if there exist $0 < \epsilon_1 < \epsilon_2$ such that for all probabilistic families of Turing machines $\{A_k\}_{k \in \mathbb{N}}$ with running time $O(2^{k^{\epsilon_2}})$ and $q \leq 2^{k^{\epsilon_1}}$, we have:*

$$\left| \Pr_x \left[A_k(g, g^x, \ldots, g^{(x^q)}, R) = 1 \mid R \leftarrow g^{1/x} \right] - \Pr_x \left[A_k(g, g^x, \ldots, g^{(x^q)}, R) = 1 \mid R \overset{\$}{\leftarrow} G_Q \right] \right| \leq poly(k)/2^{k^{\epsilon_2}},$$

where the probability is taken over the coin tosses of A_k and the random choice of $x \in \mathbb{Z}_Q^*$ and $R \in G_Q$.

By Theorem 1, the strong DDHI assumption immediately allows us to support inputs of size k^{ϵ_1} for small $\epsilon_1 > 0$.

What we need is a shrinking hash function, which maps long $l(k) = k$ bit inputs to smaller $a(k) = k^{\epsilon_1}$ bit inputs, which can be used as an input to $F_{SK}^{DY}(\cdot)$. A typical tool used for this purpose is a family of δ-universal hash functions $\mathcal{H} = \{h_i : \{0,1\}^l \mapsto \{0,1\}^a\}_{i \in \mathbb{Z}_Q^*}$.[4] The simplest such construct is

$$h_i(x) = (ix \bmod Q) \bmod 2^a,$$

where the collision probability $\delta = 1/2^a = 1/2^{k^{\epsilon_1}}$ is the best we can hope for.

We can thus define a new function $F' : \{0,1\}^l \mapsto G_Q$ as

$$F'_{SK,i}(x) = F_{SK}^{DY}(h_i(x)),$$

which is easily seen to be a secure PRF for polynomially sized inputs using a standard hybrid argument.

This new PRF can be used in the Feistel transformation. We describe how to distribute its computation in Algorithm 2.

The security of the protocol again follows by composition theorem from security of its subcomponents. Unlike Algorithm 1, this algorithm uses $O(1)$ rounds of communication. However, it relies on a rather strong complexity assumption. Line 8 dominates the running time. It requires $O((mn + m \log m)\mathcal{B})$ bit operations, which is more than l times cheaper than the Naor-Reingold distributed protocol.

---

[4] We say that a hash family is δ-universal if, for all distinct inputs $x, x' \in \{0,1\}^l$, we have $\Pr_i[h_i(x) = h_i(x')] \leq \delta$.

Algorithm 2. A protocol $\text{PRF-DY}([i]_j^Q, [SK]_j^Q, [x_1]_j^Q, \ldots, [x_l]_j^Q)$

---

1 $[x]_j^Q \leftarrow \sum_{i=0}^{l-1} 2^i \cdot [x_{i+1}]_j^Q \bmod Q$ $\triangleright$Encode input x as an element in $\mathbb{Z}_Q^*$.

2 $[r]_j^Q \leftarrow \text{MUL}([i]_j^Q, [x]_j^Q)$ $\triangleright$Then hash it to $ix \bmod Q$.

3 $a \leftarrow \lfloor l^{1/3} \rfloor$ $\triangleright$Shrinking factor $a = l^{1/3}$.

4 $([r_1]_j^Q, \ldots, [r_a]_j^Q) \leftarrow \text{BITS}([r]_j^Q, a, \mathbb{Z}_Q)$ $\triangleright$Chop all but a least significant bits.

5 $[\tilde{x}]_j^Q \leftarrow \sum_{i=0}^{a-1} 2^i \cdot [r_{i+1}]_j^Q \bmod Q$

6 $[s]_j^Q \leftarrow [\tilde{x}]_j^Q + [SK]_j^Q \bmod Q$ $\triangleright$A share of $(\tilde{x} + SK)$.

7 $[t]_j^Q \leftarrow \text{INV}([s]_j^Q)$ $\triangleright$Invert the share into $1/(\tilde{x} + SK)$.

8 $[y]_j^P \leftarrow \text{EXP}_1(g, [t]_j^Q)$ $\triangleright$Exponentiate to get shares of $g^{1/(\tilde{x}+SK)}$.

9 **return** $[y]_j^P$

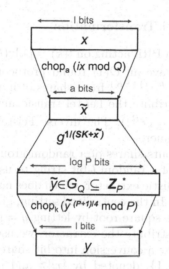

Fig. 1. Transformation of $F_{SK}^{DY}(\cdot)$ into a length-preserving PRF

4 Distributed Pseudo-Random Permutations

We now show how to construct a **threshold pseudo-random permutation** by distributing the Luby-Rackoff construction. In principle, the Luby-Rackoff construction can be used with any PRF. However, we will use it with the PRF by Dodis-Yampolskiy [25], which allows us to evaluate the threshold PRP in only $O(1)$ communication rounds.

We begin by reviewing some formal definitions in Section 4.1. In Section 4.2, we show how to distribute a single Feistel permutation. In Section 4.3, we put all of the pieces together and explain how to distribute the entire Feistel cascade. Finally, in Section 4.4, we analyze our protocol's security and sketch how to make it proactive.

4.1 Definitions

Definition 2 (Feistel transformation). *Let* $F : \{0,1\}^l \mapsto \{0,1\}^l$ *be an l-bit mapping. We denote by $\bar{F}$ the permutation on $\{0,1\}^{2l}$ defined as $\bar{F}(x) = (x^R, x^L \oplus F(x^R))$, where $x = (x^L, x^R)$. Note that $\bar{F}$ is a permutation even if F is not. Its inverse is given by $\bar{F}^{-1}(y^L, y^R) = (f(y^L) \oplus y^R, y^L)$.*

Definition 3 (Feistel network). *Let* $F_1, \ldots, F_k : \{0,1\}^l \mapsto \{0,1\}^l$ *be l-bit mappings. Then a k-round Feistel network is a composition*

$$\Psi(F_1, \ldots, F_k) = \bar{F}_1 \circ \bar{F}_2 \cdots \bar{F}_k$$

Theorem 2 (Luby-Rackoff). *The permutation $\Psi(F_1, F_2, F_3, F_4)$ on $\{0,1\}^{2l}$ cannot be distinguished from a random permutation by a PPT adversary. Here, F_i are independently keyed pseudo-random functions.*

4.2 Distributed Feistel Transformation

In Section 3.2, we defined a PRF acting on $l(k) = \text{poly}(k)$ bit inputs by $F'_{i,SK}(x) = F^{DY}_{SK}(h_i(x))$. We also gave an $O(1)$ round protocol PRF-DY that computes shares of a PRF value $g^{1/(h_i(x)+SK)}$ from shares of input's bits and secret key. We now show how to distribute the Feistel transformation $\overline{F'_{i,SK}}$, which maps (x^L, x^R) to $(x^R, x^L \oplus F'_{i,SK}(x^R))$. The inverse Feistel transformation can be computed in a similar manner.

Our PRF protocol outputs shares of a random group element in G_Q. Meanwhile, we need a sharing of a random l-bit string to use in the Feistel transformation. We use a deterministic extractor, which does not lose any entropy, to extract l bits of randomness. In the centralized setting, given PRF output $\tilde{y} \in G_Q$, we can simply compute its square root by letting $y = (\tilde{y}^{(P+1)/4} \bmod P) \bmod 2^l$ (see also Figure 1). To distribute the extractor, we use a distributed exponentiation protocol followed by a conversion into bit shares. Notice that if we have shares of bits $x_i, y_i \in \{0,1\}$, denoted by $[x_i]^Q_j$ and $[y_i]^Q_j$, we can compute a share of their exclusive-OR as $[z_i]^Q_j \leftarrow [x_i]^Q_j + [y_i]^Q_j$ and $[z_i]^Q_j \leftarrow \text{MUL}([z_i]^Q_j, 2 - [z_i]^Q_j \bmod Q)$.

We show how to compute Feistel transformation in Algorithm 3.

Security follows from composition theorem and security of its subprotocols. The protocol requires $O(1)$ rounds of communication between servers, because the for-loop is computed in parallel, and all other primitives take $O(1)$ rounds. The bit complexity is dominated by a call to the PRF-DY protocol in line 1 and by $l = o(m)$ calls to MUL in line 7, yielding $O((mn + m \log m)\mathcal{B})$ bit operations per player.

4.3 Distributed Luby-Rackoff Construction

Once we have a distributed protocol for the Feistel transformation, it is easy to distribute the Luby-Rackoff construction of PRP $g_s(x) = \Psi(F_1, F_2, F_3, F_4)(x)$.

Algorithm 3. One round of Feistel transformation $\mathtt{FEISTEL}([i]_j^Q, [SK]_j^Q,$ $[x_1]_j^Q, \ldots, [x_{2l}]_j^Q)$.

---

1 $[\tilde{y}]_j^P \leftarrow \mathtt{PRF\text{-}DY}([i]_j^Q, [SK]_j^Q, [x_{l+1}]_j^Q, \ldots, [x_{2l}]_j^Q)$ $\triangleright$PRF value at x^R.

2 $[y]_j^P \leftarrow \mathtt{EXP}_2([\tilde{y}]_j^P, (P+1)/4)$ $\triangleright$Extract square root $y^{(p+1)/4} \bmod P$.

3 $([y_1]_j^Q, \ldots, [y_l]_j^Q) \leftarrow \mathtt{BITS}([y]_j^P, l, \mathbb{Z}_Q)$ $\triangleright$Truncate to l bits.

4 **for** $i \leftarrow 1$ **to** l *(in parallel)* $\triangleright$For all bits i

5 **do**

6     $[z_i]_j^Q \leftarrow [x_i]_j^Q + [y_i]_j^Q \bmod Q$

7     $[z_i]_j^Q \leftarrow \mathtt{MUL}([z_i]_j^Q, 2 - [z_i]_j^Q \bmod Q)$ $\triangleright$We compute a share of $x_i \oplus y_i$.

8 **end**

9 **return** $([x_{l+1}]_j^Q, \ldots, [x_{2l}]_j^Q, [z_1]_j^Q, \ldots, [z_l]_j^Q)$ $\triangleright$Return shares of
     $(x^R, x^L \oplus F_{SK}^{DY}(x^R))$.

Initially, the n servers own shares of four independently chosen secret keys for the PRFs. These keys may either be jointly generated by servers or distributed to servers by a trusted party. An untrusted user, who wants to evaluate the PRP on input $x = (x^L, x^R)$, broadcasts x to the servers.[5] The servers convert x into bit shares and then run the distributed Feistel transformation for four rounds. We thus get Algorithm 4.

The round complexity is $O(1)$. Bit complexity is dominated by four calls to the Feistel protocol, which take $O((mn + m \log m)\mathcal{B})$ bit operations per player.

Similarly, we can distribute the inverse permutation $g_s^{-1}(\cdot)$ by replacing calls to Feistel transforms with calls to inverse Feistel transforms. We denote the resulting protocol by $\mathtt{LUBY\text{-}RACKOFF}^{-1}$. The round and bit complexity remain the same.

Algorithm 4. $\mathtt{LUBY\text{-}RACKOFF}(([i_1]_j^Q, [SK_1]_j^Q), \ldots, ([i_4]_j^Q, [SK_4]_j^Q), x)$

---

1 $[0]_j^Q \leftarrow \mathtt{JRPZ}(\mathbb{Z}_Q)$ $\triangleright$Shares of zero.

2 **for** $i \leftarrow 1$ **to** $2l$ *(in parallel)* $\triangleright$Locally compute shares of input's bits.

3 **do**

4    $[y_i]_j^Q \leftarrow [0]_j^Q + x_i \bmod Q$

5 **end**

6 **for** $rnd \leftarrow 1$ **to** 4 $\triangleright$Run the Feistel transformation for four rounds.

7 **do**

8    $([y_1]_j^Q, \ldots, [y_{2l}]_j^Q) \leftarrow \mathtt{FEISTEL}([i_{rnd}]_j^Q, [SK_{rnd}]_j^Q, [y_1]_j^Q, \ldots, [y_{2l}]_j^Q)$

9 **end**

10 **return** $([y_1]_j^Q, \ldots, [y_{2l}]_j^Q)$

---

[5] Alternatively, the user can split x into bit shares himself.

4.4   Security

In the stand-alone case, the security of a PRP $g_s(\cdot) : \{0,1\}^{2l} \mapsto \{0,1\}^{2l}$ is formalized via a game between an attacker and an oracle. The attacker can query the oracle for $g_s(\cdot)$ and $g_s^{-1}(\cdot)$ on messages of his choice. Roughly, the PRP is deemed secure if no attacker can tell apart $g_s(x^*)$ from random for any message x^*, which was not asked as a query.

In the distributed setting, the attacker also gets transcripts of semi-honest servers. The **security property of threshold PRP** states that these transcripts do not help the attacker in any way. Formally, for any PPT $\mathcal{A} = (\mathcal{A}_1, \mathcal{A}_2)$ that breaks the security of threshold PRP by corrupting servers $P_{i_1}, \ldots, P_{i_\tau}$, there exists a PPT $\mathcal{B} = (\mathcal{B}_1, \mathcal{B}_2)$ that breaks the security of the original PRP.

The attacker $\mathcal{A}$ learns key shares of corrupted servers. Then $\mathcal{A}_1$ runs in the first stage where it can interact with any honest servers on inputs of his choice. Attacker can ask servers either **encryption queries** where he learns shares of $g_s(x)$ or **decryption queries** for $g_s^{-1}(y)$. At the end of the phase, $\mathcal{A}_1$ outputs state information for $\mathcal{A}_2$ and a challenge input x^*, whose PRP value was not asked as a query. In the second stage, a random coin $b \in \{0,1\}$ is tossed. $\mathcal{A}_2$ receives a challenge Γ_b, which is either $\Gamma_0 \leftarrow g_s(x^*)$ or $\Gamma_1 \xleftarrow{\$} \{0,1\}^{2l}$. We let $\mathcal{A}_2$ interact with honest servers, but prohibit it from asking encryption queries on x^* or decryption queries on Γ_b. Finally, $\mathcal{A}_2$ outputs a guess b'. We say that $\mathcal{A}$ breaks the scheme if $\Pr[b = b'] > 1/2 + negl(k)$.

Theorem 3. LUBY–RACKOFF *protocol is an* $\lfloor \frac{n-1}{2} \rfloor$*-secure threshold pseudorandom permutation in the static, honest-but-curious setting.*

Proof (sketch). In the honest-but-curious setting, LUBY–RACKOFF protocol correctly computes a permutation $g_s(x) = \Psi(F_1, F_2, F_3, F_4)(x)$ for some secret key $s = ((i_1, SK_1), \ldots, (i_4, SK_4))$. We thus concentrate on the pseudorandomness property.

For sake of contradiction, suppose there exists adversary $\mathcal{A} = (\mathcal{A}_1, \mathcal{A}_2)$ that breaks the security of LUBY–RACKOFF. Since $\mathcal{A}$ is static, we assume it corrupts the maximum allowed threshold of servers before the protocol starts[6]. By symmetry, we can assume corrupt servers P_j have indices $Bad = \{1, \ldots, \tau\}$. Bad servers learn their shares of secret key s. They also observe the protocol's input x, output $y = g_s(x)$, shares of output's bits $y_1, \ldots, y_{2l}$ of both good and bad servers, and all messages Ξ exchanged during the protocol. The adversarial view $\text{VIEW}_{\text{LUBY}, \mathcal{A}}$ is thus a random variable

$$\left\langle ([i_1]_k^Q, [SK_1]_k^Q), \ldots, ([i_4]_k^Q, [SK_4]_k^Q), x, y, [y_1]_j^Q, \ldots, [y_{2l}]_j^Q, \Xi \right\rangle$$

for $j = 1, \ldots, n$ and $k \in Bad$.

We construct a simulator $\mathcal{B} = (\mathcal{B}_1, \mathcal{B}_2)$ that breaks the security of a PRP $g_s(\cdot)$. It will run $\mathcal{A}$ in a virtual distributed environment and imitate $\mathcal{A}$'s replies to distinguish $g_s(\cdot)$ from a random permutation, thereby violating Theorem 2.

---

[6] If not, we can arbitrarily fix some of the honest servers to be corrupt.

Setup. Algorithm $\mathcal{B}$ generates random shares of keys for corrupt servers. For $j \in Bad$, it picks

$$([i_1]_j^Q, [SK_1]_j^Q), \ldots, ([i_4]_j^Q, [SK_4]_j^Q) \xleftarrow{\$} \mathbb{Z}_Q^* \times \mathbb{Z}_Q^* \text{ and gives them to } \mathcal{A}.$$

Responding to Queries. When $\mathcal{A}$ initiates an honest server P_j ($j \notin Bad$) on input x, $\mathcal{B}$ in turn asks his oracle for $y = g_s(x)$. It generates random output shares $[z_1]_j^Q, \ldots, [z_{2l}]_j^Q \xleftarrow{\$} \mathbb{Z}_Q^*$ for $j \in Bad$. Then, $\mathcal{B}$ augments the set of shares of corrupted servers into a full and random sharing of y's bits. For each bit $y_i \in \{0, 1\}$ ($1 \leq i \leq 2l$), $\mathcal{B}$ picks a random polynomial $\alpha_i(x) \in \mathbb{Z}_Q[X]$ satisfying $\alpha_i(j) = [z_i]_j^Q$ and $\alpha_i(0) = y_i$. The adversary $\mathcal{A}$ receives randomized output shares $(\alpha_1(j), \ldots, \alpha_{2l}(j))$ for all servers P_j ($1 \leq j \leq n$). In the semi-honest setting, we can simulate the transcript of each subprotocol used by LUBY-RACKOFF given its input and output values. We can thus use these protocols as black-boxes and simulate messages Ξ in $\text{VIEW}_{\text{LUBY}, \mathcal{A}}$. These values provide a perfect simulation of the coalition's view. Decryption queries are handled just like encryption queries except $\mathcal{B}$ queries another oracle $g_s^{-1}(\cdot)$.

Challenge. Eventually, attacker $\mathcal{A}$ outputs a message x^* on which it wants to be challenged. It claims to be able to distinguish output of LUBY-RACKOFF(x^*) from a random $2l$-bit string. $\mathcal{B}$ sends the same challenge x^* to the trusted party and gets back Γ, which is either $g_s(x^*)$ or a random string. Finally, $\mathcal{B}$ gives Γ to $\mathcal{A}$.

Guess. Attacker $\mathcal{A}$ continues to issue queries for messages other than x^*. Simulator $\mathcal{B}$ responds to queries as before. Finally, $\mathcal{A}$ outputs a guess $b' \in \{0, 1\}$, which $\mathcal{B}$ also returns as its guess. $\square$

An adversary who controls less than $\tau = \lfloor (n - 1)/2 \rfloor$ servers cannot break the privacy of our protocol. The protocol can easily be amended to achieve **proactive security** [33] and withstand the compromise of even all servers as long as at most τ servers are corrupted during each time period. The basic idea is to have servers periodically refresh their shares of the input and the secret keys. To be exact, each server P_j will from time to time execute the JRPZ protocol to generate a random share of zero, called $[0]_j^Q$. It will then update its input share to $[x]_j^Q \leftarrow [x]_j^Q + [0]_j^Q$ and its secret keys' shares to $[SK_i]_j^Q \leftarrow [SK_i]_j^Q + [0]_j^Q$.

5   Applications of Our Construction

In the previous section, we have constructed a **threshold pseudorandom permutation**. Whenever a PRP is used as part of the construction, we can plug in our protocol instead.

Let $g_s : \{0, 1\}^{2l} \mapsto \{0, 1\}^{2l}$ be a $2l$-bit pseudo-random permutation obtained from the Luby-Rackoff construction $\Psi(F_1, F_2, F_3, F_4)$. The PRP's key s consists of four secret keys SK_i of pseudo-random functions F_i used in the construction. We denote by LUBY-RACKOFF our distributed protocol, which evaluates the $g_s(\cdot)$.

5.1 CCA-Secure Symmetric Encryption

A PRP is deterministic, so by itself it cannot be a secure encryption scheme [30]. The adversary can easily detect if the same message has been encrypted twice. Desai [18] described how a **CCA-secure symmetric encryption scheme** can be obtained from a PRP: The encryption $\mathcal{E}_s(m)$ of a $(2l - k)$-bit message m is defined as $\mathcal{E}_s(m) = g_s(m, r)$, where r is a k-bit randomly generated nonce. To decrypt, the user computes $\mathcal{D}_s(c) = g_s^{-1}(c)$ and extracts the message. To distribute the computation, the key s is split into shares among the n servers. Upon receiving a message m, the servers run the JRP protocol to generate shares of a secret random number r. They can extract shares of k bits, written $[r_1]_j^Q, \ldots, [r_k]_j^Q$, using the BITS protocol. Bit shares of m are easy to compute since m is public. Finally, the servers invoke LUBY-RACKOFF on $([m_1]_j^Q, \ldots, [m_{2l-k}]_j^Q, [r_1]_j^Q, \ldots, [r_k]_j^Q)$ to get shares of $g_s(m, r)$.

5.2 Authenticated Encryption

If we make the nonce r public and check during decryption that it matches the nonce in the ciphertext, then we get a distributed **authenticated encryption scheme** (AE). Encryption of m is given by $\mathcal{AE}_s(m) = (r, g_s(m, r))$. The decryption algorithm $\mathcal{AD}_s(r', c)$ computes $(r, m) = g_s^{-1}(c)$ and checks that $r = r'$ before returning m to the user. The message here is rather short: It is limited to $(2l - k)$ bits by the length of the PRP. For longer messages, we can use an amplification paradigm of Dodis and An [24]: We compute a **concealment** (b, h) of message m ($|b| \ll |m|$), which is a specialized publicly known transformation. In fact, we can even implement distributed **remotely keyed authenticated encryption** (RKAE) [7], where the servers do not need to perform any checks and just serve as PRP oracles for an untrusted user. The secret key s is split into shares among several computationally bounded **smartcards**, and an insecure, powerful **host** performs most of computations. The insecure host computes a concealment (b, h) of m and sends it to the smartcards, who run the LUBY-RACKOFF protocol, and return shares of $g_s(b)$.

5.3 Cipherblock Chaining Mode

We often need to encrypt messages that are longer than $2l$ bits. The message m is usually split into blocks $(m_1, \ldots, m_k)$ each of length $2l$. Then a PRP may be used in **cipher block chaining** (CBC) mode [38], which initializes c_0 with a random $2l$-bit string and sets $c_i = g_s(c_{i-1} \oplus m_i)$ for $i = 1, \ldots, k$. The encryption of m is defined to be $(c_0, c_1, \ldots, c_k)$. To decrypt, the user can compute $m_i = g_s^{-1}(c_i) \oplus c_{i-1}$. The servers own shares of secret key s. The untrusted user broadcasts message m to the servers. We must be careful to guard against the blockwise adaptive attacks [34]; hence, we require the user to send an entire message m. The servers run the JRP protocol to generate a random shared number from which shares of a $2l$-bit c_0 are extracted. For $i = 1, \ldots, k$ rounds, the servers distributively XOR shares of c_{i-1} and m_i (as in Section 4.3), and then run the LUBY-RACKOFF protocol on the result.

5.4   Variable Input Block Ciphers

Existing block ciphers operate on blocks of fixed length (FIL). Often, one needs a block cipher that can operate on inputs of variable length (VIL). There exist centralized constructions for VIL ciphers, which use a FIL block cipher as a black box: most notably, CMC [32], EME* [31] and an unbalanced Feistel network [43]. Our threshold PRP enables us to distribute the computation of these modes. Besides basic arithmetic operations, these modes XOR the ciphertexts (to distribute, we would use BITS), evaluate the fixed-length block cipher (LUBY-RACKOFF), compute the universal hash function $h_{a,b}(x) = ax + b$ (MUL), and truncate the outputs (BITS).

6   Conclusion

We gave a simple construction of a threshold PRP in the semi-honest model. Our scheme is fairly practical. PRPs are commonly used tools in protocol design. Our techniques enable distributing many protocols (using PRPs), which until now only existed in the centralized setting. In particular, we showed how to distribute the computation of a CBC encryption mode and a remotely keyed authenticated encryption scheme.

One open problem is whether we could use group multiplication to implement the distributed Feistel transformation rather than having to convert group elements into bit strings and avoid using the expensive protocol by Damgård et al. [16] altogether.

Acknowledgments

We thank Tomas Toft for his explanation of the bit conversion protocol.

References

1. J. Algesheimer, J. Camenisch, and V. Shoup. Efficient computation modulo a shared secret with applications to the generation of shared safe prime products. In *Advances in Cryptology - Proceedings of CRYPTO 2002*, volume 2442 of *Lecture Notes in Computer Science*, pages 417–432. Springer-Verlag, 2002.
2. E. Bach. *Analytic Methods in the Analysis and Design of Number-Theoretic Algorithms*. A.C.M. Distinguished Dissertations. MIT press, Cambridge, MA, 1985.
3. J. Bar-Ilan and D. Beaver. Non-cryptographic fault-tolerant computing in a constant number of rounds. In *Proceedings of the ACM Symposium on Principles of Distributed Computation*, pages 201–209, 1989.
4. D. Beaver, S. Micali, and P. Rogaway. The round complexity of secure protocols. In *Proceedings of the 22nd Annual ACM Symposium on the Theory of Computing*, pages 503–513, 1990.
5. M. Ben-or, S. Goldwasser, and A. Wigderson. Completeness theorems for non-cryptographic fault-tolerant distributed computing. In *Proceedings of the 20th Annual ACM Symposium on the Theory of Computing*, pages 1–10, 1988.

6. E. Biham. A fast new DES implementation in software. In *Fast Software Encryption - Fourth International Workshop*, volume 1267 of *Lecture Notes in Computer Science*, pages 260–272. Springer-Verlag, 1997.
7. M. Blaze, J. Feigenbaum, and M. Naor. A formal treatment of remotely keyed encryption. In *Advances in Cryptology - Proceedings of EUROCRYPT 98*, Lecture Notes in Computer Science, pages 251–265. Springer-Verlag, 1998.
8. D. Boneh and M. K. Franklin. Efficient generation of shared RSA keys. *Journal of the Association for Computing Machinery*, 48(4):702–722, 2001.
9. E. F. Brickell, G. D. Crescenzo, and Y. Frankel. Sharing block ciphers. In E. Dawson, A. Clark, and C. Boyd, editors, *ACISP*, volume 1841 of *Lecture Notes in Computer Science*, pages 457–470. Springer, 2000.
10. C. Cachin, K. Kursawe, F. Petzold, and V. Shoup. Secure and efficient asynchronous broadcast protocols. In *Advances in Cryptology - Proceedings of CRYPTO 2001*, volume 2139 of *Lecture Notes in Computer Science*, pages 524–541. Springer-Verlag, 2001.
11. R. Canetti. Universally composable security: A new paradigm for cryptographic protocols. In *Proceedings of the 42nd IEEE Symposium on Foundations of Computer Science*, pages 136–145, 2001.
12. D. Catalano, R. Gennaro, and S. Halevi. Computing inverses over a shared secret modulus. In *Advances in Cryptology - Proceedings of EUROCRYPT 2000*, volume 1807 of *Lecture Notes in Computer Science*, pages 190–206. Springer-Verlag, 2000.
13. D. Coppersmith, A. M. Odlyzko, and R. Schroeppel. Discrete logarithms in $GF(p)$. *Algorithmica*, 1(1):1–15, 1986.
14. I. Damgård. Collision free hash functions and public key signature schemes. In *Advances in Cryptology - Proceedings of EUROCRYPT 87*, Lecture Notes in Computer Science, pages 203–216. Springer-Verlag, 1987.
15. I. Damgård, M. Fitzi, E. Kiltz, J. B. Nielsen, and T. Toft. Unconditionally secure constant-rounds multi-party computation for equality, comparison, bits and exponentiation. In *Third Theory of Cryptography Conference*, 2006. To appear.
16. I. Damgård and Y. Ishai. Constant-round multiparty computation using a black-box pseudorandom generator. In *Advances in Cryptology - Proceedings of CRYPTO 2005*, volume 3621 of *Lecture Notes in Computer Science*, pages 378–394. Springer-Verlag, 2005.
17. I. Damgård and M. Jurik. A generalisation, a simplification and some applications of Paillier's probabilistic public-key system. In *Fourth International Workshop on Practice and Theory in Public Key Cryptography*, pages 119–136, 2001.
18. A. Desai. New paradigms for constructing symmetric encryption schemes secure against chosen-ciphertext attack. In *Advances in Cryptology - Proceedings of CRYPTO 2000*, pages 394–412, 2000.
19. Y. Desmedt. Society and group-oriented cryptography: a new concept. In *Advances in Cryptology - Proceedings of CRYPTO 87*, pages 120–127, 1987.
20. Y. Desmedt. Some recent research aspects of threshold cryptography. In *First International Workshop On Information Security*, pages 158–173, 1997.
21. Y. Desmedt and Y. Frankel. Threshold cryptosystems. In *Advances in Cryptology - Proceedings of CRYPTO 89*, pages 307–315, 1989.
22. Y. Desmedt and Y. Frankel. Shared generation of authenticators and signatures. In *Advances in Cryptology - Proceedings of CRYPTO 91*, pages 457–469, 1991.
23. Y. Dodis. Efficient construction of (distributed) verifiable random functions. In *Proceedings of 6th International Workshop on Theory and Practice in Public Key Cryptography*, pages 1–17, 2003.

24. Y. Dodis and J. H. An. Concealment and its applications to authenticated encryption. In *Advances in Cryptology - Proceedings of EUROCRYPT 2003*, volume 2656 of *Lecture Notes in Computer Science*, pages 312–329. Springer-Verlag, 2003.

25. Y. Dodis and A. Yampolskiy. A verifiable random function with short proofs and keys. In *Eighth International Workshop on Theory and Practice in Public Key Cryptography*, pages 416–431, 2005.

26. Y. Dodis, M. Yung, and A. Yampolskiy. Threshold and proactive pseudo-random permutations. Technical Report YALEU/DCS/TR-1325, Yale University, Nov. 2005. Available at `ftp://ftp.cs.yale.edu/pub/TR/tr1325.pdf`.

27. R. Gennaro, S. Jarecki, H. Krawczyk, and T. Rabin. Secure distributed key generation for discrete-log based cryptosystems. In *Advances in Cryptology - Proceedings of EUROCRYPT 99*, pages 295–310, 1999.

28. R. Gennaro, S. Jarecki, H. Krawczyk, and T. Rabin. Robust threshold DSS signatures. *Inf. Comput.*, 164(1):54–84, 2001.

29. O. Goldreich, S. Goldwasser, and S. Micali. How to construct random functions. *Journal of the Association for Computing Machinery*, 33:792–807, 1986.

30. S. Goldwasser and S. Micali. Probabilistic encryption and how to play mental poker keeping secret all partial information. In *Proceedings of the 14th Annual ACM Symposium on the Theory of Computing*, pages 270–299, 1982.

31. S. Halevi. EME*: Extending EME to handle arbitrary-length messages with associated data. In *Advances in Cryptology - Proceedings of INDOCRYPT 2004*, pages 315–327, 2004.

32. S. Halevi and P. Rogaway. A tweakable enciphering mode. In *Advances in Cryptology - Proceedings of CRYPTO 2003*, pages 482–499, 2003.

33. A. Herzberg, S. Jarecki, H. Krawczyk, and M. Yung. Proactive secret sharing or: How to cope with perpetual leakage. In *Advances in Cryptology - Proceedings of CRYPTO 95*, pages 339–352, 1995.

34. A. Joux, G. Martinet, and F. Valette. Blockwise-adaptive attackers. revisiting the (in)security of some provably secure encryption modes: CBC, GEM, IACBC. In *Advances in Cryptology - Proceedings of CRYPTO 2002*, volume 2442 of *Lecture Notes in Computer Science*, pages 17–30. Springer-Verlag, 2002.

35. M. Luby and C. Rackoff. How to construct pseudorandom permutations from pseudorandom functions. *SIAM Journal of Computing*, 17:373–386, 1988.

36. A. Lysyanskaya. Unique signatures and verifiable random functions from DH-DDH separation. In *Proceedings of the 22nd Annual International Cryptology Conference on Advances in Cryptology*, pages 597–612, 2002.

37. K. M. Martin, R. Safavi-Naini, H. Wang, and P. R. Wild. Distributing the encryption and decryption of a block cipher. *Designs, Codes, and Cryptography*, 2005. to appear.

38. A. J. Menezes, P. C. van Oorschot, and S. A. Vanstone. *Handbook of applied cryptography*. CRC press LLC, Boca Raton, FL, 1997.

39. S. Micali, M. O. Rabin, and S. P. Vadhan. Verifiable random functions. In *Proceedings of the 40th IEEE Symposium on Foundations of Computer Science*, pages 120–130, 1999.

40. M. Naor, B. Pinkas, and O. Reingold. Distributed pseudo-random functions and KDCs. In *Advances in Cryptology - Proceedings of EUROCRYPT 99*, volume 1592 of *Lecture Notes in Computer Science*, pages 327–346. Springer-Verlag, 1999.

41. M. Naor and O. Reingold. Number-theoretic constructions of efficient pseudo-random functions. In *Proceedings of the 38th IEEE Symposium on Foundations of Computer Science*, pages 458–467, 1997.

42. J. B. Nielsen. A threshold pseudorandom function construction and its applications. In *Advances in Cryptology - Proceedings of CRYPTO 2002*, volume 2442 of *Lecture Notes in Computer Science*, pages 401–416. Springer-Verlag, 2003.
43. S. Patel, Z. Ramzan, and G. S. Sundaram. Efficient constructions of variable-input-length block ciphers. In *Selected Areas in Cryptography 2004*, pages 326–340, 2004.
44. T. P. Pedersen. A threshold cryptosystem without a trusted party. In *Advances in Cryptology - Proceedings of EUROCRYPT 91*, pages 522–526, 1991.
45. T. Rabin. A simplified approach to threshold and proactive RSA. In *Advances in Cryptology - Proceedings of CRYPTO 98*, pages 89–104, 1998.
46. T. Rabin and M. Ben-Or. Verifiable secret sharing and multiparty protocols with honest majority. In *Proceedings of the 21th Annual ACM Symposium on the Theory of Computing*, pages 73–85, 1989.
47. A. Shamir. How to share a secret. *Communications of the ACM*, 22(11):612–613, 1979.
48. A. Yao. Protocols for secure computation (extended abstract). In *Proceedings of the 23rd IEEE Symposium on Foundations of Computer Science*, pages 160–164, 1982.

PRF Domain Extension Using DAGs

Charanjit S. Jutla

IBM T. J. Watson Research Center,
Yorktown Heights, NY 10598

Abstract. We prove a general domain extension theorem for pseudo-random functions (PRFs). Given a PRF F from n bits to n bits, it is well known that employing F in a chaining mode (CBC-MAC) yields a PRF on a bigger domain of mn bits. One can view each application of F in this chaining mode to be a node in a graph, and the chaining as edges between the node. The resulting graph is just a line graph. In this paper, we show that the underlying graph can be an arbitrary directed acyclic graph (DAG), and the resulting function on the larger domain is still a PRF. The only requirement on the graph is that it have unique source and sink nodes, and no two nodes have the same set of incident nodes. A new highly parallelizable MAC construction follows which has a critical path of only $3 + \log^* m$ applications of F.

If we allow Galois field arithmetic, we can consider edge-colored DAGs, where the colors represent multiplication in the field by the color. We prove an even more general theorem, where the only restriction on the colored DAGs is that if two nodes (u and v) have the same set of incident nodes W, then at least one w in W is incident on u and v with a different colored edge. PMAC (Parallelizable Message Authentication [6]) is a simple example of such graphs. Finally, to handle variable length domain extension, we extend our theorem to a collection of DAGs. The general theorem allows one to have further optimizations over PMAC, and many modes which deal with variable lengths.

Keywords: PRF, MAC, DAG, Partial Order, Galois Field.

1 Introduction

There is often a need to extend the domain of a given pseudo-random function (PRF). One of the most popular and well-known such schemes is the CBC-MAC [1]. In [3] it was shown that if F is a pseudo-random function from n bits to n bits, then the CBC (cipher block chaining) construction yields a PRF from mn bits to n bits. Although the construction is called a MAC (message authentication code), which is a strictly weaker notion than PRF [9], the above shows that it is indeed a PRF domain extension method. Other domain extension schemes are known as well, for example, the cascade construction [2] and the protected counter sum construction [5]. Recently, a scheme PMAC (or Parallelizable Message Authentication) [6] (also see XECB [11]) was also shown to be a domain extension scheme.

S. Halevi and T. Rabin (Eds.): TCC 2006, LNCS 3876, pp. 561–580, 2006.
© Springer-Verlag Berlin Heidelberg 2006

Despite all these results, there is no unifying theme in these results. In this paper, we attempt to remedy this situation, by proving a general theorem for domain extension. In essence, we show that arbitrary acyclic networks of the same pseudo-random function can be used to build a pseudo-random function on a larger domain. To illustrate this paradigm, consider the CBC-MAC scheme. Let F be a PRF from n bits to n bits (and which takes k bits of secret key). For example, DES [10] is usually assumed to be such a PRF on 64 bits, with 56 bits of secret key. A PRF $\tilde{F}$ from mn bits to n bits is defined as follows. The mn bit input is divided into m blocks $P_1, P_2, ..., P_m$. The function F_k (i.e. F with key k) is applied to the first block P_1 to yield an intermediate value C_1. The function F_k is next invoked on the xor of the next block P_2 and previous intermediate value C_1, to yield C_2. This chaining process is continued, and the output of $\tilde{F}_k$ is just C_m. The chaining process defines an underlying directed graph of m nodes $V_1, V_2, ..., V_m$, with an edge from V_i to V_{i+1}.

Now, consider an arbitrary directed acyclic graph (DAG) $G = (V, E)$, with m nodes V, and edges E. Assume that G has only one source node V_1, and only one sink node V_m. Given a PRF F from n bits to n bits, a composite PRF $\tilde{F}$ from mn bits to n bits is defined as follows. As before, assume that the input is a sequence $P_1, ..., P_m$. The first intermediate value is just $C_1 = F_k(P_1)$. Inductively assume that we have computed the intermediate values of all predecessors of a node V_i. Then, the intermediate value C_i for the node V_i is

$$C_i = F_k(P_i \oplus \bigoplus_{(i,j) \in E} C_j)$$

The output of the composite function $\tilde{F}_k$ is just C_m. See Figure 1 for an example.

Of course, not all DAGs are expected to yield a PRF. However, consider DAGs with the restriction that no two nodes have the same set of incident nodes

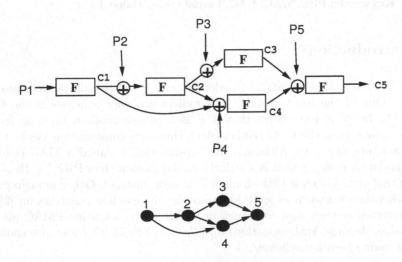

Fig. 1. A PRF Domain Extension Mode and its DAG

(u is said to be incident on v if there is an edge from u to v), and that they have unique source and sink nodes. In this paper we show that given a PRF F from n bits to n bits, the composite $\tilde{F}$ defined as above on such DAGs, is a PRF from mn bits to n bits.

An immediate application is that if a party has access to parallel hardware, then instead of simple chaining as in CBC-MAC, it can compute the PRF in parallel. For instance, if it has four processors, then it can employ the method given by the graph in Figure 2. A parallel mode with critical path of length only $3 + \log^* m$ also follows. Unlike PMAC [6], this mode does not use any Galois arithmetic.

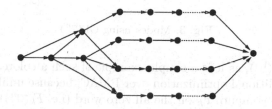

Fig. 2. A Parallel Mode for four processors

If we allow Galois Field arithmetic (in particular, fields $GF(2^n)$), we can consider edge-colored DAGs. The colors on the edges represent multiplication in the field by the color (assume that each color is mapped to a unique element in the field). For example, going back to figure 1, suppose we employ three colors, col1, col2 , and col3. Let w be a primitive element in the field. We map col1 to unity in the field, col2 to w, and col3 to w^2. Then, if we color the edge $(1, 4)$ by col2, then in the definition of the composite function, we multiply the intermediate result C_1 with w in the field, before xoring it with the plaintext P_4 and C_2, and applying F_k.

The main result of the paper can be stated as follows. Consider an edge colored DAG G with unique source and sink nodes and m total nodes, and with the condition that if two nodes (say u and v) have the same set of incident nodes (say W), then for at least one node w in W, the color on the edge (w, u) is different from the color on the edge (w, v). Given a PRF F from n bits to n bits, the composite $\tilde{F}$ built using the graph G as above, is a PRF from mn bits to n bits. The result is proven under the adaptive adversary model, which is of course the difficult case. Our proof technique is novel, and even when considered as just a proof for CBC-MAC it offers a simpler and novel proof in the adaptive adversary model. It is well known that the difficulty in analyzing the security of such schemes stems from the fact that we need to model the underlying oracle as a function, i.e. an oracle replying consistently with earlier queries. The key advance is an identity (lemma 1) which reduces the analysis to a scheme where the oracle replies randomly. The adversary remains adaptive, but the analysis in this "random game" becomes much easier.

Using the new theorem, the mode in fig 2 can now be parallelized further as in fig 3(a). The additional cost is a few $GF(2^n)$ operations. Security of PMAC [6]

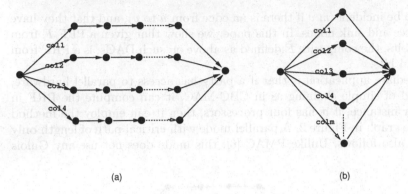

(a)   (b)

Fig. 3. Modes using $GF(2^n)$

follows (see fig 3(b)), as it is a simple example of such a colored DAG. Further, we obtain an additional optimization over PMAC, because unlike PMAC, we do not even need to compute F_k on the all zero word (i.e. $F_k(0^n)$).

In Section 5 we extend our results to *variable length domain extension*.

2   Definitions

Definition 1. For positive integers n, l, let $\mathcal{F}(n \rightarrow l)$ be the set of all functions from n bits to l bits.

Definition 2. (PRF) A *pseudo-random function* has signature

$$F : \{0,1\}^k \times \{0,1\}^n \rightarrow \{0,1\}^l.$$

Define $\mathrm{Sec}_F(q, T)$ to be the maximum advantage an adaptive adversary can obtain when trying to distinguish between $F_K(\cdot)$ (with K chosen uniformly at random) and a function chosen uniformly at random from $\mathcal{F}(n \rightarrow l)$, when given q queries and time T.

3   Domain Extension Using Arbitrary Acyclic Graphs

Definition 3. Let $G = (V, E)$, be a directed acyclic graph (DAG) [13] with a finite vertex set V and edges E. A node u is said to be **incident** on a node v, if there is an edge from u to v, i.e $E(u, v)$. Such an edge will sometimes be denoted $\langle u, v \rangle$. Define a DAG to be **non-redundant** if for every pair of nodes, the set of their incident nodes is different. For two vertices u and v, we say that $u \prec v$ if there is a directed path from u to v. Since G is a finite DAG, the relation $\prec$ is a finite partial order.

Definition 4. Given a function f from n bits to n bits, and a non-redundant DAG $G = (V, E)$ with only one source node and only one sink node, and a total of m nodes, define $f^G : \{0,1\}^{nm} \rightarrow \{0,1\}^n$ as follows:

- Let the input to f^G be an mn bit string P, which is divided into m n-bit strings $P_1, P_2, ..., P_m$.
- Since $|V| = m$, let $V_1, ..., V_m$ be an enumeration of the nodes. When it is clear from context, we will identify the index of a vertex with the vertex itself. Let the unique source node be V_1, and the unique sink node be V_m.
- For the unique source node, define $M_1 = P_1$.
- For every non-source node V_j, $j > 1$, inductively (over $\prec$) define $M_j = P_j \oplus_{u:E(u,j)} f(M_u)$
- For notational convenience, for every node V_j, let C_j denote $f(M_j)$.
- The output of the function f^G is just C_m.

It is clear that the restriction of one sink node is crucial, for if there was another sink node other than V_m, then the plaintext fed into this other sink node has no influence on C_m. It is possible that there are instances of DAGs G with *two source nodes* such that F^G is a PRF; however, a more stringent requirement than non-redundancy will definitely be required. Consider a DAG G, with two source nodes V_1 and V_2, both with only one outgoing edge and that too to the same vertex. Then, the resulting function is clearly not a PRF. A similar situation motivates the requirement of non-redundancy.

One may be tempted to weaken the non-redundancy requirement. For instance, one idea is to have a condition on the DAG that it have no non-trivial automorphism. However, such a DAG may not yield a secure PRF, as illustrated in Figure 4. The two queries $\langle p1, p2, p2, p4, p5, p6 \rangle$ and $\langle p1, p2, p2, p5, p4, p6 \rangle$ yield the same result.

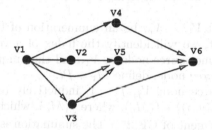

Fig. 4. A non-automorphic DAG

Theorem 1. *For a non-redundant DAG $G = (V, E)$ with unique source and sink nodes, and m total nodes, let f^G be as above. Then, no adaptive adversary, with q queries, can distinguish between (a) f^G where f is chosen uniformly at random from $\mathcal{F}(n{\rightarrow}n)$, (b) and a function chosen uniformly at random from $\mathcal{F}(nm{\rightarrow}n)$, with probability more than $(mq)^2 2^{-(n+1)}$.*

In the next section, we state and prove a more general theorem.

4   Domain Extension Using Colored DAGs and GF(2^n)

If we allow Galois field arithmetic, we get an even more general construction, and a corresponding PRF domain extension theorem. Assuming that the underlying function F has an n-bit output, we will use the Galois field $GF(2^n)$. Such fields have the property that they have exactly 2^n elements. Moreover, each element can be represented as a n bit vector, with addition in the field being just the bitwise exclusive-or ($\oplus$). Since multiplication distributes over addition in a field, it follows that if a, b and c are three elements in the field then $a * (b \oplus c) = a * (b + c) = (a * b) + (a * c) = (a * b) \oplus (a * c)$. A further useful property of finite fields is that for a fixed non-zero a in the field, if b is picked uniformly at random from the field, then $a * b$ is also uniformly distributed in the field.

Definition 5. Let $G = (V, E)$, be a directed acyclic graph (DAG). Let $|V| = m$. A coloring χ of the edges of the graph is a map $\chi : E \to [1..m]$. The triple (V, E, χ) will be called an **edge-colored** DAG. Define an edge-colored DAG to be **non-singular** if for every pair of nodes u, v, if the set of their incident nodes is same (say W), then at least for one $w \in W$, $\chi(\langle w, u \rangle) \neq \chi(\langle w, v \rangle)$. For two vertices u and v, we say that $u \prec v$ if there is a directed path from u to v. Since G is a finite DAG, the relation $\prec$ is a finite partial order.

Definition 6. Given a function f from n bits to n bits, and a non-singular edge-colored DAG $G = (V, E, \chi)$ with only one source node and only one sink node and a total of $m < 2^n$ nodes, define $f^G : \{0, 1\}^{nm} \to \{0, 1\}^n$ as follows:

- Since $m < 2^n$, we can view χ as a map from E to $GF(2^n)^*$, i.e. the non-zero elements of the field.
- Let the input to f^G be mn bit string P, which is divided into m n-bit strings $P_1, P_2, ..., P_m$.
- Since $|V| = m$, let $V_1, ..., V_m$ be an enumeration of the nodes. When it is clear from context, we will identify the index of a vertex with the vertex itself. Let the unique source node be V_1, and the unique sink node be V_m.
- For the unique source node, define $M_1 = P_1$.
- For every non-source node V_j, $j > 1$, inductively (over $\prec$) define $M_j = P_j + \sum_{u:E(u,j)} \chi(\langle u, j \rangle) * f(M_u)$, where $f(M_u)$, which is an n-bit quantity, is viewed as an element of $GF(2^n)$. The summation is addition in the field, which is the same as n-bit exclusive-or.
- For notational convenience, for every j, we denote $f(M_j)$ by C_j.
- The output of the function f^G is just C_m.

Theorem 2. : *(Main Theorem) For a non-singular edge-colored DAG $G = (V, E, \chi)$ with unique source and sink nodes, and $m < 2^n$ total nodes, let f^G be as above. Then, no adaptive adversary, with q queries, can distinguish between (a) f^G where f is chosen uniformly at random from $\mathcal{F}(n \to n)$, (b) and a function chosen uniformly at random from $\mathcal{F}(nm \to n)$, with probability more than $(mq)^2 2^{-(n+1)}$.*

Theorem 3. *Given a PRF $F : \{0, 1\}^k \times \{0, 1\}^n \to \{0, 1\}^n$, and a non-singular edge-colored DAG $G = (V, E, \chi)$ with unique source and sink nodes, and $m < 2^n$ total nodes, a function $F^G : \{0, 1\}^k \times \{0, 1\}^{mn} \to \{0, 1\}^n$ can be defined be letting for each K, $(F^G)_K$ to be $(F_K)^G$ (as in definition 6). Then,*

$$Sec_{FG}(q, T) \leq Sec_F(q, T) + (mq)^2 2^{-(n+1)}$$

The proof follows from Theorem 2 by standard techniques.

4.1 Background

Most theorems in cryptography involving PRF [3,2,5,6,11] and PRP [21,22,15,12] constructions, as well as modes of operations of block ciphers [4,16,11,24,19][1], from other primitives must tackle the issue of *collisions* in the oracle calls to the smaller primitive. Fortunately, these collision probabilities are usually low, and conditioning on distinctness of oracle calls, the target construct can be shown to behave like a random function or permutation.

Upper bounding the collision probability requires different techniques in many of these theorems, and the difficulty in the proof can depend on issues like whether there are two independent oracles (as in [22,16]) or if a fresh initial vector is used in each invocation (as in [4,16,24,19]). In particular, the proof of security of CBC-MAC [3] is more involved than that of CBC [4] for precisely this reason, i.e. in the former there is no fresh initial vector in each invocation. As our theorem generalizes CBC-MAC, we expect similar intricacies in our proof. However, as mentioned in the introduction, we prove a novel technical lemma which precisely captures this nuance of CBC-MAC.

4.2 Notation

Before we prove theorem 2, we need to fix more notation and give a general idea of the proof. We first note that we allow arbitrary functions as adversaries and not just computable functions. Then without loss of generality, we can assume that the adversary is deterministic, as every probabilistic adversary is just a probability distribution over all deterministic adversaries [18].

Fix an adaptive adversary. Since the adversary is deterministic, the first query's plaintext (say $P^1 = \langle P_1^1, ..., P_m^1 \rangle$) is fixed for that adversary. Thus, the first query's output, say C_m^1 is only a function of f. The adversary being adaptive, its second query is a function of C_m^1. But, since C_m^1 is only a function of f, the second query's plaintext can also be written just as a function of f. Thus, C_m^2 is only a function of f, and so forth.

We will denote probabilities under the first scenario, i.e. (a) in the theorem 2 statement, as Pr, and the probabilities in the second scenario, i.e. (b) in the theorem 2 statement, as $\Pr_{(b)}$. Most of the analysis will be devoted to the first scenario. So, unless otherwise mentioned, all random variables from now on are in the first scenario.

For all variables corresponding to a query, we will use superscripts to denote the query number. Subscripts will be used to denote blocks within a query. The variables will be as in Definition 6, i.e. P standing for plaintext input, M standing for the variable on which the f function is applied, and C standing for the output of f.

---

[1] These references are not meant to be exhaustive.

Thus, by the convention above C_j^i is the output of f in the ith query's j^{th} block. We will use C to denote the whole **transcript** $\{C_j^i\}_{i\in[q],j\in[m]}$ of f outputs. There will often be a need to just refer to the sequence of last blocks of each query; we will use C_m^* to denote the sequence $C_m^1,...,C_m^q$, i.e. the mth block from all the queries. More precisely, as argued earlier, these variables should be written as a function of f, e.g. $C(f)$, but we will drop the argument when it is clear from context.

Let c denote a constant mqn-bit transcript, i.e. a prospective value for $C(f)$. For a fixed c, P_j^i and M_j^i can be viewed as functions of only c (see definition 6), and we will write them as $P_j^i(c)$ and $M_j^i(c)$. Just as for C, we will use $P(c)$ to denote the whole sequence.

Definition 7. Given a constant mqn-bit transcript c, let the plaintext chosen by the adversary be $p = P(c)$. For any vertices j, j', and query indices i, i' we say that $(i, j) \equiv_c (i', j')$ if

$$(j = j') \text{ and } \forall k \preceq j : p_k^i = p_k^{i'}$$

Define

$$\mu_c(i, j) = \min\{i' | (i', j) \equiv_c (i, j)\}.$$

Not every mqn-bit constant c can be a real transcript $C(f)$ for some f. So, we define a notion of consistent c. We call c **consistent** ($\mathrm{con}(c)$) if

$$\forall j \in [1..m], \forall i : c_j^i = c_j^{\mu_c(i,j)}$$

Define the following "correcting" function ρ from mq n-bit blocks to mq n-bit blocks:

$$\rho(c) = \bar{c}, \text{ where } \bar{c}_j^i = c_j^{\mu_c(i,j)}.$$

The above definition of a correcting function is similar to the one used in the proof of the Luby-Rackoff theorem (see [20]).

Define the *core* index set of c to be $I = \{(i, j) | \mu_c(i, j) = i\}$. Informally, I is the set of indices which are not required to be "consistent" with smaller indices. Thus, $\rho(c)$ retains the values at core indices, and corrects them otherwise.

Consider the following condition **PD** (*pairwise different*).

Definition 8. For any constant c, define $\mathrm{PD}(c)$ to be

$$\forall i, i' \in [1..q], \forall j, j' \in [1..m], j \neq j' : M_j^i(c) \neq M_{j'}^{i'}(c),$$

$$\text{and} \quad \forall i, i' \in [1..q], \forall j \in [1..m] : (i, j) \not\equiv_c (i', j) \Rightarrow M_j^i(c) \neq M_j^{i'}(c).$$

4.3   Proof of Main Theorem

Lemma 1 (PRF Technical Lemma). *For every qn-bit constant $r = \langle r^i \rangle_{i\in[1..q]}$*

$$Pr_{c\in_U\{0,1\}^{mqn}, f\in_U\mathcal{F}(n\rightarrow n)}[\, C(f) = c \;\wedge\; PD(c) \mid c_m^* = r\,]$$
$$= 2^{-mqn} * Pr_{c\in_U\{0,1\}^{mqn}}[\, PD(\rho(c)) \mid c_m^* = r\,]$$

In the left hand side of the lemma, we have that c is consistent, as it is not difficult to see that $C(f)$ is consistent (as proven below in lemma 2(i)). Now for consistent c, it is also easy to see that "correcting" it leaves it unchanged, i.e. $\rho(c) = c$ (see lemma 2(f)). Hence, we can replace $PD(c)$ by $PD(\rho(c))$ in the left hand side above. Thus, the lemma can be restated as

$$\Pr_{c,f}[\, C(f) = c \mid PD(\rho(c)) \wedge c_m^* = r] = 2^{-mqn}$$

To prove this, we can try to see what constraints are imposed on f and c by $C(f) = c$. For $C(f)$ to be same as c, the transcript c must be consistent, as $C(f)$ is consistent. Let I be the *core* index set of c. Let $l = |I|$. Then for c to be consistent, there are exactly $(mq - l)$ n-bit linear constraints on c. We will also see that for every consistent c (in which case $c = \rho(c)$), such that $PD(c)$ holds, a function f_c can be defined using the M and the c values at core indices I such that $C(f_c) = c$ (see lemma 3). Moreover, such an f_c is *unique* on these l input values M (lemma 2(f)). Thus, there are exactly l n-bit constraints on f such that $C(f) = c$. Thus, there are a total of mq n-bit constraints on f and c. However, we have not addressed the issue of whether the condition $PD(\rho(c))$ perhaps influenced this count of mq total constraints. We show below rigorously that even under the condition $PD(\rho(c))$ the number of n-bit constraints on f and c is exactly mq.

So to start with, for each consistent c, we would like to define a function f_c such that $C(f_c) = c$. We also show below (lemma 2(h)) that for consistent c, $M_j^i(c) = M_j^{\mu_c(i,j)}(c)$. Thus, if we define f_c at core indices, i.e. define $f_c(M_j^i(c)) = c_j^i$ for all i in I, we might have $C(f_c) = c$. There is a slight problem however, i.e. f_c may not be well-defined, as the $M(c)$ values at core indices may not be distinct. In fact, we will need an even stronger distinctness condition, i.e. PD defined above, than just being distinct at core indices.

Definition 9. For each c, such that $PD(c)$ holds, define f_c as follows. Let $I = \{(i,j) \mid \mu_c(i,j) = i\}$ be the *core* index set. For $(i,j) \in I$, define $f_c(M_j^i(c)) = c_j^i$. This is well defined as $PD(c)$ holds. We will not need to define f_c on other values.

In Lemma 3 below we show that for every consistent c such that $PD(c)$ holds it is indeed the case that $C(f_c) = c$.

We collect all simple statements about μ, ρ and consistency and their relationships to each other in the following lemma.

Lemma 2. *For all $i, i' \in [1..q]$, $i \neq i'$, for all $j \in [1..m]$ and mqn bit constant transcript c:*

(a) $(i, m) \not\equiv_c (i', m)$, *i.e.* $\mu_c(i, m) = i$,
(b) $\equiv_c$ *is an equivalence relation,*
(c) $\mu_c(\mu_c(i, j), j) = \mu_c(i, j)$,
(d) $\mu_c = \mu_{\rho(c)}$,
(e) $\rho(c)$ *is consistent,*
(f) *Let c be consistent, and let b be such that for all i s.t. $\mu_c(i, j) = i$, $b_j^i = c_j^i$. Then $\rho(b) = c$. Also, for consistent c, $\rho(c) = c$*

(g) For $u \preceq j$, $\mu_c(i,u) = \mu_c(\mu_c(i,j),u)$.
(h) For consistent c, $M_j^i(c) = M_j^{\mu_c(i,j)}(c)$
(i) $C(f)$ is consistent,
(j) For the transcript c let $p = P(c)$ be its corresponding plaintext. If for all u s.t. $E(u,j)$, $\mu_c(i,u) = \mu_c(i',u)$, and $p_j^i = p_j^{i'}$, then $\mu_c(i,j) = \mu_c(i',j)$.

Proof: (a) As we have assumed, wlog, that the adversary does not repeat queries, it follows that i and i' ($i \neq i'$) can never be equivalent over all vertices V. In particular, it is not the case that $(i,m) \equiv_c (i',m)$. To see this, note that we have assumed that the graph has only one sink node, i.e. V_m. It follows that for every node j, $j \preceq m$, hence the claim.
(b) & (c) straightforward.
(d) Note that the adversary's choice of $p = P(c)$ depends only on c_m^*. So we first show that for all i, $\rho(c)_m^i = c_m^i$. This follows as $\mu_c(i,m) = i$ by (a). Thus p remains same for $\rho(c)$.
(e) We just note that for all i,i', $(i,j) \equiv_c (i',j)$ implies $\mu_c(i,j) = \mu_c(i',j)$. Thus, by definition of ρ, we have $\rho(c)_j^i = \rho(c)_j^{i'}$.
(f) We first note that, since by (a), $\mu_c(i,m) = i$, we have $b_m^i = c_m^i$. Thus, as in proof of (d) above, $\mu_b = \mu_c$. Now, $\rho(b)_j^i = b_j^{\mu_b(i,j)} = b_j^{\mu_{c,j}(i)} = c_j^{\mu_{c,j}(i)}$, the last equality following from (c) and condition on b. For consistent c, this is same as c_j^i.
(g) For $u \preceq j$, $(i,j) \equiv_c (i',j)$ implies $(i,u) \equiv_c (i',u)$. Now, $(i,j) \equiv_c (\mu_c(i,j),j)$. Thus, $(i,u) \equiv_c (\mu_c(i,j),u)$.
(h) $M_j^i(c) = P_j^i(c) + \sum_{u:E(u,j)} \chi(\langle u,j \rangle) * c_u^i$. First note that $P_j^i(c) = P_j^{\mu_c(i,j)}(c)$. Also, for consistent c and $u \preceq j$, $c_u^i = c_u^{\mu_c(i,u)} = c_u^{\mu_c(\mu_c(i,j),u)}$ by (g). Again by consistency of c, the latter is same as $c_u^{\mu_c(i,j)}$. This shows that $M_j^i(c) = M_j^{\mu_c(i,j)}(c)$.
(i) by induction on the finite partial order $\prec$.
(j) We just need to show that $(i,j) \equiv_c (i',j)$. But $\mu_c(i,u) = \mu_c(i',u)$ implies $(i,u) \equiv_c (i',u)$. This along with $p_j^i = p_j^{i'}$ shows that p agrees in queries i and i' over all blocks $j' \preceq j$. $\square$

Lemma 3. *For any consistent c such that $PD(c)$ holds:*

$$C(f_c) = c$$

Proof: Follows by induction. See the appendix for a full prove. $\square$

Lemma 1 (PRF Technical lemma restated). *For every qn-bit constant $r = \langle r^i \rangle_{i \in [1..q]}$*

$$\Pr_{c \in_U \{0,1\}^{mqn}, f \in_U \mathcal{F}_{(n \rightarrow n)}}[C(f) = c \wedge PD(c) \mid c_m^* = r]$$
$$= 2^{-mqn} * \Pr_{c \in_U \{0,1\}^{mqn}}[PD(\rho(c)) \mid c_m^* = r]$$

Proof: We first show that the LHS above is same as

$$\Gamma = \Pr_{c,b \in_U \{0,1\}^{mqn}}[b_j^i = c_j^i|_{(i,j):\mu_c(i,j)=i} \wedge con(c) \wedge PD(c) \mid c_m^* = r]$$

By lemma 2(i), the conjunct $con(c)$ can be added to the LHS of the lemma. We show that the two probabilities are same for every constant c. So, fix a c. As before, let $I = \{(i,j) \mid \mu_c(i,j) = i\}$ be the core index set of c. Let $S = \{M_j^i(c) \mid (i,j) \in I\}$. Since PD(c) holds, $|S| = |I|$. Let S' be an arbitrary set of n bit strings, disjoint from S, and $|S'| = mq - |I|$. Thus, $|S \cup S'| = mq$.

By lemma 3, $C(f_c) = c$. Thus, for each b agreeing with c on I, we have a function f_c defined on $|I|$ inputs S, such that $C(f_c) = c$. We can use the remaining $mq - |I|$ values of b (i.e. from indices which are not in I) to extend f_c to be defined on $S \cup S'$. This map from b to the extended f_c is 1-1.

Similarly, for any function f defined on $S \cup S'$, such that $C(f) = c$ (note that f need only be defined on S for $C(f)$ to be well defined), we can define an mqn-bit long b which agrees with c on I. For indices in $(i,j) \in I$, use $f(M_j^i(c))$ to define b_j^i, and use $f(s)$, $s \in S'$, to define the remaining part of b. This map from f to b is also 1-1. This shows that the LHS of the statement of the lemma is same as Γ.

We next show that, the RHS of the statement of the lemma is same as Γ. To this end, we show that the following two sets are equinumerous, i.e. we show a bijection between the two sets. The two sets are

$$\mathcal{C} = \{c \mid c \in \{0,1\}^{mqn}, \text{PD}(\rho(c)), \text{ and } c_m^* = r\}$$

$$\mathcal{D} = \{(c,b) \mid c, b \in \{0,1\}^{mqn}, b_j^i = c_j^i|_{(i,j) \in I}, \text{con}(c), \ PD(c), \text{ and } c_m^* = r\}$$

That they are equinumerous follows easily from lemma 2(e,f,a,d), but to be rigorous consider the following extension of ρ to a function $\hat{\rho}$ from $\mathcal{C}$ to $\mathcal{D}$.

$$\hat{\rho}(c) = (\rho(c), c)$$

It needs to be shown that the function has $\mathcal{D}$ as its range, is 1-1 and onto. The function is obviously 1-1. To prove that its range is $\mathcal{D}$, we need to prove three things:

1. $\rho(c)$ is consistent: follows by lemma 2(e).
2. $c_j^i = \rho(c)_j^i|_{\mu_{\rho(c)}(i,j)=i}$: follows directly from definition of ρ and lemma 2(d).
3. $\forall i$, $\rho(c)_m^i = r^i$: by lemma 2(a) and definition of ρ we have $\rho(c)_m^i = c_m^i$; and hence $c_m^i = r^i$ implies $\rho(c)_m^i = r^i$.

To prove that it is onto, for any (c,b) in $\mathcal{D}$, we show that b is in $\mathcal{C}$ and $\hat{\rho}(b) = (c,b)$. But for any (c,b) in $\mathcal{D}$, by lemma 2(f), $\rho(b) = c$. Thus, $\hat{\rho}(b) = (c,b)$. It also follows that PD($\rho(b)$) holds. Moreover, by lemma 2(a), $b_m^* = c_m^*$. Thus b is in $\mathcal{C}$.

The lemma follows by noting that $\Gamma = |\mathcal{D}|/2^{2mqn} = 2^{-mqn} * (|\mathcal{C}|/2^{mqn})$.

$\square$

Lower bounding the right hand side of the above lemma is a much easier task, as there is no function f involved.

We will denote by Δ the quantity $(mq)^2 2^{-(n+1)}$.

Lemma 4. *For every qn bit constant r,*

$$Pr_{c \in_U \{0,1\}^{mqn}}[\ PD(\rho(c)) \mid c_m^* = r] \geq 1 - \Delta$$

Proof: First note that for all i, $c_m^i = \rho(c)_m^i$, by lemma 2(a) and definition of ρ. Thus, once c_m^* is fixed (and hence $\rho(c)_m^*$) to r, the plaintext $p = P(c)$ is fixed, independent of other c_j^i ($i \in [1..q]$, $j < m$). We will prove the lemma by upper bounding the probability of $\neg PD$ by union bound.

For each vertex j, let V_j be its set of incident vertices, i.e. $V_j = \{u \mid E(u,j)\}$. Recall,

$$M_j^i(\rho(c)) = p_j^i + \sum_{u:E(u,j)} \chi(\langle u,j \rangle) * c_u^{\mu_c(i,u)}$$

If $j \neq j'$, and $V_j \neq V_{j'}$, wlog let $w \in V_j$ and $w \notin V_{j'}$. Then $M_j^i(\rho(c)) = M_{j'}^{i'}(\rho(c))$ iff

$$\chi(\langle w,j \rangle) * c_w^{\mu_c(i,w)}$$
$$= p_j^i + p_{j'}^{i'} + \sum_{u:E(u,j),u \neq w} \chi(\langle u,j \rangle) * c_u^{\mu_c(i,u)} + \sum_{u:E(u,j')} \chi(\langle u,j' \rangle) * c_u^{\mu_c(i',u)}$$

Since, $c_w^{\mu_c(i,w)}$ does not appear on the RHS, and $w < m$, and $\chi(\langle w,j \rangle) \neq 0$, the probability of above is 2^{-n}.

If $j \neq j'$, and $V_j = V_{j'}$, then for some $w \in V_j$, $\chi(\langle w,j \rangle) \neq \chi(\langle w,j' \rangle)$, as the underlying graph G is non-singular. Thus, similarly to the argument above, $M_j^i = M_{j'}^{i'}$ happens with probability 2^{-n}.

When j equals j' (and $i \neq i'$), we have three cases. If for some u incident on j ($E(u,j)$), $\mu_c(i,u) \neq \mu_c(i',u)$, then the probability of the two Ms being equal is at most 2^{-n}. Otherwise, if $p_j^i \neq p_j^{i'}$, then the probability is zero. If $p_j^i = p_j^{i'}$, we have $\mu_c(i,j) = \mu_c(i',j)$ by lemma 2(j), and hence the corresponding disjunct in $\neg PD$ is false.

Since all the probabilities are 2^{-n} or zero, the bound in the lemma follows. $\quad\square$

Lemma 5

$$Pr_f[\ PD(C(f))] \geq 1 - \Delta$$

Proof:

$$Pr_f[\ PD(C(f))]$$

$$= \sum_r \sum_c Pr_f[C(f) = c \wedge PD(c) \wedge c_m^* = r]$$

$$= \sum_r Pr_{c,f}[C(f) = c \wedge PD(c) \wedge c_m^* = r] * 2^{mqn}$$

$$= \sum_r Pr_{c,f}[C(f) = c \wedge PD(c) \mid c_m^* = r] * 2^{-qn} * 2^{mqn}$$

$$= \sum_r 2^{-qn} * Pr_c[PD(\rho(c)) \mid c_m^* = r] \quad \text{(by lemma 1)}$$

$$\geq 1 - \Delta \quad \text{(by lemma 4)} \quad\quad\quad\quad\quad\quad\quad\quad\quad\quad\quad\quad \square$$

Since the adversary A decides 0 or 1 based on the oracle replies, say $O = \langle O^1, O^2, ..., O^q \rangle$, we can write its output as $A(O)$. In scenario (a), O is really $C_m^*(f)$, with f chosen randomly. Since in scenario (b), the oracle is a random function with range n bits, O is just a uniformly random string of length qn.

Lemma 6

$$\Pr_{(b)}[A(O) = 0] \geq \Pr_f[A(C_m^*) = 0 \wedge \mathrm{PD}(C(f))] \geq (1 - \Delta)\Pr_{(b)}[A(O) = 0]$$

Proof: To begin with, we have

$$\Pr_f[A(C_m^*) = 0 \wedge \mathrm{PD}(C(f))] = \sum_c \Pr_f[A(c_m^*) = 0 \wedge C(f) = c \wedge \mathrm{PD}(c)]$$

$$= 2^{mqn} * \Pr_{c \in_{\mathcal{U}} \{0,1\}^{mqn}, f}[A(c_m^*) = 0 \wedge C(f) = c \wedge \mathrm{PD}(c)]$$

$$= 2^{mqn} * \Pr_{c \in_{\mathcal{U}} \{0,1\}^{mqn}, f}[C(f) = c \wedge \mathrm{PD}(c) \mid A(c_m^*) = 0] * \Pr_{(b)}[A(O) = 0]$$

The above is at least $(1 - \Delta)\Pr_{(b)}[A(O) = 0]$ by lemma 1 and lemma 4, and at most $\Pr_{(b)}[A(O) = 0]$. $\qquad\square$

Proof of Theorem 2 (Main Theorem): By lemma 6 and lemma 5 it follows that

$$|\Pr_f[A(C_m^*(f)) = 0] - \Pr_{(b)}[A(O) = 0]| \leq \Delta \qquad\square$$

5   Variable Length Domain Extension and Family of Graphs

The previous constructions were devoted to extending the domain of a function from n bits to mn bits, for a fixed m. In other words, the plaintext queries of the adversary were restricted to be exactly mn bits. We could fix m to be large enough, say $m = 2^n$, and use a canonical encoding of smaller sized plaintexts into length mn bit strings. Such an encoding exists for all plaintexts of size less than mn by appending plaintexts of size q bits, by 10^i, where $i = mn - q - 1$. In other words, 10^i acts as an end marker. However, smaller sized plaintexts have to undergo $m = 2^n$ applications of F, which is very inefficient. This problem of a really long end marker was resolved by [23] (also see [7]) by noting that the end marker can actually be of length zero, if it can be authenticated.

The simplest way to achieve this is to have two independent PRFs $F1$ and $F2$. Use $F1$ when the plaintext is not a multiple of the block size n, and use $F2$ when the plaintext is a multiple of n. In the former case, append an end marker of the kind 10^i, but now i need only be of length at most $n - 2$.

So, given a function F_k on n bits, consider a collection of graphs, one graph G_q in the family for each (plaintext) bit-length q. Then if we define $\tilde{F}_k^{G_q}$ similarly to as before, we have a composite function from all strings to n bits. We know that individually each $\tilde{F}^{G_q}$ is a PRF given F is a PRF. As explained in the previous paragraph, we need to assure that these different functions are (almost) independent. We prove that if the family of graphs satisfy certain constraints then this is indeed the case.

We consider a fixed n throughout the rest of this section. We will assume that we are only interested in domain extension up to length $2^n * n$ bits, as theorem 2 is ineffective beyond that length (this restriction is only for sake of simplicity). Each query of the adversary will be a string p of length q bits, $(0 < q < 2^n * n)$. We let the composite function answers the query as follows: If q is a multiple of n, then it returns $f^{G_q}(p)$. Otherwise, let p' be p appended with 10^i, where i is the smallest positive number to make $|p'|$ a multiple of n. The composite function then returns $f^{G_{|p'|}}(p')$.

For every $0 \leq l < 2^n$, since strings of length $ln + 1$ to $ln + n - 1$ bits get canonically encoded in the above method, we can use the same graph for all these lengths. Thus, for each l, we really need only two graphs ([7]), one for lengths $ln + 1$ to $ln + n - 1$, and one for length $ln + n$. From now on, we will assume that all plaintexts are of bit length multiples of n. Each adversarial query will be a pair: (p, z), where p is a bit string of length multiple of n, and z is in $\{0, 1\}$ (z signifies if the plaintext was of length a multiple of n or if it was padded to make it so).

Definition 10. Let S be the set of all binary strings of length non-zero multiples of n, but less than $2^n * n$. Let $\mathcal{F}$ be the set of all functions:

$$S \times \{0, 1\} \rightarrow \{0, 1\}^n$$

Let $\tilde{F}$ be a function with signature:

$$\{0, 1\}^k \times S \times \{0, 1\} \rightarrow \{0, 1\}^n$$

Given a PRF F from n bits to n bits, we need to define $\tilde{F}$ such that no adaptive adversary can distinguish between $\tilde{F}_K$, with K chosen randomly, and a function chosen uniformly at random from $\mathcal{F}$. As in the previous sections, given a function f from n bits to n bits, and given a collection of graphs $\mathcal{G}$, we first define a function $f^{\mathcal{G}}$ in $\mathcal{F}$.

Definition 11. Let $\mathcal{G}$ be a collection of edge-colored DAGs $G(l)$ (see definition 5), $l \leq (2^n - 1) * 2$. Each $G(l)$ is required to have unique source and sink nodes. Further, each $G(l)$ is required to have at least $\lceil \frac{l}{2} \rceil$ nodes. Define a function $f^{\mathcal{G}}$ as follows:

$$f^{\mathcal{G}}(p, z) = f^{G(2*|p|-z)}(p)$$

where f^G is as in definition 6. If the graph has more nodes than the length of the plaintext, then append enough zeroes to the plaintext. Usually, graphs will have exactly the required number of nodes. However, at the base cases, i.e. small length plaintexts, it may be necessary to have extra nodes. For an example, see Section 5.1.

For a theorem similar to theorem 2 to hold, we need further restrictions on $\mathcal{G}$. In particular, it will not be enough that individual graphs in $\mathcal{G}$ be non-singular. Since, we will need to extend the notion of non-singularity to the whole

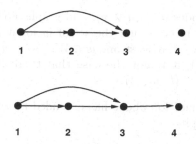

Fig. 5. An Incorrect Construction

collection of graphs, it is best to fix a set of vertices V, and just define the edges and colorings for the individual graphs. Thus, we will define $E(l)$, and $\chi(l)$. The partial order $\prec_l$ is, as before, the transitive closure of $E(l)$.

To motivate the generalized definition of non-singularity, we first consider an example where it is not enough for individual graphs to be non-singular. Let $V = [1..4]$. The graphs are identical (see fig 5), except that the second graph $G(2)$ has an extra edge from 3 to 4. The first graph $G(1)$ is used to answer queries of length 3 blocks, and the second to answer queries of length 4. Clearly, both graphs are individually non-singular. Consider two queries, one of length three, and another of length four, the latter being just an extension of the first. However, the first graph's output is C_3, and is accessible to the adversary. Thus, during the second query the internal state C_3 is available to the adversary, and it can force M_4 to be any value of its choice.

This suggests that for each graph $G(i)$, the graph $G(i)$ itself cannot be allowed to be an induced subgraph of another graph $G(i')$. We prove that this condition is sufficient for the composite function to be a PRF.

Definition 12. For any vertex j in V, let U_j^l be the set of incident vertices of j in $G(l)$.
For any vertex j in V, we say $(l,j) \cong (l',j)$ if either $(j = 1)$ or
- $U_j^l = U_j^{l'}$, and
- for all $u \in U_j^l$: $\chi_l(\langle u,j \rangle) = \chi_{l'}(\langle u,j \rangle)$, and inductively $(l,u) \cong (l',u)$.

Essentially, (l,j) is congruent to (l',j) if the two graphs $G(l)$ and $G(l')$ are identical till j.

Definition 13. Let $\mathcal{G} = \langle G(l) \rangle$, where each $G(l) = (V, E(l), \chi(l))$ is an edge-colored DAG, be a collection of graphs.

- With each $G(l)$ we associate its size $m(l)$ to be the largest numbered node in V such that there is an edge directed to it in $G(l)$.
- For each $G(l)$ we define the graph $\tilde{G}(l) = ([1..m(l)], E(l), \chi(l))$, to be the induced subgraph of $G(l)$ on vertices $[1..m(l)]$.

The collection $\mathcal{G}$ is called **PRF-preserving** if

- each $\tilde{G}(l)$ has only one source node, one sink node, has at least $\lceil \frac{l}{2} \rceil$ nodes, and

- if for any pair of nodes u, v ($u \neq v$), and graphs $G(l)$ and $G(l')$, the set of incident nodes of u in $G(l)$, and the set of incident nodes of v in $G(l')$ are same (say W), then for at least one $w \in W$, $\chi_l(\langle w, u \rangle) \neq \chi_{l'}(\langle w, v \rangle)$.
- for each graph $G(l)$, it is not the case that there is another graph $G(l')$, $l' \neq l$, s.t. $(l, m(l')) \cong (l', m(l'))$

Basically, the second condition above has extended the non-singularity requirement to be over all graphs.

Theorem 4. *For a PRF-preserving collection of $2 * (2^n - 1)$ DAGs $\mathcal{G}$, let $f^{\mathcal{G}}$ be as in definition 11. Then, no adaptive adversary, with q adaptive queries $\langle (p^i, z^i) \rangle$ ($i \in [1..q]$, and $|p^i| \leq 2^n - 1$), can distinguish between (a) $f^{\mathcal{G}}$ where f is chosen uniformly at random from $\mathcal{F}(n \rightarrow n)$, (b) and a function chosen uniformly at random from $\mathcal{F}$, with probability more than $(\sum_{i \in [1..q]} |p^i|)^2 2^{-(n+1)}$.*

Proof: To adapt the proof of theorem 2, we first need to redefine the notion of consistent transcripts c. First note that, on a fixed transcript c, the queries of the adversary are fixed, say $\langle p^i, z^i \rangle_{i \in [1..q]}$. Recall, by definition of $f^{\mathcal{G}}$, on input p^i, z^i the graph $G(2 * |p^i| - z^i)$ is used. We just denote this graph by G^i. The corresponding edge relation, coloring and partial order will be denoted E^i, χ^i, and $\prec^i$ resp. Also, for the graph G^i, its induced subgraph as per definition 13, will be denoted $\tilde{G}^i$. Similarly, the size of the graph $\tilde{G}^i$ will be denoted by m^i. Note that $m^i = |c^i| \geq |p^i|$.

Definition 14. For any vertex j in V, let V_j^i be the set of incident vertices of j in G^i.
For any vertex j in V, we say $(i, j) \cong_c (i', j)$ if either $(j = 1)$ or
- $V_j^i = V_j^{i'}$, and
- for all $u \in V_j^i$: $\chi^i(\langle u, j \rangle) = \chi^{i'}(\langle u, j \rangle)$, and inductively $(i, u) \cong_c (i', u)$.

Essentially, (i, j) is congruent (w.r.t. c) to (i', j) if the two graphs G^i and $G^{i'}$ are identical till j.

Once we generalize the definition of $\equiv_c$, rest of the definitions and proofs remain almost the same.

Definition 15. For any vertices j, j', and query indices i, i' we say that $(i, j) \equiv_c (i', j')$ if

$$(j = j') \text{ and } (i, j) \cong_c (i', j) \text{ and } \forall k \preceq^i j : p_k^i = p_k^{i'}$$

As before, define

$$\mu_c(i, j) = \min\{i' | (i', j) \equiv_c (i, j)\}.$$

We call c **consistent** ($con(c)$) if

$$\forall j \in [1..2^n - 1], \forall i : c_j^i = c_j^{\mu_c(i, j)}$$

Define the following "correcting" function ρ:

$$\rho(c) = \bar{c}, \text{ where } \bar{c}_j^i = c_j^{\mu_c(i, j)}, \text{ for } j \in [1..m^i]$$

Since the proof of theorem 10 will be adapted from the proof of theorem 2, we will denote all lemmas for theorem 10 corresponding to lemmas for theorem 2 by the prime symbol. In the proof of lemma 2(a)′, if $m^i \neq m^{i'}$, then $(i, m) \not\cong_c (i', m')$. Otherwise, if the plaintexts p^i and $p^{i'}$ are different, then again $(i, m^i) \not\cong_c (i', m^i)$. If the plaintexts are also same, then as the adversary does not repeat queries, wlog let $G^i = G(2 * m^i - 1)$, and $G^{i'} = G(2 * m^i)$. But $(i, m^i) \cong_c (i', m^i)$ is not allowed in $\mathcal{G}$ which is PRF-preserving. That proves lemma 2(a)′.

Proof of rest of lemma 2′ is similar to proof of lemma 2. In the statement and proof of lemma 2(f)′, j must be restricted to be $[1..m^i]$. Similar restrictions apply in the definition of PD (definition 8) and definition of f_c (definition 9). Proof of lemma 3′ is similar to proof of lemma 3.

Lemma 1 is now restated as (recall S from definition 10):

Lemma 1′. *For every qn bit constant $\langle r^i \rangle$ ($i \in [1..q]$)*

$$\mathrm{Pr}_{c \in_U S^q, f}[C(f) = c \wedge \ \mathrm{PD}(c) \,|\, \forall i : c_{m^i}^i = r^i]$$
$$= 2^{-mqn} * \mathrm{Pr}_{c \in_U S^q}[\mathrm{PD}(\rho(c)) \,|\, \forall i : c_{m^i}^i = r^i]$$

Proof Sketch: The proof is similar to proof of lemma 1, if we notice that we fix c in the first part of the proof. For a fixed c, let $I = \{(i, j) \,|\, \mu_c(i, j) = (i, j), j \in [1..m^i]\}$. Let $T = \{M_j^i(c) \,|\, (i, j) \in I\}$. Since PD(c) holds, $|T| = |I|$. Let T' be an arbitrary set of n bit strings, disjoint from T, and $|T'| = \sum_{i \in [1..q]} m^i - |I|$. Thus, $|T \cup T'| = \sum_{i \in [1..q]} m^i$.

By, lemma 3′, $C(f_c) = c$. Thus, for each b agreeing with c on I, we have a function f_c defined on $|I|$ inputs T, such that $C(f_c) = c$. We can use the remaining $\sum_{i \in [1..q]} m^i - |I|$ values of b (i.e. from indices which are not in I) to extend f_c to be defined on $T \cup T'$. This map from b to the extended f_c is 1-1.

The reverse direction is done as in lemma 1.

Rest of the proof is also as in proof of lemma 1.   □

Let Δ denote $(\sum_{i \in [1..q]} m^i)^2 * 2^{-(n+1)}$.

Lemma 7′. *For every qn bit constant $\langle r^i \rangle$ ($i \in [1..q]$),*

$$\mathrm{Pr}_{c \in_U S^q}[\ \mathrm{PD}(\rho(c)) \,|\, c_{m^i}^i = r^i] \geq 1 - \Delta$$

Proof: First note that for all i, $c_{m^i}^i = \rho(c)_{m^i}^i$, by lemma 2(a)′ and definition of ρ. As opposed to lemma 4, we need to show that it is not the case that a $\rho(c)_j^i$, with $j \neq m^i$, can be defined to be a c_j^i, such that $j = m^{i'}$. Suppose, there is indeed an $(i', j) \equiv_c (i, j)$, such that $j = m^{i'}$. Since, $(i', j) \equiv_c (i, j)$, we have $(i', j) \cong_c (i, j)$. Thus the graphs G^i and $G^{i'}$ are identical till $j = m^{i'}$. Thus, unless they are the same graph, this is not allowed by the condition on PRF-preserving $\mathcal{G}$. If they are the same graph, then $j = m^i$, a contradiction.

Rest of the proof is similar to proof of lemma 4.   □

Rest of the proof of theorem 10 is identical to that of theorem 2.

5.1   Applications to Variable Length Domain Extension

As an application of theorem 10, we get the variable length domain extension scheme as described in figure 6. In the figure, for each plaintext block length two graphs are given as required in definition 11. The number on the left of the graphs denotes the plaintext block lengths for which those graphs are to be employed. The $0/1$ bit signifies if the plaintext was padded to make its bit-length a multiple of n. We have only illustrated graphs up to length five, as for larger lengths, we follow similar methods as for length four and five. Note that for plaintext block length one, we have graphs which have two nodes. As remarked at the end of definition 11, this requires that plaintexts of length one block must be appended with a zero block, before employing graphs "ONE-0" or "ONE-1".

This mode has an advantage over XCBC [7], and OMAC [14] that it does not even need to employ the initial F on a constant like 0^n. Moreover, the scheme shows that if the plaintexts are restricted to be more than three blocks in length, then no Galois field arithmetic is required.

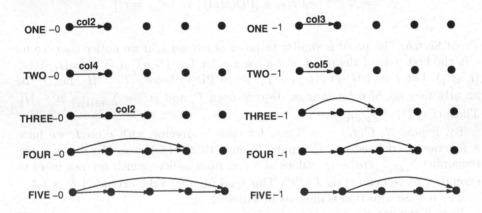

Fig. 6. A Variable Length Mode

References

1. ANSI X3.106, "American National Standard for Information Systems - Data Encryption Algorithm - Modes of Operation", *American National Standards Institute, 1983.*
2. M . Bellare, R. Canetti, H. Krawczyk, " Pseudorandom Functions Revisited: The Cascade Construction and its Concrete Security", Proc. IEEE FOCS 1996.
3. M. Bellare, J. Kilian, P. Rogaway, "The Security of Cipher Block Chaining", *JCSS*, Vol. 61, No. 3, Dec 2000, pp. 362-399
4. M. Bellare, A. Desai, E. Jokiph, P. Rogaway, "A Concrete Security Treatment of Symmetric Encryption: Analysis of the DES Modes of OPeration", 38th IEEE FOCS, 1997
5. D. Bernstein, " How to Stretch Random Functions: The security of Protected Counter Sums", J. of Cryptology, Vol 12,No. 3, (1999).

6. J. Black, P. Rogaway, " A Block Cipher Mode of Operation for Parallelizable Message Authentication", Proc. Eurocrypt 2002.
7. J. Black, P. Rogaway, "CBC MACs for arbitrary length messages: The three key constructions". CRYPTO 2000, LNCS 1880.
8. J. Carter, M. Wegman, "Universal Classes of Hash Functions", *JCSS*, Vol. 18, 1979, pp 143-154.
9. O. Goldreich, S. Goldwasser, and S. Micali, " How to construct random functions", J. ACM, vol. 33, no. 4, 1986.
10. National Bureau of Standards, Data Encryption Standard, U.S. Department of Commerce, FIPS 46 (1977)
11. V.D. Gligor, P. Donescu, "Fast Encryption Authentication: XCBC Encryption and XECB Authentication Modes",
    http://csrc.nist.gov/encryption/modes/workshop1
12. S. Halevi and P. Rogaway, "A Tweakable Enciphering Mode", CRYPTO 2003, LNCS 2729.
13. F. Harary, *Graph Theory*, Addison-Wesley 1969.
14. T. Iwata, K. Kurosawa, " OMAC: One -key CBC-MAC", FSE 2003, LNCS 2887.
15. C. S. Jutla, "Generalized Birthday Attacks on Unbalanced Feistel Networks", CRYPTO 1998, LNCS 1462.
16. C. S. Jutla, " Encryption Modes with Almost Free Message Integrity", *Proc. Eurocrypt 2001*, LNCS 2045, 2001.
17. Hugo Krawczyk, "LFSR-based Hashing and Authentication", *Proc. Crypto 94*, LNCS 839, 1994
18. H.W. Kuhn, "Extensive games and the problem of information" in *Contributions to the Theory of Games II*, H.W. Kuhn and A. W. Tucker eds., Annals of Mathematical Studies No. 28, Princeton Univ. Press, 1950.
19. M. Liskov, R. Rivest and D. Wagner, "Tweakable Block Ciphers", CRYPTO 2002, LNCS 2442.
20. M. Luby, "Pseudorandomness and Cryptographic Applications", *Princeton Computer Science Notes*, Princeton Univ. Press, 1996
21. M. Luby and C. Rackoff, "How to Construct Pseudorandom Permutations From Pseudorandom Functions", SIAM J. on Computing, Vol. 17, 1988, pp. 373-386.
22. M. Naor and O. Reingold, "On the construction of pseudo-random permutations: Luby-Rackoff revisited", *Proc. 29th ACM STOC*, 1997, pp 189-199.
23. E. Petrank, C. Rackoff, "CBC-MAC for real-time data sources", J. of Cryptology, vol 13, no. 3, nov 2000.
24. P. Rogaway, M. Bellare, J. Black and T. Krovetz, "OCB: A block-cipher mode of operation for efficient authenticated encryption", *Proc. 8th ACM Conf. Comp. and Comm. Security* (CCS), ACM, 2001.

Appendix

Lemma 3. For any consistent c such that $PD(c)$ holds:

$$C(f_c) = c$$

Proof: Let $p = P(c)$ and $M = M(c)$ be shorthands. Also, we will use $\bar{c}$ as shorthand for $C(f_c)$. Similarly, let $\bar{M} = M(\bar{c})$, $\bar{p} = P(\bar{c})$.

We do induction over the query index.

Base Case: Since the adversary is fixed, the first plaintext message is the same, i.e. $\bar{p}^1 = p^1$. Since $\bar{M}_1^1 = p_1^1$, $\bar{c}_1^1 = f_c(\bar{M}_1^1) = f_c(M_1^1) = c_1^1$, as $(1,1)$ is trivially in I. For $j > 1$, $\bar{M}_j^1 = p_j^1 + \sum_{u\,:E(u,j)} \chi(\langle u,j \rangle) * \bar{c}_u^1$ But, by induction over the partial order $\prec$, $\bar{c}_u^1 = c_u^1$, hence $\bar{M}_j^1 = M_j^1$. Moreover, $(1,j)$ is trivially in I, and hence $\bar{c}_j^1 = c_j^1$.

So, assume that for all $i' < i$, and all j, $\bar{c}_j^{i'} = c_j^{i'}$. Thus, $\bar{p}^i = p^i$. Again, $\bar{M}_1^i = p_1^i = M_1^i$. Thus, $\bar{c}_1^i = f_c(M_1^i) = f_c(M_1^{\mu_c(i,1)})$ by lemma 2(h). By definition of f_c, this is same as $c_1^{\mu_c(i,1)} = c_1^i$. For $j > 1$, $\bar{M}_j^i = p_j^i + \sum_{u\,:E(u,j)} \chi(\langle u,j \rangle) * \bar{c}_u^i$. But, by induction over the partial order $\prec$, $\bar{c}_u^i = c_u^i$, thus $\bar{M}_j^i = M_j^i$. As before, using lemma 2(h), we are done.   □

Chosen-Ciphertext Security from Tag-Based Encryption*

Eike Kiltz

CWI Amsterdam,
The Netherlands
kiltz@cwi.nl
http://kiltz.net

Abstract. One of the celebrated applications of Identity-Based Encryption (IBE) is the Canetti, Halevi, and Katz (CHK) transformation from any (selective-identity secure) IBE scheme into a full chosen-ciphertext secure encryption scheme. Since such IBE schemes in the standard model are known from previous work this immediately provides new chosen-ciphertext secure encryption schemes in the standard model.

This paper revisits the notion of Tag-Based Encryption (TBE) and provides security definitions for the selective-tag case. Even though TBE schemes belong to a more general class of cryptographic schemes than IBE, we observe that (selective-tag secure) TBE is a sufficient primitive for the CHK transformation and therefore implies chosen-ciphertext secure encryption.

We construct efficient and practical TBE schemes and give tight security reductions in the standard model from the Decisional Linear Assumption in gap-groups. In contrast to all known IBE schemes our TBE construction does not directly deploy pairings. Instantiating the CHK transformation with our TBE scheme results in an encryption scheme whose decryption can be carried out in one single multi-exponentiation.

Furthermore, we show how to apply the techniques gained from the TBE construction to directly design a new Key Encapsulation Mechanism. Since in this case we can avoid the CHK transformation the scheme results in improved efficiency.

1 Introduction

Since Diffie and Hellman proposed the idea of public key cryptography [14], one of the most active area of research in the field has been the design and analysis of public key encryption (PKE) schemes. In [16, 27] efficient primitives were suggested from which to build encryption schemes. Formal models of security were developed in [19, 23, 26] and nowadays it is widely accepted that security against chosen-ciphertext attacks provides the "right level of security" for public-key encryption schemes.

---

* The paper was written while the author was a visitor at University of California, San Diego, supported by a DAAD postdoc fellowship.

S. Halevi and T. Rabin (Eds.): TCC 2006, LNCS 3876, pp. 581–600, 2006.

There have been numerous efficient schemes that were shown to be chosen-ciphertext secure in the *random oracle model* [2]. Unfortunately a proof in the random oracle model can only serve as a heuristic argument and has proved to possibly lead to insecure schemes when the random oracles are implemented in the standard model (see, e.g., [10]).

Dolev, Dwork, and Naor [15] were the first to come up with a public-key encryption scheme provably chosen-ciphertext secure in the standard model (without random oracles). Later Cramer and Shoup [12] presented the first really practical public-key encryption scheme. Their approach was further generalized in [13] and later shown by Elkind and Sahai [17] to fit into a more general framework. The nowadays most efficient chosen-ciphertext secure encryption scheme in the standard model is the one due to Kurosawa and Desmedt [21, 1] itself being an improvement of the original Cramer-Shoup scheme. Both schemes, Cramer-Shoup and Kurosawa-Desmedt are secure under the Decisional Diffie-Hellman (DDH) assumption.

FROM IBE TO PKE. One of the recent celebrated applications of Identity-Based Encryption (IBE) is the work due to Canetti, Halevi, and Katz [11] showing an elegant black-box transformation from any IBE into a PKE scheme without giving up its efficiency. We will refer to this as the *CHK transformation*. If the IBE scheme is *selective-identity* secure then the resulting PKE scheme is chosen-ciphertext secure. Efficient constructions of IBE schemes in the standard model were recently developed by Boneh and Boyen [3] so the CHK transformation provides further alternative instances of chosen-ciphertext secure PKE schemes in the standard model.[1]

Another fact worth mentioning about the CHK transformation is that it does not seem to fall into the general framework characterized by Elkind and Sahai. Boneh and Katz [7] later improve the CHK transformation resulting in shorter ciphertexts and more efficient encryption/decryption. Since the two IBE schemes from [3] employ pairing operations the resulting schemes are still less efficient than the Kurosawa-Desmedt scheme.

TAG-BASED ENCRYPTION. MacKenzie, Reiter, and Yang [22] introduce the notion of tag-based encryption (TBE) and show (independent from [11]) that the CHK transformation also transforms any "weakly secure" TBE scheme into a chosen-ciphertext secure PKE scheme. However, the only TBE schemes in the standard model mentioned in [11] are directly derived from known PKE schemes (for example the Cramer-Shoup scheme) and the CHK transformation applied to TBE schemes does not readily give us new instantiations of chosen-ciphertext secure PKE schemes.

---

[1] The underlying computational assumptions for the security reduction of the two IBE schemes from [3] are both "pairing-assumptions", i.e. the Bilinear Decisional Diffie-Hellman (BDDH) assumption and the q-strong Decisional Bilinear Diffie-Hellman Inversion (q-strong BDDHI) assumption.

1.1 Our Contribution

FROM TBE TO PKE. As pointed out in the last two paragraphs selective-identity secure IBE (or weakly secure TBE) schemes are sufficient to construct chosen-ciphertext secure PKE schemes. The natural question that arises is if in the transformation some of the security requirements made to the IBE/TBE scheme can be dropped while still preserving security of the resulting PKE scheme. One of our contributions is to answer this question to the affirmative.

We revisit the security definitions for TBE schemes and introduce the notion of selective-tag secure TBE schemes. Selective-tag security for TBE can be seen as the selective-identity analog for IBE and is weaker than the TBE definition from [22] and the IBE definition from [11]. One of our main results is to show that selective-tag secure TBE is sufficient to build chosen-ciphertext secure PKE. Our construction uses the CHK transformations.

On the theoretical side our result underlines that for the CHK transformation, an IBE scheme is basically overkill since some of its functionality is superfluous. In particular, there is no need to have an IBE *key-derivation algorithm*, which seems to be what distinguishes IBE from all other public-key encryption primitives. The notion of TBE can be viewed as some sort of "flattened IBE scheme" (i.e., as IBE without key-derivation) and therefore exactly captures the above observation. Our contribution is to extract the best out of the afore mentioned papers: we are able to combine the known CHK transformation with a security requirement that is substantially weaker than the requirements that were believed to be necessary.

COMPARING DIFFERENT SECURITY NOTIONS OF TBE, IBE, AND PKE. What distinguishes TBE from IBE is the IBE key-derivation algorithm. Indeed, as we will point out later, it seems to be hard to transform (even particular instances of) TBE schemes into IBE schemes. The difference between selective-tag TBE and weakly secure TBE schemes seems marginal at first glance but (similar to the IBE case [3]) it turns our that the "selective-tag" property is the key to make security proofs for TBE schemes much easier to construct. An even stronger security definition of TBE schemes was already used by Shoup [29] (where the tag was called "label"). Interestingly we show that such "strongly secure" TBE schemes are equivalent to chosen-ciphertext secure PKE schemes. Since the CHK transformation is black-box, our results imply that all the afore mentioned three flavors of TBE security together with chosen-ciphertext secure PKE are in fact all *equivalent* through efficient black-box reductions.

TBE AND PKE ARE EQUIVALENT. SO WHAT IS TBE GOOD FOR? One may ask the question why to make the long detour over TBE when designing PKE schemes at all? The answer is simple. Since TBE is simpler and more general than PKE (and IBE) our hope is that TBE may prove itself useful in the future to come up with more chosen-ciphertext secure encryption in the standard model. In particular, we would like to have chosen-ciphertext secure PKE schemes based on different intractability assumptions. (Different from the BDDH or DDH assumption, hopefully even weaker or at least unrelated.)

AN EFFICIENT TBE SCHEME WITHOUT PAIRING OPERATIONS. To underline the usefulness of our TBE to PKE transformation we present an efficient TBE scheme that (in contrast to all known IBE schemes) does not directly rely on pairing operations for encryption and decryption. In particular, the decryption operation of our new TBE scheme is very efficient and (similar to the KD scheme) only performs one single multi-exponentiation. The recently introduced decisional linear (DLIN) assumption [4] states that, roughly, it should be computational infeasible to decide if $w = z^{r_1+r_2}$, given random $(g_1, g_2, z, g_1^{r_1}, g_2^{r_2}, w)$ as input. Our TBE scheme can be proved to meet the necessary security properties under the DLIN assumption in the standard model. The security reduction is tight, simple, and very intuitive. In contrast to all known efficient IBE schemes our TBE scheme does not directly use pairings. However, our proofs of security have to be carried out in *gap-groups* [25], i.e. groups in which CDH is believed to be hard even though they are equipped with an algorithm that efficiently solves the Decisional Diffie-Hellman (DDH) problem. One particular instance of such gap-groups (which is actually the only one we know at the time being) is obtained using pairings.

Instantiating the scheme with our TBE to PKE transformation we obtain a new and reasonably efficient chosen-ciphertext secure encryption scheme in the standard model based on the DLIN assumption. We remark that this is the first (practical) chosen-ciphertext secure PKE based on the DLIN assumption in the standard model.

DIRECT KEY ENCAPSULATION. A key encapsulation mechanism (KEM) is a light PKE scheme intended to encapsulate and decapsulate a random (symmetric) key. It is well known how to transform any chosen-ciphertext secure KEM into a fully fledged chosen-ciphertext secure PKE scheme using symmetric encryption (with appropriate security properties).

Surprisingly, our techniques from constructing the TBE scheme can also be exploited to directly build a chosen-ciphertext secure KEM in the standard model. Our construction avoids the CHK transformations and (similar to [12, 21]) only deploys a target collision-resistant hash function. As a result the ciphertext size of the scheme is more compact compared to the PKE scheme obtained using the above transformation. Furthermore encryption and decryption can be done more efficiently. Our KEM construction is practical and enjoys a simple proof of security with a tight reduction to the DLIN assumption in the standard model.

1.2 Related Work

Independent of our work, Boyen, Mei, and Waters [9] recently look at some specific PKE schemes obtained from the CHK transformation instantiated with the IBE schemes from [3, 30] and show how to make the resulting schemes more efficient (in terms of computation time and ciphertext length). In particular, they also come up with a practical chosen-ciphertext secure KEM (BMW-KEM) whose security is based on the BDDH assumption in the standard model.[2] Compared to our KEM, the BMW-KEM is based on bilinear pairings and therefore results in a less efficient decryption algorithm (one pairing and one exponentiation compared to one multi-exponentiation in our KEM). The BMW-KEM, however, is slightly more efficient in terms of encryption operations and comes with smaller ciphertexts. Compared to our KEM, the Kurosawa-Desmedt PKE scheme provides the same efficiency for decryption whereas it is more efficient for encryption. In Section 7.1 we discuss efficiency of all known encryption schemes in the standard model. Comparing the overall performance of all known encryption schemes in the standard model the Kurosawa-Desmedt scheme [21] can still be considered as the most efficient.

However, in contrast to the Kurosawa-Desmedt/Cramer-Shoup scheme, our KEM shares with the BMW-KEM the nice property that the validity (or consistency) of ciphertexts can be verified even without knowledge the the secret key. This observation was recently used in [9] to propose a threshold cryptosystem based on their BMW-KEM. With a similar idea and also based on the public validity test our KEM can also be used to build a threshold encryption scheme.

2 Notation

If x is a string, then $|x|$ denotes its length, while if S is a set then $|S|$ denotes its size. If $k \in \mathbb{N}$ then 1^k denotes the string of k ones. If S is a set then $s \xleftarrow{\$} S$ denotes the operation of picking an element s of S uniformly at random. Unless otherwise indicated, algorithms are randomized. "PT" stands for polynomial time and "PTA" for polynomial-time algorithm or adversary. We write $\mathcal{A}(x, y, \dots)$ to indicate that $\mathcal{A}$ is an algorithm with inputs $x, y, \dots$ and by $z \xleftarrow{\$} \mathcal{A}(x, y, \dots)$ we denote the operation of running $\mathcal{A}$ with inputs $(x, y, \dots)$ and letting z be the output. We write $\mathcal{A}^{\mathcal{O}_1, \mathcal{O}_2, \dots}(x, y, \dots)$ to indicate that $\mathcal{A}$ is an algorithm with inputs $x, y, \dots$ and access to oracles $\mathcal{O}_1, \mathcal{O}_2, \dots$ and by $z \xleftarrow{\$} \mathcal{A}^{\mathcal{O}_1, \mathcal{O}_2, \dots}(x, y, \dots)$ we denote the operation of running $\mathcal{A}$ with inputs $(x, y, \dots)$ and access to oracles $\mathcal{O}_1, \mathcal{O}_2, \dots$, and letting z be the output.

3 Definitions

In this section we formally introduce PKE and TBE schemes together with a security definition. We also give a parameter generating algorithm for bilinear groups and pairings and state our complexity assumptions.

3.1 Public-Key Encryption

An *public-key encryption* (PKE) scheme $\mathcal{PKE} = (\mathsf{PKEkg}, \mathsf{PKEenc}, \mathsf{PKEdec})$ consists of three polynomial time algorithms (PTAs). Via $(pk, sk) \xleftarrow{\$} \mathsf{PKEkg}(1^k)$

---

[2] We note that the same scheme as in [9] was independently discovered during research for this paper. Since [9] is already published at the time of writing this extended abstract we decided not to include it here.

the randomized key-generation algorithm produces keys for security parameter $k \in \mathbb{N}$; via $C \xleftarrow{\$} \mathsf{PKEenc}(pk, M)$ a sender encrypts a message M under the public key pk to get a ciphertext; via $M \leftarrow \mathsf{PKEdec}(sk, C)$ the possessor of secret key sk decrypts ciphertext C to get back a message. Associated to the scheme is a message space MsgSp. For consistency, we require that for all $k \in \mathbb{N}$ and messages $M \in \mathsf{MsgSp}(k)$ we have $\Pr[\mathsf{PKEdec}(sk, \mathsf{PKEenc}(pk, M)) = M] = 1$, where the probability is taken over the coins of all the algorithms in the expression above.

PRIVACY. Privacy follows [26]. Let $\mathcal{PKE} = (\mathsf{PKEkg}, \mathsf{PKEenc}, \mathsf{PKEdec})$ be an PKE scheme with associated message space MsgSp. To an adversary $\mathcal{A}$ we associate the following experiment:

$$\textbf{Experiment } \mathbf{Exp}^{\mathrm{pke\text{-}cca}}_{\mathcal{PKE},\mathcal{A}}(k)$$

$$(pk, sk) \xleftarrow{\$} \mathsf{PKEkg}(1^k)$$
$$(M_0, M_1, st) \xleftarrow{\$} \mathcal{A}^{\mathrm{DEC}(\cdot)}(\mathtt{find}, pk)$$
$$b \xleftarrow{\$} \{0, 1\} \; ; \; C^* \xleftarrow{\$} \mathsf{PKEenc}(pk, M_b)$$
$$b' \xleftarrow{\$} \mathcal{A}^{\mathrm{DEC}(\cdot)}(\mathtt{guess}, C^*, st)$$
$$\text{If } b \neq b' \text{ then return } 0 \text{ else return } 1$$

where the oracle $\mathrm{DEC}(C)$ returns $M \leftarrow \mathsf{PKEdec}(sk, C)$ with the restriction that in the $\mathtt{guess}$ phase adversary $\mathcal{A}$ is not allowed to query oracle $\mathrm{DEC}(\cdot)$ for the target ciphertext C^*. Both challenge messages are required to be of the same size ($|M_0| = |M_1|$) and in the message space $\mathsf{MsgSp}(k)$. We define the advantage of $\mathcal{A}$ in the above experiment as

$$\mathbf{Adv}^{\mathrm{pke\text{-}cca}}_{\mathcal{PKE},\mathcal{A}}(k) \;=\; \left| \Pr\left[\mathbf{Exp}^{\mathrm{pke\text{-}cca}}_{\mathcal{PKE},\mathcal{A}}(k) = 1 \right] - \frac{1}{2} \right| .$$

PKE scheme $\mathcal{PKE}$ is said to be *secure against chosen ciphertext attacks* (CCA-secure) if the advantage function $\mathbf{Adv}^{\mathrm{pke\text{-}cca}}_{\mathcal{PKE},\mathcal{A}}$ is a negligible function in k for all PTAs $\mathcal{A}$.

The weaker security notion of *security against chosen-plaintext attacks* (CPA-security) is obtained in the above security experiment when depriving adversary $\mathcal{A}$ of the the access to the decryption oracle.

3.2   Tag-Based Encryption

Informally, in a tag-based encryption scheme [22], the encryption and decryption operations take an additional "tag". A tag is simply a binary string of *appropriate length*, and need not have any particular internal structure. We define security for tag-based encryption in manners analogous to security for standard encryption schemes. In particular, we define selective-tag security against chosen-ciphertext attacks. The selective-tag variant is reminiscent to the selective-identity variant of IBE schemes [11] and was not considered in [22].

More formally, a *tag-based encryption* (TBE) scheme $\mathcal{TBE} = (\mathsf{TBEkg}, \mathsf{TBEenc}, \mathsf{TBEdec})$ consists of three PTAs. Via $(pk, sk) \xleftarrow{\$} \mathsf{TBEkg}(1^k)$ the randomized key-generation algorithm produces keys for security parameter $k \in \mathbb{N}$; via $C \xleftarrow{\$}$

TBEenc(pk, t, M) a sender encrypts a message M with tag t to get a ciphertext; via $M \leftarrow$ TBEdec(sk, t, C) the possessor of secret key sk decrypts ciphertext C to get back a message or the symbol *reject*. Note that the tag t must explicitly be provided as the input of the decryption algorithm and is usually not explicitly contained in the ciphertext. Associated to the scheme is a message space MsgSp. For consistency, we require that for all $k \in \mathbb{N}$, all tags t and messages $M \in$ MsgSp(k) we have $\Pr[\text{TBEdec}(sk, t, \text{TBEenc}(pk, t, M)) = M] = 1$, where the probability is taken over the choice of $(pk, sk) \stackrel{\$}{\leftarrow}$ TBEkg(1^k), and the coins of all the algorithms in the expression above.

PRIVACY. To an adversary $\mathcal{A}$ we associate the following experiment:

$$\textbf{Experiment } \textbf{Exp}_{\mathcal{TBE}, \mathcal{A}}^{\text{tbe-stag-cca}}(k)$$

$(t^*, st_0) \stackrel{\$}{\leftarrow} \mathcal{A}(1^k, \texttt{init})$

$(pk, sk) \stackrel{\$}{\leftarrow}$ TBEkg(1^k)

$(M_0, M_1, st) \stackrel{\$}{\leftarrow} \mathcal{A}^{\text{DEC}(\cdot, \cdot)}(\texttt{find}, pk, st_0)$

$b \stackrel{\$}{\leftarrow} \{0, 1\}$; $C_{tbe}^* \stackrel{\$}{\leftarrow}$ TBEenc(pk, t^*, M_b)

$b' \stackrel{\$}{\leftarrow} \mathcal{A}^{\text{DEC}(\cdot, \cdot)}(\texttt{guess}, C_{tbe}^*, st)$

If $b \neq b'$ then return 0 else return 1

where the oracle DEC(C, t) returns $M \leftarrow$ TBEdec(sk, t, C) with the restriction that $\mathcal{A}$ is not allowed to query oracle DEC for tag t^* (called *target tag*). Both messages must be of the same size ($|M_0| = |M_1|$) and in the message space MsgSp(k). We define the advantage of $\mathcal{A}$ in the above experiment as

$$\textbf{Adv}_{\mathcal{TBE}, \mathcal{A}}^{\text{tbe-stag-cca}}(k) = \left| \Pr\left[\textbf{Exp}_{\mathcal{TBE}, \mathcal{A}}^{\text{tbe-stag-cca}}(k) = 1 \right] - \frac{1}{2} \right|.$$

TBE scheme $\mathcal{TBE}$ is said to be *selective-tag weakly secure against chosen ciphertext attacks* if the advantage function is negligible for all PTAs $\mathcal{A}$.

In the security experiment adversary $\mathcal{A}$ is allowed to make decryption queries for any tag $t \neq t^*$, t^* being the tag the challenge ciphertext is created with. In particular, this includes queries for the target ciphertext C_{tbe}^* (when queried with a different tag $t \neq t^*$). In other words, the security notion offers chosen-ciphertext security for all tags $t \neq t^*$ and chosen-plaintext security for $t = t^*$. The target tag t^* has to be output by $\mathcal{A}$ before even seeing the public key. That means that a simulator may "tailor" the public-key to secure the scheme with respect to the above definition.

DISCUSSION OF DIFFERENT TBE VARIANTS. Tags in public-key encryption were already considered by Shoup [29] (and were called "labels") and later by MacKenzie, Reiter, and Yang [22]. While functionality is the same as in our definition, in terms of security there are small but crucial differences between the definitions given in the different papers. We recall the two TBE security variants from [29, 22] and point out the differences to our definition. Let C_{tbe}^* be the target ciphertext and t^* be the target tag selected by the adversary $\mathcal{A}$ in the security experiment.

- To obtain the notion of *weak CCA security* for TBE schemes (as considered in [22][3]) we modify the above security experiment in a way such that $\mathcal{A}$ does not have to commit to the target tag t^* in the beginning of the experiment. Instead, $\mathcal{A}$ is allowed to choose t^* at the end of its find stage, possibly depending on the public key and on its queries. Clearly, this is a stronger security requirement.
- To get (full) *CCA-security* (as considered in [29]), we further modify the security experiment (of weak CCA security) such that the adversary is allowed to ask any decryption query suspect to $(t, C_{tbe}) \neq (t^*, C_{tbe}^*)$. In particular this includes queries for the target tag t^* as long as $C_{tbe} \neq C_{tbe}^*$.

The differences between the different TBE security notions are summarized in the following table.

TBE security	Restriction to $\text{DEC}(t, C_{tbe})$ queries	Selective-tag?
(full) CCA [29]	$(t, C_{tbe}) \neq (t^*, C_{tbe}^*)$	no
weak CCA [22]	$t \neq t^*$	no
selective-tag weak CCA	$t \neq t^*$	yes

Clearly, the three definitions form a hierarchy of security notions, Shoup's CCA security being the strongest and our selective-tag weak CCA security being the weakest. We want to remark that selective-tag weak CCA security is strictly weaker than weak CCA security, i.e. there exists a TBE scheme that is selective-tag but not weakly CCA secure. (This can be shown by an example recently used in [18] to show a similar separation related to IBE schemes.)

RELATION BETWEEN TBE AND PKE. It is easy to see that by identifying a message/tag pair (M, t) with a message $M\|t$, any CCA-secure PKE scheme is also a CCA-secure TBE scheme. On the other hand, by identifying a message M with message/tag pair (M, t) (for an arbitrary tag t that is appended to the ciphertext in the plain) any CCA-secure TBE scheme can be used as a CCA-secure PKE scheme. Note that the same trick is not possible anymore if we weaken the security requirement to the TBE scheme to weak CCA security. (An adversary against the CCA security of the PKE scheme could query the decryption oracle for (C_{tbe}^*, t) for $t \neq t^*$ what would give it the plaintext M_b.) The above remarks show that the two notions of CCA-secure TBE and CCA-secure PKE can in fact be seen as equivalent. Fig. 1 in Section 4 is summarizing the relations between PKE and the different security flavors of TBE.

3.3 Identity Based Encryption

An identity based encryption (IBE) scheme can be viewed as a special kind of tag-based encryption scheme where the tag t is associated with an identity id.

---

[3] Note that weak CCA-security for TBE schemes was called CCA-security in [22]. But for its relation to PKE schemes we prefer to refer to it as weak CCA-security. This should become clear later.

The difference is that an IBE scheme is equipped with an additional algorithm, the key derivation algorithm KeyDer. On input of the secret key sk and an identity id, KeyDer generates a user secret key $usk[id]$ for identity id. This secret key allows the identity to decrypt all messages that were encrypted to identity id. In the terminology of TBE this means that $usk[t]$ is a "wild-card" to decrypt arbitrary ciphertexts that were encrypted with tag t, without knowing the secret key. A formal definition of IBE, together with a security model for (selective-identity) chosen-plaintext security, is given in the full version [20].

RELATION BETWEEN IBE AND TBE. By the above it is easy to see that every IBE scheme can be transformed into a TBE scheme while maintaining its security properties. In the transformation TBE tag t is identified with IBE identity id. The key generation and encryption algorithms are the same. The TBE decryption algorithm first computes the secret key $usk[t]$ for "identity" t and then uses the public IBE decryption algorithm to recover the plaintext. It is easy to verify that if the IBE scheme is (selective-identity) CPA-secure then the TBE scheme is (selective-tag) weakly CCA-secure.[4] Furthermore, a CCA-secure IBE scheme translates to a CCA-secure TBE scheme. (See full version [20] for exact IBE security definitions.)

To the best of our knowledge it is not known how to generically transform a TBE scheme into an IBE scheme. This seems particularly difficult since it is not clear how, in general, the user secret key $usk[id]$ of the IBE scheme can be defined since in TBE there is no such concept as the "user secret key".

The above observations together with the discussion from Section 3.2 indicate that the class of selective-tag weakly CCA-secure TBE schemes is more general than the class of weakly CCA-secure TBE/selective-identity CPA-secure IBE schemes and gives furthermore hope that TBE schemes in the weak selective-tag model are easier to construct. Fig. 1 in Section 4 is summarizing the relations between TBE and IBE.

4   Chosen-Ciphertext Security from Tag-Based Encryption

Canetti, Halevi, and Katz [11] demonstrate how to transform any selective-identity CPA-secure IBE scheme into a CCA-secure PKE scheme by adding a one-time signature (we will refer to this as CHK transformation). Independent of [11], MacKenzie, Reiter, and Yang [22] exploit the same construction as [11] and describe how to convert any weakly CCA-secure TBE scheme into a CCA-secure PKE scheme. In this section we combine the above three papers [11, 22, 7] and show that a selevtice-tag weakly CCA-secure TBE scheme is sufficient to construct an CCA-secure PKE scheme. More precisely, we note that the CHK transformation may as well be instantiated with any TBE scheme (the PKE decryption algorithm needs to be adapted to the TBE definition). If the TBE

---

[4] Note that CCA security for TBE schemes naturally corresponds to CPA security for IBE schemes.

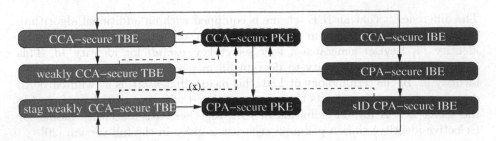

Fig. 1. Relation between IBE, TBE, and PKE with different security definitions. Solid arrows indicate direct implications, dashed lines indicate relations through a black-box reduction. All direct implications were discussed in Section 3. The upper left dashed black-box implication is due to [22], the right one due to [11], and the one with the marker (x) shows our contribution.

scheme is selective-tag weakly CCA-secure then the resulting PKE scheme is CCA-secure. We summarize the known relations among TBE, PKE, and IBE in Fig. 1. The results of this section settle the implication marked by (x).

4.1   The Transformation

Given a TBE scheme $\mathcal{TBE} = (\mathsf{TBEkg}, \mathsf{TBEenc}, \mathsf{TBEdec})$ with tag-space TagSp we construct a public-key encryption scheme $\mathcal{PKE} = (\mathsf{PKEkg}, \mathsf{PKEenc}, \mathsf{PKEdec})$. In the construction, we use a one-time signature scheme $\mathcal{OTS} = (\mathsf{SKG}, \mathsf{SIGN}, \mathsf{VFY})$ in which the verification key output by $\mathsf{SKG}(1^k)$ is an element from TagSp. We require that this scheme be secure in the sense of strong unforgeability (cf. [20]). The transformation defines the public/secret key pair of the PKE scheme to be the public/secret key pair of the TBE scheme, i.e. $\mathsf{PKEkg}(1^k)$ outputs whatever $\mathsf{TBEkg}(1^k)$ outputs. The construction proceeds as follows:

TBE to PKE transformation	
$\mathsf{PKEenc}(pk, M)$	$\mathsf{PKEdec}(sk, C)$
$\quad (vk, sigk) \xleftarrow{\$} \mathsf{SKG}(1^k)$	$\quad$ Parse C as (C_{tbe}, vk, sig)
$\quad C_{tbe} \xleftarrow{\$} \mathsf{TBEenc}(pk, vk, M)$	$\quad$ If $\mathsf{VFY}(vk, C_{tbe}, sig) = \mathtt{reject}$
$\quad sig \xleftarrow{\$} \mathsf{SIGN}(sigk, C_{tbe})$	$\quad\quad$ then return $\mathtt{reject}$.
$\quad$ Return $C \leftarrow (C_{tbe}, vk, sig)$	$\quad$ Else return $M \leftarrow \mathsf{TBEdec}(sk, vk, C_{tbe})$

It is easy to check that the above scheme satisfies correctness.

Let us now give some intuition why the PKE scheme is CCA-secure. Let (C_{tbe}^*, vk^*, sig^*) be the challenge ciphertext output by the simulator in the security experiment. It is clear that, without any decryption oracle queries, the value of the bit b remains hidden to the adversary. This is so because C_{tbe}^* is output by TBEenc which is CPA-secure, vk^* is independent of the message, and sig^* is the result of applying the one-time signing algorithm to C_{tbe}^*.

We claim that decryption oracle queries cannot further help the adversary in guessing the value of b. Consider an arbitrary ciphertext query $(C_{tbe}, vk, sig) \neq (C_{tbe}^*, vk^*, sig^*)$ made by the adversary during the experiment. If $vk = vk^*$ then $(C_{tbe}, sig) \neq (C_{tbe}^*, sig^*)$ and the decryption oracle will answer reject since the adversary is unable to forge a new valid signature sig with respect to vk^*. If $vk \neq vk^*$ then the decryption query will not help the adversary since the actual decryption using $\mathcal{TBE}$ will be done with respect to a tag vk different to the target tag vk^*. A formalization of the above arguments leads to the following:

Theorem 1. *Assuming the TBE scheme is selective-tag chosen-ciphertext secure, the $\mathcal{OTS}$ is a strong, one-time signature scheme, then the above public-key encryption scheme is chosen-ciphertext secure.*

The security reduction is tight (linear) with respect to all the public-key components. The proof follows along the lines of [11,5] and is therefore omitted here. We note that the CHK transformation can also be used to transform a (straight-forward definition of) tag-based KEM into a full KEM.

For simplicity we only described the CHK transformation in this Section. We want to remark that the more efficient BK transformation [7,5] (which basically employs a MAC insteas of a signature) works as well for TBE schemes. The use of a MAC instead of a one-time signature somewhat complicates exposition and proof. The description of the BK transformation, together with all necessary definitions, is deferred to the full version [20].

5   An Efficient TBE Scheme Based on the Linear Assumption

In this section we demonstrate the usefulness of the TBE to PKE transformation of Section 4. Whereas the only known IBE schemes are using pairings [3] we give a simple and practical TBE scheme that does not perform any pairing operation.

5.1   Parameter Generation Algorithm for Gap Groups

All schemes will be parameterized by a *gap parameter generator*. This is a PTA $\mathcal{G}$ that on input 1^k returns the description of an multiplicative cyclic group $\mathbb{G}$ of prime order p, where $2^k < p < 2^{k+1}$, and the description of a Diffie-Hellman oracle DDHvf. A tuple $(g, g^x, g^y, g^z) \in \mathbb{G}^4$ is called a *Diffie-Hellman tuple* if $xy = z \bmod p$. The oracle DDHvf is a PTA that for each input $(g, g^x, g^y, g^z) \in \mathbb{G}^4$ outputs 1 if (g, g^x, g^y, g^z) is a Diffie-Hellman tuple and 0 otherwise. More formally we require that for each $(\mathbb{G}, p, \mathsf{DDHvf}) \xleftarrow{\$} \mathcal{G}(1^k)$ and for each $(g, g^x, g^y, g^z) \in \mathbb{G}^4$,

$$\Pr[\mathsf{DDHvf}(g, g^x, g^y, g^z) = (xy = z)] \geq 1 - neg(k)$$

where the probability is taken over all internal coin tosses of DDHvf and "$xy = z$" is defined as 1 is $xy = z \bmod p$ and 0 otherwise. We use $\mathbb{G}^*$ to denote $\mathbb{G} \setminus \{0\}$, i.e. the set of all group elements except the neutral element. Throughout the paper we use $GG = (\mathbb{G}, p, \mathsf{DDHvf})$ as shorthand for the description of the gap group. See [25] for a more formal treatment of gap groups. We note that one *specific instantiation* of such gap-groups can be obtained using bilinear pairings [6].

5.2 The Decision Linear Assumption

Let GG as above and let $g_1, g_2, z \in \mathbb{G}$ be random elements from group $\mathbb{G}$. Consider the following problem introduced By Boneh, Boyen, and Shacham [4]: Given $(g_1, g_2, z, g_1^{r_1}, g_2^{r_2}, w) \in \mathbb{G}^6$ as input, output yes if $w = z^{r_1+r_2}$ and no otherwise. One can easily show that an algorithm for solving the Decision Linear Problem in $\mathbb{G}$ gives an algorithm for solving DDH in $\mathbb{G}$. The converse is believed to be false. That is, it is believed that the Decision Linear Problem is a hard problem even in gap-groups where DDH is easy. To an adversary $\mathcal{A}$ we associate the following experiment.

$$\textbf{Experiment } \mathbf{Exp}_{\mathcal{G},\mathcal{A}}^{\mathrm{dlin}}(1^k)$$
$$\mathcal{PG} \xleftarrow{\$} \mathcal{G}(1^k) \,;\; g_1, g_2, z \xleftarrow{\$} \mathbb{G}^* \,;\; r_1, r_2, r \xleftarrow{\$} \mathbb{Z}_p$$
$$\beta \xleftarrow{\$} \{0,1\} \,;\; \text{if } \beta = 1 \text{ then } w \leftarrow z^{r_1+r_2} \text{ else } w \leftarrow z^r$$
$$\beta' \xleftarrow{\$} \mathcal{A}(1^k, \mathcal{PG}, g_1, g_2, z, g_1^{r_1}, g_2^{r_2}, w)$$
$$\text{If } \beta \neq \beta' \text{ then return } 0 \text{ else return } 1$$

We define the advantage of $\mathcal{A}$ in the above experiment as

$$\mathbf{Adv}_{\mathcal{G},\mathcal{A}}^{\mathrm{dlin}}(k) = \left| \Pr\left[\mathbf{Exp}_{\mathcal{G},\mathcal{B}}^{\mathrm{dlin}}(1^k) = 1 \right] - \frac{1}{2} \right| .$$

We say that the *decision linear assumption relative to generator* $\mathcal{G}$ holds if $\mathbf{Adv}_{\mathcal{G},\mathcal{A}}^{\mathrm{dlin}}$ is a negligible function in k for all PTAs $\mathcal{A}$.

To put more confidence in the DLIN problem it was shown in [4] that the DLIN problem is hard in generic gap-groups.

A BASIC SCHEME BASED ON DLIN. Since it's introduction the DLIN assumption has already found some interesting applications (e.g., see [4, 8, 24]). As noted in [4] the DLIN assumption readily gives a CPA-secure PKE scheme (called linear encryption scheme) as follows: The public key consists of random elements $g_1, g_2, z \in \mathbb{G}$, the secret key of elements x_1, x_2 such that $g_1^{x_1} = g_2^{x_2} = z$. Encryption of a message M is given by $(C_1, C_2, E) \leftarrow (g_1^{r_1}, g_2^{r_2}, z^{r_1+r_2} \cdot M)$, where $r_1, r_2 \in \mathbb{Z}_q^*$ are random elements. The message M is recovered by the possessor of the secret key by computing M as $M \leftarrow E/(C_1^{x_1} C_2^{x_2})$.

5.3 The Scheme

The starting point of our scheme will be the (CPA-secure) linear encryption scheme from Section 5.2. By adding two additional values to the ciphertext we can update it to a selective-tag CCA-secure TBE scheme. The values contain redundant information and also depend on the tag. In the decryption algorithm the two values are used to check the ciphertext for "validity" or "consistency". We build a TBE scheme $\mathcal{TBE} = (\mathsf{TBEkg}, \mathsf{TBEenc}, \mathsf{TBEdec})$ as follows:

DLIN-based TBE
$\mathsf{TBEkg}(1^k)$
$\quad (\mathbb{G}, p, \mathsf{DDHvf}) \overset{\$}{\leftarrow} \mathcal{G}(1^k)$
$\quad g_1 \overset{\$}{\leftarrow} \mathbb{G}^* ; \ x_1, x_2, y_1, y_2 \overset{\$}{\leftarrow} \mathbb{Z}_p^*$
$\quad$ Chose $g_2, z \in \mathbb{G}$ with $g_1^{x_1} = g_2^{x_2} = z$
$\quad u_1 \leftarrow g_1^{y_1} ; \ u_2 \leftarrow g_2^{y_2}$
$\quad pk \leftarrow (\mathbb{G}, p, g_1, g_2, z, u_1, u_2) ; \ sk \leftarrow (x_1, x_2, y_1, y_2)$
$\quad$ Return (pk, sk)

$\mathsf{TBEenc}(pk, t, M)$ $\qquad\qquad\qquad$ $\mathsf{TBEdec}(sk, t, C_{tbe})$

$\quad r_1, r_2 \overset{\$}{\leftarrow} \mathbb{Z}_p^* \qquad\qquad\qquad\qquad$ Parse C_{tbe} as (C_1, C_2, D_1, D_2, E)

$\quad C_1 \leftarrow g_1^{r_1} ; \ C_2 \leftarrow g_2^{r_2} \qquad\qquad\ \ s_1, s_2 \overset{\$}{\leftarrow} \mathbb{Z}_p^*$

$\quad D_1 \leftarrow z^{tr_1} u_1^{r_1} ; \ D_2 \leftarrow z^{tr_2} u_2^{r_2} \qquad K \leftarrow \dfrac{C_1^{x_1 + s_1(tx_1 + y_1)} \cdot C_2^{x_2 + s_2(tx_2 + y_2)}}{D_1^{s_1} \cdot D_2^{s_2}}$

$\quad K \leftarrow z^{r_1 + r_2}$

$\quad E \leftarrow M \cdot K \qquad\qquad\qquad\qquad\quad M \leftarrow E \cdot K^{-1}$

$\quad C_{tbe} \leftarrow (C_1, C_2, D_1, D_2, E) \qquad\quad$ Return M

$\quad$ Return C_{tbe}

Note that the public key pk does not contain the description of the Diffie-Hellman verification oracle DDHvf.

5.4 Correctness and Alternative Decryption

Let $C_{tbe} = (C_1, C_2, D_1, D_2, E) \in \mathbb{G}^5$ be a (possibly malformed) ciphertext. C_{tbe} is called *consistent with tag* t if $C_1^{tx_1 + y_1} = D_1$ and $C_2^{tx_2 + y_2} = D_2$. Note that any ciphertext that was properly generated by the encryption algorithm for tag t is always consistent with (the same) tag t, i.e. for $i = 1, 2$ we have $(g_i^{r_i})^{tx_i + y_i} = z^{tr_i} u_i^{r_i}$ for any $r_i \in \mathbb{Z}_p$.

The key K in the decryption algorithm is computed as

$$K = \frac{C_1^{x_1 + s_1(tx_1 + y_1)} C_2^{x_2 + s_2(tx_2 + y_2)}}{D_1^{s_1} D_2^{s_2}} = C_1^{x_1} C_2^{x_2} \cdot \left(\frac{C_1^{tx_1 + y_1}}{D_1} \right)^{s_1} \cdot \left(\frac{C_2^{tx_2 + y_2}}{D_2} \right)^{s_2}$$

for uniform $s_1, s_2 \in \mathbb{Z}_q$. This can be viewed as an implicit test if the ciphertext is consistent with tag t. If so the key is computed as $K = C_1^{x_1} \cdot C_2^{x_2}$. If not then at least one of the two fractions in the above equation is different from $1 \in \mathbb{G}$ and (since $\mathbb{G}$ has prime order) a random key K is returned, completely independent of the "real key" $C_1^{x_1} \cdot C_2^{x_2}$. Hence the decryption algorithm in the above construction is equivalent to the following (less efficient) decryption algorithm:

$\mathsf{TBEdec}'(sk, t, C_{tbe})$

$\quad$ Parse C_{tbe} as (C_1, C_2, D_1, D_2, E)

$\quad$ If $C_1^{tx_1 + y_1} \neq D_1$ or $C_2^{tx_2 + y_2} \neq D_2$ then $K \overset{\$}{\leftarrow} \mathbb{G}^*$

$\quad$ Else $K \leftarrow C_1^{x_1} \cdot C_2^{x_2}$

$\quad$ Return $M \leftarrow E \cdot K^{-1}$

It leaves to verify that, in case the ciphertext is consistent, $K \leftarrow C_1^{x_1} \cdot C_2^{x_2}$ computes the correct key. Indeed we have $(g_1^{r_1})^{x_1} \cdot (g_2^{r_2})^{x_2} = z^{r_1} \cdot z^{r_2} = z^{r_1 + r_2}$. This shows correctness.

5.5 Public Verification

In this section we show that consistency (or validity) of a given TBE ciphertext can be publicly verified. The above alternative decryption procedure TBEdec' gives rise to an algorithm TBEpv(pk, t, C_{tbe}) for public verification of the ciphertext by checking if $(g_1, z^t u_1, C_1, D_1)$ and $(g_2, z^t u_2, C_2, D_2)$ are Diffie-Hellman tuples. Both checks can be carried out using the Diffie-Hellman verification algorithm DDHvf that we additionally have to provide in the public-key. To verify correctness of the above public consistency check we have to show that for $i = 1, 2$, $C_i^{t x_i + y_i} = D_i$ iff $(g_i, z^t u_i, C_i, D_i)$ is a Diffie-Hellman tuple. Let $C_i = g^{r_i}$. Then $(g_i, z^t u_i = g_i^{x_i t + y_i}, C_i = g_i^{r_i}, D_i)$ is a proper Diffie-Hellman-tuple iff $g_i^{(x_i t + y_i) \cdot r_i} = D_i$ iff $C_i^{x_i t + y_i} = D_i$.

5.6 Security and Efficiency

Theorem 2. *Under the decision linear assumption relative to generator $\mathcal{G}$, the TBE scheme from Section 5.3 is selective-tag secure against chosen-ciphertext attacks.*

Theorem 2 is proved in Appendix A. The intuition of the proof is as follows: Given an adversary $\mathcal{A}$ against the security of the TBE scheme, we can build an adversary $\mathcal{B}$ that breaks the linear assumption with the same success probability of $\mathcal{A}$. For simulating $\mathcal{A}$'s view we use two main ingredients: First, when answering the decryption queries, $\mathcal{B}$ can test for consistency using the public ciphertext verification algorithm TBEpv from Section 5.5. (This is the reason why pairings are needed for the security proof.) Second, we borrow techniques from [3] to make sure that $\mathcal{B}$ can answer the (consistent) decryption queries for all tags but for the target tag t^* output by $\mathcal{A}$ in the beginning of the security experiment.

Encryption requires three exponentiations (to compute C_1, C_2 and K) and two multi-exponentiation (to compute D_1, D_2) in $\mathbb{G}$. Encryption may as well be carried out in 7 exponentiations what is considerably faster when the receiver's public key is considered to be fixed and precomputation for fixed-base exponentiation is used. Decryption is very fast and can be done with one multi-exponentiation.

6 Key Encapsulation Based on the Linear Assumption

A *key encapsulation mechanism* [29] (KEM) $\mathcal{KEM} = ($KEMkg, KEMencaps, KEMdecaps$)$ consits of three PTAs can be seen as a light PKE scheme. Instead of encrypting messages, the encapsulation algorithm KEMencaps generates a (random) symmetric key K and a corresponding ciphertext C. The decapsulation algorithm inputs the secret key and a ciphertext and reconstructs the symmetric key K. In practice the key K is usually fed to a symmetric encryption scheme. CCA-security of a KEM can be analogously defined as CCA-security security of a PKE scheme; in the security game an adversary is given a ciphertext/key pair and has to decide if the two pairs match or if the key is random and independent from the ciphertext. A formal definition of a CCA-secure KEM can be looked up in the full version [20].

6.1   The KEM Scheme

We build a KEM scheme as follows. Let $\mathsf{KEMkg}(1^k)$ be as in the TBE scheme of Section 5.3. The public key pk additionally contains a target collision resistant hash function $\mathsf{TCR} : \mathbb{G} \times \mathbb{G} \to \mathbb{Z}_q$ (i.e. given $t = \mathsf{TCR}(g_1, g_2)$ it should be hard to find $(h_1, h_2) \in \mathbb{G} \times \mathbb{G} \setminus \{(g_1, g_2)\}$ such that $\mathsf{TCR}(h_1, h_2) = t$; we refer to [12] for a formal definition).[5] The encapsulation/decapsulation algorithms are as follows:

DLIN-based KEM	
$\mathsf{KEMencaps}(pk)$	$\mathsf{KEMdecaps}(sk, C_{kem})$
$r_1, r_2 \overset{\$}{\leftarrow} \mathbb{Z}_p^*$	Parse C_{kem} as (C_1, C_2, D_1, D_2)
$C_1 \leftarrow g_1^{r_1} \; ; \; C_2 \leftarrow g_2^{r_2}$	$t \leftarrow \mathsf{TCR}(C_1, C_2)$
$t \leftarrow \mathsf{TCR}(C_1, C_2)$	$s_1, s_2 \overset{\$}{\leftarrow} \mathbb{Z}_p^*$
$D_1 \leftarrow z^{tr_1} u_1^{r_1} \; ; \; D_2 \leftarrow z^{tr_2} u_2^{r_2}$	$K \leftarrow \dfrac{C_1^{x_1 + s_1(tx_1 + y_1)} \cdot C_2^{x_2 + s_2(tx_2 + y_2)}}{D_1^{s_1} \cdot D_2^{s_2}}$
$K \leftarrow z^{r_1 + r_2}$	Return K
$C_{kem} \leftarrow (C_1, C_2, D_1, D_2)$	
Return (C_{kem}, K)	

Analogous to the TBE construction from Section 5 consistency of a ciphertext $C_{kem} = (C_1, C_2, D_1, D_2)$ can be publicly verified by computing $t \leftarrow \mathsf{TCR}(C_1, C_2)$ and checking if $(g_i, z^t u_i, C_i, D_i)$ is a Diffie-Hellman tuple for $i = 1, 2$.

6.2   Security

Theorem 3. *Assume* TCR *is a target collision resistant hash function. Under the decision linear assumption relative to the generator* $\mathcal{G}$ *the KEM from Section 6.1 is secure against chosen-ciphertext attacks.*

The security reduction is tight and compared to the reduction from Theorem 2 there appears an additional additive factor taking into account a possible collision in the hash function TCR. The proof of Theorem 3 is similar to that of Theorem 2 and is given in the full version [20].

The way we use the target collision hash function is reminiscent to the Cramer-Shoup cryptosystem [12]. Indeed, the intuition is the same. Given an adversary $\mathcal{A}$ against the security of the KEM, we can build an adversary $\mathcal{B}$ that breaks the linear assumption with the same success probability of $\mathcal{A}$. Let $(C_1^*, C_2^*, D_1^*, D_2^*)$ be the challenge ciphertext given to adversary $\mathcal{A}$ and let $t^* = \mathsf{TCR}(C_1^*, C_2^*)$. Consider a ciphertext (C_1, C_2, D_1, D_2) queried by adversary $\mathcal{A}$ during the CCA experiment and let $t = \mathsf{TCR}(C_1, C_2)$. Similar to the proof of Theorem 2 we can setup the public-key in a way such that $\mathcal{B}$ is able to correctly simulate all such decryption queries as long as $t \neq t^*$ and the ciphertext is constentent. The latter one can be checked using the public consistency algorithm. Assume $t = t^*$. On one hand, when $(C_1, C_2) \neq (C_1^*, C_2^*)$ then $\mathcal{B}$ found a collision in the hash

---

[5] More formally we need a family of hash functions indexed by some random key c, where c is contained in the public key and the description of the hash function is included in the scheme parameters.

function. On the other hand, when $(C_1, C_2) = (C_1^*, C_2^*)$ then consistency of the ciphertext also implies $D_1 = D_1^*$ and $D_2 = D_2^*$ and hence the queried ciphertext matches the target ciphertext what is forbidden in the experiment.

6.3 From KEM to Full PKE

It is well known that if both the public-key encapsulation scheme and the underlying symmetric-key encryption scheme are CCA-secure, then the resulting hybrid public-key encryption scheme is CCA-secure [13, Sec. 7]. The security reduction is tight.

7   Discussion

7.1   Efficiency Considerations

An efficiency comparison of all previously known CCA-secure PKE schemes in the standard model is assembled in Figure 2. The Cramer-Shoup scheme [12] and the Kurosawa-Desmedt scheme [21] are listed for reference. BK/BBx refers to one of the two Boneh-Boyen IBE schemes from [3] instantiated with the MAC based BK-transformation (since the signature-based CHK transformation is less efficient we decided not to list it in our comparison).

BMW is the recent KEM from Boyen, Mei, and Waters [9]. To obtain a fair comparison we equipped the two KEM schemes (the BMW-KEM and ours from §6) with a hybrid encryption scheme to obtain a fully fledged PKE scheme.

Together with the Kurosawa-Desmedt PKE, our proposed DLIN-based KEM offers the nowadays fastest decryption algorithm. Compared to all other schemes the obvious drawbacks of our schemes are slower encryption and longer ciphertexts. Interestingly, the BK/BBx and BMW constructions tie with the KD scheme in terms of encryption but lose in terms of decryption, whereas our scheme loses in encryption but ties in decryption.

We note that the long ciphertexts are basically due to the different assumption; this is since the basic (chosen-plaintext secure) linear encryption scheme from Section 5.2 already comes with a ciphertext overhead of $2|p|$.

7.2   Remarks

We hope that by having provided weaker sufficient conditions for the CHK/BK transformations we make a step directed towards a better understanding and utilization of CCA-security in PKE schemes. From a designer's point of view the definition of selective-tag security means that the scheme only has to be "secured" with respect to the target tag. Furthermore, in the security reduction, the generated keys may depend on this tag. Having that designing concept in mind it would be interesting to come up with new CCA-secure TBE/PKE schemes based on different assumptions.

A very efficient TBE construction based on the Kurosawa-Desmedt encryption scheme [21] is obtained by removing the target collission-resistant hash function

Scheme	Origin	Assumption	Encryption	Decryption	Ciphertext	Public
			#pairings + #[multi,reg,fix]-exp		Overhead	Vfy?
KD	direct	DDH	$0 + [1, 2, 0]$	$0 + [1, 0, 0]$	$2|p|(+\text{hybrid})$	—
CS	KEM	DDH	$0 + [1, 3, 0]$	$0 + [1, 1, 0]$	$3|p|$	—
BK/BB1	BK/IBE	BDDH	$0 + [1, 2, 0]$	$1 + [1, 0, 0]$	$2|p|+\text{com}+\text{mac}$	—
BK/BB2	BK/IBE	q-BDDHI	$0 + [1, 2, 0]$	$1 + [0, 1, 1]$	$2|p|+\text{com}+\text{mac}$	—
BMW	KEM	BDDH	$0 + [1, 2, 0]$	$1 + [0, 1, 0]$	$2|p|$	yes
Ours (§5)	BK/TBE	DLIN	$0 + [2, 3, 0]$	$0 + [1, 0, 0]$	$4|p|+\text{com}+\text{mac}$	—
Ours (§6)	KEM	DLIN	$0 + [2, 3, 0]$	$0 + [1, 0, 0]$	$4|p|$	yes

Fig. 2. Efficiency comparison for CCA-secure PKE schemes. Some figures are borrowed from [7, 5, 9]. All "private-key" operations (such as hash function/MAC/KDF) are ignored. Cipher overhead represents the difference (in bits) between the ciphertext length and the message length, and $|p|$ is the length of a group element. For concreteness one can think of mac = 128 and the commitment com = 512 bits. For comparison we mention that relative timings for the various operations are as follows: bilinear pairing ≈ 5 [28], multi-exponentiation ≈ 1.5, regular exponentiation = 1, fixed-base exponentiation $\ll 0.2$.

and taking the former output of the hash function as the tag. A straightforward question is if we can somewhat modify either this KD based TBE scheme or our proposal from Section 5 to obtain an IBE scheme that does not use pairings.

Acknowledgments

We thank Mihir Bellare, Xavier Boyen, Yoshi Kohno, Gregory Neven, and the anonymous TCC referees for useful remarks.

References

1. M. Abe, R. Gennaro, K. Kurosawa, and V. Shoup. Tag-KEM/DEM: A new framework for hybrid encryption and a new analysis of Kurosawa-Desmedt KEM. In R. Cramer, editor, *EUROCRYPT 2005*, volume 3494 of *LNCS*, pages 128–146. Springer-Verlag, May 2005.
2. M. Bellare and P. Rogaway. Random oracles are practical: A paradigm for designing efficient protocols. In *ACM CCS 93*, pages 62–73. ACM Press, Nov. 1993.
3. D. Boneh and X. Boyen. Efficient selective-id secure identity based encryption without random oracles. In C. Cachin and J. Camenisch, editors, *EUROCRYPT 2004*, volume 3027 of *LNCS*, pages 223–238. Springer-Verlag, May 2004.
4. D. Boneh, X. Boyen, and H. Shacham. Short group signatures. In M. Franklin, editor, *CRYPTO 2004*, volume 3152 of *LNCS*, pages 41–55. Springer-Verlag, Aug. 2004.
5. D. Boneh, R. Canetti, S. Halevi, and J. Katz. Chosen-ciphertext security from identity-based encryption. Journal submission. Available from author's web page http://crypto.stanford.edu/~dabo/pubs.html, November 2005.
6. D. Boneh and M. K. Franklin. Identity based encryption from the Weil pairing. *SIAM Journal on Computing*, 32(3):586–615, 2003.

7. D. Boneh and J. Katz. Improved efficiency for CCA-secure cryptosystems built using identity-based encryption. In A. Menezes, editor, *CT-RSA 2005*, volume 3376 of *LNCS*, pages 87–103. Springer-Verlag, Feb. 2005.

8. D. Boneh and H. Shacham. Group signatures with verifier-local revocation. In *ACM CCS 04*, pages 168–177. ACM Press, Oct. 2004.

9. X. Boyen, Q. Mei, and B. Waters. Simple and efficient CCA2 security from IBE techniques. In *ACM Conference on Computer and Communications Security—CCS 2005*, pages 320–329. New-York: ACM Press, 2005.

10. R. Canetti, O. Goldreich, and S. Halevi. The random oracle methodology, revisited. In *30th ACM STOC*, pages 209–218. ACM Press, May 1998.

11. R. Canetti, S. Halevi, and J. Katz. Chosen-ciphertext security from identity-based encryption. In C. Cachin and J. Camenisch, editors, *EUROCRYPT 2004*, volume 3027 of *LNCS*, pages 207–222. Springer-Verlag, May 2004.

12. R. Cramer and V. Shoup. A practical public key cryptosystem provably secure against adaptive chosen ciphertext attack. In H. Krawczyk, editor, *CRYPTO'98*, volume 1462 of *LNCS*, pages 13–25. Springer-Verlag, Aug. 1998.

13. R. Cramer and V. Shoup. Design and analysis of practical public-key encryption schemes secure against adaptive chosen ciphertext attack. *SIAM Journal on Computing*, 33(1):167–226, 2003.

14. W. Diffie and M. E. Hellman. New directions in cryptography. *IEEE Transactions on Information Theory*, 22:644–654, 1978.

15. D. Dolev, C. Dwork, and M. Naor. Nonmalleable cryptography. *SIAM Journal on Computing*, 30(2):391–437, 2000.

16. T. El Gamal. A public key cryptosystem and a signature scheme based on discrete logarithms. In G. R. Blakley and D. Chaum, editors, *CRYPTO'84*, volume 196 of *LNCS*, pages 10–18. Springer-Verlag, Aug. 1985.

17. E. Elkind and A. Sahai. A unified methodology for constructing public-key encryption schemes secure against adaptive chosen-ciphertext attack. Cryptology ePrint Archive, Report 2002/042, 2002. http://eprint.iacr.org/.

18. D. Galindo and I. Hasuo. Security notions for identity based encryption. Cryptology ePrint Archive, Report 2005/253, 2005. http://eprint.iacr.org/.

19. S. Goldwasser and S. Micali. Probabilistic encryption. *Journal of Computer and System Sciences*, 28:270–299, 1984.

20. E. Kiltz. Chosen-ciphertext security from tag-based encryption. Cryptology ePrint Archive, 2005. http://eprint.iacr.org/.

21. K. Kurosawa and Y. Desmedt. A new paradigm of hybrid encryption scheme. In M. Franklin, editor, *CRYPTO 2004*, volume 3152 of *LNCS*, pages 426–442. Springer-Verlag, Aug. 2004.

22. P. D. MacKenzie, M. K. Reiter, and K. Yang. Alternatives to non-malleability: Definitions, constructions, and applications. In M. Naor, editor, *TCC 2004*, volume 2951 of *LNCS*, pages 171–190. Springer-Verlag, Feb. 2004.

23. M. Naor and M. Yung. Public-key cryptosystems provably secure against chosen ciphertext attacks. In *22nd ACM STOC*. ACM Press, May 1990.

24. L. Nguyen and R. Safavi-Naini. Efficient and provably secure trapdoor-free group signature schemes from bilinear pairings. In P. J. Lee, editor, *ASIACRYPT 2004*, volume 3329 of *LNCS*, pages 372–386. Springer-Verlag, Dec. 2004.

25. T. Okamoto and D. Pointcheval. The gap-problems: A new class of problems for the security of cryptographic schemes. In K. Kim, editor, *PKC 2001*, volume 1992 of *LNCS*, pages 104–118. Springer-Verlag, Feb. 2001.

26. C. Rackoff and D. R. Simon. Non-interactive zero-knowledge proof of knowledge and chosen ciphertext attack. In J. Feigenbaum, editor, *CRYPTO'91*, volume 576 of *LNCS*, pages 433–444. Springer-Verlag, Aug. 1991.
27. R. L. Rivest, A. Shamir, and L. M. Adleman. A method for obtaining digital signature and public-key cryptosystems. *Communications of the ACM*, 21(2):120–126, 1978.
28. M. Scott. Faster pairings using an elliptic curve with an efficient endomorphism. Cryptology ePrint Archive, Report 2005/252, 2005. `http://eprint.iacr.org/`.
29. V. Shoup. A proposal for an ISO standard for public key encryption (version 2.1). manuscript, 2001. Available on `http://shoup.net/papers/`.
30. B. R. Waters. Efficient identity-based encryption without random oracles. In R. Cramer, editor, *EUROCRYPT 2005*, volume 3494 of *LNCS*, pages 114–127. Springer-Verlag, May 2005.

A Proof of Theorem 2

Adversary $\mathcal{B}$ inputs an instance of the decisional linear problem, i.e. $\mathcal{B}$ inputs the values $(1^k, \mathbb{GG}, g_1, g_2, z, g_1^{r_1}, g_2^{r_2}, w)$. $\mathcal{B}$'s goal is to determine whether $w = z^{r_1+r_2}$ or w is a random group element.

Now suppose there exists an adversary $\mathcal{A}$ that breaks the selective-tag CCA security of the TBE scheme with (non-negligible) advantage $\mathbf{Adv}^{\text{tbe-stag-cca}}_{TBE,\mathcal{A}}(k)$. We show that adversary $\mathcal{B}$ can run adversary $\mathcal{A}$ to solve its instance of the decisional linear problem (i.e. to determine whether $w = z^{r_1+r_2}$ or if w is a random group element) with advantage

$$\mathbf{Adv}^{\text{dlin}}_{\mathcal{G},\mathcal{B}}(k) \geq \mathbf{Adv}^{\text{tbe-stag-cca}}_{TBE,\mathcal{A}}(k) . \tag{1}$$

Now Eqn. (1) proves the Theorem. Adversary $\mathcal{B}$ runs adversary $\mathcal{A}$ simulating its view as in the original TBE security experiment. We now give the description of adversary $\mathcal{B}$.

Init Stage. Adversary $\mathcal{B}$ runs adversary $\mathcal{A}$ on input 1^k and init. $\mathcal{A}$ outputs the target tag t^* that is input by $\mathcal{B}$.

Find Stage. $\mathcal{B}$ picks two random values $c_1, c_2 \in \mathbb{Z}_p$ and sets

$$u_1 \leftarrow z^{-t^*} \cdot g_1^{c_1}, \qquad u_2 \leftarrow z^{-t^*} \cdot g_2^{c_2} .$$

The public key pk is defined as $(\mathbb{G}, p, g_1, g_2, z, u_1, u_2)$ and it is identically distributed as in the original TBE scheme. Let $x_1 = \log_{g_1} z$ and $x_2 = \log_{g_2} z$, as in the original TBE scheme. This implicitly defines the values y_1, y_2 as

$$y_1 = \log_{g_1} u_1 = -t^* x_1 + c_1, \qquad y_2 = \log_{g_2} u_2 = -t^* x_2 + c_2 .$$

Note that no value of the corresponding secret key $sk = (x_1, x_2, y_1, y_2)$ is known to $\mathcal{B}$.

Now consider an arbitrary ciphertext $C_{tbe} = (C_1, C_2, D_1, D_2)$ and let $t \in \mathbb{Z}_p$ be a tag. Recall that C_{tbe} is consistent with tag t if $C_i^{x_i \cdot t + y_i} = D_i$ for $i \in \{1, 2\}$. The way the keys are setup this condition can be rewritten as

$$D_i = C_i^{t x_i + y_i} = C_i^{x_i t - t^* x_i + c_i} = (C_i^{x_i})^{t-t^*} \cdot C_i^{c_i}, \quad i \in \{1, 2\} . \tag{2}$$

By Equation (2), $D_i/C_i^{c_i} = (C_i^{x_i})^{t-t^*}$ and if $t \neq t^*$ then the session key $K = C_1^{x_1} \cdot C_2^{x_2}$ can alternatively be reconstructed as

$$K \leftarrow \left(\frac{D_1 \cdot D_2}{C_1^{c_1} \cdot C_2^{c_2}} \right)^{\frac{1}{t-t^*}} . \tag{3}$$

Now adversary $\mathcal{B}$ runs $\mathcal{A}$ on input find and pk answering to its decryption queries as follows: Let $C_{tbe} = (C_1, C_2, D_1, D_2)$ be an arbitrary ciphertext submitted to the decryption oracle $\text{DEC}(C_{tbe}, t)$ for tag $t \neq t^*$. First $\mathcal{B}$ performs a public consistency check as explained in Section 5.5 using the Diffie-Hellman verification algorithm DDHvf. If C_{tbe} is not consistent then $\mathcal{B}$ returns a random message, as in the alternative (but equivalent) decryption algorithm (Section 5.4) of the original TBE scheme. Otherwise, if the ciphertext is consistent adversary $\mathcal{B}$ computes the session key by Equation (3) as $K \leftarrow (\frac{D_1 D_2}{C_1^{c_1} C_2^{c_2}})^{\frac{1}{t-t^*}}$ and returns $M \leftarrow E \cdot K^{-1}$. This shows that as long as $t \neq t^*$ the simulation of the decryption queries is always perfect, i.e. the output of oracle $\text{DEC}(C_{tbe}, t)$ is identically distributed as the output of TBEdec(sk, C_{tbe}, t).

Guess Stage. $\mathcal{A}$ returns two distinct messages M_0, M_1 of equal length. Adversary $\mathcal{B}$ picks a random bit b and constructs the challenge ciphertext $C_{tbe}^* = (C_1^*, C_2^*, D_1^*, D_2^*, E^*)$ for message M_b as follows:

$$(C_1^* = g_1^{r_1}, \ C_2^* = g_2^{r_2}, \ D_1^* = (g_1^{r_1})^{c_1}, \ D_2^* = (g_2^{r_2})^{c_2}, \ E^* = M_b \cdot w)$$

By Equation (2), C_{tbe}^* is always consistent with target tag t^*. If $w = z^{r_1+r_2}$, then $E = M_b \cdot w$ is indeed a valid ciphertext of message M_b and tag t^* under the public key pk. On the other hand, when w is uniform and independent in $\mathbb{G}$ then $E = w \cdot M_b$ is independent of b in the adversary's view.

Adversary $\mathcal{A}$ is run with challenge ciphertext C_{tbe}^* answering to its decryption queries as in the find stage.

Eventually, $\mathcal{A}$ outputs a guess $b' \in \{0, 1\}$. Algorithm $\mathcal{B}$ concludes its own game by outputting a guess as follows: If $b = b'$ then $\mathcal{B}$ outputs 1 meaning $w = z^{r_1+r_2}$. Otherwise, it outputs 0 meaning that w is random.

This completes the description of adversary $\mathcal{B}$. We now analyze $\mathcal{B}$'s success in breaking the decisional linear problem.

When the value w input by $\mathcal{B}$ equals to $w = z^{r_1+r_2}$, then $\mathcal{A}$'s view is identical to its view in a real attack game and therefore $\mathcal{A}$ must satisfy $|\Pr[b = b'] - 1/2| \geq \mathbf{Adv}_{TBE,\mathcal{A}}^{\text{tbe-stag-cca}}(k)$. On the other hand, when w is uniform in $\mathbb{G}$ then $\Pr[b = b'] = 1/2$. Therefore $\mathbf{Adv}_{\mathcal{G},\mathcal{B}}^{\text{dlin}}(k) \geq \left| \left(\frac{1}{2} \pm \mathbf{Adv}_{TBE,\mathcal{A}}^{\text{tbe-stag-cca}}(k) \right) - \frac{1}{2} \right| = \mathbf{Adv}_{TBE,\mathcal{A}}^{\text{tbe-stag-cca}}(k)$. This proves Equation (1) and concludes the proof.

Separating Sources for Encryption and Secret Sharing

Yevgeniy Dodis[1,*], Krzysztof Pietrzak[2,**], and Bartosz Przydatek[2]

[1] Department of Computer Science, New York University
New York, NY, USA
dodis@cs.nyu.edu
[2] Department of Computer Science, ETH Zurich,
8092 Zurich, Switzerland
{pietrzak, przydatek}@inf.ethz.ch

Abstract. Most cryptographic primitives such as encryption, authentication or secret sharing require randomness. Usually one assumes that perfect randomness is available, but those primitives might also be realized under weaker assumptions. In this work we continue the study of building secure cryptographic primitives from imperfect random sources initiated by Dodis and Spencer (FOCS'02). Their main result shows that there exists a (high-entropy) source of randomness allowing for perfect encryption of a bit, and yet from which one cannot extract even a single weakly random bit, separating encryption from extraction. Our main result separates encryption from 2-out-2 secret sharing (both in the information-theoretic and in the computational settings): any source which can be used to achieve one-bit encryption also can be used for 2-out-2 secret sharing of one bit, but the converse is false, even for high-entropy sources. Therefore, possibility of extraction strictly implies encryption, which in turn strictly implies 2-out-2 secret sharing.

1 Introduction

For many important tasks, such as cryptography, randomness is indispensable. Usually one assumes that all parties have access to a perfect random source, but this assumption is at least debatable, and the question what kind of imperfect random sources can be used in various applications has attracted a lot of attention.

EXTRACTION. The easiest such class of sources consists of *extractable* sources for which one can deterministically extract nearly perfect randomness, and then use it in any application. Although examples of such non-trivial sources are known [vN51, Eli72, Blu86, LLS89, CGH+85, BBR88, AL93, CDH+00, DSS01, KZ03, TV00], most natural sources such as the so called entropy sources[1] [SV86,

---

* Supported in part by NSF career award CCR-0133806 and NSF grant CCR-0311095.
** Supported by the Swiss National Science Foundation, project No. 200020-103847/1.

[1] Informally, entropy sources guarantees that every distribution in the family has a non-trivial amount of entropy (and possibly more restrictions), but do not assume independence between different symbols of the source. In this sense they are the most general sources one would wish to tolerate.

S. Halevi and T. Rabin (Eds.): TCC 2006, LNCS 3876, pp. 601–616, 2006.

CG88, Zuc96] are easily seen to be non-extractable. One can then ask a natural question whether perfect randomness is indeed needed for the considered application. Clearly, the answer depends on the application. In particular, the natural fundamental question is to understand the extent to which a given application can be based on imperfect randomness, and also to compare the randomness requirements for different applications.

PROBABILISTIC ALGORITHMS AND INTERACTIVE PROTOCOLS. For example, a series of celebrated results [VV85, SV86, CG88, Zuc96, ACRT99] showed that entropy sources are necessary and sufficient for simulating probabilistic polynomial-time algorithms — namely, problems which do not *inherently* need randomness, but which could potentially be sped up using randomization. Thus, extremely weak imperfect sources can still be tolerated for this application domain. This result was recently extended to interactive protocols by Dodis et al. [DOPS04].

ENCRYPTION. On the other hand, McInnes and Pinkas [MP90] showed that unconditionally secure symmetric encryption cannot be based on entropy sources, even if one is restricted to encrypting a single bit. This result was recently strengthened by Dodis et al. [DOPS04], who showed that entropy sources are not sufficient even for *computationally* secure encryption (as well as essentially any other task involving "privacy"). On the opposite side, Dodis and Spencer [DS02] showed that randomness extraction is not necessary for the existence of secure encryption (at least when restricted to a single bit). Specifically, they show that there are sources which can be used to perfectly encrypt a bit but cannot be used to extract a single bit. This even holds if one additionally requires all the distributions in the imperfect source to have high min-entropy. Thus, good sources for encryption lie strictly in between extractable and entropy sources.

AUTHENTICATION. In the usual non-interactive (i.e., one-message) setting, Maurer and Wolf [MW97] show that for sufficiently high entropy rate (specifically, more than $1/2$), even general entropy sources are sufficient for unconditional one-time authentication, while Dodis and Spencer [DS02] showed that smaller rate entropy sources are indeed insufficient to authenticate even a single bit. On the other hand, [DS02] also show that for all entropy levels (in particular, below $1/2$) there exist "severely non-extractable" imperfect sources which are sufficient for non-trivial authentication. Thus good sources for authentication once again lie strictly in between extractable and entropy sources. The relation to encryption sources is currently open (see Section 5). On a related note, [DOPS04] considered the existence of computationally secure digital signature (and thus also message authentication) schemes, and show that the latter seem to be possible even with general entropy sources, at least under very strong but seemingly reasonable computational assumptions. In the interactive setting, Renner and Wolf [RW03] show (indeed, highly interactive) information-theoretic authentication protocols capable of tolerating any constant-fraction entropy rate.

SECRET SHARING? In this work we consider for the first time another crypto-graphic primitive which inherently requires randomness: secret sharing. In particular, we concentrate on the simplest case of 2-out-2 (denoted simply 2-2) secret sharing: one wants to split a message m into shares S_1 and S_2 so that neither share leaks any information about m, and yet m can be reconstructed from both shares. We first observe that (either information-theoretic or computational) encryption implies the existence of a corresponding 2-2 secret sharing: one simply sets S_1 to be the decryption key, and S_2 to be the encryption of the message M under this key. Our main technical result is to show that the converse of this statement is false, at least when restricting to one-bit message. Namely, there exist imperfect sources sufficient for perfect secret sharing of a bit, but for which any bit encryption scheme can be insecure with constant distinguishing probability (on a positive note, we show that one cannot push this probability too close to 1). Additionally, just like in the case of separation between encryption and extraction [DS02], our separation can be extended to hold even if one additionally requires all the distributions in the imperfect source to have high min-entropy.[2] Moreover, our information-theoretic separation above can be extended even to the computational setting. This means that there exist high-entropy sources for which one can build *efficient* 2-out-2 secret sharing, but any (efficient) encryption scheme can be broken by an *efficient* distinguisher on an *efficiently-samplable* distribution from our source.[3]

To summarize (see Figure 1), extraction strictly implies encryption [DS02] which in turn, as we show, strictly implies 2-2 secret sharing.

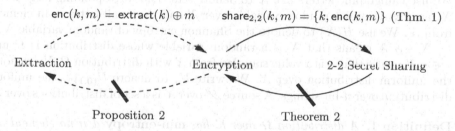

$$\mathsf{enc}(k,m) = \mathsf{extract}(k) \oplus m \qquad \mathsf{share}_{2,2}(k,m) = \{k, \mathsf{enc}(k,m)\}\ (\text{Thm. 1})$$

Extraction   Encryption   2-2 Secret Sharing

Proposition 2   Theorem 2

Fig. 1. The solid arrows indicate the implication and the separation we will prove

COMPARING CRYPTOGRAPHIC PRIMITIVES. As we see, our work continues the approach initiated by Dodis and Spencer [DS02] to compare different crypto-graphic tasks according to how they utilize randomness. Namely, given a block length n of our randomness source, and the (min-)entropy threshold $m \leq n$, we say that primitive A implies primitive B if whenever an imperfect source $\mathscr{S}$ of length n and (min-)entropy m is sufficient to implement A, then one could also

---

[2] In particular, we construct sources with min-entropy only a constant away from the maximal entropy (cf. Lemma 3 and Theorem 2).

[3] In fact, even the process of finding such an efficiently-samplable distribution can be done efficiently (with exponentially high probability), given only the oracle access to the encryption oracle.

implement B with $\mathscr{S}$. When $m = n$, we get back to the case of perfect randomness, where primitive A implies B if and only if the smallest number of truly random bits needed to implement A is at least as large as the smallest number of truly random bits to implement B. As was shown by [DS02, DOPS04] and continued here, many implications true in the perfect case simply stop being true the moment we allow for slightly imperfect random sources (i.e., allow $m < n$). In other words, these implications inherently rely on perfect randomness. On the other hand, some implications continue to hold (at least to some extent) even with imperfect randomness, implying they have more to do with the cryptographic aspect of the problem rather than the availability of true randomness. We believe that such comparison between cryptographic primitives sheds more light on how they utilize randomness, and also serves as a stepping stone toward classifying imperfect sources sufficient for different cryptographic tasks.

ORGANIZATION. We give the preliminary definitions of our primitives in Section 2. Our main technical result comparing sources for (information-theoretic) encryption and 2-2 secret sharing is given in Section 3. In Section 4 we extend our results to the computational setting. Finally, in Section 5 we take a brief look at authentication and discuss some open problems considering imperfect sources sources sufficient for various cryptographic applications.

2   Notation and Definitions

We use calligraphic letters like $\mathcal{X}$ to denote sets. The corresponding large letter X usually denotes a random variable over $\mathcal{X}$ and the small letter x an element from $\mathcal{X}$. We use $H(X)$ to denote the Shannon entropy of random variable X.

$X \in_\Omega \mathcal{X}$ means that X is a random variable whose distribution is Ω and $x \in_\Omega \mathcal{X}$ means that x is a value sampled from $\mathcal{X}$ with distribution Ω. $U_\mathcal{X}$ denotes the uniform distribution over $\mathcal{X}$. We write U_n to denote $U_{\{0,1\}^n}$, the uniform distribution over n-bit strings. A source $\mathscr{S}$ over $\mathcal{X}$ is a set of distributions over $\mathcal{X}$.

Definition 1. *A distribution Ω over $\mathcal{K}$ has* **min-entropy** *d if no element has probability more than 2^{-d}, i.e. $\max_{k \in \mathcal{K}} \Pr(k = k' | k' \in_\Omega \mathcal{K}) \leq 2^{-d}$. The largest such d is denoted $H_\infty(\Omega)$. A source $\mathscr{S}$ over $\mathcal{K}$ has* **min-entropy** *d if it only contains distributions with min-entropy at least d. The d-**weak source** over $\mathcal{K}$ is the source which contains all distributions over $\mathcal{K}$ with min-entropy at least d.*

Definition 2. *A random variable B over $\{0, 1\}$ is ϵ-**fair** if*

$$\min\{\Pr(B = 0), \Pr(B = 1)\} \geq \epsilon$$

(so a uniform random bit is $1/2$-fair and a constant bit is 0-fair). A source $\mathscr{S}$ over $\mathcal{K}$ is ϵ-fair if there exists a one-bit extractor (which is simply a function extract $: \mathcal{K} \to \{0, 1\}$*) such that* extract$(K)$*, where $K \in_\Omega \mathcal{K}$, is ϵ-fair for all $\Omega \in \mathscr{S}$.*

Definition 3. *An* **encryption scheme** *is a pair of algorithms* $\mathsf{enc} : \mathcal{K} \times \mathcal{M} \to \mathcal{C}$ *and* $\mathsf{dec} : \mathcal{K} \times \mathcal{C} \to \mathcal{M}$ *which for all keys* $k \in \mathcal{K}$ *and messages* $m \in \mathcal{M}$ *satisfies*

$$\mathsf{dec}(k, \mathsf{enc}(k, m)) = m \tag{1}$$

A source $\mathscr{S}$ *over* $\mathcal{K}$ *allows for* **perfect encryption** *of* $\mathcal{M}$ *if there is an encryption scheme such that for all distributions* $\Omega \in \mathscr{S}$ *the ciphertexts leak no information about the encrypted message* M, *i.e. for any random variable* M

$$\forall \Omega \in \mathscr{S} : H(M \mid \mathsf{enc}(K, M)) = H(M) \text{ where } K \in_\Omega \mathcal{K} \tag{2}$$

A source $\mathscr{S}$ *over* $\mathcal{K}$ *allows for* δ**-encryption** *if there is an encryption scheme such that for all distributions* $\Omega \in \mathscr{S}$ *the statistical distance of the encryption of any two distinct messages* m_1 *and* m_2 *is at most* δ, *i.e.*

$$\max_{\Omega \in \mathscr{S}, m_1 \neq m_2} \frac{1}{2} \sum_{c \in \mathcal{C}} |\Pr_{k \in_\Omega \mathcal{K}}(\mathsf{enc}(k, m_1) = c) - \Pr_{k \in_\Omega \mathcal{K}}(\mathsf{enc}(k, m_2) = c)| \leq \delta \tag{3}$$

Note that perfect encryption is 0-encryption and sending the plaintext is 1-encryption.

Definition 4. *For* $t, n \in \mathbb{Z}, t \leq n$ *a* t-n **secret sharing** *is a pair of algorithms* $\mathsf{share}_{t,n} : \mathcal{K} \times \mathcal{M} \to \mathcal{X}^n$ *and* $\mathsf{reconstruct}_{t,n} : \mathcal{X}^t \to \mathcal{M}$ *which for all keys* $k \in \mathcal{K}$ *and all* $m \in \mathcal{M}$ *satisfies*

$$\forall \mathcal{T} \subseteq \mathsf{share}_{t,n}(k, m) \text{ where } |\mathcal{T}| = t \text{ we have } \mathsf{reconstruct}_{t,n}(\mathcal{T}) = m \tag{4}$$

A source $\mathscr{S}$ *over* $\mathcal{K}$ *allows for* **perfect t-n secret sharing** *of* $\mathcal{M}$ *if any set of less than* t *shares does not reveal any information about the shared* M, *i.e. for all* $\Omega \in \mathscr{S}$ *and all* $1 \leq i_1 < i_2 < \ldots < i_{t-1} \leq n$ *we have for distributions* M

$$H(M \mid S_{i_1}, S_{i_2}, \ldots, S_{i_{t-1}}) = H(M) \text{ where } K \in_\Omega \mathcal{K}, \{S_1, \ldots, S_n\} \leftarrow \mathsf{share}_{t,n}(K, M) \tag{5}$$

Note that (4) means that from any t shares one can reconstruct m. In terms of *perfect randomness*, the uniform distribution over $\{0,1\}^n$ is necessary and sufficient to perfectly encrypt $\mathcal{M} = \{0,1\}^n$ (i.e. n-bit strings) for example by using the key k as a one time pad:

$$\mathsf{enc}(k, m) = k \oplus m \qquad \mathsf{dec}(k, c) = c \oplus m$$

where $\oplus$ denotes the bitwise XOR. U_n is also necessary and sufficient (as the dealer's randomness) to construct a perfect 2-2 secret sharing of $\{0,1\}^n$, for example as:

$$\mathsf{share}_{2,2}(k, m) = \{k, k \oplus m\} \qquad \mathsf{reconstruct}_{2,2}(s_1, s_2) = s_1 \oplus s_2$$

In the next section we will show that in terms of non-perfect randomness these two tasks are no longer equivalent. The sources which allow for perfect encryption also allow for 2-2 secret sharing (of the same message space) but not vice-versa. More precisely, we show that every source which allows for perfect 2-2 secret sharing of one bit allows for 1/2-encryption of one bit, but in general not for δ-encryption of one bit for $\delta < 1/3$. This even holds if we require the source to have high min-entropy.

3   Separating Encryption from Secret Sharing

We can now formally state the results of [MP90] and [DS02].

Proposition 1 ([MP90]). *The $(n-2)$-weak source over $\{0,1\}^n$ does not allow for δ-encryption of even 1 bit for any $\delta \neq 0$.*

So for every one-bit encryption scheme with key-space $\{0,1\}^n$ there exists a distribution for the keys with min-entropy $n-2$ such that the ciphertext always completely reveals the message.

Proposition 2 ([DS02]). *There is a source over $\{0,1\}^n$ which allows for perfect encryption of one bit, but which is not $2^{-n/2}$-fair.*
    This separation holds even if we require the source to have high min-entropy: for any $\epsilon > 2^{-n/2+1}$ there is a source $\mathscr{S}$ over $\{0,1\}^n$ with min-entropy $n - \log(1/\epsilon) - O(1)$ which allows for perfect encryption of one bit but which is not ϵ-fair.

3.1   Encryption → 2-2 Secret Sharing

Theorem 1. *Any source $\mathscr{S}$ over $\mathcal{K}$ which allows for perfect encryption of $\mathcal{M}$ allows for perfect 2-2 secret sharing of $\mathcal{M}$.*

Proof: For enc, dec which satisfy properties (1) and (2) we define for all $k \in \mathcal{K}, m \in \mathcal{M}$

$$\text{share}_{2,2}(k,m) = (k, \text{enc}(k,m)) \quad \text{and} \quad \text{reconstruct}_{2,2}(s_1, s_2) = \text{dec}(s_1, s_2).$$

Property (1) implies immediately that this scheme satisfies (4). It also satisfies property (5) as for any random variables M and $\Omega \in \mathscr{S}, K \in_\Omega \mathcal{K}$ we have that $H(M \mid K) = H(M)$ as K is independent of M and $H(M \mid \text{enc}(K,M)) = H(M)$ follows from (2).   ∎

In the following section we show that for $\mathcal{M} = \{0,1\}$, the converse is not true.

3.2   2-2 Secret Sharing ↛ Encryption

In this section we will prove our main technical result (Theorem 2 below), namely that sources which allow for 2-2 secret sharing do not allow for encryption in general. We split the proof of the theorem into the following three lemmas.

Lemma 1. *There is a source which allows for perfect 2-2 secret sharing of a bit but does not allow for δ-encryption of a bit for any $\delta < 1/3$.*

This separation is in some sense not so strong as the separation for encryption from extraction where a source was shown which allows perfect encryption but not even a weak form of extraction. The question arises if we can get something as $\delta \leq 1 - o(1)$ (and not just $\delta < 1/3$) here too. The answer is no, since already $\delta \leq 1/2$ is not achievable as shown in the next lemma.

Lemma 2. *Any source which allows for perfect 2-2 secret sharing of a bit allows for 1/2-encryption of a bit.*

We prove Lemma 1 by showing a concrete source which contains only four distributions over a domain of size six. Here the question arises whether this separation only works for such toy examples and possibly breaks down when we require the source to have high min-entropy. This is not the case: we show how one can turn such a toy-example into a high min-entropy source with the same parameters.

Lemma 3. *For any $t \in \mathbb{N}$ there is a source as in Lemma 1, where the distributions in the source have range of size $6t$ and the min-entropy of each distribution is at least $\log(6t) - \log(192)$.*

Combining Lemma 2 and Lemma 3 we get the following theorem.

Theorem 2. *There are sources over any $\mathcal{K}$ with min-entropy $\log|\mathcal{K}| - 11$ which allow for perfect 2-2 secret sharing but do not allow for δ-encryption of one bit for any $\delta < 1/3$.*
From the positive side, any source which allows for perfect 2-2 secret sharing of a bit allows for 1/2-encryption of one bit.

Theorem 2 is stated for sources over any $\mathcal{K}$ and not just for sets of size $6t$ as in Lemma 3. This is compensated for by an additional factor of $\log(6)$ in the min-entropy gap (i.e. we have a gap of $11 > \log(192) + \log(6)$).

Proof of Lemma 1. Let $\mathscr{S}$ be a source over $\mathcal{K} = \{k_1, \ldots, k_6\}$ which contains 4 distributions $\Omega_1, \ldots, \Omega_4$ where each Ω_i is the uniform distribution over $\mathcal{S}_i \subset \mathcal{K}$ with $\mathcal{S}_1 = \{k_1, k_2\}, \mathcal{S}_2 = \{k_3, k_4\}, \mathcal{S}_3 = \{k_1, k_3, k_5\}$ and $\mathcal{S}_4 = \{k_1, k_4, k_6\}$ respectively. Lemma 1 follows from the two claims below.

Claim 1. $\mathscr{S}$ *allows for perfect 2-2 secret sharing of one bit.*

Proof: We define the sharing $\mathsf{share}_{2,2} : \mathcal{K} \times \{0,1\} \to \mathcal{A} \times \mathcal{B}$, where $\mathcal{A} = \{a_1, a_2, a_3, a_4\}$ and $\mathcal{B} = \{b_1, b_2, b_3, b_4\}$ as shown in Figure 2. A key k_i is represented by a pair of directed edges, where the edge from $\mathcal{A}$ to $\mathcal{B}$ corresponds to the shares of 0, and the edge from $\mathcal{B}$ to $\mathcal{A}$ to the shares of 1. For example $\mathsf{share}_{2,2}(k_1, 0) = (a_3, b_2)$ and $\mathsf{share}_{2,2}(k_1, 1) = (a_1, b_2)$.

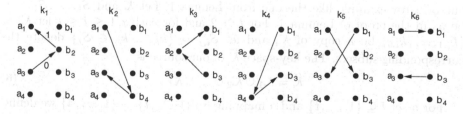

Fig. 2. The mapping $\mathsf{share}_{2,2}$ from the proof of Lemma 1

For any (a_i, b_j) there is at most one possible $m \in \{0, 1\}$ such that $(a_i, b_j) =$ $\mathsf{share}_{2,2}(k, m)$ for some $k \in \mathcal{K}$. Thus for any random variable M it always holds that $H(M \mid \mathsf{share}_{2,2}(k, M)) = 0$.

Note that for any $i, 1 \leq i \leq 4$, Ω_i is the uniform distribution over some subset of $\mathcal{K}$ whose corresponding directed edges (as shown in Figure 2) form a directed cycle, where the edges alternate between $\mathcal{A}$ and $\mathcal{B}$ (e.g. for Ω_1 we have the cycle $a_3 \to b_2 \to a_1 \to b_4 \to a_3$). So the distribution on $\mathcal{A}$ is the same no matter if we choose a random edge from $\mathcal{A}$ to $\mathcal{B}$ (a sharing of the secret 0) or from $\mathcal{B}$ to $\mathcal{A}$ (a sharing of the secret 1) on this cycle. This proves that the random variable A defined as $(A, B) = \mathsf{share}_{2,2}(k \in_{\Omega_i} \mathcal{K}, M)$ is independent of M and $H(M \mid A) = H(M)$ (and similarly for $H(M \mid B) = H(M)$). $\qquad\square$

Claim 2. $\mathscr{S}$ *does not allow for δ-encryption of a bit for any $\delta < 1/3$.*

Proof: Consider any mapping $\mathsf{enc} : \mathcal{K} \times \{0, 1\} \to \mathcal{C}$. We will prove that for our source $\mathscr{S}$ this enc cannot satisfy Definition 3 with $\delta < 1/3$. Recall that $\mathscr{S}$ contains the distributions $\Omega_i, 1 \leq i \leq 4$, where Ω_i is uniform over $\mathcal{S}_i$. Consider the graphs G_i, with $V(G_i) = \mathcal{C}$, $E(G_i) = \{(\mathsf{enc}(k, 0), \mathsf{enc}(k, 1)) \mid k \in \mathcal{S}_i\}$, where each edge is labeled with the corresponding $k \in \mathcal{K}$. We will show that $E(G_i)$ does not form a directed cycle for at least one $i, 1 \leq i \leq 4$.

Suppose this is not the case, then $E(G_1)$ forms a cycle of length 2, say $k_1 = (c_1, c_2), k_2 = (c_2, c_1)$. And similarly for $E(G_2)$, say $k_3 = (c_3, c_4), k_4 = (c_4, c_3)$. If $E(G_3)$ forms a cycle then (because of the above) either $c_1 = c_4$ or $c_2 = c_3$ but not both must hold (e.g. if $c_1 = c_4$ then we can set $k_5 = (c_2, c_3)$). Similarly if $E(G_4)$ is a cycle then either $c_1 = c_3$ or $c_2 = c_4$ but not both must hold. We can write this two conditions as $(c_1 = c_4 \oplus c_2 = c_3) \wedge (c_1 = c_3 \oplus c_2 = c_4) = true$, which cannot be satisfied and we have a contradiction.

For an i where $E(G_i)$ does not form a directed cycle (such an i exists as we just proved) there are vertices $c', c'' \in V(G_i)$ such that $indegree(c') >$ $outdegree(c')$ and $outdegree(c'') > indegree(c'')$, which using $|E(G_i)| = |\mathcal{S}_i| \leq 3$ gives

$$\frac{1}{2} \sum_{c \in \mathcal{C}} |Pr_{k \in \Omega_i} \mathcal{K}(\mathsf{enc}(k, 0) = c) - Pr_{k \in \Omega_i} \mathcal{K}(\mathsf{enc}(k, 1) = c)| \geq 1/3.$$

So because of this Ω_i property (3) cannot be satisfied with $\delta < 1/3$. $\qquad\square$

$\qquad\blacksquare$

Proof of Lemma 3. We will now show how to make a high min-entropy source out of a toy example like the one from Lemma 1. Let $\mathcal{K}$ and $\mathcal{S}_1, \ldots, \mathcal{S}_4 \subset \mathcal{K}$ be as in the proof of Lemma 1. For $t \geq 2$ and for each $i, 1 \leq i \leq t$ let $\mathcal{K}_i = \{k_{i,1}, \ldots, k_{i,6}\}$ be a copy of $\mathcal{K}$, and let $\mathcal{S}_{i,j} = \{k_{i,x} : k_x \in \mathcal{S}_j\}$ denote the corresponding subsets. The key-space $\widetilde{\mathcal{K}}$ of our source is

$$\widetilde{\mathcal{K}} = \mathcal{K}_1 \cup \mathcal{K}_2 \cup \ldots \cup \mathcal{K}_t. \tag{6}$$

For a set $I \subseteq \{1, \ldots, t\}$ and a mapping $\sigma : \{1, \ldots, t\} \to \{1, \ldots, 4\}$ we define

$$\mathcal{T}_{I,\sigma} = \bigcup_{i \in I} \mathcal{S}_{i,\sigma(i)}.$$

Our source $\mathscr{S}$ contains all uniform distributions over the sets $T_{I,\sigma}$ with $|I| \geq \lceil t/64 \rceil$. That is, our source contains all the uniform distributions over sets which are constructed by taking the union of subsets $S_{i,j} \subset \mathcal{K}_i$ from at least a $1/64$'th fraction of the $\mathcal{K}_i$'s. Since each distribution is uniform over a set of size at least $2t/64$, $\mathscr{S}$ has min-entropy at least $\log(t/32) = \log(6t) - \log(192)$.

The source $\mathscr{S}$ allows for perfect 2-2 secret sharing of one bit: On input a key $k_{i,j} \in \mathcal{K}_i$ the dealer can compute the shares as he would for the key $k_j \in \mathcal{K}$ in Claim 1 in the proof of Lemma 1. One can assume (but this is not necessary) that the dealer also publishes i, then we have the same situation as in Claim 1 and it follows from (the proof of) Claim 1 that this is indeed a perfect 2-2 secret sharing.

It only remains to show that $\mathscr{S}$ does not allow for δ-encryption of one bit with $\delta < 1/3$. For this consider any mapping $\mathrm{enc} : \widetilde{\mathcal{K}} \times \{0,1\} \to \mathcal{C}$. As shown in the proof of Lemma 1, for each $i, 1 \leq i \leq t$, there is a distribution Ω_i which is uniform over the set $S_{i,j}$ for some $j, 1 \leq j \leq 4$, satisfying

$$\frac{1}{2} \sum_{c \in \mathcal{C}} |\mathrm{Pr}_{k \in \Omega_i \widetilde{\mathcal{K}}}(\mathrm{enc}(k,0) = c) - \mathrm{Pr}_{k \in \Omega_i \widetilde{\mathcal{K}}}(\mathrm{enc}(k,1) = c)| \geq 1/3. \qquad (7)$$

Let $\mathcal{X}_i$ denote such a $S_{i,j}$. Consider a random mapping $\phi : \mathcal{C} \to \{-1, 1\}$. We say that $\mathcal{X}_i$ is *good* if

$$\forall c \in \mathcal{C} : \quad 0 \leq \phi(c) \cdot (\,|\{k \in \mathcal{X}_i : \mathrm{enc}(k,0) = c\}| - |\{k \in \mathcal{X}_i : \mathrm{enc}(k,1) = c\}|\,). \qquad (8)$$

As $|\mathcal{X}_i| \leq 3$ the rhs of eq. (8) can be nonzero for at most 6 different $c \in \mathcal{C}$ and thus $\mathcal{X}_i$ is good with probability at least 2^{-6}. This shows (as a simple application of the probabilistic method) that there is a ϕ for which at least $\lceil t2^{-6} \rceil$ of the $\mathcal{X}_i$'s are good. Fix such a ϕ and let $\mathcal{X}$ be the union of good sets. The uniform distribution over $\mathcal{X}$, $\Omega_\mathcal{X}$, is in $\mathscr{S}$, but it does not allow for δ-encryption with $\delta < 1/3$: Let γ_i be the event that $k \in \mathcal{X}_i$ (below all not explicitly labeled probabilities are over $k \in_{\Omega_\mathcal{X}} \widetilde{\mathcal{K}}$).

$$\frac{1}{2} \sum_{c \in \mathcal{C}} |\mathrm{Pr}(\mathrm{enc}(k,0) = c) - \mathrm{Pr}(\mathrm{enc}(k,1) = c)|$$

$$\overset{(8)}{=} \frac{1}{2} \sum_{c \in \mathcal{C}} \phi(c)(\mathrm{Pr}(\mathrm{enc}(k,0) = c) - \mathrm{Pr}(\mathrm{enc}(k,1) = c))$$

$$= \frac{1}{2} \sum_{c \in \mathcal{C}} \sum_{i=1}^{t} \phi(c)(\mathrm{Pr}(\gamma_i)\mathrm{Pr}(\mathrm{enc}(k,0) = c|\gamma_i) - \mathrm{Pr}(\gamma_i)\mathrm{Pr}(\mathrm{enc}(k,1) = c|\gamma_i))$$

$$\overset{(8)}{=} \sum_{i=1}^{t} \mathrm{Pr}(\gamma_i)\frac{1}{2} \sum_{c \in \mathcal{C}} |(\mathrm{Pr}(\mathrm{enc}(k,0) = c|\gamma_i) - \mathrm{Pr}(\mathrm{enc}(k,1) = c|\gamma_i)|$$

$$= \sum_{i=1}^{t} \mathrm{Pr}(\gamma_i)\frac{1}{2} \sum_{c \in \mathcal{C}} |\mathrm{Pr}_{k \in \Omega_i \widetilde{\mathcal{K}}}(\mathrm{enc}(k,0) = c) - \mathrm{Pr}_{k \in \Omega_i \widetilde{\mathcal{K}}}(\mathrm{enc}(k,1) = c)|$$

$$\overset{(7)}{\geq} \sum_{i=1}^{t} \mathrm{Pr}(\gamma_i)\frac{1}{3} = \frac{1}{3} \qquad \blacksquare$$

Proof of Lemma 2. Let $\mathscr{S}$ be a source with distributions over $\mathcal{K}$ which allows for perfect 2-2 secret sharing and $\text{share}_{2,2} : \mathcal{K} \times \{0,1\} \to \mathcal{A} \times \mathcal{B}$ be an appropriate sharing (we can wlog. assume that $\mathcal{A} \cap \mathcal{B} = \emptyset$). To prove the lemma we first define a mapping $\text{enc} : \mathcal{K} \times \{0,1\} \to \mathcal{C}$ (where $\mathcal{C} = \mathcal{A} \cup \mathcal{B}$), and then prove that it is a $1/2$-encryption (i.e. satisfies eq. (3) with $\delta = 1/2$).

For $k \in \mathcal{K}$ and $m \in \{0,1\}$ let $(a_{m,k}, b_{m,k}) = \text{share}_{2,2}(k,m)$, we set

$$\text{enc}(k,m) = a_{m,k} \text{ if } a_{0,k} \neq a_{1,k} \text{ and } b_{m,k} \text{ otherwise}$$

We cannot have $a_{0,k} = a_{1,k}$ and $b_{0,k} = b_{1,k}$ simultaneously as otherwise the share $(a_{0,k}, b_{0,k})$ could be a share of either 0 or 1 which is impossible when the secret sharing is perfect. So we always have $\text{enc}(k,0) \neq \text{enc}(k,1)$ and decryption is always possible. We will now prove that this enc satisfies eq.(3) with $\delta \leq 1/2$ (as our plaintext-domain is only one bit we can set $m_1 = 0$ and $m_2 = 1$ in (3) wlog.). For any $\Omega \in \mathscr{S}$ we have (all probabilities are over $k \in_\Omega \mathcal{K}$) using $\mathcal{C} = \mathcal{A} \cup \mathcal{B}, \mathcal{A} \cap \mathcal{B} = \emptyset$

$$\frac{1}{2} \sum_{c \in \mathcal{C}} |\Pr(\text{enc}(k,0) = c) - \Pr(\text{enc}(k,1) = c)| \tag{9}$$

$$= \frac{1}{2} \sum_{a \in \mathcal{A}} |\Pr(\text{enc}(k,0) = a) - \Pr(\text{enc}(k,1) = a)| \tag{10}$$

$$+ \frac{1}{2} \sum_{b \in \mathcal{B}} |\Pr(\text{enc}(k,0) = b) - \Pr(\text{enc}(k,1) = b)| \tag{11}$$

We will show that the term (9) is $\leq 1/2$ (which is exactly the statement of the Lemma) by showing that (10) is equal to 0 and (11) is $\leq 1/2$. Let the random variables A_m, B_m be defined as $(A_m, B_m) = \text{share}_{2,2}(k \in_\Omega \mathcal{K}, m)$, and let $\mathcal{K}^{\neq} = \{k \in \mathcal{K} | a_{0,k} \neq a_{1,k}\}$. From (5) we see that in a perfect 2-2 secret sharing the distribution of the share of each player is independent of the shared secret. This is used in the first step below (again all probabilities are over $k \in_\Omega \mathcal{K}$)

$$0 = \frac{1}{2} \sum_{a \in \mathcal{A}} |\Pr(A_0 = a) - \Pr(A_1 = a)| \tag{12}$$

$$= \frac{1}{2} \sum_{a \in \mathcal{A}} |\Pr(k \in \mathcal{K}^{\neq}) \left(\Pr(A_0 = a | k \in \mathcal{K}^{\neq}) - \Pr(A_1 = a | k \in \mathcal{K}^{\neq}) \right) \tag{13}$$

$$+ \Pr(k \notin \mathcal{K}^{\neq}) \underbrace{\left(\Pr(A_0 = a | k \notin \mathcal{K}^{\neq}) - \Pr(A_1 = a | k \notin \mathcal{K}^{\neq}) \right)}_{= 0 \text{ by the definition of } \mathcal{K}^{\neq}} | \tag{14}$$

$$= \frac{1}{2} \sum_{a \in \mathcal{A}} |\Pr(k \in \mathcal{K}^{\neq}) \left(\Pr(A_0 = a | k \in \mathcal{K}^{\neq}) - \Pr(A_1 = a | k \in \mathcal{K}^{\neq}) \right)| \tag{15}$$

Here (15) is exactly (10) so (10) is 0. We now show that (11) is $\leq 1/2$:

$$\Pr(\text{enc}(k,0) = b) = \Pr(B_0 = b \wedge k \notin \mathcal{K}^{\neq}) \leq \Pr(B_0 = b)$$

$$\Pr(\text{enc}(k,1) = b) = \Pr(B_1 = b \wedge k \notin \mathcal{K}^{\neq}) \leq \Pr(B_1 = b)$$

Perfect secret sharing implies $\Pr(B_0 = b) = \Pr(B_1 = b)$ and as the difference of two positive values cannot be larger than any those values we get

$$\frac{1}{2}\sum_{b \in \mathcal{B}} |\Pr(\text{enc}(k,0) = b) - \Pr(\text{enc}(k,1) = b)| \leq \frac{1}{2}\sum_{b \in \mathcal{B}} \Pr(B_0 = b) = \frac{1}{2}.$$

∎

4 Some Computational Aspects

Until now we have only considered an information theoretic setting. In particular, we did not care about whether the primitives, the attacks or the sampling considered can be efficiently realized. In this section, which we keep rather informal, we examine some computational aspects of the results from the previous section.

4.1 Computational Version of Theorem 1

The proof of Theorem 1, which states that any source which can be used for encryption can also be used for secret sharing, easily translates in the computational setting.

Proposition 3 (Computational version of Theorem 1). *(informal) Any source $\mathcal{S}$ over $\mathcal{K}$ which allows "computationally secure" encryption of $\mathcal{M}$ allows for "computationally secure" 2-2 secret sharing of $\mathcal{M}$.*

In the above proposition we left open what "computationally secure" exactly means. A direct translation from the information theoretic setting would advise the following security notion for encryption: the adversary can choose two messages m_0 and m_1 and then, given the encryption of m_b for a random b, should not be able to guess b (much better than with prob. $1/2$).[4] Then the security achieved for secret sharing is the following: First the adversary can choose two messages m_0 and m_1. Then, given one share of m_b for random b he cannot guess b (i.e. which message was shared). A stronger notion for encryption (e.g. semantic security) will result in a stronger security guarantee for secret sharing.

4.2 Computational Version of Theorem 2

We now take a look at Theorem 2 which follows from the Lemma 2 and Lemma 3. The proof of Lemma 2 translates into the computational setting.

Proposition 4 (Computational version of Lemma 2). *(informal) Any source which allows for "computationally secure" 2-2 secret sharing of a bit allows for "computationally secure" 1/2-encryption of a bit.*

---

[4] This notion is weaker than the notion of semantic security, where the adversary can additionally ask for encryptions of his choice except for m_0 and m_1.

As before, "computationally secure" can have several meanings (and a stronger notion for secret sharing implies a stronger notion for 1/2-encryption). Also the concept of δ-encryption has a natural meaning in the computational setting, where it means that the distinguishing advantage of any efficient adversary for the ciphertexts of two messages m_0 and m_1 is at most negligibly larger than $1/2 + \delta/2$.

Lemma 3 states that there is a source with high min-entropy which allows for 2-2 secret sharing but not for δ-encryption of one bit with $\delta < 1/3$. We can strengthen this lemma in several ways by considering computational aspects. In particular, we can require the following properties:

i. The secret sharing is efficient.
ii. For every encryption scheme the source contains an *efficiently samplable* distribution, for which the encryption-scheme is not 1/3-secure.
iii. There exists an efficient algorithm which breaks the 1/3-security of the encryption scheme under the efficiently samplable distribution from (ii).
iv. One can efficiently find the distribution from (ii).

We can achieve all four points simultaneously. However, to satisfy properties (ii) and (iv) we must be able to efficiently compute encryptions (either by getting a polynomial-size circuit or access to an oracle which computes encryptions given a key and a message). We now describe how one can adapt the proof of Lemma 3 (which we assume the reader is familiar with) to achieve these additional properties.

We can encode the key-space (see eq. (6)) as pairs of integers, i.e. $\widetilde{\mathcal{K}} \equiv [1, \ldots, t] \times [1, \ldots, 6]$. With this encoding property (i) (efficient secret sharing) is achieved: recall that the shares of $m \in \{0, 1\}$ under key (i, j) are $\mathsf{share}_{2,2}(k_j, m)$ with $\mathsf{share}_{2,2}(\cdot, \cdot)$ as defined in the proof of Lemma 1, which can be computed in constant time.

We now describe how to efficiently sample a distribution from our source which breaks the 1/3-security of enc. The distribution from the lemma is not efficiently samplable as the ϕ used to define it cannot be computed efficiently; We have only shown that a suitable ϕ exists — where suitable means that at least $\lceil t2^{-6} \rceil$ of the $\mathcal{X}_i$'s are good — using the probabilistic method. The argument used there was that a random ϕ satisfies (8) with probability at least 2^{-6}, as the rhs of (8) is nonzero for at most 6 different $c \in \mathcal{C}$. Fortunately for this argument we don't need a *random* ϕ — in fact, 6-wise independence is enough. Therefore if $\tau : \mathcal{W} \times \mathcal{C} \to \{-1, 1\}$ is an (efficiently computable) 6-wise independent function, then there is some key $w \in \mathcal{W}$ such that $\tau(w, \cdot)$ is good for $\lceil t2^{-6} \rceil$ of the $\mathcal{X}_i$'s (as $t2^{-6}$ is a lower bound for the expected number of good $\mathcal{X}_i$'s for a 6-wise independent function).

With this efficient $\phi(\cdot) = \tau(w, \cdot)$, we can now efficiently sample a key (using uniform randomness) according to the distribution of Lemma 3 (i.e. a random key from the union of all good sets) as follows:

1. Choose an integer $i, 1 \le i \le t$ uniformly at random.
2. Find a $j, 1 \le j \le 4$ (say the smallest) such that Ω_i, the uniform distribution over $\mathcal{X}_i = \mathcal{S}_{i,j}$, satisfies (7).

3. Check if this $\mathcal{X}_i$ is good, i.e. satisfies (8). If it does not, return to step 1.
4. If $|\mathcal{X}_i| = 2$ then return to step 1 with probability 1/3. (This is done to equalize the proportional weights of the $\mathcal{X}_i$'s of size 2 and 3.)
5. Output a key chosen uniformly at random from $\mathcal{X}_i$.

Note that this sampling will terminate in expected polynomial time if we can compute enc (in Step 2) and ϕ (in Step 3) efficiently.

We now describe an efficient breaking algorithm for enc, thus satisfying property (iii). Equation (8) tells us that the encryption of 0 and 1 have statistical distance at least 1/3, and from (7) we see that given a ciphertext c of a message m, $\phi(c)$ is an optimal guess on m. So if $\phi(\cdot)$ can be efficiently computed (which is the case if we set it to $\tau(w, \cdot)$, as described before), then we can efficiently break the 1/3 security of enc.

Finally we come to property (iv), which now can be stated as how to find a key w for our 6-wise independent function τ, such that $\phi(\cdot) = \tau(w, \cdot)$ is good for a 1/64 fraction of the $\mathcal{X}_i$'s. Unfortunately, for a given w one can't efficiently check if $\tau(w, \cdot)$ is good on a 1/64 fraction as for that we would have to go over all $\mathcal{X}_i$ for $i = 1, \ldots, t$, but t is exponential. But we can efficiently find a w such that $\tau(w, \cdot)$ will be good on a slightly smaller subset, say a 1/66 fraction, with probability exponentially close to 1 as follows. Choose a random w and approximate the fraction on which $\tau(w, \cdot)$ is good by randomly sampling $i \in [1, \ldots, t]$ and checking if it is good for $\mathcal{X}_i$. Accept this w if it was good on, say at least a 1/65 fraction, of the $\mathcal{X}_i$'s. By the Chernoff bound, the probability we will accept a w which is not good on at least a 1/66 fraction of all $\mathcal{X}_i$'s is exponentially small in the number of samples we have drawn for the approximation. Further, by the Markov bound we are guaranteed that we pick a w which is good on at least a 1/65 fraction after a constant number of tries.

5   Open Problems

There are many interesting open questions considering imperfect sources for various cryptographic applications. In our opinion the most dazzling one is whether the reductions from Proposition 2 and Theorem 2 generalize to larger domains. Already if we only extend the domain of the message space from two to three we cannot even show a sub-constant bound for the fairness.

Open Problem 1. *Is there an $\epsilon(n) \in o(1)$ such that there exist sources over $\{0, 1\}^n$ which allow for the encryption (or 2-2 secret sharing) of a trit[5] but cannot be used to extract an $\epsilon(n)$-fair bit (recall that for bits one can show $\epsilon(n) = 2^{-n/2}$).*

AUTHENTICATION.   Another interesting primitive we did not consider so far is authentication. Here, we will only consider the one-bit case, which already leaves several interesting open questions.

---

[5] A trit is like a bit but can take three and not just two values.

Definition 5. *We say that a source $\mathscr{S}$ with distributions over some set $\mathcal{K}$ allows for τ-authentication of one bit if there is a mapping* auth $: \mathcal{K} \times \{0,1\} \to \mathcal{A}$ *such that for all distributions $\Omega \in \mathscr{S}$ and $k \in_\Omega \mathcal{K}$ we have* $\min_k H_\infty(\text{auth}(k,0) \mid \text{auth}(k,1)) \geq -\log \tau$ *and* $\min_k H_\infty(\text{auth}(k,1) \mid \text{auth}(k,0)) \geq -\log \tau$.

Note that τ-authentication of a bit simply means that given the authenticator $\text{auth}(k,b)$ of a bit $b \in \{0,1\}$ the probability that one can guess $\text{auth}(k,1-b)$ (the authenticator of the other bit) correctly is at most $2^{-\tau}$. Authentication is very undemanding in its randomness requirements, any source whose min-entropy is large enough will do.

Proposition 5 ([MW97]). *The $(n+\tau)$-weak source over $\{0,1\}^{2n}$ allows for τ-authentication of one bit.*

The authentication which achieves the above bound is extremely simple: use the first and the last n bits to authenticate 0 and 1 respectively. Note that any half has min-entropy at least τ even when given the other half as an n-bit string has min-entropy of at most n. As such weak-sources are not enough for encryption (see Proposition 1), this already shows that sources for authentication do not allow for encryption, and one can easily show that they do not allow for secret sharing and any other cryptographic primitive requiring privacy we could think of.

But how about the other direction? Can sources which allow for encryption or secret sharing always be used for authentication? Recall, in the case of perfect randomness the result of [DS02] implies that $2n$ uniform bits are both necessary and sufficient for achieving n-authentication of 1-bit (in particular, we need at least 2 bits to do anything non-trivial at all), which means we can only hope that encryption (or 2-2 secret sharing) of at least $2n$-bits might (or might not) imply n-authentication of even a single bit. More generally,

Open Problem 2. *Find a lower bound for $\tau(n)$ and an upper bound for $\gamma(n)$ in the following statement (bounds for $n = 1$ only are already interesting):*

> *A source (possibly with some guaranteed min-entropy) which allows for the encryption (or 2-2 secret sharing) of $2n$ bits must always allow for $\tau(n)$-authentication of one bit, but in general not for $\gamma(n)$-authentication.*

As we remarked, we know that $n \geq \gamma(n) \geq \tau(n)$. Interestingly, we observe below that 3-3 secret sharing *does* imply authentication. (As a sanity check, in case of perfect randomness both n-authentication of a bit and 3-3 secret sharing of n bits need $2n$ perfectly random bits.)

Claim 3. *Any source which allows for perfect 3-3 secret sharing of n bits allows for n-authentication of a bit.*

Proof: This reduction can be achieved as follows: first compute the sharing (S_1, S_2, S_3) for some constant message, say $m = 0^n$. Now use S_1 as the authentication of 0 and S_2 as the authentication of 1. We observe that the joint distribution of shares S_1 and S_2 when $m = 0^n$ is the same as when m is uniform over $\{0,1\}^n$ as otherwise the shares S_1 and S_2 would leak information on which is the case. So let M be uniform over $\{0,1\}^n$ and K be chosen according to a distribution from our source. With the above observation we now must only prove that

$$H_\infty(S_1|S_2) \geq n \quad \text{and} \quad H_\infty(S_2|S_1) \geq n \quad \text{where} \quad (S_1, S_2, S_3) = \mathsf{share}_{3,3}(K, M).$$

Here $H_\infty(S_1|S_2) \geq n$ means that $H_\infty(S_1|S_2 = s) \geq n$ for all s in the support of S_2 (and not as sometimes used that the expectation over S_2 is at least n, i.e. not $\sum_s \Pr(S_2 = s)H_\infty(S_1|S_2 = s) \geq n$). Now by the definition of perfect secret sharing we have

$$H_\infty(M|S_2S_3) = n \quad \text{and} \quad H_\infty(M|S_1S_2S_3) = 0, \tag{16}$$

which implies $H_\infty(S_1|S_2S_3) \geq n$. To see this assume that this was not true, i.e. we have for some s_1, s_2, s_3 that $\Pr(S_1 = s_1|S_2 = s_2, S_3 = s_3) > 2^{-n}$, but then for $m = \mathsf{reconstruct}_{3,3}(s_1, s_2, s_3)$ also $\Pr(m|S_2 = s_2, S_3 = s_3) > 2^{-n}$ which contradicts $H_\infty(M|S_2S_3) = n$. The desired $H_\infty(S_1|S_2) \geq n$ now easily follows from $H_\infty(S_1|S_2S_3) \geq n$, and $H_\infty(S_2|S_1) \geq n$ can be shown similarly. $\square$

Now, as encryption implies 2-2 secret sharing and 3-3 secret sharing implies authentication, a proof that 2-2 secret sharing implies some non-trivial 3-3 secret sharing would immediately give a non-trivial bound for $\tau(n)$ from Open Problem 2. Moreover, we think that comparing 2-2 secret sharing of $2n$ bits with 3-3 secret sharing of n bits is interesting in its own right, since it would show that different t-m secret sharing schemes have (or have not) different requirements on the way they utilize randomness, even if the same amount of perfect randomness is required for them.

References

[ACRT99]   Alexander Andreev, Andrea Clementi, Jose Rolim, and Luca Trevisan. Dispersers, deterministic amplification, and weak random sources. *SIAM J. on Computing*, 28(6):2103–2116, 1999.

[AL93]   Miklós Ajtai and Nathal Linial. The influence of large coalitions. *Combinatorica*, 13(2):129–145, 1993.

[BBR88]   Charles H. Bennett, Gilles Brassard, and Jean-Marc Robert. Privacy amplification by public discussion. *SIAM J. on Computing*, 17(2):210–229, 1988.

[Blu86]   Manuel Blum. Independent unbiased coin flips from a correlated biased source — a finite state Markov chain. *Combinatorica*, 6(2):97–108, 1986.

[CDH+00]   Ran Canetti, Yevgeniy Dodis, Shai Halevi, Eyal Kushilevitz, and Amit Sahai. Exposure-resilient functions and all-or-nothing transforms. In *Proc. EUROCRYPT'00*, pages 453–469, 2000.

[CG88]   Benny Chor and Oded Goldreich. Unbiased bits from sources of weak randomness and probabilistic communication complexity. *SIAM J. on Computing*, 17(2):230–261, 1988.

[CGH+85]   Benny Chor, Oded Goldreich, Johan Håstad, Joel Friedman, Steven Rudich, and Roman Smolensky. The bit extraction problem of t-resilient functions. In *Proc. 26th IEEE FOCS*, pages 396–407, 1985.

[DOPS04] Yevgeniy Dodis, Shien Jin Ong, Manoj Prabhakaran, and Amit Sahai. On the (im)possibility of cryptography with imperfect randomness. In *Proc. 45th IEEE FOCS*, pages 196–205, 2004.

[DS02]   Yevgeniy Dodis and Joel Spencer. On the (non-)universality of the one-time pad. In *Proc. 43rd IEEE FOCS*, pages 376–388, 2002.

[DSS01]   Yevgeniy Dodis, Amit Sahai, and Adam Smith. On perfect and adaptive security in exposure-resilient cryptography. In *Proc. EUROCRYPT'01*, pages 301–324, 2001.

[Eli72]   Peter Elias. The efficient construction of an unbiased random sequence. *Ann. Math. Stat.*, 43(2):865–870, 1972.

[KZ03]   Jess Kamp and David Zuckerman. Deterministic extractors for bit-fixing sources and exposure-resilient cryptography. In *Proc. 44th IEEE FOCS*, pages 92–101, 2003.

[LLS89]   David Lichtenstein, Nathan Linial, and Michael Saks. Some extremal problems arising from discrete control processes. *Combinatorica*, 9(3):269–287, 1989.

[MP90]   James L. McInnes and Benny Pinkas. On the impossibility of private key cryptography with weakly random keys. In *Proc. CRYPTO'90*, pages 421–436, 1990.

[MW97]   Ueli Maurer and Stefan Wolf. Privacy amplification secure against active adversaries. In *Proc. CRYPTO'97*, pages 307–321, 1997.

[RW03]   Renato Renner and Stefan Wolf. Unconditional authenticity and privacy from an arbitrary weak secret. In *Proc. CRYPTO'03*, pages 78–95, 2003.

[SV86]   Miklos Santha and Umesh V. Vazirani. Generating quasi-random sequences from semi-random sources. *JCSS*, 33(1):75–87, 1986.

[TV00]   Luca Trevisan and Salil Vadhan. Extracting randomness from samplable distributions. In *Proc. 41st IEEE FOCS*, pages 32–42, 2000.

[vN51]   John von Neumann. Various techniques used in connection with random digits. *National Bureau of Standards, Applied Mathematics Series*, 12:36–38, 1951.

[VV85]   Umesh V. Vazirani and Vijay V. Vazirani. Random polynomial time is equal to slightly-random polynomial time. In *Proc. 26th IEEE FOCS*, pages 417–428, 1985.

[Zuc96]   David Zuckerman. Simulating BPP using a general weak random source. *Algorithmica*, 16(4/5):367–391, 1996.

Author Index

Lecture Notes in Computer Science

For information about Vols. 1–3787

please contact your bookseller or Springer